IAASTD

International Assessment of Agricultural Knowledge, Science and Technology for Development

North America and Europe (NAE) Report

IAASTD
International Assessment of Agricultural Knowledge, Science and Technology for Development

U N
D P

UNEP

UNESCO

THE WORLD BANK

W.H.O. O.M.S.
WHO

GLOBAL
ENVIRONMENT
FACILITY

International Assessment of Agricultural Knowledge, Science and Technology for Development

North America and Europe (NAE) Report

Edited by

Beverly D. McIntyre
IAASTD Secretariat

Hans R. Herren
Millennium Institute

Judi Wakhungu
African Centre for Technology Studies

Robert T. Watson
University of East Anglia

Library of Congress Cataloging-in-Publication data.

International assessment of agricultural knowledge, science and technology for development (IAASTD) : North America and Europe (NAE) report / edited by Beverly D. McIntyre . . . [et al.].
p. cm.
Includes bibliographical references and index.
ISBN 978-1-59726-548-5 (cloth : alk. paper) —
ISBN 978-1-59726-549-2 (pbk. : alk. paper)
1. Agriculture—North America—International cooperation.
2. Agriculture—Europe—International cooperation. 3. Sustainable development—North America. 4. Sustainable development—Europe.
I. McIntyre, Beverly D. II. Title: North America and Europe (NAE) report.
HD1428.I5454 2008
338.94′07—dc22 2008046048

British Cataloguing-in-Publication data available.

Printed on recycled, acid-free paper 

Interior and cover designs by Linda McKnight, McKnight Design, LLC.

Manufactured in the United States of America

10 9 8 7 6 5 4 3 2 1

Contents

vii Foreword
viii Statement by Governments
x Preface

1 **Chapter 1** Setting the Stage
20 **Chapter 2** Changes in Agriculture and Food Production in NAE since 1945
79 **Chapter 3** Environmental, Economic and Social Impacts of NAE Agriculture and AKST
116 **Chapter 4** Changes in the Organization and Institutions of AKST and Consequences for Development and Sustainability Goals
151 **Chapter 5** Looking into the Future for Knowledge, Science and Technology and AKST
208 **Chapter 6** Options for Action

277 **Annex A** NAE Authors and Review Editors
279 **Annex B** Peer Reviewers
281 **Annex C** Glossary
289 **Annex D** Acronyms, Abbreviations and Units
292 **Annex E** Secretariat and Cosponsor Focal Points
293 **Annex F** Steering Committee for Consultative Process and Advisory Bureau for Assessment
296 **Annex G** Reservations on NAE Report
297 **Annex H** Additional NAE Figures and Tables

299 Index

Foreword

The objective of the International Assessment of Agricultural Knowledge, Science and Technology for Development (IAASTD) was to assess the impacts of past, present and future agricultural knowledge, science and technology on the

- reduction of hunger and poverty,
- improvement of rural livelihoods and human health, and
- equitable, socially, environmentally and economically sustainable development.

The IAASTD was initiated in 2002 by the World Bank and the Food and Agriculture Organization of the United Nations (FAO) as a global consultative process to determine whether an international assessment of agricultural knowledge, science and technology was needed. Mr. Klaus Töepfer, Executive Director of the United Nations Environment Programme (UNEP) opened the first Intergovernmental Plenary (30 August-3 September 2004) in Nairobi, Kenya, during which participants initiated a detailed scoping, preparation, drafting and peer review process.

The outputs from this assessment are a global and five subglobal reports; a Global and five Sub-Global Summaries for Decision Makers; and a cross-cutting Synthesis Report with an Executive Summary. The Summaries for Decision Makers and the Synthesis Report specifically provide options for action to governments, international agencies, academia, research organizations and other decision makers around the world.

The reports draw on the work of hundreds of experts from all regions of the world who have participated in the preparation and peer review process. As has been customary in many such global assessments, success depended first and foremost on the dedication, enthusiasm and cooperation of these experts in many different but related disciplines. It is the synergy of these inter-related disciplines that permitted IAASTD to create a unique, interdisciplinary regional and global process.

We take this opportunity to express our deep gratitude to the authors and reviewers of all of the reports—their dedication and tireless efforts made the process a success. We thank the Steering Committee for distilling the outputs of the consultative process into recommendations to the Plenary, the IAASTD Bureau for their advisory role during the assessment and the work of those in the extended Secretariat. We would specifically like to thank the cosponsoring organizations of the Global Environment Facility (GEF) and the World Bank for their financial contributions as well as the FAO, UNEP, and the United Nations Educational, Scientific and Cultural Organization (UNESCO) for their continued support of this process through allocation of staff resources.

We acknowledge with gratitude the governments and organizations that contributed to the Multidonor Trust Fund (Australia, Canada, the European Commission, France, Ireland, Sweden, Switzerland, and the United Kingdom) and the United States Trust Fund. We also thank the governments who provided support to Bureau members, authors and reviewers in other ways. In addition, Finland provided direct support to the Secretariat. The IAASTD was especially successful in engaging a large number of experts from developing countries and countries with economies in transition in its work; the Trust Funds enabled financial assistance for their travel to the IAASTD meetings.

We would also like to make special mention of the regional organizations who hosted the regional coordinators and staff and provided assistance in management and time to ensure success of this enterprise: the African Center for Technology Studies (ACTS) in Kenya, the Inter-American Institute for Cooperation on Agriculture (IICA) in Costa Rica, the International Center for Agricultural Research in the Dry Areas (ICARDA) in Syria, and the WorldFish Center in Malaysia.

The final Intergovernmental Plenary in Johannesburg, South Africa was opened on 7 April 2008 by Achim Steiner, Executive Director of UNEP. This Plenary saw the acceptance of the Reports and the approval of the Summaries for Decision Makers and the Executive Summary of the Synthesis Report by an overwhelming majority of governments.

Signed:

Co-chairs
Hans H. Herren,
Judi Wakhungu

Director
Robert T. Watson

Statement by Governments

All countries present at the final intergovernmental plenary session held in Johannesburg, South Africa in April 2008 welcome the work of the IAASTD and the uniqueness of this independent multistakeholder and multidisciplinary process, and the scale of the challenge of covering a broad range of complex issues. The governments present recognize that the Global and Sub-Global Reports are the conclusions of studies by a wide range of scientific authors, experts and development specialists and while presenting an overall consensus on the importance of agricultural knowledge, science and technology for development also provide a diversity of views on some issues.

All countries see these reports as a valuable and important contribution to our understanding on agricultural knowledge, science and technology for development recognizing the need to further deepen our understanding of the challenges ahead. This assessment is a constructive initiative and important contribution that all governments need to take forward to ensure that agricultural knowledge, science and technology fulfills its potential to meet the development and sustainability goals of the reduction of hunger and poverty, the improvement of rural livelihoods and human health, and facilitating equitable, socially, environmentally and economically sustainable development.

In accordance with the above statement, the following governments accept the North America and Europe (NAE) Report.

Armenia, Finland, France, Ireland, Republic of Moldova, Poland, Romania, Sweden, Switzerland, United Kingdom of Great Britain (10 countries).

While approving the above statement the following governments did not fully accept the North America and Europe (NAE) Report and their reservations are entered in Annex G.

Canada and United States of America (2 countries).

Preface

In August 2002, the World Bank and the Food and Agriculture Organization (FAO) of the United Nations initiated a global consultative process to determine whether an international assessment of agricultural knowledge, science and technology (AKST) was needed. This was stimulated by discussions at the World Bank with the private sector and nongovernmental organizations (NGOs) on the state of scientific understanding of biotechnology and more specifically transgenics. During 2003, eleven consultations were held, overseen by an international multistakeholder steering committee and involving over 800 participants from all relevant stakeholder groups, e.g., governments, the private sector and civil society. Based on these consultations the steering committee recommended to an Intergovernmental Plenary meeting in Nairobi in September 2004 that an international assessment of the role of AKST in reducing hunger and poverty, improving rural livelihoods and facilitating environmentally, socially and economically sustainable development was needed. The concept of an International Assessment of Agricultural Knowledge, Science and Technology for Development (IAASTD) was endorsed as a multi-thematic, multi-spatial, multi-temporal intergovernmental process with a multistakeholder Bureau cosponsored by the FAO, the Global Environment Facility (GEF), United Nations Development Programme (UNDP), United Nations Environment Programme (UNEP), United Nations Educational, Scientific and Cultural Organization (UNESCO), the World Bank and World Health Organization (WHO).

The IAASTD's governance structure is a unique hybrid of the Intergovernmental Panel on Climate Change (IPCC) and the nongovernmental Millennium Ecosystem Assessment (MA). The stakeholder composition of the Bureau was agreed at the Intergovernmental Plenary meeting in Nairobi; it is geographically balanced and multistakeholder with 30 government and 30 civil society representatives (NGOs, producer and consumer groups, private sector entities and international organizations) in order to ensure ownership of the process and findings by a range of stakeholders.

About 400 of the world's experts were selected by the Bureau, following nominations by stakeholder groups, to prepare the IAASTD Report (comprised of a Global and five Sub-Global assessments). These experts worked in their own capacity and did not represent any particular stakeholder group. Additional individuals, organizations and governments were involved in the peer review process.

The IAASTD development and sustainability goals were endorsed at the first Intergovernmental Plenary and are consistent with a subset of the UN Millennium Development Goals (MDGs): the reduction of hunger and poverty, the improvement of rural livelihoods and human health, and facilitating equitable, socially, environmentally and economically sustainable development. Realizing these goals requires acknowledging the multifunctionality of agriculture: the challenge is to simultaneously meet development and sustainability goals while increasing agricultural production.

Meeting these goals has to be placed in the context of a rapidly changing world of urbanization, growing inequities, human migration, globalization, changing dietary preferences, climate change, environmental degradation, a trend toward biofuels and an increasing population. These conditions are affecting local and global food security and putting pressure on productive capacity and ecosystems. Hence there are unprecedented challenges ahead in providing food within a global trading system where there are other competing uses for agricultural and other natural resources. AKST alone cannot solve these problems, which are caused by complex political and social dynamics, but it can make a major contribution to meeting development and sustainability goals. Never before has it been more important for the world to generate and use AKST.

Given the focus on hunger, poverty and livelihoods, the IAASTD pays special attention to the current situation, issues and potential opportunities to redirect the current AKST system to improve the situation for poor rural people, especially small-scale farmers, rural laborers and others with limited resources. It addresses issues critical to formulating policy and provides information for decision makers confronting conflicting views on contentious issues such as the environmental consequences of productivity increases, environmental and human health impacts of transgenic crops, the consequences of bioenergy development on the environment and on the long-term availability and price of food, and the implications of climate change on agricultural production. The Bureau agreed that the scope of the assessment needed to go beyond the narrow confines of S&T and should encompass other types of relevant knowledge (e.g., knowledge held by agricultural producers, consumers and end users) and that it should also assess the role of institutions, organizations, governance, markets and trade.

The IAASTD is a multidisciplinary and multistakeholder enterprise requiring the use and integration of information, tools and models from different knowledge paradigms including local and traditional knowledge. The IAASTD does not advocate specific policies or practices; it assesses the major issues facing AKST and points towards a range of AKST options for action that meet development and sustainability goals. It is policy relevant, but not policy prescriptive. It integrates scientific information on a range of topics that are critically interlinked, but often addressed independently, i.e., agriculture, poverty, hunger, human health, natural resources, environment, development and innovation. It will enable decision makers to bring a richer base of knowledge to bear on policy and management decisions on issues previously viewed in isolation. Knowledge gained from historical analysis (typically the past 50 years) and an analysis of some future development alternatives to 2050 form the basis for assessing options for action on science and technology, capacity development, institutions and policies, and investments.

The IAASTD is conducted according to an open, transparent, representative and legitimate process; is evidence-based; presents options rather than recommendations; assesses different local, regional and global perspectives; presents different views, acknowledging that there can be more than one interpretation of the same evidence based on different worldviews; and identifies the key scientific uncertainties and areas on which research could be focused to advance development and sustainability goals.

The IAASTD is composed of a Global assessment and five Sub-Global assessments: Central and West Asia and North Africa (CWANA); East and South Asia and the Pacific (ESAP); Latin America and the Caribbean (LAC); North America and Europe (NAE); and Sub-Saharan Africa (SSA). It (1) assesses the generation, access, dissemination and use of public and private sector AKST in relation to the goals, using local, traditional and formal knowledge; (2) analyzes existing and emerging technologies, practices, policies and institutions and their impact on the goals; (3) provides information for decision makers in different civil society, private and public organizations on options for improving policies, practices, institutional and organizational arrangements to enable AKST to meet the goals; (4) brings together a range of stakeholders (consumers, governments, international agencies and research organizations, NGOs, private sector, producers, the scientific community) involved in the agricultural sector and rural development to share their experiences, views, understanding and vision for the future; and (5) identifies options for future public and private investments in AKST. In addition, the IAASTD will enhance local and regional capacity to design, implement and utilize similar assessments.

In this assessment agriculture is used to include production of food, feed, fuel, fiber and other products and to include all sectors from production of inputs (e.g., seeds and fertilizer) to consumption of products. However, as in all assessments, some topics were covered less extensively than others (e.g., livestock, forestry, fisheries and the agricultural sector of small island countries, and agricultural engineering), largely due to the expertise of the selected authors.

The IAASTD draft Report was subjected to two rounds of peer review by governments, organizations and individuals. These drafts were placed on an open access web site and open to comments by anyone. The authors revised the drafts based on numerous peer review comments, with the assistance of review editors who were responsible for ensuring the comments were appropriately taken into account. One of the most difficult issues authors had to address was criticisms that the report was too negative. In a scientific review based on empirical evidence, this is always a difficult comment to handle, as criteria are needed in order to say whether something is negative or positive. Another difficulty was responding to the conflicting views expressed by reviewers. The difference in views was not surprising given the range of stakeholder interests and perspectives. Thus one of the key findings of the IAASTD is that there are diverse and conflicting interpretations of past and current events, which need to be acknowledged and respected.

The Global and Sub-Global Summaries for Decision Makers and the Executive Summary of the Synthesis Report were approved at an Intergovernmental Plenary in April 2008. The Synthesis Report integrates the key findings from the Global and Sub-Global assessments, and focuses on eight Bureau-approved topics: bioenergy; biotechnology; climate change; human health; natural resource management; traditional knowledge and community based innovation; trade and markets; and women in agriculture.

The IAASTD builds on and adds value to a number of recent assessments and reports that have provided valuable information relevant to the agricultural sector, but have not specifically focused on the future role of AKST, the institutional dimensions and the multifunctionality of agriculture. These include: FAO State of Food Insecurity in the World (yearly); InterAcademy Council Report: Realizing the Promise and Potential of African Agriculture (2004); UN Millennium Project Task Force on Hunger (2005); Millennium Ecosystem Assessment (2005); CGIAR Science Council Strategy and Priority Setting Exercise (2006); Comprehensive Assessment of Water Management in Agriculture: Guiding Policy Investments in Water, Food, Livelihoods and Environment (2007); Intergovernmental Panel on Climate Change Reports (2001 and 2007); UNEP Fourth Global Environmental Outlook (2007); World Bank World Development Report: Agriculture for Development (2008); IFPRI Global Hunger Indices (yearly); and World Bank Internal Report of Investments in SSA (2007).

Financial support was provided to the IAASTD by the cosponsoring agencies, the governments of Australia, Canada, Finland, France, Ireland, Sweden, Switzerland, US and UK, and the European Commission. In addition, many organizations have provided in-kind support. The authors and review editors have given freely of their time, largely without compensation.

The Global and Sub-Global Summaries for Decision Makers and the Synthesis Report are written for a range of stakeholders, e.g., government policy makers, private sector, NGOs, producer and consumer groups, international organizations and the scientific community. There are no recommendations, only options for action. The options for action are not prioritized because different options are actionable by different stakeholders, each of whom has a different set of priorities and responsibilities and operates in different socio-economic and political circumstances.

1

Setting the Stage

Coordinating Lead Authors:
Molly Anderson (USA), Les Firbank (UK)

Lead Authors:
Sergey Alexanian (Russia), Dorota Metera (Poland), Tanja Schuler (Germany)

Review Editors:
Denis Ebodaghe (USA), Yuriy Nesterov (Ukraine)

Key Messages

1.1 Scope and Structure 3
1.1.1 Geographic scope of NAE 3
1.1.2 Structure of the report 3
1.1.2.1 Interface with Global Assessment and other Sub-Global Assessments 3
1.1.2.2 Thematic issues 3
1.1.3 Conceptual framework 4
1.1.3.1 Conceptual diagram 4
1.1.3.2 Drivers of change 4
1.1.3.3 AKST dynamics 4
1.2 Agriculture, Development and Sustainability Goals 5
1.2.1 Eradicating hunger and food insecurity; providing adequate amounts of healthy, safe food; and improving human health 5
1.2.2 Reducing extreme poverty, improving livelihoods and creating rural employment 6
1.2.3 Promoting equity across gender and social gaps 7
1.2.4 Enhancing environmental quality 8
1.3 Significance of NAE in the Generation, Use and Control of AKST 8
1.3.1 Importance within the region 8
1.3.1.1 Impacts on development and sustainability goals 8
1.3.1.2 Economic 8
1.3.1.3 Sociocultural 9
1.3.1.4 Environmental 9
1.3.2 Importance to the rest of the world 9
1.3.2.1 Impacts on development and sustainability goals 9
1.3.2.2 Export of AKST, other forms of KST and concepts of development 9
1.3.2.3 NAE's footprint 10
1.3.2.4 Wealth and political power 10
1.4 Description of the Region 10
1.4.1 Social, political and economic development 10
1.4.1.1 Prior to 1945 10
1.4.1.2 After 1945 11
1.4.2 Natural resources and their exploitation 12
1.4.2.1 Fresh water 13
1.4.2.2 Energy 13
1.4.2.3 Fisheries 13
1.4.2.4 Marginal lands 13
1.4.2.5 Forests 13
1.4.3 Agrifood systems 14
1.4.3.1 The development of agrifood systems to 1945 14
1.4.3.2 Agrifood systems post 1945 14
1.4.3.3 The development of policy 16
1.5 Challenges for AKST 16

Key Messages

1. The application of agricultural knowledge, science and technology (AKST) within NAE since 1945 has increased productivity and production substantially, such that NAE produces more than enough food overall to meet basic needs of the region. Yet its application has also undermined the achievement of development and sustainability goals within the region and in other sub-global regions by contributing to environmental degradation (e.g., habitat transformation, freshwater contamination and over-exploitation of fisheries), increasing inequity in wealth and assets in the food system, increasing the vulnerability of livelihoods dependent on agriculture and contributing to diet-related diseases, obesity and overweight.

2. NAE's agricultural activities have significant influence on the capacity of countries in other regions to meet development and sustainability goals. This is largely due to NAE's volume and variety of exports and imports and the many actors and networks based in NAE that dominate agrifood chains and AKST. For example, businesses within NAE have a powerful impact on consumer demand in the rest of the world; they obtain and profit from commodities, landraces and other valuable genetic resources and immigrant labor from other regions. NAE countries house ex situ genetic resources collections, and they have a legacy of substantial investment in AKST dating back centuries. NAE generated and initially used many advances in AKST, so this region shows the impacts of specific forms of AKST over the longest time period and can provide illustrative lessons on its application and resulting positive and negative (intended and unintended) consequences.

3. Choices about investment in, generation and control of, and access to AKST have great potential to help solve critical current and future challenges to human well-being within NAE and globally, including

- Mitigating and adapting to global climate change,
- Managing resources for human use while maintaining their ability to provide a full array of ecosystem services,
- Creating markets with fair access and compensation to participants,
- Developing renewable energy sources and other alternatives to products made from fossil fuels,
- Improving human health and reducing exposure to foodborne contaminants and disease,
- Increasing the availability of and equitable access to food and other agricultural products,
- Improving equity across gender and social divides, and
- Creating and sustaining urban and rural livelihoods.

4. AKST interacts with and is driven by knowledge and technology in non-agricultural domains such as demography, economics, international trade and cultural developments. It encompasses formal and informal education, training, research, research and innovation policy, and national and international regulations and agreements. Regulations and agreements address issues such as control, exchange and access to agrobiodiversity and natural resources; information and technology; land tenure arrangements; and Intellectual Property Rights.

5. AKST within NAE has been characterized by a paradigm emphasizing increases in production and productivity. The generation and dissemination of knowledge and development of technology have typically been fragmented and hierarchical, with some stakeholders excluded from setting and implementing AKST agendas. This paradigm is changing; continued development of a new paradigm for generation, access and use of AKST is important to meeting development and sustainability goals.

6. Different political and socioeconomic histories during the 20th century and variable access to all forms of capital (human, social, financial, physical and natural) have driven very different paths of agricultural development and AKST within NAE's subregions (North America, western Europe, eastern Europe and Israel). These are associated with widely varying attitudes about the importance of national agrifood self-sufficiency, trade and subsidies for agriculture. At the same time, geographic and political similarities across the region have important consequences for AKST and agrifood systems: most of the region is in a temperate zone; the region overall has enjoyed relative peace and stability over the last half-century compared with other sub-global regions; and many of its countries and businesses have made substantial investments in AKST.

7. Very small numbers of people (less than 2% of the population) are engaged in primary agrifood production in some NAE countries, although the proportions of small-scale subsistence or semi-subsistence growers remain quite high in other countries. Agrifood systems (including processing, distribution and sales) employ a substantial proportion of the population in all countries. In addition to providing raw materials for traditional products—food, feed, seed, fiber, fuel, paper, etc.—agricultural management in NAE is expected to deliver environmental, social and cultural goods and services. These include clean, abundant water; biodiversity and landscape quality; rural employment; recreation; and mitigation of climate change.

8. Agrifood systems have become dominated by fewer, larger actors. All sectors of agrifood systems have shown vertical and horizontal integration, although in many eastern European countries, smallholders and local outlets still raise and market most of the agrifood products (especially livestock, potatoes and other vegetables). Agrifood systems are starting to respond to consumer markets for food and other goods produced to high environmental and social standards (known as the "quality turn"). Small- and mid-scale producers and distributors through most of NAE increasingly market higher-value, differentiated goods. Vertically integrated supermarkets are attempting to expand market share and satisfy regulatory requirements for higher quality, codified in environmental and social standards and implemented in labeling and certification schemes. Concentrated enterprises and changing expectations and standards

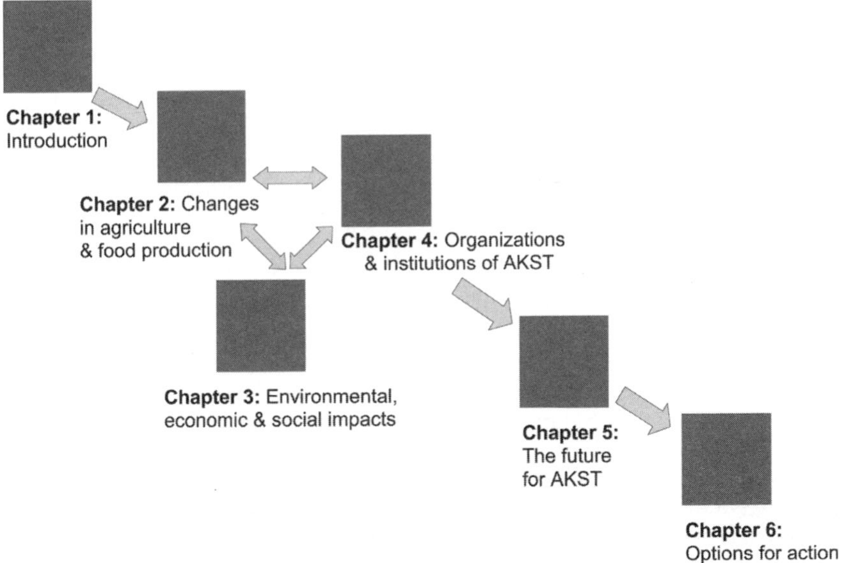

Figure 1-1. *Roadmap for North America and Europe NAE assessment*

require different forms of AKST than sufficed for previous agrifood systems. Choices about future investment in AKST and associated policies will affect who benefits from and who controls agrifood systems and their products, thus affecting who is able to meet development and sustainability goals.

1.1 Scope and structure

1.1.1 Geographic scope of NAE

For the purposes of the IAASTD, the North America and Europe (NAE) region is considered to consist of three subregions. North America comprises the US and Canada; western Europe comprises the 27 countries of the European Union[1] with Iceland, Norway, San Marino and Switzerland; while Eastern Europe is the remaining countries in the Balkans[2], Russia, and its neighboring states Belarus, Georgia, Kazakhstan, Moldova, Uzbekistan and Ukraine. Israel is also included in the region.

1.1.2 Structure of the report

This chapter introduces the NAE assessment and leads to an analysis (Chapter 2) of the changes in agrifood systems that have occurred over the past 50 years resulting from the generation, introduction and application of AKST (Figure 1-1). Chapter 3 examines environmental, economic and social impacts of these changes over the same time period; and Chapter 4 examines changes in the organization and institutions of AKST and their consequences for development and sustainability goals. Chapter 5 introduces forecasting as a method for analyzing the consequences of different options in AKST, congruent with different alternative future developments in the region, and describes major trends affecting agriculture and AKST in NAE. Chapter 6 draws from the lessons summarized in Chapters 2, 3 and 4 about the past generation and application of AKST, and significant trends noted in Chapter 5, to explore options for future investment, policies, education, training and funding.

1.1.2.1 Interface with Global Assessment and other Sub-Global Assessments

The Sub-Global Assessments (including NAE) were developed simultaneously with the Global Assessment by different working teams, which gave limited opportunities to coordinate the findings. However, representatives of the different assessments met together twice and shared their plans. Each Sub-Global Assessment has a slightly different structure to accommodate the particular issues that contributors thought needed the most attention. The Global Assessment is longer and more comprehensive than any of the Sub-Global Assessments, but does not go into detail on individual regions other than to illustrate points via case studies or vignettes. The separate Synthesis Report combines the major points of all of the reports (Global and Sub-Global) and highlights findings from the crosscutting thematic issues.

1.1.2.2 Thematic issues

Each Sub-Global Assessment deals with the past 50 years of AKST in relation to the development and sustainability goals specified in the original mandate to conduct the IAASTD. However, certain themes emerged that deserved special attention because of their importance to meeting the development and sustainability goals, their contentiousness, or the lack of adequate attention to them in previous assessments. These themes are:

[1] Countries of EU27 are: Austria, Belgium, Bulgaria, Cyprus, Czech Republic, Denmark, Estonia, Finland, France, Germany, Greece, Hungary, Ireland, Italy, Latvia, Lithuania, Luxembourg, Malta, Netherlands, Poland, Portugal, Romania, Slovakia, Slovenia, Spain, Sweden, United Kingdom.

[2] Other Balkan countries are Albania, Bosnia and Herzegovina, Croatia, Macedonia, Serbia and Montenegro

- Bioenergy
- Biotechnology
- Climate change
- Human health
- Natural resource management
- Traditional knowledge and community based innovation
- Trade and markets
- Women in agriculture

The Global and Sub-Global assessments contributed experts on these issues to develop key messages and integrate their treatment across the reports.

1.1.3 Conceptual framework

1.1.3.1 Conceptual diagram

AKST is the intersection of knowledge, science and technology (including that developed in other realms) with agricultural systems. It is influenced by and draws from other kinds of knowledge and technology in important ways not confined to food production. For example, advances in transportation and communications technology have been key to the globalization and integration of global value chains.

While the term "AKST" is often used in the IAASTD as if it were a consistent, coherent bloc, there are many different forms and permutations of AKST that have different development histories and impacts. For example, the AKST underlying traditional practices of hunting or grazing on communal lands is vastly different from the AKST leading to the patenting of goats that have been genetically engineered to express human lysozyme in their milk. Therefore, in referring to drivers of AKST or impacts in specific circumstances, it is important to clarify which forms of AKST are under consideration.

The conceptual diagram (Figure 1-2) depicts in a very general way how indirect and direct drivers of change affect development and sustainability goals through AKST. Note that AKST is a subset of science and technology, which is only one of several indirect drivers of the development and sustainability goals.

1.1.3.2 Drivers of change

The direct drivers of AKST highlighted in the conceptual diagram are food demand and consumption, the availability and management of natural resources, land use, climate change, energy and labor. These drivers of change are influenced in turn by a set of indirect drivers, including demographics; economics and international trade; the socio-political context; the broader context of science and technology; education, culture and ethics; and the biogeophysical environment. Chapter 4 examines the complexities of how AKST has interacted with these factors over the past 50 years. AKST is a driver of agrifood system changes, but it is also influenced by these changes.

1.1.3.3 AKST dynamics

Actors and networks are the agents or groups of agents that generate, disseminate, use or control AKST, e.g., public and private agricultural research organizations, universities, public extension services, independent agricultural consultants and other businesses, the International Consultative Group on International Agricultural Research (CGIAR), supply chains and civil society organizations (Figure 1-3).

Processes are the avenues by which AKST are developed, transmitted and used, and avenues that determine

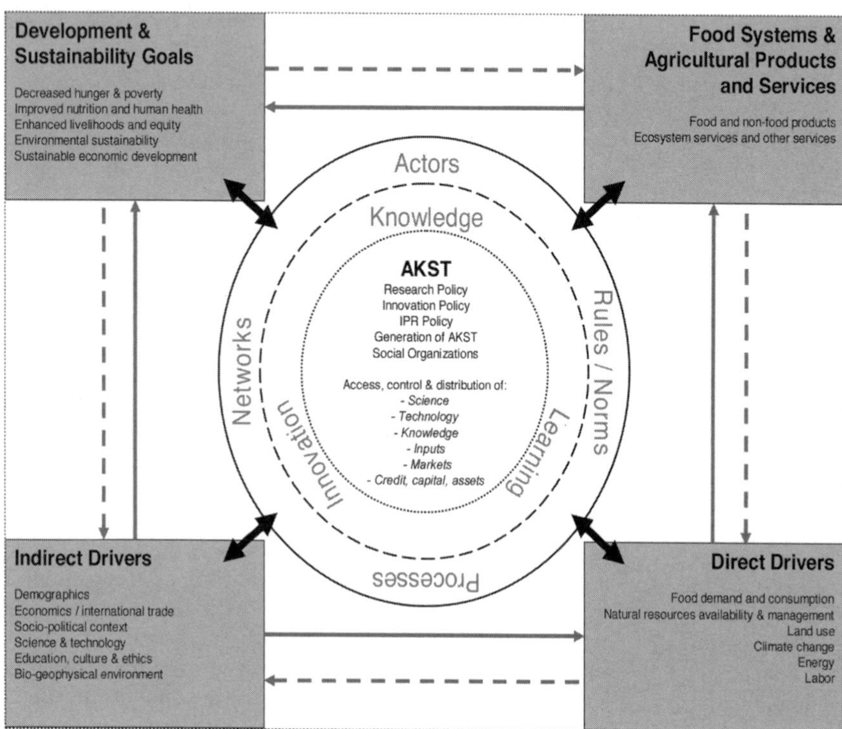

Figure 1-2. *Conceptual diagram of IAASTD*

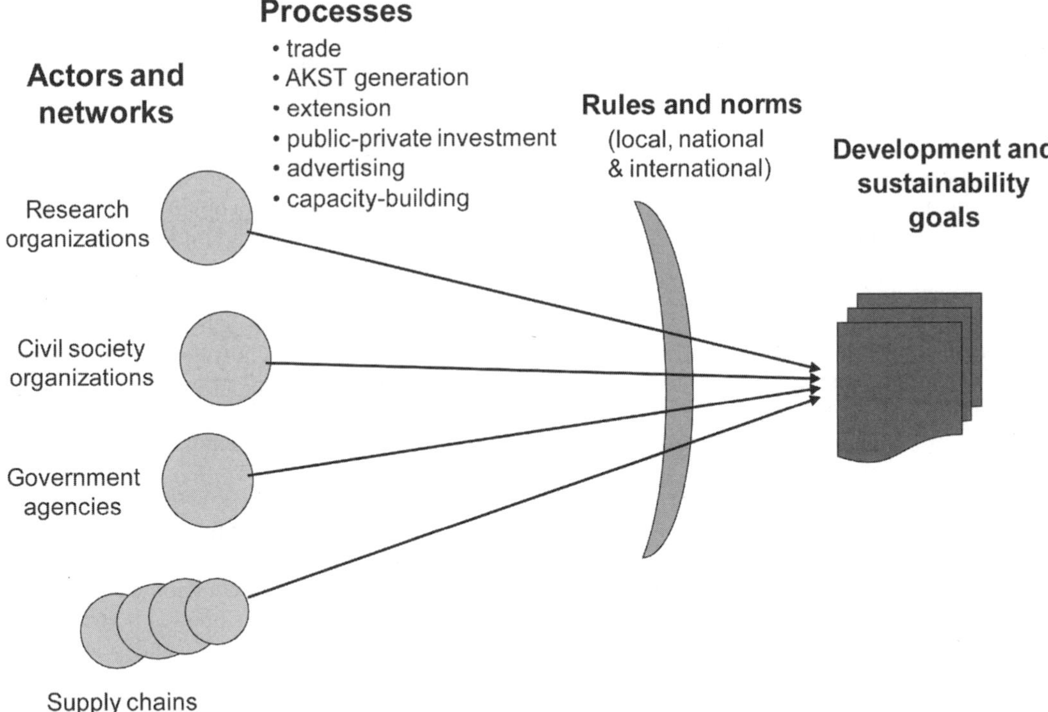

Figure 1-3. *AKST dynamics*

their development and availability, access and use. They include knowledge and technology generation, dissemination and extension, adoption, and evaluation for all sectors in supply chains; trade; public-private investment; advertising; and provision of credit and other financial resources.

Rules and norms are the sociocultural and legal conventions that manage and control AKST processes. They include international agreements and treaties, such as the International Treaty on Plant Genetic Resources for Food and Agriculture and the Convention on Biological Diversity; agreements related to trade, such as trade-related Intellectual Property Rights and World Trade Organization (WTO) settlements; international quality standards such as ISO14000 and the Codex Alimentarius; subsidies; national regulations; tax structures; and local customs.

Many of the actors, rules and norms relevant to the development, distribution and use of AKST were established by World War II Allies in the post-war years. The United Nations (UN) and the World Bank were founded in 1945 to promote economic development and the avoidance of conflict; and the UN agencies Food and Agriculture Organization (FAO), United Nations Children's Fund (UNICEF) and the World Health Organization (WHO) were created shortly afterwards. International agricultural research stations were established during the 1960s, with support from the Rockefeller Foundation, and became the first centers of the new CGIAR in 1971. Discussions among wartime allies of the United States in the late 1940s to create a multilateral agreement for the reciprocal reduction of tariffs on trade in goods led to the General Agreement on Tariffs and Trade negotiation rounds and ultimately the WTO in 1995.

1.2 Agriculture, development and sustainability goals

1.2.1 Eradicating hunger and food insecurity; providing adequate amounts of healthy, safe food; and improving human health

The development and implementation of agricultural knowledge, science and technology have delivered real benefits in food availability worldwide. Few countries in NAE have large numbers of hungry and impoverished people, yet food insecurity is still present in all of them. For example, the United States (US) has the highest Gross Domestic Product of NAE countries; but 10.9% of the US population was food-insecure in 2006, meaning that they lacked access at all times to enough food for an active, healthy life for all household members; and 4% suffered from "very low food security," indicating that the food intake of one or more adults was reduced and their eating patterns were disrupted at times during the year because the household lacked money and other resources for food (Nord et al., 2007). Food insecurity in NAE is more frequently a consequence of poverty and specific government policies that fail to ensure access to available food than due to general lack of AKST, yet such policies are elements of agrifood systems, and can draw upon AKST.

Advances in agricultural productivity have been uneven through NAE, and subsistence farming still predominates in parts of Eastern Europe, with high levels of food insecurity in some countries that have been torn by war or political instability or have been under Soviet influence. The FAO estimates that rates of food insecurity are no more than 6% in countries that have recently acceded to the European Union; but levels are higher in the Balkans and some of the countries in the Commonwealth of Independent States (Skoet and Stamoulis, 2006; Figure 1-4). Food security in Uzbekistan has deteriorated since 1993-1995; it was over 25% in 2001-2003. Food insecurity in Georgia was also high, but improved significantly during the same time period as it emerged from armed conflict (Skoet and Stamoulis, 2006). Comparing food security across NAE is difficult, as different countries use different metrics.

In NAE, policies and investment in AKST have led to frequent surplus production of many crops over the past 50 years, the overconsumption of foods that lead to poor health and rising incidences of obesity and overweight. Of course, many factors contribute to obesity and dietary choices in addition to AKST. However, the development and application of food-processing technology to meet interests other than health; the use of advertising to promote foods of low nutrient value; and policies that determine the availability, price, access and consumption of healthy foods are partly responsible for the rapid rise of obesity and diet-related diseases in the region. Average life expectancy in the US is expected to fall over the next few decades as a result of obesity and associated health problems (Olshansky et al., 2005), reversing for the first time a steady upward trend that has persisted for centuries. The three leading causes of death in the US—heart disease, cancer and stroke—are diet-related; adult-onset diabetes, which is closely correlated with obesity, ranks sixth (NCHS, 2006a). The percentage of overweight adults between 20 and 74 years was 66% in 2001-2004, with 32.1% obese. During the same time period, 17.5% of children between the ages of 2 and 17 were overweight (NCHS, 2006b). For adults and children, the proportion of the population that is overweight and obese is significantly correlated with sex, race and ethnicity (NCHS, 2006b).

Obesity and diet-related diseases are rising throughout the NAE region, although not as rapidly or to the levels currently in the US. For example, the Canadian Community Healthy Survey estimated an obesity rate of 23.1% among Canadians 18 and older in 2004 (Tjepkema, 2006); and 26% of Canadian children between the ages of 2 and 17 were overweight or obese (Shields, 2006). The Regional Office for Europe of the World Health Organization reports that the prevalence of obesity has tripled in many countries in the region since the 1980s and continues to rise, with obesity already responsible for 2-8% of health costs and 10-13% of deaths in different parts of the region.

Outbreaks of salmonella, bovine spongiform encephalitis (BSE) and its human form variant Creutzfeldt-Jakob Disease (vCJD), and foot and mouth disease (even though FMD does not transfer to humans) have raised concerns over disease transmission via food. Up until December 2006 there were 158 deaths attributable to vCJD in the UK, or about 15% of all CJD cases (Andrews, 2007). Other diseases associated with agriculture and the food chain continue to emerge. Avian flu is an example, with a total of 335 cases and 206 deaths from avian influenza A/H5N1 reported by November 2007 (WHO, 2007).

Pesticide poisoning occurs in NAE through ingesting contaminated food, through manufacture and through end use. For example, over 1,000 cases of illness were associated with aldicarb use in watermelons in California in 1985 (Goldman et al., 1990). Although reported agricultural pesticide poisonings decreased from a yearly average of 665 cases (1991-1996) to 475 (1997-2000) in California, the numbers are considered to be conservative because farmworkers face numerous barriers to reporting poisonings (Reeves and Schafer, 2003). Many people in the US carry high body burdens of pesticides, with children bearing the highest levels of some toxic pesticides such as the organophosphate chlorpyrifos (CDC, 2005). There are additional potential health risks to humans through reductions in water quality caused by runoff from fertilizers, slurries and manure that contains fecal coliforms (Mølbak, 2004); and from the conversion of nitrates in water to nitrite in the guts of small children, reacting with their hemoglobin and resulting in the "Blue Baby syndrome."

1.2.2 Reducing extreme poverty, improving livelihoods and creating rural employment

While the application of AKST has contributed to NAE's development and relatively high levels of wealth, it also has contributed to the persistence of poverty and poor prospects for livelihoods based on agriculture. Inequity is rising in some countries of NAE, such as the US; and many people are trapped in pockets of underdevelopment characterized by poverty and non-viable rural livelihoods. Poverty is most intense in eastern Europe: for example, 14% of the population of Uzbekistan lives in extreme poverty, defined as subsisting on less than US$1 per day (Skoet and Stamoulis, 2006).

Many people in production agriculture in North America face poor prospects of sustainable livelihoods when incomes are tied solely to the market; for several decades the majority of producers' income has been derived from off-farm sources and government subsidies rather than crop or product prices. For example, the average proportion of household income from farming activities across all farm types was only 18% in 2005 in the US; the remainder came from off-farm sources. The average government payment was more than half of the average income from farming activities (USDA, 2005). In Canada, farmers' Market Net Income, which subtracts out government payments, fell to negative $10,000 per farm after bottoming at negative $16,000 in 2003. Market Net Incomes dropped close to zero in the 1980s after 40 years of relative stability (NFU-Canada, 2005). While crop prices have gone up recently because of increasing demand for meat and biofuels and poor harvests, the strength and longevity of this trend is uncertain. Production subsidies have allowed farmers to stay in business through periods when product prices were below costs of production, but they are now contested by many actors because of their impacts on total commodity production and on incomes of producers in developing countries. Subsidies in the US tend to support the largest farmers disproportionately because they are based on historical production of commodity crops (MacDonald et al., 2006). Medium-scale and large-scale farms in the US received 78% of the commodity program benefits paid to farmers in 2004 (Hoppe et al., 2007). Production subsidy payments contribute to concentration of farmland among fewer farm operators and

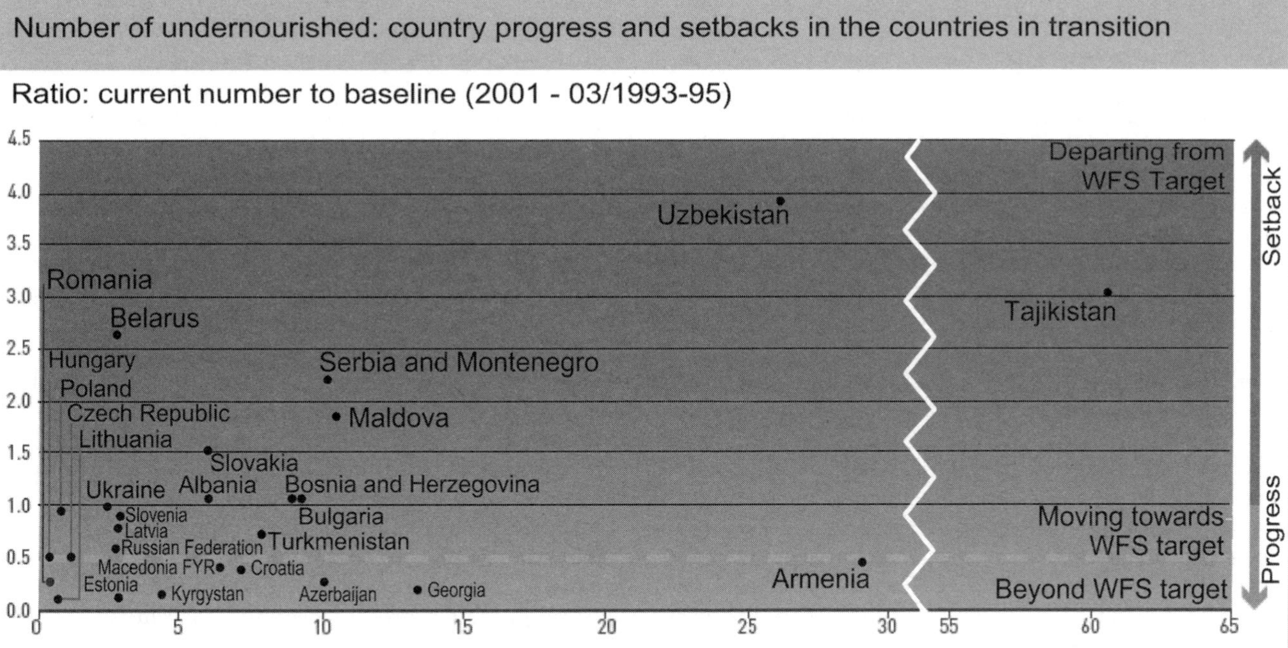

Figure 1-4. *Prevalence of undernourishment in countries in transition, 2001-2003*

raise land prices, thus creating a barrier to beginning farmers. Furthermore, patterns of discrimination in their allocation have prevented minorities from receiving subsidies at the same rate that white male farmers receive them (Oxfam America, 2007).

Wageworkers in agrifood systems (e.g., farmworkers, meat-packers, poultry processing workers, waitresses, cashiers in supermarkets) often have incomes below the official poverty threshold, and may face hazardous or substandard working conditions. Farmworkers in the US are excluded from legislation that guarantees rights such as forming trade unions and bargaining collectively. They suffer from low and stagnant wages, job instability, dangerous and unhealthy working conditions and substandard accommodations (Oxfam America, 2004). Undocumented farmworkers are the most vulnerable of all people in the food system, yet they have little recourse in the current political and legal environment. An estimated 10.3 million undocumented workers are in the US, with about 57% from Mexico (Passel, 2005). Migrants to NAE countries frequently work in low-wage agrifood system jobs, but their livelihoods are precarious because of their legal status.

In contrast, people involved in management tiers of agribusiness have seen dramatic rises in wealth and power because of policies that created environments conducive to the success of larger businesses reliant on AKST. Agricultural resources and most stages of input production and commodity processing, distribution and retail have become concentrated into fewer, much larger enterprises at an accelerating pace over the past century. Five of the inheritors of the Wal-Mart fortune were among the top thirty wealthiest people in the world in 2006 (Kroll and Fass, 2007). Wal-Mart's Supercenters sell more groceries than any other retailer in the United States, and Wal-Mart has moved rapidly into other countries in NAE and other regions.

1.2.3 Promoting equity across gender and social gaps

The proportion of women and minorities that are farmers and work in other sectors of agrifood systems varies considerably across NAE. Women and minorities constitute small proportions of the farm population in most North American and western European countries, but the numbers of women operators have been increasing. For example, in Canada, 27.8% of farm operators were women in 2006 (Statistics Canada, 2006). In the USA, women are 9% of farm operators but 16% of operators of limited resource farms. Minorities account for 5% of all principal operators and a similar percentage for each farm type except for limited-resource farms, of which about 12% have a minority operator (Hoppe et al., 2007).

In contrast, women are a large proportion of the population engaged in agricultural production in some of the countries newly acceded to the EU, eastern Europe, and transition countries in the Balkans and the Commonwealth of Independent States. Women tend to be employed at the same or even slightly higher rates than men in agriculture in these countries: 42% of employed women in Moldava were in this sector in 2003 (compared with 44% for men) and 37% in Romania (compared with 35% for men). Women also are more likely to be engaged in subsistence agriculture and agricultural jobs in the informal sector, as their employment rates in the formal sector dropped markedly after the early 1990s (Jacobs, 2006). Women constitute a small minority of migrant farmworkers, but they are particularly vulnerable to exploitation in that occupation.

AKST needs of women and minorities are often different from others working in agrifood systems. Women and minorities tend to have smaller farms in the US and Canada, although farms operated by Hispanics in the US are larger on average because more are extensive livestock operations (Effland et al., 1998). Women and minorities also tend to raise a different selection of crops and livestock than white male operators. They may face special barriers in attempting to access information and technology because of language, culture or discrimination; and they may be at greater risk of losing their farms (Effland et al., 1998; FAO, 1996; Oxfam America, 2007).

1.2.4 Enhancing environmental quality

The application of AKST in NAE has led to habitat transformation, loss of biodiversity, declining quantities of fresh water and increasing competition for what remains, degradation of the quality of groundwater and surface water, and impacts on soil quality. Transportation of agricultural products contributes to greenhouse gas emission and poor air quality due to particulates. AKST can also improve environmental quality, through practices such as no-till planting, crop rotation and sustainable management of cultural landscapes.

The consequences to agricultural production of over-exploiting natural resources are seen most vividly in the abrupt decline of marine fishery stocks (Pauly and Alder, 2005). Perhaps the most dramatic illustration of contamination is the hypoxic zone extending into oceans from the mouth of all major rivers in industrialized countries, caused by runoff of unused nitrogen applied as fertilizer to agricultural systems (Schlesinger et al., 2006). The full environmental costs associated with substantial gains in human well-being and economic development are only now becoming apparent (Tegtmeier and Duffy, 2004; MA, 2005ab; Sumelius et al., 2005; Foster et al., 2006).

The development and sustainability goals are interlinked, and ways that they intersect with agriculture and AKST are complex (Chapters 3 and 4). Awareness of environmental costs and their implications for future generations has given strength to demands for a more multifunctional agriculture, promoted through policy incentives and supporting the production of ecosystem goods and services beyond provisioning food and feed, water, fuel, fiber and forest products. Policy supporting multifunctional agricultural systems would compensate producers for maintaining supporting ecosystem services such as nutrient cycling and soil formation; cultural services such as aesthetic, spiritual and educational value; and regulating services such as climate and flood regulation and water purification (Aldington, 1998; OECD, 2001; Boody et al., 2005). The recent rise in food prices in NAE, triggered in part by the diversion of land from food and feed grains into ethanol production, adds urgency to finding the right approach to agriculture that is both sustainable and meets human needs (e.g., Cloud, 2007; Howden, 2007). There is an emerging agenda for AKST that fosters economically, environmentally and socially sustainable farming and food systems; public benefits via the food value chain; and equity between producers within the region and with those elsewhere in the world. Yet many tensions remain to be resolved in implementing this "new agenda" and ensuring that people most in need—within NAE and globally—are among the beneficiaries.

Across the region, agriculture is in flux as regulators seek to limit or reverse environmental damage caused by agriculture, migration and other demographic shifts change the complexion of rural areas, and consumers and citizens become more concerned about diet-related health issues and social externalities of agriculture. Increasingly differentiated markets are responding to new consumer desires in what has been denoted as a "quality turn" (reviewed in Wilkinson, 2006). Markets are opening up in NAE for products that promote social and environmental quality, with labels such as "Fair Trade Certified", "organic" and "dolphin-safe." The consequences of greenhouse gas emissions on global climate change have led to new attention on "food miles," or the distance that food travels from point of production to point of consumption, and interest in consuming foods producing locally. Enthusiasm for local foods is also fed by the desire to preserve unique foodways, cultures and landscapes associated with agricultural production; this has resulted in the defense of geographic indicators to demarcate foods' point of origin.

1.3 Significance of NAE in the Generation, Use and Control of AKST

1.3.1 Importance within the region

1.3.1.1 Impacts on development and sustainability goals

In most NAE countries, AKST is less relevant than public policy in other realms to meeting development and sustainability goals of reducing hunger and poverty, improving nutrition and human health, enhancing livelihoods and equity, fostering environmental sustainability and sustaining economic development. That is, AKST is not the main limiting factor in achieving these goals, although it is increasingly important to deal with emerging issues, such as building resilience and adaptive capacity to handle the consequences of global climate change, learning how to restore degraded ecosystem services and coping with new foodborne, crop and livestock diseases. In addition, uneven access to AKST in countries in transition within the NAE region constrains productivity and has serious effects on abilities to meet development and sustainability goals.

1.3.1.2 Economic

Many NAE countries and businesses have made substantial investments in AKST, which have resulted in economic gains for actors in those countries in addition to benefits in other regions. The past few decades of application of AKST resulted in the consolidation of global value chains that control the supply of most agricultural products, and have had tremendous effects on the distribution of wealth in society and prospects for making a living through agricultural production. In general, only producers with very large-scale operations are able to support a household by full-time commodity farming; other commodity producers are reliant on off-farm income because of low and unstable commodity prices. This problem is aggravated in the US by lack of universal access to health care, so that an off-farm job with associated affordable health benefits is often necessary to provide social security for a farm household. On the other hand, immense wealth is sequestered among the small number of shareholders with expansive holdings, owners, and executives of transnational corporations involved in food supply (TNCs). The application of AKST has enabled the growth of these companies, sometimes through research partially subsidized in public universities and laboratories and other forms of public support, such as the development of irrigation and water delivery systems, roads and railroads.

Analyzing the economic impacts of AKST and impacts on distribution of wealth in NAE are important in order to understand ways to meet the development and sustainability goals of more sustainable livelihoods and greater economic security for producers and wage-workers in agrifood systems, and to increase rural employment. While enhancing productivity, the application of AKST in globalized agrifood systems has increased the vulnerability of many livelihoods dependent on agriculture because production sites have become more specialized and more sensitive to sudden changes in the market. Farms usually are linked with processing, distribution and marketing enterprises in value chains that can have global reach. Therefore, producers have less control over prices and the timing and circumstances of sales; they are more likely to be price-takers than price-setters; and they are in competition with other producers globally rather than in a local or regional market.

1.3.1.3 Sociocultural

The application of AKST within NAE is associated with changing diets and health, and the disconnection of most people from food production. While AKST has improved the availability and access of many foods in NAE and eased hunger, diet-related health problems caused by excessive consumption of processed foods low in nutrient value and lack of physical activity are on the rise.

The loss of traditional knowledge is apparent in many parts of NAE. At present, indigenous populations are small and often concentrated in lands that are marginal for agriculture. In North America, 90-95% of the indigenous population died in wars or through exposure to diseases introduced by European settlers; so indigenous models of agriculture and resource use are scarce and can only be understood through laborious archaeological investigation. Traditional knowledge such as the *acequia* systems of irrigation in the southwestern US were part of agricultural systems sometimes maintained for centuries and could be valuable in the future. Assessing the value of traditional knowledge and technology can help in efforts to conserve it.

1.3.1.4 Environmental

The impacts on AKST on ecosystem goods and services and social factors in NAE may preview some of the unintended consequences of the application of contemporary AKST elsewhere in the world, and also how they can be managed. Examples of unintended environmental consequences include the vulnerability to disease that often accompanies widespread monocultures; soil erosion and fertility declines from deforestation and inappropriate methods of soil disturbance; changes in biodiversity and flood risk from draining wetlands; aquifer depletion and land subsidence from overpumping aquifers; and effects of synthetic agricultural chemicals on water quality, biodiversity, and related ecosystem services. It is possible to draw tentative conclusions about the attributes of AKST that are most likely to enhance the resilience and sustainability of global agroecosystems and ecosystem services, based on experiences in NAE.

1.3.2 Importance to the rest of the world

1.3.2.1 Impacts on development and sustainability goals

Countries and companies based in NAE control resources that are crucial for achieving development and sustainability goals, such as uncultivated arable land, the world's most extensive *ex situ* gene banks, money, scientific infrastructure and human capital. Therefore, development and sustainability goals in other countries can be met more easily with investment and material assistance from NAE. Enterprises based in NAE and governmental agencies in its countries control technology now that could make a real difference in poor regions, were it affordable to the people who most need it. However, the assistance needed most desperately may be in building capacity to educate poor regions' farmers, food system employees, teachers and researchers so that they can generate their own site-specific agricultural knowledge and technology. Resources for building such capacity are most likely to come in part from NAE, as the source of greatest wealth and AKST assets at present, although some developing countries are catching up very quickly.

Global climate change is predicted to cause less severe environmental disruption in NAE than in developing regions, even though NAE countries bear the most responsibility for the accumulation of greenhouse gases. NAE may be an essential source of emergency food assistance and resources for restoration of productive capacity in other regions following severe storms, heat waves, floods and droughts expected to result from global climate change.

Some of NAE's current policies and patterns of trading with developing countries diminish their ability to feed their own people by dumping food at below the cost of production, thereby undercutting prices of farmers in developing countries, and delivering food aid that cuts out local and regional farmers. These policies have led to demands by producers and consumers within NAE and other regions for food sovereignty, the right of peoples and sovereign states to control their own agricultural and food policies.

1.3.2.2 Export of AKST, other forms of KST and concepts of development

NAE countries produce much of the knowledge and technology used outside the region, as well as within. International development agencies, financial institutions and TNCs have exported many elements of the agricultural systems developed in NAE into developing countries through extension services, other types of training, demands to adopt certain kinds of agriculture as part of structural adjustments on which loans are conditioned, market incentives and deals set up between corporations and the governments of developing countries. This is a push/pull flow of knowledge and technology because developing countries often are eager for access to the factors that have helped NAE become a dominant force in global production and marketing of agricultural goods and services.

Although NAE no longer controls most of the developing world as colonies, it still holds sufficient political and economic power to influence the internal affairs of developing countries, including choices of development paths. Along with tangible exports and discrete clusters of AKST, NAE has the influence to ensure that concepts and ideas about economic development held by powerful entities in the region are adopted in other regions. These concepts have been especially influential when they infuse mandatory structural adjustment plans or poverty reduction plans, or when granting a loan or other aid is contingent on adopting them.

1.3.2.3 NAE's footprint

The application of AKST in NAE has expanded its ecological footprint and social impacts in other regions. This is largely due to NAE's volume and variety of exports and imports, and the many actors and networks within NAE that dominate agrifood chains. The globalized agricultural system, in which importers source raw products from the cheapest source and seek to sell processed products where they garner the highest price possible, touches every country in the world. Developing countries frequently supply the genetic resources; unprocessed food, feed and fiber; and labor to process commodities into goods that can be sold at higher prices, yet they do not garner the profits gained from adding value. Demand for cash crops, including biofuel, has trumped land use for subsistence farming from Brazil to Indonesia; demand for cheap labor for farm work or food processing results in workers finding it increasingly difficult to earn livelihoods from agriculture.

Nations do not simply use the natural resources within their own borders. For example, a country may appear to be self sufficient in water if water is not imported directly, but large quantities of "virtual water" may be imported via water used to grow, manufacture and transport agricultural and industrial produce. The US has the largest water footprint in the world, with 2480 m^3 yr^{-1} per capita, followed by the people in southern European countries such as Greece, Italy and Spain (2300-2400 m^3 yr^{-1} per capita) (Hoekstra and Chapagain, 2007). Equally, by importing food, countries import primary production also, denying it from local ecosystems. Most countries in NAE have footprints beyond the capacity of their own territories to support, except for the arctic countries of Canada, Russia and Sweden, and Romania and Belarus (Global Footprint Network, 2006).

NAE has a "consumption footprint" as well as a footprint connected with production. "Western" diets high in fats, salt, sugar and processed foods have spread rapidly into developing countries around the world. Their brandnames can be seen in the smallest and remotest villages. Advertising has helped to create demand for dietary changes, such as more processed food and bottled water, which has large impacts on material use and waste disposal. Dietary changes in the proportions of beef and other large animals consumed have dramatic effects on land use worldwide and the amount of land required to feed a population.

1.3.2.4 Wealth and political power

Countries in NAE have unique characteristics stemming from their histories and assets that make them critical to meeting development and sustainability goals. NAE countries and corporations have disproportionate power in AKST (and in science and technology more generally) compared with other world regions. The NAE region contains the wealthiest nations in the world and many of the countries with the steadiest and most sustained growth in per capita income since 1945. Much of this wealth has accrued through extracting resources and the profits of labor from other regions.

Agricultural knowledge and technology are controlled in unusual ways in NAE, compared to other regions of the world: the private sector plays a dominant role, especially in North America. The ratio of public to private investment in agricultural research has dropped steadily since 1980. As late as 1940, agricultural research represented 40% of all federal research funding, but national security concerns became preeminent with World War II (Fuglie et al., 1996). Investment in private agricultural research has grown more rapidly than public investment in research and has exceeded funds for public research since 1980 (Meeks, 2006). NAE countries are the point of origin of most TNCs that now dominate globalized food systems. Six of the top ten pesticide companies by total sales, eight of the top ten seed companies, all of the top ten global food retailers and all of the top ten beverage and food processing corporations are based in the United States or Western Europe (ETC Group 2005). NAE countries are also the source of development of most genetically modified organisms, and companies based in NAE hold more than half of the Intellectual Property Rights (IPR) relevant to agriculture.

Institutional and organizational shifts in NAE agriculture may presage similar shifts in developing countries. As the influence of traditional agricultural interest groups and state governments wanes, the power of organizations dominated by the private sector and civil society is waxing. These power shifts are related to the application of AKST: greater wealth and power allow greater access to AKST and sometimes greater capacity to create AKST to serve one's own needs. For example, knowledge of global market trends and prices is critical to the success of major agribusinesses involved in trade, and they can access this information more readily than a small-scale producer. While part of their advantage is due to economies of scale, policies that favor large businesses are influential as well. Power shifts in agriculture in North America are mirrored by rising inequity of wealth and assets.

1.4 Description of the Region

1.4.1 Social, political and economic development

1.4.1.1 Prior to 1945

It is thought that people first entered the NAE region via southwest Asia, spreading northwest into Europe and east into Asia some 40,000 years ago, and into North America across the Bering Strait land bridge between 15,000 and 9,000 years ago (Dixon, 2001). While waves of nomadic migration, conquest and trade resulted in intermittent interactions among the Eurasian cultures, sea level rise cut all but the most tenuous links with North America until European exploration and colonialism began in the 15th century. The subsequent migrations into North America were essentially economic, involving well over 12 million people from Europe (Gibson and Lennon, 1999) and an estimated 500,000 slaves imported from Africa (US Census Bureau, 2002). As the United States grew in area and economic power, the indigenous peoples were greatly reduced in numbers and largely displaced by the end of the 19th century. Indigenous food systems based largely, but not entirely, on hunter-gathering were replaced by arable agriculture and extensive grazing. Simultaneously, Russia expanded its political control from eastern Europe to the whole of mainland north Asia and Alaska (sold to the US in 1867). At the turn of the century, the global economy of the region was politically and economically dominated by NAE major powers linked by trade and diplomacy. However, the Russian Revolution, the economic and agricultural depressions of the 1920s and 1930s and World War II polarized the region into the largely communist eastern Europe and USSR and the largely democratic and capitalist western Europe and North America. This polarization, along with variable access to all forms of capital (human, social, financial, physical and natural) drove widely varying attitudes about the importance of agricultural development, agrifood self-sufficiency, trade, subsidies for multifunctional agriculture and AKST in the different subregions of NAE in the following decades.

1.4.1.2 After 1945

Post-war economic recovery in western Europe was rapid, despite the loss of cheap raw materials and captive markets as Britain, the Netherlands, France, Belgium and Portugal decolonized from around the world. Comprehensive welfare systems were developed, drawing on lessons from the depression in the interwar years. Belgium, France, Italy, Luxembourg, the Netherlands and West Germany established the European Economic Community in 1958. Denmark, Ireland and the UK joined in 1973 (a referendum in Norway rejected membership), with Greenland withdrawing in 1985. Greece, Spain and Portugal joined in the 1980s, by which time they were governed by democracies. Further expansion took place in 1994, leaving only the neutral Switzerland, Norway and Iceland as major western European countries outside what had become the European Union (EU). The early focus of the European Economic Community was on the Common Agricultural Policy (CAP) and common policies for coal and steel. Over time, a much wider range of common policies was developed, addressing domains including culture, consumer affairs, competition, the environment, energy, transport and trade.

After liberation from Nazi occupation, countries in central and eastern Europe found themselves strongly influenced politically and economically by the Soviet Union. In response to the establishment of NATO in 1949, the Soviet Union and its allies set up the Warsaw Pact in 1955. Creation of the Berlin Wall marked the final division of the communist East (including Czechoslovakia, Hungary, Bulgaria and Romania) and capitalist West by the "iron curtain." Despite also being communist, Yugoslavia was never part of the Eastern Bloc, and Albania broke away in the 1960s, aligning instead with China. In eastern Europe, private enterprises were mostly taken over by the state, as was agricultural and forest land, except in Poland and Yugoslavia. The Communist Party controlled production through rigid Five-Year Plans for required outputs by sector and by commodity. These were not uniform, nor did they provide consistent benefits. Hungary introduced limited market mechanisms, and relaxed controls on compulsory deliveries and land ownership. By the mid-1960s Hungary was relatively prosperous, as was Yugoslavia, where private ownership of land and enterprises was maintained along with freedom of international trade and travel. Elsewhere, Five-Year Plans gave the illusion of continuing quantitative success even when growth rates slowed and targets failed to be met.

By the 1980s it was obvious that the Soviet Union was lagging economically behind the West. Uncontrolled military spending (consuming over 30% of the Soviet GDP) and diminishing domestic economic returns could not be maintained politically or economically (Davies, 1996). Communist regimes started to lose power; in 1989 the Berlin Wall came down and in 1990 East Germany committed to unity with the West, while the communist federal government of Yugoslavia gave way to largely nationalist democracies in the constituent republics. A wave of establishment of independent states followed: Czech Republic, Slovakia, Serbia, Slovenia, Croatia, Bosnia-Herzegovina, Macedonia. The Soviet Union was dissolved in 1991. While most of these transformations were peaceful, many thousands of Bosnians, Croats, Serbs and Albanians were killed during the wars of 1991-2001, and disputes continue in the Caucasus. Overall, NAE has enjoyed relative peace and stability over the last half-century, compared with other sub-global regions.

The basic choice facing post-Communist governments was either to attempt a quick transformation from subsidized socialist economies into market-driven capitalism or to proceed cautiously, disposing of problematic sectors of the economy while preserving for as long as possible cheap rents, guaranteed jobs and free social services. Poland, Hungary and the Czech Republic already enjoyed relatively high standards of living (with average monthly wages approaching $400 in the late 1990s) and familiarity with western lifestyles. These countries adopted the first approach and were among the ten countries joining the EU in 2004; Bulgaria and Romania followed in 2007. By contrast, Ukraine (with monthly wages around $80 in the late 1980s) was reluctant to liberalize domestic markets or reduce the state's share in the economy and delayed change (Judt, 2005). Further east, attempts to reform the inefficient and militarized economies of the former USSR caused sharp rises in unemployment and destitution.

Foreign investment into Canada doubled during 1945-55, and discoveries of oil, gas, iron ore and other raw materials helped to expand industrial production (Sautter, 2000). The US economy was much larger, producing half of the world's goods by the 1950s. Standards of living soared; and new, consumer-based lifestyles evolved, increasingly reliant on cars. By contrast, 39.5 million people in the US lived below the poverty line, with African Americans, American Indians and farming households particularly affected. Racial segregation led to the civil rights movement in the 1950s and 60s, to be followed by campaigns promoting peace, women's rights and the environment. Their development was closely associated with pop and rock music that proclaimed a radical, English-language culture across the radio waves of much of NAE (Jones, 2005).

During the period 1913-1998 but primarily prior to 1950, populations in western Europe increased 1.5-fold and those in eastern Europe and USSR increased 1.7-fold, while the US population increased 2.8-fold, much nearer the global average of around 3.3-fold (Maddison, 2001). NAE populations have become older: the median age in the US is now 36 years, although it was 28 as recently as 1970. In Europe, 1.6% of the population was aged 80 or over in 1970, and today the figure is 3.5% (UN Population Division, 2005b). Life expectancy is 80 years for Canada, 77 for the US and 79 in western Europe, but only 65 in Russia and 66 in Ukraine (UN Population Division, 2005b). In the latter countries, there are now twice as many deaths as births; health problems include alcohol, smoking, tuberculosis and AIDS/HIV (Meier, 2006). Populations in Russia and eastern Europe are forecast to decline by over 20% by 2050; those in western Europe and North America are more likely to increase slightly (UN Population Division, 2005a). This disparity is reflected in the great variation in wealth across the region: the gross national incomes (GNI) per capita of Luxembourg, Norway and Switzerland exceed $50,000, while several countries in eastern Europe have GNI values of less than $5,000 (World Bank, 2006).

Israel was created in 1948 as a Jewish homeland in part of what had been known as Palestine, bringing people together from across many areas of the rest of NAE. Relations with the rest of Palestine and Arab states in the region have dominated Israel's politics to this date. Life expectancy at birth is nearly 80 years, with GNI per capita of $18,000 (World Bank, 2006).

The extent of urbanization varies greatly across the region. It is highest in the densely populated countries of northwest Europe, reaching over 90% in Belgium. In the US, 60% of the population now live in metropolitan areas of at least one million people, and citizens move on average ten times during their lives. Demographic change and suburbanization have been similar in Canada, where there has also been a migration westward, especially to the oil-rich state of Alberta. Urbanization is least in eastern Europe (e.g., less than 50% of the populations of Albania and Moldova), where the differences in wellbeing between urban and rural people may be the greatest. Thus in Moldava, a country where 48% still work in agriculture, half of the population earned just $19 per month in 2000 (Judt, 2005). Not surprisingly, many people in rural areas are seeking employment elsewhere, especially in western Europe, resulting in depopulation and land abandonment. It is estimated that two million Polish citizens (out of a total population of 39M) have left the country since accession to the EU, while an estimated seven million people have left Ukraine to find work since the fall of the Soviet Union (Meier, 2006). Many of these people are employed in the food and agricultural sectors; this is also true of the 14 million economic migrants to the US since 1990, mostly from Mexico and Asia (US Census Bureau, 2007).

The political and economic situation of indigenous peoples in North America has changed greatly in recent decades. In Canada, the Canadian Constitution was amended to protect aboriginal rights and the Northwest Territories was partitioned to form Nunavat, a self-governing homeland of two million square kilometers for the Inuit. Canadian aboriginal peoples account for around one million of the total population of 32 million, and they are a young and increasingly urbanized population (Statistics Canada, 2006). The proportion of reported American Indian and Alaskan Natives in the US is smaller, at around one percent of the population, and also with a lower than average median age (US Census Bureau, 2007). The purchase of the Hard Rock Café chain by the Seminole and establishment of casinos on native lands indicate the increasing wealth and power of at least some of the tribes, yet sharp disparities persist between US American Indian and white populations in most indicators of health and well-being. There are far fewer native peoples in Eurasia, with around 30,000 native speakers of the Samoyedic and other languages dispersed across northern Scandinavia and Siberia and into the Aleutian Islands.

Literacy rates are high across the region, and funding for education is at least 3% of GDP in every country. The number of women studying for a university degree has dramatically increased since 1945. In both eastern and western Europe, the proportion of women students ranges between 45 and 62%, with almost twice as many taking humanities and arts than science, mathematics or computing. Women now account for over 40% of non-agricultural jobs across most of the region, but in all EU countries women earn less on average than men. The gender pay gap ranges from less than 10% in Portugal, Belgium and Italy to 22-25% in the UK, US and Germany; in the US white women earn 76% of the wages of white men for comparable work. In most EU countries women spend about twice as much time on domestic work as men, although the ratio is considerably smaller in Sweden and Finland and much larger in Italy and Spain (EUROSTAT, 2007b).

1.4.2 Natural resources and their exploitation

Taken as a whole, the region is well endowed with land, with temperate climates and soil conditions suitable for farming and forestry. The NAE region is circumpolar, bounded in the south by mountains, deserts and the Gulf of Mexico, Mediterranean and Black Seas. The climates, and hence conditions for agriculture, are determined largely by latitude, altitudes and proximity to prevailing winds from the oceans. North-south gradients range from polar to desert: Russia is the coldest populated country in the world, with a mid-annual temperature of -5.5°C, and more than half of the country currently covered with permafrost. By contrast, California experiences the hottest temperatures recorded on the planet. Precipitation is governed more by east-west gradients. In Eurasia, the climate is milder in the northwest, which is warmed by the Gulf Stream that also carries rain from the Atlantic. Farther east, the climate becomes drier and more continental, with greater variation between winter and summer. The equivalent gradients in North America are east to west, with precipitation decreasing until deserts are reached in the southwest. Wet, warm winters and hot, dry summers characterize the climates of the Mediterranean and California.

1.4.2.1 Fresh water

The freshwater resources of the region are distributed unevenly across the region, both in terms of geography and per capita. In 1995, the region consumed around 300 km^3 of water, out of a global total of 1,800 km^3; nearly two-thirds of this was used for irrigation (Rosegrant et al., 2002). The rivers running from the region's mountains determine the water supplies to the lowlands and the potential for hydropower; the reliability of these resources is at risk because of changing climates. Water supplies in western Europe are most under pressure in Germany, France and the Mediterranean (including Israel) because of the low rainfall, high irrigation demand and high populations. Many states in Eastern Europe also have water use rates of over 20%. The Russian utilization rate is low (2%), but hides great inequalities. This is why it has been proposed to divert the Volga, Ob and Irtysh rivers to provide more water to Central Asia, with very uncertain environmental consequences. Increasing competition for water exists in the arid western sections of the US, not only to meet agricultural and hydropower needs, but also for drinking water in growing urban areas, Native American water rights, industry, recreation and natural ecosystems. As a result, many aquifers are losing water at rates far higher than recharge rates. In Canada, water consumption per capita is high by international standards (1420 m^3 per capita in 1996); but total consumption is only 2% of the available renewable supplies. The electricity sector consumed 64%, the manufacturing sector 14% and the primary-resource sector 11% (mostly for agriculture) (Gunton et al., 2005).

1.4.2.2 Energy

For much of the region's history, the major energy sources were wood and charcoal until replaced by coal. As agriculture intensified, it became increasingly reliant on fossil fuels for the production of fertilizers, the transport of materials and the processing and transport of the final product. A recent study in Sweden showed that a meal of beef, rice, tomatoes and wine required inputs of 19.0 MJ, compared with the dietary energy of a mere 2.5 MJ (Carlsson-Kanyama et al., 2003). Agriculture and forestry are increasingly seen as sources for renewable energy, in the forms of biomass, biofuels and biogas. Very large increases in production are anticipated, driven by changing policies across the region. The EU is now committed to replacing 5.75% of all transport fuels with biofuels by 2011 (EU directive 2003/30/EC); and US biodiesel production capacity is expected to increase to 9.5×10^9 liters per year by the end of 2008, from 6.5 million liters in 2000 (National Biodiesel Board, 2007). The diversion of large areas from food to biofuel production will have uncertain but very large consequences for agrifood systems, especially as land and water availability are simultaneously reduced through climate change, sea level rise and increased urbanization. These consequences are already becoming apparent in some countries as the price of bread and other staples has risen more rapidly than the rate of inflation.

1.4.2.3 Fisheries

The NAE region borders the largest marine fishery, the northwest Pacific (21.6 million tonnes in 2004), and the fourth largest, the northeast Atlantic (10 million tonnes). Most of the marine fisheries are fully or over-exploited. Catches have declined in the northern Pacific, but not as precipitously as in the northwest Atlantic, where catches are now around two million tonnes yr^{-1}, around half the levels in the early 1970s. Five species of fish caught here are now considered to be critically endangered (Devine et al., 2006). To prevent further erosion of the resource base and ensure sustainable development, Fisheries and Oceans Canada is working with a range of stakeholders to develop and implement integrated ocean management plans as part of the 1997 Oceans Act (Quigley and Harper, 2006; Rutherford et al., 2005). Reporting of inland catch fisheries is much less precise, but it seems that Europe and North America account for only around 6% of global catch, with dramatic declines in Europe. Aquaculture is increasing, but at very low levels compared with Asia (FAO, 2007).

1.4.2.4 Marginal lands

The areas north of the tree line constitute the arctic and tundra. Hunter-gatherers have long exploited this biome, and were at least partly responsible for the extinction of the megafauna of the region. Indigenous peoples still continue traditional practices; but population densities are very low, and impacts on natural populations tightly regulated. Thus while Nunavat relies heavily on hunting for its economy, it has a total population of less than 30,000 (Statistics Canada, 2006). The Sami people of northern Scandinavia herded reindeer, but this nomadic lifestyle has only been practiced by small numbers in recent centuries and has virtually ceased. It is continued by some of the Nenets people farther east. The natural resources of the area (fossil fuels, minerals and marine fisheries) are exploited more by external peoples. Climate change is already influencing this biome: polar icecaps are shrinking, glaciers retreating and permafrost beginning to thaw, releasing methane to the atmosphere, changing hydrology and transforming the region from a sink of greenhouse gases to a source (ACIA, 2005).

In the region's mountain chains, fishing and hunter-gathering is often dominated by tourists. Herders have traditionally exploited the uplands during summers, bringing cattle, sheep and horses down to lower elevations during the winter. This practice of transhumance influenced culture and biodiversity, creating and then maintaining very ecologically diverse landscapes of meadows interspersing forests. Transhumance continues on public and private lands in the US West. The conservation of transhumance in Europe is now a matter of choice more than economic necessity. The EC seeks to retain such landscapes through regional development and agri-environmental policies; but in many areas, meadows are giving way to forest as rural areas become depopulated and land is abandoned.

1.4.2.5 Forests

South of the tree line is a belt of coniferous forest, extending across Canada, Scandinavia and Siberia. These boreal and taiga forests are very extensive, accounting for much of the estimated 1.6×10^9 ha found in NAE, 40% of the world total (FAO, 2006). Russia has the largest area of forest of any country, at 809×10^6 ha, nearly twice as much as Brazil, with a vast proportion found in Siberia. Canada and the US hold the third and fourth largest areas of forest

(310 and 303 x 10^6 ha respectively). The forests continue to support indigenous cultures, including 80% of indigenous Canadians and 26 distinct peoples in Siberia (Taiga Rescue Network, 2007), along with populations of large mammals including moose, caribou and the extremely rare Siberian tiger. Further south of the coniferous forests, the climate is milder, suitable for a natural vegetation of broadleaved woodland where rainfall is high enough; scrub, grassland and desert elsewhere. In the absence of people, much of Europe west of the Black Sea and much of the US east of the Great Plains would have been forested. There remains very little European forest in its primeval state, most having been cleared or transformed by management by the 1500s. Deforestation took place much later in America. Most of the US east of the Mississippi was covered by virgin forest in 1650, with large tracts remaining 200 years later; now only fragments survive. Across NAE, most remaining forests have been transformed by management aimed at increasing productivity of timber. Conflicts between commercial, environmental and indigenous interests have sharpened in recent decades and in some areas forests are now managed to provide multiple functions, including leisure, fuel and provision of forest foods (FAO, 2006). Rates of wood removal have been more or less constant during 1990-2005 (FAO, 2006). The industry now provides livelihoods for three million people in Europe.

1.4.3 Agrifood systems

1.4.3.1 The development of agrifood systems to 1945

Eurasian agriculture began in southwest Asia around 9,000 BC with the deliberate cultivation of emmer and einkorn wheat. Crops were typically small-seeded (e.g., wheat, lentils), grown with the use of plows. Most farm animals were domesticated in central and southwest Asia, except for the horse in the Caucasus and the pig in China (Solbrig and Solbrig, 1994). As arable agriculture spread north and west across Europe, forests and scrub were cleared by felling or fire to make way for complex farming and forestry systems to provide food, fiber, fuel and other products. Crop rotation systems were developed to manage crop nutrition and diseases. Terracing, irrigation, drainage and flood plain management were used to manage water availability, while woodland edges were retained as hedgerows and lines of trees to provide barriers to livestock, animal shelter and additional food resources. The resulting mosaics of woodland, crops, grasslands and heaths created landscapes now highly valued for their biodiversity, cultural heritage and beauty. Farther east, nomadic societies developed that herded domesticated animals for meat, milk, hides and transport.

Agriculture developed independently in the Americas, with cropping of maize, squash and beans, sown with the help of a hoe and digging stick; plows were unknown until the arrival of the Spaniards. Cropping supplemented hunting and gathering for a population of around ten million in what is now the US in the late 15th century. The colonization of North America from Europe involved the import of farming systems, their crops and animals, and the introduction of some American plant species into Europe. This "Columbian exchange" resulted in the introduction of whole ecosystems to America, including pests, weeds and diseases, transforming indigenous habitats (Crosby, 1986) and subjecting the indigenous populations of the Americas to new diseases (Diamond, 1997). The much-reduced indigenous peoples were pushed to the margins of productive land or forcibly assimilated.

By the early 19th century, small-scale, mixed farming had developed in ways that had much in common across the region, providing farming families and local communities with food, fiber, animal feed and fuel. The genetic diversity of cropped species was maintained by adaptation to local conditions and selection by farmers, giving rise to many landraces of plant and animal species. Gathered (non-cropped) plants and hunted animals remained important in the diet until populations became urbanized.

Trade and exchange of agricultural produce has taken place for millennia, in both commodities (e.g., the import of grain from North Africa by the Roman Empire) and luxury goods (e.g., the medieval spice trade). The scale increased dramatically in the 19th and 20th centuries, thanks to developments in transport and refrigeration. Goods, capital and labor flowed freely between western Europe, North America and many other parts of the world as benefits of competition were considered to outweigh those of protecting markets.

In western Europe, profitability was sought through increases in production and labor efficiency, and developed through the increasing application of science and technology to breeding, fertilization and mechanization. The steppe areas of eastern Europe and the Great Plains of the US were brought into agricultural production for ranching and cereal cropping supported by irrigation. The rate of change was far slower in eastern Europe, where much land remained in the hands of peasants and former serfs. In the early years of the Soviet Union, all aspects of agricultural production and science development were influenced by the centralized administrative-command system. Collectivization began in mid-1918, and by 1940 as much as 97% of peasant holdings had been merged into kolkhozes.

Western agriculture fell into depression during the 1930s; and in the US, cropped lands recently converted from prairies were struck with drought, degrading the land and creating the "dust bowl". Resulting poverty displaced hundreds of thousands of rural families from Oklahoma.

1.4.3.2 Agrifood systems post 1945

Many traditional agrifood systems were localized; food, fuel and fiber were consumed close to the point of production. In the second half of the 20th century, these fragmented agrifood systems became increasingly integrated so that global value chains based on the international trade of commodities now dominate the region. The axes used to develop scenarios in the Millennium Ecosystem Assessment (MA, 2003) are reflected in the contrasts between agrifood systems that are globally integrated and fragmented, and between those that are responsive to multifunctional signals or primarily to economic signals.

Fragmented agrifood systems, responsive to economic signals. Dependence on hunter-gathering for food is now restricted to very small numbers of people, almost entirely in polar and forest regions, though hunting, fishing and gathering natural products from forests is of high economic value, especially because of tourism. Labor-intensive patterns of land management declined, as they were both economically inefficient and increasingly unattractive to young people who increasingly migrated to urban areas.

In many eastern European countries, smallholders and local outlets still raise and market most of the agrifood products (especially livestock, potatoes and other vegetables). For example, in the Caucasus and southern Balkans, small farmers produce cereal and oil crops for subsistence and fruit, vegetable and animal products to supplement often very low incomes, in agrifood systems that are largely independent of the rest of the economy (Dixon et al., 2001). Disproportionate numbers of limited-resource and minority producers in the US sell their products through fragmented agrifood systems.

Globalized agrifood systems, responsive to economic signals. Post-WWII Europe faced massive food shortages. While food rationing was an effective crisis-management tool, the longer-term policy in western Europe and the US was to stimulate production by economic instruments (tariffs, quotas and subsidies) and by providing AKST in the form of new varieties, synthetic fertilizers, synthetic pesticides, machinery and advice on their use. Production increased, and the successful farmers were often those who responded to market signals and produced commodity foods at competitive prices, increasing efficiency by increasing in scale and productivity. The numbers of people employed in agriculture (including forestry, fishing and hunting) fell to less than 5% in the EU in 2005 (EUROSTAT, 2007a) and 1.5% in the US (Hecker, 2004). This happened throughout the supply chain, resulting in fewer, larger corporations providing seed, fertilizers, agrochemicals and machinery to farmers, and consolidation of the supply chain from the farm. This has increased inequality of wealth and assets in the agrifood system.

In the EU, food insecurity gave way to surpluses during the 1980s. In the Soviet Union, food production increased more slowly than in the West. Food prices were kept artificially low, with rationing and inflation, periodic food shortages and long lines in shops (Patterson, 2000).

NAE now produces more than enough food to meet its basic needs, and abundant food supplies are now taken for granted across most of NAE. The share of total income devoted to food varies from 14% in the US to well over 50% in the Balkans and Ukraine (FAOSTAT, 2006). After long periods in which the food supply has been constrained by economics, technology or politics, it became increasingly driven by consumers (Ponte and Gibbon, 2005). The experience of buying food has been transformed: at the turn of the 20th century, the goods in shops were on shelves behind a counter, and were packaged and passed to the customer by a shop assistant. Supermarkets reduced costs by enabling the customers to select the produce themselves. They first appeared in the US in the 1930s, and after WWII became part of suburban culture, combining car parking, low prices and an increasingly wide choice. There is now an abundant variety of affordable native and exotic foods available in all but the poorest countries in the region. Consumption of prepared food from shops, fast food outlets and restaurants has grown rapidly. Post-production sectors of agrifood systems (processing, distribution and sales) now employ a substantial proportion of the workforce throughout NAE. Agrifood systems have become dominated by fewer, larger actors as companies have integrated both horizontally and vertically. Wal-Mart now dominates the American market, Carrefour in France and Tesco in the UK, integrating food production, distribution, preparation and supply into value chains and adding value at each step.

However, the increasing scale and productivity of agriculture became associated with increasing concerns over environment and human health. Environmental concerns included low levels of agricultural biodiversity: 80% of calories consumed worldwide (directly or through milk, eggs and meat) come from just four crops, wheat, rice, soybeans, and maize (Gressel, 2007). The loss of non-cropped biodiversity and landscape quality is regarded as an even more important issue, especially in Europe. Few unmodified habitats remain beyond the poles and high mountains, because so little potentially suitable land is not currently used for agriculture (Fischer et al., 2001). Farmland birds have declined from the 1970s across Europe, although losses of birds of prey due to bioaccumulative pesticides have now been reversed. Traditional agricultural landscapes are threatened by the twin pressures to either intensify or abandon production, resulting in landscape homogenization and further reduction in biodiversity (Petit et al., 2001). The increase in intensive agriculture has also been associated with the decline of other ecosystem functions, including resource protection, water supply and pollination (MA, 2005ab).

Fragmented value chains, responsive to multifunctional signals. Increasing numbers of consumers are concerned about the ethical, social and environmental concerns raised by intensive agriculture, new technologies and globalization (see e.g., Harvey, 1997; Pretty, 1998; Heller, 2003; Tudge, 2003). The vegetarian movement is largely a reaction to factory farming of animals and concerns over animal welfare. In Britain, for example, there were 100,000 vegetarians in 1945 and by the 1990s there were three million, the number having doubled during the 1980s. Purchasing goods certified as produced under standards of fair trade is a reaction to concerns that global trading systems and TNCs disadvantage those who are already poor. Health concerns focus on both the presence of undesirable "contaminants" in food and the overall diet. The increasing market for organic produce largely reflects the wish by some consumers to avoid pesticide residues, growth hormones, antibiotics and GMOs. Organic and locally-produced goods are also valued as ways of avoiding the perceived blandness and stereotypical nature of much modern food (Spencer, 2000). Local foods also benefit from increasing recognition that long-distance transport of food contributes to greenhouse gas emissions (Pretty et al., 2005). Each of these alternate systems has in turn promoted these concerns as a means of increasing their market share.

These trends have expanded markets for higher value, differentiated goods, not least to small farms that were not competitive in more integrated agrifood systems. Some are charging premium prices through farmers' markets, specialist retailers and on the Web. They have achieved this by adding value to the food more widely available in the supermarkets, such as by providing food that is organic, has local distinctiveness, has high standards of animal welfare or has been locally processed and packaged. Markets for food with local provenance, traditional varieties and breeds are increasing in both Europe and North America. Europe has seen a rapid growth in organic agriculture since the early 1990s. Latest figures suggest around 3.4% of EU agricultural land area is now organic, compared with around 0.3% in North America (Willer and Yussefi, 2006).

Globalized, integrated value chains, responsive to multifunctional signals. Most agricultural input industries and food processing, distribution and retail are becoming highly concentrated, with resultant shifts in power dynamics in food systems (ETC Group, 2005; MacMillan, 2005; Ollinger et al., 2005; Arda, 2006; Murphy, 2006). Interlinked networks of powerful TNCs have expanded their reach upwards and downwards in the chain of production through strategic mergers with input companies such as seed suppliers and biotechnology firms involved in seed production and through financial arrangements with global retailers. The outcome has been global value chains or networks that exert increasing control over what is to be produced, how and by whom (Gereffi et al., 2001).

Integrated agrifood chains are also adapting to changing regulations and customer demands. All levels of the industry are increasingly seeking to demonstrate their commitment to high quality, responsible production and retailing through accreditation, auditing, traceability and labeling. To help assure these new standards, companies such as Sainsbury's (UK) are establishing direct contracts with farmers around the world, bypassing systems of wholesalers. Such vertical integration of the agricultural system not only allows a more proactive approach to retailing, it allows control and auditing. The Environmental and Social Report of Unilever (Unilever, 2005), the environmental plans for Wal-Mart and the commitments to Fair Trade by Sainsbury's (Sainsbury's, 2006) are but a few of a rapidly increasing examples of industry efforts to implement and demonstrate to the public efforts to achieve greater sustainability and help meet development goals. Equally, some supermarkets are encouraging local production, exploiting the new markets in local food systems. In the UK, some ASDA stores (a Wal-Mart company) devote shelf space to local producers, while Waitrose and Booth's stores stress to customers their use of named local suppliers. In this way, agribusinesses are seeking to integrate local sensitivity within global strategies. There is competition from new companies, such as Whole Foods Markets, which preferentially sells organic and "natural" produce through a rapidly expanding network of outlets in the US, Canada and the UK, each with considerable local autonomy.

1.4.3.3 The development of policy

Inevitably, public policy addresses all of the issues raised by changing agrifood systems. In response to the food surpluses of the 1980s, the CAP moved away from simply increasing production to transfer wealth from urban to rural areas, and transformed several million peasants into relatively prosperous farmers (Davies, 1996). Subsequently, the circle between the political requirements to stop subsidizing food production and to continue to support farming communities has been squared by changing the emphasis of the CAP to support rural development and environmental goals.

Sustainable development is a major policy goal across most of the region, encompassing agriculture, forestry and fisheries. It is monitored using a wide range of social, economic and environmental indicators. The policy trend in Europe is to promote more proactive agricultural systems, both global and fragmented, with a much greater emphasis on the provision of ecosystem services such as pollution management, carbon storage and diverse habitats, and the management of natural resources such as soil, water, air and landscape quality (Miliband, 2006). Regulations such as the EU Water Framework Directive and support mechanisms such as agri-environmental schemes are helping to raise the environmental standards of agriculture. The appropriate balance between open trade and the use of barriers, tariffs and subsidies remains highly contentious, as can be seen in the Doha Round of the world trade talks. The roles of international trading blocs, such as the EU and North America, consolidated commercially through the North American Free Trade Agreement (NAFTA), have increased.

Recently, issues of energy security (National Economic Council, 2006) and climate change (Stern, 2007) have increased greatly in priority. Policies and practices are being developed to use agricultural land to mitigate climate change by carbon sequestration (Lal, 2004) and to replace some fossil fuel use by the production of biorenewables (Brown, 2003). Increasing biofuel production is already leading to rising prices for cereals, and is likely to increase competition for land, with potentially dramatic changes to farming systems, landscapes and rural economies (e.g., Firbank, 2005). Potential markets in production and management of energy, pharmaceuticals and water add to the uncertainties about the future of agriculture.

1.5 Challenges for AKST

NAE agrifood systems are now facing new challenges that involve simultaneously enhancing social, environmental *and* economic elements. The responses to these challenges within NAE will affect development and sustainability goals, both within the region and globally.

The importance of agrifood systems to human health and social change may become more pronounced. AKST is required to improve standards of nutrition; to reduce exposure to foodborne contaminants and diseases, including those transmitted from animals; and to increase the availability of and equitable access to food and other agricultural products. AKST also has a role in promoting markets with fair access and compensation to participants, improving equity across gender and social divides; and creating and sustaining urban and rural livelihoods. These goals must be met while maintaining the stability and resilience of agroecosystems, particularly as they are threatened by global environmental change.

AKST will be required if agrifood systems are to mitigate and successfully adapt to global climate change. Agriculture may need to cope with very different economic and environmental conditions, with new patterns of trade, climate, pests and diseases, while facing more stringent requirements to mitigate greenhouse gas emissions. Climate change may transform the conditions for management of land-based natural resources to ensure delivery of a full array of ecosystem services and food use. Demands are increasing for plant-based substitutes for fossil fuels for energy and industry, in competition with increasing global demands for plant and livestock products. These changes imply the generation and effective sharing, access and use of AKST to develop new genotypes, land management systems and value chains that can deliver multiple functions and are sustainable in rapidly changing social, governance, economic and environmental conditions.

References

ACIA. 2005. Arctic climate impact assessment. Cambridge Univ. Press, New York.

Aldington, T.J. 1998. Multifunctional agriculture: A brief review from developed and developing country perspectives. FAO, Rome.

Andrews, N.J. 2007. Incidence of variant Creutzfeldt-Jakob Disease in the UK. Available at http://www.cjd.ed.ac.uk/vcjdqdec06.htm. Nat. Creutzfeldt-Jakob Dis. Surveil. Unit.

Arda, M. 2006. Tracking the trend toward market concentration: The case of the agricultural input industry. UNCTAD/DITC/COM/2005/16, UN Conf. Trade Development. Geneva.

Boody, G., B. VonDracek, D.A. Andow, M. Krinke, J. Westra, J. Zimmerman and P. Welle. 2005. Multifunctional agriculture in the United States. BioScience 55(1):27-38.

Brown, R.C. 2003. Biorenewable resources: Engineering new products from agriculture. Iowa State Press, Ames.

Carlsson-Kanyama, A., M.P. Ekstrom, and H. Shanahan. 2003. Food and life cycle energy inputs: consequences of diet and ways to increase efficiency. Ecol. Econ. 44:293-307.

CDC (Centers for Disease Control and Prevention). 2005. Third national report on human exposure to environmental chemicals. Available at http://www.cdc.gov/exposurereport/pdf/thirdreport.pdf. CDC, National Center for Environ. Health, Atlanta.

Cloud, J. 2007. The rising costs of food. Time Magazine. 21 June.

Crosby, A.W. 1986. Ecological imperialism: The biological expansion of Europe. Cambridge Univ. Press, Cambridge.

Davies, N. 1996. Europe, a history. Oxford Univ. Press, New York.

Devine, J.A., K.D. Baker, and R.L. Haedrich. 2006. Fisheries: deep-sea fishes qualify as endangered. Nature (London) 439:29.

Diamond, J. 1997. Guns, germs and steel: The fate of human societies. W.W. Norton, New York.

Dixon, E.J. 2001. Human colonization of the Americas: Timing, technology and process. Quatern. Sci. Rev. 20:277-299.

Dixon, J., A. Gulliver, and D. Gibbon. 2001. Farming systems and poverty: Improving livelihoods in a changing world. FAO and World Bank, Rome and Washington, DC.

Effland, A.B.W., R.A. Hoppe, and P.R. Cook. 1998. Minority and women farmers in the U.S. Agric. Outlook (May):16-21.

ETC Group. 2005. Oligopoly, Inc. 2005. Communiqué 91 (Nov/Dec).

EUROSTAT. 2007a. Agricultural statistics: Data 1995-2005. European Communities, Luxembourg.

EUROSTAT. 2007b. Population and social conditions [Online]. Available at http://epp.eurostat.ec.europa.eu/.

FAO. 1996. Overview of the socio-economic position of rural women in selected central and eastern European countries—Bulgaria, Croatia, the Czech Republic, Estonia, Hungary, Latvia, Lithuania, Poland, Slovakia and Slovenia. FAO, Rome.

FAO. 2006. Global forest resources assessment 2005: Progress towards sustainable forest management. FAO, Rome.

FAO. 2007. State of the world's fisheries and aquaculture 2006. FAO, Rome.

FAOSTAT. 2006. Food security statistics. Available at http://www.fao.org/faostat/foodsecurity/index_en.htm. FAO, Rome.

Firbank, L. 2005. Striking the balance between agricultural production and biodiversity. Ann. Appl, Biol. 146:163-175.

Fischer, G., M. Shah, H. van Velthuizen, and F.O. Nachtergaele. 2001. Global agro-ecological assessment for agriculture in the 21st century. IIASA, Laxenburg.

Foster, C., K. Green, M. Bleda, P. Dewick, D. Evans, A. Flynn, J. Mylan. 2006. Environmental impacts of food production and consumption: A report to the Dep. Environ. Food Rural Affairs. Manchester School of Business. DEFRA, London.

Fuglie, K., N. Ballenger, K. Day, C. Klotz, M. Ollinger, J. Reilly, et al. 1996. Agricultural research and development, public and private investments under alternatives markets and institutions. ERS, US Dep. Agric., Washington DC.

Gereffi, G., J. Humphrey, R. Kaplinsky and T.J. Sturgeon. 2001. Introduction: Globalisation, value chains and development. IDS Bulletin 32(3), Inst. Dev. Studies, Univ. Sussex.

Gibson, C.J., and E. Lennon. 1999. Historical census statistics on the foreign-born population of the United States: 1850-1990. US Census Bureau, Washington DC.

Global Footprint Network. 2006. Ecological footprint and biocapacity. Available at http://www.footprintnetwork.org/gfn_sub.php?content=national_footprints.

Goldman, L.R., D.F. Smith, R.R. Neutra, L.D. Saunders, E.M. Pond, J. Stratton et al. 1990. Pesticide food poisoning from contaminated watermelons in California, 1985. Archives Environ. Health 45:229-236.

Gressel, J. 2007. Genetic glass ceilings: Transgenics for crop biodiversity. Johns Hopkins Univ. Press, Baltimore.

Gunton, T.I., K.S. Calbick, A. Bedo, E. Chamberlain, A. Cullen, K. Englund, et al. 2005. The maple leaf in the OECD—comparing progress toward sustainability. The David Suzuki Foundation, Vancouver, BC.

Harvey, G. 1997. The killing of the countryside. Jonathan Cape, London.

Hecker, D.E. 2004. Occupational employment projections to 2012. Mon. Labor Rev. 127:80-105.

Heller, R. 2003. GM nation? The findings of the public debate. Dep. Trade Industry, London, UK.

Hoekstra, A.Y. and A.K. Chapagain. 2007. Water footprints of nations: Water use by people as a function of their consumption pattern. Water Resour. Manage. 21:35-48.

Hoppe, R.A., P. Korb, E.J. O'Donoghue and D.E. Banker. 2007. Structure and finances of US farms: Family farm report. EIB-24, ERS, USDA, Washington, DC.

Howden, D. 2007. The fight for the world's food: Population is growing. Supply is falling. Prices are rising. What will be the cost to the planet's poorest? The Independent. 23 June.

Jacobs, G. (ed). 2006. The story behind the numbers: Women and employment in central and eastern Europe and in the Commonwealth of Independent States. UN Dev. Fund Women (UNIFEM), Bratislava, Slovak Republic.

Jones, C.B. 2005. Twentieth century USA. Contemporary Books, Chicago.

Judt, T. 2005. Post war. William Heinemann, London.

Kroll, L., and A. Fass. 2007. The world's richest people. Available at http://www.forbes.com/lists/2007/10/07billionaires_The-Worlds-Billionaires_Rank.html?boxes=custom. Forbes Magazine.

Lal, R. 2004. Soil carbon sequestration impacts on global climate change and food security. Science 304:1623-1627.

MA (Millennium Ecosystem Assessment). 2003. Ecosystems and human well being: a framework for assessment. Island Press, Washington, DC.

MA (Millennium Ecosystem Assessment). 2005a. Ecosystems and human well-being: Current states and trends. Vol. 1. Island Press, Washington, DC.

MA (Millennium Ecosystem Assessment. 2005b. Synthesis report. Island Press, Washington, DC.

MacDonald, J., R. Hoppe and D. Banker. 2006. Growing farm size and the distribution of farm payments. EB-6, ERS, USDA, Washington, DC.

MacMillan, T. 2005. Food power: Understanding trends and improving accountability. Background Paper for UK Food Ethics Council. Available at http://www.agribusinessaccountability.org/pdfs/323_Power-in-the-Food-System.pdf.

Maddison, A. 2001. The world economy: A millennial perspective. OECD, Paris.

Meeks, R.L. (ed). 2006. Federal research and development funding by budget function: Fiscal years 2004-2006 (Table 38). Div. Sci. Resources Stat., Nat. Sci. Foundation, Washington, DC.

Meier, A. 2006. Die ertrotzte Demokratie. Nat. Geogr. (Deutschland):82-108.

Miliband, D. 2006. One planet farming. DEFRA, London.

Mølbak, K. 2004. Spread of resistant bacteria and resistance genes from animals to humans—The public health consequences. J. Vet. Med. Series B 51:364-369.

Murphy, S. 2006. Concentrated market power and agricultural trade. Available at http://www.ecofair-trade.org/pics/de/EcoFair_Trade_Paper_No1__Murphy.pdf. Disc. Pap., EcoFair Trade Dialog.

National Biodiesel Board. 2007. US biodiesel production capacity. Available at http://www.biodiesel.org/pdf_files/fuelfactsheets/Production_Capacity.pdf. Nat. Biodiesel Board.

National Economic Council. 2006. Advanced energy initiative. The White House/Nat. Econ. Council, Washington, DC.

NFU-Canada. 2005. The farm crisis and corporate profits. Available at http://www.nfu.ca/new/corporate_profits.pdf. National Farmers Union, Canada.

Nord, M., M. Andrews, and S. Carlson. 2007. Household food security in the United States, 2006. Econ. Res. Rep. No. (ERR-49). USDA, Washington, DC.

NCHS. 2006a. Deaths—Leading causes. Available at http://www.cdc.gov/nchs/fastats/lcod.htm. Nat. Center for Health Statistics, Centers for Disease Control, Hyattsville.

NCHS. 2006b. Health, United States, 2006, with chartbook on trends in the health of Americans. National Center for Health Statistics, Hyattsville.

OECD. 2001. Multifunctionality: A framework for policy analysis. OECD, Paris.

Ollinger, M., S.V. Nguyen, D, Blayney, B. Chambers, and K. Nelson. 2005. Structural change in the meat, poultry, dairy, and grain processing industries. ERR-3, Econ. Res. Serv. USDA, Washington, DC.

Olshansky, J., D.J. Passaro, R.C. Hershow, J. Layden, B.A. Carnes, J. Brody, et al. 2005. A potential decline in life expectancy in the United States in the 21st century. New Engl. J. Med. 352:1138-1145.

Oxfam America. 2004. Like machines in the fields: Workers without rights in American agriculture. Available at http://www.oxfamamerica.org/newsandpublications/publications/research_reports/art7011.html.

Oxfam America. 2007. Shut out: How US farm programs fail minority farmers. Available at http://www.oxfamamerica.org/newsandpublications/publications/research_reports/shut-out.

Passel, J.S. 2005. Unauthorized migrants: Numbers and characteristics. Available at http://pewhispanic.org/files/reports/46.pdf. Pew Hispanic Center, Washington DC.

Patterson, K.D. 2000. Russia. p.1240-1247. *In* The Cambridge world history of food. Vol. 2 K.F. Kiple and K.C. Ornelas (ed) Cambridge Univ. Press, Cambridge.

Pauly, D. and J. Alder. 2005. Marine fisheries systems. p. 477-511. *In* Millennium Ecosystem Assessment. Ecosystems and human well-being: Current states and trends. Vol. 1. Island Press, Washington, DC.

Petit, S., L. Firbank, B. Wyatt, and D. Howard. 2001. MIRABEL: Models for integrated review and assessment of biodiversity in European landscapes. Ambio 30:81-88.

Ponte, S. and P. Gibbon. 2005. Quality standards, conventions and the governance of global value chains. Econ. Soc. 34(1): 1-31.

Pretty, J. 1998. The living land. Earthscan, London.

Pretty, J., A. Ball, T. Lang, and J. Morison. 2005. Farm costs and food miles: an assessment of the full cost of the UK weekly food basket. Food Policy 30:1-19.

Quigley, J.T., and D.J. Harper. 2006. Effectiveness of fish habitat compensation in Canada in achieving no net loss. Environ. Manage. 37:351-366.

Reeves, M. and K.S. Schafer. 2003. Greater risks, fewer rights: US farmworkers and pesticides. Int. J. Occup. Environ. Health 9:30-39.

Rosegrant, M.W., X. Cai, and S.A. Cline. 2002. Global water outlook to 2025: Averting an impending crisis. IFPRI, Washington, DC, and IWMI, Columbo.

Rutherford, R.J., G.J. Herbert, and S.S. Coffen-Smout. 2005. Integrated ocean management and the collaborative planning process: the Eastern Scotian Shelf Integrated Management (ESSIM) Initiative. Marine Policy 29:75-83.

Sainsbury's. 2006. Corporate responsibility. Available at http://www.sainsburys.co.uk/aboutus/policies/policies.htm?WT.seg_1=nav_secondary.

Sautter, U. 2000. Geschichte Kanadas. Beck, München.

Schlesinger, W.H., K.H. Reckhow and E.S. Bernhardt. 2006. Global change: the nitrogen cycle and rivers. Water Resour. Res. 42:W03S06.

Shields, M. 2006. Overweight and obesity among children and youth. Statistics Canada Health Rep. 17(3):27-42.

Skoet, J. and K. Stamoulis. 2006. The state of food insecurity in the world 2006. FAO, Rome.

Solbrig, O.T., and D.J. Solbrig. 1994. So shall you reap: Farming and crops in human affairs. Island Press, Washington, DC.

Spencer, C. 2000. The British Isles. p. 1217-1226. *In* K.F. Kiple and K.C. Ornelas (ed) The Cambridge world history of food. Vol. 2. Cambridge Univ. Press, Cambridge.

Statistics Canada. 2006. Snapshot of Canadian agriculture. Available at http://www.statcan.ca/english/agcensus2006/articles/snapshot.htm.

Stern, N. 2007. The economics of climate change; The Stern review. Cambridge Univ. Press, Cambridge.

Sumelius, J., S. Bäckman and T. Sipiläinen. 2005. Agri-environmental problems in central and eastern European countries before and during the transition. Sociol. Rural. 45(3):153-170.

Taiga Rescue Network. 2007. Indigenous people. Available at http://www.taigarescue.org/en//index.php?sub=2&cat=86. Taiga Rescue Network, Sweden.

Tegtmeier, E.M. and M.D. Duffy. 2004. External costs of agricultural production in the United States. Int. J. Agric. Sustainability 2(1):1-19.

Tjepkema, M. 2006. Adult obesity. Statistics Canada Health Reports 17(3):9-26.

Tudge, C. 2003. So shall we reap. Penguin Books, London.

Unilever. 2005. Unilever Environmental and Social Report. Unilever, Rotterdam.

UN Population Division. 2005a. World population prospects; the 2004 population revision database [Online]. Available at http://esa.un.org/unpp. UN, New York.

UN Population Division. 2005b. World population prospects; the 2004 population revision. Vol. III. Analytical Report. UN, New York.

US Census Bureau. 2002. Table 1. United States—Race and Hispanic origin: 1790 to 1990 [Online] http://www.census.gov/population/documentation/twps0056/tab01.xls.

US Census Bureau. 2007. Statistical abstract: 2007 ed. US Census Bureau.

USDA. 2005. Agricultural Resource Management Survey. Available at http://www.ers.usda.gov/Briefing/ARMS. USDA, Washington, DC.

WHO. 2007. Cumulative number of confirmed human cases of avian influenza A/(H5N1) reported to WHO [Online]. Available at http://www.who.int/csr/disease/avian_influenza/country/cases_table_2007_03_12/en/index.html. WHO, Geneva.

Wilkinson, J. 2006. The mingling of markets, movements and menus: The renegotiation of rural space by NGOs, social movements and traditional actors. Paper presented at Int. Workshop, Globalisation: Social and cultural dynamics. Rio de Janeiro, 23 Mar 2006. Available at http://www.minds.org.br/arquivoswilkinsontheminglingofmarkets.pdf.

Willer, H., and M. Yussefi. 2006. The world of organic agriculture. Statistics and Emerging Trends 2006. IFOAM and Research Institute of Organic Agriculture FiBL, Bonn and Frick UK.

World Bank. 2006. World development indicators database [Online]. Available at http://web.worldbank.org/WBSITE/EXTERNAL/DATASTATISTICS/0,,contentMDK:20535285~menuPK:1192694~pagePK:64133150~piPK:64133175~theSitePK:239419,00.html. World Bank, Washington, DC.

2

Changes in Agriculture and Food Production in NAE Since 1945

Coordinating Lead Authors:
Mary Hendrickson (USA), Mara Miele (Italy)

Lead Authors:
Rebecca Burt (USA), Joanna Chataway (UK), Janet Cotter (UK), Béatrice Darcy-Vrillon (France), Guy Debailleul (Canada), Andrea Grundy (UK), Kenneth Hinga (USA), Brian Johnson (UK), Helena Kahiluoto (Finland), Peter Lutman (UK), Uford Madden (USA), Miloslava Navrátilová (Czech Republic)

Contributing Authors:
Dave Bjorneberg (USA), Randy L. Davis (USA), William Heffernan (US), Susanne Johansson (Sweden), Veli-Matti Loiske (Sweden), Luciano Mateos (Spain), Selyf Morgan (UK), Jyrki Niemi (Finland), Fred Saunders (Australia), Paresh Shah (UK), Gerard Porter (UK), Elisabeth Ransom (USA), Peter Reich (USA), Leonid Sharashkin (Russia), Timo Sipiläinen (Finland), Joyce Tait (UK), K.J. Thomson (UK), Francesco Vanni (Italy), Bill Vorley (UK), Markku Yli-Halla (Finland)

Review Editors:
Barbara Dinham (UK), Maria Fonte (Italy)

Key Messages
2.1 Agriculture and Food System Specialization in NAE 22
2.1.1 Changes in farming and rural population in North America 26
2.1.2 Changes in European farming and rural populations 27
2.2 Farm Policies and the Development of NAE Agriculture 30
2.2.1 US farm policy: A legacy of the Great Depression 30
2.2.2 Canada: A bipolar farm policy 31
2.2.3 Common Agricultural Policy and the building of a single market 31
2.2.4 Agricultural policies in CEE countries 33
2.3 Changes in Market Structure 34
2.4 Changes in NAE Cropping Systems since 1945 37
2.4.1 Changes in soil AKST and use since 1945 37
2.4.2 Changes in cropping systems in NAE 38
2.4.3 Increasing cropping systems productivity through inputs 40
2.4.3.1 Mechanization 40
2.4.3.2 Plant breeding, seeds and genetics 40
2.4.3.3 Nutrients in cropping systems 43
2.4.3.4 Pesticide usage in NAE cropping systems 44
2.4.3.5 Water control in NAE cropping systems 46
2.4.4 Agricultural products for energy and fuels 47
2.4.5 Organic cropping systems 47
2.4.6 Key changes in cropping systems and drivers 48
2.5 Changes in Livestock Systems in NAE 48
2.5.1 Trends in output and productivity since 1945 49
2.5.2 Drivers of increased livestock output and productivity 50
2.5.3 Key changes in the NAE livestock sector 53
2.6 Changes in Forestry Systems 53
2.6.1 Main trends in NAE forests and forestry production 53
2.6.2 Forest ownership and control 54
2.6.3 Forestry as an industry 55
2.6.4 AKST in forestry 55
2.6.5 Forest institutions 55
2.6.6 Drivers of changes in forestry 56
2.6.7 Trends in NAE forestry 56
2.7 Changes in Aquaculture Production 57
2.7.1 North American aquaculture 57
2.7.2 European aquaculture 59
2.7.3 Science and technology in aquaculture 60
2.7.4 Key changes in aquaculture 61
2.8 Key Changes in Post-Harvest and Consumption Systems 61
2.8.1 Changes in the food retail sector in NAE 61
2.8.2 Concentration and trends at national levels 63
2.8.3 Changes in food manufacturing and processing 65
2.8.4 Market segmentation 65
2.8.5 Food safety, quality regulation and food market niches 66
2.8.6 Changes in diet/consumption 69
2.8.7 Key changes in consumption systems 70

Key Messages

1. Following WWII rapid advances in the understanding of plant and animal biology fueled productivity increases and provided new tools for identifying and addressing agricultural problems. In this period, agricultural production and productivity increased significantly, especially in Western Europe and North America, but more slowly in Central and Eastern Europe. The increased productivity of agriculture was supported by technological development and food supply policies.

An increased range of technologies and tools has been available to agriculture primarily through advances in AKST. Farmers have accessed AKST to enhance crop and livestock productivity and quality. Efficient knowledge transfer systems developed in the governmental and private sectors have facilitated the dissemination of these new tools. Information technology (IT) has revolutionized AKST as well as food manufacturing, transportation and distribution and has allowed efficient dissemination of AKST.

The broad range of new technologies, some of them controversial, has had and is having significant impacts for all NAE societies. The impacts of scientific and technological advances have been and are being felt in both conventional plant and animal breeding programs and those involving biotechnology. Biotechnology, including genetic engineering, has greatly expanded the speed at which traits critical to agriculture can be identified and manipulated.

Crop production has increased considerably over the last 50 years in the NAE particularly in output per unit area. These increases have been due to improved soil management, increased fertilizer use, including new synthetic fertilizers, greater technological sophistication and scale of agricultural mechanization and development of agrochemicals for pest and disease control. Wider adoption of irrigation coupled with the conversion of pasture to permanent cultivation has contributed to production increases. The development of plant breeding technologies, including hybrids and genetically engineered varieties, have changed the way most North American and Western European farmers obtain seed to annual purchases rather than saving seed. Uptake of genetic engineered crops has differed markedly in the region. They form a part of just a few cropping systems (predominantly soybeans and cotton, but also maize and canola) in North America.

Overall, livestock productivity and output in NAE has increased since 1945 with beef, pig meat and milk production almost doubling and a four-fold increase in numbers of poultry. This has been driven by increasing demand from a growing and wealthier population and by production-oriented policies. Increases in productivity are due to animal breeding developments, intensive rearing systems, antibiotic use and high-yielding pastures. Technical advances in fish breeding and rearing have led to considerable increases in production in both saltwater and freshwater fish farming.

NAE is the only region where there has been an increase in forest area since the 1960s, partly as a result of increased plantations and partly resulting from re-growth following abandonment of agricultural land. Demand for forest products in NAE has increased dramatically because of a larger and wealthier population. New management and processing technologies have been introduced to meet these demands resulting in increased efficiency and better access to remote areas. The environmental quality of forests in NAE has declined somewhat over the last 50 years. This has been caused by a variety of factors, including a significant increase in forest fires across NAE; it is a complex issue still not fully understood.

In North America and Western Europe, agricultural policies were adopted and implemented to improve farm income, to promote use of technology and to sustain productivity. In terms of increasing productivity and total production, these policies were largely successful. They also helped improve average farm income, ameliorated poverty in rural populations in some regions and contributed to overall economic development.

2. These increases in total food production addressed much of the problem of hunger and food shortages across NAE. The increase in food supply in NAE has progressively led to a greater availability of food both in quantity and variety and more recently to an overabundance of calories. Despite the absolute quantity of calories available, poor households across the region often do not have access to an adequate nutritious diet.

The increase in agricultural productivity has led to a decrease in real prices of agricultural products in North America and Western Europe over the last 40 years. This situation has led to more affordable food and ensured food security for the majority of the NAE population. Nevertheless, increased food availability and changes in human behavior and lifestyle have favored the development of nutrition-related chronic diseases. Over the last 15 years, these chronic diseases, including obesity, have had a heavy economic, public and social cost throughout the region.

In Eastern Europe and the Soviet Union, the degree of food self-sufficiency increased from the late 1940s until the 1970s; however, in the USSR, food and agricultural shortages from the 1960s to the 1980s led to increased agricultural imports. In the 1990s a transition period occurred in Central and Eastern European countries characterized by falling output. Household allotments have been particularly important in the former Soviet Union and now Russia, for food security where small household producers account for 25-50% of agricultural output (e.g., potatoes, key vegetables and meats).

3. Knowledge systems used for breeding new plant and animal varieties and for agrochemicals have been partially protected as intellectual property and increasingly privatized. The emergence of technologies protected as intellectual property has created synergies that have favored industry consolidation and has facilitated the creation of NAE-based transnational agribusinesses. These transnationals now account for almost a third of commercial seeds worldwide and a significant share of livestock genetics.

4. The structure of the food system has changed with time in NAE. The agricultural and food system has become more vertically integrated from agricultural in-

puts to food retailing. Improved productivity and food security led to mature markets for staple foodstuffs and limited the opportunities for further growth. Food suppliers responded by increased differentiation and food innovation. The largest actors, including large-scale food retailing and food catering/service businesses have increasing influence over the production of food. Food suppliers sought to expand the market initially by increasing the range of available foodstuffs through trade in "exotic" foods, through yearround supply of fruits and vegetables, through the development of the processed food market and through the development of "quality" food products. Crop and livestock enterprises have become fewer and larger due to economies of scale; this trend is likely to continue. Changes in agricultural labor have been uneven across the region and across agricultural systems. The need for farm labor has generally decreased in conventional cropping and livestock system, but some farming systems, particularly fruit and vegetable production, have intensive demands for farm labor. Increases in sizes of farm and food processing entities have often led to reliance on immigrant labor.

5. Biofuels have always been a component of energy production in NAE, especially for heat, although biomass is generally less important as a fuel source in NAE than in other regions. In the past several years, biofuel production has dramatically increased in importance and application. Policy directives across much of the NAE have led to the subsidization of the use of biofuels to replace fossil fuels, which has spurred the production of bioethanol and biodiesel, mostly from maize and oilseed rape. There is active research to generate "second-generation" biofuels from other more energy-rich plant source materials, especially biomass.

6. The concerns over the application of new tools and technologies and the changed production systems resulting from them have contributed to a growing environmental, social and health awareness in NAE. Crop and livestock production in the NAE is among the most intensive in the world and this has had serious adverse impacts on the environment. Increased awareness of these adverse effects has resulted in regulatory frameworks for the use of agrochemicals, the use of new tools and technologies and the development of alternative production systems, including organic agriculture. This awareness has led to changes away from production-oriented policies toward those that are market-driven or environmentally led. The recognition of the multiple roles of agriculture has emerged in political and economic agendas.

In these agendas, agriculture is now seen as delivering not only food but services that meet emerging social demands such as environmental protection (including the management of resources such as water and land, landscape, biodiversity and natural habitat); environmentally-friendly production of food; use of land for residential needs and recreational activities; protection of local cultures and knowledge; protection of cultural heritage through the production of traditional foods; ethical dimensions of food production such as positive contributions to food security and social justice (e.g., fair trade); and animal welfare considerations. These developments have been concurrent with an increasing demand for variety, including increased demand for foods that are high quality; locally produced; regionally specialized; organic; fairly traded; humanely produced; and ethnic.

The relative peace and stability in NAE has been an important component in securing food security.

2.1 Agriculture and Food System Specialization in NAE

In the past few decades agriculture in North America and Europe has gone through dramatic structural change. There has been a decrease in the number of farms, reduction in the agricultural labor force, increased specialization geographically and at the farm level, and a loss of self-sufficiency at the farm level.

Technological change has been rapid in NAE and the introduction of any new agricultural technology has implications for markets, producers and consumers (Hayami and Ruttan, 1985; Kislev and Peterson, 1986). In most of NAE, technological change has favored capital intensive technologies and economies of scale. Mechanization has increased, generally allowing for larger average farm sizes although there is considerable heterogeneity in farm size and scale in NAE. Most NAE farmers have attempted a scale of operation characterized by the lowest cost per unit of output. The average unit cost follows an L-shape function; the unit cost at first decreases sharply with size but then reaches a plateau (Hall and Leveen, 1978; Nehring, 2005). The evidence for diseconomies of size is weak or non-existent. In spite of the fact that the average size of farms has increased in most of NAE, they are mainly managed by private farm families, most of which rely on off-farm income in addition to income from farming activities (Hoppe and Banker, 2006).

The decreasing number of farms, combined with increasing total output has led to concentration of production (Figure 2-1). The number of farms necessary to produce a particular share of output has fallen; for example, from 1989 to 2003 the fraction of US farm production by large scale family and non-family farms increased from 57.7 to 72.8%. In Western Europe the farm size in terms of land area is only one tenth of that in the US; the number of farms is much higher but rapidly decreasing. From 1983 to 2001 the number of farms decreased in EU-12 from about 9 million to 6.5 million, but farms grew larger, especially in the livestock sector. A larger percentage of the farms in Europe compared to farms in North America operate on a part-time basis because of the smaller farm size.

Economic growth also contributes to farm structure (Heady, 1962). Other things being equal, including the labor share of inputs, the scale of farm businesses must increase in proportion to the increase in non-farm labor earnings. The growth of other sectors of the economy has driven labor from agriculture to more productive sectors in most parts of the NAE.

Specialization, an important aspect of productivity growth in NAE agriculture, has improved the spatial organization of the food chain and lowered production and transportation costs (Chavas, 2001). In Western Europe and

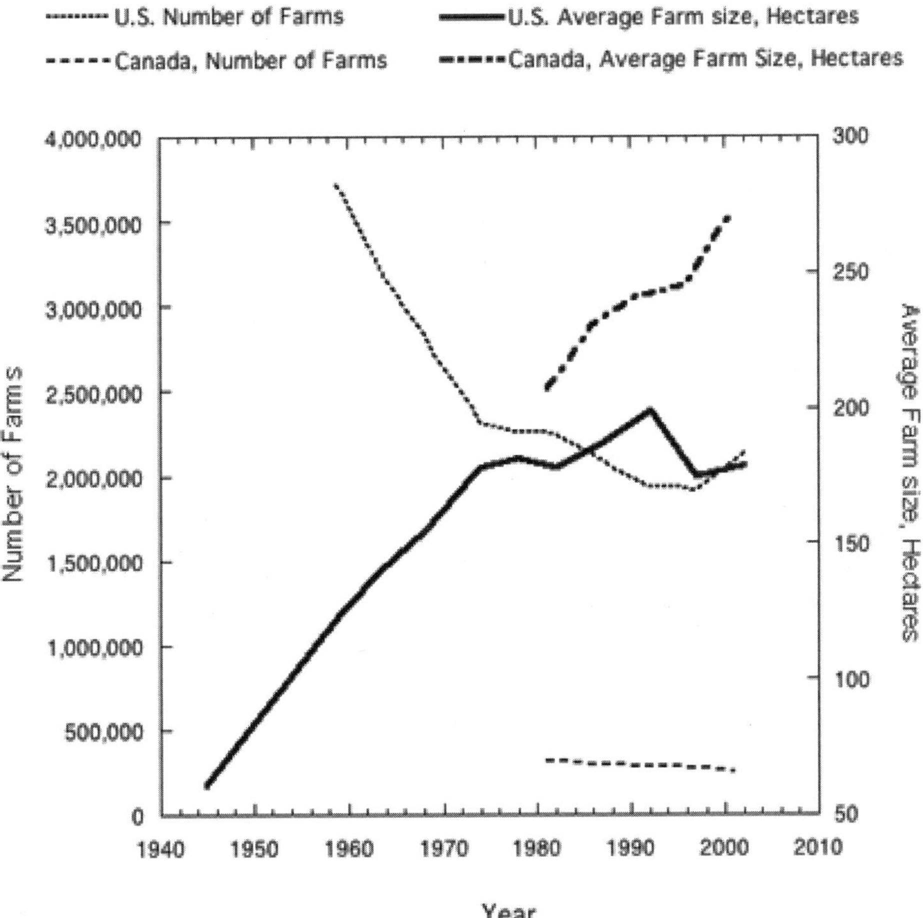

Figure 2-1. *Change in farm size and number of farms in North America from 1940-2000.* Source: USDA data; author elaboration.

North America, specialization occurred largely because of economies of scale, larger economic forces and technological change. When economies of scale (the unit cost decreases with size) prevails over economies of scope (synergies between products and by-products), specialization increases which is followed by an increased size in production units. This leads to regional specialization and concentration. Government policies may also influence farm size and numbers. Agricultural policies after World War II directly promoted specialization through incentives (e.g., Pirog et al., 2001; for a fuller discussion of policies see 2.2). Yet larger trends have usually overshadowed the impact of policy programs on farm structure.

Farm specialization is particularly pronounced in North America (see Table 2-1) and in central and eastern European areas that experienced collectivization. Specialization differs by farm size with smaller farms the most likely to produce one commodity (Cash, 2002) (Figure 2-2). The average number of commodities produced per farm has fallen from 4.6 in 1945 to 1.3 in 2002 (Dimitri and Effland, 2005) even though financially successful farms have tended to be more diversified (Hoppe, 2001). Farms in the United States now have a bimodal distribution, with the number of farms in the middle declining (Kirschenmann et al., 2003). More than 25% of very large family farms are specialized in hog

Table 2-1. **100 years of structural change in U.S. agriculture.**

	1945*	1970	2000/02
Number of farms (millions)	5.9	2.9	2.1
Average farm size (acres)	195.0	376.0	441.0
Average number of commodities produced per farm	4.6	2.7	1.3
Farm share of population (percent)	17.0	5.0	1.0
Rural share of population (percent)	36.0	26.0	21.0
		percent	
Off-farm labor*	27	54	93

*1945 = percent of farmers working off-farm; 1970 and 2000/02 = percent of households with off-farm income

Source: Dimitri and Effland, 2005.

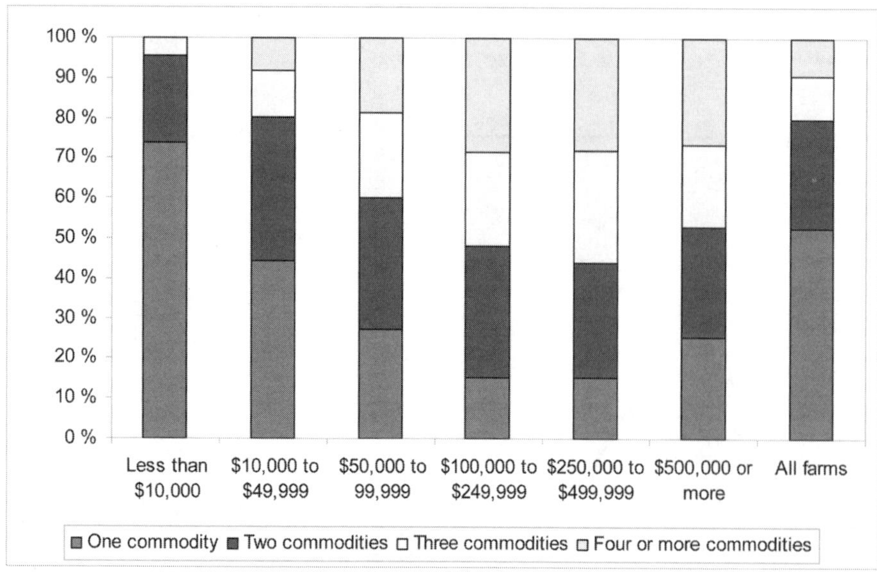

Figure 2-2. *Distribution of number of commodities by sales class in U.S.* Source: USDA, 1999

and poultry and closely linked to processors (Hoppe and Korb, 2005).

While agricultural production is now highly concentrated in large farms, there still are a large number of more diverse small farms coexisting with a small number of very large farms that capture most of the markets for agricultural commodities (Miljkovic, 2005). Crop diversity declined between the 1930s and 1980s; the area sown to grain crops increased and woodland on farms declined (Medley et al., 1995). During this period the number of farms decreased by 60% and farm size increased from 37 ha in 1925 to 72 ha in 1987.

An examination of farm by type of ownership-operation provides a useful look at the diversity of farm types currently in the US. Land is distributed fairly evenly among different types of farms, ranging from part-time farmers to very large scale operations (Figure 2-3). The large-scale, very-large-scale and non-family farms represent a disproportionately large fraction of the total US farm production (73% of the production from 38% of the farm area). Yet the majority of farms (98%) in the US as of 2003 are family-owned farms, though they may be organized as proprietorships, partnerships, or family corporations (Hoppe and Banker, 2006).

Specialization in the eastern part of NAE has followed a different path due to collectivization after World War II. The collectivization of agriculture was intended to exploit economies of scale, particularly in respect to mechanization and the use of agrichemicals. These were more obvious in large-scale crop production and possibly in intensive livestock production; they were less clearly applicable to farming in mountainous areas, or with labor-intensive crops. Collectivization led to the establishment of large collective or state farms which were highly mechanized and specialized but often inefficient in their use and allocation of resources. In the former Soviet Union the collectivized sector of agriculture (99.6% of agricultural producers were collectivized by 1955) grew significantly during the post-war decades (Matskevich, 1967). After World War II, the Central and Eastern Europe (CEE) countries were major suppliers of agricultural products to the Soviet Union. Compared to the more arid regions of the Soviet Union, soils were relatively productive and a system of large collective farms were developed in the 1930s (Wheatcroft and Davies, 1994). This system was only economically viable under the centralized agricultural economies of the Socialist era.

As in the rest of NAE, the farm structure was dualistic in many CEE countries with numerous small self-subsistence plots and large-scale farms producing most of the gross output. Soviet agriculture essentially branched into two sectors. The collectivized sector was characterized by state-controlled, large-scale reliance on off-farm inputs, mechanization and hired labor and centralized processing and distribution of outputs. This sector was capital intensive and emphasized the management of quantities rather than qualities, because of the lack of price signals for quality, whether judged by processing enterprises or final consumers (Sharashkin and Barham, 2005). Moreover, there was widespread use of agronomic and veterinary expertise (sometimes located within individual farms), which led to the provision of improved varieties of crops and livestock. Collectivized farms were linked to centralized input-supply and product-processing facilities. The other branch of Soviet agriculture was the household-managed sector, characterized by micro-scale, lack of state support or inputs, manual labor provided by the household and self-provisioning goals (Sharashkin and Barham, 2005). The latter was authorized by Soviet authorities at the beginning of WWII to fight impending food shortages and quickly spread throughout the country (Lovell, 2003). This household-based sector continued to grow and by the mid-1950s accounted for 25% of the country's agricultural output (Wadekin, 1973). Throughout the Socialist period, the authorities maintained an ambivalent attitude toward household producers; their importance to food security was tacitly recognized, yet the government

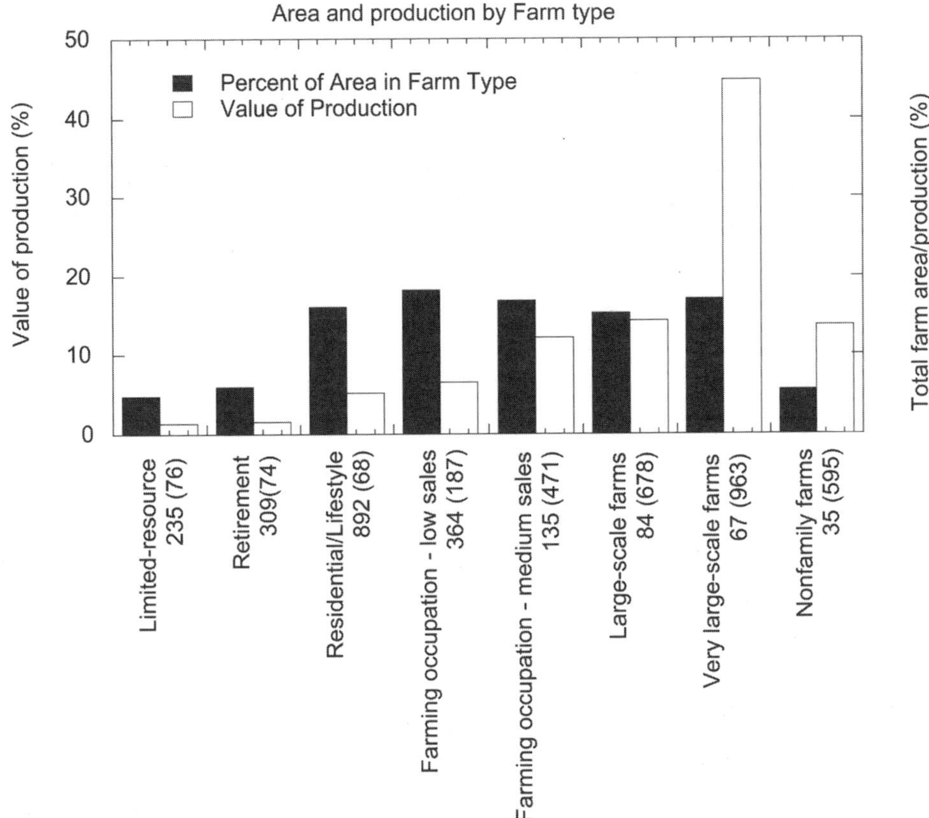

Figure 2-3. *U.S. farmland area and total U.S. farm production by type of farm in 2005.* Source: Hoppe and Banker, 2006

refrained from providing any support to household production so as not to encourage any "capitalistic", private ownership tendencies (Lovell, 2003).

Specialization was less pronounced in other parts of CEE. For instance, most farming production remained small-scale in Poland although some state farms were initiated when collectivization after 1956 affected the supply of inputs and, crucially, the distribution of most output. Unlike Yugoslavia, where a similar semi-collectivization was enforced, the Polish government continued to exercise strenuous but erratic central control over agriculture in an effort to improve performance by balancing and linking the 85% of agriculture that was privatized with socialized sectors of farming. These efforts, which usually favored the larger-scale collectivized sector, were seldom successful and led to inefficient use of new technologies and inputs in this sector, while private producers were starved of both funds and technology and resultant stagnation.

After transition of the CEE countries to democracy in the early 1990s, the collective and State farm system rapidly broke down, partly because the system became uncompetitive when forced to compete in world markets and partly because the Soviet markets were no longer easily available to the transition countries. State farms were broken into smaller units and/or sold to private investors, which led to a rapid fall in agricultural output in many countries. Large farms remain a feature of many CEE countries, although many of these are now owned by corporations (Lerman et al., 2004). Production stabilized at the lower level but has started to recover in connection to the EU membership. In places like Russia, where household enterprises have been particularly important, small household producers produce nearly all vegetables and potatoes and over 50% of meat and milk products (O'Brien and Patsiorkovsky, 2006) (Table 2-2).

Privatization of agricultural land, as well as upstream and downstream parts of the agrifood chain, was largely completed by 2001, although it is very much an ongoing process in some areas (e.g., Poland and Czech Republic). Land privatization has created a highly fragmented ownership structure across the region—less so in the Czech Republic and Hungary due to the restitution of land title exchangeable for investment vouchers or cash and more so in Bulgaria, Lithuania and Romania where the operational structure allows land to be farmed in large viable units. In Poland and Slovenia most of the land continues to be farmed as family type units as in the pre-transition period. The process of privatization has resulted in a bimodal structure in the region with both small and large scale farms especially important in Bulgaria, Estonia and Hungary. Large scale farms are dominant in Czech Republic and Slovakia and small and medium size farms in Latvia, Lithuania, Poland, Romania and Slovenia. In general, however, policies promote consolidation of holdings (OECD, 2001).

In Albania, the almost complete breakdown of the preexisting system left the countryside open to fragmentation

Table 2-2. **Agricultural output by product and enterprise in Russia.**

Structure of agricultural output by type of product and type of enterprise in Russia 1990-2004									
Type of agricultural product	Type of enterprise (%)								
	Large enterprise			**Private farmer**			**Household**		
	1990	1995	2004	1990	1995	2004	1990	1995	2004
Grain	99.7	94.4	81.2	0.01	4.7	17.4	0.3	0.9	1.4
Sugar Beets	99.9	95.9	88.6	0.01	3.5	10.3	0.0	0.6	1.1
Sunflower	98.6	86.3	74.4	0.0	12.3	24.5	1.4	1.4	1.1
Potatoes	33.9	9.2	6.2	0.0	0.9	2.0	66.1	89.9	91.8
Vegetables	69.9	25.3	14.9	0.0	1.3	4.9	30.1	70.4	80.2
Meat	75.2	49.9	45.1	0.0	1.5	2.4	24.8	48.6	52.5
Milk	76.2	57.1	45.0	0.0	1.5	2.8	23.8	41.4	52.2
Eggs	78.4	69.4	72.8	0.0	0.4	0.5	21.6	30.2	26.7
Share of total agricultural output			43.1			5.9			51.0

Source: O'Brien and Patsiorkovsky, 2006.

and a shift to household self-sufficiency in food. This process was evident in many CEE countries during the 1990s as a substantial proportion of the population, often older, newly unemployed and unskilled, retreated from the cities and towns to rural housing where an older, poorer but more secure way of life could be pursued.

Farm restructuring involved the reallocation of land, labor and capital and included organizational reform such as a move from cooperatives to family farms. In CEE there is now a wide range in the type of farm organization from family farms, private cooperatives, joint stock companies and part-time farmers. The restructuring has led to production efficiency gains but also contributed to the short term production declines seen in the early 1990s. Restructuring was complicated by conditions in the industry pre-reform including the type of farm organization, the degree of capital intensity, the extent of technology use and the degree and speed by which these initial conditions were reformed.

Crop production in the former USSR increased at about the same rate from 1961 to 1980 as world production. However, production levels remained stagnant in the 1980s, before falling about 30% in the 1990s to where production levels in 2000 were the same as in 1961 (Lerman et al., 2003).

2.1.1. Changes in farming and rural population in North America

In NA, the proportion of farm and rural populations as part of the total population has declined significantly since 1945 (Figure 2-4). Mirroring these changes in population have been changes in the agricultural workforce. In 1945, 16% of the total labor force in the United States was employed in agriculture, but this dropped to 4% by 1970 and 1.9% by 2002 (Dmitri et al., 2005). Primary farm operators also begin to work more off-farm jobs during this time period. In 2002, 93% of farm households had off-farm income, a threefold increase since 1945, when 27% of farmers worked off-farm (Table 2-1). The decade of the 1950s saw the largest exodus from farming (Lobao and Meyer, 2000). During the "Farm Crisis," 600,000 farmers exited farming between 1979 and 1985 (Heffernan and Heffernan, 1986); this exit particularly affected the economic base of rural communities in the Midwestern states.

The shift in the relative percentage of urban to rural dwellers is often perceived as an exodus from rural areas, but during this time the total rural population has held relatively constant (Figure 2-4). It is important to look at the geographical consequences of changes in the farming population. For instance, farm size in the US heartland rose by 18% between 1980 and 2000 (Paul and Nehring, 2005). Similar strong growth in farm size occurred in the Lake and Northern Plains states but slower growth was evident in some other areas. Farming-dependent counties were sprinkled throughout much of the US in 1950. By 2000, more than two-thirds of farming-dependent counties were concentrated in the Great Plains of the United States, a giant swath in the middle of the country stretching from the Prairie Provinces of Canada to the panhandle of Texas (Barkema and Drabenstott, 1996; Dimitri et al., 2005).

Agricultural workers in NA. Since WWII the characteristics of hired farm labor supply have fluctuated widely both in North America and in Europe, with labor supply and demand being dependent on changes in farm structures, changes in consumer preferences, the growing power of retailers and the changing importance of the agricultural sector relative to other industrial sectors in the economies of NAE countries. The agricultural sector has been insulated from some of these changes because of its particular labor force structure, being largely dependent on farmer and farmer-family labor. It is estimated that 70% of the US agricultural production workforce in 2003 were farm operators, partners and their unpaid family members. Hired workers make up the remaining third of the workforce (Runyan, 2000;

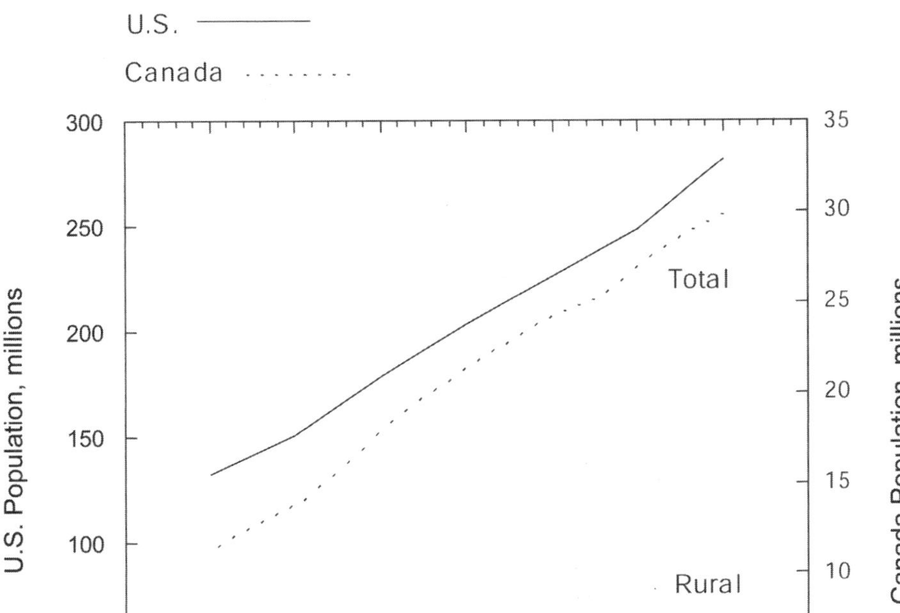

Figure 2-4. *Total change in rural and farm populations in North America from 1930-2000.* Source: USDA data; author elaboration.

Vogel, 2003). Along with the variation of the size of the hired workforce among countries, there is also considerable variation between different sectors of agriculture and a concentration of demand for hired workers in vegetable, fruit and horticulture systems (Frances et al., 2005). The seasonality in these sub-sectors has encouraged the use of temporary workforces, but the nature of this workforce has itself been in flux for the last fifty years as conditions in the industry have changed. Temporary work in agriculture continues to require minimal skills and be physically demanding with poor pay and poor work conditions.

In the US this employment has traditionally been taken up by immigrant populations, which in the past have included Chinese, Japanese, Indian, Pakistani, Mexican and Dust Bowl migrants. The racial division between farm owners and farm workers has persisted; the 1997 US Census of Agriculture found 98% of US farmers were white and 1.5% Hispanic, but 90% of the hired farm workers were Hispanic (Martin, 2002). Hispanics living in rural areas are more likely to be working in lower skilled sectors such as agriculture and because of low wage levels are more likely to live in poverty than non-Hispanic whites (USDA-ERS, 2005a).

Immigrant agricultural populations in the US have been regulated with varying levels of success by means of a number of laws, recruitment schemes and immigration policies including the Immigration Reform and Control Act (1986), which instituted the Special Agricultural Worker program and two guest worker programs (H-2A and Replenishment Agricultural Worker). These were intended to provide a legal work force that could join unions and result with better border control in reducing illegal immigration and creating better working conditions for the legal agricultural labor force (Martin, 2002). These objectives have not been realized given that in the first part of the decade an estimated 50% of all hired workers in crops and livestock farming, 25% in meat processing and 17% in food services are undocumented or unauthorized workers (Wells and Villarejo, 2004; Passel, 2005; Simonetta, 2006). These changes have also been happening in the context of the influence on international migration of the North American Free Trade Agreement (NAFTA), which came into force in 1994, although it did not formally include labor mobility as part of the framework agreement. The economic changes wrought by NAFTA have shifted relative economic power between the signatories to the agreement with differential effects on migration and on relative wage levels occurring among Mexico, USA and Canada (Canales, 2000; Aydemir and Borjas, 2007).

2.1.2 Changes in European farming and rural populations

Europe emerged from the 1940s with the sector predominantly consisting of small "mixed" farms. As technology

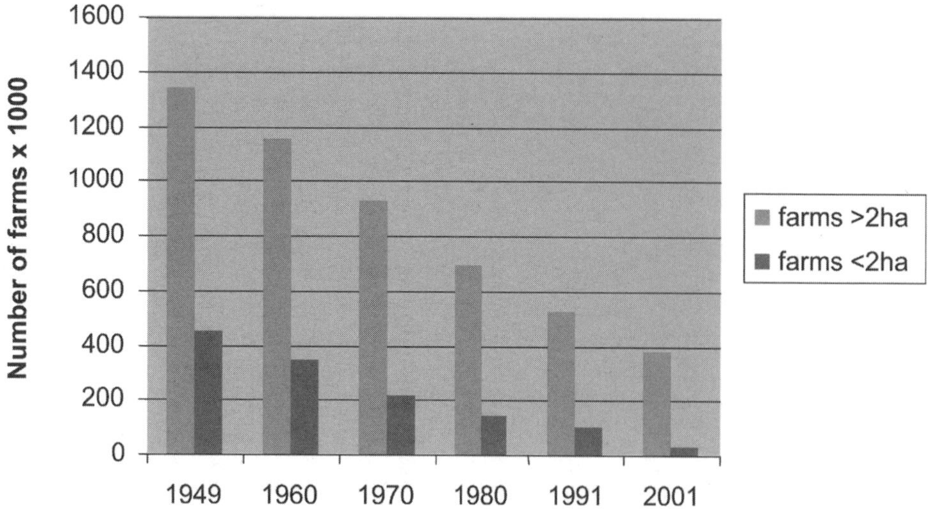

Figure 2-5. *Changes in the number of farms in West Germany 1949-2001.* Source: Government of Germany, 2006.

advanced during the following 50 years the number of farms and the number of farmers and farm workers has declined dramatically. In West Germany, for example, large farms (i.e., those over 2 ha) have declined from over 1,000,000 to less than 400,000, while the number of "small farms", mainly run by part-time farmers has declined even more dramatically. At the same time the area of farmed land has only declined from 12.8 million ha in 1949 to 11.4 million ha in 2001 (Gov. Germany, 2006), indicating that there has been a dramatic increase in average farm sizes (Figure 2-5). In France the agricultural workforce declined from 8% to about 4% of the total working population in between 1977 to 1997. However, since the reform to the Common Agricultural Policy (CAP) in 1992 this decline in Europe, both in agricultural employment and the number of farms, has slowed down as can be seen in the annual percentage changes in labor force. Different countries and different areas in those countries have followed this pattern since 1990 to varying extents.

The changes in the agricultural labor force differed greatly throughout Europe with a noticeable North-South divide. Southern European countries such as Spain and Portugal lost more than a third of their labor force between 1987 to 1997, while the average for the European Community was a 25% reduction. This more dramatic decline reflects the fact that these southern Member States traditionally have a more labor-intensive Mediterranean style of agricultural production; approximately 9% of jobs in countries with Mediterranean production systems were associated with farming. Greece has a particularly high rate agricultural employment (about 20%). Northern European countries such as Denmark and the UK showed average agricultural employment figures closer to 3% for 1997.

In western Europe, individual national migration policy has been gradually subsumed under general EU agreements, although as exemplified by the expansion of the Union to 27 members, full legal labor mobility for citizens of EU states may be delayed and circumscribed by a number of local national regulations. The UK, for example, has developed regulations (e.g., the Seasonal Agricultural Workers Scheme: SAWS) that respond to the need to attract farm workers for seasonal and temporary employment, building on a long history of dependence on migrant workers both from within and outside the UK (e.g., Collins, 1976). This demand has continued and a preference for migrant workers in agriculture remains strong (Dench et al., 2006).

The structure of demand for migrant workers in UK agriculture has been described as dependent on the relationship between growers and retailers; recent changes in favor of retailers has meant a decline in margins for growers. Worsening terms of trade for the growers has been reflected in changing demands made of the workforce, which include more demanding working practices and lower wage rates. The characteristics of the workforce desired by the growers changed accordingly with greater premium put on reliability and the capacity and willingness to accept hard work and lower wages. Growers report that foreign nationals have provided these characteristics more readily, possibly due to their relative lack of security and greater vulnerability and the attraction of high earnings relative to home-country wages and immigration/work permit status (Rogaly, 2006; Frances et al., 2005).

The accession of CEE countries from 2004 to 2007 has changed the supply and character of migrant labor to western European agriculture and to southern EU states such as Greece (Kasimis and Papadopoulos, 2005). Progressive opening of labor markets in western Europe for workers from the new EU states has offered migrants a greater range of work and increasing confidence in asserting employment rights and some evidence has been forthcoming of possible shortages in the supply of seasonal agricultural workers from these sources (e.g., Topping, 2007). These changes have their own cascading effects illustrated by the re-focusing of the SAWS scheme in the UK to relate primarily to workers from Romania and Bulgaria who can only obtain work permits in the UK for agricultural labor. These two countries are the latest to join the EU; most western EU member states (including the UK) have imposed transitional restrictions on the movement of workers to their economies. In turn, there is some evidence that improvement in the economies of new EU member states, in addition to the movement of workers from those states to more developed EU states, has created opportunities for migrants from Russia, Ukraine and Moldova and other former USSR states, some of whom are available for work in the agricultural sector (Patzwaldt, 2004).

The changes in CEE are more complex as collectivization greatly reduced the number of farming units in some countries (e.g., E. Germany and Czechoslovakia) but not others (e.g., Poland). Following the demise of collectivization, there has been a variable re-allocation of land to former owners resulting in fragmentation of the farming units, which has been followed by a re-amalgamation of the small units to create more financially viable enterprises (Bouma et al., 1998). An underlying factor in most transitions was the situation of the land and credit sectors, which together determined the ability—and sometimes the identity—of new landowners and farmworkers during the processes of land restitution and business privatization. In some countries, such as the Czech Republic, Slovenia and much of Poland, viable private farming businesses emerged quickly in the hands of families or companies. In Russia, Belarus and Ukraine, with their much longer period under communist leadership and only partial acceptance of market-oriented systems, structural transformation in the countryside was slow and patchy, despite harsher economic conditions.

Despite the general trend observed across Europe for a decline in farm numbers, increase in farm size and laying off of farm workers, some countries have seen a recent change in emphasis towards developing new on-farm enterprises, expansion into higher value-added crops and engagement in environmental schemes. These activities have actually resulted in an increase in agricultural labor in countries such as Denmark and Greece. Similarly, the recent rise in consumer demand for organic produce has seen an increase in labor in this part of the farming sector to meet needs of labor intensive operations and provide the necessary technical support. For example, data for Denmark has shown that conversion to organic farming has led to a 38% increase in labor costs. A small increase in job creation in the agricultural sector is also resulting from the increase in agrienvironment schemes such as those being implemented in the UK.

The contribution of women to the agricultural workforce largely reflects the overall declining trend in farm employment in the European region. Overall, women make up more than one-third of the European agricultural workforce. However, women make a greater contribution to the agricultural labor force in Southern European countries than Northern, with the exception of Finland. In France, fewer farmers' wives now work on the farm, approximately half in 1997, as opposed to three-quarters in 1979. Part-time work is also less widespread in Northern European countries compared with southern Europe. This high level of part-time employment in southern Europe is associated with the greater number of seasonal activities in this region and is reflected in the employment of both men and women, but is generally more common among women.

Across the EU, women have lower overall labor force participation rates compared to men, higher levels of participation in part-time work, higher rates of unemployment and lower wages (nearly 25% below those of men). Part-time work is by and large a female phenomenon; 85% of the part-time workforce in the EU is female. Non-standard employment (zero hour contracts, casual and seasonal work, temporary work, home working and unpaid family work) accounts for a disproportionately high share of women's employment. In a majority of EU member countries, at least 10% of the female labor force is in temporary employment with the highest rates in the Iberian countries and Greece. Outwork and homework are almost exclusively performed by women. In the more marginalized areas of the EU, two different developments are affecting farm women. On the one hand there is noticeable out-migration, especially of young women, particularly in areas where a strong patriarchal culture coexists with difficult working and living conditions, e.g., Spain and Italy; on the other hand, there is also an increase in the number of female-headed farms (Spain, Portugal and Italy) (Van der Plas and Fonte, 1994). Women provide safety nets where male out-migration has become a dominant feature. In these areas, women adjust farming to reflect the reduced availability of labor (e.g., smaller areas farmed, conversions to extensive farming, greater emphasis on subsistence, cooperatives, and agrotourism) and receive remittances from their spouses.

In CEE countries women are mainly employed as low-skilled workers. As in North America, farm household income in Europe is increasingly from off-farm salaries. The reduction in agricultural employment has, therefore, had a generally greater negative effect on female employment.

Rural women and poverty in the EU. Since a key trend in Europe is concentration (regional, sectoral and among firms), the division between the richer and poorer countries and the more and less prosperous regions is expected to deepen, as are the divisions between women. Within the EU, large regional imbalances occur. Portugal had the highest incidence of poverty followed by Spain, Ireland, Greece and the United Kingdom. In four out of the six countries where poverty rates are reported by economic activity of the household head, they are higher for farmers than for any other group (Denmark, Germany, the Netherlands and Portugal).

With the exception of the Netherlands, female-headed households have higher poverty rates than male-headed households, with the highest incidences of poverty among female-headed households occurring in the UK, Ireland, France and Spain. Several countries also have an unequal ratio of poor men to poor women. For example, in Germany and the UK there are 120 to 130 poor women per 100 poor men. In Italy and the Netherlands the ratios are nearly equal, while in Sweden the ratio is reversed, with fewer poor women (90-93 women per 100 men). The existence of strong family ties (Italy), high rates of female employment (Sweden) and a strong system of social assistance (the Netherlands) appear to influence these ratios positively. In general, rural women constitute one of the major groups most vulnerable to poverty in the Western European population—as members of poor farm families, as female heads of household and as off-farm workers (Borjas and de Rooij, 1998).

2.2 Farm Policies and the Development of NAE Agriculture

Farm policies have played a major role in the transformation of the agricultural sectors in Western countries during the last six decades and clearly contributed to the rapid adoption of new technologies and to dramatic increases in output and productivity. The agricultural legislation and policies of most Western countries during the past fifty years have had two underlying themes. One is to provide farm families with incomes equivalent to those in other segments of society; the second is to ensure an adequate and safe food supply for all the people in the country. To these ends a complex combination of measures has been produced, which at one end of the spectrum has attempted to keep small-scale farmers on the land and at the other has encouraged the consolidation of holdings into efficient mechanized units. Quotas and tariffs barriers have been used to protect local production from foreign competition. Price supports, production subsidies and supply controls have all been used to raise minimum family incomes while meeting some government budget constraints (Stanton, 1985).

2.2.1 US farm policy: A legacy of the Great Depression

The US farm policies implemented after WWII were designed and tried during the Great Depression. As part of the Great Depression, falling prices of agricultural products gripped all the rural areas, prompting the federal government to intervene into agricultural markets to support farmers' incomes, stabilize prices and guarantee cheap food to low income populations (Dmitri et al., 2005). The most important instruments were production controls and government loans.

Beginning with Franklin Roosevelt's New Deal in 1933, the solution to rapidly falling farm incomes was primarily price supports, achieved through dramatic reductions in supply. Supply controls for staple commodities included payments for reduced planting and government storage of market-depressing surpluses when prices fell below a predetermined level. For perishable commodities, supply control worked through a system of marketing orders that provided negative incentives for producing beyond specified levels. In these farm programs were the seeds of later food programs, including food stamps, commodity foods and school lunch programs. The combination of price supports and supply management functioned as the general outline of Federal farm policy from 1933 until the present and continues to figure in current debates, although the mechanisms and relative weights of the policies' components were modified by successive farm legislation. In some years, notably during World War II and postwar reconstruction and again during the early 1970s and mid-1990s, global supplies tightened sharply, sending demand and prices soaring above farm price supports and rendering acreage reduction programs unnecessary. But for most of the period, repeated cycles of above-average production and/or reduced global demand put downward pressure on prices, keeping the programs popular and well funded. Continued public support for direct intervention after World War II arose for different reasons. The low prices and consequent low farm incomes of the 1920s and early 1930s resulted from surpluses created by sharply reduced global and domestic demand, beginning with Europe's return to normal production after World War I and followed by the international economic depression of the 1930s. In contrast, surpluses following World War II resulted from rapidly increasing productivity, exacerbated by continuing high price supports that kept production above demand.

The apparent success of production controls and price supports in raising and maintaining farm incomes by the mid-1930s, made a continuation of these policies publicly acceptable. Nonetheless, intense debate between proponents of high price supports and those who believed farm prices should be allowed to fluctuate according to market demand continued from the mid-1950s to the mid-1960s. The debate was set in the context of large surpluses, low prices and efforts led by the Eisenhower administration to return the US economy and government bureaucracy to pre-New Deal, pre-World War II structures. Out of the debate—between advocates of very high price supports and mandatory production controls and those who wished to end direct government market intervention—came a compromise for farm policy. The Food and Agriculture Act of 1965 made most production controls voluntary and set price supports in relation to world market prices, abandoning the "parity" levels intended to support farm income at levels comparable to the high levels achieved during the 1910s. A system of direct income support ("deficiency") payments compensated farmers for lower support prices. Some exports programs aimed at concessional prices and food aid programs (PL 480) were implemented during the 1950s and 1960s in addition to programs already in place to promote exportations in order to deal with a part of excess output.

The debate over price supports and supply control recurred with enough intensity to divert the direction of policy in the mid-1980s. The new setting was the farm financial crisis and its aftermath, along with efforts by the Reagan presidency to place the American farm economy on a free-market footing. This time, with steadily increasing government stocks of program commodities and Federal budget deficits at record levels, the argument against continuing expensive government support of the farm economy gained support. At the same time, the farm crisis began to undermine some of the farm sector's confidence that domestic price supports and production controls were a very effective way to secure US farm income in a global economy. Supported US prices reduced international marketing opportunities and increasing global supplies undercut domestic production control efforts. Farm legislation passed in 1985 and 1990 maintained the traditional combination of price supports, supply controls and income support payments, but introduced changes that moved farmers toward greater market orientation i.e., lower price supports, greater planting flexibility and more attention to developing export opportunities for farm products. In the 1985 Farm Bill, environmental cross compliance measures were also introduced in order to address specific issues such as soil erosion and conservation of humid areas. This Farm Bill also reintroduced direct subsidies to farm exports: Export Enhancement Program (EEP) and Targeted Export Assistance (TEA).

The stable economic development provided by farm programs in conjunction with rapid technological development resulted in rapid adoption of new and improved technologies on farms, relatively heavy investments in non-farm produced inputs, increased production efficiency and a rapid rate of growth in aggregate production capacity which exceeded aggregate demand (Cochrane, 1987).

There are several shortcomings of these farm programs. First is the failure to understand the structural excess capacity problem confronting commercial agriculture during the period between the end of the Korean War and the increase of the demand for agricultural exports at the beginning of the 1970s. This problem was largely understood as a temporal one. That led to various weaknesses in the farm programs: for instance, unwillingness to impose strict production controls and the tendency to impose production controls over only the commodity in most serious oversupply while permitting the released resources to shift into the production of other commodities. This last weakness was not seriously addressed until the 1980s. Another important shortcoming of the farm programs was the almost complete reliance on acreage controls as a means of controlling supply which induced the substitution of fertilizer, pesticides, machinery and power for land and labor, contributing to the land and water pollution of modern agriculture (Debailleul and Deleage, 2000). In addition, while acreage diversion was also considered as a means to reduce the soil erosion, farmers tended to divert the less productive parts of their land and to intensify the agricultural practices on the most fertile part of their land, often the most vulnerable to the erosion. The farm policy was supposed to protect farmers against sharp declines in agricultural prices and in the same time to contribute to provide consumers with declining prices for food, what was possible due to the improvement in farm productivity. But experience shows that in periods of rapidly increasing farm prices, such as occurred during 1972 to 1975, consumers were not protected against the rise of food prices.

2.2.2 Canada: A bipolar farm policy

In the five decades following WWII, a highly complex set of programs and institutions were implemented as Canadian farm policy. This uncommon situation was due to two reasons. First, the federal government as well as provincial governments have the jurisdiction to intervene in the agricultural field, so some provinces, like Quebec, have adopted a set of farm programs in the last few decades. The second major reason was the bipolar structure of Canadian agriculture: an export-oriented western agriculture devoted to grain and oil-seed crops and a domestic-market oriented agriculture in Ontario and Quebec specialized in dairy, poultry and egg production. In these latter systems, supply management and border protection have been implemented as instruments to adjust the supply to the domestic demand. Beginning in the 1930s, marketing boards were implemented in the western provinces; their monopoly on marketing grain outside of the country was considered the best way to assure good prices for farmers. However, during the 1950s and 1970s some other programs were implemented, including a program to subsidize the transportation of grain from prairies to the central and eastern provinces and the implementation of minimum prices for several crops.

During the 1990s, the federal government undertook a drastic reform of its farm programs. Because of budgetary deficits, combined with trade liberalization and free-trade agreements, the legitimacy of such programs was questioned. Due to budgetary constraints, some programs were phased out and the direct support of farm price programs was abandoned in favor of programs which supported the net average farm income, thereby decoupling farm payments. The supply management programs have been maintained but the future for these programs is still uncertain.

2.2.3 Common Agricultural Policy and the building of a single market

As with North American agriculture, European agriculture was greatly affected by the economic crisis of the 1930s. After WWII, most Western European countries pursued protectionist policies in order to increase self-sufficiency and reduce their agricultural trade deficits. As a consequence, food prices were maintained at a high level. Production responses to high food prices differed from country to country. In several countries, the agricultural sector began to modernize and become more competitive, while in other countries, agricultural structures were still inefficient, leading to greatly different agricultural systems among those countries working to form the European Community.

The implementation of Common Agricultural Policy (CAP) was supposed to be divided in two periods; the period from 1958 to 1970, the "transitional period," was supposed to experiment with new instruments and the "permanent period" beginning in 1970 was devoted to the achievement of a single agricultural market. Actually, the transition to the permanent phase was completed in 1968.

The CAP was designed with several different objectives, including increasing agricultural production through the development of technological progress as well the efficient use of factors of production, in particular labor; ensuring equitable standards in living for farm people particularly through an increase of personal income; stabilizing markets; securing the food supply and ensuring reasonable prices for consumers. This domestically oriented farm policy was based on three major principles:

- A unified market in which there is a free flow of agricultural commodities within the EEC;
- Product preference in the internal market over foreign imports through common customs tariffs; and
- Financial solidarity through common financing of agricultural programs.

Thus, individual nations were supposed to gradually leave their decision-making power in agricultural matters, both at the domestic and international levels, in the hands of the Community. Decisions made in Brussels were to be applicable equally to all member states. Today the CAP's main instruments include agricultural price supports, direct payments to farmers, supply controls and border measures. Major reform packages have significantly modified the CAP over the last decade. The first reform, adopted in 1992, began the process of shifting farm support from prices to direct payments. The 1992 reforms reduced support prices and created direct payments based on historical yields and introduced new supply control measures. These reforms affected the grain, oilseed, protein crop (field peas and beans), tobacco, beef and sheep meat markets. The second reform,

"Agenda 2000" began in 2000 in preparation for EU enlargement. Similar to the first CAP reform, Agenda 2000 used direct payments to compensate farmers for half of the loss from new support price cuts. Agenda 2000 reforms focused on the grain, oilseed, dairy and beef markets.

The most recent reforms (begun in 2003 and 2004) represent a degree of re-nationalization of farm policy, as each member state will have discretion over the timing and method of implementation. The 2003 reforms allow for decoupled payments—payments that do not affect production decisions—that vary by commodity. Called single farm payments (SFP), these decoupled payments will be based on 2000-02 historical payments and replace the compensation payments begun by the 1992 reform.

When member states implement the reforms, compliance with EU regulations regarding environment, animal welfare and food quality and safety will be required to receive SFPs. Moreover, land not farmed must be maintained in good agricultural condition. Coupled payments, which can differ by commodity and require planting of a crop, are allowed to continue to reinforce environmental and economic goals in marginal areas. The CAP budget ceiling has been fixed from 2006-13; if market support plus direct payments fall within 300 million euros of the budget ceiling SFPs will be reduced to stay within budget limits.

Domestic price support

Prices for major commodities such as grains, oilseeds, dairy products, beef, veal and sugar depend on the EU price support system, although price support has become less important for maintaining grain and beef farmers' incomes under the CAP reforms. The major method of maintaining domestic agricultural prices is through price intervention and high external tariffs. Farmers are guaranteed intervention prices for unlimited quantities of eligible agricultural products. This means that EU authorities will purchase at the intervention price unlimited excess products meeting minimum quality requirements that cannot be sold on the market, which are then stored or sold for export with subsidies.

Other mechanisms, such as subsidies to assist with surplus storage and consumer subsidies paid to encourage domestic consumption of products like butter and skimmed milk powder, also support domestic prices. The 2003 reforms, however, cut storage subsidies by 50%. Some fruits and vegetables are withdrawn from the market in limited quantities by authorized producer organizations when market prices fall to specified levels. Reforms have lowered the cost of the CAP to consumers as intervention prices have been reduced. However, taxpayers now bear a larger share of the cost because more support is provided through direct payments.

Direct payments

While price supports remain a principal means of maintaining farm income, payments made directly to producers provide substantial income support. Compensation payments for price cuts generated by the 1992 reform began in 1994 and were increased for the Agenda 2000 reform. These compensation payments were established on a historical-yield basis for arable crops by farm and required planting to receive a payment. Production requirements have been eliminated in the 2003 reform for both crops and livestock, with payments made to farmers based on the average level of payments received during 2000-02. Direct payments currently account for about 35% of EU producer receipts and for an even higher percentage of net farmer income (once input costs are subtracted from receipts).

Supply control

The 1992 reforms instituted a system of supply control that has been maintained through subsequent reforms. To be eligible for direct payments, producers of grains, oilseeds, or protein crops must remove a specified percentage of their area from production. Small producers are exempt from the set-aside requirement. Supply-control quotas have been in effect for the dairy and sugar sectors for nearly two decades.

Border measures

The CAP maintains domestic agricultural prices above world prices for most commodities. In preferential trade agreements, such as those with former colonies and neighboring countries, the EU satisfies consumer demand while protecting high domestic prices through import quotas and minimum import price requirements. The CAP also applies tariffs at EU borders so that imports cannot be sold domestically below the internal market prices set by the CAP. Although the Uruguay Round of Agreement on Agriculture called for more access to the EU market, market access to the EU's agricultural sector remains highly restricted in practice. In addition, the EU subsidizes the agricultural exports to make domestic agricultural products competitive in world markets.

Additional aspects of 2003 reform

Important components of the 2003 reform reflect a philosophical change in the approach to EU agricultural policy. For the first time, much of the pressure to reform the CAP came from environmentalists and consumers. The requirement to comply with environmental and animal welfare standards to qualify for the SFP reflects these pressures. Moreover, farmers must meet food quality and food safety regulations for payments to continue. Another important feature of the 2003 reforms is the move from a price support policy to an income support policy through decoupled payments. EU farmers will have more choices in their planting decisions because of decoupled payments. Commodity support prices continue to exist but at lower levels, while direct payments to farmers without requirements to plant a crop are more widespread.

There is also a marked shift in the way rural development is treated. The 2003 CAP reforms established two pillars in the budget: Pillar I for market and price support policies and Pillar II for rural development policies. In the reforms, a ceiling was imposed on Pillar I spending, whereas Pillar II spending seems open-ended. The intended budget for rural development will more than double over the next 10 years, while the CAP budget for Pillar I may only increase by 1% per year in nominal terms from 2006-13. Moreover, in a concept called modulation, SFP payments greater than 5,000 Euros are reduced by 5%, while farmers whose SFP is less than that are not penalized. The budget funds saved through modulation are transferred to the Pillar II rural development fund. At least 80% of the funds from the penalties will remain in the country where the SFPs were reduced and are to be used for rural development purposes.

The increase in agricultural productivity within the EC was very rapid. While increases in the rate of agricultural productivity in the United States appeared in the 1930s, this trend didn't began until the 1950s in the EC and continued in the subsequent decades primarily due to the implementation of CAP. While protectionist policies were employed by EC member countries before the CAP was established in 1962, it has played a fundamental role in increasing the size of supply and the agricultural productivity.

Benefits and shortcomings of farm policies

Consumer benefits from price stabilization are lower probabilities of shortages and extremely high prices. A large part of gains in agricultural productivity have also been transmitted to the consumer through a long-term tendency of declining real farm prices. Food processing firms benefited from more stable supplies and prices that resulted in more efficient use of processing facilities and improved management decisions. The agricultural supply industry also benefited as farm programs constituted great incentive for investment and adoption of new technologies. For the same reasons, livestock producers also gain from grain price stabilization and government storage policies.

Despite the underlying theme of support for the family farm in both NA and the EU policies, long run effects promoted larger farms. For instance, higher price supports, benefits, deficiency payments, disaster payments and direct aids are generally proportional to output or to acreages. Between 20% and 30% of the farmers are able to capture between 60% and 80% of government payments in either the US or the EU. For instance, 70% of the direct payments of CAP during the financial year 2000 went to 16% of EU eligible farmers.

The results of US and European attempts to dispose of surplus commodities have been particularly damaging to the agricultural sectors of the developing countries. The availability of cheap surplus food from Europe and the US has made it possible for some nations to maintain urban food prices at relatively low levels. This discouraged production by their own farmers and encouraged rural people to migrate to the cities. In addition it made poor nations dependent upon American and European willingness to continue to overproduce agricultural commodities (Bonnano et al., 1990). Moreover, the modernization and intensification of agriculture that have been promoted by these policies has had damaging environmental and social consequences that have not been entirely addressed by reforms.

2.2.4 Agricultural policies in CEE countries

Three broad stages can be identified in agricultural price policy reforms in CEE countries. These began in the early 1990s with the dismantling of administered pricing, production targets and the state monopoly on trade as well as the adoption of price and trade liberalization and limited intervention in agricultural markets. This was followed by an *ad hoc* reapplication of controls on price and market support and on trade restrictions. By the late 1990s and continuing up to EU accession by many countries in 2004, agricultural policy was dominated by the alignment of their agricultural sectors with that of the European Union, particularly to the CAP and to food hygiene and welfare standards (OECD, 2001). Structural reform was directed to improve overall performance of the agrofood sector such as investment to improve market infrastructure, to modernize plants and equipment and eliminate management inertia, as well as consolidation of holdings to ensure viable farming units which depend on a functioning land and land lease market (Cochrane, 2002)

EU support was provided to certain CEE countries for pre-accession restructuring through various programs, with the Special Accession Programme for Agriculture and Rural Development (SAPARD) being important in agriculture. SAPARD is a 7 year program which started in 2000 and allocated two-thirds of its funding program to Poland, Romania and Bulgaria.

In Russia and the NIS, reforms were required in farm-level organization and management and in the development of the physical and institutional infrastructure. Private farming had not developed during the 1990s to any substantial degree and land and rural credit markets remained ineffective as a credible commercial legal system to protect property and enforce contracts remained undeveloped (Virolainen, 2006). However in Russia, there were signs by the 21st century that vertically integrated forms of organizations were emerging. It has been suggested that any productivity gains in Russia in the short to medium term might come more from strengthening vertical ties for production and distribution rather than from real technological or systemic change because of the increasing attractiveness for investment that would result (Liefert et al., 2002).

In Russia in particular there has been "a rapid, quite fundamental change in the principles for developing agricultural production" (Virolainen, 2006). The emphasis has shifted from the family farm to supporting large, commercial farm enterprises. These enterprises form so-called agroholding companies, consisting of either a single farm enterprise or a collection of individuals. These agroholdings may also be part of a larger industrial-economic grouping, such as the Alfa group, Interros, Lukoil, Metalinvest or Rusagro. These enterprises perform as vertically integrated enterprises ensuring raw material supply to group member companies and may be used to ensure the supply of foodstuffs for the core company's employees.

The political reforms that began in 1989 shifted the emphasis in agricultural policy toward developing an efficient, productive, export oriented agriculture based on comparative advantage instead of a focus on responding to basic production targets formulated by national plans with their goal of achieving self sufficiency. At the same time the role of agriculture in the post communist era declined relative to other sectors that began to achieve a relatively faster rate of development (OECD, 2001).

The reforms led to a substantial decline in agricultural production in the Central and Eastern European countries

(CEECs).[3] The gross agricultural output fell by between 15 and 30% for these countries between 1989 and 1992, although for both the Czech Republic and Slovenia, that followed a brief initial increase of some 10%. The decline subsequently moderated for these countries during the remainder of the 1990s and even reversed for the Czech Republic, Poland and Hungary. For Albania, by 1998, output had even reached higher than the 1989 level by over 10% annually (Macours and Swinnen, 2000).

Political and economic reform in Russia, republics of the Soviet Union and the Newly Independent States (NIS) of the 1990s produced similar consequences for agricultural *productivity*. Estimates for Russian crop production indicate a drop of 8% in productivity overall between 1993 and 1998, while overall agricultural productivity rose in Russia and Ukraine between 1992 and 1997 but only by 7% and 2% respectively (Liefert et al., 2002). The major changes in Russian agricultural production and trade following transition included a halving of the livestock inventory resulting from a reduction in imports of animal feed. Fertilizer, machinery and fuel use also fell substantially, resulting in cuts in domestic grain yields and harvest levels. The same applied to Ukraine as fertilizer output was switched to export supply (Liefert et al., 2002).

2.3 Changes in Market Structure

Specialization in agricultural production has been accompanied by significant changes in market structure for both agricultural inputs and outputs. Economic power in food and agriculture and thus the power to make decisions about what to produce and where to produce it, has moved toward fewer and fewer transnational firms which are embedded in a web of relationships in food production, from genetics to food retailing (Yoon, 2006). Some view these changes positively as a way to increase efficiency in the food system (Barkema et al., 2001) while others point toward increased marginalization of farmer and rural livelihoods and negative impacts on communities (Goldschmidt, 1978; Lobao, 2000; Stofferhan, 2006)

In Europe, concentration in the food system started at the retail stage, becoming most obvious during the 1980s and 1990s (Vorley, 2003). In the US, concentration of ownership and control became most visible in the production and processing stages, especially in the poultry sector in the mid-twentieth century. Contrary to European trends, in the US and Canada increased market share by fewer firms occurred in the agricultural input sectors and the food processing stage much earlier than in the food retailing sector.

Horizontal integration is occurring at all stages of the food system from the genetics to raw agricultural commodities to food retailing. The concentration ratio (CR4), which is a measure of the market share of the top four firms in a particular commodity, has continued to increase during the past decade in the US The largest four processors for all the major commodities now have from 50 to 80% of the market share (Table 2-3 and Figure 2-6) which can indicate decreased competition in the marketplace forcing farmers into a relatively powerless position vis-à-vis suppliers or buyers. Others argue that competition is sufficient for farmers to obtain a fair price (Tweeten, 1992; MacDonald et al., 2000). Nevertheless, farmers across the NAE faced with decreasing choices buying agricultural inputs and selling outputs can face a cost-price squeeze that affects their ability to earn a livelihood from agriculture.

The structure of the market for agricultural inputs has changed markedly in the last 50 years. For instance, two firms provide most of the fertilizer used today in North America while one firm has a 25% market share for fertilizers in Europe. The seed industry is even more instructive for other inputs. Globally, the seed industry is increasingly driven by NAE based transnational agrifood businesses (UNCTAD, 2006). Four NAE-based transnational companies provide almost 30% of the world's commercially available seeds while NAE accounts for 43% of the commercial seed market globally (Table 2-4).

Many of the changes in NAE were anticipated by the changing nature of the US seed industry, the most heavily commercialized in the world. In the 1930s, over 150 companies formed to sell hybrid maize, but by the mid-1960s, American farmers had essentially abandoned open-pollinated maize varieties with nearly all maize acreage planted to hybrid maize (Fernandez-Cornejo, 2004). Maize provided the kernel of transformation for the seed industry in general. Between 1970 and 2000, small private seed firms essentially vanished, with more than 50 acquisitions of seed firms by pharmaceutical and chemical firms (Fernandez-Cornejo, 2004). By the 1980s, the maize seed market was dominated by two firms and by the late 1990s, over 90% of cotton seed, 69% of maize seed and nearly half of soybean seeds were sold by the four largest firms in each crop. The same privatization trends are seen in Europe and as a consequence, the private sector is becoming increasingly important.

One of the more striking features of industry changes in the last two decades has been the convergence of ownership between agrochemical and seed/genomic firms. This strategy has worked well "to better control and market proprietary lines of chemicals, genetic technologies and seeds, often sold in a single-bundled package" (UNCTAD, 2006). These bundles can be attractive to farmers and farmer managers as a purchased management tool. However, such packaged bundles can reduce flexibility of on-farm management strategies for pests and weeds, as well as implementation of novel consumer-driven production systems and increase reliance on purchased inputs (c.f. Hendrickson and James, 2005).

When farmers sell their products, they also face highly concentrated markets. In the US less than 10 firms slaughter and process most of the broilers, turkeys, cattle (heifers and steers) and pork in the United States. Many of these are the same firms that operate in Canada. Moreover, the CR4 ratio has been increasing for all livestock processing—particularly steers and heifers and hogs—since 1980 in the US.

[3] The countries that are included under the rubric of the CEEs differ. Some authors restrict the definition to the ten countries that underwent accession to the EU between 2004 and 2007, namely Estonia, Latvia, Lithuania, Poland, Romania, Slovakia, Czech Republic, Hungary, Bulgaria and Slovenia. Others include Albania and the remaining Balkan states, but these are also referred to as the South East European Countries (SEEC). Most of the material here regards the CEEs as to the ten accession countries, unless other countries are referred to specifically.

Table 2-3. **Concentration in the U.S. and Canadian food industry.**

Commodity Market and Top Firms	2007 Concentration Ratio*	Historical CR4
Beef packing (Tyson, Cargill Excel, Swift & Co, National Beef)	CR4=83.5%	CR4=72% (1990)
Pork packing (Smithfield, Tyson, Swift & Co, Hormel)	CR4=66%	CR4=37% (1987)
Broilers (Pilgrims' Pride, Tyson, Perdue, Sanderson Farms)	CR4=58.5%	CR4=35% (1986)
Turkeys (Smithfield/Maxwell Foods, Hormel, Cargill, Sara Lee)	CR4=55%	CR4=31% (1988)
Flour milling (Cargill/CHS, ADM, ConAgra)	CR3=55%	CR4=40% (1982)
Soybean crushing (ADM, Bunge, Cargill)	CR3=71%	CR4=54% (1977)
Food retailing (Wal-Mart, Kroger, Albertson's, Safeway, Ahold USA)	CR5=48%	CR5=24% (1997)
Selected information about concentration in the Canadian agriculture and food industry		
Commodity market and top firms	Concentration Ratio, 2006	
Beef packing (Cargill, Lakeside Packers [owned by Tyson], XL Foods)	CR3=75%	
Durum milling (ADM, Robin Hood Foods [owned by J.M. Smucker Co])	CR2=57%	
Flour milling (ADM, Robin Hood Foods [owned by J.M. Smucker Co])	CR2=57%	

*Concentration Ratio refers to the market share that the top four firms (or three as in the case of soybean crushing, and five in the case of food retailing) control. Concentration Ratios are calculated using statistics reported in trade journals.

Source: Hendrickson and Heffernan, 2006, 2007.

Livestock production in Europe is less consolidated than in North America. For instance, the top 10 integrated broiler producers in Europe account for only 36% of production compared with 66% in the US.

The grain trading sector worldwide is dominated by three NAE based firms. These three players are in the process of rationalizing crushing capacity, closing down some factories and increasing the utilization rate of others.

During the 1990s, intensive mergers among farmer dairy cooperatives left only two major US cooperatives, one of which currently produces 33% of the US milk supply. Two of the largest private companies merged to become the largest dairy processor, controlling 30% of the US milk supply (Hendrickson and Heffernan, 2005). Retail consolidation in dairy increased prices for consumers, yet decreased farm gate prices (Cotterill and Franklin, 2001). Across Europe, there has been a process of international consolidation in dairy processing, led by farmer-owned businesses, in the race to remain competitive with multinational companies. Concentration in dairy is also a trend in Central and Eastern Europe (Csaki et al., 2004).

It is estimated that 60% of retail food purchases in the United States go to the ten largest global food corporations (Lyson and Raymer, 2000). The major food manufacturing

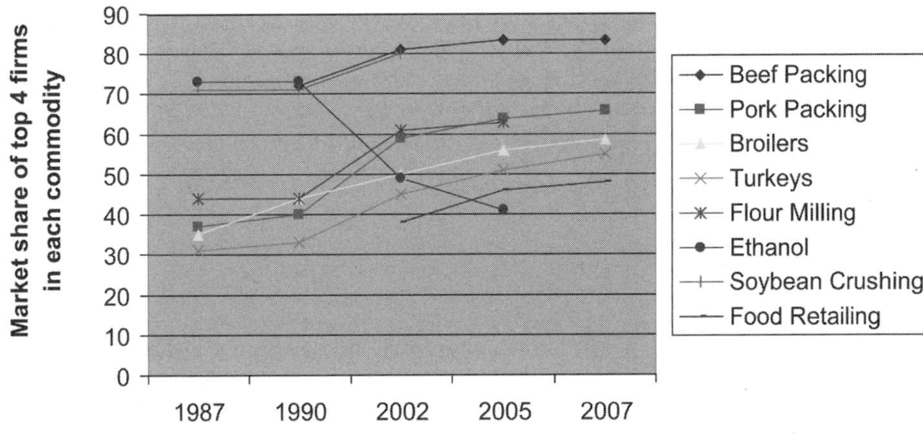

Figure 2-6. *Trends in consolidation in the US food industry from 1990 to 2007.* Source: Hendrickson and Heffernan, 2007

Table 2-4. **Global seed sales by NAE based companies.**

Company	2004 Seed Sales (million US$)	Market Share (%)
DuPont/ Pioneer	2,624	10
Monsanto	2,277	9
Syngenta	1,239	5
Limagrain	1,239	5
Others	17,821	71
Total	25,200	100

Source: UNCTAD, 2006.

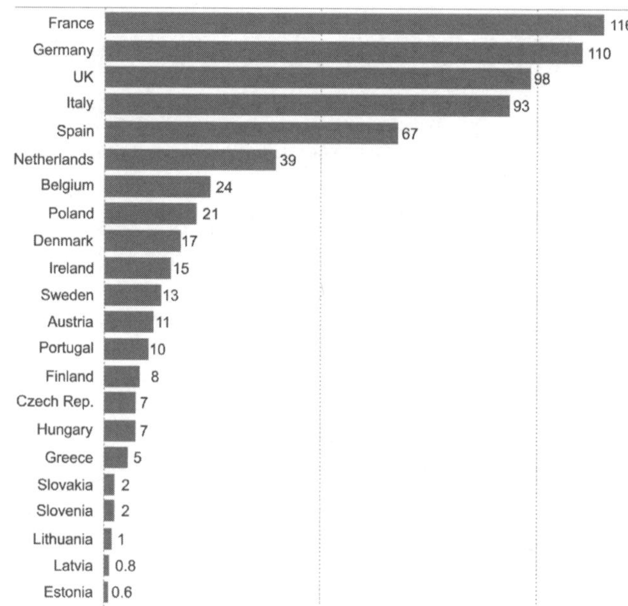

Figure 2-7. *EU-25 food and drink sector 2001, value of production (billion Euro).* Source: USDA-FAS, 2005a

countries in Western Europe are France, Germany the UK and Italy (Figure 2-7). Meat, beverages and dairy are the biggest sectors, comprising 20, 15 and 15% respectively of over EUR 600 billion production value in 2001. It is Europe's leading industrial sector and third-largest industrial employer and concentration in the sector is relatively low (Table 2-5).

Another striking feature of the food system in NAE is that the same firms appear in different sectors of the food system, from genetics to processing because of vertical integration. While not a new term, or process, vertical integration has accelerated rapidly in NAE since 1945. Mostly, this process combines the management (but historically ownership) of a series of stages in the food system. Vertical integration leads to supply chain management, which when exercised in non-competitive markets resulting from horizontal integration, replaces the competitive market providing the coordinating function in a competitive system (Hildred and Pinto, 2002).

We can look to NAE, particularly the US, to see some early examples of vertical integration, e.g., poultry. The poultry industry has now become the prototypical model of industrialized agriculture and is often referred to as a model of the structure that may come to characterize much of US farming in the future (Perry et al., 1999; Hendrickson et al., 2001). Before the 1950s, chickens were raised on more farms in more regions of the US than any other farm animal. The chicken farmer was supported by thousands of local hatcheries, feed mills and processors where chicks, feed and other supplies could be purchased and the birds could be sold. Following the WWII, large feed companies recognized the broiler industry's potential for growth and moved quickly into the production of broilers (Heffernan, 1998; Martinez, 1999; Ollinger et al., 2000). These companies began buying up hatcheries and developing relationships with retailers. By 1960, 286 firms were selling broilers (Heffernan, 1972) and the top four firms controlled 12% of the market. By 1998, only 52 firms remained and in 2007 the top four firms accounted for over 58% of the market (Hendrickson and Heffernan, 2007). Today, a typical broiler complex includes breeder farms, hatcheries, feed mills, grow-out farms, processing plants and retail markets. Commercial feed firms became the major consolidators in the broiler industry, traveling out 25 to 30 miles in a circle from the processing plant to the growers' buildings (Heffernan, 1984). The geographical layout is much the same today except the number of integrating firms and the number of processing facilities are greatly reduced. These firms have about 250 sets of processing facilities across the country producing broilers. Very few growers live in an area where two circles of competing integrating firms overlap. As a result, most growers live in places where they have access to only one integrating firm.

Vertical integration has been manifested through the development of food system clusters or integrated food supply chains; both terms connote a direct line of control for a firm from one stage of the food system to another (Barkema and Drabenstott, 1996; Drabenstott and Smith, 1996). In 1999 three emerging food system clusters appeared to be dominant forces in the food system from genetic material to food manufacturing (Heffernan et al., 1999; Hendrickson and Heffernan 2002). These food chain clusters are still major entities in the agrifood system, but have significantly evolved, including mergers and divestments. Other strong firms remain that have likely formed, or will form, new clusters. It is important to note that much movement to reorganize supply chains in the early 21st century, particularly in the fruit and vegetable sector, has come from large, global retailers, all of whom are based in the NAE, especially in Europe.

One form of vertical integration is the agricultural contract, manifested either as a production or marketing contract. In the US, agricultural contracting covers nearly 40% of the value of agricultural production, up from 11% in 1969 (MacDonald and Korb, 2006). Production contracts exist when an integrating company retains ownership of the commodity as it moves through the chain, with growers receiving a fee for providing labor and/or capital (Sommer et

Table 2-5. **Top European food manufacturers, ranked by turnover in 2007.**

Company	Headquarters	Year end	Sales in € billion	Growth from previous year (%)	Main sectors
Nestlé	CH	Dec. 2006	22.7	5.5	Multi-product
Heinken N.V.	NL	Dec. 2006	8.8	7.3	Beer
Groupe Danone	FR	Dec. 2006	8.6	6.2	Dairy, Multi-product
Unilever Plc/Unilever NV	NL/UK	Dec. 2006	8.6	0.7	Multi-product
Danish Crown Amba	DK	Oct. 2006	6.5	0.5	Meat products
Groupe Lactalis	FR	Dec. 2006	6.4	30.6	Dairy products
Associated British Food	UK	Sept. 2006	5.7	9.5	Sugar, Starch,
Sudzucker	DE	Feb. 2007	5.8	7.8	Prepared foods, Sugar, Multi-product
InBev SA	BE	Dec. 2006	5.5	7.2	Beer
Carlsberg	DK	Dec. 2006	5.2	6.1	Beer
Scottish&Newcaslte	UK	Dec. 2006	4.9	2.0	Beer, Beverages
Royal Friesland Foods N.V.	NL	Dec. 2006	4.7	5.8	Dairy products
Ferrero	IT	Dec. 2006	4.6	0.0	Confectionery
Campina	NL	Dec. 2006	3.6	1.5	Dairy products
Oetker-Group	DE	Dec. 2006	3.6	-1.1	Multi-products

Source: CIAA, 2007.

al., 1998). In marketing contracts, farmers retain ownership and use the contract to specify price, quantity and quality of product to be delivered. About 10% of all US farms use a contract of some sort, with almost 50% of large commercial farms involved in contract production (MacDonald and Korb, 2006). Contract usage varies among commodities. In 2003, nearly 60% of hogs and almost 90% of poultry and eggs in the U.S. were sold through contract production, primarily production contracts. Crops like vegetables, fruit and rice tend to have higher rates of contracting than corn, soybeans, wheat and sugar beets. Marketing contracts are much more prevalent in crop production while production contracts predominate in livestock production. While contracting can provide risk management for producers, contract farming can also pose risks to social structure when it creates the structural equivalent of factory or piece-rate workers who lose control over decision-making or assets; and to family well-being given the contractor grower's asymmetrical bargaining power relationship with integrating firms (Hendrickson and James, 2005; Stofferahn, 2006; Hendrickson et al., 2008).

2.4 Changes in NAE Cropping Systems Since 1945

2.4.1 Changes in soil AKST and use since 1945

Soil is one of the basic natural resources and is vital for agricultural productivity across NAE, a region with extensive amounts of productive soils. Knowledge of soil is critical to agriculture, especially in low input agricultural systems, such as organic agriculture. Traditionally, knowledge of soil type on a particular farm passed from one generation of farmer to the next and traditional practices of manure application were followed to improved soil productivity. Since the end of the WWII, development and availability of soil analytical techniques has led to a more science-based approach for increasing and conserving soil productivity.

Soil testing facilities have been developed largely in response to issues related to agricultural productivity. Concerns such as nutrient depletion and acidification led to the establishment of soil testing programs at publicly-funded institutes in the late 1930s to the early 1950s. These provided services to help farmers make decisions about fertilizer and lime applications (e.g., Olsen et al., 1954; Mehlich, 1984). In recognition that saline and alkali soil conditions reduced the value and productivity of considerable areas of land in the US, the United States Salinity Laboratory was created in 1947 (Richards, 1954). During the 1970s soil testing expanded, providing additional tests and services in response to renewed emphasis on the efficient use of agricultural inputs such as fertilizers, largely due to the energy crisis and an increased public concern for the protection of water quality and prevention of pollution from chemical fertilizers. Similarly, increased ability to analyze trace elements allowed recommendations to be given to farmers concerning shortages, excesses or trace elements.

Until the 1980s, there was substantial investment by governments in soil science research, predominantly focused on soil productivity and aimed to increase agronomic yields. However, since the 1980s, this investment has decreased and the institutional knowledge about analytical methods for soils, water and plant material is lodged more and more in the US private sector (Prunty, 2004). In contrast, following shrinkage in the 1980s, soil science is re-emerging as vital component of agricultural and environmental sciences in Europe, with a current EC strategy and publicly-funded research program to protect Europe's soils from erosion and degradation and ensure sustainable use (EC, 2006).

Extensive and detailed mapping of US and European soils was initiated following World War II and today has evolved into comprehensive, digital national maps of soils in many countries across NAE. This has resulted in more appropriate land use based on soil classification (e.g., rough pasture, arable land). Over the last three decades, there has been an evolution to, assemblage and development of long-term soil resource assessment technologies that are land or ecological. This is especially applicable to forestry management in both the US and Canada (Hills, 1952; Smalley, 1986; O'Neil et al., 2005). Since 1945, there has been development and refinement of soil and water conservation technologies (USDA-SCS, 1955; USDA, 1957; Troeh et al., 1980; USDA-NRCS, 1996; Weesies et al., 2002).

There is greater appreciation of the value of manures and sludge for providing both nutrients and organic matter to soils used for crop production. Proper application rates have been increasingly understood to minimize movement of nutrients off site, which could cause adverse ecological effects, e.g., eutrophication elsewhere. Organic systems are sometimes thought to lead to increased manure runoff, due to their increased reliance on organic fertilizers (Stolze et al., 2000). However, studies from the UK at least (Shepherd et al., 2003) indicate that awareness of the problem has largely alleviated it. In addition, the reduced excess of nutrients on organic farms can have beneficial effects on water quality via reduced nutrient runoff (Shepherd et al., 2003).

2.4.2 Changes in cropping systems in NAE

Increased productivity is the key change in NAE cropping systems. Arable crops, especially the major commodity small grain crops, such as wheat, barley and maize along with the oilseed crops (soybeans, oilseed rape, sunflower), the legumes (peas, beans) and root crops (sugar beet, potatoes) have formed the backbone of crop production in the NAE while fruits and vegetables, with their great range of crops, from lettuces to apple trees, make up the remaining production sector. Over the last 50 years there has been some change in the proportions of different crops grown, such as the increase in oilseed production, but the overall area of agricultural land has not increased during this period. In fact, data from FAOSTAT indicates an approximately 10% reduction in agricultural lands for the EU(15) and for the USA between 1961 and 2003, with a lesser decline in Canada. In the CEE, the amount of land in agricultural use initially remained constant after the end of the Socialist era, although today there seems to be increase in the amount of uncultivated land across the region, which by certain estimates amount up to 30% in some countries (OECD, 2001).

Despite stable or declining arable land, production of virtually all crops has increased significantly (Figure 2-8), in some cases more than doubling, in NAE during this time. The increases in production, particularly in Western Europe and North America, have been stimulated by the increasing demands for food from the rising NAE population during the last 50 years. This was particularly important in the 1950s, as there were real food shortages in many countries in the years following World War II. Post-war agricultural conditions in the Soviet Union were dire, with famine conditions in 1946-47 (Medvedev, 1987) and per capita production of grain and meat below 1913 levels. These conditions were due to the direct destruction of farming and food distribution resources in CEE. In the western NAE, the continued momentum to increase production was encouraged by the politically driven agricultural financial support systems in Western Europe and USA (see 2.2), aimed at ensuring the continued viability of the rural economy. The Soviet Union turned to centralized planning, collectivization and ultimately the Virgin Land Program, when 36 million ha in dry areas were ploughed and sowed in the late 1950s to increase grain production.

Although production lagged behind that in Western Europe and most of the world, CEE farms steadily increased arable production from 1945 to 1980 (Lerman et al., 2003). In the Soviet Union, by the mid-1950s cereal production exceeded the 1913 level and between 1950 and 1970 rose by more than 2.3 times to 186.8 million tonnes (Goskonstat, 1971). After the breakdown of the collective farm system, there was a rapid decline in productivity starting in 1991, with large areas of arable land essentially left unfarmed. For instance, up to 40% of arable land in the Baltic States was abandoned in the 1990s, with a similar decrease in agricultural output (Lerman et al., 2003). This in turn led to a 38% decrease in per capita income in rural areas. Far less land was abandoned in Hungary and Poland where markets were more robust. There has been a recovery in production in most CEE countries, but production levels in the smaller countries are still only at 1960s levels (Lerman et al., 2003). Farmland in the larger countries, especially in eastern Germany, Hungary and Poland, was seen by investors from Western Europe as having good potential for further increases in production by applying modern technology and having relatively low labor costs. Some areas of arable farmland in these countries are increasingly owned by Western consortia. Most CEE countries are now members of EU-27 and EU management of CEE grain production is expected to increase it by around 25%, an increase of some 50 million tonnes. This increase has already become apparent in places like eastern Germany where yields of all grains now equal or exceed those in Western Germany.

Another factor in increased crop production in NAE has been the increasing demand for meat, (see discussion in 2.5) coupled with the increasing intensification of meat production often resulting in intensive housed systems, requiring large quantities of grain, protein and oilseeds. Increased crop production was facilitated by, and to some extent stimulated by, the development of new cultivars and technologies aimed at increasing yields and decreasing yield threats from biotic and abiotic factors (e.g., pest and disease attack, weather impacts on crop growth and harvesting). Research on crop production inputs and the dissemination of the information to farmers has played a key role in providing tools for farmers to increase their production. The major contributors to these yield increases are:

1. Breeding of higher yielding cultivars and the adoption of high-yielding hybrid seeds for planting;

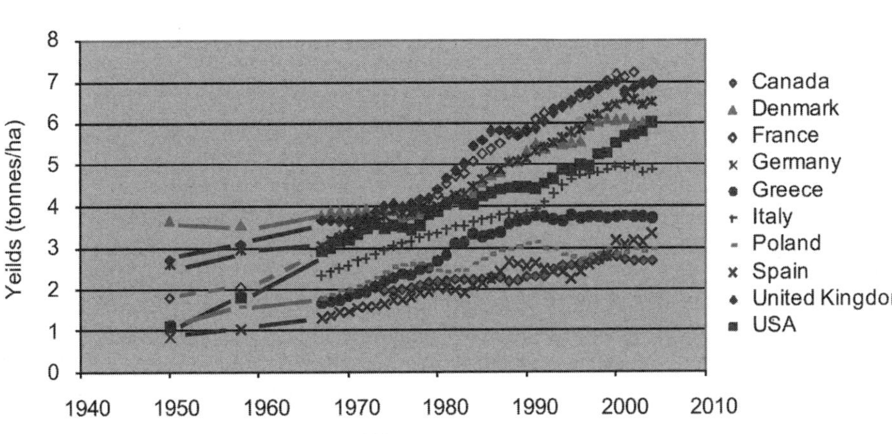

Figure 2-8. *Wheat yields in ten NAE countries since 1950.* Source: FAOSTAT, 2006; FAO Yearbooks 1950 and 1958.

2. Increased availability of fertilizers and increased knowledge of how to use them;
3. Development of new pesticides to control weeds, pests and diseases;
4. Better understanding of the biotic and abiotic factors constraining yields, leading to optimizing agronomic practices (e.g., sowing dates, plant densities, fertilizer timing);
5. Improvement in machinery design and range to assist optimization of crop production;
6. Increased use of irrigation;
7. Enhanced mechanisms for technology transfer, such as development of national agricultural advisory systems; and
8. The delivery of information by the private sector, e.g., on the use of their products. It is as important a source of information to farmers as the public sector extension services and related public sector support.

These advances are summarized in data from the long-term Rothamsted wheat experiment (Figure 2-9), which clearly shows the role played by a number of different inputs in delivering higher yields.

While increasing productivity has been the main goal of

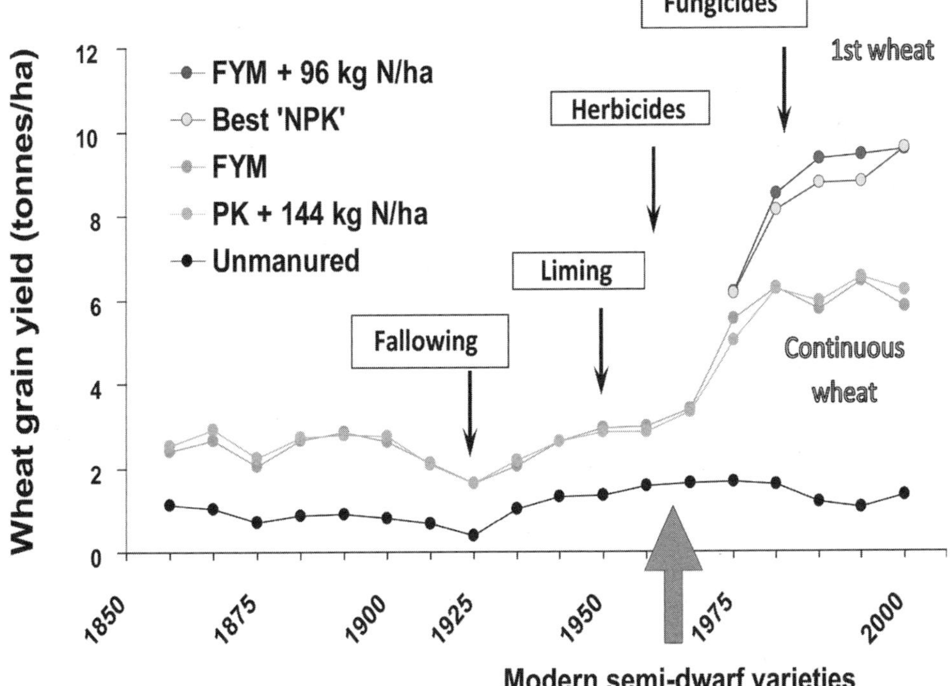

Figure 2-9. *Yield responses on the Broadbalk winter wheat experiment at Rothamsted Research (UK) in relation to the introduction of novel agronomic practices.* Source: Updated from Poulton, 1995.

the last 60 years, there is evidence of little increase in yields since 2000, suggesting that farmers may have reached economically optimal yield achievable with the cultivars available at the present time and in the current economic and policy atmosphere. Similar responses can be identified for other major arable crops.

As well as the direct contribution of science and technology to increases in yields, the establishment of effective technology transfer systems to ensure that the "new" advice was conveyed to the farmer users was also of great importance. Such advisory systems have sometimes involved the public sector (government sponsored advice) and sometimes the private sector. In the US, development of an extensive public knowledge transfer system through the cooperative extension service of land-grant universities contributed greatly to agricultural productivity (Hildreth and Armbruster, 1981). However, today there is a transition from publicly supported technology transfer systems to private technology transfer systems (see Chapter 4). The former tended to be more holistic in approach while the latter has primarily been associated with commercially viable products, whether new agrochemicals or new cultivars (c.f. Fuglie et al., 1996).

2.4.3 Increasing cropping systems productivity through inputs

As noted above, changes in outputs of cropping systems across the NAE reflect changes in production and management systems that utilize inputs such as mechanization, labor, seeds, genetics, nutrients and irrigation, in new and different ways.

2.4.3.1 Mechanization

The last half of the 20th century saw dramatic changes in farming operations because of increased mechanization. The introduction of the diesel engine, compact combine harvesters and sophisticated hydraulic and transmission equipment has reduced labor requirements in weeding, harvesting and threshing (Park et al., 2005).

Improved efficiency and increase in machine scale may explain some of the decline in the number of harvesters and threshers observed in the USA in the 1960s, which has maintained a plateau since the mid-1970s. In contrast, data for Europe showed a large increase in uptake during the 1960s and 1970s showing a continued investment in this machinery and reaching a peak in the number of machines during the mid-1980s.

New developments in mechanization also relate to precision agriculture, which seeks to improve performance by mapping the specific nutrient needs or levels of pest damage to growing crops in such a way that differing treatments may be provided within the same field (e.g., McBratney et al., 2005). By providing precise information about variable field conditions, precision agriculture can substitute knowledge for chemical inputs such as fertilizer and pesticides (Bongiovanni and Lowenberg-DeBoer, 2005), while improving management techniques for environmental and economic goals. It is often—but not necessarily, associated with the incorporation of new technologies (e.g., global positioning service or electronic sensors) into varying agricultural machinery (McBratney et al., 2005). Precision agriculture can benefit the environment by reducing excess applications of inputs and reducing losses due to nutrient imbalances or pest damage, but the necessary technology is at present best suited to relatively large farms so that the capital cost of investment can be spread over a large output, primarily in places like the United States and Canada (Natural Resources Canada, 2006).

In some CEE countries, the collectivization of agriculture tried to exploit economies of scale, particularly in the fields of mechanization and in the use of agrichemicals. In the Soviet Union, productivity advances were largely achieved by government-mandated and government-sponsored industrialization of agriculture. Thus, between 1950 and 1974 the production of plough-tractors increased by 79% to 218,000 units per year and the production of cereal harvesters increased by 91% to 88,400 units per year. However, investment in machinery was limited by lack of state resources for collectivized farms and lack of access to credit for private landowners (Kovách, 1999).

Another agricultural sector that has seen significant mechanization advances is glasshouse production, which is used for high value crops such as tomatoes and ornamentals. The use of glasshouses and other structures enable horticultural crops to be protected from frost, irrigated as required, protected from pests and disease and brought to market out of normal season in first class conditions. Since 1950 growing sophistication resulting from the use of automatic temperature, humidity and ventilation controls has improved performance and reduced the labor requirement. However, as transport becomes cheaper, protected crops face growing competition from imports grown in climates that are more favorable. One response has been to devise cheaper ways of protecting crops, notably the use of plastic and polytunnels.

Mechanization of agriculture allows more timely completion of tasks and reduces labor requirements, thereby increasing productivity, avoiding labor shortages and eliminating unpleasant jobs. It also allows cropping of lands previously too difficult to cultivate. But mechanization also has disadvantages; including loss of jobs, costs of maintenance and fuel as well as elimination of hedges and expanded field size to accommodate larger equipment (Wilson and King, 2003).

The main drivers of mechanization have been the desire for greater productivity in the 1950-60s (EEA, 2003), the reduction of the labor leading to an increased quality of life and increased economic needs. Moreover, AKST has provided mechanisms for the achievement of engineering improvements for agricultural and forestry equipment and more sophisticated handling of milking, as well as allowing for the development of computer management in animal feeding. Thus, mechanization is correlated with field size across NAE, changed management systems and increased flexibility of land use and management. All of these changes have had very important economic, environmental and social implications.

2.4.3.2 Plant breeding, seeds and genetics

A key contributor to productivity increases in crops has been the major advances in crop breeding since the late 1930s, including the development of hybrid crops, cell fusion, embryo rescue and genetic engineering. Many of these new techniques derived from new discoveries in biological sciences and major advances in the fields of genetics (e.g., the discovery of the structure of DNA and the understanding, at the molecular level, of genes as physical entities that could give rise to Mendelian-style inheritance). Post WWII, the study of genetics led to the development of new techniques to introduce inheritable traits into organisms, a subset of the broad set of methods known as biotechnologies designed to adapt living things for the production of useful products. These new techniques include genetic engineering (where a genetic "cassette" manipulated in vitro and containing a recombinant DNA gene for a desired trait is inserted into the organism) and marker assisted breeding (where the use of known "marker" sequences associated with a desired trait are used to determine if the desired trait is inherited in offspring from conventional breeding).

The new techniques of genetic engineering and marker assisted selection have yet to result in improved cultivars with higher yields and other quantitative traits controlled by many genes simultaneously. The current seed varieties available in NAE for most crops, including those for increased yield, have been developed largely through conventional breeding where plants with desired traits are cross-bred and the resultant offspring contain the desired trait. Commercial hybrids are produced by the conventional breeding of two carefully chosen different high-quality true-breeding parental lines to yield progeny that themselves do not breed true, but that in combination give good yield (show vigor) and exhibit superior qualities, above those of traditional (open pollinated) varieties.

Hybrid varieties generally have increased vigor over their open-pollinated counterparts. With the growth of mechanization of agriculture, hybrids could provide uniform characteristics amenable to mechanical harvesting such as uniform maturity, concentrated fruit set, etc., thereby increasing their attractiveness to and profitability for farmers. At the end of World War II, the emphasis was almost solely on yield, rather than nutritional quality because of food shortages in Europe. Later this trend continued because of the rise of processed food where uniform standards were required. This emphasis has remained until very recently with the advent of foods with additional or extra vitamins or minerals.

Between 1940 and 1960, new maize hybrids were developed by private companies such as the forerunners to Pioneer Hi-Bred (Troyer, 1999) that were suited to the application of nitrogen fertilizers. Between 1950 and 1980, the amount of nitrogen fertilizer applied to corn in the USA increased by a factor of 17 (Kloppenburg, 2004). Changes in plant architecture brought about by hybridization allowed these plants to be grown more densely with higher rates of fertilizer application and they were typically managed with the use of insecticides, fungicides and herbicides. Indeed, developments in crop protection have tended to parallel those in fertilizers.

Breeding with conventional techniques and biotechnologies has made considerable contributions to the development of non-cereal crops. The main targets for breeding have been agronomic properties such as crop pest and disease resistance and tolerances to biotic stresses (e.g., cold, heat, salt). Extending crop flavor, quality, nutritional characteristics, shelf life and seasonality are increasingly of importance in breeding programs for high value crops. Some breeding programs are even targeted at improving harvesting and transport. For vegetable cropping, quality has been the main driver of different breeding. There is currently renewed interest in breeding for resistance against pest and diseases in order to decrease pesticide inputs.

Mutagenesis

Radiation (usually gamma or x-ray) and certain chemicals have been used to induce mutations in plants as part of plant breeding for the past 50-60 years. Induced mutations are used to provide a general increase in genetic variation for use in plant breeding, or for the direct production of a variety with a certain characteristic. The techniques have been applied to almost all crops. Seed producing crops form the majority of new varieties produced through mutagenesis, but varieties of crops that can be reproduced vegetatively (e.g., the banana, trees, ornamental flowers) have also been developed (Ahloowalia et al., 2004). Mutagenesis has unpredictable effects and after exposure, plants must be grown to see if any useful mutants result that can be multiplied and developed as distinct varieties or used in plant breeding.

Mutagenesis is reported to have resulted in the production of 2,252 varieties according to the FAO/IAEA mutation varieties database up to the end of 2000 (Maluszynski et al., 2000). It has been increasingly applied to ornamental plants and flowers. One factor favoring the use of induced mutants has been the lack of intellectual property restrictions on access for use in cross breeding programs. One of the highest profile uses of mutagensis in plant breeding in recent years has been in the production of non-GE herbicide tolerant crops, e.g., for imidazolinone tolerance.

Marker assisted selection

DNA knowledge-based techniques, such as marker-assisted selection (MAS) and genetic engineering, rely on genomic characteristics and mapping and have shown great promise over the past few years (Asíns, 2002). This is especially true for complex characteristics such as drought resistance that tend to be controlled by multiple genes and hence are not amenable to straightforward genetic engineering strategies. Furthermore, plants produced using MAS are considered conventionally bred in the US and Europe and are not subject to the same consumer and safety concerns raised with respect to GE crops, although in Canada they are regulated in the same manner. Marker assisted selection can be performed by private companies or public institutes as varieties would be protected by plant breeders rights.

Genetic Engineering

In NAE, only North America has embraced genetically engineered crops since 1996 (Figure 2-10). Predominantly herbicide tolerant (HT) and/or insect resistant (IR) GE varieties of soybean, maize, cotton and canola are grown. For the most part, European acreage is limited to field trials of GE crops (ISAAA, 2005). GE crops producing novel compounds not intended for food use (industrial and pharmaceutical crops) are currently grown only in the US in small quantities and under strict management systems.

According to surveys conducted in 2001-2003, the majority of US farmers adopting GE corn, cotton and soybeans

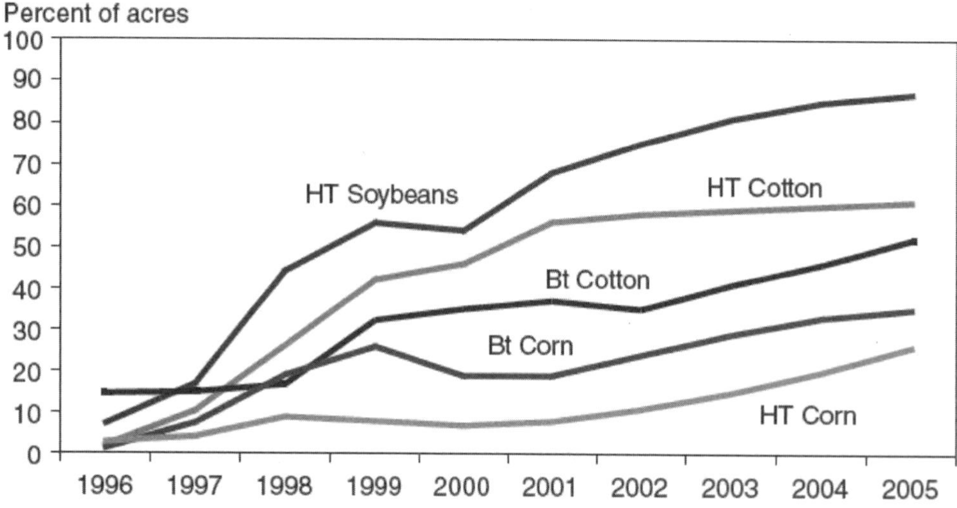

Figure 2-10. *Uptake of genetically engineered crops in the U.S.* Source: Fernandez-Cornejo and Caswell, 2006.

indicated that they did so mainly because of improved weed or pest control. Other reasons for adopting these varieties were to save management time, to make other practices easier and to decrease pesticide costs. The actual impact on farm income appears to vary from crop to crop; in some instances, management time savings have offered farm families the opportunity to generate more off-farm income (Fernandez-Cornejo and Caswell, 2006). In the EU, the total area of commercially grown GE crops is much less, accounting for only a few percent of the total maize harvest which is only grown for animal feed (GMO Compass, 2007). Regulatory differences and differences in public attitudes towards GE are the keys to understanding the different patterns of growth and are discussed later in this section.

Changes in the organizational arrangements of seeds and genetics

Plant breeders turned to new genetic techniques for a variety of reasons, including major emphasis on increased production and productivity in the political arena across the whole of NAE as well as because of market demand—c.f. discussion of political emphasis on demand in Eastern Europe (Medvedev, 1987). Moreover, efficient and well-financed knowledge transfer systems (e.g., extension and private consultants) moved these new plant breeding technologies and techniques into widespread use. In addition, plant breeders were responding to the larger scientific arena that was pushing knowledge boundaries.

Such major transformations in technologies and techniques were accompanied by significant changes in the organizational arrangements of seeds and genetics. Even as hybrid maize was developed by public institutes such as USDA in the 1920s, it became clear that there was an economic dimension to their development (Kloppenburg, 1991). Because the grain harvested from hybrid plants cannot produce economically viable seed, the seed has to be bought each year by the farmer. This contrasts with open pollinated varieties where seeds can be saved from year to year. Thus, the seed business developed from a public service to a profitable industry (Fernandez-Cornejo, 2004). At the same time, the number of varieties researched, developed and produced by public institutes waned.

A major driver of the shift from public to private research was the establishment of Plant Breeders' Rights (PBR). PBR are granted to the breeder of a new variety of plant to grant the control of the seed of a new variety and the right to collect royalties for a number of years. For several of the main commodity crops, farmers cannot sell the seed they produce but can use their own crops as seed. In 1961, the International Convention for the Protection of New Varieties of Plants, which restricted the sale of propagated protected varieties, was signed. Within Western Europe and the United States, national legislation was passed in the 1960s and early 1970s in accordance with the Convention. The WTO's Agreement on Trade-Related Aspects of Intellectual Property Rights (TRIPs) and The International Union for the Protection of New Varieties of Plants regulate plant breeders' rights internationally.

The legislation concerning plant breeders' rights was intended to stimulate private investment in producing new varieties. It certainly has done this but some maintain that there is conflict between these international agreements on plant breeders' rights and the Convention on Biological Diversity which advocates "fair and equitable sharing of the benefits arising out of the utilization of genetic resources. This has led to considerable continued discussion in a number of international forums. These uncertainties affect farmer practice and farmer profitability; clarification will be important for farmers in poor parts of the world to maintain profitability. The latter is important as the development of plant breeding techniques within NAE has had significant impacts on the rest of the world, particularly as many of these techniques and their resulting products have been transferred globally.

The way GE crops have been introduced into farming in NAE has in part depended on changes within agrochemical and seed industries. In the mid-1980s a new "technological trajectory" based on biotechnology began to emerge for the agrochemical and seed industries (Parayil, 2003; Chataway et al., 2004). Regulatory pressures, which made it more challenging and costly to bring new chemical-based products to market, and the existence of new science and a willingness on the part of industries to engage in large-scale change meant that biotechnology was adopted in research and development in a radical way (Chataway et al., 2004). However, the nature of change was such that adoption of new biotechnology based techniques (predominantly genetic manipulation) initially contributed to strengthening firms' abilities to produce chemicals rather than biotechnology-based alternatives to chemicals. Most multinational agrochemical companies used biotechnology to speed up the screening process for agrochemicals and to improve its efficiency and targeting (Steinrucken and Hermann, 2000). Biotechnology is closely related to changed developments in pharmaceuticals (Malerba and Orsenigo, 2002) and relates to three main areas:

- Using genomics to validate targets for new pesticides;
- Using combinatorial chemistry to generate large numbers of new chemicals for screening; and
- Using high throughput screening to test very large numbers of chemicals, rapidly on a range of living targets.

These new methods are unlikely to increase the number of new chemical products reaching the market but they are expected to allow companies to meet increasingly stringent regulatory requirements while still launching one or two major new products a year (Tait et al., 2000).

The development of genetically engineered crops is not entirely within the private sector in NAE; two examples thus far of publicly developed GE crops that have been commercialized or are undergoing regulatory review are virus resistant papaya and virus resistant plum (AGBIOS, 2008).

A key feature of the early evolution of biotechnology were efforts to create a "life sciences" based industrial sector. Negative public opinion is one factor that affected these plans. The concept of life science synergies played an important part in agrochemical and biotechnology industry managers' strategic planning (Tait et al., 2000). Early interpretations of the term "life science" assumed that, by using biotechnology to gain a better understanding of the functioning of cells across a wide spectrum of species, there would be useful cross-fertilization of ideas between the development of new drugs and of new crop protection products for agriculture. The vision was one of synergy at "discovery" level, where a better understanding of genomics and cell processes, made possible by fundamental knowledge gained in the life sciences can lead to new drugs, new pesticides, GE crops and genetic treatments for disease.

These assumptions were accepted without much questioning until the very early years of the 21st century, partly to justify the continued retention within the same multinational company of two sectors with markedly different profit potentials, pharmaceuticals and agrochemicals. However, the original conception of a life science sector is now being reinterpreted. The synergy worked well where both partners are interested in sources of *chemical* novelty, but not in the *gene* area. The large scale marketing of genetically engineered organisms is not a significant factor in the strategies of pharmaceutical companies. Although experience in the USA and other countries has indicated that GE crop development is potentially very profitable, the negative public reaction in Europe has created potential conflicts of interest between the two industry sectors (Tait et al., 2000).

Over a medium and longer term timescale useful synergies between pharmaceutical and agricultural areas of biotechnology may again emerge, for example genetically engineered pharmaceutical crops. However, it is not clear that a link between the agrochemical and pharmaceutical divisions of companies will be maintained (Tait et al., 2000) and this could influence the direction on agriculture related science, technology and innovation. GE crops producing novel compounds not intended for food use (industrial and pharmaceutical crops) are currently grown only in the United States in small quantities and under strict management systems. Under these conditions, no ecological impacts have been detected.

It is clear that the development of important new technologies in plant breeding (i.e., hybridization, embryo transfer, genetic engineering, etc.) has significantly increased productivity of cropping systems in NAE. Moreover, the shift from public institutions to private industry in the development of new varieties and technologies in plant breeding has had considerable impact on the development of cropping systems across the region. Where new technologies and products were developed that could be protected through IPR, industry consolidation has tended to occur. Many firms combined to take advantage of strong demand complementarities between products (Just and Hueth, 1993). This industrial concentration may create efficiencies but it may also limit the technological options as smaller firms which often bring dynamism to a sector find it harder to compete at the level of bringing products to market. However, they often arrange collaborations with larger firms in which they bring initial innovative research to a company with greater resources for product development and deployment. Similar arrangements are increasingly common between researchers in academia and large firms as well.

2.4.3.3 Nutrients in cropping systems

The productivity of agricultural crops draws on three primary sources: carbon dioxide from the atmosphere, and water and nutrients from the soil. While carbon is replenished by the atmosphere, continuous harvest of plant material can eventually strip reactive nitrogen (N), potassium (K) and phosphorus (P) from the soils impeding further plant growth. Agricultural production can also be limited by minor nutrient deficiencies, but N, P and K are the main limiting factors for production. Hence these are the main nutrients that are augmented through synthetic fertilization.

Traditional fertilizers were organic manures, but by the early to mid 1900s the use of inorganic sources of P, mined from phosphate rocks, and reactive N produced by industrial processes came into agricultural use as a result of the development of the Haber-Bosch process in 1910. After the end of World War II the use of synthetic fertilizers increased dramatically as a result of the breeding of new varieties able to respond to the increased fertilizer levels. The trends for NAE are similar to the world as a whole. Between 1950 and 1972 the supply of NPK fertilizers to Soviet agriculture increased almost 10 times and the rate of NPK application increased from 7.3 to 55.9 kg/ha per year (Goskonstat, 1975) but there was a significant temporary decrease in fertilizer use in the CEE and CIS countries in the late 1980s due to the collapse of the former Soviet Union. While P use leveled off in North America around 1980, N use is still increasing, though at a slower rate than pre-1980 (Figures 2-11 and 2-12). Fertilizer use in the intensive cropping systems of the NAE is partly responsible for the considerable gains in agricultural productivity in NAE since the 1950s. Until recently, fertilizer has been relatively cheap for farmers and the profits from yield increases achieved far exceeded the costs of the additional fertilizers.

2.4.3.4 Pesticide usage in NAE cropping systems

Synthetic chemical pesticides were developed and introduced after 1945 and have since become the major form of pest management in agriculture and stored products in NAE. The term pesticide refers to herbicides, insecticides and fungicides, as well as products that control rodents, nematodes and other pests and treat or preserve timber. Over 1000 chemicals are marketed worldwide, sold in tens of thousands of formulations (Tomlin, 2006).

A program for registration of pesticides was initiated in 1947 by the US Department of Agriculture and is currently under the authority of the US Environment Protection Agency (Pierzynski et al., 2000). All NAE countries now have stringent requirements for the registration of pesticides, which authorize specific formulations for each crop and require evidence of tests on non-target organisms, fate and transport of pesticides. Data requirements have progressively increased to address environmental and health concerns. The organochlorine pesticides which represented the first generation of insecticides were bioaccumulative and environmentally persistent. This led to a series of bans and withdrawals in NAE and worldwide. In 1960, chlorinated pesticides had represented about 75% of insecticide use in the US, but by 1997 these were less than 3% (see Aspelin, 2003). Nine of these insecticides are now scheduled to be withdrawn from production and use under the Stockholm Convention on Persistent Organic Pollutants. Since 1992, discussions have taken place to globally harmonize the classification and labeling requirements for pesticides worldwide (OECD, 2004).

Synthetic chemical pesticides did not become available until after 1945; massive increases in use were recorded in NAE from 1950 onwards. Trends in use by volume in the USA (Figure 2-13) are also similar to Western Europe, showing a peak in the 1980s. Measurement by volume use is a limited indicator of pesticide use and change, as it amalgamates information on products used in undiluted form, reflects neither their toxicity to different organisms nor their persistence in the environment and masks the fact that newer pesticides are developed to be more active at lower rates of application. In 1997 approximately 350,000 tonnes (USA), 32,000 tonnes (UK) and 100,000 tonnes (France) of pesticides were used on agricultural crops (FAOSTAT, 2006).

Detailed changes at country level are difficult to access, but an example from UK national pesticide survey data demonstrates large increases in land area treated with fungicides and herbicides between 1974 and 2002. Increases

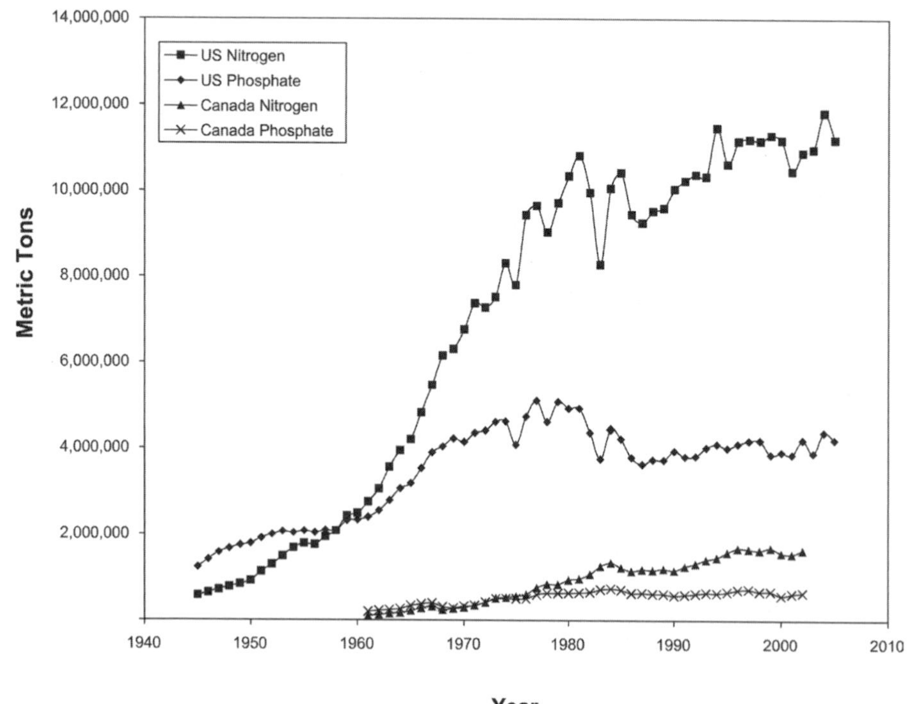

Figure 2-11. *Fertilizer use in North America.* Sources: U.S. data: USDA-ERS; Canada: FAO statistics.

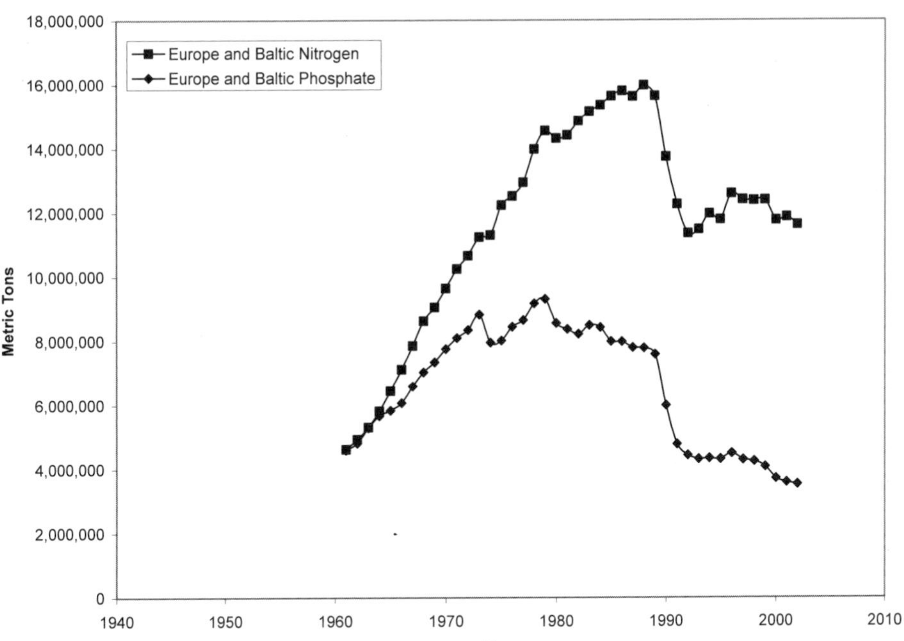

Figure 2-12. *Nitrogen and phosphorus fertilizer use in Europe and the Baltic States.* Source: FAO statistics

arise from multiple treatments on cropped areas as the area sown remained relatively static. The number of pesticide treatments applied per hectare per year increased from two to nearly nine (Chapman et al., 1977; Davis et al., 1990; Garthwaite et al., 1996, 2000, 2004; Sly, 1977, 1986). In the US, where agriculture typically encompasses 75 to 80% of total use of conventional pesticides, the growth of pesticide use through the 1950s and 1960s was primarily due to the greater application of herbicides (Kiely et al., 2004). Herbicide use peaked around 1980, with atrazine being the most used active ingredient for many years, but by 2001 it was overtaken by glyphosate as a result of the wide adoption of glyphosate-tolerant crops. Most US producers of major crops now scout for damaging insects (NASS, 2006) and only apply insecticides when the defined thresholds are exceeded and when the projected savings from yield loss will outweigh the costs of the insecticide application. Some of the decrease since 1995 is due to the use of genetically-engineered insect resistant varieties of maize and cotton (Fernandez-Cornejo and Caswell, 2006). Integrated pest management techniques are increasingly adopted and can make a significant contribution in the general reduction of insecticide use (Kogan, 1998)

A number of NAE governments have promoted programs to reduce pesticide use. A Canadian government program, Food Systems 2002, was launched in 1987 to reduce the use of pesticides in agriculture by 50% by the year 2002 (Gallivan et al., 2001) and achieved a 38.5% reduction 1983-1998. The decrease came partly from smaller cropping areas, but principally from reduction in mean application rates. In the EU, a number of countries, including in Denmark, Germany, the Netherlands and Sweden, have adopted legislation to reduce pesticide use and reductions have been achieved, partly by the use of newer products with lower environmental footprint. The European Commission is now requiring countries to develop pesticide reduction strategies.

Role of AKST

The development of pesticides has depended almost totally on scientific advances in the private sector. The majority of pesticides have been produced by multinational agrochemical companies. Research by universities and public agencies (such as US Geological Survey) has improved understanding of the fate and transport of agricultural pesticides and the impacts on drinking and groundwater (Schraer et al., 2000; Thurman and Aga, 2001; Spaulding et al., 2003).

The public sector has played a greater role in the regulatory approval for pesticides for minor or specialty crops where the small markets are not large enough to warrant conducting the necessary field tests. Science and technology have also played a role in governmental regulation, as new tools and techniques, coupled with increased understanding of environmental consequences, have led to increasingly rigorous evaluation of new products. In the US the number of new pesticides being registered that are classified as low-risk and biopesticides (naturally occurring compounds) are now greater than the number of new conventional pesticides, but they remain a small proportion of the available pesticides (EPA, 2005). Agricultural science has provided tools to develop biological control agents and other non-chemical methods.

The drivers of pesticide use in the NAE have been:

- The objective of increasing crop yield and quality;
- The demand from NAE markets for pest- and disease-free products leading to greater use of pesticides in

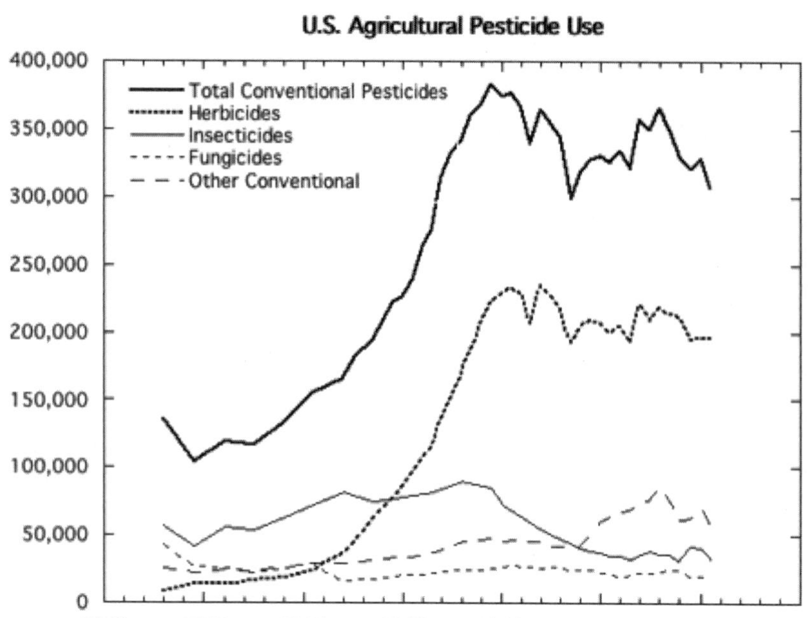

Figure 2-13. *Trends in U.S. pesticide use.* Source: Kiely et al., 2004; Aspelin, 1997, 2003.

almost all crops and horticultural crops in particular; and

- The rise of related environmental concerns among regulators and the general public resulting in greater regulation of pesticides and restrictions on use. A specific aspect of this has been the need to reduce levels of pesticides in both ground and surface waters and to minimize residue levels in food.

2.4.3.5 Water control in NAE cropping systems

Soil moisture in agriculture has a large impact on yield and plant health. Root growth and function is impaired if soils are either waterlogged or droughted and this in turn affects the vigor of the plant above ground.

Because many lowland soils are naturally waterlogged, especially during spring and fall, farmers have often drained their land with subsurface drains that are highly effective at removing water from large areas of land. The fired clay pipes were expensive and installation was labor-intensive so 1940s era drainage pipes were replaced with machine laid plastic pipes in the 1950s (Spoor and Leeds-Harrison, 1997). Large subsidies were made available to farmers to encourage soil drainage and from 1950 to 1990 vast areas were drained (c.f. Robinson and Armstrong, 1988 for a UK example), improving crop yields and increasing access to land for spring planting and harvesting at the end of the season. Access to land in fall also opened up the potential for winter cropping, which is now common over large parts of Europe and the US. While drainage on this scale certainly improved yields it also gave rise to serious water pollution problems due to oxidation of iron and sulphur compounds in soils and increased nutrient and pesticide runoff to rivers and streams (Sagardoy, 1993; EEA, 1994; Ongley, 1996; FAO, 1997). (See Chapter 3 for discussion of environmental impacts of irrigation.)

In NAE, irrigation is used extensively in southern Europe and the western United States. Much of this use focuses on high value horticultural crops, although there is also appreciable usage in some of the major arable crops such as maize, soybeans and potatoes. Overall, within the EU (15), there has been a rise in the percentage of irrigated crops from 4 to 9% over the last forty years (FAOSTAT, AQUASTAT). This average value disguises the greater areas irrigated in the hotter southern countries and the much lower usage farther north. In the United States, the area under irrigation doubled between 1949 and 1979 to 21 million hectares and by 1987 had more than doubled again (Rhoades, 1990). Although irrigated land is only 18% of the total harvested cropland, farms with irrigated land receive 60% of the total market value of crops in the United States. Irrigation not only increases crop value, it can also increase water use efficiency (Howell, 2001) by increasing the mass of crop produced per volume of water.

A major challenge for irrigated agriculture is increasing competition for water, primarily due to population increase (NRC, 1996). As a result of this, irrigation cost will increase (CAST, 1996); already the average irrigation application rate has declined from 1080 ha-mm per ha (3.55 acre-ft per acre) in 1950 to 756 ha-mm per ha (2.48 acre-ft per acre) in 2000.

The desire for increased productivity has been a major driver increasing the use of irrigation in the NAE, along with an increasing demand for products outside their normal production period (especially for fruits and vegetables) and the increased profitability of crop production using irrigation methods.

2.4.4 Agricultural products for energy and fuels

Due to a rapidly growing interest in developing alternate fuels for transportation, expectations are high for agriculture to produce liquid biofuels. The US Energy Policy Act of 2005 calls for the use of 7.5 billion gallons per year (equivalent to 2% of the US gasoline consumption) of biofuel (primarily ethanol) to be mixed into the US fuel supply by 2012. The European Union biofuels directive of 2003 sets a reference value of 5.75% for the market share of biofuels in 2010.

In the US, ethanol production capacity has increased from 1.6 billion gallons per year in 2000 to about 5 billion gallons per year in 2006, with an additional 6 billion gallon capacity under construction (Renewable Fuels Association, 2006). Biodiesel production (primarily using soybean as a feedstock) is currently much lower than ethanol, but rapidly expanding. As of 2005, there were 53 biodiesel plants with a capacity of 354 million gallons per year. Biodiesel capacity is expected to reach 1.2 billion gallons per year.

As in North America, production of biofuels is increasing in some parts of Europe. Little was produced prior to 2000 but by 2004 biofuel production had reached 2.4 million tonnes and the aim is to produce 18 million tonnes by 2010. Unlike the USA, most biofuel in Europe is biodiesel from oilseed rape and in 2004 2 million tonnes were produced. Assuming an average yield of 2.5 tonnes ha^{-1} this amount of biodiesel would have been produced by about 300,000 ha of oilseed rape. The remainder of the biofuel production was bioethanol, much of it derived from excess wine production in the EU.

Increased biofuel production can increase the price for the crops at the farm gate and provide more price stability. In addition, the biofuel industry can provide off-farm, rural employment opportunities while the byproducts of biofuel production (distilled grains and residue after oil is recovered) are considered quality feed supplements.

However, there are clearly limits as to how much biofuel can be produced, at least with current and foreseeable technologies. For example, in 2005, 14% of the US corn crop was used to produce the equivalent of 2% of gasoline use in the US (by energy content). By comparison, the US exports about 16% of its corn production. Using the same corn use to ethanol ratio, utilization of 100% of the US corn crop for ethanol would produce fuel to replace only about 14% of the US (2005) gasoline use.

While at least at a modest scale, biofuels production should benefit the NAE agricultural community, questions remain whether greatly increased production and use of biofuels will have detrimental environmental effects, or even meet the projected environmental benefits. To the extent that mandates to meet certain biofuel use targets cannot be met by domestic production, biofuels will need to be imported. This may negate some of the savings expected from import of petroleum products. Further, it may prompt increases in agricultural production elsewhere at detriment to the environment (e.g., Pearce, 2005).

One incentive for the use of biofuels is their replacement for fossil fuels. There are some estimates that the current production of biofuels is actually carbon negative in that it takes more fossil fuel to produce biofuel than the petroleum it is intended to replace (e.g., Pimentel and Patzek, 2005) though others point to a positive net carbon balance in the production and use of biofuels (e.g., Farrell et al., 2006; Worldwatch, 2006). Biofuels could be used to replace the fossil fuels in the agricultural practices to produce biofuels.

Other agricultural-related energy sources

Agricultural lands may make a contribution to energy in ways other than through production agriculture. For example, in the US the richest wind energy resource, available in wide areas, stretches from the upper Midwestern plains states to Texas (Elliot et al., 1986). Farmers have leased the land for turbines, or have invested directly in their ownership. The potential of the Midwest wind resource has been recognized and the number of installed wind turbines and overall electricity production capacity is expanding (c.f. American Wind Energy Association, n.d.; US Dep. Energy, 2007).

Forestry and other sources of plant material (e.g., biomass crops) are being increasingly used in Europe as a source of heat and energy, driven by the rising price of oil. In 2004 52.4 million tonnes (oil equivalent) were produced from these sources. A huge proportion of this was from forestry waste, especially in the well forested EU states, such as those in Scandinavia. However, the EU proposes to greatly increase the 2% of energy from biomass crops such as coppice willow and Miscanthus grass, so that it makes an appreciable contribution to the EU energy budget in the future (EC, 2005, 2007). As in the USA there are also considerable developments in the utilization of wind power. In 2004 the EU contributed 73% of the world's total capacity of 48 thousand MW. There is much debate as to the location of these wind farms and of their environmental impact, but they do offer an alternative source of income to farmers and other land owners.

2.4.5 Organic cropping systems

Largely unidentified as organic before the advent of synthetic fertilizers and pesticides, organic agriculture has been one response to public concern over the environmental and health impacts of industrialized agriculture. Since the beginning of the 1990s, organic farming has rapidly developed in almost all European countries. Growth has slowed recently. In 2004 in Europe, 6.5 million hectares were managed organically on about 167,000 farms. In the European Union more than 5.8 million hectares are under organic management and there are almost 140,000 organic farms. The country with the highest number of farms and the largest organic area is Italy. In most countries of Europe and particularly the European Union organic farming is supported with legislation and direct payments. In terms of the share of organic farmland to total agricultural area, Austria, Switzerland and Scandinavian countries lead the way. In Switzerland, for example, more than 10% of the agricultural land is managed organically (Willer and Yussefi, 2007). In fact the land under the organic certification has been largely increasing since 1994, i.e., when financial support was first introduced by the EU-Regulation 92/2078.

The support for organic production granted by the reform of the CAP, i.e., enforcement of the EU Regulation 2078/92 (mis.A3+A4), constituted a fundamental step in this evolution and largely promoted the conversion to organic farming in the Southern regions of the EU, even though the pioneers of organic agriculture were in North and in Central

Europe. In the 1990s, regions in the south of Italy recorded the highest rates of growth of farms in conversion to organic farming. In the European Union, the European Organic Action Plan implementation process is now getting under way (Miele and Pinducciu, 2001).

In North America almost 1.4 million ha are managed organically, representing approximately a 0.3% share of the total agricultural area. Currently, the number of organic farms is almost 12,000 (Willer and Yussefi, 2007). With the adoption of national standards in 2002 in the United States, the organic sector has been able to provide a guarantee to consumers that organic products using the labeling followed specific practices. The US market has been growing rapidly, estimated by the Organic Trade Association at 20% or more per year, with a growing number of certification agencies accredited by USDA and talks progressing to expedite international trade of organic products. Since 1999, the Canadian industry has had a voluntary Canada Organic Standard that is not supported by regulation. The organic industry continues to devote its energies toward implementation of a mandatory national organic regulation to help expedite trade relations with such major trading partners as the United States, European Union and Japan.

2.4.6 Key changes in cropping systems and drivers

In summary, production of arable crops has doubled and in some cases tripled over the last 50 years in the NAE. These production increases have been mainly due to increases in output per unit area, as the area of arable land in the NAE has not increased and in many countries has decreased slightly. Production increases have been facilitated by the contribution of AKST, providing farmers with new tools to enhance crop production. These have primarily been more efficient use of fertilizers, mechanization and development of novel more effective agrochemicals and the breeding of new higher yielding cultivars.

Dissemination of this new knowledge has depended on the development of efficient knowledge transfer systems, both governmental and private sector. Moreover, there has been increased technological sophistication in agricultural mechanization. The increased productivity/efficiency of cropping systems has left more time for off-farm employment and decreased labor employment in agriculture. Despite the labor savings brought about by mechanization in many agricultural systems, some production systems remain labor-intensive (e.g., horticultural crops).

New tools enabled change or extension of farming practice. For example, larger field sizes to accommodate machinery, new areas under cultivation because of improved plough/cultivation capability, increased capability for minimum tillage, increased ability to cope with management and feeding of livestock at higher densities, and a shift from silage to hay. However, there are also negative aspects associated with soil compaction and structural damage resulting from frequent passes of large heavy machinery. Still, mechanization has increased the practicality of the production of some organic crops (e.g., innovations in mechanical weeders).

2.5 Changes in Livestock Systems in NAE

As in cropping systems, the key change in livestock systems in NAE has been significant increase in both productivity and production of meat and dairy products driven by an increased demand for these products among NAE consumers. This has been made possible by improved genetics and widespread access to superior genotypes, changes in livestock feeding regimes, development of specialized production units for livestock and improvements in food safety. Consumer demand for humanely treated livestock and increased concern about environmental impacts of intensive livestock production have started to change production practices across NAE, especially in Western Europe.

Because of World War II's disruptions to production, distribution and storage, the postwar livestock industry could not meet European consumer demand until the late 1950s. Meat consumption per capita has generally increased since post-war rationing ended (Aumaitre and Boyazoglu, 2000). During the post-war years most European governments used subsidies to increase livestock production (Hodges, 1999).

Mixed farms such as those in Europe where livestock was fed mainly by grazing or cereals produced on the same farm predominated after WWII. In this period, the US had a geographically dispersed livestock sector. On the uplands in Europe, pastoralism was a way of life using summer grazing and winter stock movements ("transhumance") developed in mediaeval times.

In Europe, the mixed farms of the 1940s have today almost completely changed to either specialist arable or milk and livestock production units, using high intensity production methods promoted by the CAP and state subsidies of capital investment and/or productivity-related payments (de Haan et al., 1997). Half of all EU farms still have livestock, with 90% now specialist livestock producers, buying feed from global commodity markets (European Commission). Europe now has one of the highest livestock densities in the world (FAOSTAT), with a mixture of intensive grazing and fattening/rearing units where livestock are fed on both home-grown and imported feed. The overall result has been increased livestock numbers (although the livestock density (LU/ha) in Europe has fallen some 10% in the past decade (FAOSTAT) and increased productivity of all livestock and dairy products, leading to large-scale over-production in the cattle, pig and poultry sectors over the past twenty years.

US and Canadian livestock sectors have also undergone extensive restructuring since 1945, but in different ways (Table 2-6). One of the key developments has been the integration of the US, Canadian and Mexican livestock sector, accelerated by the adoption of NAFTA in 1994. This is particularly true in the beef and pork sectors (Young and Marsh, 1998; Farm Foundation, 2004; Haley, 2004). Prices for beef and pork tend to move together in both wholesale and live animal markets, particularly in Canada and the US (Vollrath and Hallahan, 2006) (e.g., 8% of pork slaughtered in the US now originates in Canada, a large increase over the last decade [Hahn et al., 2005]). Poultry is the exception as it is not as well-integrated because it is a managed sector in Canada.

As in Europe, the number of farms in North America with livestock has decreased (McBride, 1997), while production of red meat, poultry products and dairy products has continued to increase. In the US, there have been significant geographic concentrations in beef and broiler produc-

Table 2-6. **Changes in livestock farming operations.**

Animal Production on Farms, U.S. and Canada						
Year	Number of Farms	Farms Producing (%)				
		Beef	Dairy	Swine	Chicken
United States						
2002	2,128,982	37.4	4.3	3.7	1.5
1974	2,314,013	**44.3**	17.4	20.3	1.5
1920	6,118,956	29.7	**74.60**	**79.3**	
Canada						
2001	230,540	52.9	9.5	6.7	11.5
1971	258,716	**96.1**	56.2	47.3	46.2
1921	711,090	84.2		**63.4**	**82.4**

Source: Farm Foundation, 2004.

tion. Large feedlot operations for beef are concentrated in the Great Plains, while broiler production is heavily concentrated in the Southeast. In the 1980s hog production shifted from the Midwest to large operations in the Southeast (Figure 2-14; Welsh et al., 2003). At the same time, dairy production expanded in Western states away from the Northeast and Upper Midwest (McBride, 1997). Canada has seen similar geographic concentrations of livestock production with hog production shifting from Quebec and Ontario to the west, particularly Manitoba, while cattle production has become concentrated in Alberta (USDA-FAS, 1996).

2.5.1 Trends in output and productivity since 1945

Four groups of animals produce over 90% of Europe's meat and dairy products; cattle for milk, beef and veal, pigs for meat, poultry for meat and eggs, and sheep and goats for meat, milk and wool. Meat, dairy products and eggs account for over one-third of the total value of agricultural production in Europe. Beef sales declined during the BSE crisis from 1996 to 2001 but have now begun to recover (Morgan, 2001; USDA-FAS, 2005). Pig and poultry meat consumption increased due to the BSE-induced dip in beef demand, but have increased even further since the 1990s due to greater competitiveness with other meat production, partly as a result of CAP reforms that made cheaper cereals available for animal feed. Sheep meat production and consumption declined during the 2001 UK foot-and-mouth disease outbreak, but have now almost recovered (Eurostat Agriculture, 2007a).

In response to growing demand from a larger and richer population, production of all livestock increased very rapidly in the EU-15 from 1961 to 2000, while production of meat and dairy products has fallen in CEE from 1990, mainly as a result of the transition from a centrally planned to a market economy. However in Hungary, Slovenia, Croatia and Romania production has either remained stable or increased slightly from 1993 to 2004 (EU, 2004). Europe (EU-25) produces over three times as much meat per head of human population as the world average of 36 kg per capita (FAOSTAT, 2007).

This productivity has led to over-production. Europe is more than self-sufficient in meat, with a current net balance of around 105% for all meats (Eurostat Agriculture, 2007a). As a result of rigorous CAP reforms in the 1990s, European production of beef and veal has fallen rapidly from around 50% over-production (EU-15) in the 1990s to around 96% self-sufficiency in 2004. Beef and veal consumption has risen in the past 4 years, with the European production deficit being made up by imports of around 250,000 tonnes per year from South America. Pig meat is still being over-produced in EU-25 by about 8%, making the EU-25 a net exporter of pig meat products, mainly to Russia and Japan.

The EU is a net importer of sheep meat (EU-25 is only 78% self-sufficient in sheep and goat meat) and dairy products, mostly from New Zealand and also imports large quantities of poultry meat from Brazil and Thailand, where production costs are much lower than in Europe. Somewhat perversely the EU also exports large quantities of poultry meat and offal to Russia and the Ukraine and parts of the Middle East (Eurostat Agriculture, 2007b).

North America accounts for 16% of the world's total number of beef cows, 8% of the world's pig crop, nearly one-third of the world's poultry meat production and nearly 15% of the world's milk (Farm Foundation, 2004; Adcock et al., 2006). In the swine sector, productivity in breeding herds has increased significantly, with 3.2 million fewer sows in 2004 than in 1980 producing roughly the same amount of pigs. The US and Canada have been able to increase milk output 19% (Figure 2-15) and 6% respectively, even with fewer cows, due to significant improvements in milk productivity related to improved genetics (Farm Foundation, 2004). In the US the value of livestock production increased nearly by a factor of eight between 1948 and 2005, while the production of red meat increased nearly 50% from 1963 to 2006 (even though lamb and mutton production has declined sharply due to cheaper imports). Poultry production has also significantly increased.

In Canada, pig slaughter has nearly tripled since 1976, while cattle slaughter declined and then started to increase in the last 15 years, due to the opening of new processing

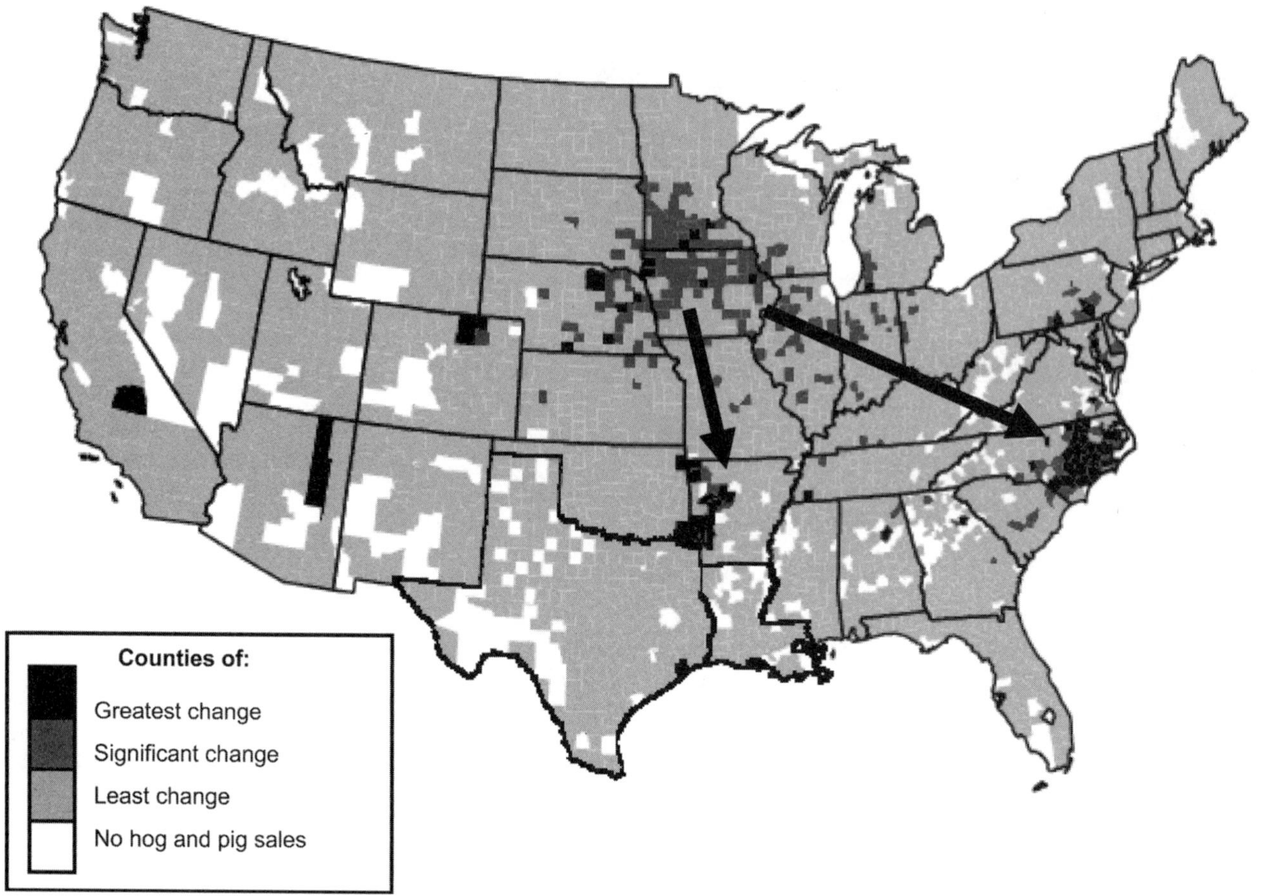

Figure 2-14. *Geographic changes in hog and pig production in the U.S.* Source: McBride, 1997 (arrows added)

facilities by US based firms, Cargill and Tyson. Sheep and lamb slaughter, while still very small has managed to almost double since 1976—a very different trend than the US.

Overall, livestock productivity and output in NAE has increased enormously since 1945 with beef, pig meat and milk production almost doubling and a four-fold increase in numbers of poultry. Sheep and goat numbers and production of meats and other products from this animal stock have remained comparatively stable (data compiled from FAO, Eurostat and USDA).

2.5.2 Drivers of increased livestock output and productivity

The spectacular rises in livestock numbers and productivity seen in NAE over the past 50 years result from six major drivers:

- Growth in population numbers and wealth, creating strong market demand for meat and dairy products (for example in dairy products a 1% growth in income gives almost the same increase in consumption in low income countries and about 0.35% increase in wealthy countries (Agra/CEAS, 2004);
- Strong policies and strategic frameworks within NAE aimed at increasing livestock production;
- Rules and regulations determining husbandry methods and processing of livestock products;
- Production-led subsidies that funded output and productivity increases (Starmer and Wise, 2007);
- The application of knowledge, science and technology to animal genetics and nutrition, including grassland management and feed formulation; and
- Improvements in animal and livestock product transport systems allowing animal production and slaughter to be situated more closely to major supplies of feed.

The most important contributors of AKST to increased productivity have been changes in livestock genetics, livestock feeding and stock management systems. For example, selection involved in animal breeding took place at the farm level until the end of the 19th Century resulting in the adaptation of cattle, pigs, sheep, goats and poultry to specific (usually regional) farming and market situations (Hodges, 1999). Yield goals were blended with emphasis on selecting livestock that would thrive on particular types of land, climate and feed (CIV website).

By contrast, in the 20th Century livestock breeding was increasingly done in either state-owned or private institutions using genetic science. Coupled with advances in land and management practices such as drainage, fertilizer use and better harvest and storage techniques, these breeding programs began to be more yield-oriented to cope with increased demands for food from a rapidly expanding urban

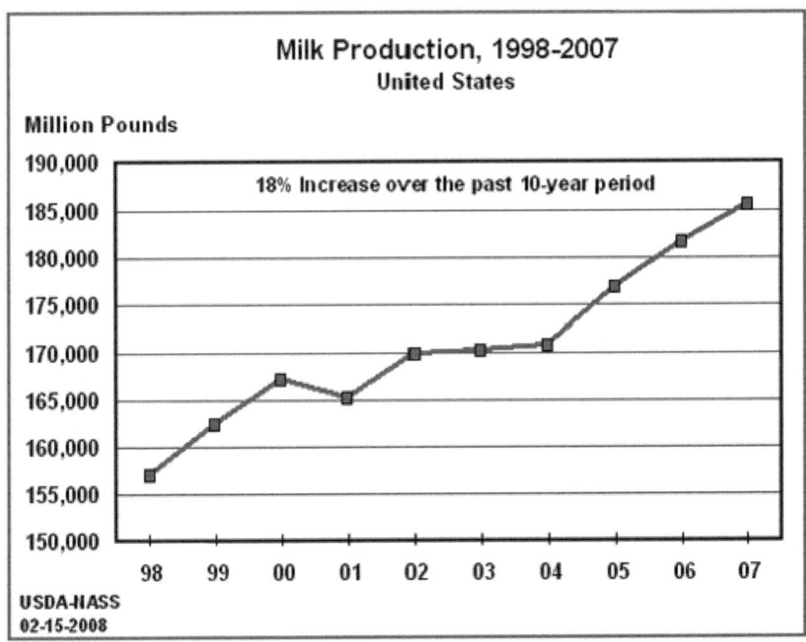

Figure 2-15. *Trends in productivity per cow in US from 1998-2007.* Source: USDA-NASS, 2008

population in Europe and desire for more meat consumption among North Americans. This drive for greater productivity accelerated in the 1950s as a response to the need to rebuild the food supply chain after World War II. Science-based livestock breeding typically produced annual genetic changes of around 2% of the mean of a trait (or trait-related index), especially in species with high reproductive rates like pigs and poultry (Simm et al., 2005). Not only were yields from livestock varieties substantially increased, but standardized livestock systems were also developed, where in cattle (and to some extent sheep), the landscape was adapted to the system. In pigs and poultry, the whole enterprise was taken off the land and into intensive housing and feeding systems. Varieties that maximized food conversion ratios were quickly developed, especially in pigs and poultry (Simm et al., 2005), with cattle breeding focused almost entirely on high milk and meat production. In N. America in the 1960s and 1970s, the so-called "British Breeds" of cattle were replaced in much of the beef sector by "Continental Breeds" that introduced size and leanness, in response to consumer desire for leaner beef. The genetic techniques used to achieve these productivity gains include:

- Better statistical methods of estimating the breeding value of animals
- The use of artificial insemination that allowed producers at any level to access superior genetics
- Better techniques for measuring performance of new breeds
- Selection focused on quantitative traits, such as weight gain and disease resistance

Despite the undoubted success of these science-based breeding programs, it is generally agreed that the maximum genetic potential of cattle, pigs and poultry has still not been reached and intensive breeding programs are still maintained in Europe although the focus is now shifting away from continued productivity increases towards animal health and welfare traits (Garnsworthy, 2005).

As a result of these breeding and husbandry techniques, the wide variety of landraces in 1945 was quickly replaced by a few high yielding varieties, such as Holstein/Friesian milking cattle (e.g., this breed comprised more than 85% of the Canadian dairy herd in 1999 [Kemp, 2001]) or white lines of pigs used in intensive production facilities. Most livestock landraces have survived in small numbers either by the activities of "rare breed societies" who try to maintain the genetic base of the "old" livestock breeds, or by being used to produce niche market high quality products, mainly meat and cheeses.

The latest developments in animal breeding include genetic engineering. Its use is unpopular in Europe, but in North America, ancillary uses of GE technology, e.g., to increase milk production through the administration of recombinant Bovine Somatotropin (rBST), has been widely adopted. Whether transgenic animals in the food supply are accepted by NAE consumers remains to be seen. In Europe, rBST use for milk production raised concerns about animal suffering and potential negative impacts on small farmers. In the context of surplus milk production in NAE, the benefits of this application continue to be debated.

Simultaneously with breeding for improved productivity, NAE scientists also focused on improving livestock feeding and management. For example, the weight gain for broilers at 56 days in 1957 was around 800g, compared to a 3900g weight gain in 2001 (Havenstein et al., 2003). Similar trends can be found for weight gain in pigs and for milk yields in cattle (Simm, 1998). Breeding and nutrition technologies for sheep and goats have not been subjected to such intensive scientific attention as they are still mainly raised on marginal land throughout Europe and the market is smaller.

Grassland-based cattle systems have changed radically throughout most of Northern and Western Europe from haymaking to silaging using the highest fertilizer inputs in the world (FAOSTAT), with great loss of non-grass biodiversity in pastures and meadows since WW2 (Johnson and Hope, 2005). Haymaking with low fertilizer use still survives in upland and marginal areas in N and W Europe and in many parts of the CEE countries, especially where traditional breeds of livestock are used. Grasslands, particularly in western regions, have been the predominant system for cow-calf and sheep production in N. America. Because of increased environmental concerns about management of federal lands in the US, there has been renewed interest in range management. Intensive grazing systems are increasingly used in beef cow and dairy herds across NA, where the focus is on increasing profitability per animal, rather than maximizing productivity (Gerrish, 2004). In the future, increased demand for grain for biofuel production may increase costs of animal production, potentially increasing consumer prices to the point where supply of cheap livestock products is reduced in less wealthy parts of NAE.

A major change contributing to increased production and better storage has been in the vertical integration of the livestock chain through standardizing genetics, feeding systems and housing units while increasing communication throughout the sector. In N. America this is particularly apparent in the poultry and pork sectors and in Europe, high throughput automated housing, feeding, slaughtering and processing facilities have grown larger, replacing smaller family-owned businesses (EC, 2001). Across NAE, many animal production, slaughter and processing units are operated by large consortia which control large parts of the food chain, increasingly outcompeting family farms by means of their economies of scale and ability to influence market prices for livestock and products.

Changes in Livestock and Labor

Changing consumer preferences and the meat industry's increased emphasis on pre-cut and pre-packaged meat and growing export levels increased the demand for labor. Between 1972 and 2001 employment in the poultry processing industry increased by 150% in the US, with jobs being on offer mainly as low-skilled manual labor. During this period re-structuring in the industry had led to a re-location of processing plants to rural areas, largely to areas that lacked a unionized tradition. With greater technological innovation, meat processing has become increasingly de-skilled and, in addition to stable or declining real wages, meat processing employment became less appealing for increasingly well-educated native born workforce (Stull, 1994). The industry had undergone a gradual change from unionized urban skilled workforce to rural based mostly non-unionized and low skilled workforce concentrated in manufacturing plants by the 1980s and these characteristics have remained the same since that time (Kandel, 2006) Hispanic workers are over-represented within the food processing industries. Between 1980 and 2000 the proportion of Hispanic meat-processing workers increased from under 10% to almost 30% of the total. Whereas previously about half of the Hispanic workforce in the meat-processing industry was born in the US, by the year 2000, 82% were foreign born.

In the second part of the 20th century, major changes also took place in animal production facilities and investment in buildings and their use have become issues of growing importance for farmers and growers (Gay and Grisso, 2002). Traditional buildings associated with livestock production were general purpose, small scale and reflected production systems relying heavily on manual labor. Faced by rising labor costs, facilitated by a variety of technological developments in machinery, building materials and methods of controlling the environment, a major transformation has taken place where modern high throughput facilities, such as dairy parlors and pig and poultry production units, have largely replaced traditional multipurpose buildings. These provide controlled environments with measured use of feed and prophylactic treatments to prevent disease. Such facilities are also very important in the vertical integration of the meat supply chain.

A less evident contributor to productivity in the NAE livestock sector is the development of effective transport systems that allowed animal feeding and slaughter to be concentrated more closely to feed sources, particularly in beef production. A parallel process was the introduction of vacuum packaging in the late 1960s. This significantly altered the value chain for beef and other protein, since retailers could sell particular cuts of meat without an on-site butchery (Duewer, 1984).

These developments in genetics, management systems and meat handling, combined with the geographical shifts in production, allowed significant restructuring in the beef, pork and poultry sectors leading to the development of confined animal feeding operations, contractual relationships in marketing and specialization in livestock agriculture. These changes have been controversial because intensive livestock production raises ethical and environmental issues. Treating animals as items on a production line offends many NAE citizens who feel this is an unacceptable relationship between humans and other species. Farm animal welfare has become an important area for policy makers, especially in Europe (Webster, 2005). The mass production of animals to specification, while producing cheap and nutritious products, also undermines traditional livestock businesses, reducing local employment and undermining the economic survival of some communities. In an area in which emotions often play an important part in determining attitudes there are a wide range of pressure groups and consumers who criticize intensive livestock production. For example, the development of confined animal feeding operations in NA have resulted in significant conflicts over air and water quality, land use issues (zoning) and regulatory control (Bonanno and Constance, 2006; Donham et al., 2007; Heederik et al., 2007).

Livestock kept in intensive systems can be prone to outbreaks of disease, illustrated by the periodic outbreaks of foot and mouth disease and encephalopathies such as BSE and scrapie; viral diseases in cattle, sheep and pigs and epidemics of viral and bacterial poultry diseases. While epidemic disease has always been part of livestock production, the larger groups of animals and widespread transport to and from markets associated with intensive systems have increased risks of large epidemics, even though biosecurity at individual units has been improved (e.g., Defra, 2006). It has been argued that intensive systems have also produced new and dangerous diseases such as *E.coli 0157:H57* and BSE (Walker et al., 2005). These epidemics have sometimes devastated livestock sectors in Europe and have largely been controlled by a slaughter policy, although for some pig and poultry diseases vaccination and the routine use of antibiotics have become common practice since the 1950s. The use of antibiotics as growth promoters and disease control agents in NAE livestock production has caused serious concern because of the rise of antibiotic resistant bacteria in humans (Khachatourians, 1998; Mellon et al., 2001).

Food safety issues are also important in the meat industry in North America. In 1995, an outbreak of *E coli 0157:H57* killed several children who had eaten fast food hamburgers in Washington state. This event led to a revolution in food safety procedures in red meat, seafood and poultry in the US with the creation and adoption of new food safety rules (see 2.8.4). Food safety concerns about *Salmonella* and *Listeria* continue to be of concern throughout the NAE livestock sector (Johnston, 2000; Raijaic et al., 2007).

Advances in productivity in the NAE livestock sector would not have been possible without public investments in AKST. In particular, many new genetic selection techniques were developed through public universities and disseminated through extension services. Today, much of the actual genetics has been privatized and is now maintained primarily in the private sector, although performance measures for stud selection are still provided in the public realm. In the same way, the research that developed the HACCP approach to food safety was performed by public entities like USDA-Agricultural Research Service and enforcement is still performed through USDA. Finally, many of the engineering advances that allowed the development of large-scale climate controlled buildings for poultry and swine and for handling wastes from these systems were developed in the public sector and disseminated widely.

2.5.3 Key changes in the NAE livestock sector

Livestock productivity and output in NAE has increased enormously since 1945 driven by policy (especially the CAP), government subsidies (Starmer and Wise, 2007) and increasing population and wealth. AKST has been a key driver of growth in the livestock sector and is likely to remain so in the future. Europe and North America have been exporters of livestock sector AKST to the rest of the world.

For the past 30 years much of NAE has been producing far more meat and dairy products than it needs with the EU and NAFTA blocks becoming some of the world's leading exporters, particularly in pork (EU), chicken and beef (NA). The search for more market sector has led to dumping of these products in less wealthy countries with consequent damage to the economic status of their agricultural producers. In common with the rest of the developed world, milk, beef, pig meat and poultry are among the most valuable agricultural commodities produced by European farmers.

Much of European lowland and landless livestock production is the most intensive in the world and this has had serious adverse impacts on the European environment. Similar situations exist in N. America because of the increased geographical concentration of livestock production. Across NAE, livestock enterprises have become fewer and larger due to economies of scale and this trend is likely to continue especially in the CEE region of EU-25.

Developments in genetics, management systems and meat handling in NAE, combined with the geographical shifts in production, allowed significant restructuring in the beef, pork and poultry sectors leading to the development of confined animal feeding operations, contractual relationships in marketing and specialization in livestock agriculture.

Subsidy-led policies are moving away from production-led subsidies towards a more market-led and environmentally friendly system, but there are still substantial direct and indirect subsidies paid to most livestock sectors that reduces the competitiveness of developing countries.

2.6 Changes in Forestry Systems

In North America and Europe, forests and woodlands have always been the dominant vegetation cover. NAE forests are largely derived from natural vegetation dominated by deciduous trees in the south and west and vast areas of conifers towards the north and east that make up over 50% of total forest cover.

NAE forests have been exploited by humans for timber supplies, fuel, food (e.g., nuts, fungi and berries), for cork (the EU is the largest producer of cork with over 80% of the world market) and for paper fiber, while still providing a significant proportion of the renewable energy used by both industrial and domestic consumers. Forests also provide valuable and irreplaceable ecosystem services such as water resource protection, biodiversity and carbon dioxide fixation (MA, 2005). For example, approximately 140,000 species of plants, animal and micro-organisms are estimated to occur in Canada of which approximately 66% are found in the forests (Canadian Forest Service, 2003).

2.6.1 Main trends in NAE forests and forestry production

NAE is the only world region where there has been an increase in forest area since the 1960s. In 1630, when conversion of North American forests to agricultural land began, 50% of US lands were forests. Today, forests are approximately 33%, but since the 1980s have been increasing by 0.3% per annum. The US growing stock volume increased 39% from 1953 to 2002. The 415 million ha of Canadian forests represent 10% of the world's forests, with 20% of the world's fresh water flowing from its watersheds. Forests cover 45% of the land mass of Canada (Lowe et al., 1996) although it is unclear whether forest cover in Canada is stable or contracting (CANFI, 2004).

Forests in Europe have been expanding over the past 40 years by around 0.8% p.a., about 880,000 ha per year. This has been mainly due to an increase in plantations, reversion of agricultural land and decreased harvesting activity especially in the Russian Federation. The Russian Federation accounts for over 90% of an estimated 1.5m ha per year natural re-colonization of non-forest land in Europe (Kuusela 1994; TBFRA, 2000; UNEP, 2002). It has more than seven times more forest cover than the European Union and almost double the combined forest area of Canada and the United States while containing the greatest area of natural forest (UNECE, 2003).

There has been a decrease in other wooded land (OWL—woodlands not dense or contiguous enough to be classified as forest) of approximately 0.2% p.a. in Europe, similar to that of North America (TBFRA, 2000). Europe (not including the Russian Federation) now has forest cover of around 35% (FAO statistics), similar to that of the US, after having reached a low of 25% during the 19th Century. Since the 1950s, there have been proportionately fewer fellings compared to the increasing forest growth and this has made it possible to supply more wood, while simultaneously increasing the growing stock.

Throughout NAE there been a steady increase in both deciduous and coniferous plantations since early in the 20th century. This is now accelerating as planting technologies have improved and more agricultural land has become available for conversion to forest (Figure 2-16) There is a distinct trend towards a greater proportion of coniferous wood (now 69% in W. Europe, 66% in CIS) being planted. European plantations make up 17% of world plantations with the Russian Federation having the greatest area in Europe. (FAO, 2000; TBFRA, 2000; UNECE/FAO, 2000).

Overall, European and Russian forests sequester around 540 million tonnes of carbon per year, some 14% of the world's total sequestration, with US and Canadian forests sequestering about 200 million tonnes of carbon per year (UNECE/FAO, 2000). There has been an increasing trend for forests to be planted specifically for carbon sequestration, funded by schemes set up as a response to the Kyoto Protocol. (Bowyer and Rametsteiner, 2004; MA, 2005)

2.6.2 Forest ownership and control

Over the past twenty years there has been a strong trend away from public towards private ownership of forests in W and S Europe, but almost all forest land remains in state ownership in the CEE countries, although this is changing towards private ownership in former Soviet states now in the EU-25.

Fifty-seven percent of all US forest land is privately owned, but 94% of Canada's forests are publicly owned. Approximately 10% of US forestland is legally protected from commercial forestry, more than double that protected in 1953. Around 66% of US forest land is classed as timberland (forest capable of producing in excess of 20 cubic feet per acre per year and not legally protected). Since 1953 the area of timberland has had a net loss of one percent primarily because it has become legally protected. Seventy-one percent of US timberland is privately owned.

In general forest growth rates exceed exploitation levels throughout NAE. Net growth rates have not been increasing as rapidly as in the past, while harvest levels have remained relatively stable since 1986. Increased imports have addressed the additional resource demands. Since 1960 the US forest resources have continued to improve in condition and quality as measured by increased average size and volume of trees; however, if quality is measured as a function of optimum stand density, i.e., optimum number of trees per acres for stands of a given age, then the overall quality of many stands has deteriorated (Smith et al., 2002).

Canada is the largest exporter of forest products with total exports valuing $44.1 billion (Natural Resources Canada, 2000). In 2002, one in 17 jobs was directly or indirectly linked to forests.

Less than two-thirds of annual forest growth in Europe (excluding Russia) is harvested, so the volume of standing wood in forests is growing. In Russia only 14% of annual growth is currently being harvested, less than the proportion being harvested in the 1970s (TBFRA, 2000).

The past thirty years have seen an increase in forest accessibility through construction of new logging access roads into remote areas. Conservation protection legislation has also been applied to many inaccessible areas over the past thirty years. In W Europe over 85% of forest is now available for wood supply; in CEE, where more forest is protected, 64% is available for wood supply (TBFRA, 2000).

Biodiversity

In both North America and Europe there has been an overall decrease in forest biodiversity due to reductions in areas

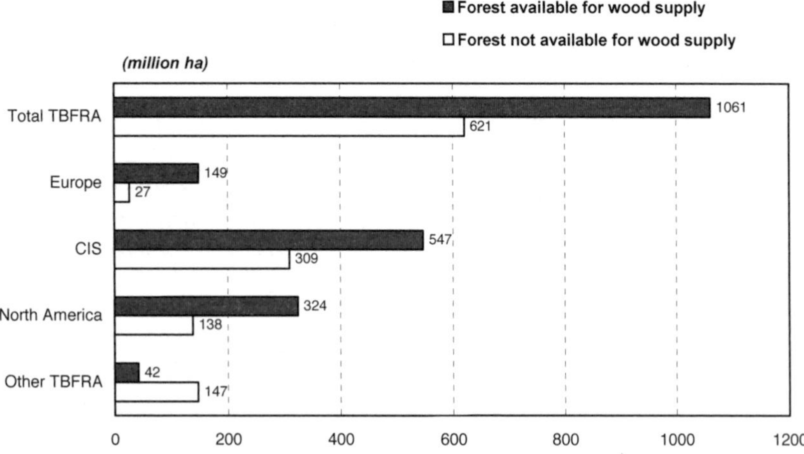

Figure 2-16. *Estimated average annual changes in area of forest and other wooded land (FOWL) in TBFRA area 1993 to 2004.* Source: TBFRA, 2000.

of natural forest, illegal felling, increases in monocultural plantations, increases in serious fires and hunting activity in some countries, adverse effects of air pollution and more urban access into forest areas. Around 60% of Europe's forests are now degraded by the factors listed above (UNEP, 2002). This degradation trend may be reversing in some more developed countries (UK, Germany, some CEE countries) with higher levels of legal protection than the rest of Europe (> 10% of area protected) and development of new plantations that alleviate pressure on natural forest. In the NAE region, Canada and CEE has the highest proportion of forest undisturbed by humans.

2.6.3 Forestry as an industry

Demand for forest products in NAE has dramatically increased since the World War II, especially for industrial wood, with consumption and production more than doubling between 1961 and 2004 (UNECE/FAO, 2003b). Demand and production of fuel wood has increased from 1990 and now exceeds 1960s levels, but is still only 20% of industrial wood production.

Because of this, the forestry industry has steadily grown over the past 50 years from a rural activity supplying urban areas with timber products to a major industry producing a wide range of added value products, especially wood-based boards where Europe is one of the world's major exporters. Not only has there been a significant rise in consumption of, and demand for, wood-based products derived from Europe, but there has also been a significant increase in the import of timber, especially fashionable tropical hardwoods, from other parts of the world, especially from Canada, S. America and the Far East. This import market has had an increasing impact on the forests of other continents and is an important factor driving forest loss in those areas. (FAO, Europa, UNECE).

2.6.4 AKST in forestry

In Western and Southern Europe the main focus of forestry science has changed recently from the traditional productivist paradigm towards a scientific approach to sustainable multifunctional use, including the conservation of species associated with forests and the impacts of climate change. This trend is also found in parts of North America. Since the classification of American forests into ecoregions in the 1970s and 1980s (Bailey, 1980; Bockheim, 1984; McNab and Avers, 1994), there has been a change in forest management away from exploitation towards multifunctional sustainability (Johnson et al., 1999; Bosworth, 2004) focusing on four objectives; watershed health and restoration (USDA-USFS, 1999), sustainable forest management, public access and recreation. These topics form the framework for most forest research in NAE.

Since 1945 many new technologies have been increasingly applied to forest production, harvesting and processing. Increased pesticides use, especially on conifer plantation monocultures, has led to less insect and disease damage to forests. Drainage and ground preparation techniques have been adapted and scaled up from agriculture, resulting in conversion of more open uplands and wetlands to forest.

Even using native tree varieties and labor-intensive forestry systems, foresters in Europe and North America have significantly increased productivity and production per unit area by employing new technologies for ground preparation (better drainage, fertilization and tree protection using physical and chemical means), planting technology using mechanical planters, improved management of plantations, advanced rapid timber harvesting and extraction machinery and high throughput processing (for paper, timber and board production). New harvesting technologies have increased harvest rates and result in a higher proportion of felled wood being processed, with less waste. For example, in Sweden the introduction of the chainsaw and mechanization of logging operations resulted in total forest work productivity increasing between 2.3 and 12.5 m^3 per person-day between 1960 and 1990 (Axelsson, 1998). Between 1970 and 1990, the degree of mechanization in final fellings increased from 25% to 85% and in thinning from zero to 60% (Frej and Tosterud, 1989).

The NAE timber industry also makes better use of fiber by-products (for board manufacture, insulation materials and fuel) than before 1945, when many of these products were simply burnt in the open on site. Much of this development was initiated from the state forest services, both in terms of funding and technical expertise. State services continue to have a major input into technology development, especially in the CEE countries, but in West and South Europe, forest technologies are dominated by a viable industry that exports machinery and knowledge for timber production and processing worldwide. In common with other manufacturing industries, production of machinery used in forestry and wood processing is increasingly shifting to the Far East, a trend that is set to continue.

The negative impact has been that the larger scale mechanization has lead to a major decline in the number of forest workers. Another negative consequence is that in systems such as short rotation forestry, soil compaction can be an important issue when considering the mechanization. This can have a particular impact where the crop is harvested in the winter months on wet soils, as can be the case in soils of Northern Europe. In these regions the crops are frequently grown on soil that is saturated during the winter months and soil damage is more likely to be significant (Culshaw and Stokes, 1995).

Unlike in agriculture, crop varieties used in plantations for commercial forestry are largely derived from selected wild stocks of trees, but not necessarily grown in their native region. Some of these are taken from stands known to grow well in the prevailing conditions and to produce good quality timber. Domestication of trees is still at a very early stage largely because selective breeding is more difficult with plants that have long generation times and that only exhibit desirable traits close to maturity, typically after several decades. Biotechnology and genomic knowledge is beginning to open up the possibility of true domestication of trees, partly by producing varieties with shorter generation times, but mainly through increasing knowledge of the genes responsible for desirable traits.

2.6.5 Forest institutions

Forest management in the United States and Canada has changed dramatically since 1945. In the United States, the Forest Service was formally established in 1905, assisting private forest land owners with management. The limited applicability of European management models to the US context, especially in the area of forest fires, provided impetus for forestry research (Williams, 2000). The US and Canada collaborate over research on forest health, sustainability and soils (Lal et al., 1997; O'Neill et al., 2005; Powers et al., 2005)

Europe has a large number of institutions that underpin the development of forestry as an industry and a social resource (UNECE, 2001). There are at least 150 forest research organizations and learned societies in Europe ranging from industry-sponsored research facilities, to academic departments (and entire "Forestry Universities" in the CEE countries) and state-funded research institutions. These include at least 30 state forest services in Europe, some of them also responsible for wider land use issues such as agriculture, biodiversity conservation and water resources. They are often powerful and influential organizations, with substantial funding, human and capital resources. Besides the training available through the organizations above, forestry is included in the general higher educational curriculum of many NAE countries and there are dedicated training establishments for forestry and wood-based processing.

Throughout NAE forestry NGOs promote sustainable use of forests and campaign for better protection of natural forests. They include forest product consumers who question the ways in which their countries' forests are being managed and exploited. Consumer organizations are increasingly involved in lobbying for more sustainable forestry, both within and outside NAE. This has led to the establishment and expansion of certification schemes throughout NAE, which although controversial, are aimed at assuring consumers that the forests from which their products are derived from are forests managed according to a published set of management rules and objectives.

Although many NAE forestry societies, state forest services and research organizations were established over 100 years ago, these institutions have developed rapidly over the past 50 years, largely driven by the post-war need to increase timber and paper supplies to an expanding and increasingly wealthy public. They hold considerable political power and continue to be a key influence on the success of the forestry industry (World Bank, 2005).

2.6.6 Drivers of changes in forestry

Markets have always played an important part in forestry production, driven by demand for structural timber for rebuilding NAE infrastructure needed after World War II, meeting demand for increased timber and paper pulp due to an increasing population and demand for fuel wood that is now increasing after a decline from 1950 to 1980. There has been a steady increase in global demand for wood-based boards used in construction and fitments and this is expected to continue in the 21st century.

State ownership and subsidies have also played an important role in the development of NAE forestry science and technology, especially the increased use of modern soil preparation, planting and harvesting technologies and processing equipment, and has enabled the increases in forest output seen in the past fifty years. Rules and regulations have become increasingly important as drivers of forest management and protection, especially enabled by conservation legislation driven by EU Directives and North American statutes.

In NA, the main drivers of change in forestry have been the decreased demand for conversion of forestland to agriculture; increased demand and market pressures in North America and globally for wood and wood products; increased emphasis on non-timber products of forests, e.g., wildlife, range, water, outdoor recreation; and the increased recognition of the role of forests in climate change and protecting biodiversity.

European Forests and Livelihoods

Within the EU-15 area, some 2.7 million people are employed in forestry and forest-based industries such as woodworking, the cork industry, pulp and paper manufacture and board production. The industry produces an annual value of at least EUR 335 billion (UNECE/FAO, 2003a; http://europa.eu/). The EU is one of the world's largest traders and consumers of forest products, with a net income in this sector. The EU also imports large quantities of forest products, primarily roundwood from the Russian Federation and wood pulp from the Americas, where higher growth and lower production costs make forest products from this region very competitive. The EU excels in the production of high value wood products such as boards, cork and specialist papers and is a key exporter in this sector. (Bowyer and Rametsteiner, 2004; http://europa.eu/).

At least 12 million people own forest holdings within the EU-15, mostly small scale owners with an average holding of 13 ha, with most owning around 3 ha, contrasting with the average area of 1,000 ha for public holdings. Private owners occupy around 65% of Europe's forested land. Since enlargement of the EU large areas of previously state-owned forest holdings have been restored to private ownership. There is an increasing trend for private owners to supplement their incomes from urban-based incomes, with less dependence on income from forestry (http://europa.eu/).

European forests are also economically and socially important because, besides providing the wood for industry, they also provide services such as leisure use (tourism, general recreation and hunting) and provide casual income for rural people from collecting valuable products such as fungi, berries and nuts. In Europe forests give many communities and individuals a strong sense of identity that is deeply ingrained in culture and societal values in many parts of Europe (e.g., rights to fuelwood, hunting and the collection of forest foods).

2.6.7 Trends in NAE forestry

NAE is the only world region where forest cover is increasing. Throughout NAE there been a steady increase in both deciduous and coniferous plantations since early in the 20th century. Timber productivity has increased since 1945 to meet increased demand, but NAE continues to import large quantities of wood, including hardwoods from tropical forests. This has been partly responsible for reductions in cover and quality of forests in other world regions.

Since 1945 there has been a shift from private to state forest ownership in the US. This trend was also apparent in Europe, but here ownership is increasingly being privatized.

Forestry research and development has increased significantly since 1945. Technologies, especially mechanization, have been developed to achieve faster and more efficient harvests and to access and harvest timber in areas previously considered too fragile for harvest.

Across NAE, there has been an overall decrease in forest biodiversity. However, adoption of ecosystem-based approaches to manage national forests and grassland has changed the way US and Canadian public/federal land managers administer natural resources. Forest management for multifunctionality is an increasing trend in Europe, with the exception of Russia where productivity is still the key driver of management.

Forestry management continues to provide livelihoods and a cultural focus for large numbers of people in NAE and the forestry product industry has grown rapidly to accommodate increased demand for timber and other forestry-derived products.

2.7 Changes in Aquaculture Production

2.7.1 North American aquaculture

It is useful to divide aquaculture into two distinct types, freshwater and salt water (Figures 2-17, 2-18, 2-19 and 2-20). As a whole, Canadian aquaculture between 1986 and 2004 has grown at an annual rate of 20%.

In the US modest amounts of fresh water aquaculture, dominated by catfish culture, have been practiced since at least the 1940s. In 2003, there were some 300 tonnes of catfish raised, representing 71% of all US aquaculture, fresh and salt water by weight; trout, talapia, crawfish and baitfish comprised the remainder. Canadian freshwater aquaculture consists primarily of the rainbow trout and secondarily brook trout.

In Canada, the major aquaculture crop is salmon. The majority of the cultured salmon, 55 to 60%, is exported to the United States, with the other two largest export markets, Japan and Taiwan, each representing less than 2% of production. Steelhead trout is the other seawater finfish aquaculture, but is produced in much lower amounts (Figure 2-19). Through the late 1980s and 1900s there was a rapid expansion of clam and especially mussel aquaculture such that mussel is now the major shellfish aquaculture product by weight and by value (Figure 2-20).

By contrast, before the 1990s US saltwater aquaculture was dominated by oyster culture. However, starting in the mid 1980s and continuing through the 1990s there has been a very large expansion of salmon aquaculture to become the dominant saltwater product. Although, salmon is the currently largest saltwater aquaculture harvest by weight, the dollar value of oyster production ($63 million in 2003) is greater than that of salmon ($54 million).

Aquaculture products are growing in importance in both the US and Canada, although they are less than 15% of wild fishery landings. Aquaculture in 2003 represented about 10% of US wild fishery landings. The total Canadian com-

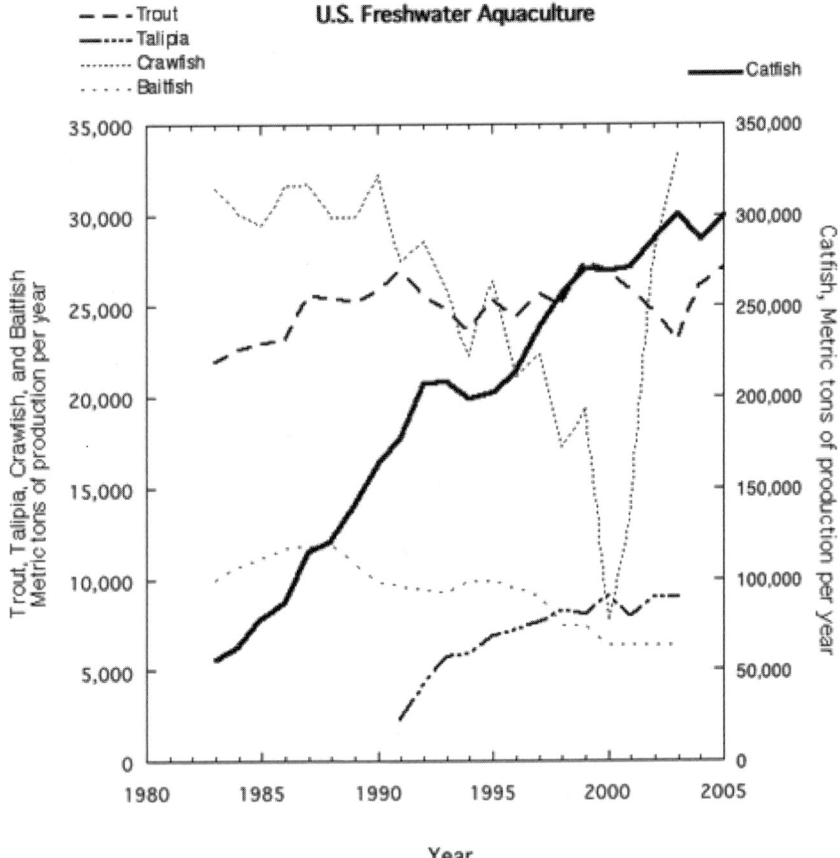

Figure 2-17. *Production of major aquaculture species in the U.S.* Source: Author elaboration of USDA and Canada STAT data.

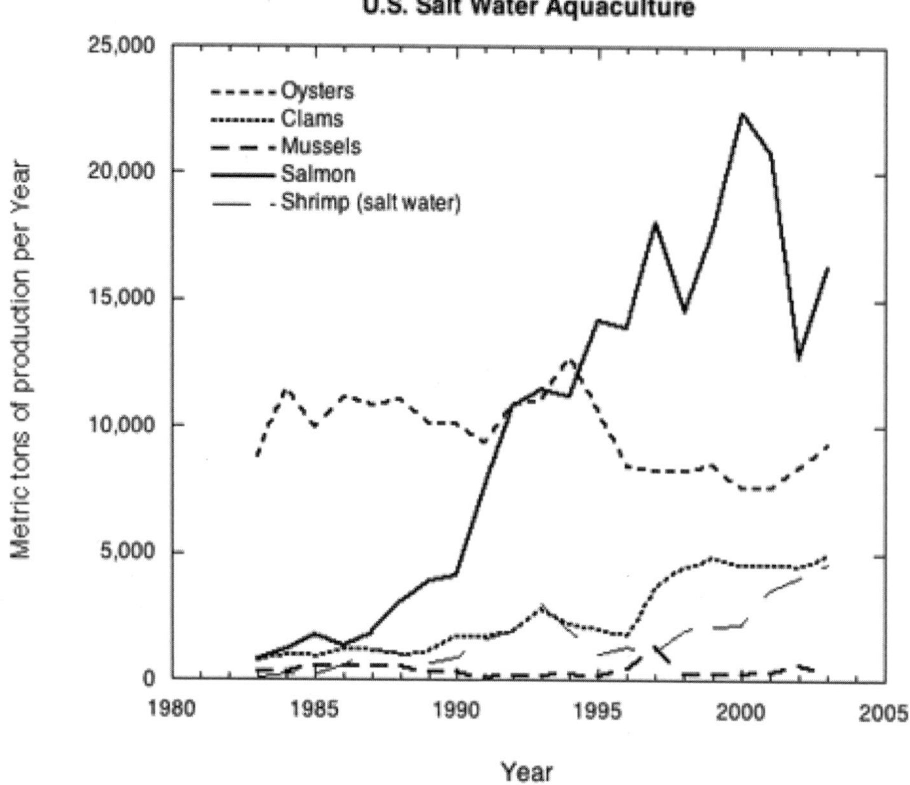

Figure 2-18. *Production of major salt water aquaculture species in the U.S.* Source: Author elaboration of USDA and Canada STAT data.

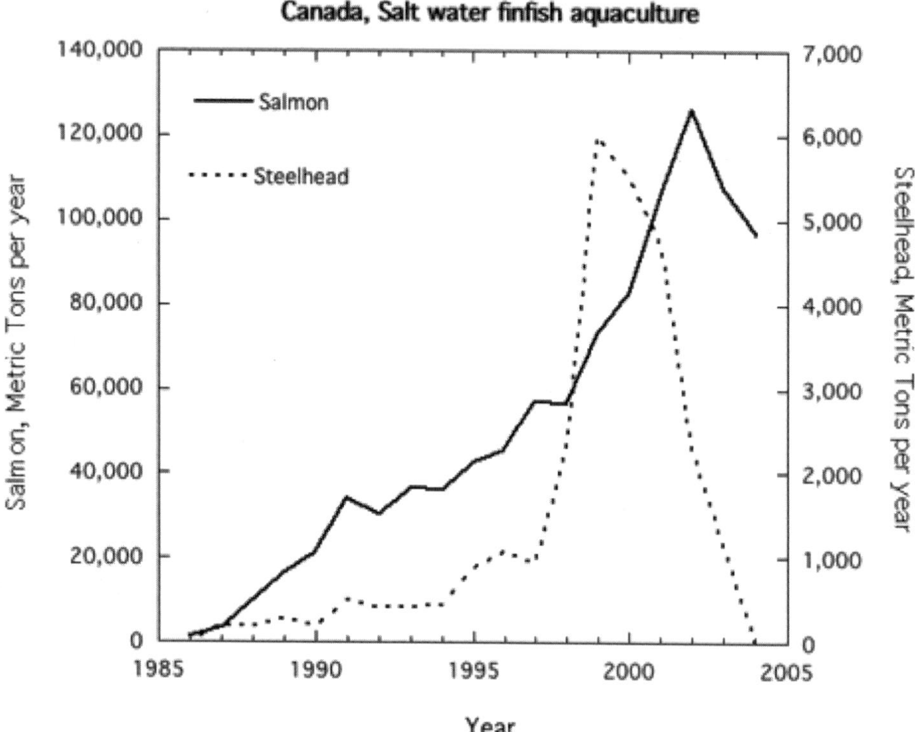

Figure 2-19 Canadian saltwater finfish aquaculture production. Source: Fisheries and Oceans Canada, 2007.

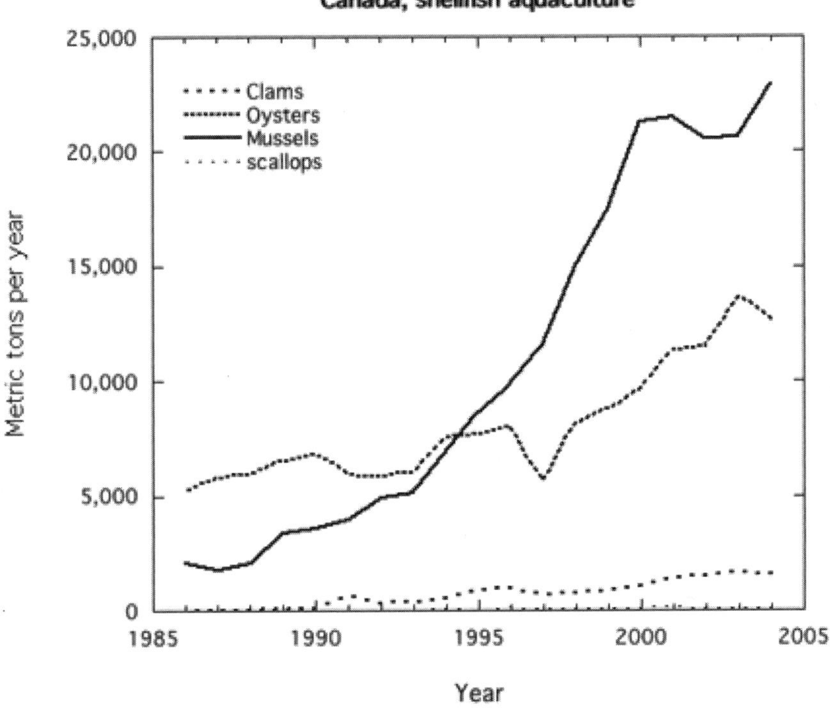

Figure 2-20. *Canadian shellfish aquaculture.* Source: Fisheries and Oceans Canada, 2007.

mercial landings of wild fisheries in 2004 were 1,071,182 tonnes, while aquaculture production was 145,840 tonnes, or 13.6% of the wild harvest. However, for salmon in Canada the wild fishery landed just over 25% of aquaculture production in 2004. The US is a net importer of seafood primarily from Asia.

2.7.2 European aquaculture

The aquaculture sector in Europe has a very diverse production, processing and marketing structure, ranging from small traditional enterprises, through medium-sized family fish farms, to the large-scale intensive businesses dominated by multinational companies (Fédération Européenne de Salmoniculture, 1990; MacAlister Elliott and Partners Ltd., 1999; Varadi et. al., 2001). Although there are structural differences between aquaculture sectors in different European regions, markets are now the determining factors of success and therefore the major driver in the aquaculture business with consumer demands, international competitiveness, health and environment issues and product quality all driving demand and price (Stirling Aquaculture, 1996ab).

The total output from European aquaculture has increased steadily since 1945 (Tacon, 1997). From the 1960s to the present the broad pattern of aquaculture development has been (FAO, 1996, 2000; Tacon and Barg, 1998):

- High growth in Northern Europe and medium growth in Western Europe fuelled by the development of salmonid mariculture;
- Low growth in Southern Europe with a focus on mariculture of sea fish; and
- Decline in CEE due to general post-transition economic decline and changing consumer habits (Staykov, 1994; Szczerbowski, 1996).

Increases in the production of finfish and molluscs have almost always led to value reduction as the price falls. This has become a serious issue for the viability of salmonid farming in Northern and Western Europe, where ex-farm prices have dropped from 3.5 Euro kg^{-1} in 1997 to 2.4 Euro kg^{-1} in 2005. In Southern Europe the value of farmed sea fish has remained relatively steady. Overall production increases in European aquaculture have slightly outpaced falls in price, leading to an increase in total value from 3.4 million Euro in 1999 to 3.9 million Euro in 2005.

Subsidies from the EU have contributed to the development of the salmonid sector, but withdrawal of state support in CEE may have contributed to the decline in cyprinid production. Other challenges for aquaculture include increasing concern from the public and from governments about the quality of fish produced in intensive systems and about the environmental impacts of fish farming and the competition for resources such as high quality water, high protein feed based on fish meal and labor.

Freshwater production has grown since 1945, but remained almost static in the 1980s, largely because output from the CIS countries and Russia declined (FAO, 1996). Increased fish consumption is expected, especially in CEE, where per capita fish consumption still remains far below that of the EU-15 (Tacon, 1997). Overall production from freshwater aquaculture is now increasing, albeit at a much slower rate than production from saltwater (FAO, 1996).

Aquaculture in saltwater has seen a spectacular rise in output since the mid 1970s, when farming salmon in sea cages began to develop in Norway, Scotland and Ireland. Salmonid finfish production now dominates the saltwater sector, overtaking mollusc production in 1995. The success of increasing output from the salmon industry has been tempered by a collapse in prices in the early 1990s, in turn leading to government intervention such as the destruction of smolts and feed quota systems introduced in Norway in the mid 1990s (Anon., 1996). Besides salmonid production, other higher value species of saltwater finfish such as bass, turbot, sea bream, cod and halibut are now being intensively farmed in European seas, lagoons and purpose built tanks in coastal waters of the warmer southern European countries such as Greece, Italy and Spain (Tacon, 1997). The industry is still developing from a low base in the 1980s but production has risen rapidly, with for example sea bream and bass production growing annually by over 40% (315 tonnes to 17,000 tonnes) from 1984 to 1995 (FAO, 2000; www.fao.org/fi/statist/FISOFT/FISHPLUS.asp). Production rose to 120,000 tonnes in 2001, most of which was exported from Greece to Italy and Spain, but the market for these fish has now expanded to other European countries.

The main finfish species groups cultivated in the region are salmon and rainbow trout, with about 85% of total farmed finfish production (Eurostat and FAO). Salmonids freshwater cyprinids (mostly carp and eels) constitute the second major finfish species group cultivated in the region at around 12% of total farmed finfish production (Voronin and Gavrilov, 1990; Dushkina, 1994; Zaitsev, 1996). Production of mussels and oysters and other molluscs is still a major part of total aquaculture output in Europe. There has been a slow decline in output of molluscs since the mid 1980s driven by a combination of disease problems (Figueras et.al., 1996), changing consumer habits and competition from other aquaculture sectors. Europe is the leading world producer of farmed turbot (100%), eels (99%), mussels (70%), sea bass and bream (68%), salmon (60%) and trout (54%).

From a low base at the end of World War II, European mollusc production increased rapidly until the 1970s and then output has remained relatively static, with some evidence for a decline of about 4% in the past twenty years. Blue Mussel production in France illustrates this trend with output at 8,500 tonnes in 1950 rising to 47,000 tonnes in 1977, a level that is the average maintained since then (FAO FISHPLUS website). Mussels remain the dominant species in this sector (60% of total output), with oysters making up around 25% output and several species of clams the rest. The main mollusc production regions are in France (35% of total), Italy (26%), Spain (17%) and the Netherlands (13%). Mollusc production makes up around 25% of the total monetary value of aquaculture in Europe (Tacon, 1997; FAO, 2000).

Institutions in aquaculture production in Europe

National organizations representing the aquaculture industry have grown rapidly since the 1960s in the Northwestern European countries, handling policy, advice, marketing and research. Some of these, like the Fiskeoppdretternes Salgslag in Norway are effectively production and marketing monopolies, but most others are NGOs independent of the industry. For producers there is a European wide organization, the Federation of European Aquaculture Producers (FEAP), representing all national associations at EU level. In most Eastern European countries, aquaculture is usually organized and advised by the Ministries of Agriculture and Food, with the exception of the USSR where it is in a separate Ministry of Fisheries. This state intervention is rapidly changing as private companies are beginning to gain market share within the Central and Eastern parts of EU-25.

Public investment in fish farming has been and remains a major factor in the development of European aquaculture. In CEE, public funding has come via state intervention, whereas in other parts of Europe, state and EU subsidies and development programs have played a significant role in developing both the fresh and saltwater aquaculture industries. Thus, although policy has historically been a driver of aquaculture development, state intervention is declining and markets are becoming more important drivers.

Fish farming is now strictly regulated in Europe with a number of Directives and domestic legislation covering water use and pollution control, the use of disease control measure (including pesticides) and feed regulations. There are also rules and regulations relating to the processing and marketing of aquaculture products. There is a trend towards stricter regulation and monitoring that adversely affects small family-owned enterprises (Varadi et al., 2001).

2.7.3 Science and technology in aquaculture

Since 1945 major breakthroughs have been made in fish farming techniques, including:

- The intensive hatching and rearing of sea fish in the southern countries
- Control of density dependent fungal and bacterial diseases in finfish
- Techniques for rearing salmonids in salt water
- The development of fish food processing and supply, including better formulation, the development of specialized feed and automatic feeding

These developments have enabled the spectacular increases in production seen in Europe over the past thirty years, especially in farmed salmonid and sea fish output (FAO, 2000). Most of this research and development has focused on high value finfish production, with far less work being done on mollusc and carp production, where production is mostly from units using traditional methods developed over centuries.

However, now research in aquaculture has changed to helping production systems address environmental issues including:

- Pollution of the sea caused intensive cage systems in coastal waters
- Pollution of rivers and streams caused by trout farming units
- Pesticide residues in fish flesh and the impacts of pesticide use in the marine and freshwater environment
- The impact on marine ecosystems of large-scale supply

of sea fish for aquaculture feed, for example the 1990s near-collapse of food webs dependent on sandeels in parts of the Northwest Atlantic.

2.7.4 Key changes in aquaculture

Aquaculture, while practiced for centuries across NAE, has grown in importance since the 1940s, in most parts of the region except for CEE. In Canada, for example, the industry is growing at 20% per year. There have been very large increases in aquaculture—both freshwater and saltwater—across NAE, propelled in part by explosive growth in salmon production. Despite this growth, North American aquaculture represents 15% or less of wild fishery landings by weight.

In the US, salmon has overtaken oysters as the major saltwater aquaculture and is the most important aquaculture crop in Canada. Salmon production is very important in Northern Europe, fuelled by good prices in the 1970s and 1980s. However, by the late 1990s, prices had dropped precipitously.

Due to developments in AKST, intensive rearing methods came to dominate aquaculture production. These production systems required the development of specialized feeds and control of fungal and bacterial diseases. Increases in salmon production were possible because of new techniques for saltwater production. However, the environmental impacts of these intensive production systems has caused aquaculture research to shift to addressing pollution concerns, pesticide residues and impacts on ecosystems.

2.8 Key Changes in Post-Harvest and Consumption Systems

Postwar consumer desire for adequate and safe food at modest prices has driven some of the changes described in the last few subchapters. We now turn our attention to changes in the consumption systems that exist across NAE. In line with trends across the OECD, the share of overall consumer spending on essentials (food, clothing, energy) has declined in Europe; in the UK, it has halved in 40 years. In the UK, one pound in three spent on food is spent away from home and in Ireland it is estimated that one Euro in every four is spent away from home (Henchion and McIntyre, 2004). Declining relative expenditure on food and even food price deflation is a major factor in the level of competition in food retail.

2.8.1 Changes in the food retail sector in NAE

Food retailing has experienced significant changes since 1945. Today, the giants of European food retail are Germany, France and the UK, based on their high populations and mature markets. The ownership structure of the largest companies in European food retail is varied. Carrefour (the world's second largest retailer) and Tesco are publicly held. Metro is publicly held, but with a large proportion owned by founder Otto Beisheim, the Haniel group and the Schmidt-Ruthebeck family. Rewe is a cooperative owned by its 3000 retail members, while ITM Intermarché is a consortium of independent merchants. Food accounts for around three-quarters of sales for these companies, except Metro where the figure is closer to 50%.

In 2003, European food retailers accounted for 46% of all European retail sales. The food retail market in Europe is very mature, but the food retail sector has increased its share of the wider retail market in all but four of 19 countries (France, Spain, Sweden and Denmark) by 19% to €870bn between 1999 and 2003. Tesco's sales rose by 54% and Wal-Mart Europe by 32% thanks entirely to the Asda operation in the UK. Non-food is the driver of this supermarket growth, since food sales are relatively stagnant.

There is a close relationship between per capita GDP and the penetration of "modern" retail (Figure 2-21). But what is interesting from a European perspective are the outriders, such as Italy with about 20% below that predicted and the UK, which is about 15% above that predicted by this relationship. Whether this phenomenon points to durable exceptions to the rule based on cultural or policy differences, or simply to time lags in some countries, is not currently clear.

In CEE countries, the penetration of large supermarket chains in the national food retail markets is quickly approaching saturation. The EU average is 15 hypermarkets per one million inhabitants. Hungary has 10 million inhabitants and by the end of 2005 there will be 98 hypermarkets in the country. Hypermarkets in Hungary now account for around a quarter of the market. Modern retailing already has an 18% share of the Russian market. This trend towards supermarket penetration in food retail has decreased the number of farmers' markets in many CEE countries.

While there is a general trend toward concentration in Europe, the emerging structures of food retail are not always the same (Dobson et al., 2001). These authors use a typology of the *dominant firm* (when the market share of the top firm is >25% and at least twice as high as the second rated firm), the *duopoly*, the *asymmetric oligopoly*, the *symmetric oligopoly* and *unconcentrated* structure (when no firm has a market share >10%) (Table 2-7). In 1999 Italy was the only country ranked as "unconcentrated", though this no longer applies now that Coop Italia has a 12.5% share.

The internationalization of retail in Europe has been, by comparison with other sectors, a recent phenomenon. There is still quite a strong national characteristic to food retailing in many Western European countries (Table 2-8) though this (1) hides high levels of international collaboration between firms in pan-European sourcing to increase buying power, with buying groups especially strong in Scandinavia and (2) the rise of the deep discounters such as Aldi up the ranks of national players. Food retail in most CEE countries is dominated by the multinational chains. The top 10 retailers in the Czech Republic, for example, are all multinationals. Nevertheless, some domestic cooperatives, trade associations and retail chains (such as COOP, CBA and Reál in Hungary, or VP Market in the Baltic countries) have been able to hold their own against competition by international retailers.

Internationalization allows retailers to use their distribution systems for pan-European procurement. Tesco, for instance, exports Hungarian products under its private labels; the firm announced last year it aimed to export HUF 1 billion in Hungarian goods in 2005, with increases of Hungarian goods to the Czech Republic, Slovakia and Poland. French-owned hypermarket Auchan also said recently it will increase the sale of Hungarian products outside Hungary's borders to HUF 5 billion in several years' time.

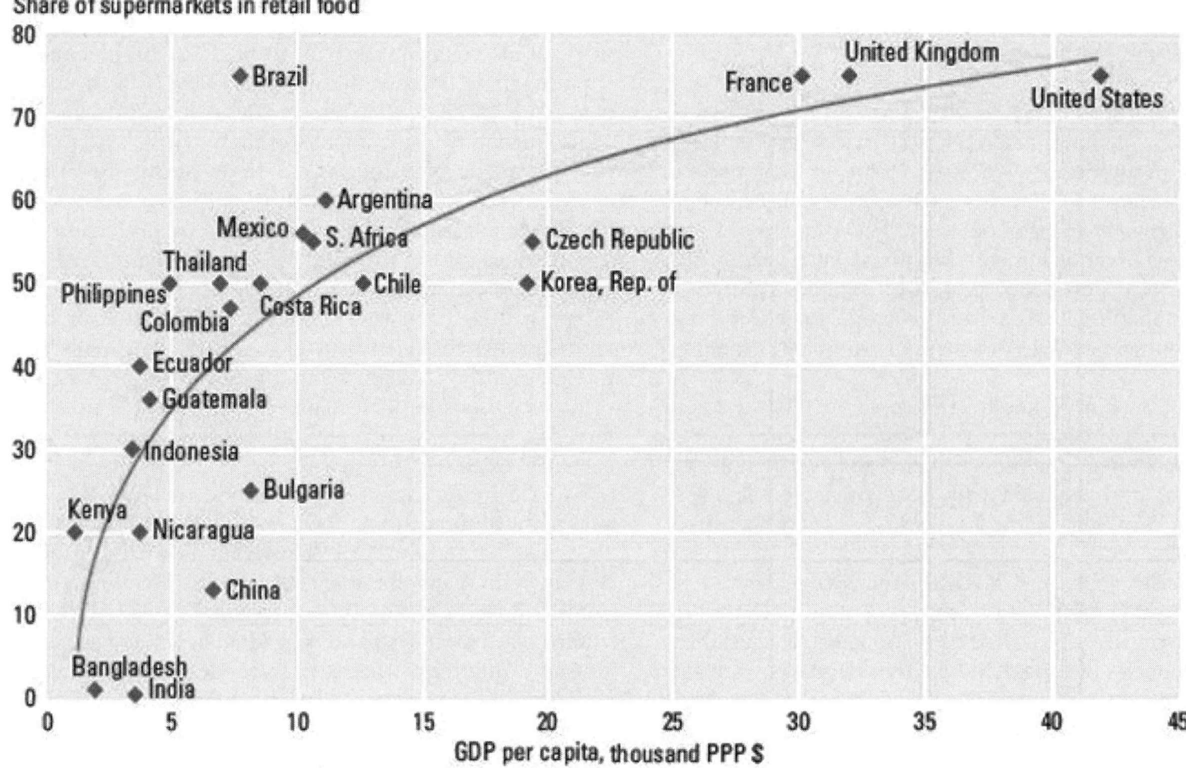

Figure 2-21. *Large supermarket penetration vs GDP per capita*

Own brand (private labels) are still rising in the European supermarket scene with an average 26% market share in Western Europe (Table 2-9). Growth is strong in parts of CEE—the share of private label products in Hungary was 15% in 2003 and own-brand goods account for around 25% of the total Tesco revenue in Hungary. The tight price squeeze forced by supermarkets has been responsible for own brand manufacturers such as Northern Foods struggling with profitability.

Table 2-7. **Market structure of retail in Western Europe, based on market shares of top 5 retailers, based on 1999 data.**

Country	Market structure
Austria	Asymmetric oligopoly
Belgium/Lux	Asymmetric oligopoly
Denmark	Duopoly
Finland	Duopoly
France	Asymmetric oligopoly
Germany	Symmetric oligopoly
Ireland	Asymmetric oligopoly
Italy	Unconcentrated
The Netherlands	Dominant firm
Portugal	Duopoly
Spain	Asymmetric oligopoly
Sweden	Dominant firm
UK	Asymmetric oligopoly

Source: Dobson et al., 2001.

"Trade spend" is another important feature of European retail, also known as *marges arrières* (back margins). Supermarkets have been able to use their gatekeeper position to make money on the buy side. This "trade spend" for suppliers to secure business with supermarkets comprises reimbursements to the retailer for the range of products it carries and promotions it carries out and includes supplier rebates, overriders (a discount or rebate related to the performance of the customer, paid in retrospective), unilateral deductions from money due or even demands for ad hoc cash payments. "A typical big European retailer might extract the equivalent of 10% of its total revenues via trade spending" (Economist, 2003).

Discounters are a growing part of the European food retail landscape with some notable exceptions such as the UK and Ireland. Discounters are a huge part of the market in Germany—in 2003, Germany accounted for 43% of Western Europe's 32,500 discount stores. But deep discounting is also growing fast in France, where there is a growing emphasis on price.

Buying groups or "international purchasing and marketing organizations" are means by which supermarket companies and consortia can increase their buyer power especially when negotiating with the big brand manufacturers. This is demonstrated by the GNX platform offering for

Table 2-8. **Top retailers across Europe—summary.**

Country	CR3	CR4	Top 3-4 firms
	percent		
Austria	n/a		n/a
Belgium/Lux	n/a		Carrefour, Delhaize Group, Colruyt, Aldi
Czech rep	30.1		Metro, Ahold, Schwartz
Denmark	78		FDB, Dansk Supermarkt, Supergros
Finland	79		Kesko, S Group
France	50.8	63.2	Carrefour, Intermarché, Leclerc, Casino
Germany	44.3	56.1-66.7	Metro, Rewe, Edeka/AVA, Aldi
Hungary	48.2	51	CBA, Tesco, Co-op Hungary, Metro, Reál Hungária
Ireland	54.7		Tesco, Dunnes Stores, Superquinn,
Italy	29.1	36.0	Coop Italia, Auchan, Carrefour, Conad
The Netherlands	62.6	82.6	Ahold, Casino, Sperwer, Makro
Norway	83		Norgesgruppen, Coop, Hakon
Poland	17.3		Metro, Jerónimo Martins, Tesco, Auchan
Portugal	n/a	n/a	n/a
Romania	n/a	27.0	Metro, Rewe, Carrefour, Delhaize
Slovakia	24.4		Tesco, Metro, Rewe
Spain	53.8	62.5	El Corte Inglés, Carrefour, Marcadona Eroski,
Sweden	95		ICA/Ahold, Coop, Axfood
UK	42.3	49.3-76.5	Tesco, Asda-Wal-Mart, Sainsbury's, Morrisons

Note: CR3 and CR4 refer to concentration ratios of the market share of top 3 (CR3) and top 4 (CR4) firms

Source: Planet Retail, 2007, 2006; Nielson, 2005.

auction contracts worth $8bn. Associations between buying groups and the top 30 retailers in Europe are common. The largest, EMD, has a 10.6% market share in Europe and a sales volume of EUR 950 million. Buying groups can have a significant impact on actual industry concentration. For instance in Hungary, from the Top-10 list SPAR and Metro form the buyer group METSPA with more than USD 1,800 million sales and Cora (Delhaize group) and Csemege are part of the PROVERA buyer group. Because of the buying groups, in western Europe only around 110 buying desks account for about 85% of the total retail food (not food-service) sales of the western European countries (Grievink, 2003) (see Figure 2-22).

Consolidation of retailers' supply base is creating conditions in which competition between suppliers creates its own pressure on producer prices. For example, between May and August 2004, the big three UK supermarket companies all announced rationalization of their milk supply, to two suppliers in the cases of Tesco and Sainsbury's and one in the case of Asda.

2.8.2 Concentration and trends at national levels

Germany is famously the toughest market in Europe. Deep discounters have a huge share of the market, accounting for 27% of modern grocery distribution sales, with that share around 50% for some product areas such as milk. The position of discounters is supported by strict planning laws for "big box" retailing, consumer perceptions of discounter private labels as good quality and popularity across income groups. The rate of growth of the UK food market has slowed and competition at the consumer side is very intense, with a permanent price war. Many firms have struggled to remain competitive and build critical mass in a market where market share is perceived to be key to success, including Morrison's (following the acquisition of Safeway), Marks and Spencer, Sainsbury's (only just starting to reverse a decline) and even Asda (part of Wal-Mart group) which has recently reported disappointing figures. This turmoil is not limited to publicly owned companies. The Cooperative Group is now searching for "efficiencies" after poor sales figures following a series of acquisitions. Only Tesco seems

Table 2-9. **Outlook for private label in Europe (% sales).**

	2000	2005	2010
Western Europe (of which):	20	26	30
Northern	25	29	32
Southern	12	18	25
Nordic	15	20	25
Central & Eastern Europe	1	4	7
World	15	19	23

Source: Planet Retail, 2006, 2007; Nielsen, 2005.

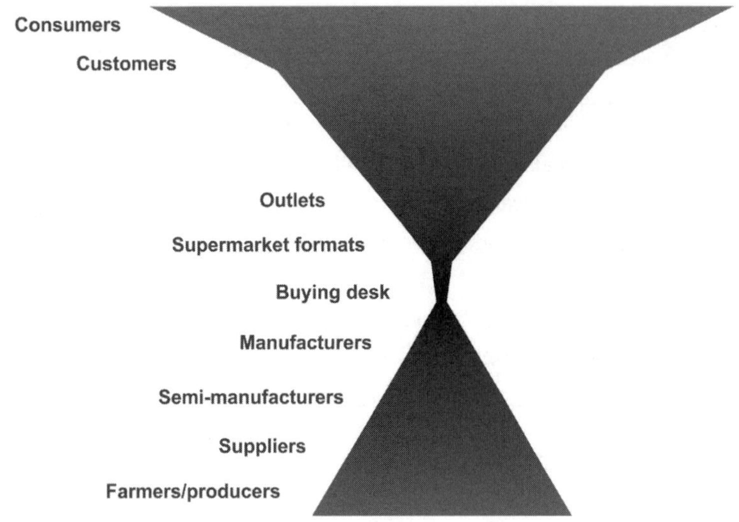

Figure 2-22. *The supply chain funnel in Europe.* Source: Grievink, 2003

to have managed consistently strong growth in market share at home and abroad (half of that shelf space is now overseas), profits and shareholder value in this period of consolidation of the UK retail sector, while taking massive chunks of business from clothing, electronics, financial service and other non-food sectors. The craft retailer Waitrose has also prospered.

Primary producers and suppliers are feeling the squeeze on prices. In a recent survey of farmers by Farmers Weekly magazine, a massive 95% of those questioned were concerned about power imbalance between buyers and suppliers, saying that the government must find ways to make trading relationships between retailers, processors and producers more equitable. Caribbean banana producers have called the price war "perverse transfer of wealth, by some of the supermarkets, from farmers and farm workers of developing countries to the consumers of developed countries" and "anti-development and regressive" (Eurofruit, 2004).

Despite investigations by the Competition Commission in 2000 and again in 2003 (around the Safeway takeover by Morrison's) and the resulting Supermarkets Code of Practice and subsequent review by the Office of Fair Trading, it is clear that consumer interests remain dominant over those of suppliers in the eyes of the Office of Fair Trading. Indeed, the situation in the UK around producer-supermarket trading can only be described as policy paralysis.

The UK independent retail sector is in steep decline, with a 7.4% decline in the number of corner shops in the last year alone. Industry watchers say 30,000 local shops—including specialists such as butchers, bakers and greengrocers—will be lost in a decade.

In North America, food retailing had a relatively slow pace of consolidation. A major wave of consolidation happened in the late 1990s, when Albertson's Kroger became the first coast-to-coast supermarket chains. By 2001, Kroger and Albertson's were the largest US grocers. However, Wal-Mart, which until the early 1990s had never sold any groceries, became the largest grocery retailer in 2004, with about 15% of the US grocery market. In Canada, Loblaw's is the dominant grocer in the Canadian market, with Sobey's competing for the number two position.

Today, the top five supermarket chains (Wal-Mart, Kroger, SuperValu, Safeway and Ahold) account for almost 50% of food retail sales in the United States (Table 2-10). By comparison, the top five food retailers accounted for only 20% of food sales in 1993.

When Wal-Mart entered the supermarket business in the mid-1990s, other stores were wary because of the incredible logistics system and supplier pricing that Wal-Mart brought to the business. More importantly, Wal-Mart's large size and market power caused concern as it integrated backward in the food system by creating relationships with the dominant food chain clusters. Wal-Mart is one of the first supermarkets to use case-ready meat in its stores.

The end of the 20th century saw the emergence of truly global food retailers like Carrefour, Wal-Mart and Tesco. Considering the rapid consolidation of the Latin American supermarket industry by transnational firms, development policy will need to respond to the resulting exclusion of

Table 2-10. **Food retailing in the USA.**

Supermarket*	Grocery Sales (billion $)		
Wal-Mart Stores	66.5		
Kroger Co.	46.3		
Albertsons, Inc.	32.0		
Safeway, Inc.	30.0		
Ahold USA, Inc.	25.1		
Historical CR5	**1997**	**2001**	**2005**
	24%	38%	46%

*Progressive Grocer reports only grocery sales from supermarkets, and does not report general merchandise, drug or convenience sales. In the 4/15/04 issue, it reported that total 2003 supermarket sales were $432.8 billion in the U.S.
Note: CR5 refers to concentration ratios of the market share of top 5 firms
Source: Planet Retail, 2007.

small farmers from regional agrifood markets (Reardon and Berdegué, 2002).

The significance of the changes in food retailing for production is in the restructuring of supply and distribution networks and in the development of standards enforced by retailers (Reardon and Berdegué, 2002). While food manufacturers have sometimes embraced consolidation because it decreases transaction costs, it also distorts power in the chain and puts the food retailers in a more powerful position (Stanton, 1999). Another result of restructuring is increasing retailer fees, some of which cover real costs but which are also used to generate an income stream that creates more gross profit for retailers (FTC, 2000). Manufacturers attributed the rising use of fees to greater retailer influence, while retailers attributed it to the increased cost of handling products (FTC, 2000).

In this arena of negotiated power between manufacturers and retailers, US retailers seem to have an edge, with bigger chains charging higher retailer fees (FTC, 2000). As power shifts to the largest retailers, evidence from the UK indicates that profitability does also (Wrigley, 1997). However, retailers are at the mercy of those manufacturers who have successful brands because branding is one way to create leverage with retailers. Retailers begin to develop one-on-one relationships with dominant food manufacturers who can service their far-flung systems. Moreover, retailers can start dictating terms to food manufacturers from their position of power at the point of consumption (Mehegan, 1999). Increasing consolidation of the retail sector has essentially constrained the way that farmers can respond to the changing nature of the global food system (Burch and Goss, 1999).

The point is that there exist dynamic social relationships within the channel from production to consumption although the trend seems to be that it is more and more difficult for smaller entities in any one sector of the chain to compete effectively. The development of these anti-competitive practices in supply chain management concerns many observers, including those from business schools (Hildred and Pinto, 2002).

2.8.3 Changes in food manufacturing and processing

The major food manufacturing countries in Western Europe are France, Germany the UK and Italy. Meat, beverages and dairy are the biggest sectors, comprising 20, 15 and 15% respectively of the value of production in 2001 totaling over EUR 600 billion. It is Europe's leading industrial sector and third-largest industrial employer (Table 2-11). Concentration in the food manufacturing sector is relatively low.

2.8.4 Market segmentation

One of the main changes occurring in the last 50 years in NAE can be described as a growing segmentation of the food markets and the emergence of food niche markets, such as PDO/PGI and TSG products in Europe and organic and fair trade production both in Europe and in North America. The process of market segmentation has been facilitated by the development of an increasing number of food standards and an articulated system of food labeling and certification.

Rise of uniform quality standards for food manufacturing/retailing

In recent years there has been a great increase of a all types of standards[4] in the agrifood system (e.g., food safety, food quality, environmental standards). The prominence of stan-

[4] "Standards are documented criteria or specifications, used as rules, guidelines or definitions of characteristics, to ensure consistency and compatibility in materials, products and services. In use standards become measures by which products, processes and producers are judged" (Bain et al., 2005). Standards for animal agriculture tend to focus either on food safety or product attributes, which generally encompass quality concerns like meat tenderness or animal welfare issues (Ransom, 2006, 2007).

Table 2-11. **Top European food manufacturers, ranked by turnover in 2002.**

Manufacturer	Country	Sales (EUR billion)	
Nestlé	Switz	52.6	Cereal, dairy, beverages, confectionery
Unilever	NL/UK	32.1	Dairy, beverages, dressings, frozen foods, cooking products
Diageo	UK	19.0	Alcoholic beverages, dough products
Danone	France	14.5	Dairy, beverages, biscuits and cereals
Cadbury Schweppes	UK	8.9	Beverages, confectionary
Heineken	NL	8.1	Alcoholic beverages
Parmalat	Italy	7.8	Dairy, gourmet, biscuits, beverages
Interbrew	Belgium	7.3	Alcoholic beverages
ABF	UK	7.1	Sugar, starches, baking products, meat, dairy
Tate & Lyle	UK	6.4	Sweeteners, starches
Lactilis	France	5.5	Dairy
Arla Foods	Denmark	5.0	Dairy
Sudzucher	Germany	4.8	Sugar

Source: CIAA, 2007.

dards has started as regulation of agrifood systems has shifted from nation-states to a broader set of organizations and institutions of the agrifood systems that also include global governance organizations, (e.g., World Trade Organization), multilateral and regional regulatory schemes (e.g., the EU) and private sector organizations, including transnational corporations (e.g., Cargill and Wal-Mart) (Scholte, 2000; McMichael, 2004; Higgins and Lawrence, 2005).

As the organization of agrifood systems has shifted, standards have become one of the most significant emerging practices for governing food (Bain et al., 2005; Higgins and Lawrence, 2005). Economists have typically highlighted the role standards play in helping to reduce transaction costs, increasing the predictability of a product and in general, simplifying what could be a very tedious and complicated process. With the increasing importance of standards, however, a shift has occurred from the use of standards as technical tools for market homogeneity to the use of standards as strategic tools for accessing markets, coordinating systems, enhancing quality and safety assurance, product branding and creating niche markets (Giovannucci and Reardon, 2000; Reardon et al., 2001).

The importance of standards has been recognized especially as the way in which the globalization of agriculture and food has been operationalized. Many authors have pointed to the growing concerns surrounding the distributional benefits of standards, especially for poor countries, small scale producers (both in poor and rich countries) and farmers utilizing alternative production systems (Dolan and Humphrey, 2000; Reardon and Farina, 2002; Dunn, 2003; Freidberg, 2004; Unnevehr and Roberts, 2004; Bain et al., 2005). In particular, this growing body of research has highlighted the rise of different types of standards, the lack of opportunity for specific groups to participate in standard setting, the high costs associated with standards adoption and the elevation of standards that require adherence to specific forms of production and processing in agrifood systems.

Historically, standards in most national food sectors have focused on what are called product (or performance) standards—that is, the composition (e.g., shape, color, etc.) of the final product and/or health features of the product (e.g., pesticide residues, contaminants, etc.) all of which are easily measured in the end product (Hannin et al., 2006). In much of the recent standards literature, the explanation for the emergence of food safety (or product) standards has to do with the decline of nation-state regulation combined with the many well-publicized food safety scares that have occurred in various countries (e.g., BSE—bovine spongiform encephalopathy, E-Coli contaminated meats and vegetables and dioxin-contaminated chicken). Thus, in order to reassure consumers of the safety of food products, countries and companies have imposed more stringent food safety standards. In Europe, NGOs pressure activities and consumers demand are often mentioned as the explanation for the increase in animal welfare standards and more broadly quality standards (Murdoch and Miele, 2004; Miele et al., 2005). Quality standards, (i.e., organics, fair trade, animal welfare) as opposed to food safety standards (i.e., pesticides residues, contaminants), are processed based standards, which means that the focus is on how the product is produced, with definitions of quality revolving around shared, socially constructed values (such as environmental conservation or regional characteristics) (Renard, 2005). Moreover, quality standards are voluntary standards and it is argued that industry leaders adopt voluntary quality standards due to consumer demand, or at the very least, to allow retailers to differentiate products along lines that appeal to consumers, such as animal welfare, environmental sustainability and worker welfare (Hatanaka et al., 2005).

2.8.5 Food safety, quality regulation and food market niches

Created by FAO and WHO, the Codex Alimentarius Commission has elaborated many international standards. According to the Codex Alimentarius definition, food safety is the assurance that food will not cause harm to the consumer when it is prepared and/or eaten according to its intended use (Codex Alimentarius, 1997).

Recent food scares in NAE have stimulated public concerns about food and farming. Consumers find it difficult to know where their food comes from, how it is produced and how far it has traveled. Food provision is increasingly organized through complex supply chains, often on a global scale. This has implications for consumer confidence, food safety and public health. In order to address this problem at the global level a number of international standards for food have been elaborated. For food safety the most widespread standard is HACCP which stands for "Hazard Analysis at Critical Control Point". The Codex Alimentarius Commission has adopted HACCP as the international standard for food safety. Under the EU food hygiene legislation, there are over a dozen measures covering specific products, an initiative to consolidate all hygiene legislation into one single text led to the implementation of EU Hygiene of Foodstuffs Regulations, 1998. While HACCP had its origin in the USA, it has now been introduced by the Hygiene Rules 93/43/EWG in the production line of food in Europe. It bears the main ideas from the worldwide-accepted HACCP-System of the FAO/WHO Codex Alimentarius (Wilm, 2005).

Chronology of HACCP development (Wilm, 2005)

1959—Development of the HACCP concept to assure one hundred percent safety of food to be used in space.

1971—The HACCP system was published and documented in the USA.

1985—The National Academy of Science (NAS) recommended the use of the system. Worldwide the system became used and the FAO/WHO Codex Alimentarius (Food and Agriculture Organisation/World Health Organisation) cited the system in the Codex.

1993—The European regulation 93/43 EG since 1993 provides the use of the system for the production of food.

The International Organization for Standardization (ISO) has developed the ISO-9001:2000 quality system that aims to enhance customer satisfaction. This includes the processes for continual improvement of the quality system and the assurance of conformity to the customer and applicable regulatory requirements. In global business the certification according ISO 9000 turned out to be an imperative duty. Certification to an ISO 9000 standard does not guarantee the compliance (and therefore the quality) of end products and services; rather, it certifies that consistent business processes are being applied.[5] Although the standards originated in manufacturing, they are now employed across a wide range of other types of organizations, including colleges and universities. A "product," in ISO vocabulary, can mean a physical object, services, or software. ISO 9000 and ISO 14000 standards are implemented by 760,900 organizations in 154 countries (Table 2-12) (ISO, 2005).

ISO 22000:2005 Food Safety Management Systems Standard is an international standard that defines the requirements of a food safety management system covering all organizations in the food chain from "farm to fork," including catering and packaging companies. This standard has been developed to harmonize the growing number of national standards for food safety management. The standard combines generally recognized key elements to ensure food safety along the food chain including interactive communication; system management; control of food safety hazards through pre-requisite programs and HACCP plans; and continual improvement and updating of the management system.

Niche markets. Product differentiation has provided special niches in food markets. These markets have been developed by granting protected trade marks/names so that consumers can easily distinguish the special flavor or quality of niche products among similar commodities. These schemes are increasing important for rural development across Europe. Their implementation in the US is a relatively new phenomenon with such regions now being delineated ecologically rather than politically, culturally or economically.

The market for organic products

In 2004, the market value of organic products worldwide reached 23.5 billion EUR (27.8 billion USD), with a market growth of about 9%. The leading regions were Europe, with a share of 49% and North America with a share of 47%. The three largest country markets were USA ($12.2 billion); Germany ($4.2 billion) and the UK ($1.9 billion) (Willer and Yussefi, 2006). In 2005, the global market for organic products reached a value of 25.5 billion Euros, with the vast majority of products being consumed in North America and Europe. For 2006, the value of global markets is estimated to be at more than 30 billion Euros.

The distribution of the European organic market continues to broaden and deepen as more consumers are attracted in more sectors and in more countries. In Germany a growing number of conventional supermarkets are offering organic products and the number of organic supermarkets continues to increase with 40 new organic supermarkets opening in 2004 alone. The UK market continues to show healthy growth, with much of the growth occurring in non-supermarket channels like organic food shops, box schemes and farmers' markets. A growing number of catering and food service companies are also offering organic food. The

Table 2-12. **Top ten countries for ISO 14001 certificates.**

Country	Certificates
Japan	13,416
UK	5,460
China	5,064
Spain	4,860
Germany	4,144
USA	3,553
Sweden	3,404
Italy	3,006
France	2,344
Korea, Rep. of	1,495

Source: ISO, 2003.

Italian and French markets are the next most important in Europe, however growth rates have slowed in these countries. A smaller market for organic food is found in CEE countries with the region comprising less than 3% of European revenues. Demand for organic products is growing through all CEE countries including Russia, particularly in metropolitan areas.

The data for the European market is fragmented and reliable detailed country comparisons are difficult to make because of the differences in data collection methods. However, FiBL have estimated the data which contribute to the profile of the European market reflected in the following tables for 2003 in which year the European market for organic food and beverages amounted to €11 billion (Table 2-13).

The North American market for organic products has reported the highest growth worldwide. Organic food and drink sales in the US were estimated to have totaled approximately 14.5 billion USD in 2005. With healthy growth rates continuing, the region is expected to overtake Europe and represent most global revenues in 2006. The driver for growth is the increase in marketing and distribution channels, with traditional, dedicated organic retailers like Whole Food Market and Wild Oats being joined by mainstream food multiples. Mainstream grocery retailers now comprise most organic food sales and the range of products is expanding in supermarkets such as Safeway, Albertson's, Wal-Mart and Kroger. The Canadian market is also reporting high market growth.

Demand in North America has become so high that local producers are having difficulty in matching supply and organic products are being imported from across the world e.g., organic seeds and grains are coming in from Europe and Asia; organic herbs and spices from Latin America and Asia; organic beef is imported from Australia and Latin America. Large food companies dominate almost every sector with companies such as Dean Food and General Mills active in the market. North America has organic food companies such as Hain Celestial, Sun Opta, Whole Food Market and Planet Organic listed on the stock exchange.

[5] Certification body is URS Certification Ltd in India and Europe which accredited by NABCB and UKAS (http://www.ursindia.com)

Table 2-13. **EU market countries clustered by stage of organic market development, 2001.**

Mature market countries	Growth market countries	Emerging market countries
Austria	Finland	Czech Republic
Denmark	Italy	Greece
Germany	The Netherlands	Ireland
Switzerland	Sweden	Slovenia
	France	Spain
	Belgium	Norway
	United Kingdom	Portugal

Source: OMIaRD, 2004.

Fair trade

In 2003, the global Fair Trade sales were over $895m and sales could increase by a factor of 20 or more in the next few years (Nicholls and Opal, 2004). Half the UK population is now aware of Fair Trade and there are similar figures for other European countries. Sales of fair trade products in Europe are growing remarkably well in several countries, largely stagnant in other countries and are not prominent in CEE countries. In 2004 sales grew of 102% in France, 50% in Belgium and 60% in Italy (Wills, 2005) (Table 2-14).

The findings of the 2005 Fair Trade Trends Report (Fair Trade Foundation, 2005) clearly demonstrate that the Fair Trade movement has continued to grow rapidly over the past five years. In 2003, total Fair Trade sales in North America including Mexico reached $291.75 million, a 53% increase over 2002. The US Fair Trade sales currently represent a potentially huge market for the initiative. US Fair Trade market is the largest single national market in the world after UK and the sales are increasing remarkably (Table 2-15).

Fair Trade Coffee. In 2002, FLO estimated the income benefit to Fairtrade producers at £21m, of which £17m was attributable to sales of Fairtrade certified coffee. TransFair USA estimated that, in five years of activity in the USA, Fair Trade has returned over £16.8m to coffee farmers in developing countries above what they would have received in the conventional market (TransFair USA, 2004). Fair Trade coffee sales vary considerably among different European countries. While coffee sales keep increasing in some countries, in general in Europe are largely stagnant.

Table 2-14. **Fair trade in Europe (data 2003-2004).**

Importing Organizations= 200	
Sales outlets	**Number**
World shops	2,845
Supermarkets	56,700
Others	19,300
Total sales outlets	**78,900**
Paid staff	
Importing organizations	851
World shop associations	107
Labeling organizations	113
Total paid staff	**1,071**
Turnover (in 000 €)	
Importing organizations	243,300
World shops, net retail value	103,100
Labeling organizations, net retail value	597,000
Education/PR/Marketing (in 000 €)	
Importing organizations	11,400
World shops associations	1,700
Labeling organization	5,100
All world shops, net retail value, estimate (in 000 €)	**120,000**
All Fair Trade products net retail value, estimate (in 000 €)	**660,000**

Source: Fine, 2006.

Table 2-15. **Total gross sales in North America (US and Mexico) 2001-2003.**

Year	Total gross sales Fair Trade (million US$)
2001	125.2
2002	180
2003	276.1

Source: Fair Trade Foundation, 2005.

By contrast, in North America, strong national campaigns have allowed a significant growth and it is likely that in the US and Canada, fair trade coffee sales will reach a market ceiling similar to that in Europe (Murray et al., 2003). Fair Trade Certified coffee is now the fastest-growing segment of the US specialty coffee market. The retail value of TransFair USA certified coffee increased by 59% in 2003 for a total of $208 million and by 77% in 2004 for a total of $369 million.

Fair Trade bananas

Fair Trade bananas were introduced in Europe by Max Havelaar in 1996. Since then, Fair Trade bananas had grown 14,655 tonnes by 1998 (Murray and Raynolds, 2000). They have captured unprecedented market shares; sales have risen by over 25% per year since 1999, reaching a market share of over 45% in Switzerland (Fine, 2006).

Alternatively, traded bananas have emerged in US in different ways compared to Europe. In US the NGO Rainforest Alliance has certified bananas under its ECO-OK and "Better bananas" program in 1999. Instead of building an alternative trade that challenges the power of bananas multinational corporations, this NGO has fostered a close collaboration with those companies (Murray and Raynolds, 2000). Trainsfair USA began certifying Fair trade bananas only in January 2004. Data of market shares for FT labeled bananas are not available.

2.8.6 Changes in diet/consumption

The general context in NAE is that of a contrasted situation between the food shortage post WWII, especially in Europe and the present situation of affluence and surplus in North America and Europe. This trend is attested by a number of key indicators of food provision (c.f. Wood et al., 2005). The average food production per capita in the world increased from 1961 to 2003 by around 25%. There were huge inequities between industrial and developing countries. This was accompanied by falling food prices, as there was a strong decline in the relative importance of food within total consumption expenditure from above 40% after WWII to 12-20% in Europe in 1999 (Eurostat, 2001) and to 10% in the United States in 1996.

According to 2001 estimates, 13% of the household budget in the EU15 was spent on food and non-alcoholic beverages, but the share of the budget spent on food fell between 1995 and 2001, mainly as a result of increasing available household income. Logically the share varies with GDP per head: the lower GDP per head of a country, the higher the share of money spent on food.

In 2005, the consumption of food and drink represented on average 16% of total consumption expenditure per person in the EU-15 countries and 27% in the new Member States (EEA, 2005). Food and drink used to account for the largest share of household consumption, before being gradually overtaken by other necessities such as housing, transport and leisure (Table 2-16). Consumer patterns across the enlarged EU reflect income differences but also the availability of goods and services.

Significant differences persist among member states (Tables 2-17 and 2-18). The lowest share of expenditure is found in the United Kingdom (9.7%) and the highest in Portugal (18.5%). The share of food and drinks in household expenses remains important in the new member states with an average of 22% against 12% in the EU 15 (Eurostat, 2005). Consumers' habits vary substantially among the 25 Member States. In addition to income, factors such as culture, tradition, household composition and degree of urbanization can influence habits in each country. The accession of the 10 new Member States has made the differences even more apparent than before. The share of citizens' total expenditure on food is projected to continue decreasing. Indeed, food consumption expenditure in the EU is projected to increase by 17% between 2000 and 2020, while in the same period total household expenditure could increase by 57% (EEA, 2005).

Changes in food provision and food nutrients. Increased food availability was made possible by increases in production and labor productivity in all sectors of the agricultural and food chains (see data in previous parts of chapter 2). AKST has played a major role in this phenomenon, as intensive livestock and crop systems were developed in order to meet quantitative food demand. These changes in food provision resulted in increased amounts of food calories, as well as protein and fats available for consumption in Europe and North America (Table 2-19).

Table 2-16.
Household consumption expenditure in the EU-25 in 2003.

Expenditures	%
Food and non-alcoholic beverages	13.1
Alcoholic beverages, tobacco and narcotics	3.8
Clothing and footwear	6.1
Housing, water, electricity, gas and other fuels	21.5
Furnishing, household equipment and routine maintenance of the house	6.6
Health	3.5
Transport	13.5
Communications	2.8
Recreation and culture	9.4
Education	1.0
Restaurants and hotels	9.0
Miscellaneous goods and services	9.9

Source: Eurostat, 2005.

Table 2-17. **Proportions of expenditures in real values (average of 1995 and 1999).**

	Housing	Food	Furnishings	Education and leisure	Transport and communications	Clothing	Health
Western EU	19.24	18.45	6.81	12.03	12.73	5.45	8.54
Central and Eastern EU	24.66	22.06	3.43	17.42	8.61	3.46	10.89
Total	21.02	19.66	5.69	13.83	11.36	4.79	9.32

Source: Schenkel et al., 2005.

Table 2-18. **Index of relative price.**

	Food	Clothing	Housing	Furnishings	Transport and communications	Education and leisure	Health
Western EU	86.94	107.9	97.5	93.06	109.6	107.6	99.28
Central and Eastern EU	139.6	183.2	75.38	157.1	175.5	51.85	66.62

Note: GDP index for each country, 100.

Source: Schenkel et al., 2005.

Table 2-19. **NAE food supply: Energy, protein and fats per capita per day.**

	Western Europe			Eastern Europe			USA		
	Calories (kcal)	Protein (g)	Fats (g)	Calories (kcal)	Protein (g)	Fats (g)	Calories (kcal)	Protein (g)	Fats (g)
1961	3,001	87	106	3,118	91	79	3,100	92	138
2003	3,535	109	149	3,227	95	109	3,900	112	178

Source: FAOSTAT, 2006 and USDA-ERS, 2005b.

Available food calories have increased in the range of 18-26% in Western Europe and USA between 1961 and 2003, presently reaching values of 3500 to 3900 calories per capita per day. During the same period, protein supply has increased by 22-25% and fat supply by 29-41%. Increases were much more modest in Eastern Europe, as food calories increased by only 3% and protein by 4% between 1961 and 2003. In contrast, total fat supply increased considerably, i.e., by 37% in the same period.

Noteworthy is the amount of calories provided by lipids in the diet, which is presently around 40% in Western Europe and America, but 30% in Eastern Europe (derived from data presented in Table 2-20). Another feature is the change in the percentage of calories or nutrients derived from animal vs. plant products for Western and Eastern Europe (Table 2-20). Whereas the percentage of calories from animal origin slightly increased between 1961 and 2003, the percentage of proteins from animal origin increased more dramatically (reaching 60% in 2003 for Western Europe). In contrast, the percentage of animal fats in the diet actually decreased over the same period, especially in Eastern Europe where it was quite high in the 1960s.

2.8.7 Key Changes in consumption systems

Across NAE, the amount that consumers spend on food provisioning has significantly decreased, reflecting the decline in real prices for food. However, this change has been accompanied by an increasingly differentiated food marketplace. Consumers across NAE are spending more on food eaten away from home. Strong markets for organic, fair

Table 2-20. **NAE food supply: Percentage of energy, protein and fats from animal sources.**

	Western Europe			Eastern Europe		
	Calories	Protein	Fats	Calories	Protein	Fats
	Percent from animal sources					
1961	29	51	64	23	36	73
2003	31	60	55	26	50	59

Source: FAOSTAT, 2006.

trade and other nice food products have developed in NA and Western Europe, with less interest in these markets in most CEE countries.

The food retail market has become increasingly consolidated across the entire region, resulting in a shift in power away from farmers to food retailers. The increase in standards, some resulting from concerns about food safety and others from demand for quality, has also created some market barriers for farmers. In addition, the widespread availability of so much food has affected diets and diet related diseases across the region.

References

Adcock, F., D. Hudson, P. Rosson, H. Harris, and C. Herndon. 2006. The global competitiveness of the North American livestock industry. Choices 21(3):171-176.

AGBIOS. 2008. GM database. Available at http://www.agbios.com/dbase.php. AGBIOS, Ontario.

Agra/CEAS. 2004. Study on the socio-economic implications of the various systems to keep laying hens. Agra/CEAS Consulting Ltd., Report for the European Commission, Brussels.

Ahloowalia, B.S., M. Maluszynski, and A.K. Nicherlein. 2004. Global impact of mutation-derives varieties. Euphytica 135:187-204.

Anon. 1996. Norwegians destroy 40 million smolts. Scottish Fish Farmer 87:7.

Asíns, M.J. 2002. Present and future of quantitative trait locus analysis in plant breeding. Plant Breeding 121(4):281-291.

Aspelin, A.L. 1997. Pesticides industry sales and usage 1994 and 1995 market estimates. Rep. 733-R-97-0002. U.S. EPA, Washington, DC.

Aspelin, A.L. 2003. Pesticide usage in the United States: Trends during the 20th Century. Available at http://cipm.ncsu.edu/cipmpubs/index.cfm. Center for IPM, North Carolina State Univ., Raleigh.

Aumaitre, A.L., and J.G. Boyazoglu. 2000. A note on livestock production and consumption in Europe. Available at http://bsas.org.uk/downloads/mexico/007.pdf. Eur. Assoc. Animal Production, Rome.

Axelsson, A. 1998. The mechanisation of logging operations in Sweden and its effect on occupational safety and health. Int. J. For. Engineer. 9(2):25-31.

Aydemir, A., and G. Borjas. 2007. Cross-country variation in the impact of international migration: Canada, Mexico, and the United States. J. Eur. Econ. Assoc. 5(4):663-708.

Bailey, R.G. 1980. Descriptions of the ecoregions of the United States. Misc. Publ. 1391. USDA-USFS, Washington, DC.

Bain, C.B., J. Deaton, and L. Busch. 2005. Reshaping the agrifood system: The role of standards, standard makers, and third-party certifiers. p. 71-83. *In* V. Higgins and G. Lawrence (ed) Agricultural governance: Globalization and the new politics of regulation. Routledge, NY.

Barkema, A., and M. Drabenstott. 1996. Consolidation and change in heartland agriculture. p. 61-77 *In* Economic forces shaping the rural heartland. Federal Reserve Bank, Kansas City.

Barkema, A., M. Drabentstott, and N. Novack. 2001. The new U.S. meat industry. Federal Reserve Bank of Kansas City. Econ. Rev. (2nd Q):33-56.

Bockheim, J.G. 1984. Proceedings of the Symposium, forest land classification: Experiences, problems, perspectives. Misc. Publ. USDA-USFS, Washington, DC.

Bonanno, A., and D. Constance. 2006. Corporations and the state in the global era: The case of Seaboard Farms and Texas. Rural Sociol. 71(1):59-84.

Bonnano, A., D. Fernandez Navarrete, and J.L. Pilles. 1990. Agrarian policy in the US and EC: A comparative analysis. p. 331. *In* A. Bonnano (ed) Agrarian policies and agricultural systems. Westview Spec. Studies Agric. Sci. Policy. Westview Press, Boulder.

Bongiovanni, R. and J. Lowenberg-Deboer. 2005. Precision agriculture and sustainability. Precision Agric. 5(4):359-387.

Borjas, H., and S. de Rooij. 1998. Rural women and food security: Current situation and perspectives Available at http://www.fao.org/docrep/003/w8376e/w8376e06.htm#3.%20women%20in%20the%20farming%20sector. FAO, Rome.

Bosworth, D.N. 2004. Four threats to the Nation's forests and grasslands speech. Idaho Environ. Forum, Boise, ID.

Bouma, J., G. Varallyay, and N.H. Batjes. 1998. Principal land use changes anticipated in Europe. Agric. Ecosyst. Environ. 67:103-119.

Bowyer, J., and E. Rametsteiner. 2004. Policy issues related to forest products markets in 2003 and 2004. *In* UNECE/FAO forest products annual market review 2003-2004. Timber Bull: 57(3):11-18. Available at www.unece.org/trade/timber/docs/fpama/2004/2004_fpamr.pdf.

Burch, D., and J. Goss. 1999. An end to fordist food? Economic crisis and the fast food sector in Southeast Asia. p. 87-110. *In* D. Burch et al. (ed) Restructuring global and regional agricultures: Transformations in Australasian agri-food economies and spaces. Ashgate, London.

Canadian Forest Service. 2003. Available at http://www.pfc.cfs.nrcan.gc.ca/canforest/.

Canales, A. 2000. International migration and labour flexibility in the context of NAFTA. Int. Migration 63:221-252.

CANFI. 2004. Available on line at http://nfi.cfs.nrcan.gc.ca/canfi/index_e.html. Canada Nat. Forest Inventory, Ottawa.

Cash, J.A. 2002. Where's the beef? Small Farms Produce Majority of Cattle. Agric. Outlook 297:21-24.

CAST. 1996. Future of irrigated agriculture. Council Agric. Sci. Tech., Task Force Rep. 127. Ames IA.

Chapman, P.J., J.M.A. Sly and J.R. Cutler. 1977. Pesticide usage survey report 11—Arable farm crops 1974. MAFF, London.

Chataway, J., J. Tait, and D. Wield. 2004. Understanding company R&D strategies in agro-biotechnology: Trajectories and blindspots. Res. Policy 33(6-7):1041-1057.

Chavas, J.P. 2001. Structural change in agricultural production. p. 263-285. *In* B.L. Gardner and G.C. Rausser (ed) Handbook of agricultural economics 1A.

CIAA. 2007. Data and trends of the European food and drink industry [Online]. Available at http://www.ciaa.be/documents/brochures/dataandtrends_2007.pdf. Confederation of the food and drink industry of the EU. Brussels.

CIV. 2007. Available at: http://www.civ-viande.org/uk/ebn.ebn?pid=57&rubrik=1&item=1&page=1. Centre d'Information des Viandes.

Cochrane, N. 2002. Pressures for change in Eastern Europe's livestock sectors. Agric. Outlook Jan-Feb.

Cochrane, W.W. 1987. Saving the modest-sized farm or the case for part-time farming. Choices 2(2):4-7.

Codex Alimentarius. 1997. Available at: http://www.haccphelp.com/Documents/Codex.pdf.

Collins, E. 1976. Migrant labour in British agriculture in the nineteenth century. Econ. Hist. Rev. 29(1):38-59.

Cotterill, R.W., and A.W. Franklin. 2001. The public interest and private economic power: A case study of the northeast dairy compact. Food Policy Marketing Center, Univ. Connecticut.

Csaki, C., C, Forgacs, and B. Kovacs. 2004. CEE regioinal report: Regoverning markets in food and agricultural. CEE, Budapest.

Culshaw, D., and B. Stokes. 1995. Mechanisation of short rotation forestry. Biomass Bioenergy 9:127-140.

Davis, R.P., D.G. Garthwaite, and M.R. Thomas. 1990. Pesticide usage survey report 78—
arable farm crops 1988. MAFF, London.

Debailleul, G., and E. Deleage. 2000. Les agriculteurs et la Conditionnalite Environnementale en France et aux Etats-Unis. Rapport D'etudes pour le Ministere de L'amenagement du Territoire et de l'environnement.

De Haan, C., H. Steinfeld and H. Blackburn. 1997. Livestock and the environment: Finding the balance. Chapter II. Wrenmedia, UK.

Dench, S., J. Hurstfield, D. Hill, and K. Akroyd. 2006. Employers use of migrant labour main report. Available at http://www.homeoffice.gov.uk/rds/pdfs06/rdsolr0406.pdf. Home Office, London.

Dmitri, C., A. Effland and N. Conklin. 2005. The 20th century transformation of U.S. agriculture and farm policy. EIB 3. USDA, Washington, DC.

Dobson P.W., M. Waterson, and S.W. Davies. 2001. The patterns and implications of increasing concentration in European food retailing. J. Agric. Econ. 54:111-125.

Dolan, C., and J. Humphrey. 2000. Governance and trade in fresh vegetables: The impact of UK supermarkets on the African horticulture industry. J. Dev. Studies 37:147-176.

Donham, K.J., S. Wing, D., Osterberg, J.L. Flora, C. Hodne, K.M. Thu, et al. 2007. Community health and socioeconomic issues surrounding concentrated animal feeding operations. Environ. Health Perspect. 115(2):317-310.

Drabenstott, M. and T.R. Smith. 1996. The changing economy of the rural heartland. p. 1-11. *In* Economic forces shaping the rural heartland. Federal Reserve Bank, Kansas City.

Duewer, L., 1984. Changing trends in the red meat distribution system. AER 509. ERS, USDA, Washington, DC.

Dunn, E.C. 2003. Trojan pig: Paradoxes of food safety regulation. Environ. Planning A 35(8):1493-1511.

Dushkina, L.A. 1994. Farming of salmonids in Russia. Aquacult. Fisheries Manage. 25:121-26.

EC. 2001. Agriculture in the European Union—Statistical and economic information 2001. Available at: http://ec.europa.eu/comm/agriculture/agrista/2001/table_en/en353.htm and for imports/exports data at http://ec.europa.eu/comm/agriculture/agrista/tradestats/index_graph.htm#part1.

EC. 2005. Biomass action plan. Report {SEC(2005)1573}. Brussels. Available at http://ec.europa.eu/energy/res/biomass_action_plan/doc/2005_12_07_comm_biomass_action_plan_en.pdf, accessed 3/27/07.

EC. 2006. A strategy to keep Europe's soils robust and healthy. Available at http://ec.europa.eu/environment/soil/index.htm.

EC. 2007. Innovation and technological development in energy. Available at http://ec.europa.eu/energy/res/sector/bioenergy_en.html. EC, Brussels.

Economist. 2003. Trouble in store: the murkiness of retailing. May 17, p. 74. http://ec.europa.eu/energy/res/sectors/bioenergy-en.html. EC, Brussels.

EEA. 2003. Europe's environment: the third assessment. State of environment Rep. No 1/2003. Available at http://reports.eea.europa.eu/environmental_assessment_report_2003_10/en/kiev_chapt_02_3.pdf. European Environ. Agency.

EEA. 2005. Household consumption and the environment. EAA report 11/2005 available at http://reports.eea.europa.eu. European Environmental Agency.

EEA. 1994. European River and Lakes: Assessment of their environmental state. Mono. 1, Eur. Environ. Agency, Copenhagen.

Elliot, D.L., C.G. Holladay, W.R. Barchet, H.P. Foote, and W.F. Sandusky, 1986. Wind energy resource atlas of the United States. Washington Pacific Northwest Lab., Richland.

EPA. 2005. Taking care of business: Protecting public health and the environment. EPA's Pesticide Program, FY 2004 annual report. EPA-735-R-05-001. Available at http://www.epa.gov/oppfead1/annual/2004/04annualrpt.pdf. U.S. Environ. Prot. Agency, Washington, DC.

EU. 2004. Consumption trends for dairy and livestock products and the use of feeds in production, in the CEEC accession and candidate countries. Available at http://ec.europa.eu/comm/agriculture/publi/reports/ccconsumption/fullrep_en.pdf.

EU Commission. 2006.**<B>[[FILL IN]]<B>**

Eurofruit. 2004. Interview with Bernard Comibert. April.

EUROPA database on forestry. Available at: http://ec.europa.eu/enterprise/forest_based/forestry_en.html, accessed 19 June 2006

Europa website on forest-based industries 2006 available at: http://ec.europa.eu/enterprise/forest_based/indusfed_en.html

Eurostat Agriculture. 2007a. Livestock production statistics. Available at http://epp.eurostat.ec.europa.eu/portal/page?_pageid=1996,39140985&_dad=portal&_schema=PORTAL&screen=detailref&language=en&product=Yearlies_new_agriculture&root=Yearlies_new_agriculture/E/E1/E12/eda33040.

Eurostat Agriculture: 2007b. Trade Statistics. Available at http://ec.europa.eu/comm/agriculture/agrista/tradestats/index_sem.htm.

Eurostat. 2001. Consumers in Europe-facts and figures. Available at http://epp.eurostat.ec.europa.eu.

Eurostat. 2005. Consumers in Europe-facts and figures. Available at http://bokkshop.europa.eu.

Fair Trade Foundation. 2005. Available at http://www.fairtrade.org.uk/what_is_fairtrade/facts_and_figures.aspx.

FAO. 1996. Fisheries and aquaculture in Europe: situation and outlook in 1996. Fisheries Circ. 911. Rome, FAO.

FAO. 1997 Management of agricultural drainage water quality. *In* C.A. Madramootoo et al. (ed) Water Rep. 13. Int. Comm. Irrig. Drainage. FAO, Rome.

FAO. 2000. Global forests resource assessment 2000. Main Report Available at www.fao.org/forestry/site/19155/en. FAO, Rome.

FAOSTAT. 2006. Available at http://faostat.fao.org/faostat/.

FAOSTAT. 2007. Chartroom. Available at http://www.fao.org/es/ess/chartroom/default.asp.

Farm Foundation. 2004. The future of animal agriculture in North America: An overview of issues. http://www.farmfoundation.org/projects/documents/InitialWhitePaperNovember04.pdf.

Farm Foundation. 2006. The future of animal agriculture in North America. Available at http://www.farmfoundation.org/projects/04-32ReportTranslations.htm.

Farrell, A.E., R.J. Plevin, B.T. Turner, A.D. Jones, M. O'Hare, and D.M. Kammen. 2006. Ethanol can contribute to energy and environmental goals. Science 311:506-508.

Fédération Européenne de Salmoniculture. 1990. A market study of the portion-sized trout in Europe.

Fernandez-Cornejo, J. 2004. The seed industry in U.S. Agriculture: An exploration of data and information on crop seed markets, regulation, industry structure, and research and development. AIB 786. ERS, USDA, Washington, DC.

Fernandez-Cornejo, J., and M. Caswell. 2006. The first decade of genetically engineered crops in the United States. EIB-11. USDA-ERS, Washington, DC.

Figueras, A., J.A.F. Robledo, and B. Novoa. 1996. Brown ring disease and parasites in clams (*Ruditapes decussatus* and *R. philippinarum*) from Spain and Portugal. J. Shellfish Res. 15(2):363-368

Fine, B. 2006. The new development economics. p. 1-20. *In* B. Fine et al. (ed) The new development economics: After the Washington consensus. Tulika, India.

Fisheries and Oceans Canada. 2007. Statistical services [Online]. Available at http://www.dfo-mpo.gc.ca/communic/statistics/aqua/index_e.htm. Fisheries and Oceans Canada, Ottowa.

Frances, J., S. Barrientos, B. Rogaly. 2005.

Temporary workers in UK agriculture and horticulture: A study of employment practices in the agriculture and horticulture industries and co-located packhorses and primary food processing sectors. Precision Prospecting, Suffolk for DEFRA, London.

Freidberg, S. 2004. French beans and food scares: Culture and commerce in an anxious age. Oxford Univ. Press, New York.

Frej, J., and A. Tosterud. 1989. Systems and methods used in large scale forestry. Rep. 6. Forest Oper. Inst., Sweden.

FTC. 2000. Public Workshop on Slotting Allowances and other Grocery Marketing Practices. Transcript: June 1. Fed. Trade Commission, Washington, DC.

Fuglie, K., N. Ballenger, K. Day, C. Klotz, M. Ollinger, J. Reilly et al. 1996. Agricultural research and development: Public and private investments under alternative markets and institutions. AER735. USDA, Washington, DC.

Gallivan, G.J., G.A. Surgeoner, and J. Kovach. 2001. Pesticide risk reduction on crops in the province of Ontario. J. Environ. Qual. 30:798-813.

Garnsworthy, P.C. 2005. Livestock yield trends: Implications for animal welfare and environmental impact. p. 379-401. *In* R. Sylvester-Bradley and J. Wiseman (ed) Yields of farmed species: Constraints and opportunities in the 21st century. Nottingham Univ. Press, UK.

Garthwaite, D.G., and M.R. Thomas. 2000. Pesticide usage survey report 159—arable farm crops in Great Britain 1998. MAFF, London.

Garthwaite, D.G., M.R. Thomas, A. Dawson, and H. Stoddart. 2004. Pesticide usage survey report 187—arable farm crops in Great Britain 2002. MAFF, London.

Garthwaite, D.G., M.R. Thomas, and M. Hart. 1996. Pesticide usage survey report 127—arable farm crops in Great Britain 199. MAFF, London.

Gay, S.W., and R. Grisso. 2002. Planning for a farm storage building. Publ. 442-760. Biol. Syst. Engineer. Dep., Virginia State Univ.

Germany (Government of). 2006. Statistik und Berichte. (in German) [Online]. Available at http://www.bmelv-statistik.de. German Gov. Statistics, Berlin.

Gerrish. J. 2004. Management-intensive grazing: The grassroots of grass farming. Green Park Press, Purvis, MS.

Giovannucci, D., and T. Reardon. 2000. Understanding grades and standards and how to apply them. *In* D. Giovannucci (ed) A guide to developing agricultural markets and agro-enterprises. World Bank, Washington, DC.

GMO Compass. 2007. GM maize growing in five EU member states. Available at http://www.gmo-compass.org/eng/agri_biotechnology/gmo_planting/191.gm_maize_110000_hectares_under_cultivation.html.

Goldschmidt, W. 1978. Large-scale farming and the rural social structure. Rural Sociol. 43:362-366.

Grievink, J.W. 2003. The changing face of the global food industry. OECD Conf. Changing Dimensions of the Food Economy: Exploring the policy issues, The Hague, 6 Feb 2003. Available at webdomino1.oecd.org/comnet/agr/foodeco.nsf/viewHtml/index/$FILE/GrievinkPPT.pdf.

Hahn, W.F., H. Mildred, D. Leuck, J. Miller, J. Perry, F. Taha, et al. 2005. Market-integration of the North American animal products complex. Outlook Rep. LDP-M-13101. USDA-ERS, Washington, DC.

Hall, B.F. and E.P. Leveen. 1978. Farm size and economic efficiency: The case of California. J. Agric. Econ. 60:589-600.

Haley, M.M. 2004. Market integration in the North American hog industries. Outlook Rep. LDP-M-125-01. ERS, USDA, Washington, DC.

Hannin, H., J.-M. Codron, and S. Thoyer. 2006. Standardization issues in the wine sector. p. 73-92. *In* J. Bingen and L. Busch (ed), Agricultural standards: The shape of the global food and fiber system. OIV and WTO, Springer, Dordrecht.

Hatanaka, M., C. Bain, and L. Busch. 2005. Third-party certification in the global agrifood system. Food Policy 30:354-369.

Havenstein, G.B., P.R. Ferket, and M.A. Qureshi. 2003 Carcass composition and yield of 1957 vs. 2001 broilers when fed representative 1957 and 1991 broiler diets. Poultry Sci. 82:1509-1518.

Hayami, Y., and V.W. Ruttan. 1971. Agricultural development: An international perspective. Johns Hopkins Univ. Press, Baltimore.

Hayami, Y., and V.W. Ruttan. 1985. Agricultural development. 2nd ed. Johns Hopkins Univ. Press, Baltimore.

Heady, E.O. 1962. Agricultural policy under economic development. Iowa State Univ. Press, Ames.

Heady, H.F., and D. Child. 1994. Rangeland ecology and management. Westview Press, Boulder.

Heederick, D. et al. 2007. Health effects of airborne exposures from concentrated animal feeding operations. Environ. Health Perspect. 115(2):298-302.

Heffernan, J.B., and W.D. Heffernan. 1986. When families have to give up farming. Rural Dev. Perspect. 2(3):28-31.

Heffernan, W.D. 1972. Sociological dimensions of poultry production in the United States. Sociol. Ruralis 12(3/4):481-499.

Heffernan, W.D. 1984. Constraints in the U.S. poultry industry. p. 237-260. *In* Research in rural sociology and development. Vol.1. JAI Press Inc.

Heffernan, W.D. 1998. Agriculture and monopoly capital. Monthly Rev. 50:46-59.

Heffernan, W.D, M. Hendrickson, and R. Gronski. 1999. Consolidation in the food and agriculture system. Report to National Farmers Union. Washington, DC.

Henchion, M., and B. McIntyre. 2004. Developments in the Irish food supply chain: Impacts and responses by SMEs. J. Int. Food Agribusiness Market. 16:103-122.

Hendrickson, M., and H. James, Jr. 2005. The ethics of constrained choice: How the industrialization of agriculture impacts farming and farmer behavior. J. Agric. Environ. Ethics 18:269-291.

Hendrickson, M., and W. Heffernan. 2002. Concentration of agriculture markets table. Dep. Rural Sociology, Univ. Missouri.

Hendrickson, M., and W. Heffernan. 2005. Concentration of agriculture markets table. Dep. Rural Sociology, Univ. Missouri.

Hendrickson, M., and W. Heffernan. 2006. Concentration of agriculture markets table. Dep. Rural Sociology, Univ. Missouri.

Hendrickson, M., and W. Heffernan. 2007. Concentration of Agricultural Markets. Available at http://www.nfu.org/wp-content/2007-heffernanreport.pdf. Univ. Missouri.

Hendrickson, M., W.D. Heffernan, P.H. Howard, and J.B. Heffernan. 2001. Consolidation in food retailing and dairy. Brit. Food J. 3(10):715-728.

Hendrickson, M., W.D. Heffernan, D. Lind and E. Barham. 2008. Contractual integration in agriculture: Is there a bright side for agriculture of the middle? *In* T.A. Lyson et al. (ed) Better food choices: Strategies for renewing an agriculture of the middle and transforming its supply chains. MIT Press, Cambridge, MA.

Higgins, V., and G. Lawrence. 2005. Agricultural governance: Globalization and the new politics of regulation. Routledge, NY.

Hildred, W., and J. Pinto. 2002. Impacts of supply chain management on competition. Working Pap. Ser. 02-10. Coll. Business Admin., Northern Arizona Univ.

Hildreth, R.J., and W.J. Armbruster. 1981. Extension program delivery: Past, present and future—An overview. Am. J. Agric. Econ. 63(5):853-858.

Hills, G.A. 1952. The classification and evaluation of site for forestry. Ontario Dept. Lands and Forests. Resource Div. Report 24.

Hodges. J. 1999. Jubilee history of the European Association for Animal Production. Livestock Production Science p. 105-168

Hoppe, R.A. 2001. Structural and financial characteristics of U.S. farms: 2001 family farm report. Agric. Inform. Bull. 768. USDA, Washington, DC.

Hoppe, R.A., and D. Banker. 2006. Structure and finances of U.S. Farms: 2005 family farm report. Available at http://ssrn.com/abstract=923592. EIB 12, ERS, USDA, Washington, DC.

Hoppe, R.A., and P. Korb. 2005. Large and small farms: Trends and characteristics. *In* D.E. Banker and J.M. MacDonald (ed) Structural and financial characteristics of U.S. farms: 2004 family farm report. AIB 797. USDA, Washington, DC.

Howell, T.A. 2001. Enhancing water use efficiency in irrigated agriculture. Agronomy J. 93:281-289.

ISAAA. 2005. Briefs. No.34, ISAAA: Ithaca, NY.

Johnson, B., and A. Hope, 2005. Productivity, biodiversity and sustainability. p. 335-350. *In* R. Sylvester-Bradley and J. Wiseman. Yields of farmed species: Constraints and opportunities in the 21st century. Nottingham Univ. Press, UK.

Johnson, N.C., A.J. Malk, R.C., Szaro, W.T. Sexton (ed) 1999. Ecological stewardship: A common reference for ecosystem management. Elsevier, UK.

Johnston, A.M. 2000. Animal health and food safety. Brit. Med. Bull. 56:51-61.

Just, R.E., and D.L. Hueth. 1993. Multimarket exploitation: The case of biotechnology and chemicals. Am. J. Agric. Econ. 75(4): 936-945.

Kandel, W. 2006. Meat-processing firms attract Hispanic workers to rural America. Amber Waves 4(3).

Kasimis, C., and A. Papadopoulos. 2005. The multifunctional role of migrants in the Greek countryside: Implications for the rural economy and society. J. Ethnic Migration Studies 31(1).

Kemp, R. 2001. Innovation in the livestock industry. Report Prepared Canadian Biotechnology Advisory Committee. Available at http://strategis.ic.gc.ca/epic/internet/incbac-cccb.nsf/vwapj/LivestockInnov_Kemp_f.pdf/$FILE/LivestockInnov_Kemp_f.pdf.

Khachatourians, G.G. 1998. Agricultural use of antibiotics and the evolution and transfer of antibiotic-resistant bacteria. Canadian Med. Assoc. J. 159(9):1129-1136.

Kiely, T., D. Donaldson, and A. Grube. 2004. Pesticides industry sales and usage, 2000 and 2001 market estimates. Available at http://www.epa.gov/oppbead1/pestsales/. UE EPA, Washington, DC.

Kirschenmann, K., S. Stevenson, F. Buttel, T. Lyson, and M. Duffy. 2003. Why worry about agriculture of the middle? White paper prepared for the Agriculture of the Middle Project. Available at http://www.agofthemiddle.org/papers/whitepaper2.pdf

Kislev, Y., and W. Perterson. 1986. Economies of scale in agriculture: A survey of the evidence. Report DRD 203. World Bank, Washington, DC.

Kloppenburg, J. 1991. Social theory and the de/reconstruction of agricultural science: local knowledge for an alternative agriculture. Rur. Sociol. 56(4):519-548.

Kloppenburg, J.R. 2004. First the Seed: The political economy of plant biotechnology 1492-2000. 2nd ed. Univ. Wisconsin Press, Madison.

Kogan, M. 1998. Integrated pest management: Historical perspectives and contemporary developments. Ann. Rev. Entomol. 43:243-270.

Kremnev, I. 1920. Puteshestvie moego brata Alekseia v stranu krest'ianskoi utopii. Moskva: Gosudarstvennoe izdatel'stvo.

Kuusela, K. 1994 Forest resources in Europe 1950-90. EFI Res. Rep. 1.Cambridge Univ. Press, Cambridge.

Lal, R., W.H. Blum, C.Valentinin, and B.A. Stewart. 1997. Methods for assessment of soil degradation. Adv. Soil Sci. Lewis Publ., Boca Raton, FL.

Lerman, Z., C. Csaki, and G. Feder. 2004. Evolving farm structures and land use patterns in former socialist countries. Q. J. Int. Agric. 43(4):309-335.

Lerman, Z., Y. Kislev, A. Kriss, and D. Biton. 2003. Agricultural output and productivity in the former Soviet Republics. Econ. Dev. Cultural Change 51(4):999-1018.

Liefert, W., S. Osborne, M. Trueblood, and O. Liefert. 2002. Could the NIS Region become a major grain exporter? Agric. Outlook, May.

Lobao, L. and K. Meyer. 2000. Institutional sources of marginality: Employment change and economic decline in midwestern family farming. p. 23-49. *In* R. Hodson (ed), Research in the sociology of work. JAI Press, Greenwich, CT.

Lowe, J.J., K. Power, and M.W. Marsan. 1996. Canada's forest inventory 1991: Summary by terrestrial ecozones and ecoregions. Inform. Rep. BC-X-364E. Nat. Resourc. Canada, Canadian Forest Serv., Victoria, B.C.

Lovell, S. 2003. Summerfolk: A history of the dacha, 1710-2000. Cornell Univ. Press. Ithaca and London.

Lyson, T. and A.L. Raymer. 2000. Stalking the wily multinational: Power and control in the U.S. food system. Agric. Human Values 17:199-208.

MA. 2005. Millennium ecosystem assessment synthesis report. Millennium Ecosystem Assessment, Island Press, Washington, DC.

MacAlister Elliott and Partners Ltd. 1999. Forward study of community aquaculture. Summary report for the European Commission, Fisheries Directorate General.

MacDonald, J.M., and P. Korb. 2006. Agricultural Contracting Update: Contracts in 2003. United States Department of Agriculture, Economic Research Service. Economic Information Bulletin No. 9.

MacDonald, J.M., M.E. Ollinger, K.E. Nelson, and C.R. Handy. 2000. Consolidation in U.S. meatpacking. AER Rep. 785. ERS, USDA, Washington, DC.

Malerba, F., and L. Orsenigo. 2002. Innovation and market structure in the dynamic of the pharmaceutical industry and biotechnology: Towards a history friendly model. Industrial and Corporate Change, Vol. 11(4):667-703.

Maluszynski, M., K. Nichterlein, B. Van Zanten, and S. Ahloowalia. 2000. Officially released mutant varieties—the FAO/IAEA database. Mutation Breeding Rev. 12:1-88.

Mancours, K., and J.F.M. Swinnen. 2000. Causes of output decline in economic transition: The case of Central and Eastern European agriculture. J. Compar. Econ. 28:172-206.

Martin, P. 2002. Mexican workers and U.S. agriculture: The revolving door. Int. Migration Rev. 36(4):1124-1142.

Martinez, S.W. 1999. Vertical coordination in the pork and broiler industries: Implications for pork and chicken products. Agric. Econ. Rep. 777. ERS, USDA, Washington, DC.

Matskevich, V.V. 1967. Sotsialisticheskoe pereustroistvo selskogo khoziaistva. Moskva. [Socialist restructuring of agriculture].

McBratney, A., Whelan, B., A. Tihomir, and J. Bouma. 2005. Future directions of precision agriculture. Precision Agric. 6:7-23.

McBride, W.D. 1997. Changes in U.S. livestock production, 1969-92. Agric. Econ. Rep. AER754. ERS, USDA, Washington, DC.

McMichael, P. 2004. Development and social change: A global perspective. Pine Forge Press, Thousand Oaks, CA.

McNab, W.H., and P. Avers. 1994. Ecological subregions of the United States: Section descriptions. WO-WSA-5. USDA-USFS, Washington, DC.

Medley, K.E., B.W. Okey, G.W. Barrett, M.F. Lucas, and W.H. Renwick. 1995. Landscape change with agricultural intensification in a rural watershed, southwestern Ohio, USA. Landscape Ecol. 10:161-176.

Medvedev, Z.A. 1987. Soviet agriculture. W.W. Norton, New York.

Mehegan, S. 1999. When worlds collide: Part I, the supermarket universe today. Baking Buyer. March. Available at www.bakingbuyer.com.

Mehlich, A. 1984. Mehlich 3 soil test extractant: A modification of Mehlich 2 extractant. Commun. Soil Sci. Plant Anal. 15:1409-1416.

Mellon, M., C. Benbrook, and K.L. Benbrook. 2001. Hogging it: Estimates of antimicrobial abuse in livestock. Available at http://go.ucsusa.org/publications/report.cfm?publicationID=308#food. Union Concerned Scientists, Boston.

Miele, M., and D. Pinducciu. 2001. A market for nature: Linking the production and consumption of organics in Tuscany. J. Environ. Policy Plan. 3:149-162.

Miele, M., J. Murdoch, and E. Roe. 2005. Animals and ambivalence: Governing farm animal welfare in the European food sector. p. 169-85. *In* V. Higgins and G. Lawrence (ed), Agricultural governance: Globalization and the new politics of regulation. Routledge, NY.

Miljkovic, D. 2005. Measuring and causes of inequality in farm sizes in the United States. Agric. Econ. 33(1):21-27.

Morgan, N. 2001. Repercussions of BSE on international meat trade: Global market analysis. Available at: www.fao.org/ag/aga/agap/frg/Feedsafety/pub/morgan%20 bse.doc. Commodities and Trade Division, FAO, Rome.

Murdoch, J., and M. Miele. 2004. A new aesthetic of food? Relational reflexivity in the "alternative" food movement. *In* M. Harvey et al. (ed) Qualities of food. Manchester Univ. Press, UK.

Murray, D.L., and L.T. Raynolds. 2000. Alternative trade in bananas: Obstacles and opportunities for progressive social change in the global economy. Agric. Human Values 17:65-74.

Murray, D.L., L.T. Raynolds, and P.L. Taylor. 2003. One cup at a time: Poverty alleviation and fair trade coffee in Latin America. Available at http://www.colostate.edu/Depts/Sociology/FairTradeResearchGroup/doc/fairtrade.pdf. Fair Trade research group, Colorado State Univ.

Narodnoe khoziaistvo. 1971. SSSR v 1970. Goskomstat, Moskva.

Narodnoe khoziaistvo. 1975. SSSR v 1974. Goskomstat, Moskva.

NASS. 2006. Agricultural chemical usage 2005 field crops summary. Rep. Ag Ch 1(06), Nat. Agric. Statistics Board. USDA, Washington, DC.

Natural Resources Canada. 2000. The state of Canada's forests—Forests in the new millennium 1999-2000. Natural Resources Canada, Canadian Forest Service, Ottawa, ON.

Natural Resources Canada. 2006. Precision farming Canada. Available at http://ccrs.nrcan.gc.ca/optic/hyper/farming_e.php. Centre for Remote Sensing. Canadian Forest Service, Ottawa, ON.

Nehring, R. 2005. Farm size, efficiency and off-farm work. *In* D.E. Banker, and J.M. MacDonald (ed) Structural and financial characteristics of U.S. farms: 2004 family farm report. Agric. Infor. Bull. 797. USDA, Washington, DC.

Nicholls, A., and C. Opal. 2004. Fair trade: Market-driven ethical consumption. Sage, London.

NRC. 1996. A new era for irrigation. Nat. Res. Council. Nat. Academy Press, Washington, DC.

O'Brien, D., and V. Patsiorkovsky. 2006. Measuring social and economic change in rural Russia: Surveys from 1991 to 2003. Lexington Books, Lanham MD.

O'Neill, K.P., M.C. Amacher, and C.H. Perry. 2005. Soils as an indicator of soil health: A guide to the collection, analysis, and interpretation of soil indicator data in the forest inventory and analysis program. Gen. Tech. Rep. NC-258. USDA-USFS, North Centr. Res. Stat., St. Paul MN.

OCED. 20XX. <B>[[FILL IN]]<B>

OECD. 2001. Challenges for the agro-food sector in European transition countries. OECD Observer, Paris.

Ollinger, M., J. MacDonald, and M. Madison. 2000. Structural change in U.S. chicken and turkey slaughter. Agric. Econ. Rep. 787. ERS, USDA, Washington, DC.

Olsen, S.R., and C.V. Cole, F.S. Watanabe, and L.A. Dean. 1954. Estimation of available phosphorus in soils by extraction with sodium bicarbonate. USDA Circ. 939. U.S. Gov. Print. Office, Washington, DC.

Ongley, E.D. 1996. Control of water pollution from agriculture. FAO Irrig. Drain. Pap. 55. FAO, Rome.

OURFOOD. 2005. http://www.ourfood.com/Introduction.html.

Park, J., J. Finn, and D. Cooke. 2005. Environmental challenges in farm management course outline. Available at http://www.ecifm.rdg.ec.uk. Reading Univ., UK.

Parayil, G. 2003. Mapping technological trajectories of the Green Revolution and the Gene Revolution from modernization to globalization. Res. Policy 32(6):971-990.

Paul, C.J., and R. Nehring. 2005. Product diversification, production systems, and economic performance in US agricultural production. J. Econometrics 126:525-548.

Passel, J. 2005. Unauthorized migrants: Numbers and characteristics. Pew Hispanic Center, Washington, DC.

Patzwaldt, K. 2004. Labour Migration in Eastern Europe and Central Asia: Current issues and next political steps. UNESCO Series of Country Reports on the Ratification of the UN Convention on Migrants. UNESCO, Paris.

Pearce, F. 2005. Forests paying the price for biofuels. New Scientist 2526:19.

Perry, J., D. Banker, and R. Green. 1999. broiler farms' organization, management, and performance. Res. Econ. Div. ERS, USDA, Washington, DC.

Pierzynski, G.M., J.T. Sims, and G.F. Vance. 2000. Soils and environmental quality. 2nd ed. CRC Press, Boca Raton, FL.

Pimentel, D., and T.W. Patzek, 2005. Ethanol production using corn, switchgrass, and wood; Biodiesel production using soybean and sunflower. Natural Resourc. Res. 14:65-76.

Pirog, R., T. van Pelt, K. Enshayan, and E. Cook. 2001. Food, fuel and freeways. Leopold Center Sustainable Agric., Iowa State Univ., Ames.

Poulton, P.R. 1995. The importance of long-term trials in understanding sustainable farming systems: the Rothamsted experience. Aust. J. Expl. Agric. 35:825-34.

Powers, R.F., D.A. Scott, F.G. Sanchez, R.A. Voldseth, D. Page-Dumroese, J.D. Elliott, and D.M. Stone. 2005. The North American long-term soil productivity experiment: Findings from the first decade. Forest Ecol. Manage. 220:31-50.

Prunty, L. 2004. Soil science and pedology disciplines in the North Dakota 1862 Land Grant College of Agriculture: A review. North Dakota State Univ. White Paper, Fargo.

Raijaic, A., L.A. Waddell, J.M. Sargeant, S. Read, J. Farber, M.J. Firth, et al. 2007. An overview of microbial food safety programs in beef, pork and poultry from farm to processing in Canada. J. Food Protection 70(5):1286-1294.

Ransom, E. 2006. Defining a good steak: Global constructions of what is considered the best red meat. p. 159-175. *In* J. Bingen and L. Busch (ed), Agricultural standards: The shape of the global food and fiber system. Int. Library Environmental, Agricultural and Food Ethics. Springer, NY.

Ransom, E. 2007. The rise of agricultural animal welfare standards as understood through a neoinstitutional lens. *In* M. Miele and B. Bock (ed) Competing discourses of farm animal welfare and agri-food restructuring. Int. J. Sociol. Agric. Food, Spec. Vol. 15(3).

Reardon, T., and J.A. Berdegué. 2002. The rapid rise of supermarkets in Latin America: Challenges and opportunities for development. Dev. Policy Rev. 20(4):317-334.

Reardon, T., J.M. Codron, L. Busch, J. Bingen and C. Harris. 2001. Global change in agrifood grades and standards: Agribusiness strategic responses in developing countries. Int. Food Agribusiness Manage. Rev. 2:421-35.

Reardon, T., and E. Farina. 2002. The Rise of Private Food Quality and Safety Standards: Illustrations from Brazil. Int. Food Agric. Manage. Rev. 4:413-421.

Renard, M.C. 2005. Quality certification, regulation and power in fair trade. J. Rural Studies 21:419-431.

Renewable Fuels Association. 2006. Ethanol industry outlook 2006: From niche to nation. Available at http://www.ethanolrfa.org/objects/pdf/outlook/outlook_2006.pdf. RFA, Washington, DC.

Rhoades, J.D. 1990. Principal effects of salts on soils and plants. *In* A. Kandiah (ed), Water, soil, and crop management relating to the use of saline water. AGL Misc. Series Publ. 16/90. FAO, Rome.

Richards, L.A. (ed). 1954. Diagnosis and improvement of saline and alkali soils. US Salinity Laboratory. USDA Handbook 60. U.S. Govt. Print. Office, Washington, DC.

Rogaly, B. 2006. Intensification of work-place regimes in British agriculture: The role of migrant workers. Available at http://www.sussex.ac.uk/migration/documents/mwp36.pdf. Migration Working Pap. 36. Univ. Sussex, UK.

Robinson, M., and A.C. Armstrong. 1988. The extent of agricultural field drainage in

England and Wales, 1971-80. Trans. Inst. Brit. Geographers (New Ser.) 13(1):19-28.

Rossiia v tsifrah. 2003. 2004. Goskomstat, Moskva.

Rossiia v tsifrah. 2005. 2006. Goskomstat, Moskva.

Runyan, J. 2000. Profile of hired farm workers. USDA-ERS, Washington, DC.

Sagardoy, J.A. 1993. An overview of pollution of water by agriculture. p. 19-26. *In* Prevention of water pollution by agriculture and related activities. Proc. FAO Expert Consultation, Santiago, Chile, 20-23 Oct. 1992. Water Rep. 1. FAO, Rome.

Schenkel, M., D. Sturam, and F. Occari. 2005. Between transition and enlargement: The composition of consumption in European households. Transition Studies Rev. 12: 58-73.

Scholte, J.A. 2000. Globalization: A critical introduction. Palgrave, Basingstoke.

Schraer, S.M., D.R. Shaw, M. Boyette, R.H. Coupe, and E.M. Thurman. 2000. Comparison of enzyme linked immunosorbent assay and gas chromatography procedures for the detection of cyanazine and metolachlor in surface water samples. J. Agric. and Food Chem. 48:5881-5886.

Sharashkin, L., and E. Barham. 2005. From peasantry to dachas to ringing cedars kin estates: Subsistence growing as a social institution in Russia. Paper presented at the Rural Sociological Society meeting in Tampa, Florida, 9-12 Aug 2005.

Shepherd, M., B. Pearce, B. Cormack, L. Philipps, S. Cuttle, A. Bhogal, et al. 2003. An assessment of the environmental impacts of organic farming. A review for Defra-funded project OF0405. Available at http://www.defra.gov.uk/farm/organic/policy/research/pdf/env-impacts2.pdf. Defra, London.

Simm, G. 1998 genetic improvement of cattle and sheep. CABI Publ., UK.

Simm, G., L. Bünger, B. Villanueva, and W.G. Hill. 2005. Limits to yield of farm species: Genetic improvement of livestock. p. 123-141. *In* R. Sylvester-Bradley and J. Wiseman (ed)Yields of farmed species: Constraints and opportunities in the 21st century. Nottingham Univ. Press, UK.

Simonetta, J. 2006. National agricultural workers survey findings, 1989-2004. Available at http://www.doleta.gov/reports/pdf/National_Agriculture_Survey_Findings.pdf. US Dep. Labor, Washington, DC.

Sly, J.M.A. 1977. Review of usage of pesticides in agriculture and horticulture in England and Wales 1965-1974. Survey Rep. 8. MAFF, London.

Sly, J.M.A. 1986. Arable farm crops and grass 1982. Pest. Usage Survey Rep. 35. MAFF, London.

Smalley, G.W. 1986. Site classification and evaluation for the Interior Upland. Tech. Pub. R8-TP9. Southern Reg., USDA, USFS, Atlanta GA.

Smith, W.B., P.D. Miles, J.S. Vissage, and S.A. Pugh. 2002. Forest resources of the United States, 2002. Available at http://www:ncrs.fs.fed.us. USDA-USFS, North Central Res. Station, St. Paul MN.

Sommer, J.E., R.A. Hoppe, R.C. Green, and P.J. Korb. 1998. Structural and financial characteristics of U.S. farms, 1995. 20th Annual Family Farm Report to the Congress. AIB 746. ERS, USDA, Washington, DC.

Spalding, R.F., D.G. Watts, D.D. Snow, D.A. Cassada, M.E. Exner and J.S. Schepers. 2003. Herbicide loading to shallow ground water beneath Nebraska's management systems evaluation area. J. Environ. Qual. 32:84-91.

Spoor, G., and P.B. Leeds-Harrison.1997. Drainage of heavy soils and mole drainage. Chapter 33. *In* J. van Schilfgaarde and R.W. Skaggs (ed), Agricultural drainage. ASA, Madison.

Stanton, B.F. 1985. Commentary (to the Sumner article). p. 321-328. *In* B. Gardner (ed) U.S. agricultural policy: The 1985 Farm Legislation. Am. Enterprise Inst. Public Policy Res., Washington, DC.

Stanton, J. 1999. Support the independent grocer—or else. Food Process. 60(2):36.

Starmer, E., and T. Wise. 2007. Living high on the hog: Factory farms, federal policy and the structural transformation of swine production. Working Pap. 07-04. Global Dev. Environ. Inst., Tufts Univ., Boston.

Staykov, Y. 1994. Economic problems of aquaculture development in Bulgaria during the transition period into a market economy. p. 91-96. *In* Y.C. Shang et al. (ed) Socioeconomics of aquaculture. Tungkang Marine Laboratory Conf. Proc. 4. Taiwan Fisheries Res. Inst., Taiwan.

Steinrucken, H.C.M., and D. Hermann. 2000. Speeding the search for crop chemicals. Chem. Industry 7:246-249.

Stirling Aquaculture Ltd. 1996a. The present state of aquaculture in the EU Member States and its future up to 2005. Aquaculture: development, environmental impact, product quality improvements. Vol. 2. European Parliament, Direct. Gen. Res., Directorate A, the STOA Prog., Luxembourg.

Stirling Aquaculture Ltd. 1996b. Technical aspects and prospects for a sustained development of the aquaculture sector, Aquaculture: development, environmental impact, product quality improvements. Vol. 3. European Parliament, Direct. Gen. Res., Directorate A, the STOA Prog., Luxembourg.

Stofferahn, C. 2006. Industrialized farming and its relationship to community well-being: An update of a 2000 Report by L. Lobao. Prepared for the State of North Dakota, Office of the Attorney General for the case *State of North Dakota vs. Crosslands*. September. Available at www.und.nodak.edu/org/ndrural/Lobao%20&%20Stofferahn.pdf.

Stolze, M., A. Piorr, A. Häring, and S. Dabbert, 2000. The environmental impacts of organic farming in Europe. Organic farming in Europe: Economics and policy. Vol. 6. Univ. Hohenheim, Stuttgart-Hohenheim.

Stull, D. 1994. Knock 'em dead: Work on the killfloor of a modern beefpacking plant. p. 44-77. *In* L. Lamphere et al. (ed) Newcomers in the workplace: Immigrants and the restructuring of the U.S. economy. Temple Univ. Press, Philadelphia.

Szczerbowski, J. 1996. European aquaculture production trends and outlook—carp. p. 157-168. *In* A.G.J. Tacon (ed) European aquaculture trends and outlook. FAO/GLOBEFISH Res. Prog. Vol. 46. FAO, Rome.

Tacon, A.G.J. and U.C. Barg. 1998. Major challenges to feed development for marine and diadromous finfish and crustacean species. p. 171-207. S.S. De Silva (ed) Tropical mariculture. Academic Press, London.

Tacon, A.G.J. 1997. Review of the state of world aquaculture. p. 120-125. Regional reviews: Europe. Fish. Circ. No. 886 Rev. 1. FAO, Rome.

Tait, J., J. Chataway, and D. Wield. 2000. Final Report, PITA Project. Available at www.technology.open.ac.uk/cts/projects.htm#biotechnology.

TBFRA. 2000. UNECE/FAO database [Online]. Available at http://www.unece.org/trade/timber/fra/welcome.htm. UN Econ. Comm. Europe, Geneva.

Thurman, E.M., and D.S. Aga. 2001. Detection of pesticides and pesticide metabolites using the cross reactivity in immunoassays. J. Assoc. Off. Analy. Chemists Int. 84(1):163-167.

Tomlin, C. 2006. The pesticide manual: A world compendium. British Crop Prot. Council, Alton, Hampshire.

Topping, A. 2007. Strawberries rot as migrants shun agricultural work. The Guardian Newspaper, London, 16 June 2007.

TransFair USA. 2005. Fair trade coffee facts and Figures. Available at http://www.transfairusa.org/content/Downloads/2005Q2FactsandFigures.pdf.

Troeh, F.R., J.A. Hobbs, and R.L. Donahue. 1980. Soil and water conservation. For productivity and environmental protection. Prentice-Hall, NJ.

Troyer, A.F. 1999. Background of U.S. hybrid corn. Crop Sci. 39(3):601-626.

Tweeten, L. 1992. Productivity, competitiveness and the future of U.S. agriculture. Res. Domes. Int. Agribusiness Manage. 10: 127-147.

UNCTAD. 2006. Tracking the trend towards market concentration: The case of the agricultural input industry. UNCTAD/DITC/COM/2005/16. Available at http://

www.unctad.org/en/docs/ditccom200516_en.pdf. UN Conf. Trade Development, Geneva.

UNECE. 2001. Forest policies and institutions in Europe 1998-2000. Geneva Timber and Forest Disc. Pap. 19. UN Econ. Comm. Europe, Geneva.

UNECE. 2003. The statistical yearbook of the Economic Commission for Europe 2003. UN Econ. Comm. Europe, Geneva. Available at http://www.unece.org/trade/timber/fra/screen/chp3_tot.pdf. UN Econ. Comm. Europe, Geneva.

UNECE/FAO. 2000. Forest resources of Europe, CIS, North America, Australia, Japan and New Zealand (industrialized temperate/boreal countries), Contribution to the Global Forest Resources Assessment 2000, Main Report. Geneva Timber and Forest Study Papers, No. 17. Available at http://www.unece.org/trade/timber/fra/screen/chp3_tot.pdf. UN Econ. Comm. Europe, Geneva.

UNECE/FAO. 2003a. Employment trends and prospects in the European forest sector. Geneva Timber and Forest Disc. Pap. ECE/TIM/DP/29. UN Econ. Comm. Europe, Geneva.

UNECE/FAO. 2003b. The development of European forest resources, 1950 To 2000: A better information base. Geneva Timber and Forest Disc. Pap. 31, ECE/TIM/DP/31. UN Econ. Comm. Europe, Geneva.

UNEP. 2002. Global Environmental Outlook 3. http://www.unep.org/GEO/geo3/. UNEP, Nairobi.

Unnevehr, L., and D. Roberts. 2004. Food safety and quality: Regulations, trade and the WTO. p. 512-30 *In* G. Anania et al. (ed), Agricultural policy reform and the WTO: Where are we heading? G. Anania et al. (ed) Edward Elgar Publ., Northampton, MA.

US Dep. Energy. 2007. Wind powering America. Energy efficiency and renewable energy program. Available at http://www.eere.energy.gov/windandhydro/windpoweringamerica/. DOE, Washington, DC.

USDA. 1957. The yearbook of agriculture. 85th Congress, 1st Session, House Doc. No. 30. USDA, U.S. Govt. Print. Office, Washington, DC.

USDA. 1999. Agricultural resource management survey. Economic Res. Serv., USDA, Washington, DC.

USDA. 2005a. EU-25 food processing ingredients sector. The EU's food and drink industry 2005. GAIN Rep. E35067, USDA Foreign Agricultural Service, Washington, DC.

USDA. 2005b. Farm and ranch irrigation survey (2003). USDA Nat. Agric. Statistics Service. Vol. 3, Special Studies Part 1. http://www.nass.usda.gov/census/census02/fris/fris03.htm. USDA, Washington, DC.

USDA-ERS. 2005a. Rural Hispanics at a glance. EIB No. 8. USDA, Washington, DC.

USDA-ERS. 2005b. Food consumption data system. Available at http://www.ers.usda.gov/Data/FoodConsumption/. USDA, Washington, DC.

USDA-ERS. 2006. Food budget shares for 114 countries. Available at http://www.ers.usda.gov/Data/InternationalFoodDemand/

USDA-FAS. 1996. Survey of the Canadian livestock and meat economy. World Markets and Trade, October. Available at http://www.fas.usda.gov/dlp2/circular/1996/96-11/canada.html. USDA, Washington, DC.

USDA-FAS. 2005. EU-25 Food processing Ingredients Sector, the EU's Food and drink industry. GAIN Report number E35067. Available at http://fas.usda.gov. USDA, Washington, DC.

USDA-NASS. 2008. National agricultural statistics service. Available at http://www.nass.usda.gov/Charts_and_Maps/Milk_Production_and_Milk_Cows/). USDA, Washington, DC.

USDA-NRCS. 1996. America's private land. A geography of hope. U.S. Govt. Print. Office, Washington, DC.

USDA-SCS. 1955. Recommendations of the soil conservation service to the departmental committee on land use problems in the Great Plains, May 12, 1955. Historical SCS Reports File, Great Plains Conserv. Program Files, Soil Conserv. Serv., Washington, DC.

USDA-USFS. 1999. Issues and examples of forest ecosystem health concerns. 1999 Health Update. USDA-USFS. (Online). Available at http://www.fs.fed.us/foresthealth/.

Varadi, L., I. Szucs, F. Pekar, S. Blokhin, and I. Csavas. 2001. Aquaculture development trends in Europe. p. 397-416. *In* R.P. Subasinghe et al. (ed), Aquaculture in the third millennium. Tech. Proc. Conf. Aquaculture in the Third Millennium, Bangkok, Thailand, 20-25 Feb 2000. NACA, Bangkok and FAO, Rome.

Virolainen, M. 2006. Working Paper 06/06. Available at http://tradeag.vitamib.com.

Vollrath, T., and C. Hallahan. 2006. Testing the integration of the U.S.-Canadian meat and livestock markets. Can. J. Agric. Econ. 54:55-79.

Vogel, F. 2003. Agricultural sustainability—the human dimension. Stat. J. UNECE 20:1-8.

Vorley, B. 2003. Food, Inc.: Corporate concentration from farm to consumer. Available at www.ukfg.org.uk/docs/UKFG-Foodinc-Nov03.pdf. IIED/UK Food Group, London.

Voronin, V.M., and V.S. Gavrilov. 1990. Inland fisheries of the USSR, today and in prospect. p. 505-510. *In* W.L.T. van Densen et al. (ed), Management of freshwater fisheries. Proc. Symp. organized by the European Inland Fisheries Advisory Commission. Goteborg, Sweden, 31 May-3 June 1988. Pudoc, Wageningen.

Wadekin, K.E. 1973. The private sector in Soviet agriculture. Univ. California Press, Berkeley.

Walker P., P. Rhubart-Berg, S. McKenzie, K. Kelling, and R. Lawrence. 2005. Public health implications of meat production and consumption. Publ. Health Nutr. 8:348-356.

Webster, J. 2005. Animal welfare limping towards Eden. Blackwell, New York.

Weesies, G.A., D.L. Schertz, and W.F. Kuenstler. 2002. Erosion control by agronomic practices. p. 402-406. *In* R. Lal (ed), Encyclopedia of soil science. Marcel Dekker, New York.

Wells, M.J., D. Villarejo. 2004. State structures and social movement strategies: The shaping of farm labor protections in California. Politics Society 32(3):2291-2326.

Welsh, R., B. Hubbell, and C.L. Carpentier. 2003. Agro-food system restructuring and the geographic concentration of US swine production. Environ. Planning A 35(2): 215-229.

Wheatcroft, S.G., and R.W. Davis. 1994. Agriculture. *In* R.W. Davis et al. (ed), The economic transformation of the Soviet Union, 1913-1945. Cambridge Univ. Press, Cambridge.

Willer, H., and M. Yussefi (ed). 2006. The world of organic agriculture. Statistics and emerging trends 2006. 8th ed. Available at http://orgprints.org/5161/01/yussefi-2006-overview.pdf. IFOAM, Bonn, and Res. Inst. Organic Agric. FiBL, Frick, Switzerland.

Willer, H., and M. Yussefi (ed). 2007. The world of organic agriculture. Statistics and emerging trends 2007. 9th ed. Available at http://www.orgprints.org/10506. IFOAM, Bonn, and Res. Inst. Organic Agric. FiBL, Frick, Switzerland.

Williams, G.W. 2000. The USDA Forest Service—First century. FS-650 July. USDA-USFS, Washington, DC.

Wills, C. 2005. Fair Trade works. PES Fair Trade Conf., Brussels, 22 June. Available at http://www.ifat.org/downloads/general/EP%20Carol%20speech%20June%2005.doc.

Wilson, P., and M. King. 2003. Arable plants—a field guide. Available at http://www.arableplants.fieldguide.co.uk. Wild Guides and English Nature, London.

Wood, S., S. Ehui, J. Alder, S. Benin, K.G. Cassman, H.D. Cooper et al. 2005. Food. p. 209-241. Chapter 8. *In* R. Hassan et al. (ed), Ecosystems and human well-being. Vol. 1. Current state and trends. Available at http://www.maweb.org/en/Products.Global.Condition.aspx. A report of the Millennium Ecosystem Assessment. Island Press, Washington, DC.

World Bank. 2005. Forest institutions in transition: Experiences and lessons from Eastern Europe http://www.profor.info/pdf/FITfinal.pdf. World Bank, Washington, DC.

World Bank. 2007. World development report, 2008: Agriculture for development. World Bank, Washington, DC.

Worldwatch. 2006. Biofuels for transportation. Global potential and implications for sustainable agriculture and energy in the 21st Century. Extended Summary. Worldwatch Inst., Washington, DC.

Yoon, B. 2006. Who is threatening our dinner table? The power of transnational agribusiness. Monthly Rev. 58(6):56-64.

Young, L.M., and J. Marsh. 1998. Integration and interdependence in the U.S. and Canadian live cattle and beef sectors. Am. Rev. Canadian Studies 28:335-354.

Zaitsev, G., 1996. The fishery industries in Russia. FAO/GLOBEFISH Res. Prog. 43. FAO, Rome.

3

Environmental, Economic and Social Impacts of NAE Agriculture and AKST

Coordinating Lead Authors:
Peter Lutman (UK); John Marsh (UK)

Lead Authors:
Rebecca Burt (USA), Joanna Chataway (UK), Janet Cotter (UK), Béatrice Darcy-Vrillon (France), Guy Debailleul (France), Andrea Grundy (UK), Mary Hendrickson (USA), Kenneth Hinga (USA), Brian Johnson (UK), Helena Kahiluoto (Finland), Uford Madden (USA), Mara Miele (Italy), Miloslava Navrátilová (Czech Republic) and Tanja Schuler (Germany)

Contributing Authors:
Riina Antikainen (Finland), Dave Bjorneberg (USA), Henrik Bruun (Finland), Randy L. Davis (USA), William Heffernan (USA), Susanne Johansson (Sweden), Richard Langlais (Canada), Veli-Matti Loiske (Sweden), Luciano Mateos (Spain), Jyrki Niemi (Finland), Fred Saunders (Australia), Paresh Shah (UK), Gerard Porter (UK), Timo Sipiläinen (Finland), Joyce Tait (UK), K.J. Thomson (UK), Francesco Vanni (Italy), Markku Yli-Halla (Finland)

Review Editors:
Barbara Dinham (UK), Maria Fonte (Italy), Michael Schechtman (USA), Dariusz Szwed (Poland)

Key Messages

3.1 Environmental Impacts of Agriculture and AKST within NAE 81
3.1.1 Environmental consequences of changes in crop production 82
3.1.1.1 Environmental effects of soil management 82
3.1.1.2 Environmental consequences of pesticides and other agricultural chemical use 84
3.1.1.3 Environmental consequences of increased field drainage 84
3.1.1.4 Environmental consequences of irrigation 85
3.1.1.5 Environmental consequences of the adoption of genetically engineered crops 85
3.1.1.6 Environmental consequences of increased mechanization 86
3.1.1.7 Environmental consequences of changes in farm size and structure 87
3.1.1.8 Environmental consequences of growing more bioenergy crops 87
3.1.2 Environmental consequences of changes in animal production 87
3.1.2.1 Environmental impacts of differing animal husbandry systems 87
3.1.2.2 Environmental effects of manures produced by animal production 88
3.1.2.3 Animal husbandry and methane 88
3.1.2.4 Environmental consequences of the use of veterinary medicines 88
3.1.3 Environmental impacts of a larger aquaculture sector 88
3.1.4 Environmental consequences of changes in forest management 90
3.1.5 Overall environmental consequences of changes in the agricultural industry 90
3.1.5.1 Overall environmental consequences of increased intensity of agriculture 90
3.1.5.2 Environmental consequences of the increase in food miles 92
3.2 Economic Impacts of Agriculture and AKST within NAE 92
3.2.1 Economic context linking advances in AKST to production 92
3.2.2 Impact of AKST on supply and demand 93
3.2.3 Impacts of advances in AKST on the growth of output and on farm businesses 94
3.2.4 Impacts of AKST driven growth in output on processors and distributors 95
3.2.5 Impacts on market power 95
3.2.6 Structural change induced by AKST 97
3.2.7 Impacts on trade of changes in production driven by AKST 98
3.2.8 External economic impacts of the application of AKST 99
3.3 Social Impacts of Agriculture and AKST within NAE 99
3.3.1 Impacts of changes in agriculture on community well-being 100
3.3.2 Consumer concerns about the food system 100
3.3.3 Social impact of increased mechanization 101
3.3.4 Migration from rural areas 101
3.3.5 Equity (benefits, control and access to resources) 102
3.3.5.1 Equity in terms of economic benefits and value added 102
3.3.5.2 Equity in access to resources 103
3.3.5.3 Equity in control and influence 103
3.3.5.4 Rise of alternative food systems 104
3.3.6 Distancing consumers from production 104
3.3.7 Nutritional consequences of NAE food systems 104
3.4 Impacts of NAE AKST through International Trade 104

Key Messages

Environmental Impacts

1. The relatively intensive and highly productive types of agriculture practiced extensively in NAE have had undesirable impacts on the environment in NAE. However, there is considerable potential for reduction, or in some cases reversal, of these impacts by application of knowledge to identify and select improved practices. Increased fertilizer use has resulted in raised levels in nitrogen and phosphorus in rivers and coastal waters causing changes in aquatic populations and contributing to eutrophication. Pesticide and sediment runoff from erosion can also damage aquatic populations. Adoption of farming practices to prevent overfertilization has helped to reduce environmental damage (e.g., controlled timing of treatments, more precise rates, creation of buffer zones). Reduction in pesticide use through methods such as integrated pest management and switching to less persistent and harmful products has reduced impacts but problems from non-target effects of pesticides remain. Soil quality in parts of NAE has been degraded by a variety of intensive land use and irrigation practices.

2. The adoption of mechanization in NAE has contributed to substantially larger fields and farm units. In some regions, this has resulted in loss of traditional landscapes and hedgerows with a subsequent loss of wildlife habitat and biodiversity. Policies and programs, especially financial payments, are available in some areas of NAE, to restore farmland habitats and increase wildlife populations.

3. Greater intensity of animal production systems, combined with the increased spatial segregation of crop and animal production units, has led to concerns over water and air pollution, development of antibiotic resistance and animal welfare. These changes in production systems have created areas where the amount of wastes cannot easily be returned as soil amendments, leading to water pollution in many parts of the NAE. Concerns over impacts have led to stronger regulatory frameworks, especially in the EU.

4. Aquaculture production in NAE, especially salmon, has been growing rapidly over the last few decades. Feeding these farmed fish with fishmeal has put further pressure on fish stocks. Also waste from such operations may overload the capacity of local waters to absorb or process these nutrients, leading to environmental degradation. Further, caged aquatic livestock can incubate diseases that may infect wild populations and escaped fish bred for fast growth in aquaculture may outcompete native wild populations.

5. Agriculture is a sizable contributor to greenhouse gas emissions, especially of methane and nitrous oxide. Greenhouse gas emissions from agriculture are in the range of 7-20% of total country emission inventories (by radiative effect) for NAE. Approximately 30% of global methane is thought to originate from agriculture, of which digestive fermentation from ruminant livestock is by far the greatest contributor. Agriculture in NAE contributes at least one third of global emissions of nitrous oxide and it is the primary contributor to increases in reactive nitrogen.

6. The evidence for the presence of direct environmental impacts arising from the current genetically engineered (GE) crops grown on a large scale compared with conventional agriculture remains controversial. Conclusions that the production of GE crops in North America have not led to adverse environmental effects are not accepted by some stakeholders. It must be pointed out that the agricultural system chosen as comparator is important in the evaluation of GE crops. Measurable reductions of insecticide use have been observed with insect resistant GE crops but not eliminated and vary with crop type. Herbicide tolerant GE crops have facilitated conservation tillage resulting in environmental benefits. Weed populations tolerant to herbicides used in conjunction with certain GE herbicide tolerant crops have become an issue in some parts of North America, but options exist for their management.

7. Bioenergy crops. The use of crops for the production of biomass and liquid biofuels is increasing rapidly. Their use is already having an impact on food crop surpluses, crop production patterns and prices. There is concern that high levels of production of biofuels from food crops could encourage crop production on lands presently reserved for conservation purposes with undesirable effects on the environment.

8. Reorganization of supermarket supply chains and consumer demand in NAE for varied fresh food products and counter-seasonal food products have caused an increase in the long-distance transport of food (food miles). Agricultural policies have encouraged the production of high-value horticultural crops in developing countries which must be shipped in high-energy cool chains. While this trend has had negative effects on the environment, primarily because of increased energy use, it has given some farmers in developing countries access to export markets. In contrast, another trend towards sourcing local food whenever possible may reduce food transport miles in the future.

Economic impacts

9. The application of AKST in a dynamic economic and political environment has allowed consumers to purchase food at relatively low prices, but the technologies that have developed from AKST have encouraged concentration at all levels of the agriculture and food sectors. Declines in prices have forced farmers to adopt more productive practices or increase production and landholdings, reducing the number of farmers and, in many cases, necessitating dependence on off-farm incomes to maintain living standards.

10. Across much of the NAE, large-scale food retailers and processors have a dominant role in determining what people can buy and farmer profits. This has given rise to concern about the impact on competition across the chain and the relatively weak position of farm and food businesses that supply those companies. The development of standardized products which can be processed intensively, as well as the imposition of quality/safety standards by retailers and processors, can increase monopoly power. However, there is an increasing desire among certain consumers to source foods they perceive to have improved quality/safety (e.g., organics, fair-trade), which is providing new opportunities for some farmers.

11. In the last 30 years, a number of food safety breakdowns and animal health issues (e.g., Salmonella, E-coli 0157:H7 and BSE) have occurred and have had extensive impacts, given the increased scale of agricultural and food production. In response to these breakdowns, most of the NAE region has developed far-reaching regulatory mechanisms (e.g., tools for traceability and biosecurity) to detect and prevent the spread of pathogens, weeds and pests and for the detection of pesticide and chemical residues. Some vertically integrated food chains have developed new forms of governance by setting up articulated systems of quality standards, including those aimed at increasing food safety and animal welfare. These forms of governance have been used by major food retailers in some parts of the NAE as a way to regain consumer confidence after food safety scandals. Some retailers have required farmers to comply with specific farm assurance schemes for quality standards in order to sell their products. This can potentially increase costs and raise barriers for farmers.

12. Many of the applications of AKST in agriculture and food systems have created significant waste streams across the food chain, from post-harvest wastage of raw product to end-consumer packaging. Disposable packaging and creation of uniform products have increased commercial appeal of food products and have contributed to food hygiene, but have also increased costs to local communities for disposal.

Social impacts

13. Since 1945, food insecurity across the NAE region has largely been resolved, due to an increasingly wealthy population, decreases in the real prices for food and the substantial increases in food production and productivity. But some sectors of the population across the region remain food insecure (e.g., one in ten households in the US).

14. The needs of labor intensive agricultural systems (such as fruits, vegetables and meat processing) are being met by migrant (largely immigrant) workers. While this has allowed the survival of these labor intensive agricultural systems within NAE and provided workers with a foothold into richer host countries, it has left these workers vulnerable to exploitation across the NAE. They typically have poor working and living conditions, low wages and lack rights to organize. In many cases, they have high levels of poverty and in some regions (especially North America), high levels of food insecurity.

15. Despite gains in agricultural productivity, food security and overall wealth, inequities remain in much of the food system. Within NAE populations there are large variances in the degree of rural poverty, access to affordable, nutritious diets and the sharing of benefits from the reorganization of the food system and global trade. There has been a growing interest in much of the NAE in "alternative" food systems, in which participants seek to incorporate principles of social, environmental and economic sustainability. These systems are currently still small in scale but are increasing.

16. Obesity and associated diseases (diabetes, cardiovascular diseases and metabolic syndrome) have become an increasing concern across the NAE, partly as a result of inadequate nutrition. This is due to the interaction of various factors: general abundance of food and a high degree of food marketing, lifestyle and dietary choice. Some nutritional and educational policy changes have recently been instituted, particularly in schools, to ameliorate these trends, but their impact is yet to be evaluated. Despite a situation of overabundance of food, some sections of the population cannot access a sufficiently healthy diet, mostly due to poverty. Some countries are now facing the double burden of food insecurity and nutrition-related diseases.

Impacts outside NAE

17. NAE has had a major impact on agriculture in the rest of the world, both directly by importing food and raw materials and indirectly, through the impact of NAE AKST. This impact of NAE import requirements has had environmental and economic consequences for the rest of the world. Research undertaken in NAE has also had a global impact. While other countries have derived some benefit, the focus of NAE research has not been on their problems. The development of international research capacity, via the CGIAR institutes, has sought to balance this by stimulating research relevant to the needs of developing countries. The intellectual paradigm that determines the conduct and direction of this research remains powerfully influenced by the model of research in NAE countries and this may sometimes have diminished the usefulness and applicability or research results.

3.1 Environmental Impacts of Agriculture and AKST within NAE

Farming practices have a considerable impact on the environment. Cultivation agriculture has replaced natural forest or grassland ecosystems with species and varieties of plants that have been adapted to cultivation and planted in near-monoculture, such that the original native ecosystem and its native biodiversity have been severely modified or lost altogether. Grazed lands may be similarly altered by the grazing of cultivated livestock and the deliberate planting of forage.

Agroforestry has often replaced the native mix of trees with species selected for a desirable eventual harvest creating a different and likely less diverse forest.

In NAE, where most available arable land has been under cultivation for decades, if not centuries, the farmlands, grazed rangelands and forest plantations may be viewed as a normal or accepted state, even though these systems are far from natural ecosystems. Relative to urban and peri-urban environments, the agricultural landscapes provide valuable habitats for wildlife, non-cultivated plants and animals, open space, catchments for watersheds and recreation areas.

Given the general acceptance of agricultural lands in NAE, the changing environmental impacts of evolving agricultural practices over the last 50 years on the off farm environment and on the agricultural lands themselves, can best be viewed relative to farming in the early parts of the 20th Century, rather than to pre-existing non-farming environments in NAE. The trends in agriculture over the last 50 years, increased mechanization, larger average farms, increased use of fertilizers and pesticides, are documented elsewhere in this report (see Chapter 2). This section addresses the environmental impacts of agriculture as practiced in recent years, recognizing that agriculture practices are continuing to change, and is organized into sub-sections by different agricultural practices.

3.1.1 Environmental consequences of changes in crop production

While certain natural processes can damage soil quality, human activity in agriculture can initiate or accelerate soil degradation. The major threats to soil functions have been identified as erosion, a decline in organic matter and overall soil nutrition status, local and diffuse contamination, sealing and crusting, compaction, a decline in biodiversity and salinization (Van Lynden, 2000; CEC, 2002).

3.1.1.1. Environmental effects of soil management

In both Europe and North America agriculturally induced soil degradation has been a major concern over the last 50 years and, indeed, was of considerable importance in the earlier decades of the 20th century (e.g., the Dust Bowl in the Great Plains of the USA in the 1930s). Soil erosion, by both wind and rain, is arguably the most serious issue (Kirkby et al., 2004). In general, soil erosion is more severe in North America than in much of Europe, due to in part to differences in climate, e.g., higher intensity rains and climatic extremes (hot summers, cold winters) increasing the soil's susceptibility to water erosion (Lal, 1990). Other reasons are related to intensive land use, monocropping without frequent use of soil-conserving cover crops, continuous cropping and the excessive and often unnecessary use of heavy machinery (Lal, 1990). According to expert estimates based on non-standardized data (GLASOD, 1992), 26 million ha in the EU suffer from water erosion and at least 1 million ha from wind erosion. Erosion particularly affects the Mediterranean region but problems also arise in other parts of Europe (GLASOD, 1992; CEC, 2002). USDA data on soil erosion on US cropland indicated soil losses of 1.75 billion tonnes, with sheet and rill erosion of 971 million tonnes and wind erosion of 776 million tonnes (Figure 3-1) (USDA-NRCS, 2003a). However, these figures also demonstrated a dramatic decline of 43% since 1982.

Intensive agriculture can also have great effects on soil fertility. This can manifest itself in loss of nutrients and organic matter and in soil acidification. Many practices can cause these effects, including intensive cropping with inadequate or no return of crop residues, heavy tillage systems which accelerate organic matter decomposition and increase nutrient release, excessive or inappropriate application of fertilizers and lime and irrigation. According to the European Soil Bureau nearly 75% of the total area analyzed in Southern Europe has a low (3.4%) or very low (1.7%) soil organic matter content. Land use changes from forest or grassland to arable agriculture have been and still are a significant source for the release of former plant and soil carbon into the atmosphere (Sauerbeck, 2001), thus increasing atmospheric levels of CO_2. With increased AKST, considerable advances have been made over the last 30 years in resolving these issues, but problems remain both in North America and in Europe. For example, conservation tillage has been a major part of the US conservation program since the 1970s and is being used to sustain or increase soil organic matter (SOM) (Bruce et al., 1990; Havlin et al., 1990; Wood et al., 1991; Franzluebbers et al., 1994; Reeves and Wood, 1994; Aase and Pikul, 1995). Similarly, the introduction of no-till and reduced till techniques is reported to have increased the carbon content of arable soils in Europe (Arrouays et al., 2002). This increase results in a net transfer of CO_2 from the atmosphere to the soil. While it is clear that conservation tillage increases SOM in surface soils (up to 0.2-0.3 m), consideration of SOM in deeper soils (which is much less often measured) indicates that reduced tillage may not promote carbon sequestration as much as earlier studies based on samples from surface layers of the soil indicated (Baker et al., 2007). So, although reduced cultivation may have other benefits (e.g., reduced energy use, less impact on soil invertebrates), its effects on total profile soil carbon levels are not clear.

Human activities have also greatly increased the amount of soil compaction, largely related to mechanical stress caused by off-road wheel traffic and machinery traffic (Hakansson and Voorhees, 1998). Heavy metals and other industrial pollutants, together with synthetic organic and inorganic chemicals used in agriculture have all had a negative impact on soil fertility and can end up in surface and groundwaters (Thurman et al., 1992). These issues are discussed in more detail below, in relation to pesticide use.

The highly productive agriculture in much of NAE has been supported by increased inputs of fertilizers, especially synthetically produced inorganic fertilizers (see Chapter 2). Not all the nitrogen and phosphorus applied to agricultural fields ends up in the target crops. For example, it is estimated that for the US only 65% of the nitrogen applied to fields is harvested (NRC, 2000) and 20% leached to water. A small, portion of the nitrogen is volatilized to the atmosphere (2%) and the remainder is either building up in soils or is denitrified. Nutrients that are lost from fields often become large sources of nitrogen and phosphorus that can severely pollute aquatic and marine ecosystems. Manure used as organic fertilizer also contains nutrients, which can run off fields after application. Where livestock is finished in

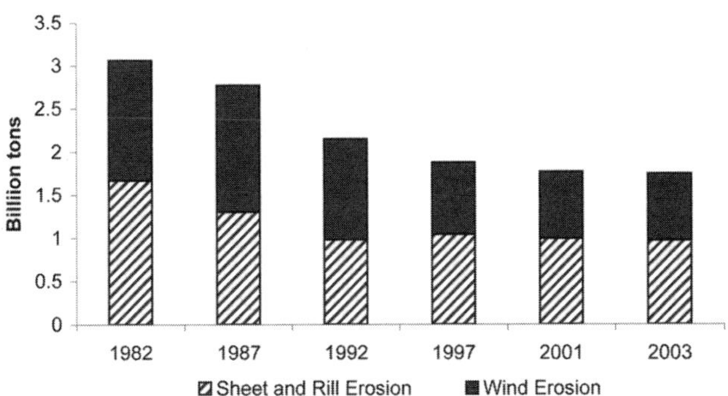

Figure 3-1. *Erosion on cropland by year in the US.* Source: USDA-NRCS, 2003.

intensive feeding operations which produce large amounts of manure, a local oversupply of fertilizer may be created which if not properly managed can also cause pollution (see 3.1.2).

Phosphorus from agriculture can contribute to eutrophication of fresh waters and agricultural nitrogen to eutrophication of coastal marine waters (Lavelle et al., 2005). In recent decades concern over eutrophication has been focused on effects in coastal waters, as there are numerous hypoxic zones in the coastal waters of North America and Europe (UNEP, 2004). The contribution of agricultural nitrogen to coastal eutrophication in different watersheds is quite variable (NRC, 2000) and depends upon the relative amount of atmospheric deposition of nitrogen from combustion sources and point sources in the watershed. Nevertheless, it is clear that agricultural nitrogen is often a significant, if not the major source.

Ammonia emissions to the atmosphere from manure and ammonia-based fertilizers can contribute to local odor problems. The ammonia can be converted to nitrate in the atmosphere, contributing to acid rain and the nitrogen will be redeposited, contributing to eutrophication. Another volatilization path results in production of nitrous oxide (N_2O), a greenhouse gas of importance secondary only to carbon dioxide and methane. The increased soil nitrogen availability from agricultural fertilization has led to greater N_2O production.

Use of appropriate on-field farming practices can make major reductions in fertilizer runoff and emissions without significant reductions in agricultural productivity (Table 3-1). Significant runoff reductions can be achieved through use of uncropped "set-aside" areas as buffer zones and wet-

Table 3-1. **Examples of the magnitude of benefit of different on-field agricultural practices.**

Practice	Contaminant	Example	Reduction in runoff or inputs	Citation
In-season optimization of nitrogen application	nitrogen	North Carolina wheat fields. Nitrogen needs evaluated on fields or sub-fields based upon plant growth properties	Average 15% (range 0 to 51%)	Flowers et al., 2004
Polymer use in furrow irrigation systems	sediments, phosphorus	Pacific Northwest wheat and bean fields. Added supplements to irrigation water to bind sediments & phosphorus	90% for sediments, 50% for phosphorus	Lentz and Sojka, 1994; Lentz et al., 1998
Changing chemical form of fertilizer	phosphorus	Fertilized New Zealand pasture, slow release fertilizer vs. single superphosphate	90%	Nguyen et al., 2002
		Arkansas pasture, organic vs. inorganic fertilizer	41%	Nichols et al., 1994; Hart et al., 2004
Optimization of applied irrigation water	nitrate	Lettuce irrigation, Salinas Valley	75% for nitrate	Tanji et al., 1994
Budgeting to reduce excess fertilizer application	nitrogen, phosphorus	Netherlands	25% for nitrogen 15% for phosphorus	Oenema et al., 2005
Controlled drainage in tile-drained fields	nitrogen	Ohio	45% for nitrate	Fausey, 2005
		Ontario, maze with ryegrass intercrop	46% for nitrogen 49% when used with conservation tillage	Drury et al., 1996
		Ontario, maize	36% for nitrate	Ng et al., 2002
Hay mulching	nitrogen, phosphorus	New Brunswick potato field	72-82%	Rees et al., 2002

Table 3-2. **Examples of the magnitude of the benefit of different off-field management practices.**

Type of Control	Runoff reduction	Citation
Vegetated Buffer 7 meter grass buffer 7 meter grass buffer plus 9 meter wooded riparian zone Iowa	95% sediment 60% nitrogen and phosphorus 97% sediment 80% nitrogen and phosphorus	Schultz, 2004
Three-zone buffer grass to wooded riparian zone, Georgia	78% nitrate 52% ammonium 66% phosphorus	Vellidis et al., 2003
Constructed wetlands to receive water from tile-drained fields Illinois 3% to 6% of drained area	46% nitrogen, 2% phosphorus	Kovacic et al., 2000

lands, or pastures can be used to process runoff from croplands adjacent to surface waters (Table 3-2).

While it is known that adopting different farm practices can make substantial reductions in nutrient runoff, the challenge is in having sufficient numbers of farmers adopt the practices to make widespread improvements in the environment.

3.1.1.2 Environmental consequences of pesticides and other agricultural chemical use

Pesticides are chemicals that target pests, weeds, or disease organisms and include veterinary products (see 3.1.2). Their potential toxic or other adverse effects on farm workers, persons handling pesticide containers, members of the public exposed to spray drift near farms and the issues of residues in food and drinking water are important topics, but are not addressed here.

While pesticides are intended to control organisms that adversely affect crop and animal production, they can also affect non-target organisms, including beneficial ones (e.g., Somerville and Walker, 1990). For example, certain insecticides are toxic to honeybees and other pollinators of cultivated and wild plants and so their usage can result in both environmental and economic losses. Insecticide and herbicide run-off from farmers' fields may have direct toxic effects on aquatic organisms.

Low-level exposure to pesticides through the food chain may affect certain organisms (Hinga et al., 2006). The case of the chlorinated, persistent pesticide DDT being concentrated in predatory birds and leading to reproductive failure is well known. Research is revealing other unpredicted effects from low-dose exposure. For example, the herbicide atrazine has been shown to feminize amphibians, with implications for reproduction in other species (Hayes et al., 2002, 2006). Endocrine disrupting and chronic effects of pesticides have been traced in mammals (Choi et al., 2004). The potential for effects that are not easy to predict or to identify is a continuing concern.

Pesticides may change the availability of food sources for higher level organisms. For example, the control of insect pests can reduce insect prey populations, which in turn limits the size of a bird population feeding on the insect. Similarly, herbicides may change habitats or limit plants that are the foundation for specific food chains. Specific research projects have demonstrated that herbicide, insecticide and fungicide use has decreased the breeding success of several farmland bird species, including grey partridge and yellowhammer (Rands, 1986; Boatman et al., 2004; Hart et al., 2006).

The unwanted effects of pesticides can be mitigated in a number of ways, including decreasing the intrinsic toxicity of the pesticides themselves. Modern pesticides are generally more environmentally benign than the older products that they have replaced. Good farming practices can also reduce unwanted exposures to pesticides. These practices include adoption of Integrated Pest Management (Kogan, 1998); treating pests when needed rather than as a preventive measure; timing spraying to avoid winds and rain; using appropriate and well-maintained machinery; training operators to reduce poor spray practices and disposing safely of waste. Use of biological controls agents, biopesticides and integrated pest management techniques, such as traps with chemical lures, may reduce pest damage sufficiently to avoid general treatment of the whole field, greatly reducing the amount of pesticide used.

3.1.1.3. Environmental consequences of increased field drainage

The land in many parts of NAE, especially the US and western Europe, has been drained with sub-surface tile drains or ditches, to allow lands that were wetlands (with standing water), or were frequently wet enough to preclude tillage, to provide suitable conditions for successful crop growth. However, artificial drainage also facilitates the transport of sediments (especially in the ditches), nutrients and pesticides from agricultural fields. Drainage also affects the hydrology of watersheds as the creation of drains and ditches results in less local water retention and increasing peak flows leading to increased risk of downstream flooding. In removing wetlands, where water may be retained, there is a loss of function of the wetland to act as a site of nutrient removal (see 3.1.1.1) and the degradation of agricultural chemicals. In the UK, over 300,000 ha of wet grassland were lost between 1970 and 1985 (Bradbury and Kirby, 2006). In the US, the conversion of wetlands, primarily for agricultural use, has resulted in the loss of approximately half of the original inventory. In recent decades US conservation policies have acted to reduce agricultural wetland loss and the total amount of wetlands on agricultural lands in the US has increased since the early 1990s (Wiebe and Gollehon, 2006).

There are a number of practices which help mitigate the undesirable loss of sediment, nutrients and pesticides. Control structures may be installed on tile drains to manage the flow of water to both reduce runoff and help provide water for growth when needed by plants. Drainage ditches may be vegetated to help prevent erosion, catch eroded sediments and take up nutrients. Both drainage ditches and tile drains may be directed into constructed or re-established wetlands to process nutrients and agrochemicals. However such practices require a significant investment and to establish wetlands some land has to be taken out of production, presenting barriers to adoption of these mitigation measures.

3.1.1.4 Environmental consequences of irrigation

Although irrigation has had tremendously beneficial effects on crop yields, irrigation systems can have detrimental environmental, economic and social effects upstream of the system, at the site of the irrigation system and downstream (Hillel and Vlek, 2005). Poorly managed irrigation can cause problems of salinization (buildup of salts), water-logging, erosion and soil crusting.

Irrigation water applies water-borne salts to the soil surface and if there is not sufficient drainage, salts accumulate and they can markedly reduce the fertility of the soil. Irrigation in naturally saline soils, without careful management of drainage, may mobilize salts to the root zone, impairing plant growth. The water drained from agricultural fields where salinization is an issue, whether from the buildup of salts delivered by irrigation or through the mobilization of native salts, may have a high salt content which can cause environmental problems in the receiving waters and associated wildlife, e.g., bird deformities resulting from selenium in the drainage water (Letey et al., 1986). Soil salinization affects an estimated 1 million ha in the EU, mainly in irrigated fields of Mediterranean countries and is a major cause of desertification. Similarly, there are approximately 10 million ha in the western US affected by salinity-related yield reductions (Barrow, 1994; Kapur and Akca, 2002).

In the last half of the last century, extensive work had been carried out in the US and globally, to research, diagnose, improve and manage salt-affected soils on irrigated agricultural lands (Miles, 1977; Moore and Hefner, 1977; Ayers and Westcot, 1985; Hoffman et al., 1990; Rhoades, 1990ab; Tanji, 1990; Rhoades et al., 1992; Umali, 1993; Sinclair, 1994; Rangasamy, 1997; Rhoades, 1998, 1999; Gratan and Grieve, 1999). Modern management techniques are being deployed to improve water use efficiency to overcome these problems, by targeting the water more accurately and by using the most appropriate application technologies. Productivity can often be maintained in salt affected areas through careful application of appropriate practices (Miles, 1977; Hoffman et al., 1990).

Soil crusting can be caused by the use of certain irrigation systems. For example, center-pivot sprinkling irrigation in the Coastal Plain area of the U.S has caused soil crusting arising from the sprinkler drop impact energy (Miller and Radcliffe, 1992). The water application rates of this high energy impact irrigation system are often limited by low infiltration rates due to crust formation. Changes in application practices can reduce this problem (Singer and Warrington, 1992; Rhoades, 2002). Erosion can also be caused by inappropriate irrigation practices (e.g., Carter et al., 1985; Carter, 1986).

Irrigation can create problems resulting from the removal of water from other locations. Abstraction of water from rivers can cause major reductions in water flow with consequent negative impacts on river and associated wetland habitats. The drying and salination of the Aral Sea as a result of abstraction of water for irrigation from the main rivers feeding the sea is a particularly stark example of off-site impacts (Micklin, 1994, 2006). Similarly, abstraction of water for irrigation from boreholes can cause a lowering of the water table with adverse effects on neighboring natural wetland areas. Society needs to assess the overall impact of irrigations schemes, not just the agricultural cost and benefits (Lemly et al., 2000). Various strategies are needed to ensure long-term sustainability of irrigation including restricting irrigation to high value crops and using the best equipment and soundest management practices (Hillel and Vlek, 2005.

3.1.1.5 Environmental consequences of the adoption of genetically engineered crops

Transgenic crops are those created through the techniques of biotechnology to select a gene from one species and incorporate it to the same or different species (also called genetically modified, GM, genetically modified organisms, GMOs, or genetically engineered, GE). These new cultivars will have new properties. Accordingly, the environmental effects of each new transgenic variety may differ and regulatory systems have to evaluate each new variety individually. Current GE crops have to undergo an extensive environmental risk assessment throughout NAE (see e.g., Directive 2001/18/EC for EU requirements (www. europa.eu.int/eur-lex/pri/en/oj/dat/2001/l_106/l_10620010417en00010038.pdf) and http://usbiotechreg.nbii.gov/lawsregsguidance.asp for EU and US requirements).

Currently, most transgenic crops are either classified as insect resistant (IR) or herbicide tolerant (HT). Cultivars with other characteristics have been approved for use in parts of NAE, or are in development, including disease resistance, pharmaceutical chemical production, abiotic stress tolerance (drought or salinity), nutritional characteristics (e.g., fatty acid composition) and storage characteristics (e.g., increased shelf life after harvest). (AGBIOS data base lists crops and traits that have been approved by nation: http://www.agbios.com/main.php).

A general review of the 10-year history of cultivation and testing of the currently planted genetically engineered crops concludes that there is no scientific evidence that the commercial cultivation of GE crops has caused environmental harm (Sanvido et al., 2006) though they note that there are no requirements to monitor for potential effects where GE crop varieties have been approved for unregulated use. This conclusion is not accepted by some stakeholders. Because of the nature of the technology it has raised greater public and governmental concerns than "conventional" plant breeding, resulting in closer scrutiny of potential environmental effects. Recommendations exist for further study of the environmental effects of transgenic crops (FAO, 2003).

Insect resistant crops are based upon the inclusion of a gene derived form bacteria resulting in production of a protein (Bt) that is toxic to certain groups of insects (moths and butterflies). As the toxicity is limited to particular groups of insects, farmers will often also treat with pesticides to control other insect pests. The primary concern is that insect resistant crops may have toxic effects on non-target or beneficial organisms (Sears et al., 2001, Dively et al., 2004). Another concern is the persistence of insecticidal proteins in the soil ecosystems, particularly over cold winter periods, although no negative impacts on non-target soil organisms have been found so far (Stotzky, 2004). As IR crops have been in use since 1996, there has been significant experience with their use. A significant reduction in total pesticide use has been found for IR crops relative to comparable non-IR varieties, especially in cotton (Brookes and Barfoot, 2005; Fernandez-Cornejo et al., 2006). Different pesticides have very different toxicities and persistence, so the total amount of pesticide used is a rather poor measure of environmental impact. A more direct measure of effect is on non-target populations. Non-target insects are generally more abundant in Bt fields than in non-transgenic maize and cotton fields managed with insecticides, although not as abundant as in pesticide free fields (Marvier et al., 2007). Concern remains about non-target effects, e.g., indications that pollen from Bt corn can affect aquatic Lepidoptera (Rosi-Marshall et al., 2007).

The planting of herbicide tolerant (HT) crops allows the farmer to control weeds by treating with a broad-spectrum herbicide because the crop will not be affected. As HT crops are intended to be used with herbicides, there is little or no reduction chemical use. However, the herbicides used in HT crops tend to be less persistent and toxic than the herbicides they have replaced (e.g., Fernandez-Cornejo and McBride, 2002; Brimner et al., 2005). HT crops can facilitate the use of conservation tillage, which provides a number of environmental benefits. A major environmental concern with HT crops is the potential development of herbicide tolerant persistent weed species through cross-pollination of transgenic crops to wild relatives or to other (non GE) varieties of the crop. The risk of gene-flow to wild relatives needs to be assessed for each new GE event and the particular geographical region, before release. Where the risk of cross-pollination to wild relatives is considered too high, restrictions have been applied. Also, it has been predicted that continued herbicide use, associated with HT crops, could lead to a reduction in the broad-leaf weed flora (Heard et al. 2003) and could potentially have toxic effects on ecosystems, including soil microflora (Lerat et al., 2005).

There is considerably less experience of potential environmental effects with the other traits that may come into use. One exception is virus resistant papaya, which was approved for use in 1996 in the US and now represents over 50% of the Hawaii papaya plantings. There is probably little environmental cause for concern in a reduction of transmission of a disease virus specific to papaya. This may also be true for bacterial and fungal diseases, provided the method of protection does not introduce properties detrimental to non-target organisms. Alteration of agronomic traits, to increase salinity and drought tolerance, which determine the conditions under which a plant can survive and grow, have greater potential for creation of varieties that could become feral and a problem either directly, or through cross breeding.

It is anticipated that, as is the case with conventional insecticides and herbicides, that insects will develop resistance to the transgenic toxin proteins and that weeds will develop resistance to the herbicides used in combination with transgenic HT crops. Weed resistance to Roundup (glyphosate) is now a serious concern in the US and other places where Roundup Ready crops are grown on a large scale (Baucom and Mauricio, 2004; Roy, 2004; Vitta et al., 2004). The development of weeds resistant to the herbicides used for transgenic crops will require farmers to switch (return) to other herbicides, potentially with consequent environmental changes.

If insects were to develop resistance to the toxic proteins used in IR crops this would cause the loss of effectiveness of the IR crops but also pose a threat to cultivation of organic crops on which the same insects are controlled by topical applications of Bt protein. The Bt protein itself and certain formulations of it, being natural products, are permitted as treatments on organic crops. As Bt is one of a very few such treatments available to organic growers, the loss of effectiveness of Bt would be a serious loss in such instances. Accordingly, growers of IR crops are required to create no-IR refuges in order to decrease the chances of development of resistant insects.

The evidence for the presence of direct environmental impacts arising from the current genetically engineered (GE) crops grown on a large scale, compared with conventional agriculture, remains controversial. Conclusions that the production of GE crops in N. America have not led to adverse environmental effects are not accepted by some stakeholders.

3.1.1.6 Environmental consequences of increased mechanization

The introduction of powerful engine driven plows opened up areas for crop production that were previously difficult to work due to less tractable soil conditions. One consequence has been large-scale removal of hedges to create larger fields to assist maneuverability of the large machinery (Wilson and King, 2003). Deep plowing can increase soil erosion, but mechanization has also increased the potential for less environmentally damaging minimum tillage soil cultivation practices. The ability to spread more fertilizers or pesticides because of increased mechanization may pose dangers of runoff into streams and rivers resulting in water and air pollution beyond the farm gate. However, the greater precision of modern machines has tended to reduce some environmental hazards (e.g., reduced spray drift, more precise fertilizer application). Frequent passes of heavy machinery in fields causes damaging soil compaction which is exacerbated when the crop is harvested in the winter months on wet ground, as can be the case in Northern Europe (Culshaw and Stokes, 1995). Thus, increased mechanization can have both positive and negative effects on the environment.

Agriculture, is a contributor to global CO_2 emissions from the burning of fossil fuels used in farm machinery, energy use for irrigation pumps, temperature control in indoor and glasshouse units, the burning of agricultural waste and drying of agricultural crops for storage. Since the mid 1960s the primary direct energy use on US farms has shifted from gasoline (petrol) to diesel powered engines. Farm energy use in the USA has been estimated to be 9.2 and 3.5 Tg

CO_2-C equivalent for diesel and gasoline respectively (Lal et al., 1998). However, relative to other sources of CO_2, these sources are small. Estimated CO_2 emission directly from agricultural energy use in the USA in 2001 is only 2% of total CO_2 emissions (USDA, 2004). Similarly, UK statistics suggest that emissions due to use of agricultural fossil fuel and lime accounted for less than 1% of total CO_2 emissions in the UK (MAFF, 2000).

3.1.1.7 Environmental consequences of changes in farm size and structure

One of the changes in farm structure over the last 50 years has been the increase in sizes of fields and farms and the simplification (in the number of products per farm) of cropping systems. In Europe changes in farm sizes are often associated with other changes in agricultural practice, which in turn can have environmental impacts. The fine grained nature of traditional European landscapes, with small fields separated by hedges, trees, walls and ditches and with small seminatural areas between fields, has become coarser with the loss of many of the traditional boundary features that are often the key to the success of indigenous plants, invertebrates, mammals and birds. (Roschewitz et al, 2005; Herzog et al., 2006)

Intensification of production in eastern Europe during the socialist era has resulted in greater negative environmental effects than has occurred in western Europe. Although crop yields were increased, politically driven, central management has resulted in greater erosion, salination and chemical pollution (Bouma et al., 1998). Changes since 1990 are now endeavoring to limit adverse side effects from agriculture.

3.1.1.8 Environmental consequences of growing more bioenergy crops

One incentive for the use of biofuels and biomass crops is their replacement of fossil fuels. While any burning of fossil fuels (without sequestration) contributes to increases in carbon dioxide in the atmosphere, power produced from bioenergy appears neutral at the point of use as the carbon in the bioenergy crops came from the atmosphere. However, much of NAE agriculture is energy intensive and the emissions saved by use of biofuels and biomass crops is significantly reduced by the fossil fuels used directly (e.g., running farm machinery) or indirectly (energy used in the production of fertilizer and agrochemicals) during the production of the crop. There are some estimates that the current production of biofuels (maize-based ethanol) is actually carbon negative in that it takes more fossil fuel to produce biofuel than the petroleum it is intended to replace (e.g., Pimentel and Patzek, 2005) though the consensus seems to be that there is a positive net carbon balance in the production and use of biofuels (e.g., Farrell et al., 2006; Worldwatch, 2006).

Two concerns associated with the expansion of biofuel and biomass production are that there is likely to be competition for land between requirements to grow crops for food or for bioenergy, with associated impacts of food prices and that there could be pressure to put uncropped land into energy crop productions, especially highly erodible lands, wetlands, buffer areas and mature forests. Many of these areas are currently providing environmental benefit and their loss would increase environmental impacts. Production of energy crops with irrigation would put increasing demands on water use. Putting or returning uncropped lands into agricultural production may (depending upon the clearing and agricultural systems used) also release the carbon in biomass and soil organic carbon into the atmosphere.

The prospects for greater production of biofuels without greater effects on the environment rely on a second generation of biofuel sources. It is expected that in the relatively near future that it will be possible to produce ethanol from the non-starch and non-sugar components of plants, expanding the amount of carbon that can be converted from food crops and making non-food plants suitable for biofuel production (Gray et al., 2006; Tilman et al., 2006). However, agricultural practices will have to assure that sufficient plant materials remain in the soil to maintain soil health and soil organic carbon and maintain other benefits (e.g., Lal and Pimentel, 2007). Losses of soil organic carbon would tend to negate benefits from use of non-fossil fuels.

Future developments may also entail breeding of food crop varieties and non-food plants specifically to increase their utility for energy production. Non-food crops may include hardwood species such as poplar and willow, switchgrass and even algae. It should be noted that ethanol and biodiesel are not the only prospects for second generation fuels. Butanol can also be produced by (bacterial) fermentation of sugars and may have significant advantages over ethanol as a gasoline replacement (Ramsey and Yang, 2004). Biogas may also be produced from plant materials.

3.1.2 Environmental consequences of changes in animal production

3.1.2.1 Environmental impacts of differing animal husbandry systems

There are three distinct animal production systems in the NAE (Seré et al., 1996): grazing, mixed farming and industrial systems. Each has potential environmental impacts, especially the latter. The increased specialization that has occurred in the last 50 years has resulted in many areas in separation of production into "crop production areas" and "animal production areas". As a result the number of mixed farms has declined.

Grazing systems feed animals mostly on native grassland, with little or no amounts of other plant material and rarely including imported inputs, resulting in low calorific output per unit land area (Jahnke, 1982). However, if too many livestock are kept on the grazed area, the desirable forage plants may be reduced too severely, creating opportunity for invasive species.

Mixed farming systems integrate livestock and crop activities and have traditionally been the dominant approach to agriculture. By-products (crop-residues, manure) from one enterprise can serve as inputs for the other, resulting in environmentally friendly systems. Thus, the detrimental environmental effects from fertilizers can be minimized by efficient use and recycling of nitrogen and phosphorus. However, even in mixed farming systems, animal by-products can cause environmental damage, if they are not recycled efficiently. The shift from haymaking to silaging for feeding grassland-based cattle in mixed (and intensive) farming systems, assisted by increasing mechanization, has led to reductions in non-grass biodiversity in pastures and meadows (Johnson and Hope, 2005).

Intensive, industrial production systems have evolved from the less intensive mixed farming systems in response to increased demand for meat, resulting in animal concentrations that are greater than the waste absorptive and feed supply capacity of nearby available land and which can cause major pollution problems and human health risks. Indoor production systems are now predominant for pigs, poultry and veal cattle. These agricultural systems have become increasingly controversial because of the amount of waste produced, odor problems, the potential for surface and groundwater contamination and animal welfare concerns. In intensive livestock farming areas excessive loss of nutrients and farm effluents in surface runoff and/or leaching, are the principal causes of degradation of water quality (Hooda et al., 2000; Tamminga, 2003).

3.1.2.2 Environmental effects of manures produced by animal production

Awareness of the environmental impacts of some animal production systems, especially in relation to phosphorus and nitrogen pollution of water and the presence of antibiotics, pesticides and micro-organisms in manures, has resulted in the development of more sustainable management practices. Increased mechanization has enhanced efficiency of management of animal waste, resulting in reduced potential for negative affects on the environment, but the use of mechanization to increase intensity of production can counteract these benefits, by producing much greater quantities. In some European countries changes in management have been supported by legislation restricting the way manures are processed. An evaluation in 2003 of the Danish National Action Plan for the Aquatic Environment showed that nitrogen leaching (primarily from intensive pig farms) had declined by 50% since 1989 (Grant et al., 2006). A range of measures have also been introduced in The Netherlands, including a manure phosphorus quota which has been allocated to every farm, limiting the amount of P that can be applied to the land (Kuipers and Mandersloot, 1999). In the UK a range of management options have been introduced to encourage reductions in water pollution from livestock farms (Hooda et al., 2000). Further legislation on the impact of nutrients on water is included in the EU's Water Framework Directive (http//ec.europa.eu), currently being promulgated across Europe. All countries in the NAE are endeavoring to reduce the effects of animal manures on the wider environment. A range of new technologies are also being developed and adopted, especially in the USA, to minimize the environmental impact of animal production, such as optimized feeding strategies and the identification of feed additives that could improve the efficiency of utilization forages and crop residues, while reducing methane emission (Makker and Viljoen, 2006). However, manure from industrial livestock systems and its impact on water systems remains a significant concern in some areas of NAE.

3.1.2.3 Animal husbandry and methane

Husbandry of ruminant animals is the major source of increased agricultural emissions of CH_4 (including lagooning and management of waste) (Prather et al., 2001). It is estimated that ruminant livestock production (including cattle and sheep) accounts for 90% of agricultural methane because of their unique digestive system allowing them to digest coarse plant material. The most recent UK estimates are that 80% of emissions are from enteric fermentation and 20% from animal waste (Anon, 2006). Beef and dairy cattle combined account for over 90% of the CH_4 enteric emissions in the USA (Table 3-3). In the UK cattle alone account for 75% of these enteric emissions. Manipulation of the diet in these concentrated animal feeding operations (CAFO's) is one of the major methods available to manage these emissions (MAFF, 2000).

Whereas methane can be collected from manure, the methane can be used as an energy source to generate heat and electricity (e.g., Williams and Gould-Wells, 2004). Extraction energy from the conversion of methane to CO_2 reduces the greenhouse effect, as CO_2 is not as strong a greenhouse gas as is methane. Such manure management also reduces potential for runoff pollution from manure wastes and may also reduce odor problems.

3.1.2.4 Environmental consequences of the use of veterinary medicines

Animal husbandry in industrialized systems often requires the use drugs to keep animals healthy or stimulate growth. Residues of such pharmaceuticals are excreted and may escape through runoff to be dispersed in the environment. Of particular concern is the routine use of antibiotics for growth promotion or prophylaxis rather than disease control. It is a near certainty that microbes will develop a tolerance if given steady exposure to low levels of antibiotics, eventually rendering the antibiotics ineffective for treatment of disease (Cohen and Tauxe, 1986). Administered hormones may be excreted by livestock, especially those held in dense populations and can affect other organisms at very low concentrations. Estrogenic compounds may affect growth, behavior and sexual development and hence breeding ability. Practices that control agricultural runoff, such as buffer zones and wetlands, are effective in retaining and degrading agricultural pharmaceuticals to prevent release into the wider environment (Lorenzen et al., 2005; Shappell et al., 2007).

Current FAO studies of the influence of livestock development practices on the natural resource base will provide information to predict and prevent possible negative affects of intensified production and enhance positive ones. These livestock studies involve feed quality, use of biomass for animal fodder, avoidance of overgrazing, manure management, animal waste disposal, domestic animal genetic diversity, plant and animal wildlife diversity and integration of cropping and livestock systems (FAO/IAEA, 2006).

3.1.3 Environmental impacts of a larger aquaculture sector

The different types of aquaculture have very different potentials for impacts on the environment and it is useful to divide aquaculture into three major categories in order to address their risks.

Substantial increases in the production of caged aquaculture in open ecosystems (e.g., salmon culture in coastal ecosystems or tilapia in caged cultures in parts of fresh wa-

Table 3-3. **Methane emissions from enteric fermentation (Gg).**

Livestock Type	1990	1995	2000	2001	2002	2003	2004	2005	2006
Beef Cattle	4,281	4,616	4,304	4,257	4,251	4,260	4,155	4,198	4,249
Dairy Cattle	1,488	1,422	1,377	1,374	1,381	1,393	1,377	1,411	1,441
Horses	91	92	94	99	108	126	144	166	166
Sheep	91	72	56	55	53	51	49	49	50
Swine	81	88	88	88	90	90	91	92	93
Goats	13	12	12	12	13	13	13	13	13
Total	**6,044**	**6,302**	**5,933**	**5,886**	**5,896**	**5,931**	**5,828**	**5,928**	**6,010**

Note: Totals may not sum due to independent rounding.

Source: US-EPA. 2008.

ter lakes) have affected wild fish populations. Aquaculture's substantial demand for fish meal is driving a large wild capture of small fishes (that are the base of food chains) (Naylor et al., 2000). In part, the over-fishing of some fish populations is to support the aquaculture industry. Recognition of this has lead to research and efforts to replace fish protein and lipids in fish meal with vegetable sources and byproducts from livestock processing (e.g., Glencross et al., 2003, Montero et al., 2003; Higgs et al, 2006).

A second issue with caged cultures in natural waters is habitat degradation in the areas of the cages due to the large inputs of organic matter and nutrients (nitrogen and phosphorus) in the feed for the aquatic livestock. These inputs can lead to reduced water quality, undesirable algal blooms and alteration in benthic communities in the near vicinity of the aquaculture operations (e.g., Gyllenhammar and Hakanson, 2005).

Caged aquaculture inevitably loses some of the cultured fish, through small accidental escapes and through occasional large losses in storms, to the wild. The escapees may interfere with native populations (Canonico et al., 2005). While the number of escaped fish are small relative to native populations, the impacts of the escapees are probably minor. However, in the case of Atlantic salmon (*Salmo salar*), escaped populations may be relatively large compared to native populations. Although aquaculture salmon may be more aggressive and may outcompete native populations they are less reproductively viable and may cross-breed, with native populations leading to reduced viability of offspring, which threatens the survival of the native gene pool (Naylor et al., 2005).

A final concern of caged populations is that dense aquaculture populations are incubators for diseases and parasites (e.g., Heuch et al., 2005), which can then spread to wild populations. Because fish diseases have led to major economic losses in aquaculture, there is increased use of veterinary drugs and vaccines in intensive production systems. The use of antibiotics in aquaculture can rapidly lead to the adaptation of disease microbes and loss of effectiveness of the antibiotic (Garcia and Massam, 2005). However, antibiotics are not used either as prophylactic (before disease occurs) agents or as growth promoters in temperate water aquaculture production in Europe and North America (Alderman and Hastings, 1998). In recent years the use of antibiotics has fallen dramatically in the farmed salmon industry in Norway from about 50 to less than one tonne annually (Figure 3-2). This is largely as a result of the successful development and use of vaccines against the principle fish pathogens (Alderman and Hastings, 1998).

Closed-system aquaculture, such as in farm-based cat-

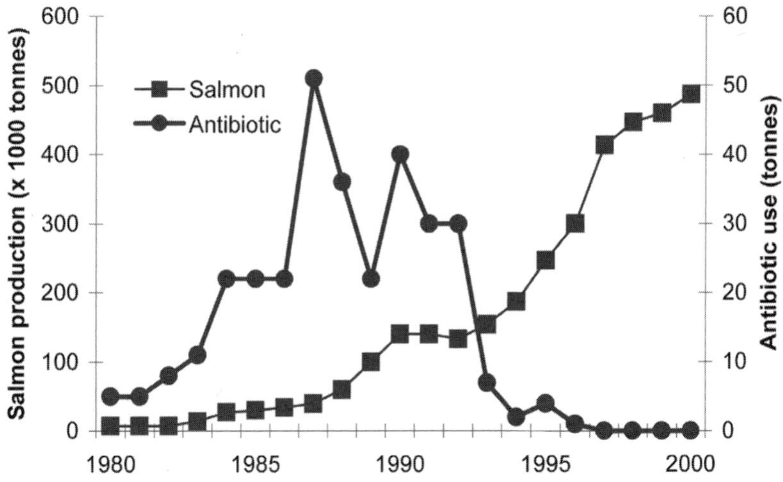

Figure 3-2. *Increase in production of farmed salmon and decrease in use of antibiotics in Norway from 1984 to 2000.* Source: FAO, 2006b.

fish ponds, trout farms and some seawater closed systems, avoid many of the problems of caged aquaculture as the possibility of escape of the livestock or transmission of diseases to native populations is greatly reduced. However, the effluent from such systems may be rich in organic matter and plant nutrients. Unrestricted discharge of these waters could impair receiving water quality. Use of systems that used the discharge from farm ponds to directly irrigate and fertilize farm fields, or use additional ponds to grow algae, which in turn is used as a fertilizer or livestock feed supplement, can eliminate or reduce the impacts on receiving waters.

Filter feeding molluscs (clams, mussels, oysters, scallops) in aquaculture rely on natural suspended particulates (i.e., phytoplankton and detritus) rather than external food sources. Such systems do not add new materials to the ecosystem and are unlikely to create the eutrophication problems of finfish caged aquaculture. However, these systems may redistribute organic matter and concentrate organic materials in sediments below the structures holding the cultured organisms.

Similarly, seaweed culture does not rely on external inputs and therefore does not have the eutrophication impacts that can occur in caged, externally fed organisms. Indeed, it has been suggested that carefully placed mollusk or seaweed culture, used in an integrated system with caged culture, could help cleanup the organic residue and algal growth promoted by externally fed aquaculture (Lindahl et al., 2005; Troell et al., 2005).

3.1.4 Environmental consequences of changes in forest management

Forests cover an appreciable proportion of the land surface of the NAE, especially in parts of N. America and in Russia, so changes in forest management have the potential to have appreciable environmental impacts. Forests provide environmental benefits of wildlife habitat, plant and animal biodiversity, timber, provision of clean water and carbon storage. High-quality riparian areas trap sediments, slow runoff, provide habitat for wildlife, fish and plants (USDA-USFS, 1999).

The quality of forests may be affected by clearing, but also can be damaged by air pollution, e.g., acid rain and ground-level ozone (USDA-USFS, 1999). Forests may also be damaged by fire, invasive species and unmanaged recreation (Bosworth, 2004). In addition, nitrogen deposition from the atmosphere may potentially cause a shift in composition of some forests. The USDA Forest Service has also identified how ozone damages trees and has screened tree varieties for those less susceptible to this gas. Studies are ongoing to identify ozone-sensitive trees in areas of ozone exposure, increasing our understanding of how to manage forest resources (USDA-USFS, 1999).

In Europe, the replacement in the last century of mixed aged stands of often deciduous woodlands with uniform age conifer plantations has had negative effects on biodiversity, especially ground flora and mammalian fauna and sometimes on soils and surface waters (Hartley, 2002; Humphrey et al., 2002; Spiecker, 2003; MA, 2005). Bird populations may also be adversely affected but in some cases, conversion and intensive management has boosted populations of birds and some mammals that were previously rare in primary forest (such as crossbills (*Loxia curvirostra*) red squirrels and pine martens) in Scotland, where 90% of woods are plantations (Marquiss and Rae, 1994). About 40% of the hundred European "priority" forest bird species are in unfavorable conservation status, mainly due to declines in old-growth forest (BirdLife International, 2007). Coniferous plantations also appear to increase the acidity of precipitation falling on them, leading to reductions in pH of streams, rivers and lakes within forested areas (Spiecker, 2003). Although the area of forested land in Europe is increasing, most of the increase is made up of plantations and secondary woodland and this does not necessarily offset the reductions in flora and fauna caused by conversion of natural forests to intensively managed plantations. Awareness of the negative impacts of uniform age conifer plantations has resulted in much debate in Europe as to the economic viability of replacing them with mixed species stands, with both conifer and deciduous species (Spiecker, 2003). Despite declines in natural forest quantity and quality in W. and some E. Europe countries, European forests remain one of the most important refuges for wildlife on the continent. Additionally, the increase in forested timber volume within the NAE increases carbon sequestration and is of value in reducing atmospheric levels of CO_2.

Environmental concerns about forestry have resulted in changes in approaches to tree production and to management in the USA since 1970. In the 1970s public concern in the USA about the effect of current clear-felling and reforestation practices led to the 1976 National Forest Management Act (NFMA). One of the important developments following the passage of this Act was the establishment of the Long-Term Soil Productivity (LTSP) research program (Williams, 2000) to explore and reduce the environmental effects of forestry practices (e.g., see Powers et al., 2005). Changes in practices arising from AKST have had some success in the last 30 years in ameliorating some of the negative environmental effects of forestry in the USA. However it must also be noted that new technologies developed since the second world war allow faster and more efficient harvests and access to timber in areas previously considered too fragile for harvest, thus expanding the potential managed forest areas.

3.1.5 Overall environmental consequences of changes in the agricultural industry

The previous sections of this chapter have highlighted the major issues associated with specific changes in crop and animal production and forestry. However there are also issues that transcend these individual components, as there are environmental consequences arising from overall changes in agriculture and which cannot easily be attributed to individual components. Two issues are highlighted here, the impacts of changes in the intensity of agricultural production on the natural ecosystem and the issue of "food miles".

3.1.5.1 Overall environmental consequences of increased intensity of agriculture

As the dominant land use throughout much of Europe, agriculture (including forestry), has a huge footprint on the overall ecosystem, especially in intensively farmed countries such as France, The Netherlands and UK. There have been widespread declines in the populations of many groups of organisms associated with farmland (e.g., arable plants, invertebrates, farmland birds) since the 1940s in Britain and North-West Europe. A review of 18 studies investigating changes in wildlife in arable farmland in Great Britain confirmed the decline of many taxa. In only two studies (on butterflies) was there evidence of an increase over the survey periods (Robinson and Sutherland 2002). Similar results have been found in Portugal (Stoate et al., 2001).

At a wider European level, decline in farmland bird populations have been related to agricultural "intensity" (Donald et al., 2002). At its simplest there is a link between average cereal yields (FAOSTAT) and the rate of bird decline (Figure 3-3). A similar study on invertebrates has reported on changes in bees and hoverfly populations in Britain and the Netherlands pre and post 1980, concluding that there has been a decline in bee diversity in most of the assessed areas in both countries since 1980 (Biesmeijer et al., 2006). This decline seemed to be linked to declines in pollinator plants, which may well have become less common as a result of agricultural intensification (Preston et al., 2002). The overall conclusion for Europe, east and west, is that increased farming intensity over the last 50 years, although leading to appreciable increases in production per unit area, has had a negative impact on the environment and ecosystem services (Tilman, 1999). A further complicating issue relates to the impact of land abandonment in some areas of East and Southern Europe on biodiversity. Economic pressures have resulted in fields not being farmed and as a consequence scrub has started to invade, degrading the habitats' suitability for many farmland species, though it does increase its suitability for others.

Concerns about the impact of food production on ecosystem services loom less large in North America, although American-based ecologists are as concerned as European scientists about the impact of agriculture on the ecosystem (Tilman, 1999). Agriculture has a much smaller "footprint" in North America, as in the USA it uses less that 50% of the land surface and in Canada less than 10% (FAOSTAT, 2006). In general, management strategies of US natural resources have moved toward land or ecological-based systems which recognize the important role of the soil (Robertson et al., 1999). There has also been a changing philosophy to rangeland management in the US over the last 50 years (Orr, 2006) with management evolving from purely grazing objectives, to a more scientific approach, recognizing the need for "resource rehabilitation, protection and management for multiple objectives including biological diversity, preservation and sustainable development for people" (Stoddart et al., 1975; Heady and Child, 1994). Despite this changed philosophy more than one-half of all US rangeland ecosystems have lost 98% of pre-settlement flora, to agricultural use. The amount of US grazing land and rangeland is expected to continue to decline slowly over the next 50 years, as the land use shifts away from grazing use but there is no indication that endangered rangeland ecosystem types are being lost except for desert grasslands.

The decline of biodiversity can be at least partly attributable to the changes in farming systems which advances in agricultural technology have made possible. These include:

- The widespread use of pesticides has affected non-target species
- The development of machinery capable of establishing crops on soils not previously amenable to crop production has caused a decline in natural and semi-natural habitats
- The increased size of machinery, aimed at increasing efficiency, has resulted in field amalgamations and losses of hedges and other semi-natural wildlife habitats
- Simplification of rotations so that only a limited number of crops are grown, has decreased the planting of those with different biology and planting times, that formerly provided a greater range of habitats for wildlife
- The replacement of hay crops by the earlier harvested silage, for intensive animal production has reduced the environmental value of grasslands

Such technologies have typically been adopted by farmers after weighing the complex tradeoffs, economic and environmental, inherent in each. However, AKST is also continuing to provide newer and better tools and expertise to assess impacts of agricultural changes on wider biodiversity

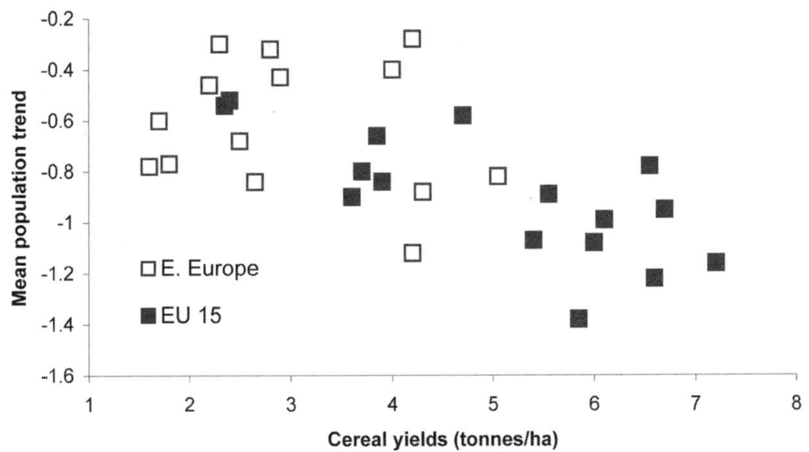

Figure 3-3. *The relationship between mean farmland bird population trend and cereal yield across Europe.* Source: Donald et al., 2002.

and thus provide guidance on how to reduce biodiversity effects. Reduction in the overall intensity of agriculture has been proposed as a technique to help restore agricultural ecosystems and retain ecosystem services. For example, less intensive organic production systems have been identified by some as more environmentally benign. The "ecological" emphasis implicit in ecosystem service approaches has been questioned by those who favor increasing intensity of production in some areas and thus conserving other areas for off-farm biodiversity (land sparing) (Green et al., 2005; Vandermeer and Perfecto, 2005). This debate may miss important opportunities for achieving win-win solutions incorporating productivity and ecosystem services (Pretty et al., 2006). The debate continues.

3.1.5.2 Environmental consequences of the increase in food miles

Increased geographical distance between producer and consumer, together with the regional specialization of agriculture has resulted in the availability of a wider selection of apparently cheap food for consumers, but at the cost of longer transport with the attendant consequences of greater energy use and deleterious effect on global climate. Distancing and regional specialization has encouraged less diverse production systems, complicating recycling of nutrients and carbon from animal husbandry back to crop production and from demand chains back to agriculture. Further, distancing consumption from production hinders feedback from the ecosystem to the human community, affecting the land use, thus impeding adaptive management (Vergunst, 2003; Deutsch, 2004; Sundkvist et al., 2005).

The increase in food transportation has a significant impact on energy use, climate change, pollution, traffic congestion and accidents. Road transport generates six times more CO_2 emissions compared with shipping and airfreight 50 times more (Jones, 2001). The dramatic increase in transportation has resulted in a rise in the amount of CO_2 emitted by food transport (Smith et al., 2005). The cost of food miles is £9bn a year to the UK. This is greater than the total contribution of the agricultural sector to GDP (£6.4bn). Several studies show that shorter supply chains would be less detrimental to the environment. Transportation, especially for fresh products, is responsible for a considerable proportion of the total energy consumption, exceeding the energy consumed for cultivation of apples, for example (Jones, 2002). The use of fossil energy and climatic effects of transportation of more local food were smaller, even when taking into account the smaller amounts transported at a time (Carlsson-Kanyama, 2004; Poikolainen, 2004; Granstedt et al., 2005). The external cost of transportation in local food systems (food basket sourced from within 20 km of retail outlet) would be less than one tenth of the current one in the UK, depending on transport vehicles (Pretty et al., 2005). In the USA, depending on the system and truck type, the conventional food system used 4 to 17 times more fuel and released 5 to 17 times more CO_2 than the Iowa-based regional and local systems (Pirog et al., 2001).

The environmental consequences of distancing are complex. If food supply chains are identical except for transportation distance, reducing transportation increases sustainability (Smith et al., 2005). However, differences in food supply systems often imply tradeoffs among various ecological, economic or social sustainability concerns. Transport mode, transport efficiency (vehicle size and loading), differences in food production systems and food storage, all affect the final outcome. The total effect depends, for example, on the energy input to production and post-harvest processes. If production is clearly less energy-intensive when performed outside the region (Cowell and Parkinson, 2003), as it can be for greenhouse vegetables (Poikolainen, 2004) and for cereals with higher yields and lower energy need for drying in warmer regions (Sinkkonen, 2002), the benefits of reduced transportation may be more than offset by the increased energy costs for production. Therefore, a simple calculation of food miles is not a valid indicator for sustainability (Seppälä et al., 2002).

3.2 Economic Impacts of Agriculture and AKST within NAE

All changes in agricultural production in the NAE over the last 50 years have economic drivers and consequences, from the field to the "plate". This sub-chapter looks at the changes that have occurred in production systems, partly as a result of advances in AKST but also due to other technological and societal changes that have occurred during this period.

3.2.1 Economic context linking advances in AKST to production

In the past 50 years agricultural output in NAE has grown more rapidly than demand. (See Chapter 1 and Chapter 2) One result has been a trend for real prices for farm products to fall. (See EU, 2003; FAO, 2005; UK, 2005a) The driving force has been improvements in technology. Farmers who did not initially use the new methods have had to adopt them, find a new niche market for their products or face falling real income. Income earned outside farming may cushion this or even make it of no great importance but where these strategies cannot be used, many working farmers and their children have had to leave farming. Although rural populations have started to stabilize and more recently to grow in some areas, the decline in the farm labor force in the second part of the 20th Century has been dramatic (Figure. 3-4).

The pressure upon the centrally planned economies of the eastern European states after the Second World War to adopt technical innovations was enormous. Failure to supply sufficient and reliable food was a major problem for the Soviet Government. Some countries in eastern Europe, such as Poland, retained many very small farm holdings. Here it was more difficult to apply the larger scale investments associated with new farm technology. In contrast, as in Hungary where private holdings were merged into collective farms, large scale farming businesses looked for innovation and invested in production related research. A failure to keep pace with AKST technology across the food industry as a whole weakened the relative position of the centrally planned economies to those of the West. Consumers had fewer choices, products were often of lower quality and the centrally planned economies became less able to compete in global markets except by cutting prices. Although substantial investments in new technology were made these did not

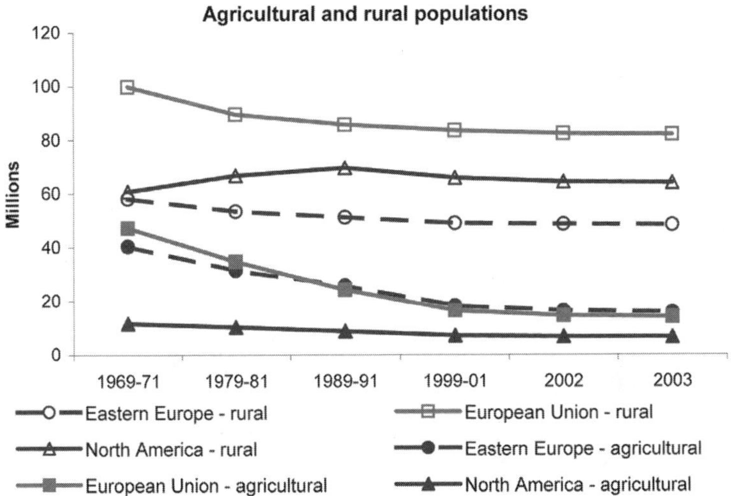

Figure 3-4. *Agricultural and Rural Population in North America and Europe.*
Source: FAO, 2008b.

overcome the relative lack of competitiveness. Compared with market driven economies the intensity of production and the levels of productivity usually remained lower although output continued to grow.

The effect of new technology is seen in the sustained and substantial improvements in productivity that were achieved (see Chapter 2). Measurements of this are complex. Yields per unit of land of major crop products are a first and very rough proxy for productivity (Chapter 2). Aggregated data of this nature conceals a good deal of variation but the overall message is clear. Yields have increased in every area and while the rate of improvement slowed in the 1980s it has recovered. The substantial gap between the former USSR and other areas has not been removed. This reflects underlying natural conditions. However, even here cereal yields have doubled over the 40 year period (see Chapter 2).

In contrast to many assumptions, GDP per person engaged in agriculture tends to be higher than in the economy as a whole in most NAE countries. Improved technology made possible rises in GDP per worker. In Europe and North America GDP (Gross Domestic Product) per person seems to have risen faster in agriculture than in the economy as a whole although the share of agriculture in the overall economy has declined (Figure 3-5).

3.2.2 Impact of AKST on supply and demand

The tendency for real prices to fall has led to demands for protection. Agricultural policies have mitigated but not prevented falling prices in markets such as the EU and USA. External markets, which have absorbed varying levels of surplus from these protected markets, have been volatile and experienced the full impact of the tendency for real prices to fall.

The EU is the largest agricultural trader (Figure 3-6). Even when intra EU trade is excluded, it remains a major player in the market for many important commodities (Table 3-4). Price support combined with rising productivity led to a situation in which substantial export subsidies were needed to enable domestic production to compete in world markets. Since 2003 many subsidies have been decoupled from production allowing the prices farmers receive to reflect market realities. Income support has been provided by direct payments fixed on the basis of production in 2002-2003.

Export subsidies mean that relatively modest shifts in consumption or production spill over into the world market where they may influence world prices. The effect of growing productivity within NAE countries, driven by AKST and price support, has thus been to depress world prices. The impact of improving productivity, combined with subsidies on exports is illustrated in the falling trend of commodity real prices shown below (Figure 3-7).

Falling prices can benefit consumers, especially poorer consumers who spend a relatively large share of their income on food. They also benefit net importing countries but may give rise to increased dependence on foreign supplies and reduced investment in local agriculture and its support services. This has had the effect of making import prices low and volatile for importing countries. For developing countries low import prices benefit consumers but reduce returns to domestic producers. Because imported food prices are also volatile, they can give rise to unpredictable and unaffordable trade deficits.

Changed technology has also led to a transformation in the way in which food reaches the consumer (Regmi and Gehlhar, 2005; UK, 2005b; USDA, 2005) and has resulted in the production of anonymous, cheap and highly processed and packaged food. Some consumers have reacted to this by seeking alternatives that represent for them higher quality. The response is multidimensional. It includes a growing demand for organic products (Dimitri and Greene, 2002); growing requirements for farmers to increase the welfare of their farm animals in order to be able to sell their products to the European retailers (Defra, 2004a); a growing market for locally produced and fairly traded products (F3, 2007; http://www.fairtrade.net/.).

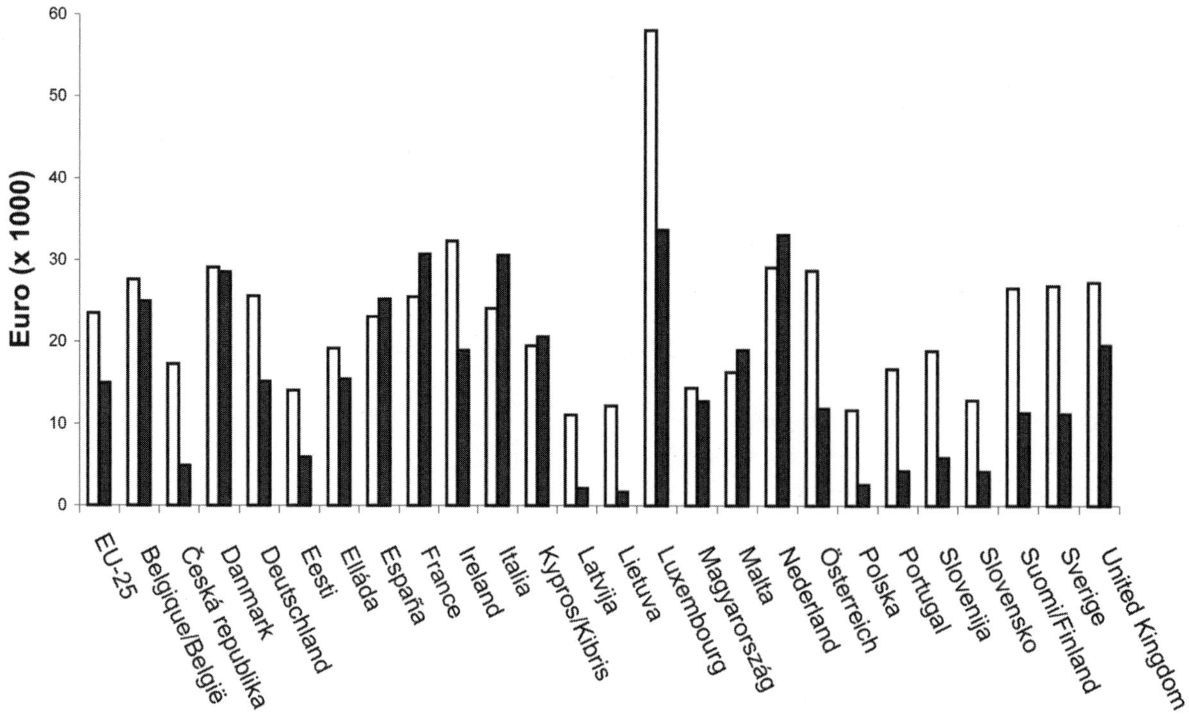

Figure 3-5. *Gross Domestic Product (GDP) per capita ($) and agricultural GDP per economically active person in agriculture (2005).* Source: European Commission, 2006b.

3.2.3 Impacts of advances in AKST on the growth of output and on farm businesses

The application of AKST has enabled farmers to increase yields, but It has also resulted in a fundamental restructuring of the industry. Many farmers now sell directly to large scale retailers or processors using a variety of contractual relationships. Small and part-time farms accounted for 86% of all farms in North America and almost half the farmers had full time jobs elsewhere (Thompson, 1986; Miljkovic, 2005). A growing number of farmers will have to get second jobs when subsidies from the CAP are slashed in 2013 (Barthelemy, 2007; Fischer-Boel, 2007).

Farmers have also sought to secure their position by diversifying their businesses to include activities that are not limited to agricultural production: these include direct selling (e.g., farmers' markets), agritourism or outdoor leisure activities. For many of these pluriactive farms, a minority of income now comes from farming. For instance, UK data shows that more than 50% of farms have income from di-

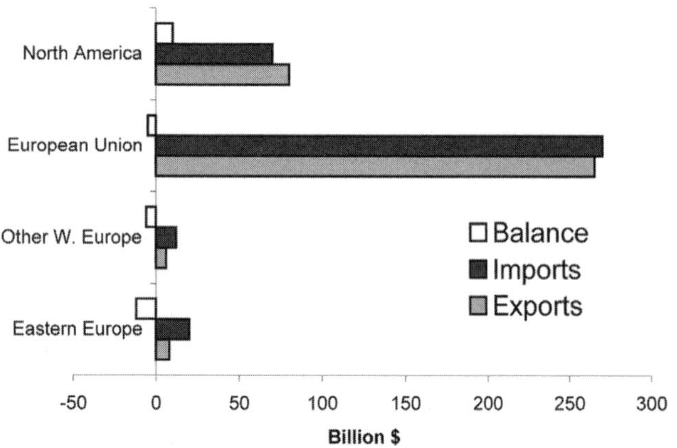

Figure 3-6. *Agricultural imports and exports in Europe and North America.* Source: FAO, 2008c.

Table 3-4. **External trade of EU 15 in 2002 in eight selected products.**

	% of world trade 2002	
	Imported by EU	Exported by EU
Total cereals	6.9	6.7
Oil seeds	29.2	1.9
Wine	15.0	29.8
Refined sugar	3.9	18.0
Fresh milk	1.1	5.6
Total meat	8.2	8.1
Eggs	3.1	14.5

Source: FAO, 2008c.

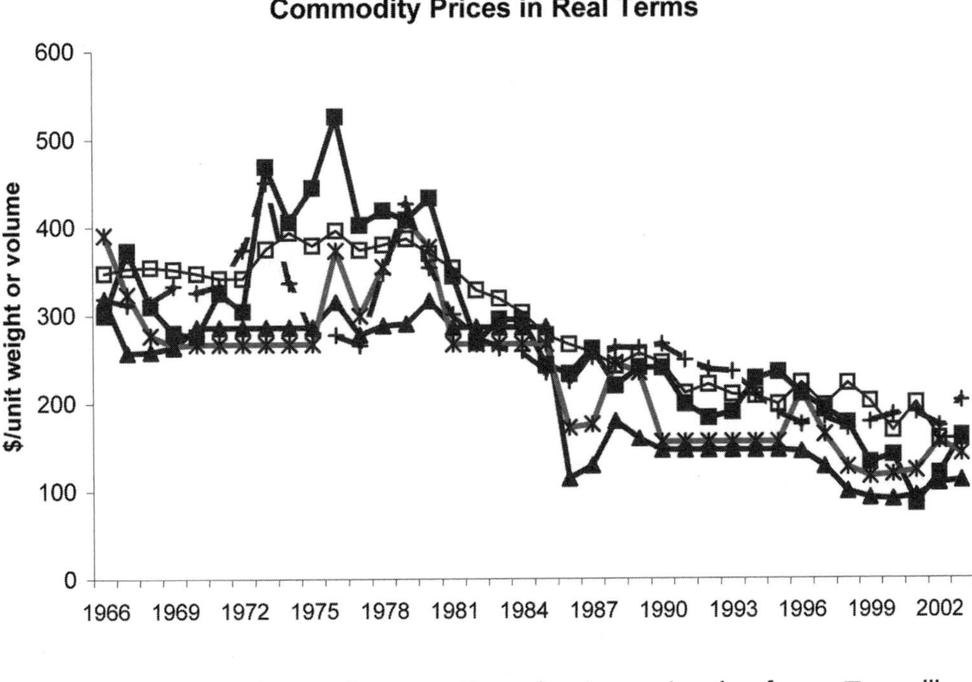

Figure 3-7. *Changes in Real Commodity Prices in the USA.* Source: FAO, 2008a

versified activity and income from these sources accounts for more than 50% of total income for 43% of the farms concerned (Defra, 2007).

In the years between 1945 and 1989 many farms in eastern Europe were collectivized and in some countries state farms were established. Where this was not the case very small scale farming persisted, often using old technologies (European Commission, 2006a) (Figure 3-8). On the large collective and state farms modern methods were used although the number of workers employed did not decline as rapidly as in the West. In the post 1989 period, as central planning gave way at varying paces to competitive markets, adjustments are taking place in the structure of farming, the level of agricultural employment, and the relationship between producers and consumers (Borzutzky and Kranidis, 2005).

Farmers in countries that have recently become members of the EU now have to compete in a wider market. This will lead to application of more AKST both on the farm and in the processing sector in order to reach the levels of quality and productivity that market demands. EU data shows that there is still a relatively low level of participation in further education and training in agriculture in the new member countries (European Commission, 2006a).

3.2.4 Impacts of AKST driven growth in output on processors and distributors

AKST has been also critical in ensuring the safety and quality of food. Food borne diseases are a matter of alarm where mass distribution increases the number of people who may suffer if products are infected or contaminated. On the farm this means attention to issues such as biosecurity and the use of pesticides. In food processing, preparation and presentation, rules of hygiene and the provision of information about ingredients to which some customers may be allergic are essential. The most valuable asset for retail distributors is their reputation. Safeguards are needed to ensure that products are consistent and safe, can be branded and can be traced to ensure that any failure is rapidly identified (Hornibrook and Fearn, 2005). Changes in the role of processors and distributors have altered the supply chain.

Traditional arms-length markets have been or are being replaced by coordinated plans for production and delivery. These minimize some elements of market risk and are a channel through which new technologies may be encouraged and supported on farm (Duffy, 2005). This development has been closely linked to progress in transport and the use of information technology to monitor performance at all stages of the food supply chain.

3.2.5 Impacts on market power

The technologies that have developed from AKST tend to encourage concentration at all levels of the agriculture and food sectors (see chapter 2). Although farms in general remain small businesses, a high proportion of output comes from the largest units (McAuley, 2004). Beyond the farm gate the concentration of the industry has advanced much more considerably (Hendrickson and Heffernan, 2007; Wiel 2007). An important repercussion of this has been a sense among both farmers and consumers that they are helpless in the face of the businesses with which they deal. This has enhanced the importance of farmer cooperatives and of direct marketing to consumers. Direct marketing includes tradi-

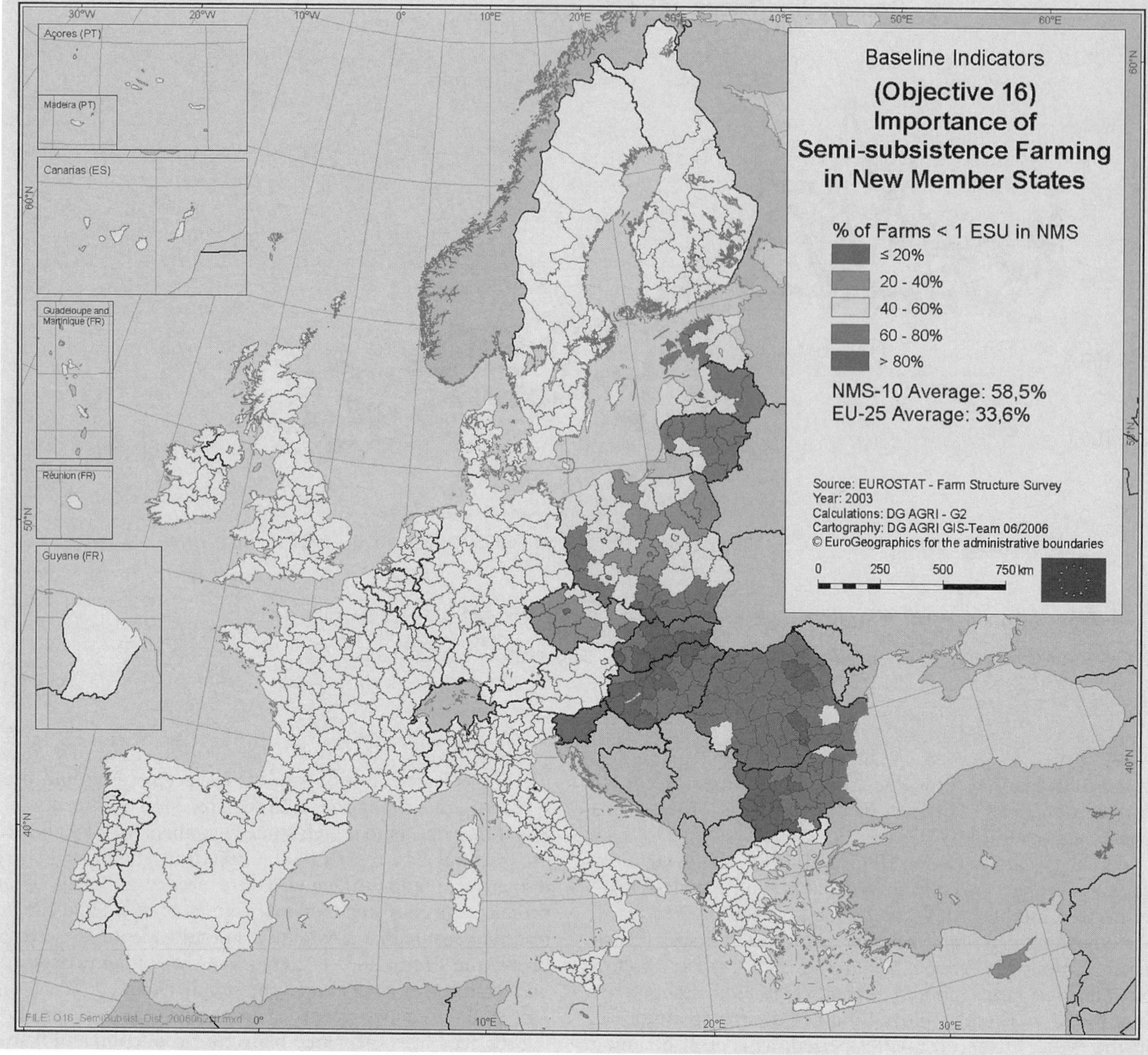

Figure 3-8. *Semi-subsistence farming among New Member States.* Source: European Commission, 2006a.

tional open markets in local towns, still a major avenue of distribution in France and the South of Europe, or farmers' markets that may take place on farms, or sometimes within open spaces in towns. Farm shops that may have started to sell the produce of the farm often develop to sell a diversity of products and services not produced on the farm itself but offering to the urban customer an attractive shopping experience.

The dominant position of multiple supermarkets in the UK led the Competition Commission to examine the food supply chain, pricing and the land banks owned by these companies (Wardell, 2007). They expressed concern about the extent to which owning but not developing sites impeded competition from other retailers. Overall, they concluded that consumers had benefited from the emergence of strong supermarket chains.

Cooperative responses to market power

Farmers have used cooperative buying and selling power to challenge the increasing power of transnational agricultural businesses. In the US and Europe, the agricultural cooperative movement flourished from the beginning to the mid-20th century. Farmers joined cooperatives to market agricultural products, as well as to obtain farming inputs and services. In Canada, the establishment of state marketing boards was a way to help farmers obtain fair prices for their products.

For example, after WW II, farmer cooperatives thrived. The total number of farm cooperatives in the US declined from a peak of 12,000 1930 to 6,293 in 1980 to 3,140 in 2002. Today less than 3 million farmers belong to cooperatives in the US. In Europe, cooperatives are very important and powerful organizations in the marketing and processing of agricultural products and in the supply of credit to farmers (Table 3-5). Farmer cooperatives are more important in some countries than others (Figure 3-9) and are also more important in some sectors than others. The dairy sector in the US, for instance, relies heavily on marketing cooperatives with 87% of US milk purchased at the first handler level through cooperatives (Kraenzle, 1998). In northern Europe and Ireland, agricultural cooperatives have captured almost majorities (or the entirety) of the dairy market and have significant shares of the markets for inputs in many western European countries. A majority of Canadian grain has been and continues to be marketed through marketing boards. However, cooperatives are less important in the livestock marketing sector in the US and Canada, while accounting for a larger portion of sales in northern Europe.

Traditional marketing and supply cooperatives have confronted the increased pressure from the consolidation of investor-owned firms and their increasing market share. Many cooperatives merged with other cooperatives, particularly in the dairy sector (Hendrickson et al., 2001) and those marketing grains and oilseeds (Crooks, 2000). Others developed joint ventures and alliances with investor-owned firms.

The agricultural and food system, that AKST has made possible, requires substantial packaging, temperature control, processing and has appreciable delivery costs. Additional costs may also occur when food is discarded because temperature control has failed, or where the "sell-by" date has been passed. For packaged goods supermarkets sell products in predetermined pack sizes. These may not match the requirements of small households who find they do not fully use all the items in a package before its "use-by" date has passed. These costs have to be absorbed within the supply chain and borne by the consumer. They may lead to environmental costs as a result of excessive packaging and problems of waste disposal. While such waste is of concern, it should be noted that substantial wastage occurred before modern AKST systems were used, as seasonal surpluses could not be safely preserved.

3.2.6 Structural change induced by AKST

The way in which resources are organized into businesses is determined by many factors including the competitiveness of different technologies. Among the other factors affecting the food and agricultural sector are rising labor costs, the development of communication systems, the operation of banking systems and the availability of transport systems. Even without changes in the state of AKST, changes in these areas would lead to changes in the sort of technology that was used in the sector.

At the farm level, the most obvious structural effects have included fewer workers, increased specialization and a tendency for full time farms to become bigger, while smaller farms become part time (see chapter 2). In some cases the statistics may not fully represent the degree to which decisions have been concentrated, as farmers share resources such as machinery or labor and in some cases run a single large enterprise on more than one "farm". The decline in the farm labor force has profound implications for rural communities. In areas where agriculture was the major source of employment the rural economy can be undermined. Community services such as schools, medical facilities and transport are no longer able to operate at an economic level. Business districts may disappear and the informal, voluntary activities that often form a crucial part of the social support system for community residents may decline. In regions close to urban centers this impact may be diminished.

Table 3-5. **Percentage of agricultural products sold through cooperatives in the EU-15 (1997).**

	Pig meat	Beef/ veal	Poultry meat	Eggs	Milk	Sugar beet	Cereals	All fruit	All vegetables
Belgium	20	0	—	—	53	—	30	75	85
Denmark	91	66	0	52	94	0	60	70-80	70-80
Germany	27	28	—	—	52	80	45-50	40	28
Greece	3	2	15	2	20	—	49	57	3
Spain	8	9	25	28	30	23	22	45	20
France	85	30	30	25	47	16	68	40	25
Ireland	66	15-20	20	—	99	—	57	14	18
Italy	13	12	35	8	40	6	20	43	8
Luxembourg	37	38	—	—	81	—	79	—	—
Netherlands	34	16	8	14	83	63	65	76	73
Austria	15	5	70		90	100	60	18	28
Portugal	—	—	—	—	—	—	—	—	—
Finland	68	65	81	54	97	—	46	—	—
Sweden	78	73	0	32	100	0	75	20	50
UK	28	—	25	—	67	—	24	67	26

Source: European Commission, 2000a.

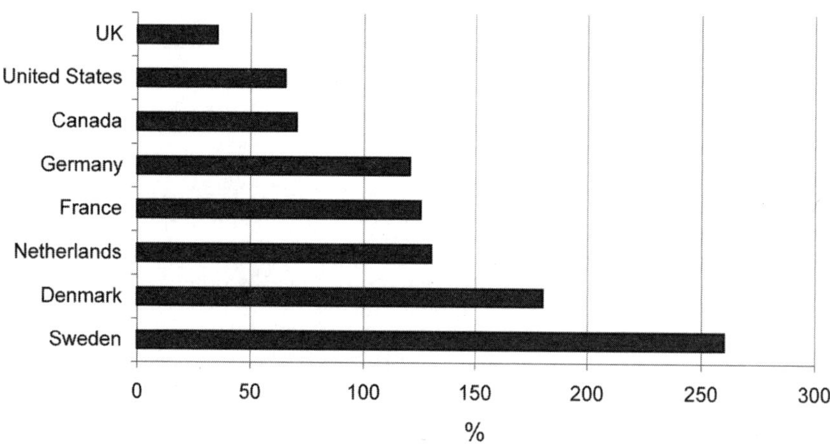

Figure 3-9. *Turnover of farmer-controlled businesses as percentage of agricultural output.* Source: English Food and Farming Partnership, 2006

Instead of working on farms, former farm workers may commute to towns. Where the urban economy is buoyant, city dwellers may move into villages, raising the price of village houses and creating new and different communities. In this type of situation impacts measured in average data tend to show these communities as relatively affluent, although they contain many poorer people who once depended on farming for their incomes.

3.2.7 Impacts on trade of changes in production driven by AKST

Fluctuation in farm incomes at times presented a major problem for most NAE governments. Powerful farm lobby groups demanded support for farm incomes. In response, policies provided subsidies that prevented declines in farmer income despite excess levels of production resulting from greater productivity. In effect, the EU and the USA subsidized farmers, limited imports and subsidized exports (Figure 3-10).

Dramatic changes in the level of support took place after the break up of the Soviet Union (Figure 3-11). From the mid-1990s support had declined to levels below those of most other developed countries.

Producers in other countries faced depressed prices and in some cases total loss of markets at least in part as a result of subsidies in NAE. This has become the major issue in international trade negotiations. Its impact extended far beyond agricultural trade itself because countries refused to make progress on trading issues without an agreement on agriculture. The debate included the level of domestic subsidies, the demand to remove export subsidies and to reduce all sorts of barriers to market access. In return for progress in these areas the NAE countries sought tariff reductions on manufactured goods; trade in services and agreements relating to intellectual property (WTO, 2005).

Agricultural issues remain critical in the current Doha round of trade negotiations. The Secretary General reported on 18 December 2005 that, after protracted negotiations, significant progress had been made on agriculture including an agreement to end export subsidies by 2013.

However, he announced the suspension of those negotiations seven months later because of lack of progress. At

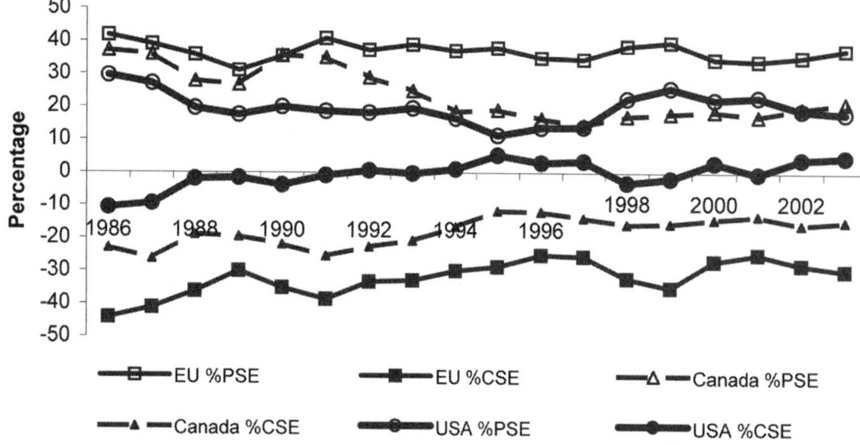

Figure 3-10. *Producer and Consumer Support Estimates (PSE & CSE) as measures of support for agriculture.* Source: OECD, 2004

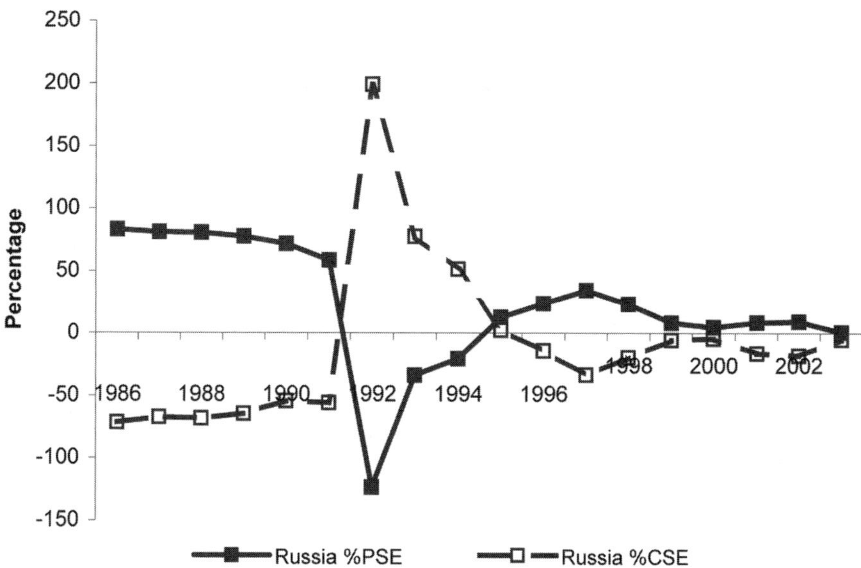

Figure 3-11. *Producer and Consumer Support Estimates (PSE & CSE) for farming in Russia.* Source: OECD, 2004

the heart of the debate were failures to agree terms for access for developing country exports to developed country markets and to reach a settlement on domestic support (WTO, 2006). According to a review of more than 200 theoretical and empirical studies about *the effect of trade liberalization* on sustainability, the effects on economic welfare and overall sustainability depend on the nature and extent of the flanking and other supporting measures that are taken (Kirkpatrick et al., 2004). The potential, aggregate economic welfare gains to be made from free trade and increased foreign investment inflows, are not necessarily shared by all countries or by all socioeconomic groups within these countries. In many examples the social (and environmental) impacts are negative if protective measures are insufficiently effective.

The trends in global demand for food safety and processed products under the conditions of free trade raise concerns about the long-term viability of small farms in developing countries (Lipton, 2005). These have often already felt the disproportionately negative impacts of structural adjustment policies on smallholders during the 1980s and 1990s. The impact of trade liberalization on distribution of income within developing countries varies, however, according to country-specific policy conditions and socioeconomic structures. In Latin America, for example, the effects on equality in income have been positive in nine countries and negative in five countries (von Braun, 2003).

3.2.8 External economic impacts of the application of AKST

Negative impacts of AKST on the environment have been discussed in sub-chapter 3.1. These environmental and social costs generally do not figure in the accounts of the businesses concerned but do represent real economic benefits or costs to other individuals. These externalities may be positive or negative and their incidence is diverse. Some, such as the costs of restoring adequate water quality that has suffered as a result of farming practices, can be calculated with relative ease. Less easily assessed are environmental losses occurring where plant nutrients or pesticides contaminate water courses (see 3.1). The use of AKST in devising and using veterinary medicines, pesticides, herbicides and in the management of more intensive stocking of livestock can raise public health issues. Food-borne diseases represent costs to affected individuals and to medical services. For the industry, market collapses as a result of food scares can destroy the value of goods already produced. Governments seek to minimize risks to human health but the costs can be very large. For example the gross total cost to the UK and the EU budgets of measures to combat BSE between 1996 and 2006 are reported below (Table 3-6) (Defra, 2006a). Similarly, UK government costs to manage bovine tuberculosis between 1997 and 2007 are presented in Table 3-7.

The cost of introducing a new medicine or pesticide involves substantial expenditure by the company concerned on testing to the approved standards. Increased public concern has led to a progressive tightening up of standards across the NAE, although particularly pronounced in western Europe and North America (Clark and Tait, 2001).

3.3 Social Impacts of Agriculture and AKST within NAE

The increase in productivity achieved by NAE agriculture over the last 60 years with the help of AKST has contributed to providing people in NAE with more wealth, choice and mobility. In NAE there is today more food and a wider range of affordable food items available than ever before. People have also more choice in where they want to live and work than in the past. Rural regions have increasingly specialized in producing and exporting natural resource-based raw materials. This development has given rise to out-migration and to major changes in social structures in rural regions.

Table 3-6. **Net UK costs of managing the outbreak of BSE 1996-2005.**

£ million									
1996/7	1997/8	1998/9	1999/00	2000/01	2001/02	2002/03	2003/04	2004/05	2005/06
1496	963	568	425	495	492	–	335.9	340.1	265.7

Source: Defra, 2004c.

3.3.1 Impacts of changes in agriculture on community well-being

The social impacts of specialization in agriculture and increased scale of agricultural production are primarily related to well-being of communities and farm families. A great deal of evidence produced using at least five different methodologies, involving a number of different researchers and looking at different regions of the US showed detrimental impacts for community well-being from industrialized farming. These studies also showed that industrialized farming involved a tradeoff effect, as it did not consistently produce detrimental effects for all time periods or for all regions, but resulted in beneficial impacts for some groups and detrimental ones for others (Goldschmidt, 1978; Lobao, 1990; Stofferahn, 2006).

3.3.2 Consumer concerns about the food system

There are different attitudes in North America and Europe with regard to GE-derived foodstuffs. While foods from GE crops are available and do not require labeling in North America, in Europe foods derived from GE crops are generally not available and where sold are required to be labeled as containing GE ingredients. This situation is viewed in Europe as a clear reflection of consumer concerns. Some in US industry and government, however, take the view that consumers have not yet been offered an adequate opportunity to accept or reject these products, because food manufacturers, out of a desire to preserve brand equity have reformulated products so they do not trigger mandatory European labeling requirements (Larson, 2002; USTR, 2003; Yoder, 2003; USDA, 2005a).

However, some experts have argued that the potential benefits of improved nutrition and increased yields from genetic engineering are so important, especially for developing countries, that GE crops should be readily and economically available (Nuffield Council on Bioethics, 1999). Early development of the technology has not been with poorer countries in mind (Kinderlerler and Adcock, 2003). Rather it has been aimed at securing profits for firms in industrialized country contexts selling products to relatively wealthy farmers. While public private partnerships and international agriculture research centers may be developing crops more appropriate to developing countries, general welfare, justice and access should also be considered (Kinderlerler and Adcock, 2003). A position that allows each country the right to accept or refuse GE crops, based solely on ethics, is not consistent with the science-based regulatory approach of the World Trade Organization, although as a matter of policy, countries are allowed to set their own level of SPS protection (Kinderlerler and Adcock, 2003).

Ethical issues are a major consideration in discussions about biotechnology and animals. A distinction is made between "intrinsic concerns" (genetic engineering as wrong or morally dubious due to the mode of production or the source of the genetic material or "it is unnatural to genetically engineer plants, animals and foods") and "extrinsic concerns" based on animal welfare perspectives (Kaiser, 2005) and environmental impacts. Reviews such as those published by the Netherlands Advisory Committee on Ethics and Biotechnology in Animals and the UK Royal Society (2001) stress the need to consider a range of health and risk implications of genetically engineered animals to humans but also our responsibility to the animals themselves.

Intensive livestock production raises several other significant ethical issues. Treating animals as items on a production line offends many who feel this is an unacceptable relationship between humans and other species. In western Europe and North America the welfare of farm animals has become an area of increased significance for policy makers (USDA, 2003; Defra, 2004b; Webster, 2005). The mass production of animals to specification undermines traditional livestock businesses, reducing local employment and

Table 3-7. **Breakdown of the net cost of managing bovine tuberculosis in Great Britain 1997-2006/7 (£m).**

	1997	1998/99	1999/00	2000/01	2001/02	2002/03	2003/04	2004/05	2005/06	2006/07 (provisional)
Cattle testing	5.5	7.3	17.6	13.3	5.4	24.7	33.2	36.4	36.7	37.8
Compensation	1.4	3.5	5.3	6.6	9.2	31.9	34.4	35.0	40.4	24.5
RBCT	1.7	2.9	4.6	6.6	6.0	6.6	7.3	7.2	6.2	1.6
Surveillance activity by the VLA	1.6	1.9	2.4	3.5	3.7	4.1	5.3	4.9	7.5	6.4
Other research	1.7	2.5	3.8	5.3	6.1	6.5	7.0	5.7	6.5	7.8
Q/Overheads	4.1	6.7	4.5	0.9	0.1	0.7	1.0	1.3	1.8	1.7
Totals	16.0	24.8	38.2	36.2	30.5	74.5	88.2	90.5	99.1	79.8

Source: DEFRA, 2007b.

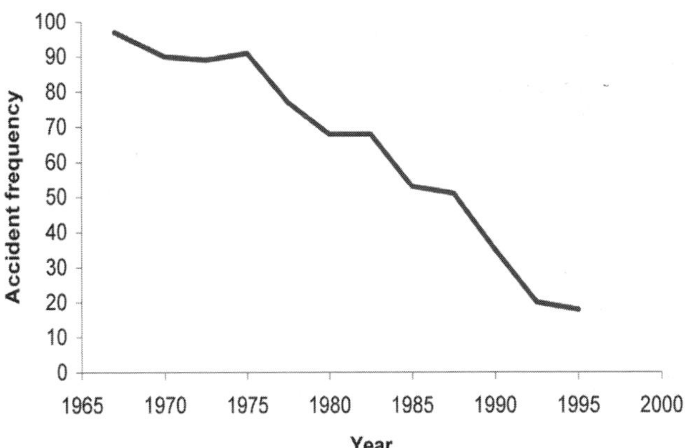

Figure 3-12. *Accident frequency rate, i.e., number of accidents per one million hours worked in Swedish forestry (1967 to 1995).* Source: Axelsson, 1998

jeopardizes the economic survival of some communities. In an area in which emotions often play an important part in determining attitudes there are a wide range of pressure groups who criticize many aspects of intensive livestock production (Compassion in World Farming, 2007).

Livestock kept in intensive systems are prone to outbreaks of disease which have been controlled by slaughter policy in some cases (e.g., foot and mouth disease) or more often through the use of antibiotics. However some production systems routinely utilize antibiotics both for disease prevention and growth promotion. This has raised serious concerns within some parts of the NAE because of the rise of antibiotic resistant bacteria in humans which some argue is linked to livestock production (Mellon, 2000).

3.3.3 Social impact of increased mechanization

In all sections of agriculture increases in mechanization have resulted in redundancy in the farm labor force but the increased productivity/efficiency has also left more time for other work and enhanced worker environment by eliminating repetitive, dangerous and disliked tasks (Culshaw and Stokes, 1995; Wilson and King, 2003). Its ability to secure lower costs implies growing pressures on small farms that cannot, or fail to, apply similar methods. Where communities depend on traditional agriculture, as in many areas of Europe, it is likely to increase pressure on farmers and farm workers to seek employment off the farm and accelerate the continuing decline of the farm labor force. The social and political consequences of this are likely to remain at the centre of agricultural policy thinking into the 21st Century.

In forestry one of the greatest impacts of the increase in mechanization has been on a reduction in accidents (Figure 3-12). Forestry is an innately dangerous operation and in Sweden between 1970 and 1990 the number of accidents decreased from 8656 to 1469. The accident risk, expressed as accident frequency rate, was reduced from 90 to 35 accidents per one million man-hours worked (Axelsson, 1998).

3.3.4 Migration from rural areas

In North America and western Europe today the population working in agriculture is only a small share of each country's overall population. In contrast, in some countries in eastern Europe the proportion of the population in agriculture is still very significant (Table 3-8). The rural population is still declining in terms of percentage of the total population in most NAE countries (Table 3-9). The rural population is still declining in terms of percentage of the total population in most NAE countries.

While overall trends are similar, different regions in NAE have different conditions impacting these changes. The farm population in the United States has decreased as a percentage of the US total population, falling to 1% in 2002 from 17% in 1945 and the rural population to 21% in 2000 from 36% in 1950, respectively (Dimitri et al., 2005). The decade of the 1950s saw the largest exodus from farming (Lobao, 1990) while 600,000 farmers exited farming between 1979 and 1985 (Heffernan and Heffernan, 1986), the latter characterized as the "Farm Crisis" of the 1980s that particularly affected the economic base of rural communities in the Midwestern states. Still, while the portion of rural dwellers in the US dropped from 50% of the population in 1945 to about 21% in 2005, this does not signal an exodus from rural areas, as the actual rural population has held relatively constant over this time.

In western Europe, as technology advanced during the 50 years following the Second World War, the number of

Table 3-8. **Employment by major economic sectors in 2003 in a selection of countries in the NAE (as a % of total employment).**

	Agriculture	Industry	Services
Austria	14.2	24.7	70.3
Canada	2.7	20.9	76.4
Czech Republic	4.2	38.3	57.5
Denmark	3.2	21.7	75.1
Estonia	6.1	32.4	61.6
Finland	5.3	26.3	68.4
France	3.7	21.1	75.2
Georgia	54.9	6.1	39.0
Germany	2.3	27.0	70.7
Greece	14.5	23.9	60.8
Hungary	5.4	33.4	61.3
Italy	4.2	29.0	66.8
Netherlands	3.4	18.1	78.5
Poland	18.4	28.6	53.0
Romania	35.7	29.8	34.5
Russia	10.9	24.2	64.9
Spain	5.7	29.7	64.7
Sweden	2.5	23.6	73.9
Turkey	33.9	22.7	43.4
Ukraine	19.5	24.6	55.9
United Kingdom	0.9	18.3	80.8
United States	1.6	21.1	77.2

Source: UNECE, 2005.

Table 3-9. **Urban and rural populations in NAE.**

Country	Population distribution (%), 2004		Average annual rate of change in population (%), 2000-2004	
	% Urban	% Rural	Urban	Rural
Austria	66	34	0.05	0.06
Czech Republic	74	26	-0.04	-0.30
Denmark	85	15	0.35	-0.38
Estonia	69	31	-1.30	-0.74
Finland	61	39	0.07	0.37
Georgia	52	48	-1.43	-0.39
Germany	88	12	0.27	-1.38
Greece	61	39	0.68	-0.64
Hungary	65	35	0.08	-1.49
Italy	68	32	0.04	-0.33
Netherlands	66	34	1.25	-0.96
Poland	62	38	0.03	-0.25
Romania	54	46	-0.34	-0.09
Spain	77	23	0.34	-0.12
Sweden	83	17	0.09	0.03
Turkey	67	33	2.07	0.06
United Kingdom	89	11	0.38	-0.27
United States	81	19	1.44	-0.74

Source: FAO, 2008b.

farms and the number of farmers and farm workers has also declined dramatically. In 1950 England had a farm labor force of 687,000 people. By 2000, the labor force on farms had declined to 375,000 (Defra, 2006b). Similar trends are apparent in other western European countries.

The changes in eastern Europe are more complex as collectivization during the communist era greatly reduced the number of farming units in most countries. For example, in East Germany in 1945, all large farms were reduced to 100ha and the rest of the land allocated to farm workers. Some of these private farms survived until 1955, but after the German Democratic Republic was established in 1949, pressure to collectivize them increased. The collectivization was completed in 1955 and after that no private ownership of land was permitted. Many of farmers left the land to work in the new factories. Then, following the demise of this system of land management in c. 1990 there has been a variable re-allocation of land to the former owners, resulting in fragmentation of the farming units. In turn there has now been re-amalgamation of the small units to create more financially viable enterprises (Bouma et al., 1998).

3.3.5 Equity (benefits, control and access to resources)

Food production per capita has been increasing in the NAE and globally, but major distributional inequalities exist. Current directions in the development of food systems have fundamentally changed the internal interaction and share of benefits in the food chains, disempowering local rural actors, such as farmers and small-scale processors. The share of retail in the control and benefits in the food chains has increased.

3.3.5.1 Equity in terms of economic benefits and value added

AKST has been a factor in enabling rural regions to specialize in producing and exporting natural resource-based raw materials for, e.g., food industry (Siegel et al., 1995) and enabling local demand to be met with imported food. The value added in production, food processing and food distribution has been transferred to urban areas and, increasingly, beyond national borders. Despite this, food production has played a central role in rural vitality and will do for a long time to come (OECD, 1996). The reduction in the number of farms and farm workers has led to out-migration and the break down of some social structures in the rural regions of all industrialized countries in Europe.

The transformation to a more advanced stage of industrialized farming over the past 60 years has led to significant increases in productivity with concomitant benefits to many consumers, but it has simultaneously, in many rural areas, had an adverse effect on economic and social vitality and arguably reduced the somewhat idealized independence of farmers (Goldschmidt, 1978; Ikerd, 2002; Stofferhan, 2006). The above description of events may be too sweeping because changes in social and economic structure of rural communities have differential effects, creating opportunities for some and disadvantaging others (Buttel, 1983). Such reasoning suggests that socioeconomic effects of industrialization and globalization are variable and fluctuate in response to local and non-local forces. Concern may also be indicative of a nostalgic worldview that idealizes how rural farming communities once were.

The rise of retail concentration (see Chapter 2) has led to the concern that retailers may abuse their market power vis-à-vis other actors with smaller market shares, in particular farmers and consumers (Hendrickson et al., 2001; Morgan et al., 2006). Farmers have for a long time noted how small a share of consumer prices for food and fiber products comes from what farmers receive for the raw commodities at the farm gate. The declining share of the consumer food Euro/dollar allocated to producers is reflected in rising retail-farm price margins. A factor contributing to this decline is the increase in consumer demand for off-farm or marketing services for food. Farmers' ever-increasing productivity has made agricultural products steadily cheaper in real terms; this alone would cut the farmer's share of retail prices if the margins for processing and retail distribution just kept up with inflation. But growing farm productivity is only half of the story. In some markets the farm-to-retail margins have risen significantly faster than overall food marketing costs. Growing retail margins may be variously explained in different markets (Reed et al., 2002). Reduced competition among retailers or (for some products) processors may produce monopoly profits, stifle cost saving innovation and dull the efficiency of management; alternatively, consumers may choose products to which more value has been added, fewer competitors may increase the importance of competition on things other than price. There may be more value-added at the retail level, including better service and a greater variety within the category. All farmers are facing a shrinking share of the retail dollar/Euro. With the ever-growing efficiency of production agriculture and the continuing tendency of the marketing system to add more value for wealthier consumers, this trend is expected to continue (Kinsey and Senauer, 1996).

3.3.5.2 *Equity in access to resources*

The development of agricultural technology in NAE, based on external, purchased inputs has affected global equity. Poor farmers especially in developing countries often do not have the option of introducing modern methods because of the lack of market integration and infrastructure, the heterogeneity of the environment, or because they cannot afford purchased inputs. The nutrient case illustrates the more general consequences. Large field areas of the NAE, especially in Europe, have been enriched with phosphorus and nitrogen but only a proportion of the industrially fixed nutrients is retained in food products. This leads to eutrophication and biodiversity decline in both aquatic and terrestrial systems (see Chapter 3.1). Conversely, the soils of several cultivated systems especially in sub-Saharan Africa are nutrient-depleted. This is especially problematic where fruits, vegetables and other crops are exported on a large scale from rural areas to urban centers, or from regions with nutrient-poor field soils to nutrient-enriched NAE. In fact, NAE increasingly relies on food, feed and resources originating beyond its borders (Deutsch, 2004). For example, only a third of African phosphate fertilizer production was used in Africa in 2002 (FAOSTAT, 2006).

3.3.5.3 *Equity in control and influence*

Critics concerned with the global equity of agri-business assert that powerful food retailers situated in the North, whose success has been partly driven by NAE AKST, largely dictate the social relations of production in the South and provide little opportunity to encourage local value capture (Marsden, 1997). Such processes are seen to be powerful drivers for divergence and marginalization in traditional farming communities. Further, it is contended that the only way forward is for these localities to disengage and reintegrate into local and regional settings. Paradoxically, in some regions (e.g., Tuscany), these same phenomena described above have been the catalyst for stimulating vibrant new livelihood strategies (such as tourism) in traditional farming communities, as they have endeavored to innovate and adapt to rapidly changing circumstances (Miele and Pinducciu, 2001; Morgan et al., 2006).

Historically, some of the effects of the trends described above have been mitigated in Europe and the US by costly market intervention to support prices, often under the policy guise of rural poverty mitigation, rural development programs or more recently nature conservation (Petit, 1997; Dimitri et al., 2005). The impacts of these policies are in the decline in the US, but due to effective lobbying and public support the agricultural sector in the EU was largely exempted from trade liberalization agendas until the Uruguay Round in 1992.

Understanding the wants and demands of consumers within highly differentiated food markets has become a source of power within food systems. Related to this point, consumers are demanding more transparency and information (essentially control) about food production methods and labor relations on which to base purchasing decisions (Miele and Parisi, 2001; Blokhuis et al., 2003). Thus the role of knowledge and information is assuming more and more importance as a point of influence and control in food systems, especially in NAE. Supermarkets and fast food outlets with their proximity to customers have a unique capacity to influence the rest of the production and food distribution chain. These powerful retailers continue to strive to meet consumer welfare concerns (price, quality and variety), often to the detriment of producer welfare. A recent spate of food controversies in North America and Europe has re-stimulated the continuing debate and concern about human and environmental health risks (the so-called food anxieties) associated with food production and consumption (Holloway and Kneafsey, 2004). The response is tougher more restrictive food quality criteria managed through resource intensive, producer responsible, certification processes to manage risk and quality. Clearly it is larger scale producers who are in a better position to meet such demands.

3.3.5.4 Rise of alternative food systems

Partly in response to the numerous concerns related to industrialized agribusiness there has been a growing interest in "alternative" food systems. Some of these reject aspects of NAE AKST provided. Local food systems with their focus on their social and economic embeddedness can overcome high costs and reduce risk for farmers and consumers by adding value locally, thereby supporting rural development (Sage, 2003; Winter, 2003). Although there are many benefits attributed to locally-oriented food systems, these models have also been criticized as benefiting primarily those who can choose based on education or income (Allen, 1999; Hinrichs and Kremer, 2002; Hinrichs, 2003).

Conceptualizing the equity of food systems at different spatial scales generates different perspectives and responses. Projects based on regional identity (e.g., Tuscany) or branding (e.g., organics, Slow Food) have been promoted as rural development alternatives in NAE (Barham, 2003; Murdoch and Miele, 2004). However, they may also serve the privileged at the expense of the poor (Allen, 1999), through the decreasing affordability of products—perhaps even magnifying existing unequal relations of consumption locally (Bellows and Hamm, 2001; Allen and Sachs, 1991). Furthermore a focus on the local may well direct attention from global-scale inequities surrounding issues of food security and material welfare, although it may reduce local communities' (implicit) involvement as consumers in exploitative labor and environmental commodity chains. The local concentration of production and consumption may also restrict opportunities to import Fair Trade goods, thus limiting market access for developing country growers.

3.3.6 Distancing consumers from production

The increasing emergence of vertical food chains (see Chapter 2) has increased spatial and social distancing between sectors in the food chain (Sumelius and Vesala, 2005). Social distancing has helped to lessen consumers' understanding of the production system and the food chain, thereby decreasing their ability to fully participate in a food system dominated by market logic. Issues of ethical, social and environmental concern are typically shielded from consumer view and may only be revealed if there are dramatic and direct societal consequences. The environmental effects of conventional agriculture and their social implications tend to be spatially bounded (rather than atmospheric or global) and often are remote from the end consumer (Marsden et al., 1999). In these circumstances, price and convenience, which are still visible, have been the predominant determinant for consumers, while adverse social and environmental effects can be isolated from consumer view.

3.3.7 Nutritional consequences of NAE food systems

The most direct and tangible benefit of food is its role in enabling individuals to pursue active, healthy, productive lives as a consequence of adequate nutrition (MA, 2005). For these reasons access to adequate, safe food has been recognized as a basic human right. Decreased hunger and poverty and improved nutrition and human health are two of the Millennium Development Goals.

Although the food insecurity and prevalence of undernourishment and hunger has been reduced worldwide, there were still 9 million undernourished people in industrialized countries and 28 million in countries in transition in 2001-2003 (FAO, 2006). These data include 21 million people in the Commonwealth of Independent States (7% of the population), 3 million people in eastern Europe (former socialist states within and without the EU) (4% of the population) and 0.1 in Baltic States (1% of the population).

An increase in consumer purchasing power, progress in food production methods and changes in the marketing of food products have dramatically improved the food situation in many countries of the European Union and in the USA. Food has been generally available, although some sections of the population do not consume a sufficiently healthy diet. For example, the consumption of fruits and vegetables has declined in the US in the last 100 years. People on a low income spend a greater proportion of their income on food, but eat a diet of lower nutritional quality than those on a high income (European Commission, 2002a).

The emerging challenges in relation to nutrition and health are thus different than those of some decades ago. North America and Europe are currently experiencing a high prevalence of noncommunicable diseases, such as cancer, cardiovascular disease, diabetes, certain allergies and osteoporosis. These are the result of the interaction of various genetic, environmental and lifestyle factors (including smoking, diet and a lack of physical activity). Numerous studies suggest nutrition is important in maintaining health and preventing many of these major diseases (Ferro-Luzzi and James, 1997).

For the European Union, estimates have been made of the total burden of ill health, disability and premature death from all causes experienced by the population and the factors most responsible for this disease burden (European Commission, 2002b). Of a broad range of causes, diet-related factors are believed to be responsible for nearly 10% of the total disease burden—including overweight (3.7%), low fruit and vegetable consumption (3.5%) and high saturated fat consumption (1.1%). Together with lack of physical exercise (1.4%), these factors account for a greater proportion of ill health than tobacco smoking (9.0%).

In recent years, overweight and obesity have been growing at a very fast rate. Today obesity represents a real threat to the public health of certain groups in North America and Europe, as shown by data from IOTF and OECD (Tables 3-10 and 3-11). A particular concern is the rapid rise in childhood obesity (Figure 3-13).

In the next 5 to 10 years obesity in the European Union will probably reach the high level of prevalence in the United States today, where one third of people are estimated to be obese and one third to be overweight. In many countries there is a 10-15 year lag behind the USA, but nevertheless European countries are narrowing this gap (Figure 3-13).

3.4 Impacts of NAE AKST through International Trade

NAE accounts for more than a quarter of global trade in agricultural products. The European Union and the United

Table 3-10. **Obesity and overweight among adults in a sample of countries within European Union.**

Country	Year of Data Collection	Males			Females		
		% BMI 25-29.9	%BMI ≥30	Combined BMI≥25	% BMI 25-29.9	%BMI ≥30	Combined BMI>25
Czech Republic	1997/8	48.5	24.7	73.2	31.4	26.2	57.6
Denmark	1992	39.7	12.5	52.2	26	11.3	37.3
England	2003	43.2	22.2	65.4	32.6	23	55.6
Finland	1997	48	19.8	67.8	33	19.4	52.4
France (self report)	2003	37.4	11.4	48.8	23.7	11.3	35
Germany	2002	52.9	22.5	75.4	35.6	23.3	58.9
Greece	1994-8	51.1	27.5	78.6	36.6	38.1	74.7
Hungary	1992-4	41.9	21	62.9	27.9	21.2	49.1
Italy (self report)	1999	41	9.5	50.5	25.7	9.9	35.6
Latvia	1997	41	9.5	50.5	33	17.4	50.4
Netherlands	1998-2002	43.5	10.4	53.9	28.5	10.1	38.6
Poland (self report)	1996	n/a	10.3	n/a	n/a	12.4	n/a
Spain	1990-4	47.4	11.5	58.9	31.6	15.3	46.9
Sweden (adjusted)	1996-7	41.2	10	51.2	29.8	11.9	41.7

BMI = body mass index. Age range and year of data in surveys may differ. With the limited data available, prevalences are not standardised. Self reported surveys may underestimate true prevalence. Sources and references are from the IOTF database.

Source: International Obesity Task Force, 2005.

Table 3-11. **Change in obesity (percentage of adult population with a BMI>30 kg/m2) from 1980-2003 in the NAE.**

	1980	1990	2000	2001	2002	2003
Canada	—	—	—	13.9	13.9	14.3
Czech Republic	—	11.2	14.2	14.2	14.8	14.8
Denmark	—	5.5	9.5	9.5	9.5	9.5
Finland	7.4	8.4	11.2	11.4	11.8	12.8
France	—	5.8	9.0	9.0	9.4	9.4
Germany	—	—	11.5	11.5	12.9	12.9
Greece	—	—	21.9	21.9	21.9	21.9
Hungary	—	—	18.2	18.2	18.8	18.8
Italy	—	—	8.6	8.5	8.5	9.0
Netherlands	5.1	6.1	9.4	9.3	9.7	10.7
Norway	—	—	6.4	8.3	8.3	8.3
Spain	—	6.8	12.6	12.6	12.6	13.1
Sweden	—	5.5	9.2	9.2	10.2	9.7
United Kingdom	7.0	14.0	21.0	22.0	22.0	23.0
United States	15.0	23.3	30.5	30.5	30.6	30.6

Source: OECD, 2006.

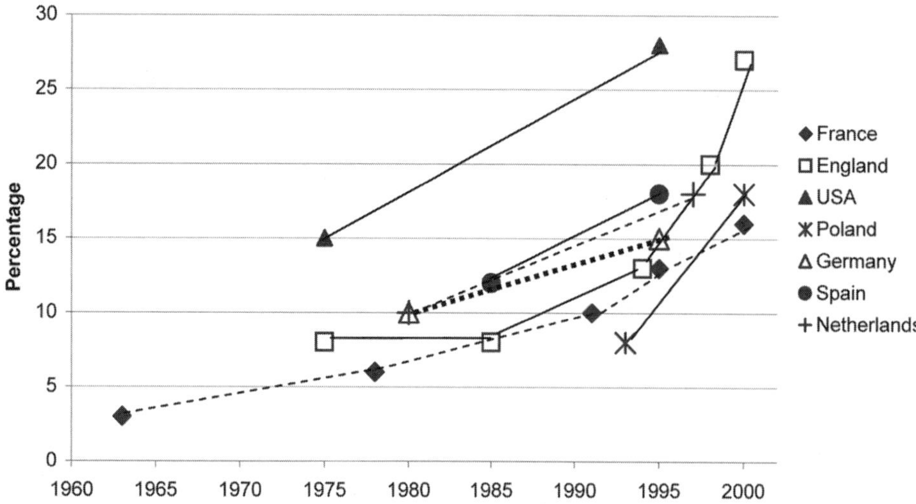

Figure 3-13. *Prevalence of overweight children in NAE (ages 5-11).* Source: International Obesity Task Force, 2005

States are major players while trade flows with the Russian Federation are much smaller (Table 3-12). Trade has also been growing. Between 1986 and 2003, substantial changes in trade flows were associated with the breakup of the system of centralized command systems in the USSR and other parts of CEE. Beyond 2004 the EU became 25 countries rather than 15.

The US has been a net exporter while the EU has been a net importer (Figure 3-14). The EU has subsidized agricultural exports while the US support system for farmers, combined with Food Aid programs, has made their farm exports competitive. Subsidized exports damage low cost producers in both developed and developing countries who face lower prices and may even lose markets to products that are effectively dumped into the world market. The damage done by export subsidies and policies that have similar effect has played a major role in trade negotiations. With the creation of WTO, agriculture was brought within the multilateral trade negotiating scene and pressure has grown for export subsidies to be reduced and eventually removed and for there to be greater access to developed country markets for produce from developing countries.

The largest volume of agricultural trade in the EU is between its member countries. Much of the external trade takes place between the US and the EU (Figure 3-15). Many EU agricultural imports, particularly from the US and Brazil, are feedstuffs for the livestock industry rather than finished products.

For the US the most important destinations for exports are its neighboring countries Mexico and Canada within the North American free trade area. Outside this free trade area Japan and the EU represent the major destinations for North American exports. China has markedly increased imports since 2002 and is expected to continue to do so in the future. A major development that may change the flow of exports from North America is the use of an increasing share of the US maize crop to produce bioethanol rather than entering the food chain.

There is a similar concentrated pattern for US imports (see figures 3-16 and 3-17) but here the EU has recently overtaken Canada as the largest supplier. Imports from Mexico have risen relatively rapidly as a result of the North American Free Trade Agreement. Among the four largest suppliers only Australia secures its market without subsidies or preferential access to the market.

Agricultural trade flows can act as catalysts for the diffusion of AKST to exporting countries. Importers may invest in production and processing activities that employ technologies developed within their own countries to meet market needs. As markets are established imported technol-

Table 3-12. **Trade in agricultural products (2003) (1000$US).**

	Imports	% World	Exports	% World
World	550,134,581	100	523,884,525	100
Russian Federation	10,993,983	2.0	2,339,450	0.4
North America Developed	67,686,614	12.2	79,902,492	15.3
EU (15) Excluding Intra-Trade	68,197,006	12.4	62,648,810	12.0

Source: FAO, 2008c.

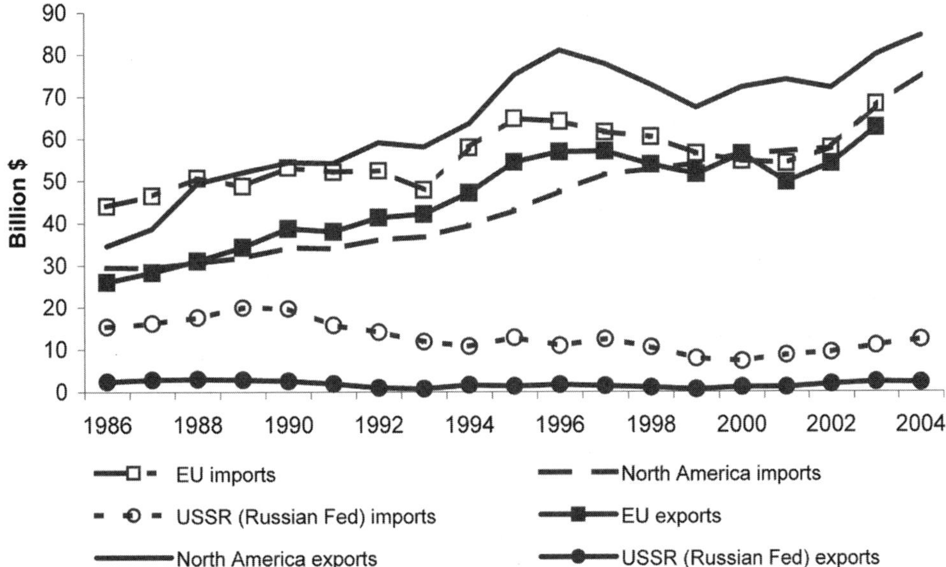

Figure 3-14. *Trade (imports and exports) in NAE from 1986-2004.* Source: FAO, 2008c

ogies can be adapted to local circumstances developing skills within the local community.

Trade also plays an important role in putting into practice public and private initiatives to encourage the development of agricultural knowledge, science and technology in the developing world. Private initiatives through the Ford, Rockefeller and Gates Foundations, for example, have supported research directed specifically at the problems of production in low-income countries. Many aid agencies such as Christian Aid, Oxfam, Farm Africa and World Vision have supported the development of education and the application of new technologies in farming. While the focus of much of this activity has been to improve the productivity of traditional farming activities in developing countries as production moves from local self-sufficiency to meet market needs whether at home or abroad, there is a need to employ technologies that cope both with the needs of storage and transport.

Much of the final value of agricultural products is embodied in processing. Imports of processed products have been increasing and this provides new opportunities for developing exporting countries that are able to access and use appropriate technology to meet the safety requirements of importing countries and respond to the needs of their retail-

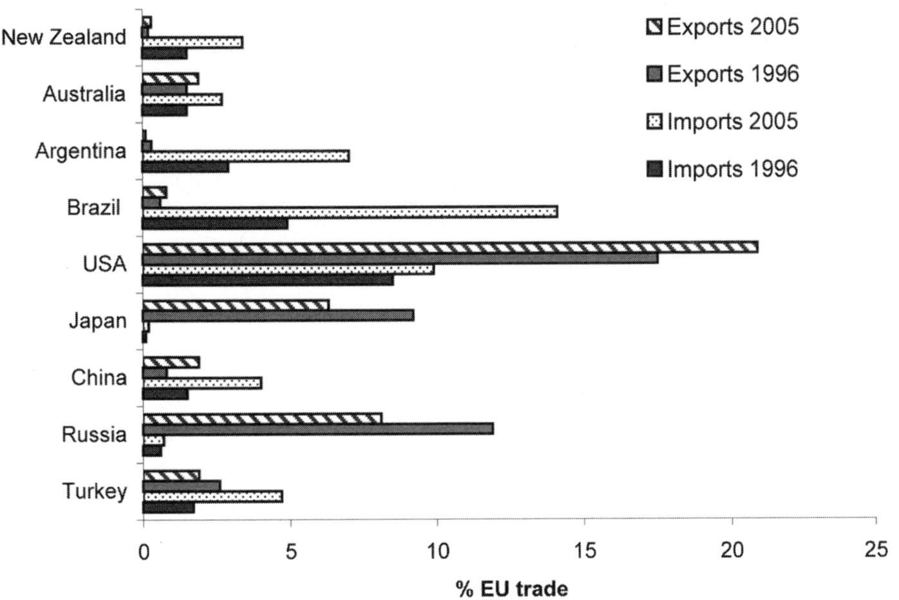

Figure 3-15. *EU Agricultural imports and exports in 1996 and 2005 as % total trade (extra EU trade only).* Source: European Commission 2000b, 2006b.

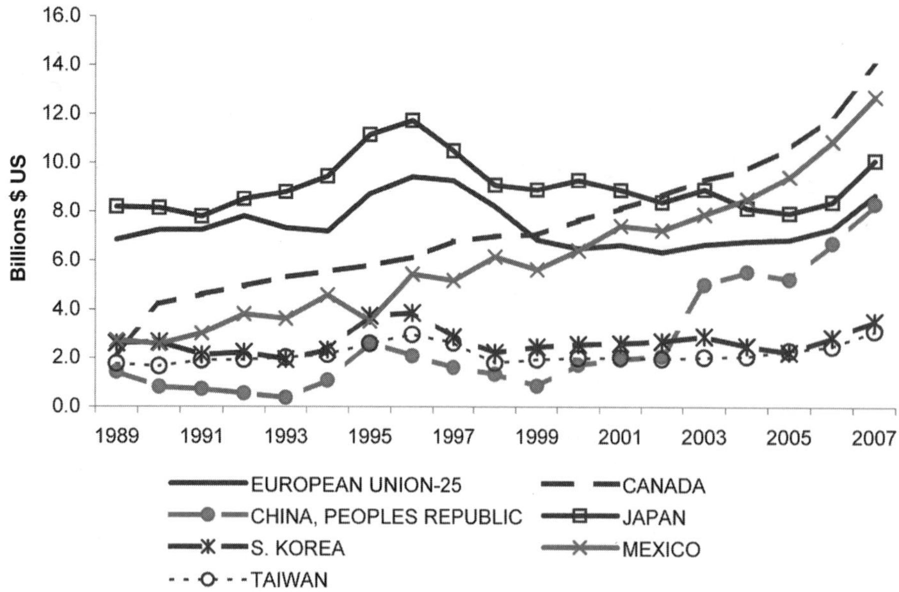

Figure 3-16. *US Exports Destinations from 1989-2007.* Source: USDA, 2008

ers and caterers. Production and transport is often organized by developed country suppliers who oversee production, handling and transport through to their final customers.

European livestock production and trade

For the past 30 years Europe has been producing far more meat and dairy products than it needs, becoming one of the world's leading exporters. Previously, the search for increased market share led to dumping of these products in less wealthy countries with consequent damage to the economic status of their agricultural producers. There are several well documented cases of disruption of and damage to, developing country agricultural markets as a result of this European strategy. As a result of rigorous CAP reforms in the 1990s, European production of beef and veal has fallen rapidly from around 50% overproduction (EU-15) in the 1990s to around 96% self-sufficiency in 2004 (Table 3-13).

The large increases in European livestock production between 1960 and 1990 relied heavily on animal feed imported from Brazil, Argentina, North America and Ukraine. In 2005 the EU 25 imported 30 million tonnes of animal feed, over half coming from Brazil and Argentina (data from Eurostat). Animal feed is the largest imported (aggregated) product for the EU-25 (European Commission, 2006b).

Pig meat is still being overproduced in EU-25 by about 8%, making Europe a net exporter of pig meat products, mainly to Russia and Japan. The EU is a net importer of sheep meat (EU-25 is only 78% self-sufficient in sheep and goat meat) and dairy products, mostly from New Zealand

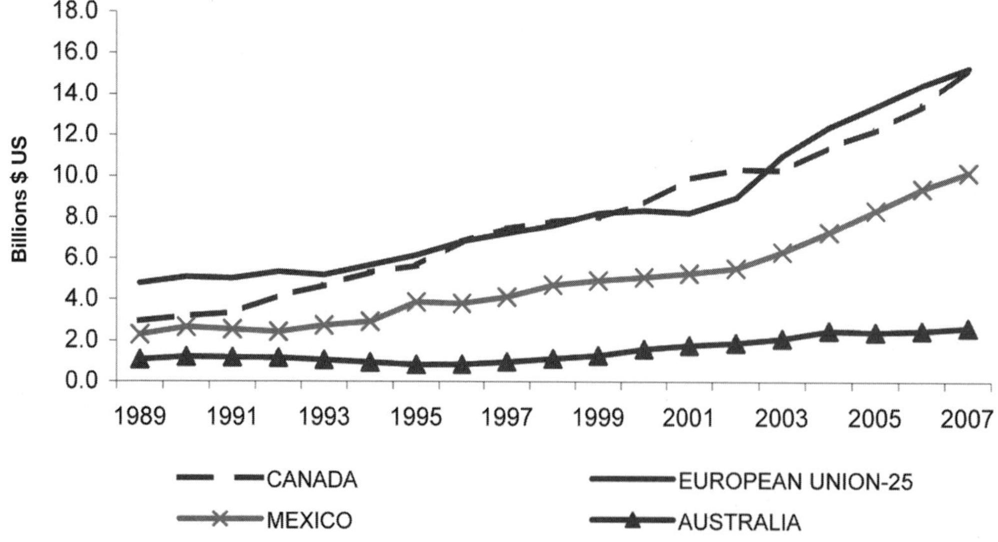

Figure 3-17. *US Imports of Agricultural Products 1989-2007.* Source: USDA, 2008

Table 3-13. **Net balance of external trade (EU) in meat products.**

Meat	Net balance (x 1000t)				% Self sufficiency			
	2000	2001	2002	2003	2000	2001	2002	2003
Pig meat	1,211	980	1,113	1,130	108.5	107.4	108.6	106.7
Beef/Veal	252	157	52	-89	102.4	112.4	99.9	96.2
Poultry meat	612	331	555	192	106.8	104.5	106.0	102.2
Sheep and goat meat	-271	-276	-280	-289	80.8	78.4	78.9	78.1
Equine meat	-43	-107	-83	-79	54.9	33.4	38.0	38.4
Other	-51	-79	-17	-8	95.0	92.1	98.2	99.1
Total	1,710	1,003	1,340	857	105.1	104.5	104.3	102.1
Edible Offals	349	331	413	420	118.5	117.3	122.2	122.5
Total	2,059	1,335	1,753	1,277	105.8	105.2	105.2	103.1

Source: European Commission, 2004.

and also imports large quantities of poultry meat from Brazil and Thailand, where production costs are much lower than in Europe. Somewhat perversely the EU also exports large quantities of poultry meat and offal to Russia and Ukraine and parts of the Middle East.

Next to India, the EU is the second largest producer of milk and milk products, exporting around 800,000 tonnes per year to a variety of global markets, including Africa (mainly Nigeria and Algeria), China, Russia and parts of the Middle East, especially Saudi Arabia. Exports of cheese and curd currently run at around 300,000 tonnes per year, going mainly to the US, Russia and Japan (Eurostat Agricultural Trade Statistics data).

The Common Agricultural Policy is moving away from production-led subsidies towards a more market-led and environmentally friendly system, but there is still a substantial subsidy paid to most participants in the livestock sector that reduces the competitiveness of developing countries.

References

Aase, J.K., and J.L. Pikul. 1995. Crop and soil response to long term tillage practices in the northern great plains. Agron. J. 87:652-656.

Alderman, D. J., and T. S. Hastings. 1998. Antibiotic use in aquaculture. Int. J. Food Sci. Tech. 33:139-155.

Allen, P. 1999. Reweaving the food security safety net: Mediating entitlement and entrepreneurship. Agric. Human Values 16:117-129.

Allen, P., and C.E. Sachs. 1991. The social side of sustainability: Class, and gender and race. Science Culture 13:569-590.

Arrouays, D., J. Balesdent, J.C. German, P.A. Jaouet, J.F. Soussana, and P. Stengel. 2002. Stocker du carbone dans les soles agricoles des France? (Increasing carbon stocks in French agricultural soils?). INRA, Paris.

Axelsson, A. 1998. The mechanisation of logging operations in Sweden and its effect on occupational safety and health. Int. J. Forest. Eng. 9(2):25-31.

Ayers, R.S., and D.W. Westcot. 1985. Water quality for agriculture. FAO Irrig. Drainage Pap. 29. FAO, Rome.

Baker, J.M., T.E. Ochsner, R.T. Venterea, and T.J. Griffis. 2007. Tillage and soil carbon sequestration—what do we really know. Agric. Ecosyst. Environ. 118:1-5.

Barham, E. 2003. Translating terroir: The global challenge of French AOC labeling. J. Rural Studies 19(1):127-138.

Barrow, C.J. 1994. Land degradation. Cambridge Univ. Press, Cambridge.

Barthelemy, P.A. 2007. Agriculture and the rural environment: Changes in agricultural employment. Available at http://ec.europa.eu/agriculture/envir/report/en/emplo_en/report_en.htm. EC, Brussels.

Baucom, R.S., and R. Mauricio. 2004. Fitness costs and benefits of novel herbicide tolerance in a noxious weed. PNAS 101(36): 13386-13390.

Bellows, A., and W.B. Hamm. 2001. Local autonomy and sustainable development: Testing import substitution in more localized food systems. Agric. Human Values 18: 71-84.

Biesmeijer, J.C., S.P.M. Roberts, M. Reemer, R. Ohlenmuller, M. Edwards, T. Peeters, et al. 2006. Parallel declines in pollinators and insect-pollinated plants in Britain and the Netherlands. Science 313:351-354.

Birdlife International. 2007. How do we know that a forest is doing FINE? Forest Indicators: What do they tell us? The European Forest Task Force Annual Workshop, Poland, 2007. Available at http://www.birdlife.org/action/change/europe/forest_task_force/index.html.

Blokhuis, H., B. Jones, R. Geers, M. Miele, and I. Veissier. 2003. Measuring and monitoring animal welfare: Transparency in the product quality chain. Animal Welfare 12:445-455.

Boatman, N.D., N.W. Brickle, J.D. Hart, T.P. Milsom, A.J. Morris, A.W.A. Murray, et al. 2004. Evidence for the indirect effects of pesticides on farmland birds. IBIS 146 (suppl 2):131-143.

Borzutzky, S., and E. Kranidis. 2005. A struggle for survival: The Polish agricultural sector from Communism to EU Accession East Europe. Politics Societies 19:614-654.

Bosworth, D.N. 2004. Four threats to the Nation's forests and grasslands speech. Idaho Environ. Forum, Boise, ID.

Bouma, J., G. Varallyay, and N.H. Batjes. 1998. Principal land use changes anticipated in

Europe. Agric. Ecosyst. Environ. 67: 103-119.
Bradbury, R.B., and W.B. Kirby. 2006. Farmland birds and resource protection in the UK: cross-cutting solutions for multi-functional farming. Biological Conserv. 129:530-542.
Brimner, T.A., G.J. Gallivan, and G.R. Stephenson. 2005. Influence of herbicide-resistant canola on the environmental impact of weed management. Pest Manage. Sci. 61:47-52.
Brookes, G., and P. Barfoot. 2005. GM crops: the global economic and environmental impact—the first nine years. AgBioForum 8:187-196.
Bruce, R.R., G.W. Langdale, and A.L. Dillard. 1990. Tillage and crop rotation effect on characteristics of a sandy surface soil. Soil Sci. Soc. Am. J. 54:1744-1747.
Canonico, G.C., A. Arthington, et al. 2005. The effects of introduced tilapias on native biodiversity. Aquatic Conserv.: Marine Freshwater Ecosyst. 15(5):463-483.
Carlsson-Kanyama, A., Å. Sundkvist, and C. Wallgren. 2004. Lokala livsmedelsmarknader—en fallstudie. (Local food markets—a case study). Rapport nr. 2. Centrum för miljöstrategisk forskning, KTH, Stockholm.
Carter, D.L. 1986. Effects of erosion on soil productivity. p. 1131-1138. *In* Proc. Water Forum '86. Long Beach, 4-6 Aug 1986. Vol. 2 Am. Soc. Civil. Eng. New York.
Carter, D.L., R.B. Berg, and B.J. Sanders. 1985. The effect of furrow irrigation erosion on crop productivity. Soil Sci. Soc. Am. J. 49:207-211.
CEC. 2002. Towards a thematic strategy for soil protection. Communication from the Commission to the Council, the European Parliament, the Economic and Social Committee and the Committee of the Regions. Brussels 16.4.2002. COM(2002)179 final. Available at http://europa.eu.int/eur-lex/en/com/pdf/2002/com2002_0179en01.pdf.
Choi, S.M., S.D. Yoo, and B.M. Lee. 2004. Toxicological characteristics of endocrine-disrupting chemicals: Developing toxicity, carcinogenicity and mutagenicity. J. Toxicol. Environ. Health Part B 7:1-32.
Cohen, M.L., and R.V. Tauxe. 1986. Drug-resistant Salmonella in the United States: An epidemiologic perspective: Science 234:964-969.
Compassion in World Farming. 2007. Intensive farming and the welfare of farm animals. Available at http://www.ciwf.org.uk/education/resources/intensive_farming.html.
Cowell, S.J., and S. Parkinson. 2003. Localisation of UK food production: an analysis using land area and energy indicators. Agric. Ecosyst. Environ. 94(2):221-236.
Crooks, A. 2000. Consolidation in the heartland. Rural Cooperatives 67(6):14-19.
Culshaw, D., and B.Stokes. 1995. Mechanisation of short rotation forestry. Biomass Bioenergy 9:1,127-140.
Defra. 2004a. Family food in 2002-2003. Defra, London.
Defra. 2004b. New Animal Welfare Bill. Available at http://www.defra.gov.uk/news/issues/2004/animal07a.htm. Defra, London.
Defra 2004c. Animal Health and welfare: Bovine spongiform encephalopathy, financial costs. Available at http://www.defra.gov.uk/animalh/bse/general/qa/section9.html.
Defra. 2006a. The costs to the UK of BSE measures. Available at http://www.defra.gov.uk/animalh/bse/general/qa/section9.html. Defra, London.
Defra. 2006b. Labour statistics for England. Available at http://farmstats.defra.gov.uk/cs/farmstats_data/DATA/historical_data/hist_pub_results.asp. Defra, London.
Defra. 2007a. Farm diversification. Jan 2007. Available at http://statistics.defra.gov.uk/esg/statnot/Diver07.pdf. Defra, London.
Defra. 2007b. Animal health and welfare: breakdown of bovine TB expenditure in Great Britain: 1997-2006/07 (£m). Available at http://www.defra.gov.uk/animalh/tb/stats/expenditure.htm. Defra, London.
Deutsch, L. 2004. Global trade, food production and ecosystem support: Making the interactions visible. PhD thesis. Dep. Systems Ecol., Stockholm Univ., Sweden.
Dimitri, C., A. Effland, and N. Conklin. 2005. The 20th century transformation of US agriculture and farm policy. EIB3. June 2005. USDA, Washington, DC.
Dimitri, C., and C. Greene. 2002. Recent growth patterns in the US organic foods market. AIB777. USDA, Washington, DC.
Dively, G.P., R. Rose, M.K. Sears, R.L. Hellmich, D.E. Stanley-Horn et al. 2004. Effects on monarch butterfly larvae (Lepidoptera: Danaidae) after continuous exposure to Cry1Ab-expressing corn during anthesis. Environ. Entomol. 33(4):1116-1125.
Donald, P.F., G. Pisano, M.D. Rayment, and D.J. Pain. 2002. The Common Agricultural Policy, EU enlargement and the conservation of Europe's farmland birds. Agric. Ecosyst. Environ. 89(3):167-182.
Drury, C.C, C.S. Tan, J.D. Gaynor, T.O. Oloya, and T.W. Welacky. 1996. Influence of controlled drainage-subirrigation on surface and tile drainage nitrate loss. J. Environ. Qual. 25:317-324.
Duffy, R. 2005. Meeting consumer demands through effective supply chain linkages. Stewart Postharvest Rev. 1(1) June 2005.
English Food and Farming Partnership. 2006. Farmer controlled businesses and agricultural output. Available at www.effp.com.
European Commission. 2000a. Agriculture in the European Union—Economic information 2000: agricultural products sold through cooperatives (1997). Available at http://ec.europa.eu/agriculture/agrista/2000/table_en/en3561.pdf. EC, Brussels.
European Commission. 2000b. Agricultural in the European Union—Economic information 2000: EU imports and exports of agricultural products to and from various groups of countries. Available at http://ec.europa.eu/agriculture/agrista/2000/table_en/en37.htm. EC, Brussels.
European Commission. 2002a. Status report on the European Commission's work in the field of nutrition in Europe, http://ec.europa.eu/health/ph-determinants/life-style/nutrition/documents/nutrition-report-en.pdf
European Commission. 2002b. European Commission farm animal welfare—Current research and future directions. Available at http://ec.europa.eu/research/agriculture/pdf/animal-welfare_en.pdf. EC, Brussels.
European Commission. 2004. Agricultural in the European Union—Economic information 2004: meat in general—net balance of external trade (EU) in meat products. Available at http://ec.europa.eu/agriculture/agrista/2004/table_en/41411.pdf. EC, Brussels.
European Commission. 2006a. Rural development in the European Union: Statistical and economic information 2006. EU Directorate for agriculture and rural development. Available at http://ec.europa.eu/agriculture/agrista/rurdev2006/index_en.htm. EC, Brussels.
European Commission. 2006b. Agricultural in the European Union: Economic information 2006. Available at http://ec.europa.eu/agriculture/agrista/2006/table_en/index.htm. EC, Brussels.
F3. 2007. F3: Enabling local and sustainable food initiatives. Available at http://www.localfood.org.uk/.
FAO 2003. Report of the FAO expert consultation on environmental effects of genetically modified crops, Rome, 16-18 June 2003. Available at http://www.fao.org/DOCREP/FIELD/006/AD690E/AD690E00.HTM.
FAO. 2004. FAO statistical yearbook 2004. Available at http://www.fao.org/es/ess/yearbook/vol_1_1/site_en.asp?page=welfare. FAO, Rome.
FAO. 2005. Agriculture commodity prices continue long-term decline. Available at http://www.fao.org/newsroom/en/news/2005/89721/index.html. FAO, Rome.
FAO. 2006a. The state of food insecurity in the world 2006. Available at www.fao.org/icatalog/inter-e.htm. FAO, Rome.
FAO. 2006b. State of world aquaculture: chapter 5: Resource use and environment. Fisheries Tech. Pap. 500. FAO, Rome.
FAO. 2008a. World produce consumer prices: crops and livestock. Available at http://faostat.fao.org/site/426/default.aspx (crops) http://faostat.fao.org/site/427/default.aspx (livestock). FAO, Rome.
FAO. 2008b. World population estimates: Available at http://faostat.fao.org/site/429/default.aspx. FAO, Rome.
FAO. 2008c. World trade data: Crops & livestock primary and processed. Available at http://faostat.fao.org/site/412/default.aspx. FAO, Rome.

FAO/IAEA. 2006. Animal Production and Health (APH). Nuclear techniques in food and agriculture. Available at http://www-naweb .iaea.org/nafa/questions/1quest-aph.html. FAO, Rome.

FAOSTAT. 2006 data. Available at http:// faostat.fao.org. FAO, Rome.

Farrell, A.E., R.J. Plevin, B.T. Turner, A.D. Jones, M.O'Hare, and D.M. Kammen. 2006. Ethanol can contribute to energy and environmental goals. Science 311:506-508.

Fausey, N. 2005. Drainage management for humid regions. Int. Agric. Engineer. J. 14:209-214.

Fernandez-Cornejo, J., M. Caswell, L. Mitchell, E. Golan, and F. Kuchler. 2006. The first decade of genetically engineered crops in the United States Available at http://www.ers. usda.gov/Publications/EIB11/. ERS, USDA, Washington, DC.

Fernandez-Cornejo, J., and W.D. McBride. 2002. Genetically engineered crops: U.S. adoption and impacts. Agric. Outlook 294:24-27.

Ferro-Luzzi, A., and P. James. 1997. Diet and health. p. 2-38. *In* Nutrition in Europe. European Parliament Scientific and Technological Options Assessment. Directorate-general for Research, PE n° 166-481.

Fischer-Boel, M. 2007. Farmers will have to get second jobs. EU Commission Jan 2007 Available at http://www.eubusiness.com/ Agri/eu-farmers.73/.

Flowers, M., R. Weisz, R. Heiniger, D. Osmond, and C. Crozier. 2004. In-season optimization and site-specific nitrogen management for soft red winter wheat. Agronomy J. 96: 124-134.

Franzluebbers, A.J., F.M. Hons, and D.A. Zuberer. 1994. Long-term changes in soil carbon and nitrogen pools in wheat management systems. Soil Sci. Soc. Am. J. 58:1639-1645.

Garcia, A., and J.P. Massam. 2005. Elimination of antibiotics in hatcheries while improving production by use of probiotics. World Aquacult. 36(1):57-60.

GLASOD. 1992. World map of the status of human-induced soil degradation. UNEP and Winand Staring Centre, Wageningen.

Glencross, B., W. Hawkins, and J.G. Curnow. 2003. Evaluation of canola oils as alternative lipid resources in diets for juvenile red seabream, *Pagrus auratus*. Aquacult. Nutr. 9(5):305-315.

Goldschmidt, W. 1978. Large-scale farming and the rural social structure. Rural Sociol. 43:362-366.

Granstedt, A., O. Thomsson, and T. Schneider. 2005. Environmental impacts of eco-local food systems. Final Rep. BERAS Work Package 2. Baltic Ecological Recycling Agriculture and Society (BERAS) Nr. 5. Ecol. Agric. nr 46. Centre for Sustainable Agric., Swedish Univ. Agric. Sciences.

Grant, R., K. Nielsen, and J. Waagpetersen, 2006. Reducing nitrogen loading of inland and marine waters—evaluation of Danish policy measure to reduce nitrogen loss from farmland. Ambio 35:117-123.

Gratan, S.R., and C.M. Grieve. 1999. Salinity-mineral nutrient relations in horticultural crops. Sci. Hort. 78:127-157.

Gray, K.A., L. Zhao, and M. Emptage. 2006. Bioethanol: Current opinion. Chem. Biol. 10:141-146.

Green, R.E., S.J. Cornell, J.P.W. Scharlemann, and A. Balmford. 2005. Farming and the fate of wild nature. Science 307:550-555.

Gyllenhammar, A., and L. Hakanson. 2005. Environmental consequence analysis of fish farm emissions related to different scales and exemplified by data from the Baltic—A review. Marine Environ. Res. 60(2):211-243.

Hakansson, I., and W.B. Voorhees. 1998. Soil compaction. p.167-179. *In* R. Lal et al. (ed), Methods for assessment of soil degradation. Adv. Soil Sci. CRC Press, Madison.

Hart J.D., T.P. Milsom, G. Fisher, V. Wilkins, S.J. Moreby, A.W.A. Murray, and P.A. Robertson. 2006. The relationship between yellowhammer breeding performance, arthropod abundance and insecticide applications on arable farmland. J. Appl. Ecol. 43(1):81-91.

Hart, M.R., B.F. Quinn, and M.L. Nguyen. 2004. Phosphorus runoff from agricultural land and direct fertilizer effects: a review: J. Environ. Qual. 33:1954-1972.

Hartley, M.J., 2002. Rationale and methods for conserving biodiversity in plantation forests. Forest Ecol. Manage. 155:81-95.

Havlin, J.L., D.E. Kissel, LD. Maddux, M.M. Claasen, and J.H. Long. 1990. Crop rotation and tillage effects on soil organic carbon and nitrogen. Soil Sci. Soc. Am. J. 54:448-452.

Hayes, T., K. Haston, M. Tsui, A. Hoang, C. Haeffele and A. Vonk. 2002. Feminisation of male frogs in the wild. Nature 419:895-896.

Hayes, T.B., P. Case, S. Chui, D. Chung, C.K. Haeffele, et al. 2006. Pesticide mixtures, endocrine disruption and amphibian declines: are we underestimating the impact? Environ. Health Perspect. 114:40-50.

Heady, H.F., and D. Child. 1994. Rangeland ecology and management. Westview Press, Boulder.

Heard, M.S., C. Hawes, G.T. Champion, S.J. Clark, L.G. Firbank, A.J. Haughton, et al. 2003. Weeds in fields with contrasting conventional and genetically modified herbicide-tolerant crops. I. Effects on abundance and diversity Phil. Trans. R. Soc. London B 358(1439):1819-1832.

Heffernan, J.B., and W.D. Heffernan. 1986. When families have to give up farming. Rural Dev. Perspect. 2(3):28-31.

Hendrickson, M., W.D. Heffernan, P.H. Howard, and J.B. Heffernan. 2001. Consolidation in food retailing and dairy. Brit. Food J. 3(10):715-728.

Herzog, F., B. Steiner, D. Bailey, J. Baudry, R. Billeter, R. Bukasek, et al. 2006. Assessing the intensity of temperate European agriculture at the landscape scale. Eur. J. Agron. 24(2):165-181.

Heuch, P.A., P.A. Bjorn, B. Finstad, J.C. Holst, L. Asplin, and F. Nilsen. 2005. A review of the Norwegian national action plan on salmon lice on salmonids: The effect on wild salmonids. Aquaculture 246(1-4):79-92.

Hillel, D., and P. Vlek. 2005. The sustainability of irrigation. Adv. Agron. 87:55-84.

Hinga, K.R., A. Batchelor, M.T. Ahmed, O. Osibanjo, N. Lewis, and M. Pilson. 2006. Waste processing and detoxification. Chapter 15. p. 417-439. *In* R. Hassan et al. (ed), Ecosystems and human well-being: Current state and trends. Millennium Ecosystem Assessment, Vol. 1. Island Press, Washington, DC.

Hinrichs, C. 2003. The practice and politics of food system localization. J. Rural Studies 19:33-45.

Hinrichs, C., and K. Kremer. 2002. Social inclusion in a Midwestern local food system project. J. Poverty 6(1):65-90.

Higgs, D.A., S.K. Balfry, J. Oakes, M. Rowshandeli, and G. Deacon. 2006. Efficacy of an equal blend of canola oil and poultry fat as an alternate dietary lipid source for Atlantic salmon (*Salmo salar* L.) in sea water. I: Effects on growth performance, and whole body and fillet proximate and lipid composition. Aquacult. Res. 37(2):180-191.

Hoffman, G.J., T.A. Howell, and K.H. Solomon. 1990. Management of farm irrigation systems. p. 667-715. *In* G.J. Hoffman et al. (ed), Salinity management. ASCE, St. Joseph MI.

Holloway, L., and M. Kneafsey (ed). 2004. Geographies of rural cultures and societies. Ashgate, Aldershot.

Hooda, P.S., A.C .Edwards, H.A. Anderson and A. Miller. 2000. A review of water quality concerns in livestock production. Sci. Total Environ. 250:143-167.

Hornibrook, S., and A. Fearn. 2005 Demand driven supply chains: Contractual relationships and the management of perceived risks. 2nd Eur. Forum Market Driven Supply Chains. Milan, April 2005. Available at http://www.kent.ac.uk/kbs/cscr/ pdf/Demand%20driven%20chains%20 EIASM%20conference%202005.pdf.

Humphrey, J.W., R. Ferris, M.R. Jukes, and A.J. Peace. 2002. The potential contribution of conifers plantations to the UK Biodiversity Action Plan. Bot. J. Scotland 54:49-62.

Ikerd, J. 2002. Sustainable Farming: Reconnecting with Consumers. Agric. Conf. 2002—Changing times: Creating opportunities in agriculture. Agric. Leadership Foundation of Hawaii. Honolulu, Hawaii, 24 Oct 2002. Available at http://www.ssu.missouri.edu/faculty/ jikerd/papers/HawaiiSA.html#_ftn1 (15-Mar-2004).

International Obesity Task Force. 2005. Obesity in Europe. EU Platform Briefing Paper Mar 2005. Available at http://www.iaso.org/popout.asp?linkto=http%3A//www.iotf.org/media/euobesity3.pdf.

Jahnke, H.E. 1982. Livestock production systems and livestock development in tropical Africa. Kieler Wissenschaftsverlag Vauk, Kiel.

Johnson, B., and A. Hope. 2005. Productivity, biodiversity and sustainability. p. 335-350. *In* R. Sylvester-Bradley and J. Wiseman (ed), Yields of farmed species: Constraints and opportunities in the 21st century. Nottingham Univ. Press, UK

Jones, A. 2001. Eating oil. Food supply in a changing climate. Sustain, London.

Jones, A. 2002. An environmental assessment of food supply chains: a case study on dessert apples. Environ. Manage. 30(4):560-576.

Kaiser, M. 2005. Assessing ethics and animal welfare in animal biotechnology for farm production. Rev. Sci. Tech. Off. Int. des Epizooties 24(1):75-87.

Kapur, S., and E. Akca. 2002. Global assessment of degradation. p. 296-306. *In* R. Lal (ed), Encyclopedia of soil science. Marcel Dekker, NY.

Kinderlerer, J., and M. Adcock. 2003. Agricultural biotechnology, politics, ethics and policy, Working Pap. 3. Regional policy dialogue on biotechnology, agriculture and food security in Southern Africa. FANRPAN/IFPRI, Johannesburg.

Kinsey, J., and B. Senauer. 1996. Consumer trends and changing food retailing formats. Am. J. Agric. Econ. 78:1187-1191.

Kirkby, M.J., R.J.A. Jones, B. Irvine, A. Gobin, G. Govers, O. Cerdan, et al. 2004. Pan-European soil erosion risk assessment: The Pesera map, Vers. 1 Oct 2003. Explanation of Spec. Publ. Ispra 2004 No. 73. (S.P.I. 04.73). Eur. Soil Bureau Res. Rep. 16. EUR 21176. Official Publ. Eur. Communities, Luxembourg.

Kirkpatrick, C., C. George, and S.S. Scrieciu. 2004. The implications of trade and investment liberalization for sustainable development: review of literature. Final report. DEFRA, London.

Kogan, M. 1998. Integrated pest management: Historical perspectives and contemporary developments. Ann. Rev. Entomol. 43: 243-270.

Kovacic, D.A., M.B. David, L.E. Gentry, K.M. Starks, and R.A. Cooke. 2000. Effectiveness of constructed wetlands in reducing nitrogen and phosphorus export from agricultural tile drainage. J. Environ. Qual. 29:1262-1274.

Kraenzle, C.A. 1998. Co-ops break supply sales record. Rural Cooperatives 65(6).

Kuipers, A., and F. Mandersloot. 1999. Reducing nutrient losses on dairy farms in The Netherlands. Livestock Production Sci. 61:139-144.

Lal, R. 1990. Soil erosion and land degradation: Their global risks. p. 129-172. *In* R. Lal and B.A. Stewart (ed), Soil degradation. Adv. Soil Sci. Springer-Verlag, NY.

Lal, R., J.M. Kimble, R.F. Follett, and C.V. Cole. 1998. The potential of U.S. cropland to sequester carbon and mitigate the greenhouse effect. Ann Arbor Press, Chelsea.

Lal, R., and D. Pimentel. 2007. Biofuels from crop residues: Soil Tillage Res. 93:237-238.

Larson, A. 2002. Remarks of U.S. Secretary of State Alan Larson before the CATO Institute. 25 Sep 2002.

Lavelle, P., R. Dugdale, R. Scholes, A.S. Berhe, E. Carpenter, L. Codispoti et al. 2005. Nutrient cycling. p. 331-353. *In* R. Hassan et al. (ed), Ecosystems and human well-being. Current state and trends. Millennium Ecosystem Assessment, v. 1. Island Press, Washington, DC.

Lemly, A.D., R.T. Kingsford, and J.R. Thompson. 2000. Irrigated agriculture and wildlife conservation: conflict on a global scale. Environ. Manage. 25:485-512.

Lentz, R.D., and R.E. Sojka. 1994. Field results using polyacrylamide to manage furrow erosion and infiltration. Soil Sci. 158: 274-282.

Lentz, R.D., R.E. Sojka, and C.W. Robbins. 1998. Reducing phosphorus losses from surface-irrigated fields: Emerging polyacrylamide technology. J. Environ. Qual. 27:305-312.

Lerat, S., L.S. England, M. Vincent, K.P. Pauls, C.J. Swanton, J.N. Klironomos, and J.T. Trevors. 2005. Real-time polymerase Lemly chain reaction (PCR) quantification of transgenes for Roundup Ready corn and Roundup Ready soybean in soil samples. J. Agric. Food Chem. 53:1337-1342.

Letey, J. Jr., C. Roberts, M. Penberth, and C. Vasek. 1986. An agricultural dilemma: Drainage water and toxics disposal in the San Joaquin Valley. Univ. Calif. Agric. Exp. Stn. Spec. Publ. 3319.

Lindahl, O., B. Hart, B. Hernroth, S. Kollberg, L-O. Loo, F. Noren et al. 2005. Improving marine water quality by mussel farming: A profitable solution for Swedish society. Ambio 34(2):131-138.

Lipton, M. 2005. The family farm in a globalizing world: The role of crop science in alleviating poverty. 2020 Disc Pap. 40. IFPRI, Washington, DC.

Lobao, L. 1990. Locality and inequality. State Univ. New York Press, Albany.

Lorenzen, A., R. Chapman, J.G. Hendel, and E.R. Topp. 2005. Persistence and pathways of testosterone dissipation in agricultural soil. J. Environ. Qual. 34:854-860.

MA. 2005. Millennium Ecosystem Assessment 2005. Available at http://www.maweb.org/en/index.aspx. Island Press, Washington, DC.

MAFF. 2000. Climate change and agriculture in the United Kingdom. Available at http://www.defra.gov.uk/environ/climate/climatechange/index.htm. Defra, London.

Makker, H.P.S.,and G.J. Viljoen (ed). 2006. Applications of gene-based technologies for improving animal production and health in developing countries. Springer, NY.

Marquiss, M., and R. Rae. 1994. Seasonal trends in abundance, diet and breeding of common crossbills (*Loxia curvirostra*) in an area of mixed species conifer plantation following the 1990 crossbill "irruption". Forestry 67(1):31-47.

Marsden, T. 1997. Creating space for food: The distinctiveness of recent agrarian development. *In* D. Goodman and M. Watts (ed), Globalising food: Agrarian questions and global reconstructing. Routledge, NY.

Marsden, T., J. Murdoch, and K. Morgan. 1999. Sustainable agriculture, food supply, chains and regional development. Editorial introduction. Int. Planning Studies 4:295.

Marvier, M., C. McCreedy, J. Regetz, and P. Kareiva. 2007. A meta-analysis of effects of Bt cotton and maize on nontarget invertebrates. Science 316:1475-1477.

McAuley, T. 2004. The food chain. Available at http://www.cfo.com/article.cfm/3014279/c_2984419/?f=archives. CFO Europe. Aug 2004.

Mellon, M. 2000. Europe just says no. Nucleus 21(4). Available at http://go.ucsusa.org/publications/nucleus.cfm?publicationID=188. Union Concerned Scientists, Boston.

Micklin, P. 1994. The Aral sea problem. Proc. Inst. Civil Engineers 102:114-121.

Micklin, P. 2006. The Aral sea crisis and its future: an assessment in 2006. Eurasian Geograph. Econ. 47:546-567.

Miele, M., and V. Parisi. 2001. L'Etica del Mangiare, i valori e le preoccupazioni dei consumatori per il benessere animale negli allevamenti: un'applicazione dell'analisi Means-end Chain. Riv. Econ Agraria LVI(1):81-103.

Miele, M., and D. Pinducciu. 2001. A market for nature: Linking the production and consumption of organics in Tuscany. J. Environ. Policy Planning 3:149-162.

Miles, D.L. 1977. Salinity in the Arkansas Valley of Colorado. EPA-AIC-D4-0544. USEPA, Denver.

Miljkovic, D. 2005. Measuring and causes of inequality in farm sizes in the United States. Agric. Econ. 33:21-27.

Miller, W.P., and D.E. Radcliffe. 1992. Soil crusting in the Southeastern United States. p. 233-266. *In* M.E. Sumner and B.A. Stewart (ed), Soil crusting, chemical and physical processes. Adv. Soil Sci. Lewis Publ., Madison.

Montero, D., T. Kalinowski, A. Obach, L. Robaina, L. Tort, M.J. Caballero et al. 2003. Vegetable lipid sources for gilthead seabream (Sparus aurata): Effects on fish health. Aquaculture 225:353-370.

Moore, J., and J.V. Hefner. 1977. Irrigation with saline water in the Pecos Valley of West Texas. p. 339-394. *In* Proc. Int. Salinity Conf. Managing Saline Water for Irrigation. Texas Tech. Univ., Lubbock.

Morgan, K., T. Marsden, and J. Murdoch. 2006. Worlds of food: Place, power and provenance in the food chain. Oxford Univ. Press, UK.

Murdoch, J., and M. Miele. 2004. A new aesthetic of food? Relational reflexivity in the "alternative" food movement. *In* M. Harvey et al. (ed), Qualities of food. Manchester Univ. Press, UK.

Naylor, R.L., R.J. Goldburg, J.H. Primavera, N. Kautsky, M. Beveridge, J. Clay, et al. 2000. Effect of aquaculture on world fish supplies. Nature 405:1017-1024.

Naylor, R.L., K. Hindar, A. Cooper, J. Eagle, I. Fleming, R. Goldburg, et al. 2005. Fugitive salmon: Assessing the risks of escaped fish from net-pen aquaculture. BioScience 55(5):427-437.

Ng, H.Y.F., C.C. Tan, C.F. Drury, and J.D. Gaynor. 2002. Controlled drainage and subirrigation influences tile nitrate loss and corn yields in a sandy loam soil in southwestern Ontario: Agric. Ecosyst. Environ. 90:81-88.

Nguyen, M.L., B.F. Quinn, and J.P.S. Sukias. 2002. Potential losses of phosphorus and nitrogen in runoff and drainage from pastoral soils applied with superphosphate and reactive phosphate rock. p. 137-153. *In* L.D. Currie and P. Loganathan (ed), Dairy farm soil management. Occas. Rep. 15. Fertilizer and Lime Res. Centre., Massey Univ., Palmerston North, New Zealand.

Nichols, D.J., T.C. Daniel, and D.R. Edwards, 1994, Nutrient runoff from pasture after incorporation of poultry litter or inorganic fertilizer: Soil Sci. Soc. Am. J. 58:1224-1228.

NRC. 2000, Clean coastal waters: Understanding and reducing the effects of nutrient pollution. Nat. Res. Council, Ocean Studies Board, Washington, DC.

Nuffield Council on Biotethics. 1999. Genetically modified crops: The ethical and social issues. Nuffield Council on Bioethics, London.

OECD. 1996. Better policies for rural development. OECD, Paris.

OECD. 2004. Agricultural policies in OECD Countries: At a glance. Producer and consumer support estimates, OECD database 1986-2003. Available from http://www.oecd .org/document/58/0,3343,en_2649_37401_32 264698_1_1_1_37401,00.html. OCED, Paris.

OECD. 2006. Health data: Obesity, percentage of adult population with a BMI>30 kg/m[2]. Available at http://www.oecd.org/dataoecd/7/38/35530193.xls. OECD, Paris.

Oenema, O., L. van Liere, and O. Schoumans. 2005. Effects of lowering nitrogen and phosphorus surpluses in agriculture on the quality of groundwater and surface water in the Netherlands. J. Hydrol. 304:289-301.

Orr, B. 2006. Defining rangeland management. A comparison of three textbooks. Rangelands West, www.rangelandswest.org.

Pardey, P.G., and N.M. Beintema. 2001. Slow magic. Agricultural R&D a century after Mendel. IFPRI, Washington, DC.

Petit, M. 1997. Western European beliefs and values regarding government intervention in agriculture and food prices. Searching for common ground. EU Enlargement and Agricultural Policy. Agric. Policy and Econ. Dev. Ser. FAO Rome.

Pimentel, D., and T.W. Patzek. 2005. Ethanol production using corn, switchgrass, and wood; Biodiesel production using soybean and sunflower: Natural Resourc. Res. 14:65-76.

Pirog, R., T. van Pelt, K. Enshayan, and E. Cook. 2001. Food, fuel and freeways. Leopold Center Sustainable Agric., Iowa State Univ., Ames.

Poikolainen, K. 2004. Vihannesomavaraisuuden vaikutus kuljetusten energiankulutukseen ja päästöihin. (The impact of self-sufficiency of vegetables on energy use and emissions of transport). Master's thesis. Univ. Helsinki, Finland.

Powers, R.F., D.A. Scott, F.G. Sanchez, R.A. Voldseth, D. Page-Dumroese, J.D. Elliott, and D.M. Stone. 2005. The N. American long-term soil productivity experiment: Findings from the first decade. Forest Ecol. Manage. 220:31-50.

Prather, M., D. Ehhalt, F. Dentener, R. Derwent, et al. 2001. Atmospheric chemistry and greenhouse gases. p. 239-287. *In* J.T. Houghton et al. (ed), Climate change 2001: The scientific basis. Contribution of working group I to the third assessment report of the Intergovernmental Panel on Climate Change. Cambridge Univ. Press, Cambridge.

Preston C.D., D.A. Pearman, and T.D. Dines. 2002. New atlas of the British and Irish flora. Oxford Univ. Press, Oxford.

Pretty, J., A.S. Ball, T. Lang, and N. Morison. 2005. Farm costs and food miles: An assessment of the full cost of the UK weekly food basket. Food Policy 30:1-19.

Pretty, J.N., A.D. Noble, D. Bossio, J. Dixon, et al. 2006. Resource-conserving agriculture increases yields in developing countries. Environ. Sci. Tech. 40(4):1114-1119

Ramsey, D., and S.-T. Yang. 2004. Production of butyric acid and butanol from biomass. US Dep. Energy, Washington, DC.

Rands, M.R.W. 1986. The survival of gamebird (*Galliformes*) chicks in relation to pesticide use on cereals. IBIS 128(1):57-64.

Rangasamy, P. 1997. Sodic soils. p. 265-277. *In* R. Lal et al. (ed), Methods of assessment of soil degradation. CRC Press, New York.

Reed, A., H. Elitzak, and M. Wohlgenant. 2002. Retail-farm price margins and consumer product diversity. Tech. Bull. 1899. ERS, USDA, Washington, DC.

Rees, H.W., T.L. Chow, R.J. Long, J. Lavoie, and J.O. Monteith. 2002. Hay mulching to reduce hay mulching and soil loss under intensive potato production in northwestern New Brunswick, Canada. Can. J. Soil Sci. 82:249-258.

Reeves, D.W., and C.W. Wood. 1994. A sustainable winter-legume conservation tillage system for maize: Effects on soil quality. p. 1011-1016. *In* Proc. 13th Int. Conf. Int. Soil Tillage Res. Org. (ISTRO). Vol. II. Roy. Vet. Agric. Univ. and the Danish Inst. Plant Soil Sci., Aalborg, Denmark, 24-29 Jul 1994.

Regmi, A., and M. Gehlhar. 2005. New directions in global food markets. AIB794. Available at www.ers.usda.gov/publications/aib794/. USDA, Washington, DC.

Rhoades, J.D. 1990a. Principal effects of salts on soils and plants. *In* A. Kandiah (ed), Water, soil, and crop management relating to the use of saline water. AGL Misc. Ser. Publ. 16/90. FAO, Rome.

Rhoades, J.D. 1990b. Soil salinity—causes and controls. p. 109-134. *In* A.S. Goude (ed), Techniques for desert reclamation. Wiley, New York.

Rhoades, J.D. 1998. Use of saline and brackish waters for irrigation: implications and role in increasing food production, conserving water, sustaining irrigation, and controlling soil and water degradation. p. 261-304. *In* R. Ragab, and G. Pearce (ed), Proc. Int. Workshop: The use of saline and brackish waters for irrigation: Implications for the management of irrigation, drainage, and crops. 10th Afro-Asian Conf. Int. Comm. Irrigation Drainage, Bali, 23-24 Jul 1998.

Rhoades, J.D. 1999. Use of saline drainage water for irrigation. p. 619-657. *In* R.W. Skaggs and J. van Schilfgaarde (ed), Agricultural drainage. ASA Mono. 38. ASA, Madison.

Rhoades, J.D., A. Kandiah, and A.M. Mashali. 1992. The use of saline waters for crop production. FAO Irrig. Drainage Pap. 48. FAO, Rome.

Rhoades, J.D. 2002. Irrigation and soil salinity. p. 750-753. *In* R. Lal (ed), Encyclopedia of soil science. Marcel Dekker, New York.

Robertson, E.G., D.C. Coleman, C.S. Bledsoe, and P. Sollins (ed). 1999. Standard soil methods for long-term ecological research. Oxford Univ. Press, Oxford.

Robinson, R.A., and W.J. Sutherland. 2002. Post-war changes in arable farming and biodiversity in Great Britain. J.Appl. Ecol. 39(1):157-176.

Roschewitz, I., C. Thies, and T. Tscharnte. 2005. Are landscape complexity and farm specialisation related to land-use intensity of annual crop fields? Agric. Ecosyst. Environ. 105:87-90.

Rosi-Marshall, E.J., J.L. Tank, T.V. Royer, M.R. Whiles, et al. 2007. Toxins in transgenic crop byproducts may affect headwater stream ecosystyems. PNAS 104:16204-16208.

Roy, B.A. 2004. Rounding up the costs and benefits of herbicide use. PNAS 101(39):13974-13975.

Royal Society. 2001. The use of genetically modified animals. May 2001. Available at www.royalsoc.ac.uk/document.asp?tip=0&id=2430. Royal Society, London.

Sage, C. 2003. Social embeddedness and relations of regard: Alternative "good food" networks in South West Ireland. J. Rural Studies 19:47-60.

Sanvido, O., M. Stark, J. Romeis, and F. Bigler. 2006. Ecological impacts of genetically modified crops: Experiences from ten years of experimental field research and commercial cultivation. ART-Schriftenreihe 1. Agroscope Reckenholz-Tanikon Res. Station, Zurich.

Sauerbeck, D.R. 2001. CO_2 emissions and C sequestration by agriculture—perspectives and limitations. Nutr. Cycl. Agroecosyst. 60:253-266.

Schultz, R. C., T.M. Isenhart, W.W. Simpkins, and J.P. Colletti. 2004. Riparian forest buffers in agroecosystems—lessons learned from the Bear Creek Watershed, central Iowa, USA. Agroforest. Syst. 61:35-50.

Sears, M.K., R.L. Hellmich, D.E. Stanley-Horn et al. 2001. Impact of Bt corn pollen on monarch butterfly populations: A risk assessment. PNAS 98(21):11937-11942

Seppälä, A., P. Voutilainen, M. Mikkola, A. Mäki-Tanila, et al. 2002. Ympäristö ja eettisyys elintarviketuotannossa—todentamisen ja tuotteistamisen haasteet. (Environment and ethics in food production—challenges of verification and product development.) Selvityksiä 11. MTT, Finland.

Seré, C., H. Steinfeld, and J. Groenewold. 1996. World livestock production systems: current status, issues and trends. Animal Production Health Pap. 127. FAO, Rome.

Shappell, N.W., L.O. Billey, D. Forbes, M.E. Paoch, T.A. Matheny, G.B. Reddy, et al. 2007. Estrogenic activity and steroid hormones in swine wastewater through a lagoon constructed-wetland system. Environ. Sci. Tech. 41:444-450.

Siegel, P., T. Johnson, and J. Alwang. 1995. Regional economic diversity and diversification. Growth and Change 21.

Sinclair, T.R. 1994. Limits to crop yield? p. 509-532. *In* K.J. Boone (ed), Physiology and determination of crop yield. ASA, Madison.

Singer, M.J., and D.N. Warrington. 1992. Crusting in the Western United States. p. 179-204. *In* M.E. Sumner and B.A. Stewart (ed), Soil crusting, chemical and physical processes. Adv. Soil Sci. Lewis Publ., Madison.

Sinkkonen, M. 2002. Impact of production method and production area on energy balance of rye consumption in Helsinki. p. 33-42. *In* J. Magid et al. (ed), Urban areas—Rural areas and recycling—The organic way forward? Proc. NJF-seminar No. 327. Copenhagen.

Smith, A., P. Watkiss, G. Tweddle, A. McKinnon, et al. 2005. The validity of food miles as an indicator of sustainable development. Final Report. DEFRA. ED50254 Issue 7. Available at http://statistics.defra.gov.uk/esg/reports/foodmiles/default.asp. Defra, London.

Somerville, L., and C.H. Walker (ed). 1990. Pesticide effects on terrestrial wildlife. Proc. Int. Workshop Terrestrial Field Testing of Pesticides, Cambridge. Taylor & Francis, London.

Spiecker, H. 2003. Silviculture management in maintaining biodiversity and resistance of forests in Europe—temperate zone. J. Environ. Manage. 76:55-65.

Stoate, C., N.D. Boatman, R.J. Borralho, C. Rio Carvalho, G.R. de Snoo, and P. Eden. 2001. Ecological impacts of arable intensification in Europe. J. Environ. Manage. 63(4): 337-365.

Stoddart, L.A., A.D. Smith, and T.W. Box. 1975. Range management. 3rd ed. McGraw Hill, New York.

Stofferahn, C. 2006. Industrialized farming and its relationship to community well-being. Update of a 2000 Report by L. Lobao for State of North Dakota, Office of the Attorney General for the case State of North Dakota vs. Crosslands. September. Available at www.und.nodak.edu/org/ndrural/Lobao%20&%20Stofferahn.pdf.

Stotzky, G, 2004 . Persistence and biological activity in soil of the insecticidal proteins from *Bacillus thuringiensis*, especially from transgenic plants. Plant Soil 266(1-2):77-89

Sumelius, J., and K.M. Vesala (ed). 2005. Approaches to Social Sustainability in Alternative Food Systems. Rep. Ecol. Agric. 47. Centre Sustainable Agric., Swedish Univ. Agric. Sciences, Uppsala.

Sundkvist, Å., R. Milestad, and A. Jansson. 2005. On the importance of tightening feedback loops for sustainable development of food systems. Food Policy 30(2):224-239.

Tait, J. 2001. Pesticide regulation, product innovation and public attitudes. J. Environ. Monitor. 3:64n-69n.

Tamminga, S. 2003. Pollution due to nutrient loss and its control in European animal production. Livestock Production Sci. 84:101-111.

Tanji, K.K. (ed). 1990. Agricultural salinity assessment and management. ASCE Manuals and Rep. Engineering No. 71. ASCE, New York.

Thompson, E. 1986. Small is bountiful: The importance of small farms in America. American Farmland Trust, USDA, Washington, DC.

Thurman, E.M., D.A. Goolsby, M.T. Meyer, M.S. Mills, M.L. Pomes, and D.W. Koplin. 1992. A reconnaissance study of herbicides and their metabolites in surface-water of the Midwestern United States using immunoassay and gas chromatography/mass spectrometry. Environ. Sci. Tech. 26:2440-2447.

Tilman, D. 1999. Global environmental impacts of agricultural expansion: the need for sustainable and efficient practices.PNAS 96(11):5995-6000.

Tilman, D., J. Hill, and C. Lehman. 2006. Carbon-negative biofuels from low-input high-diversity grassland biomass. Science 314:1598-1600.

Troell, M., A. Neori, T, Chopin, and A.H. Buschman. 2005. Biological wastewater treatment in aquaculture—more than just bacteria. World Aquacult. 36(1):27-29.

Umali, D.L. 1993. Irrigation-induced salinity: A growing problem for development and environment. World Bank, Washington, DC.

UK. 2005a. Agriculture in the United Kingdom 2005—Table 4.1. UK price indices for products and inputs. The Stationery Office, London.

UK. 2005b. Agriculture in the United Kingdom 2005—Table 6.1. UK farmers' share of the value of a basket of food items. The Stationery Office, London.

UNECE. 2005. United Nations Economic Commission for Europe: Employment by Country, Measurement, Activity and Year. Available at http://w3.unece.org/pxweb/DATABASE/STAT/20-ME/3-MELF/3-MELF.asp. UNECE, Brussels.

UNEP. 2004. GEO yearbook 2003. UNEP, Nairobi.

USDA. 2003. Animal welfare act and regulations. Available at http://www.nal.usda.gov/awic/legislat/usdaleg1.htm. Animal Welfare Information Center. USDA, Washington, DC.

USDA. 2004. Agriculture and forestry greenhouse gas inventory: 1990-2001. Tech. Bull. 1907. Available at http://www.usda.gov/oce/global_change/gg_inventory.htm. USDA, Washington, DC.

USDA. 2005. Processed food trade pressured by evolving global supply chains. Available at http://www.ers.usda.gov/AmberWaves/February05/Features/ProcessedFood.htm. USDA, Washington, DC.

USDA. 2005a. Global traceability and labeling requirements for agricultural biotechnology-derived products: Implications for the United States. Advisory Comm. Biotech. 21st century agriculture consensus report. June 2005. USDA, Washington, DC.

USDA. 2008. U.S. trade exports & imports—FATUS commodity aggregations. Available at http://www.fas.usda.gov/ustrade/USTExFatus.asp?QI=. USDA, Washington DC.

USDA-NRCS. 2003. National range and pasture handbook. Rev. 1. USDA-NRCS. Available at http://www.glti.nrcs.usda.gov/. USDA, Washington, DC.

USDA-USFS. 1999. Issues and examples of forest ecosystem health concerns. Update. Available at http://www.fs.fed.us/foresthealth/. USDA-USFS, Washington, DC.

US-EPA. 2008. Inventory of US Greenhouse gas emissions and sinks 1990-2004. Chapter 6 Agriculture. Available at http://epa.gov/climatechange/emissions/downloads/08_Agriculture.pdf. EPA, Washington, DC.

USTR. 2003. Public statement by the Office of the United States Trade Representative, May 13, 2003.

Vandermeer, J., and I. Perfecto. 2005. The future of farming and conservation. Science 308:1257.

Van Lynden, G.W.J. 2000. Soil degradation in Central and Eastern Europe: The assessment of the status of human-induced soil degradation. Rep. 2000/05. FAO and ISRIC, Rome.

Vellidis, G., R. Lowrance, P. Gay, and R.K. Hubbard. 2003. Nutrient transport in a restored riparian wetland. J. Environ. Qual. 32:711-726.

Vergunst, P. 2003. Liveability and ecological land use. The challenge of localisation. Acta Universitatis Agriculturae Sueciae. Agraria 373. PhD thesis. Swedish Univ. Agric. Sci., Uppsala.

Vitta J.I., D. Tuesca, and E. Puricelli. 2004. Widespread use of glyphosate tolerant soybean and weed community richness in Argentina. Agric. Ecosyst. Environ. 103(3):621-624.

Von Braun, J. 2003. Agricultural economics and distributional effects. 25th Conf. Int. Assoc. Agric. Economists. Durban, 16-22 Aug 2003.

Wardell, J. 2007. U.K. Watchdog to analyze supermarkets. Available at http://finance.myway.com/jsp/nw/nwdt_ge.jsp?news_id=cmt-023w3570&feed=cmt&date=20070123. Assoc. Press Copyright AFX News Limited 2006.

Webster, J. 2005. Animal welfare limping towards Eden. Feb 2005. Blackwell, NY.

Wiebe, K., and N. Gollehon (ed). 2006. Agricultural resources and environmental indicators, 2006 Ed. EIB-16. Available at http://www.ers.usda.gov/publications/arei/eib16/. ERS, USDA, Washington, DC.

Wiel, van der A. 2007. Spain: 40% market share in produce sales for supermarkets. Available at http://www.freshplaza.com/2007/0123/1-2_es_marketshare.htm. Fresh Plaza.

Williams, D.S., and D. Gould-Wells. 2004. Biogas production: Engineering and technology for sustainable world 11:11-12.

Williams, G.W. 2000. The USDA Forest Service—First century. FS-650—Jul 2000. USDA Forest Service, Washington, DC.

Wilson, P. and M. King. 2003. Arable Plants—a field guide. Available at http://www.arableplants.fieldguide.co.uk. Wild Guides and English Nature, London.

Winter, M. 2003. Embeddedness, the new food economy and defensive localism. J. Rural Studies 19:23-32.

Wood, C.W., J.H. Edwards, and C.G. Cummins. 1991. Tillage and crop rotation effects on soil organic matter in a Typic Hapludult of Northern Alabama. J. Sustainable Agric. 2:31-41.

Worldwatch. 2006. Biofuels for transportation; Global potential and implications for sustainable agriculture and energy in the 21st Century. (Extended Summary). Worldwatch Institute, Washington, DC.

WTO. 2005. Understanding the WTO: The agreements. Agriculture: Fairer markets for farmers. Available at http://www.wto.org/english/thewto_e/whatis_e/tif_e/agrm3_e.htm. WTO, Geneva.

WTO. 2006. News Items—DG Lamy: Time out needed to review options and positions. Available at http://www.wto.org/english/news_e/news06_e/tnc_dg_stat_24july06_e.htm. WTO, Geneva.

Yoder, F. 2003. Testimony of Fred Yoder, President, National Corn Growers Association before the US Senate Foreign Relations Committee, June 24, 2003.

4

Changes in the Organization and Institutions of AKST and Consequences for Development and Sustainability Goals

Coordinating Lead Authors:
Helena Kahiluoto (Finland), Tanja Schuler (Germany)

Lead Authors:
Rebecca Burt (USA), Joanna Chataway (UK), Janet Cotter (UK), Guy Debailleul (Canada), John Marsh (UK), Miloslava Navrátilová (Czech Republic)

Contributing Authors:
Henrik Bruun (Finland), Richard Langlais (Canada), Paresh Shah (UK), Gerard Porter (UK), Joyce Tait (UK)

Review Editors:
Dariusz Szwed (Poland)

Key messages
4.1 Lessons Learned—A Synthesis 118
4.2 Historical Trends in the Organization of Scientific Knowledge Generation 120
4.3 General Trends of Paradigms in Societal Context 122
4.3.1 Paradigms in NAE AKST during the first half of the period 122
4.3.2 Impacts of paradigms in NAE AKST on low-income countries 123
4.3.3 Paradigms in NAE AKST in recent decades 124
4.4 Changes in the Integration of Perspectives within AKST 126
4.4.1 Evolution 126
4.4.2 Alternatives in integration 128
4.4.3 Barriers faced by integration 129
4.4.4 Risks associated with integration 130
4.5 Development of Structures, Funding and Agenda of AKST 130
4.5.1 Establishment of structures 130
4.5.2 Drivers of change 132
4.5.3 Development of funding and agenda in NAE in the global context 133
4.5.3.1 Development in NAE 133
4.5.3.2 NAE in the global context 136
4.5.3.3 International AKST 136
4.5.4 Changes in structures and management 137
4.5.5 Influence of beneficiaries 138
4.5.6 Consequences of the changes in structures and funding 139
4.6 Development of Public Control of Agrifood Systems 141
4.6.1 Development of risk regulation 142
4.6.2 Intellectual property rights 143
4.6.3 Changes in policy goals 144

Key Messages

1. Following the Second World War, agricultural knowledge, science and technology (AKST) had a major role in developing agriculture such that food security was achieved in most parts of NAE. Higher levels of food security were achieved in western regions of NAE compared to the eastern regions partly due to a more decentralized approach to decision-making in AKST and more integration among research, education and extension.

2. Application of solely production-focused AKST in NAE has been associated with positive consequences but also major negative socioeconomic and environmental externalities, not just within but beyond the NAE borders. These externalities have been increasingly recognized and attempts are being made to address them, e.g., by addressing and quantifying them through research and reducing them through different policy instruments.

3. AKST approaches integrating different perspectives are increasingly considered to be fruitful and have been applied to varying degrees by different countries in NAE.

- Many negative externalities of AKST would likely have been less significant in the past had different disciplines and stakeholders interacted in development and application of AKST more extensively. The development in such integration has proceeded mainly in approaches (e.g., research programs, research methods, or educational programs) rather than in organizational structures. Integration has not always proceeded smoothly as a number of barriers have been encountered. On the other hand, some erosion of important established disciplinary expertise has recently occurred as public financial resources for AKST had to cover a wider range of disciplines.
- Integration amongst research, education and extension was from the beginning built into American AKST in contrast to AKST in many European countries. Such integration, as well as integration of AKST and KST and of relevant policies and administrative sectors, has to some extent proceeded recently at the governmental level in Western Europe, increasing the potential to effectively enhance interrelated development and sustainability goals.
- Food systems approaches, as an example of integration, have since the 1990s shown great potential as a way for AKST to address more comprehensively development and sustainability goals.

4. Between 1945 and the mid-1970s there was a period of rapid growth rates in public agricultural research and development expenditures in NAE. The growth rates then declined. The 1990s saw a slight increase but the growth rates stagnated thereafter despite the by then much broader scope of agricultural R&D. Even if the share of public agricultural R&D expenditure from the total R&D expenditure declined, agricultural R&D expenditure relative to the value of agricultural output increased more than the corresponding figures for science and technology research in general. The share of public agricultural research funds given to universities increased considerably from the 1970s onwards in parts of NAE, leading to a shift towards basic research. The economic returns of investments into agricultural R&D have been high with no evidence for a decline, thus offering an argument for ensuring the public funding to meet development and sustainability goals.

5. The proportion of private funding of AKST in North America and Western Europe has increased since the Second World War, a change that influenced the type of agriculture-related research conducted as well as the allocation of public funding for research, training and extension. Thus the focus of NAE AKST shifted more towards market-driven goals and away from public goods.

6. There have been efforts to streamline public agricultural research in the last quarter of the 20th century in some parts of NAE, which had positive as well as negative impacts on AKST. Competition and short-term contracts were increasingly built into the public sector funding system for AKST in NAE. The aim of this change was to ensure quality, transparency and efficiency. However, there is some evidence that this development reduces rather than increases efficiency. In addition, short-term approaches are not necessarily appropriate for all areas of AKST relevant to the development goals (e.g., integrated approaches, research aimed at sustainability and ecosystem management). Where rationalization of facilities took place in response to changes in priorities and scientific methods and to take advantage of new economies of size and scope, this has been beneficial. However, where the aim has been solely to reduce costs, this has also contributed to a fragmentation and weakening of the disciplinary research base and to loss of crucial scientific expertise and facilities.

7. NAE AKST had a major direct and indirect role in the development of the world's agrifood systems. It contributed to successfully reducing hunger in some regions beyond NAE, but had also adverse ecological and socioeconomic effects. In some areas the technology transfer approach was far from successful.

- Agricultural R&D has become increasingly spatially concentrated, increasing inequity. OECD countries and transition economies use most of the resources. This was contributed to by the increase in private funding in NAE. Spending on international R&D (CGIAR) grew in the 1970s but subsequently real spending started to stagnate and decline while the share of restricted funds increased. Expenditures have increased again since 2001 but only represent 1.5% of the global public sector investments in agricultural R&D and 0.9% of all public and private agricultural R&D spending.
- Factors that increasingly limit spillovers from NAE to developing countries include regulatory policies like IPR, biosafety protocols and trading regimes and the fact that technologies developed in NAE are increasingly less appropriate for poor farming communities.
- Indirect effects of NAE AKST on other areas of the

world—through changes in agriculture, diet and food systems in NAE—have increased.

8. The main drivers of NAE AKST in relation to development and sustainability goals were advancements in KST and changes in societal circumstances and interlinked shifts in paradigms. Societal demand, markets and policies (and consequently AKS) evolved under the influence of these developments.

- Throughout NAE, AKST made a higher degree of industrialization and technological development as well as urbanization possible, but were also crucially affected by these changes. Following the Second World War there was a strong focus in Europe on increasing food supply to ensure food sufficiency and one characteristic of the rebuilding period was a faith in technology throughout NAE. This led to the narrow focus of AKST during this time and further to the adverse environmental and social impacts, which started to gain attention from the 1960/70s onwards.
- In North America and Western Europe in the 1970s, the food crisis had been largely solved, a shift had occurred towards increasing economic liberalization and agriculture by then played a less significant role in the economy. As a result AKST in this region experienced budget cuts. Since the 1990s policies increasingly took into account the multiple interdependent roles of agriculture. Thus AKST started to cover more comprehensively issues relevant to development and sustainability goals.
- In Central and Eastern Europe, the societal restructuring in the late 1980s and 1990s had a dramatic effect on AKST. At the start of the 21st century the fulfillment of the accessional requirements became a main driver of AKST in the countries which joined the EU during this time.
- The wealth differences between NAE and the developing world as well as conflicts outside NAE have contributed to the continued inequity in AKST between NAE and other parts of the world.

4.1 Lessons Learned: A Synthesis

Paradigm shifts seem to have been major drivers for the changes that have taken place in NAE AKST after the Second World War. The main lesson learned, based on the changes in organization and institutions of AKST in NAE and their consequences, is that the dominant paradigms can substantially influence meeting development and sustainability goals of reduced hunger and poverty, improved nutrition and human health, enhanced rural livelihoods and equity, environmental sustainability and sustainable economic development. Institutional and organizational changes in AKST seem to be important factors in helping to meet these goals.

Goals and scope of AKST

The goal of food sufficiency was successfully met in North America and Western Europe through focusing AKST on the productivity of land and labor and on farmer profits. This goal was not achieved to the same extent in Eastern Europe largely due to the socioeconomic and political conditions, a centralized approach to AKST, restructuring and instability. However, the food systems developed in NAE do not provide full food security within all parts of NAE itself due to societal circumstances. They also rely to a large extent on resources outside NAE, which has not only resulted in inequity but also hinders meeting development and sustainability goals outside NAE (see Chapters 2 and 3 of this assessment). Negative consequences of this development of AKST and agriculture in NAE were great environmental, animal welfare and social costs, which did not remain within the NAE borders (see Chapter 3). Many of these costs are difficult to quantify and were initially largely ignored. Such negative externalities are increasingly being addressed but impacts can be difficult and sometimes impossible to recover (e.g., species loss, soil erosion). As discussed in the following subchapters, the attempts of NAE to assist by means of AKST in reducing hunger outside NAE were only partially successful.

The potential of AKST to contribute to meeting development and sustainability goals might have been considerably greater if the scope of AKST had broadened earlier and not only since the 1990s, to embrace whole food systems integrating all its dimensions (social, economic and ecological), levels (including e.g., inputs such as financing, agriculture, processing, transportation, trade, consumption, waste, public goods and costs) and scales (from local to global) with varied perspectives of their actors and of multiple disciplines. This broadening of the view helped AKST on a new track of providing knowledge of the kinds of food systems, which would help to meet the goals and how such food systems might be achieved. AKST has now more potential to cope with the varied societal contexts and preconditions and strive for diverse systems with synergy among the different dimensions of sustainable development.

Approaches and tools of AKST

The increasing deficits in integration of the scientific communities and varied voices (especially of the most vulnerable beneficiaries) in the AKST processes after the Second World War contributed to the partial failure of AKST and agriculture in terms of development and sustainability goals. Since the 1970s, the problems caused by these structural changes in AKST were relieved through a gradual emergence of more systems oriented approaches, more participation of varied stakeholders in AKST and increased interaction between the agricultural, environmental and social sciences. This process started in international development research and similar approaches have been increasingly adopted within NAE.

Interdisciplinarity is more widely accepted as the preferred approach for AKST rather than continuous emergence of new disciplines by unifying old ones. Interdisciplinarity still has a variety of barriers to overcome. Communication across disciplinary borders seems to be the most crucial barrier to achieve true interdisciplinarity. Organizational structures based on the basic sciences as well as disciplinary traditions in funding and merit systems have created disincentives to interdisciplinarity. Demands have increased for appropriate education and training to understand diverse science philosophic approaches, for conceptual tools to facilitate the process and for the development of interdisciplinary review systems and publication channels. Transdisciplinarity and participation (see 4.4) towards governance by the relevant agrifood system actors (including those relating to rural livelihoods, environment and the poor) have been found to require some degree of reconsideration of employees' reward systems. The progressive move from the linear technology generation and transfer to farmers, towards *knowledge networks* crossing horizontal and vertical borders implies collective learning (societal learning) with repeated feedback loops, for co-innovation processes that can successfully meet the goals of the IAASTD in dynamic and complex environments.

Structures and funding of AKST in NAE

The degree in integration of education, research and extension varied among the countries of North America, Western and Eastern Europe. The integrated model applied in the US was particularly successful, especially in contrast to the centralized approach applied in much of Eastern Europe in the past. Decentralized decision-making seems to foster diversity, adaptation to local circumstances and innovation and is thus likely to help meeting development and sustainability goals.

Private funding of AKST in NAE has increased since the Second World War, a change that influenced the type of agriculture-related research conducted as well as allocation of public funding for research, training and extension. AKST has focused increasingly on value addition through industrial, high-technology input development and food processing. Health and food safety concerns and consumers' seeking of comfort and pleasure have been increasingly addressed within the industrial framework. Distributional issues have received less attention.

Public funding in AKST-related research tended to stagnate since the mid-1970s in many parts of NAE. In recent decades increasing recognition of environmental and social problems has gradually caused a shift in allocation of public funds towards reducing negative externalities. However, the move towards diverting more funding to universities at the expense of more applied AKST institutions further emphasized the role of basic sciences and increased the gap between basic and applied research and between research and non-academic stakeholders (especially rural ones). This development emphasizes the need to develop integrative approaches, to avoid decline of chances of NAE AKST to help meeting development and sustainability goals.

Throughout much of NAE, competition and a short-term outlook were increasingly built into the public funding system for AKST on different levels, a change that continues to the present day. This change in approach was meant to ensure quality, transparency, efficiency and value for money for taxpayers. Although this approach has favored certain aspects of scientific performance and international collaboration and increased transparency, it has been suggested that at worst it also has resulted in extreme competitiveness, sub-optimal use of public resources (including increased bureaucracy) and loss of scientific commitment to public goods and long term goals. These changes might hinder the evolution of partnership-based knowledge networks which would help with achieving development and sustainability goals addressed by the present assessment. Short-term contracts also disadvantage time demanding integration and favor laboratory research at the expense of more field based agricultural and sustainability oriented R&D. Therefore, also new forms of review practices and contract arrangements have been sought. In Europe, competitive grants and a merit system based on quantification of publications increasingly encouraged method based R&D at the expense of problem oriented agricultural R&D. The latter trend was supported by the rise of method orientation over problem oriented R&D encouraged by competitive grants and a merit system increasingly based on quantification of publication outputs. In the EU the 5th and 6th Framework Programs have in recent years sought to counteract these trends and promote more integrated R&D focused on public goods, although the focus of these programs was different from that of the IAASTD.

Interaction of NAE AKST with the rest of the world

Initially the contribution of NAE AKST to international research was implemented through technology transfer with considerable success but over time the limits of this approach became apparent with some severe consequences for the achievements of development and sustainability goals. New ways forward were found through development of integrated approaches but old structures and attitudes continued to cause friction. In recent years NAE AKST has increasingly focused on applications within the developed world at the expense of applications appropriate for poor rural developing countries.

Financial resources of the world AKST have further concentrated spatially in NAE and in a few large transition economies. International R&D increasingly faced restrictions set by donors for use of funds. Part of the expert community claims that the international significance of NAE AKST in terms of meeting development and sustainability goals has declined in the latter half of the period. Others suggest that new technology developed by AKST in recent years has been very significant for developing countries although uptake has been uneven. World AKST has further concentrated spatially in NAE and in a few large developing countries. NAE AKST sciences have focused increasingly on basic sciences, high-tech approaches, industrial applications and consumer concerns and a higher proportion of food system relevant R&D has been funded by companies. Spillover of AKST from NAE to developing countries is thus declining. The introduction of regulatory policies such as intellectual property rights, biosafety protocols and trading regimes are seen by many as further endangering equity between NAE and the rest of the world. Conventions on access and benefit sharing have been, however, introduced to balance some of the perceived inequities. The dependence and adverse ecological impacts of NAE agriculture and food supply on areas outside NAE have also increased. These developments hinder meeting development and sustainability goals.

Drivers of NAE AKST in relation to development and sustainability goals

In the period since the Second World War the main direct drivers of NAE AKST in relation to development and sustainability goals were new KST and shifts in paradigms and societal demand. As a consequence, there was evolution in policies, regulations and markets. KST (incl. AKST) made industrialization and technological development as well as urbanization possible, but was also crucially influenced by these changes. The indirect drivers of NAE AKST were predominantly societal circumstances. In Europe the Second World War resulted in loss of infrastructure and in food insecurity but it also promoted industrialization and technological development throughout NAE. The war was followed by a rebuilding period characterized by a faith in technology. The establishment of larger economic and/or political structures in Europe (the EU and its predecessors) has had a marked effect on AKST in ever larger parts of Europe (as membership increases). In Eastern Europe the drastic societal restructuring in the late 1980s and the 1990s increased the risks of poverty, hunger and malnutrition for parts of the population in many of the affected countries. However, opening to the West also provided opportunities for AKST (e.g., in increasing environmental sustainability) even though positive impacts may take time to take effects. The wealth differences in the developed and developing world, intensified by wars and conflicts outside NAE, directly hindered meeting the goals: Lower production costs in developing countries resulted in cheap resources for processing in NAE, thus further enhancing inequity between NAE and developing countries. Policies and search for short term returns for AKST in NAE, together with development of IPR protection, were the main drivers behind increases in privatization and introduction of increased competition to AKST management. This again created barriers for meeting the goals.

Policy makers and governments will have a key role in developing measures that help meeting the goals. To contribute to meeting development and sustainability goals of the present assessment a stronger focus on a wider range of public goods and thus a paradigm shift towards a bigger public role in AKST seems to be required alongside the emerged shift towards more comprehensive adoption, application and institutionalization of horizontal (sciences), vertical (food system actors) and contextual (societal and ecological circumstances) integration as well as collective learning (societal learning) within NAE and in the international context.

4.2 Historical Trends in the Organization of Scientific Knowledge Generation

Much of the extraordinary increase in agricultural productivity in comparison with other industries during the last fifty to sixty years was achieved by rapid technological change. Agricultural knowledge, science and technology (AKST) was a major direct driver of this change (Evenson, 1983). These advances helped greatly to overcome the food insecurity in Europe following the Second World War.

Four decades ago, global goals were expressed such as "in ten years, no one child shall go to bed hungry" or in terms of "increasing the pile of rice on the plates of the food-short consumers" (Falcon and Naylor, 2005). World cereal production has indeed almost doubled since 1970 based on essentially the same cropping area as of 40 years ago (Falcon and Naylor, 2005) (see Chapter 2). Despite this increase in cereal production, 5 million children die from hunger-related causes per year and there are still 850 million people worldwide suffering from undernutrition today. Even though there has been a considerable decline in the proportion of people undernourished in the developing world, there has not been a big change in the absolute numbers of the undernourished since the late 1970s (Falcon and Naylor, 2005). Productivity of labor and land in NAE has increased partly at the expense of limited resources (e.g., land use for fodder export) from other regions. The carrying capacity of some ecosystems was seriously exploited and rural livelihoods in some regions injured. NAE AKST had a key role in this development and needs to learn from its successes and failures.

AKST is not formed or conducted in isolation from the rest of science. There is a long history of agricultural scientists drawing on and adapting findings from the basic biological, chemical and other sciences (Pardey and Beintema, 2001). Moreover, contemporary findings (especially in genetics and information sciences) serve to blur the boundaries between AKST and other sciences (CGIAR Science Council, 2005). The societal context and trends in research and development (R&D) often apply and interact across disciplinary boundaries. Therefore, the development in organizations and institutions related to AKST should be seen in the context of trends in the organization of scientific knowledge overall.

The contemporary organization of scientific knowledge production has its origin in the education centered scientific academies of the 17th and 18th centuries and in the invention of the research university in Prussia in the early 19th century (Rhodes, 2001). European universities had close connections to the state as codifiers of national identity, while American universities had a more pragmatic orientation towards civil society, particularly those established as land-grant universities under the 1862 Morrill Act. By 1950 the public agricultural research system of the US had developed from very small beginnings into the world's largest system, a feat made possible by the expansion of public funding for research and by the decentralized state funded land-grant system (Buttel, 2005). The disciplinary organization of education and research emerged during the latter part of the 19th century and early 20th century through a reorganization of universities and establishment of national and international scientific societies and journals. Academic development before Second World War was characterized by growth, specialization and fragmentation.

After the Second World War, spending on higher education and research increased dramatically in the industrial countries. In the 1960s many new universities were established. Science policy was based on the so-called linear model, which assumed that investments in basic science would lead automatically to technological innovations (Stokes, 1997). In the early 1970s awareness of environmental pollution and a range of societal problems surfaced (Klein, 1996) and the disciplinary structure of science was criticized as not adequate for solving real world problems. Concerns were already expressed in those years that the fragmentation of scientific knowledge had a negative impact on the capacity of people and societies to act in a coherent way (Apostel et al., 1972). Up to the mid 1970s, corporate research was characterized by a relatively high degree of self-sufficiency and secrecy. Increased globalization has since led to a streamlining of industrial R&D, with a stronger emphasis on getting products to market for short term financial returns. At the same time, corporate research started to be geared more towards interactions with R&D and business outside the mother company. Partnerships, licensing and internal venture activities became increasingly important (Chesbrough, 2003).

The growing importance of R&D for commercial opportunity also affected publicly funded organizations. With the growth in the venture capital sector in the 1970s university science could be commercialized directly, without the need to transfer a new technology first to a company. Research became a business opportunity for the researchers. Universities were encouraged to make use of this development through legislation that made it possible to assert IPRs for the output of their researchers (Slaughter and Leslie, 1997; Buttel, 2005). International organizations, such as the OECD and the EU, set up projects in collaboration with national authorities and researchers, to develop a new approach to policy in the fields of science and technology (Miettinen, 2002). The establishment of university-industry networks and the commercialization of university research was promoted by governments in a number of countries in NAE and university research was increasingly seen as an important contributor to regional and national economic competitiveness (Cooke and Morgan, 1998). The focus had shifted from basic research to a stronger emphasis on research that can be commercialized (Schienstock, 2004).

Another aspect in the recent history of the organization of scientific research is the emphasis on value creation and accountability. Since the 1960s, the growth of public research funding in Western Europe and the US has been largely in form of competitive grants rather than budget funding for universities or research institutes. The overall share of external grants has increased. Although funding systems vary from country to country within NAE, there has been a general trend to include peer review as part of the funding decision. The aim of peer review for the assessment of grant applications is to allocate the limited funds to the best projects and that investments produce scientific value. A further development arose in the 1990s as the funding of universities, research institutes, departments, groups and individual employees became increasingly based on performance according to quantitative measures such as the number of articles in journals with a high citation index, the number of citations of one's work, the number of degrees awarded and so on. Managerial systems were also introduced, in some countries, to monitor the activities of individual scientists and to create incentives for scholarly activity. The British Research Assessment Exercise is a well-known and much-debated example.[6]

The gender imbalance in science has also received increasing attention since the 1970s. Although considerable progress has been made, women are still underrepresented (Box 4-1; Figure 4-1; Table 4-1).

The organization of scientific knowledge production has thus undergone constant change. The sites of knowledge generation have become more diverse, with an increasing role for civil society organizations as they have become more professional, with increasing capacities for knowledge generation and policy input. In addition, the emphasis on the application context of research has increased (Gibbons et al., 1994). The problem oriented nature of research has led to a crossing of disciplinary boundaries in academia (in industry they were never respected) and multi- and interdisciplinary research is becoming increasingly common (Klein, 1996). Research is also more and more collective in nature. The number of copublications has increased in virtually all fields and in some areas experiments can involve tens or even hundreds of researchers (Galison and Hevly, 1992). For most industries, science provides an important stock of knowledge and basis for innovations (Klevorick et al.,

Box 4-1. Women in science in NAE.

The presence of women in science has increased in NAE since the Second World War but they are still under-represented (ETAN, 2000). In the US women in academia began to make considerable progress in the 1970s through concerted protests, appropriate legislation and class action suits. Canada has also devoted considerable attention to the issue (ETAN, 2000). In Europe the issue of under-representation of women in science was taken up first in the Nordic countries in the early 1980s, particularly in Finland and Sweden (ETAN, 2000). More attention was paid to this issue at EU level in the late 1980s. For example, the European Parliament's Resolution on Women and Research from 1988 stated that "the under-representation of women in academic life is a highly topical problem and calls for practical incentives" and called on Member States to "promote positive measures to further the presence of women at the highest levels in universities and research institutes" (ETAN, 2000). However, although women now constitute about half the undergraduate population they still play a minor role in decision-making concerning scientific policies and priorities in many NAE countries (Table 4-1, Figure 4-1) (ETAN, 2000). The proportion of women in senior scientific positions is small as there is a continuous drop in the numbers of women at each level of the academic ladder and many highly trained women are lost to science. In 2004, the proportions of females in the highest senior grade in some AKST-relevant fields of science in EU25 were 15% in agricultural sciences, 11% in natural sciences and 17% in social sciences (European Commission, 2006).

Working patterns of women vary between NAE countries. While career breaks and part-time working are common in some Northern European countries such as the UK and the Netherlands, in other parts of Europe, for example in Spain, France and Italy, women are much more likely to work full-time and throughout their adult lives. Systems of support and cultural expectations reflect and partly create these differences (ETAN, 2000).

Source: ETAN, 2000.

[6] http://www.rae.ac.uk/.

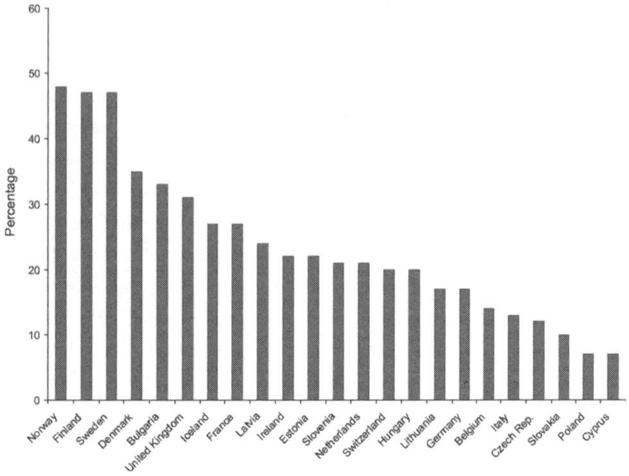

Figure 4-1. *Women on scientific boards in EU countries in 2004.*
Source: adapted from EC, 2006

Table 4-1. **Percentage of female professors in university faculties (different ranks, all disciplines).**

Country[a]	Full professor	Associate professor	Assistant professor	Year
Turkey	21.5	30.7	28.0	1996/7
Finland	18.4	–	–	1998
Portugal	17.0	36.0	44.0	1997
France	13.8	34.2	–	1997/8
USA	13.8	30.0	43.1	1998
Spain	13.2	34.9	30.9	1995/6
Canada	12.0	–	–	1998
Norway	11.7	27.7	37.6	1997
Sweden	11.0	22.0	45.0	1997/8
Italy	11.0	27.0	40.0	1997
Greece	9.5	20.3	30.6	1997/8
UK	8.5	18.4	33.3	1996/7
Iceland	8.0	22.0	45.0	1996
Israel	7.8	16.0	30.8	1996
Belgium (Fr.)	7.0	7.0	18.0	1997
Denmark	7.0	19.0	32.0	1997
Ireland	6.8	7.5	16.3	1997/8
Austria	6.0	7.0	12.0	1999
Germany	5.9	11.3	23.8	1998
Switzerland	5.7	19.2	25.6	1996
Belgium (Fl.)	5.1	10.0	13.1	1998
Netherlands	5.0	7.0	20.0	1998

[a] Figures for Portugal include only academic staff performing R&D activities. The French-speaking and the Flemish-speaking parts of Belgium keep separate statistics.

Source: adapted from ETAN, 2000.

1995). In turn public science depends on industry for its instrumentation and research materials. Countries in NAE today spend up to 4% (Sweden) of their GDP on research and development (US, 2.7%; EU15, 1.9%; Russian Federation, 1.2%; Canada, 1.9%) (OECD, 2006a).

The increasing importance of knowledge, innovation and technology development for the economy together with globalization have made the world economy more dynamic. Diffusion of knowledge relevant to innovations throughout the economy is extremely important and here the traditional linear innovation model has shown weaknesses (Stokes, 1997). The systemic or interactive model of innovation, currently broadly accepted as a representative picture of how the innovation-driven economy works, postulates the need for dynamic and flexible structures and processes (OECD, 2002). While non-economic institutions often continue to develop along the earlier path (OECD, 2005c), a third generation of an innovation policy (going beyond the linear and interactive models) is emerging. It calls attention to the process of accommodation, especially in the area of governmental science, technology and innovation policy (OECD, 2005a, b and c). This horizontal process requires governments and institutions to be more flexible and to integrate policy formulation and implementation among ministries and across other institutional boundaries to improve coherence. Despite the challenges associated with expanding knowledge and science policy into a broader innovation policy, there seems to be both a need and an opportunity for such a change, especially in the context of sustainable development. Key barriers, based on case studies in different OECD countries, are lack of recognition of innovation policy as a key driver of sustainable development, separate "missions" and lack of understanding of related policies between different ministries (OECD, 2005c). Countries in NAE have faced different obstacles in this context and have proceeded on this path to different degrees.

4.3 General Trends of Paradigms in Societal Context

4.3.1 Paradigms in NAE AKST during the first half of the period

During the past century, agriculture in NAE faced two persistent challenges linked with industrialization: technology development and rising real wage rates in the non-farm sector. The agricultural sector has undergone a major economic and social change (see Chapter 2) as it has adjusted to these forces and become more integrated into the national and world economies. The wages available in non-farm employment represent an opportunity cost to farm labor when the two labor markets are integrated. Before 1933 farm input markets were poorly integrated with non-farm input markets but by the 1970s they had become well integrated (Huffman, 1996). In the US, real manufacturing wage rates rose by a factor of 5 from 1890 to 1990; real compensation rose faster, by a factor of 7.6. These large increases represent a powerful force for drawing labor away from agriculture, made on the other hand possible by, but also causing, labor saving technical change in agriculture (for opposing views see Hayami and Ruttan, 1971; Busch et al., 1984; Olmstead and Rhodes, 1994; Huffman, 1998a; Huffman and Evenson, 2001). (For further details on changes in labor see Chapter 2.) Within agriculture specialization of tasks increased through industrialization. The 1920s saw expansion of the ammonia industry for fertilizers, development of the crop hybridization technique on a commercial scale (Buhler et al., 2002) as well as mechanization. With the number of farms declining and aggregate output growing, average output per farm grew rapidly (see Chapter 2 for changes in farm size and modernization of farms).

The main driver for the development of AKST in NAE after the Second World War has been technology development based on industrialization, globalization, policies and demand. The main direct driver of AKST during the early part of the period after the Second World War was a policy directed towards food sufficiency in NAE, to address the situation of food insecurity especially in Europe. Policies that led to a decline in real food prices greatly aided the growth of cities and allowed the rising living standards in North America and Western Europe. In Central and Eastern Europe industrialization of agriculture took place only after the Second World War as part of a planned economy and was more variable. This period was characterized by spectacular production gains (de Wit, 1986), through: (1) rapid integration of mechanization into farming activities, (2) increased use of inputs, e.g., fertilizers and other agrochemicals, adoption of hybrid seeds and crop varieties that could utilize these inputs (see Chapter 2 of this assessment) and (3) increased levels of publicly funded R&D, particularly in plant and animal genetics and farm management. The discovery of the role and structure of DNA led to advances in genetics and the development of molecular biology. Legislation on intellectual property protection applied to living organisms was developed. Together these developments fundamentally changed the nature of agricultural sciences, public and private roles as well as the roles of locally provided and internationally traded agricultural goods and services (Alston et al., 1998).

Public AKST and AKST more generally, contributed to the industrialization and development of productivity. Jorgenson and Gollop (1992) showed that the average annual total factor productivity (TFP)[7] growth in the agricultural sector over the 1947-1985 period exceeded the corresponding rate for the US private non-farm economy by more than 3.5 times and was more than double the rate of TFP growth for the manufacturing sector. For agriculture, productivity growth accounted for 82% of the growth of output, while for the rest of economy, productivity accounted for only 13% of the growth. Although there are some problems with correctly identifying causal relationships (Griliches, 1979), the evidence above and adopted from cross-sectional and over-time variation of TFP in agriculture (Evenson, 1983) indicates that investments in public and private agricultural research, public agricultural extension and farmers' schools are a major part of the explanation for the growth in productivity. Public research and education have been at least as important as private R&D and market forces for change in livestock specialization, farm size and farmers' off-farm work participation (Busch et al., 1984; Huffman and Evenson, 2001). The strength of the relationship between public research and farm growth increased from about the early 1970s to the early 1980s. Private R&D and market forces have been relatively more important than public research and education for changing crop specialization. As profitability is influenced by local geoclimatic as well as economic conditions, good adoption decisions depend to a large extent on appropriate training (see Huffman, 1998b, for a summary of the evidence), which increases the profits of early adopters (OTA, 1992; Huffman and Evenson, 1993).

Following the restoration of the food supply after the Second World War, government concern in North America and Western Europe shifted towards supporting farmers' standards of living. Technological innovation remained important, as the new technologies generally used less labor to produce a given quantity of output at any given relative input price. However, the social welfare of rural communities and income parity for primary producers became dominant drivers of change in agricultural policies, with stabilization of prices being used as the main tool (James, 1971). The Common Agricultural Policy (CAP), as formulated in the Treaty of Rome (1958), aimed to (1) guarantee food supplies at stable and reasonable prices, (2) ensure a fair standard of living for farmers and (3) improve agricultural productivity through technical progress and rational production systems that would employ labor more efficiently (see Chapter 2 for further information on CAP, trade and tariffs).

4.3.2 Impacts of paradigms in NAE AKST on low-income countries

In many developing countries, the basis for the agricultural development after the Second World War was built during colonialism, when the focus of agricultural research and extension was not on staple foods but on cash crops (such as sugar cane, tea, coffee, tobacco, spices, oil palm, cotton and rubber) (Masefield, 1972). Following independence (e.g., in Africa in the late 1950s and 1960s), the structures and methods left behind formed the basis of the R&D system of the new governments. The emphasis, especially in Africa, remained on cash crops (Roy, 1990). Although more attention was then paid to food-crop research in the subsistence livelihood context, there was little interaction with resource-poor farmers (Buhler et al., 2002).

The NAE strategy to ensure food sufficiency was reflected in the development of the Green Revolution for developing countries which started with Cooperative Wheat Production Program in 1944 to increase wheat yield in Mexico. This program involved the Rockefeller Foundation and the Mexican Ministry of Agriculture. It involved breeding high yielding, disease resistant wheat varieties and combined them with the use of artificial fertilizers, irrigation and pesticides. As a result of the program Mexico became a net exporter of wheat by 1963. A similar approach was

[7] Productivity analysis is an economist's attempt to approximate the "ultimate" impact of technical change on useful output without trying to identify "intermediate" successful technologies or count innovations. To accomplish this, total factor productivity (TFP) expresses aggregate output per unit of aggregate input—rather than per unit of one input, say labor or land. The growth of aggregate output that cannot be explained by aggregate input—under the control of producers—is defined as TFP (Griliches, 1979; Jorgenson et al., 1987).

applied in Asia with wheat and rice, which also led to impressive yield increases. This strategy was institutionalized in the 1960s with the establishment of the international and tropical research centers and with their union, the CGIAR, in the 1970s (see 4.5.1). While some of the research centers were commodity oriented, since the 1970s most have concentrated on farming systems and often promoted input intensive farming schemes (Van Keulen, 2008). The strategy was to concentrate inputs and services on a few major crops (like wheat, rice and corn) on the best arable lands and for the better-off farmers, to reduce food scarcity and to establish markets for farm inputs. Overall the Green Revolution is credited with saving over a billion people from starvation (Buhler et al., 2002; Evenson and Collins, 2003). Initially there were high hopes in translating the Green Revolution to Africa, but these attempts failed, possibly due to challenging socio-ecological conditions and because farmer's goals are different than those in Asia (Conway, 1997).

After the initial enthusiasm about the successes of the Green Revolution a whole catalogue of criticisms emerged from the late 1960s onwards. Social concerns included that the practices introduced were often not appropriate or accessible for small-scale farmers, that there was little R&D of the staple crops of the most food insecure and that the reliance on external inputs led to indebtedness of a proportion of the farmers. Environmental concerns pointed to that the building of big dams required for the new irrigation schemes resulted often in flooding of farmland, excessive use of chemical inputs leading to water pollution, soil degradation due to agricultural intensification and more extensive use of non-renewable energy sources. Mixed cropping was replaced with monocultures of single varieties and landraces of crops were lost. Other means of yield improvement tended to be ignored by farmers and crops grown for subsistence gave way to the production of cash crops (Van Keulen, 2008; Falcon and Naylor, 2005; Buttel, 2005).

Subsequently approaches that emphasized the multidimensional effects of technologies aiming to reduce negative social and/or environmental consequences while increasing positive impacts became more common in the 1970s and 1980s (Mann, 1997). Examples of such approaches are Integrated Pest Management (IPM), on-farm conservation, farming systems research (FSR), farmer-oriented approaches and participatory research, sustainable agriculture and integrated rural development (see 4.4.1). CGIAR, in collaboration with national research centers and universities, was extensively involved in IPM programs and habitat management strategies for parasitic weed and pest control (Cook et al., 2007). FSR approaches, which relate to the whole farm rather than individual elements and take into account traditional farming expertise, household goals and constraints (Stephens and Hess, 1999), rapidly became popular and supported by many donor agencies (Brown et al., 1988). As the limitations of the FSR approach became apparent, the agroecosystem analysis (AEA) approach was promoted. It broadened the perspective to take into account the long term health of the wider ecosystem (Stephens and Hess, 1999). The new approach of the Doubly Green Revolution (introduced by Conway in 1997) aims at sustainable use of resources and/or adaptive management in agriculture (Pretty, 1995; Conway, 1997; von Braun, 2000; Ashley and Maxwell, 2001).

In recent decades, accelerated by the end of the Cold War, agricultural trade has been increasingly liberalized. Developing countries, in which the agricultural sector occupied a large share of the economy and employment, sought to switch from self-sufficient agriculture to commercial agriculture. One side effect of this strategy was an increase in the number of poor people and in the gap between rich and poor. Small farmers increasingly started contract production under large farm owners. In some cases farmers lost their land, being unable to pay off credits used to finance external inputs, turning into tenant farmers or farm laborers. In the face of reduced development aid, programs and policies were outlined for poverty reduction and remedies for poor areas to reduce the regional disparities (Van Keulen, 2008). Developing countries have also responded to the increase in demand for food produced without chemical inputs, exporting organic produce to serve NAE markets, a development of interest to poor and remote farmers. In recent years, the use of genetic engineering techniques to accelerate plant breeding has resulted in some successes. The introduction of insect resistant Bt cotton in China has been reported to improve yields and yield security as well as reducing insecticide use and cases of pesticide poisoning in farmers (Pray et al., 2002). The transgenic techniques have also raised a lot of criticism due to inequity in terms of access and feared environmental and health risks.

Global insecurity, civil conflicts and lack of democracy have continued to be major problems causing food insecurity (e.g., Falcon and Naylor, 2005). During the 1990s, 1 million lives were lost annually in civil wars. The combined number of annual hunger-related deaths was 8 million people, of which 60% occurred in Africa and 25% in Asia (UN, 2004; Hunger Project, 2005). Global food supply problems for several major commodities were largely solved, but the problem of access to food was not conquered (e.g., Lappeé and Collins, 1988; Falcon and Naylor, 2005).

4.3.3 Paradigms in NAE AKST in recent decades

Negative side effects of an AKST approach focused solely on increasing the food sufficiency and farm productivity became gradually more apparent and raised concern about the externalities of agricultural technologies, in particular in terms of environment and health (e.g., effects of DDT and eutrophication). The energy crisis in the 1970s, publication of the Global 2000 report (Barney, 1981) and the Chernobyl accident in the 1980s raised concern about resource limitations. These various concerns gave rise to the concept of sustainable development, a concept brought to the fore by the Brundtland report (WCED, 1987). Declines in biodiversity and climate change also received increasing attention. The biodiversity issue in particular raised discussions in Europe about the multifunctionality and sustainability of agriculture, emphasizing the role of diverse cultural landscapes and the role of biodiversity in maintaining ecosystem functions. It led to the adoption of an ecosystem approach in World Summit on Sustainable Development in 2002 for conserving biodiversity (Plan of Implementation, 44e) (UN, 2002).

One example of an ecosystem approach is organic food

and farming (OF). Organic farming is based on the principles of health, ecology, fairness and care (IFOAM, 2005), emphasizing animal welfare, which in the 1990s raised wide concern in the society (see Chapter 2). By the mid-1980s organic farming was an established alternative to conventional farming and during the 1990s its share of field area increased considerably in NAE. In Europe the area under organic farming increased from <0.1 million ha in 1985 to 7 million ha in 2006, representing about 3.2% of the European field area (and 4% of that in the EU) (Institute Rural Sciences, 2007). Another example is Integrated Farming Systems (IFS) (also known as Integrated Crop Management). The objectives of IFS approach are a holistic pattern of land use which integrates natural regulation processes into farming activities to achieve maximum replace of off-farm inputs and to sustain farm income (El Titi, 1992; Wibberley, 1995; IOBC, 1999; Morris and Winter, 1999). Research on various aspects relating to OF and IFS has been taking place in NAE since the late 1970s in response to the environmental side effects of intensive farming practices (see Chapter 2).

The inherent conflicts that occur among environmental, economic and social costs and benefits of agriculture (ACRE, 2006) were increasingly understood. Approaches taking into account the whole food chain started to be developed in the 1980s. In the 1990s food systems approaches emerged, particularly within NAE, in an interaction with the emergence of alternative food systems initiatives (see Chapter 2). These approaches aimed not only to take into account environmental, economic and social aspects but also covered the whole food chain, from inputs to waste management and to support systems related to food, including institutions such as values and norms (see e.g., Dahlberg, 1993; Tansey and Worsley, 1995). Proceeding simultaneously on all the dimensions of sustainability remains a challenge.

The concern for rural communities and their vitality received increasing attention, which was reflected in EU policy schemes and attempts to integrate agricultural and rural policy (Figure 4-2 [in Annex H]). Abandonment of farm land, e.g., in the Mediterranean region, not only had negative social and economic consequences but often also undesirable effects on a range of environmental parameters (MacDonald et al., 2000; Suarez-Seone et al., 2002), illustrating again the multifunctionality of agriculture.

Farm animal welfare became a concern in Western Europe and North America as animal production intensified and the population became more affluent and less in touch with farming. Voices questioning whether welfare concerns are compatible with animal husbandry or meat eating increased and in the 1990s radicalism proliferated within the animal welfare movement (Buller and Morris, 2003). The farm animal welfare debate has gradually penetrated farm policy within the EU and is becoming increasingly institutionalized as a result of EU and national legislation (Buller and Morris, 2003). In parallel, renewed academic interest developed in human-animal relations, fuelled by a re-examination of society-nature relationships (Buller and Morris, 2003).

The central role of AKST as a driver of industrialization and structural change, especially but not solely of agriculture, has raised debate about whether even publicly funded agricultural research is equally accessible to all users and whether it is targeted to the full range of user and citizens' groups (BANR, 2002).

Over the past thirty years the agricultural component of developmental economics has declined in academia in parts of NAE, such as the US, rather than increased in response to continuing food security problems (Falcon and Naylor,

Box 4-2. An introduction to the evolution of the ecosystem approach.

The ecosystem approach is a strategy for the integrated management of land, water and living resources that promotes conservation and sustainable use in an equitable way. It is based on the application of appropriate scientific methodologies focused on levels of biological organization, which encompass the essential processes, functions and interactions among organisms and their environment. It recognizes that humans, with their cultural diversity, are an integral component of ecosystems. Therefore, the ecosystem approach is a crucial step towards acknowledging, conserving and relying on the ecosystem functions and structure in the development of agri-food systems, compared to the earlier approach of sustainable use, which takes nature as a source of resources and sink of wastes for agriculture and calls for stewardship (Douglass, 1984). An even more narrow approach is that of food sufficiency, which lacks long-term perspective or consideration of environmental and social impacts of food production. The environmental, social and economic consequences of the latter approach, which has dominated the development of agri-food systems for the first decades after the WWII, are described in Chapter 3.

The ecosystem approach has its critics. Wood and Lenne (2005), for example, used the CBD as a framework to reject the three "received wisdoms" in the agri-environmental policy over the past ten years: the ecosystem approach, the premise that agricultural expansion damages wild biodiversity and the premise that agricultural biodiversity ensures agricultural sustainability (c.f. MA, 2005). They proposed development of intensive agriculture to save off-farm biodiversity. Other recent contributors to this longstanding debate about intensive vs extensive agriculture include e.g., Green et al. (2005), Balmford et al. (2005) and Vandermeer and Perfecto (2005). One argument is that intensification (through increased yield per hectare), although causing declines of biodiversity on agricultural land, may help reduce the need for habitat reduction elsewhere (including natural pristine habitats). Pretty et al. (2006) suggest to exploit win-win situations that can be achieved in combining high productivity and ecosystem services. Another factor to be considered is that intensive agriculture often relies on inputs from beyond national borders (the so called "hidden hectares") to produce, e.g., feed (Deutsch, 2004; Johansson, 2005). Another view is that although the ecosystem approach may be appropriate in Europe, developing countries need the development of more intensive, highly productive agriculture, even if it has to rely on external inputs.

2005). Major US private universities that historically have trained large numbers of agricultural policy analysts have closed key academic units. The Land-Grant universities tended to focus on state agricultural interests rather than international agricultural R&D. Also, several states have made funding foreign graduate students more difficult (Falcon and Naylor, 2005).

In addition to the environmental concerns and the development of the concept of multifunctional agriculture, market-based economic liberalization and globalization were dominant drivers from 1986 until the early 2000s. These market forces contributed to large-scale agricultural industrialization. The main consequences were a shift from producing commodities to manufacturing products, emphasis on the entire food chain with increasing specialization, re-alignment and increasing power of retail and flexible system adjustment to changes in consumer demand, economic conditions and technological improvements (Van Keulen, 2008). Further, information technology was increasingly utilized to enhance the value chain's competitive ability. Development of new products was aided through new technologies: improved logistics brought about by integration of transport and storage systems, improved preservation systems, the communication "revolution" (through electronic data exchange as well as investments on efficient consumer response), biotechnology, active packaging, precision farming and an increased use of integrated pest management (Van Keulen, 2008).

These trends in AKST approaches after the Second World War were more prominent in research, extension and training than in higher education. In higher education the general trends were similar but changes proceeded more slowly and met with more resistance.

4.4 Changes in the Integration of Perspectives within AKST

Integration of perspectives within AKST has several dimensions, integration among scientific disciplines and actors representing multiple interrelated goals (e.g., different dimensions of sustainability, different policy goals), system levels (e.g., loops in the food chain, rural development), as well as spatial (local, national, global) and temporal scales (short- and long-term dimension of sustainability). Integration within AKST refers also to integration among education, research and extension. Integration between AKST and KST is also of interest. Integration and disintegration may take place in terms of approaches, methods and conventions of science and innovation, as well as through development of organizational structures.

4.4.1 Evolution

In the past AKST was well integrated, if informally, with practical agriculture and beneficiaries as well as among the emerging disciplines. This changed at the time when the disciplinary basis of universities and research institutes was established. Distancing occurred both in relation to the practitioners and among emerging disciplines (i.e., vertical and horizontal disintegration). This distancing was more extensive in Europe than in the US as the higher education, agricultural research and extension systems of the latter were established in a more integrated way. More recently AKST has moved towards re-integration.

The integration of the early days was biased towards (1) farmers and rural populations at the cost of consumers and other interest groups and (2) soil, crop and animal sciences as well as farm economics at the cost of human nutrition, ecological and social sciences. The re-integration has mainly proceeded in the form of specific integrative research approaches without this earlier bias. The latter were often first adopted in developing countries, simultaneously with still continuing disciplinary fragmentation. Thus, in most places, integration has been a functional rather than a structural, organizational phenomenon. In Europe, the strongest formal incentive to integration has been provided by recent EU Framework Programs, conceived to respond to the major socio-economic challenges facing Europe (Buhler et al., 2002).

Up until the middle of the 19th century, training of agricultural scientists did not advance rapidly. Advancement required the introduction of a new science system for agriculture, which occurred largely between 1860 and 1920. To establish this system, research methods were borrowed from the general sciences (e.g., chemistry, botany, physics) (Huffman, 1998ab). Even though the historical ideals of unity and synthesis of knowledge in natural sciences served as the first models for agricultural sciences, a fragmentary tendency dominated the infrastructure of science until the mid 20th century. This tendency was characterized by the splitting of disciplines into new subspecialties (Klein, 1990) and by focusing on separate topics, increasingly ignoring their interrelations. Thus agricultural science structures—both in education and research—rewarded a narrow orientation as a sign of a truly scientific approach. However, science and technology developed bidirectionally, facilitated by the agricultural roots of most agricultural scientists (Huffman, 1998ab). Additional methodologies were developed to meet the special circumstances associated with agriculture (Huffman, 1998a) and much applied research became multidisciplinary. While the earliest documented use of the term "interdisciplinary" in research appeared in general education and in the social sciences in the 1920s, the first problem-oriented interdisciplinary research was conducted in the 1940s in agriculture and defense (Bruun et al., 2005). In many comparative studies agriculture has turned out to be one of the most interdisciplinary science fields (Clayton, 1985; Qin et al., 1997; Song, 2003). However, these studies often used the term "interdisciplinarity" meaning multidisciplinarity with no requirement of interaction of sciences. Also combinations of closely related fields were much more common than interactions between natural and social sciences (Bruun et al., 2005).

As described above (and in more detail in Chapter 3) the narrow focus in AKST and agriculture after the Second World War on productivity of labor and land caused negative externalities which gradually become more apparent. These unintended consequences raised concern about fragmentation and overspecialization in agricultural and food sciences (Carson, 1962; White, 1967). The recognition that ecological, economic and social dimensions needed to be taken into account simultaneously led to the introduction of the concept of sustainable development (WCED, 1987; Buttel, 2005). As knowledge about agroecosystems has increased, past uses of environments and the potential for their sustainable management in the future has attracted particular integrative or interdisciplinary efforts (Pawson and Dovers, 2003). Interdisciplinarity is now increasingly claimed and practiced (Bruun et al., 2005).

Integration of perspectives representing different system levels, spatial and temporal scales, scientific disciplines and stakeholders in agricultural research and extension (and later also in education) has thus come into focus as a way to overcome the main barriers towards achieving sustainable development. Examples include hard and soft systems approaches (Box 4-3), participation (Table 4-2), interdisciplinarity and transdisciplinarity (Visser, 2001; Klein, 2004). In the mid-1960s, there was little interaction between traditional agricultural and social scientists. Although the Green Revolution (an approach relaying on natural sciences alone) was successful in reducing hunger for millions, the lack of success in using a similar approach with resource poor farmers led in the 1960s and early 1970s to the evolution of a number of new foci in international agricultural R&D (see 4.3.2) (Table 4-2).

For example, during the 1980s the CGIAR centers were encouraged to use multidisciplinary approaches, to increase inter-center cooperation and to collaborate with others (CGIAR, 2006), even if strong friction occurred due to the existing structures and management (Buhler et al., 2002).

For integration of different dimensions of farming and for participation of resource-poor farmers (and later other stakeholders) in R&D, several approaches with different coverage, emphasis and procedures were developed (see 4.3.2). Examples of farmer oriented approaches include "farmer-back-to-farmer" and "farmer-first," Rapid Rural Appraisal (RRA) (Chambers, 1983), Participatory Rural Appraisal (PRA), Participatory Poverty Assessment (PPA) (Robb, 1998), Sustainable Rural Livelihoods approach (SRL) (Carney, 1998) and Farmer Field Schools (FFSs) (Way and van Emden, 2000). The concept of participation has more and more evolved towards governance (Table 4-2; Ashley and Maxwell, 2001). Participation is also a way to introduce experiential and local/indigenous knowledge (Sillitoe et al., 2002) as well as knowledge about the locally adapted, traditional systems and practices to contribute to system development in interaction with science-based knowledge (Sumberg and Okali, 1997).

Food systems approaches (see 4.3.3) often comprehensively involve food system actors to contribute to AKST. The US academic literature on food systems echoes alternative social norms, where "local" becomes the context in which these norms can be realized, while in the European literature dealing with alternative food networks, localism is seen as a way to maintain rural livelihoods (DuPuis and Goodman, 2005). Irrespective of the scale, food system AKST is relevant to food policy. Food systems approaches make it possible to address and take into account societal preconditions when developing food systems and thus have great potential to contribute knowledge and tools to reduce hunger and poverty and increase sustainability.

The paradigmatic change towards sustainability, food chain approach and systems orientation created a demand for integrated, educational programs taking into account the multiple roles of agriculture and more problem- and improvement-oriented pedagogical solutions (Delgado and Ramos, 2006). Student-centered and experiential approaches started to emerge in higher education in food and agriculture-related subjects during the last decades. Such

Box 4-3. An introduction to systems approaches.

Beginning with Einstein's theory of relativity (1905), a more systemic approach has evolved within science (Jantsch, 1975; Ackoff, 1983), and been formulated into a general theory of systems, for example by Bertalanffy (1973). According to the systems view, useful information about a phenomenon is not obtained by studying its components in isolation, because their interrelations determine the function of both the part and the whole (Bunge, 1985). A system is seen always to be embedded in a larger system, thus implying the aspect of hierarchy, and the interrelations among system levels are important to consider. The soft systems approach (e.g., Checkland, 1981) further assumes that every system can be described in several ways depending on the underlying worldview. This shift from a hard systems methodology (an ontological systems orientation) to a soft systems methodology (an epistemological systems orientation) implies that not only is the phenomenon studied interpreted as a system but also the inquiry into it (Checkland, 1988; Bawden, 1991). This approach, participatory in its very nature (Laszlo and Laszlo, 1997), introduces the researcher as a responsible actor in the human activity system (also Aløre and Kristensen, 1998). Attempts to construct research methodologies, especially for agriculture using hard or soft systems approach, were made starting with FSR in low-income countries and by Spedding (1979), Bawden et al. (1985), Odum (1983, 1988) and others. This approach is often seen as an articulation for a plea for holism in science. The danger is in interpreting a systems approach as a need to focus solely at a certain, often relatively high level, which can lead to "upward reductionism" (see Bunge, 1985).

Soft system research has been promoted for situations where there is uncertainty about what constitutes the problem and what represents an acceptable solution as they depend on the perspective of the individuals involved (Stephens and Hess, 1999). A key feature of the soft system approach is that it aims to avoid formulating problems from one perspective to the exclusion of others. Stephens and Hess (1999) suggested that "an idealised pathway may be to adopt soft systems approaches to problem identification, hard systems methods to researching acceptable and sustainable solutions, and then to develop bilateral projects . . . [to] facilitate the uptake of outputs" although they were concerned that that the current short term funding situation does not allow the necessary time or the freedom of thought.

Table 4-2. **Types of participation in development.**

Type of Participation		Characteristic
1.	Manipulative	A pretence (no real power). For example, the presence of "people's" representatives on a board or committee, but who are outnumbered by external agents.
2.	Passive	People told about a decision or what has already happened, with no ability to change it.
3.	Consultative	People answer questions. The form of the questions and analysis of results is done by external agents.
4.	Material Incentive	People contribute resources (e.g., land, labor) in return for some incentive.
5.	Functional	Participation seen by external agents as a means to achieve goals (e.g., reduce costs) usually after major decisions have already been made.
6.	Interactive	People involved in analysis and development of action plans, for example. Participation is seen as a right and not just as a mechanical function.
7.	Self-mobilization	People mobilize themselves and initiate actions without the involvement of any external agency, although the latter can help with an enabling framework.

Source: Pretty, 1995; Buhler et al., 2002.

ideas as lifelong learning, communicative learning (Leeuwis, 2004) and collective learning (societal learning) as well as participatory approaches have led to the development of innovation systems and processes within AKST. Inclusion of multiple knowledge bases, feedback loops and learning processes now aim to enable those involved to respond to emerging unpredictable circumstances. The concept is still evolving and requires more analysis of the agents involved, their behavior, the diverse interactions that characterize it (Spielman, 2005) as well as techniques and procedures to include actors to create knowledge for use and diffusion.

Many analysts conclude from the experiences with international AKST that the constraints faced by agricultural organizations and systems are often institutional in nature (Byerlee and Alex, 1998) and that formal and informal organizations need to closely interact. Consequently, science for agricultural development has become more inclusive, consultative and participatory. It reveals new opportunities but also new challenges, such as of responding to and engaging with a widening range of interest groups, agendas, priorities and opportunities. According to the CGIAR Science Council (2005) (in accordance with OECD, 2005abc) "such a systems perspective on agricultural innovation offers the potential of realizing the promise of science and technology in the context of socio-economic development and merits increased investment in future."

4.4.2 Alternatives in integration

There are two dominant types of disciplinary integration, both appearing increasingly within agricultural sciences. The first is integration of two or more disciplinary traditions to form a new discipline involving formulation of new theoretical grounds and methodologies. Ecological economics is one example. The second type is constructive interaction among separate disciplines.

Historical evidence suggests that interdisciplinary communication and interaction often plays a key role in the emergence of new research fields, i.e., in scientific renewal and development. Thinking collectively about complex problems requires crossing boundaries both horizontally (across disciplines) and vertically (involving policy-makers, experts, practitioners, public) (Klein, 2004). This leads to participatory approaches and transdisciplinarity and thus problem solving that crosses both disciplinary boundaries and sectors of society (Scholz and Marks, 2001). It can also involve efforts towards a new unifying theory. For example, it has been proposed that agroecology could be developed and defined as an embracing discipline for studies on the entire agrifood system in all its ecological, economic and social dimensions (e.g., Dalgaard et al., 2003; Francis et al., 2003).

Constructive interaction among disciplines does not, however, necessarily imply a genesis of a new discipline. In fact, the continuous emergence of new disciplines would merely result in the continuous reconstruction of new boundaries to be overcome.[8] The greatest value of any emergent, integrating discipline would be in establishing a common language and concepts for the participating researchers. On the other hand, interdisciplinary studies benefit from the accumulated knowledge, methodologies and traditions of the contributing disciplines. In many cases an interdisciplinary orientation would supply a broader and more flexible selection of the expertise and methods required for a sound result than would reliance on the creation of new disciplinary approaches (Heemskerk et al., 2003; Lele and Norgaard, 2005; Kahiluoto et al., 2006). The short time frame of one study and the continuously evolving research needs and objectives underline this conclusion.

[8] This would be the case even if the development of the new disciplines would be based on the unifying and expanding "rhizome model" rather than the more commonly used hierarchical model, which involves branching into distinct, semiautonomous fields of enquiry (Bruun et al., 2005).

Indeed, disciplines can be interpreted as just administrative academic artifacts, which have lost their significance as an organizing principle of science during the last quarter century (Lele and Norgaard, 2005). For example, the biological sciences have dropped the historic disciplinary distinctions, e.g., between the plant and animal worlds and are organizing more according to the level of analysis from genes to organisms to ecosystems. The diversity of approaches within a discipline and the possible relatedness with an approach of another discipline suggest forgetting disciplines and thinking in terms of scientific community (Lele and Norgaard, 2005). A scientific community is a group of scholars who share a characteristic. The characteristic can be (1) a subject, (2) assumptions about the underlying characteristics of the factors they study, (3) assumptions about the larger world they do not study and about how what they do study relates to the larger world, (4) the models they use, (5) the methods they use and (6) the audience they strive to inform through their research. Crucial, according to them, is recognizing that organizational charts of universities do not coincide with the most important markers of difference and similarity found on different dimensions and scales. This recognition facilitates crossing boundaries between scientific communities.

4.4.3 Barriers faced by integration

Interdisciplinarity is increasingly considered the ideal of research but it relies heavily on high-quality disciplinary research (Lockeretz and Anderson, 1993; Bruun et al., 2005; Kahiluoto et al., 2006). In applied sciences, such as agricultural and food sciences, integrative approaches are becoming more widely accepted in education, research and extension and in some contexts are increasingly demanded by funding organizations. Participation is also an approach increasingly demanded by donors of international research funding.

Although disciplinary borders have always been crossed in research, integrative approaches are difficult to handle, not yet well understood and their adoption and wide application still face major constraints (Duncker, 2001). Seven major barriers for interdisciplinarity exist: structural, knowledge, cultural, epistemological (i.e., relating to the theory of knowledge), methodological, psychological and reception barriers (Bruun et al., 2005).

The structure of organizational decision-making and the organizational norms affect the character of research and education. The current disciplinary organization of science has been criticized as hampering interdisciplinary research and educational programs (Bruun et al., 2005), though obviously there are numerous such ongoing programs and projects. Fragmentation starts with the structure of governments, is present in the disciplinary organization of universities and research institutes and is present in the contents of education and training programs.

An important obstacle for interdisciplinarity is that scholars who review interdisciplinary project proposals have no training in the quality criteria for interdisciplinary research and that boards of reviewers often don't cover the breadth of knowledge required to give full justice to interdisciplinary research proposals. On the basis of an empirical study interviewing experimental researchers at major interdisciplinary research institutes, main quality criteria include: (1) Consistency with multiple separate disciplinary antecedents' (i.e., the way in which the work stands vis-à-vis what researchers know and find tenable in the disciplines involved); (2) Balance in weaving together perspectives (i.e., the way in which the work stands together as a generative and coherent whole); and (3) Effectiveness in advancing understanding (i.e., the way in which the integration advances the goals that researchers set for their pursuits and the methods they use) (Mansilla and Gardner, 2003). Scientists throughout much of NAE are primarily based on their refereed publication output and its impact (measured in terms of impact factors and citations). Scientific journals with high impact factors tend to have little interest in applied interdisciplinary research and often have a disciplinary orientation.

Cultural barriers include language problems (such as different technical terminology) and differences in methodologies. Problems with communication and understanding across disciplines are seen by many as the main barrier for successful multi- and interdisciplinary settings (Bärmark and Wallen, 1980; Porter and Rossini, 1984; Bauer, 1990; Duncker, 2001; Pawson and Dovers, 2003; Helenius et al., 2006; Kahiluoto et al., 2006; Mäkelä, 2006).

Epistemological problems occur when disciplines fundamentally interact. Reception barriers appear when issues and assumptions that are dealt with are unfamiliar to the established disciplines and thus not easily accepted. Problems in paradigms, communication, organization and cognitive development are often faced in interdisciplinary research (e.g., Bärmark and Wallen, 1980). The creation of "trading zones" for exchange and "interlanguages" (more or less elaborate) may be required for successful cooperation across disciplinary borders (Duncker, 2001). Many efforts failed partly because the representatives of the separate intellectual communities did not recognize the barriers created by their separate ways to understand and approach the problems (Bärmark and Wallen, 1980; Lele and Norgaard, 2005).

Institutions that have a history of interdisciplinary orientation typically can move more quickly to adopt new initiatives along these lines than those that do not (Feller, 2005). And a number of studies have indicated that the barriers for interdisciplinarity and participation can be overcome. Conceptual tools to overcome the most prominent barrier in interdisciplinary studies—communicating and understanding across the disciplinary borders—have been developed (e.g., Duncker, 2001; Heemskerk et al., 2003). It is an important challenge for science education to improve proficiency in interdisciplinarity through a better understanding of the philosophy and theory of alternative approaches and methodologies in science. This can be achieved through development and adoption of appropriate procedures and tools for communicating and through practicing interdisciplinarity (Venkula, 2006).

Barriers faced by participatory approaches are largely similar to the barriers faced by interdisciplinary approaches but are often even higher for the former and more diverse as participatory approaches usually cover integration both horizontally among disciplines and vertically among different actors. For participatory approaches involving non-academics from different parts of food systems and fields of life, communication is more challenging than in integrated approaches involving solely academics. Tools to facilitate dialogues involving different values of stakeholders have been developed (e.g., Wolfe et al., 2002). Another major barrier for participatory approaches are the limited appreciation, rewards and career opportunities for researchers, a limitation which is more pronounced than in the case of interdisciplinarity. A barrier of growing significance, specific for participatory approaches is the "digital divide" (i.e., the difference in access to information technology) between the developed and the developing world and between the rich and the poor (Rao, 2005; Britz et al., 2006; Chetty et al., 2006). It has contributed to inequity and inefficient use of AKST (Bouma et al., 2007).

The expectations for integrative scientific approaches and the practical preconditions offered by the performance of the knowledge, science and technology generation system often seem to be in conflict and it has been suggested that for integrated approaches to be feasible and to become more commonplace, institution-level changes in curricula, incentives, evaluation criteria and accountability would be required (Lele and Norgaards, 2005).

4.4.4 Risks associated with integration

Although interdisciplinarity has been increasingly considered the ideal of research, increasing the level of integration has so far been a rocky path in some countries in NAE. The barriers described above are not the only challenges as there are also risks associated with increasing the extent of integration. These risks need to be minimized and managed carefully to ensure that integrative approaches help rather than hinder achievement of the goals of this assessment.

Interdisciplinary research relies heavily on high-quality disciplinary research. However, many of the changes implemented in recent years particularly in Western Europe in the name of integration, streamlining and quality control have resulted in cuts in funding of disciplinary research. This research has long provided essential knowledge for AKST and gradual cuts have caused confusion and disillusionment of scientists involved in such research. This development has resulted in fragmentation and loss of continuity of the science base, weaker links between science and application and less security for the future (OSI, 2006). It might also limit the capacity to respond adequately to current as well as future challenges facing agrifood systems. It has been recommended that the costs and time needed for rebuilding expertise be included in evaluations of area of research considered for discontinuation. Finding the optimal balance between integrated approaches and disciplinary approaches has been (and will continue to be) an important challenge. The strategic planning of public sector funding organizations needs to be better joined up at a national level to help maintain crucial scientific expertise and facilities (OSI, 2006). There are also initiatives to improve strategic planning at an international level to avoid duplication of effort at a time of increasing funding constraints (EURAGRI, 2005).

Balancing the influence of stakeholders in the development of AKST agendas to ensure that funds are focused on the areas most relevant to society and the environment, has been a challenge (see also 4.5.3.3 and 4.5.5). Despite much progress in theoretical work there is still little agreement amongst social scientists regarding the best methodologies to be used for citizen participation (Pidgeon et al., 2005). Analytic-deliberative processes that can accommodate a very wide plurality of views in public policy discourses and decisions have been recommended (Pidgeon et al., 2005). New technologies represent particular challenges in terms of citizen participation. The problems the general public faces in judging the potential risks and benefits associated with biotechnology are one recent example. Research suggests that in general, people rely on the judgment of trusted others rather than making choices vis-à-vis technologically complex new products in a rational fashion (Grove-White et al., 2000). It is, however, noteworthy that choices of citizens are also contributed to by their value systems, where scientists are no experts.

Media have so far preferred to exploit and heighten public fears of certain new technologies although hope has been expressed that they can change "to encourage mature discussion of the implications of uncertainties and unknowns surrounding new technologies and their insertion into everyday life—as necessary for constructive public debate" (Grove-White et al., 2000). The same encouragement can be addressed at other organisations the general public uses as trustworthy sources of information. An important aspect is also thought to be the need to pay more attention at the earliest development stages to the social constitutions (i.e., the particular social values and assumptions) new technologies are perceived to have (Grove-White et al. 2000).

Following 15 to 20 years of evolution, participatory techniques are now accepted as part of the mainstream science for agricultural development, especially in developing countries. Participation is an inherent part in "innovation systems". The difference between one-directional mediation of information and creation of multidirectional, interactive knowledge networks is fundamental (Table 4-2) (Buhler et al., 2002). On the other hand, it has been argued that the more traditional approaches (e.g., technology transfer) have in places been very successful, providing the appropriate infrastructure was present and that increased use of participation techniques as a research tool has not had a clear impact (Bentley, 1994). Real impact would require more than short-term technology development efforts (Humphries et al. 2000). Seeing farmer participation in research primarily as a route to the empowerment of local populations and almost independent of any eventual research outputs has been questioned (Sumberg et al., 2003).

A more integrated approach and multi-disciplinary research programs should not lead to less disciplinary research and a depletion of agricultural research but should be seen as a reinforcement of agricultural research. The integration of different structures carries the risk of increasing the administrative burden and wasting funds where it has led lead to an additional layer of bureaucracy. Approaches in integration that do not increase the layers of bureaucracy may be a challenge but would be a more efficient use of limited resources.

4.5 Development of Structures, Funding and Agenda of AKST

4.5.1 Establishment of structures

Much of the invention and technological improvement in NAE agriculture before 1840 and to a lesser extent up to

1900, came about through the activities of private individuals such as innovative farmers, blacksmiths and estate owners. Accordingly, a large share of the technical advances from this informal system was realized in the form of mechanical innovations rather than biological advances (Hayami and Ruttan, 1971; Evenson, 1983). Agricultural societies provided early support to teaching and research institutions. Both the performance and the funding of agricultural research in the U.S. has since then been shared between private and public interests.

In most countries in NAE formalized agricultural research organizations were established from the 1840s onwards. The first experimental stations staffed with professional scientists were established in the UK, France and Germany, followed soon by most other European countries. By 1875 there were 90 national experimental stations in Europe (Grantham, 1984). In the US, acts of Congress assisted the states in establishing land-grant colleges to teach agriculture and applied sciences (in 1862), carry out agricultural research, establish the land-grant experiment stations (in 1887 and 1890) and authorize statewide informal education at colleges (in 1914). In contrast to the German model, the US experimental stations were established under the direction of a state land-grant college or university. In order to assure the dissemination of the knowledge produced by these investments, the Cooperative Agricultural Extension Service was created in the US as a partnership between federal, state and county governments. In Europe higher education in agriculture was in most cases arranged as an activity of existing universities. In further contrast to the US, distribution of their results to farmers was not a major focus of the activity of the experimental research stations in Europe. Farmers' institutes, traveling agricultural-college short courses and field demonstration activities were turn-of-the-century precursors to extension.

The second wave of public commitment to expansion of agricultural R&D in NAE took place in the first half of the 1900s, based on crucial developments in the basic and applied sciences, e.g., in chemistry, mechanization and genetics. These developments fundamentally changed the roles of private and public actors (organizations and their personnel, etc.) in science. This change coincided with the end of the Second World War, a period when science (and agricultural R&D in particular) was widely considered a potential source of major improvements in social welfare. This perception fostered a strong third wave of development of structures for agricultural R&D.

The governmental responsibility for AKST is divided in many different ways in NAE, but the responsibility is often shared among different ministries. In Russia and the now independent former Soviet states a highly centralized AKST was established. In contrast, in the US decision-making was decentralized and occurred largely at the regional level (Table 4-3), a situation that has fostered diversity, innovation and local adaptation (Miller et al., 2000). In countries in Western Europe, levels of decentralization vary. Germany is an example where decision making in agricultural research and education also occurs to a great extent at the regional ("Laender") level.

As outlined above, the higher education, agricultural research and extension systems of the US were established in a relatively integrated way. In contrast, in Russia and in the CEE countries which followed the Russian model, AKST organization have been highly divided and research, education and training were not integrated. In Russia, AKST is still divided into science academies that also provide the highest education to universities, research institutes and training systems. The public extension service is still poorly developed. The decentralization and integration of US AKST is considered an important part of the US's success in increasing productivity of agriculture (Huffman and Evenson, 1993). In a comparative analysis of the development in productivity of agriculture in relation to the organization of AKST and of the development of US public education in relation to the

Table 4-3. **A comparison of agricultural higher education in the US and Russia.**

Issue	Institution	Russia
Curriculum	Determined by faculty at each institution	Approximately 75% set by federal government
Course content	Set by faculty at each institution	Centrally determined
Enrollment	Determined by market and campus	Quota determined centrally
Tuition	Set by individual campuses	Quota students free; above quota set by campus
Student/faculty ratio	Individual campus	System
Entrance requirements	Campus determined	Centrally determined
Greatest fiscal support	State government and tuition	Federal government
Links to research and extension	Inherent in land-grant system	No extension system and only weak links to research
Quality and applicability of education	Quality comparable, applied aspects greater	Quality comparable, lacking in application
Years of education	Comparable	Comparable
Senior project	Not required in most cases	Required

Source: Miller et al., 2000.

organization of schooling research, easy access to important advances in related sciences and scientific methods seemed to be of major importance for success (Huffman, 1998). In contrast, the inefficiencies in Russian agriculture were a major factor in several changes in Soviet leadership and finally the collapse of the Soviet socialism (Miller et al., 2000). In the rest of Europe, the integration of universities, agrifood research and extension varies significantly among countries. For example, the Swedish structure is similar to that in the US while in France, Denmark and Finland the higher education and strategic R&D are organizationally separated.

International agricultural R&D (see also 4.2.2) represents in comparison a relatively recent institutional innovation as it was only initiated in 1943 with the Mexican government—Rockefeller wheat research program. This initial program became a model for many subsequent international agricultural research initiations in the 1960s, including the four international agricultural centers CIAT (tropical agriculture, Colombia, established in 1967), CIMMYT (maize and wheat, Mexico, 1966), IITA (tropical agriculture, Nigeria, 1967) and IRRI (rice, Philippines, 1960). The subsequent development of the international agricultural research centers took place mostly under CGIAR, established in 1971 to mobilize science and financial support to serve the needs of the poor. CGIAR is a strategic alliance of countries, international and regional organizations as well as private foundations supporting international research centers, which work with the national agricultural research systems and civil society organizations including the private sector. CGIAR is funded mainly through the development aid funds of developed countries, either directly to the centers or through contributions to agencies such as the World Bank, the Asian Development Bank and the European Union. CGIAR established a Technical Advisory Committee (TAC) to ensure the relevance of CGIAR-supported research and the quality of science at the centers. The expansion phase of the international AKST was in the 1970s.

In many developing countries, the National Agricultural Research Systems (NARS) started to develop based on the inherited colonial export-oriented R&D structures, which were built with the "top-down" principle. Not surprisingly the structures in the developed and developing countries were therefore closely related. It is estimated that approximately 90% of agricultural researchers in Africa were still expatriate in the early 1960s but this proportion had declined to 20% by the early 1980s (Buhler, 2002).

4.5.2 Drivers of change

Following decades of government service expansion, the mid-1970s to the late 1980s became an era of less government. However, a new paradigm emerged for the 90s: not less government, but better government, involving a shift to more enlightened regulation, improved service delivery, devolution of responsibility, openness, transparency, accountability, partnership and "new public management" (OECD, 1999).

In many developed and developing countries, public agricultural R&D policy changed dramatically between the early 1980s and the end of the 1990s. The long period of sustained growth had ended (see 4.5.3) due to general fiscal constraints and a more skeptical view of the social benefits of R&D. Clearer justification and accountability for R&D funds was requested. In Eastern Europe, the drastic changes in the socio-political system led to a re-orientation towards a market economy from about 1990, although not to the same extent in all affected countries. These changes were associated with a period of disturbance and restructuring of agrifood systems and AKST. The large budget deficits in the 1980s forced also US agricultural R&D into a contracting mode (Huffman and Just, 2000; Alston et al., 1998), while individual states largely resisted pressure to shift to peer-reviewed competitive grants (Huffman, 2005) (see 4.5.4).

On the other hand, new participants emerged in the private research sector in NAE following the introduction of incentives such as periodic strengthening of intellectual property rights (IPRs) (e.g., in the 1930s, 1970s and 1980s) and the subsequent shift of the boundary between publicly- and privately-funded research (Fuglie et al., 1996). This development was intentionally fostered by governmental science policies. During the 1990s, the shortcomings of the public research model then also contributed to the gradual emergence of private sector/broadly market-oriented reforms in agricultural R&D investments (see IAASTD Global report). The transition was facilitated by structural adjustment policies imposed in many NAE countries, the global changes in trade regime as well as developments in biotechnologies. Governmental science policies were also modified to broaden the scope of agricultural R&D and increase its efficiency (van der Meer, 1999; Huffman and Just, 2000; Rubenstein et al., 2003). This has made the agricultural R&D environment increasingly competitive and proprietary.

During the last decade, many OECD countries have adopted the explicit goal to change the structure and function of their agricultural R&D organizations. They tended to bring AKST policies closer to the general public KST policies. Also, there was a shift from the unidirectional paradigm of knowledge transfer to a paradigm of interactive knowledge networks involving multiple stakeholders, which led to various forms of peer review and merit review (OECD, 1999) of research, educational and extension programs.

In a study (Alston et al., 1998) of public agricultural R&D during the last quarter of the 20th century in developed countries (using the five OECD countries US, Netherlands, UK, Australia and New Zealand as case studies) the following major institutional changes were identified: (1) a shift towards using public funds for more basic research rather than applied or near-market research, (2) a trend towards joined funding of near-market research using different mechanisms, (3) strengthening of oversight and accountability mechanisms, (4) measures to increase competition among researchers for productivity and resource allocation, (5) measures to privatize public agricultural research institutions and (6) increasing the cost effectiveness of public agricultural research facilities.

The similarities between the countries are derived from a common set of "vectors for change," which include (1) the more market-oriented "laissez-faire" role of the government in the management of the national economy, (2) the changing nature of the scientific and agricultural research, (3) the development of a more skeptical view of the potential benefits of agricultural R&D due to the decrease in the share of agriculture in the national economy and (4) the growing influence of the "non-traditional" interest groups such as agri-business, food industry, NGO's (like environmental and consumer associations), food-safety lobbies and in the international AKST also farmer organizations (Alston et al., 1998).

4.5.3 Development of funding and agenda in NAE in the global context

4.5.3.1 Development in NAE

Public agricultural R&D expenditures

Between 1945 and the mid-1970s there was a period of rapid growth rates in public agricultural R&D expenditures in NAE. Many NAE countries financed large-scale expansions in their national science research-education systems. Alston et al. (1998) analyzed the data available for 22 OECD countries,[9] which show that agricultural R&D spending in the OECD grew on average by 7 to 8% per year during the 1950s and 60s. Alston et al. (1998) suggested that such growth rates were probably not sustainable in the long term and that by the 1970s in many OECD countries publicly funded agricultural research had become a mature industry characterized by modest rather than rapid expansion. The 1970s saw a growth rate of 2.7% per year on average for the OECD analyzed. Some NAE countries had higher growth rates, e.g., the annual growth rate was 4.2% in the Netherlands for the period 1971-81. However, in the 1980s occurred a further decline in real public agricultural R&D expenditure growth rates in many regions of NAE (Figure 4-3, Table 4-4 [both in Annex H]). While the annual growth rate in the US remained relatively stable (2.3% for the period 1981-93), the growth rate in the Netherlands was only 0.9% and expenditure even declined in the UK by 0.2% over the same period.

The dramatic declines in growth rate in the Netherlands and the UK were associated with relatively radical changes in the institutional organization and management of public agricultural research during the 1980s compared to other countries in NAE (Alston et al., 1998). In the 1990s public expenditures in NAE recovered somewhat but growth rates remained modest compared to the 1960s and 1970s. Despite minimal funding increases, demands on the public system grew increasingly complex due to increasing awareness of food safety issues, environmental externalities and increasing food consumption (Rubenstein and Heisey, 2005). This has led to efforts within the EU in recent years to coordinate funding for AKST to minimize duplication of research between member states. Such efforts have attracted criticism by the farming community concerned that they may reduce national competitiveness.

U.S. federal funding for extension has been declining in scope for more than a decade and support for agricultural experiment stations is also now under attack, partly because of an increasing preference for competitive grants over formula-based funding and also because of fragmentation of the constituency for such funds (Busch, 2005). The key niche occupied by colleges of agriculture has shrunk in scope and there has been a tendency to shed specialists dealing with minor crops while maintaining competence in major crops. These crops are, however, increasingly controlled by the private sector, while minor crops are of little interest to the private sector as they lack the potential for significant profit in input supply. These and other factors contributed to weaken the once strong links between farmers and especially farm commodity groups and colleges of agriculture (Busch, 2005). For more than a century, the colleges of agriculture were at the center of the research agenda in the US. They had few competitors as private biological research was mostly unprofitable.

[9] OECD totals reported by Alston et al. (1998) included the following NAE countries: Austria, Belgium, Canada, Denmark, Finland, France, Germany, Greece, Iceland, Ireland, Italy, the Netherlands, Norway, Portugal, Spain, Sweden, Switzerland, the UK and the US. Non-NAE countries included in their data were Australia, Japan and New Zealand.

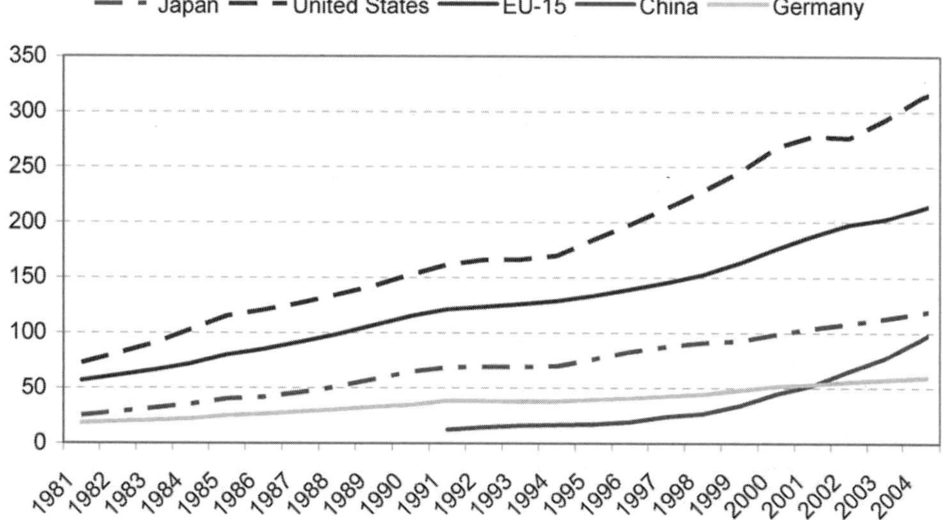

Figure 4-4. *Gross domestic expenditure on R&D (billion current parity purchase price dollars).* Source: OECD, 2006b.

The share of public agricultural research funds given to universities increased considerably from the 1970s onwards in parts of NAE, particularly the US, UK and the Netherlands, indicating a shift towards more basic research. Between 1971 and 1993, the university share of public agricultural R&D spending increased in the UK from 2.3% to 14.7%, in the Netherlands from 14.9% to 31.9% and in the US from 67.3% to 74.1%. In contrast, in the other countries analyzed by Alston et al. (1998) the average share of public agricultural R&D given to universities remained about 28% over the same period.

Public R&D expenditure relative to the value of agricultural output

The public agricultural R&D intensity ratio (ARI; public agricultural R&D expenditure relative to the value of agricultural output) increased throughout the period 1971-1992 in most NAE countries analyzed (Alston et al., 1998). The average science and technology research intensity ratio for the countries increased by a much smaller proportion than the ARIs.

Although these research intensity ratios suggest that agriculture has been treated relatively favorably in many NAE countries in terms of public R&D funds, a different picture emerges when trends in agriculture's share of total publicly performed science and technology are examined (Alston et al., 1998). In fact, the share of agricultural R&D out of the overall R&D funding declined in the 1980s to the early 1990s in analyzed countries (Figure 4-4) (Alston et al., 1998). In the Netherlands, for example, agriculture's share of the total public R&D budget declined from 14.5% in 1981 to 12.4% in 1993. In the US it declined from 6.2 to 5.6% and in the UK from 7.1 to 6.6% over the same time span. Across the 22 OECD countries analyzed, agriculture's share of the total public science and technology R&D budget declined on average from 8.9% in 1981 to 7.4% in 1993, a proportional decrease of close to 17%. A likely cause for the changes was pressure to reallocate funds to other science R&D programs (such as health) (Alston et al., 1998). This shift was also reflected in the declining space devoted to agriculture and natural resources in major journals (e.g., economic journals) while coverage of issues such as manpower, labor, population developments, welfare programs, consumer economics as well as urban and regional economics increased (Ryan, 2001). A meta-analysis of all the available studies of the impact (in terms of rates of return) of agricultural R&D between 1953 and 1998 found no evidence of a decline in returns to investments throughout these decades (Alston et al., 2000). These results imply that equally large returns to current spending on agricultural R&D will also be feasible in the future (CGIAR Science Council, 2005). During the 1990s agricultural R&D spending in the US increased again, from 3216 million in 1991 to 3828 million in 2000 (in 2000 international dollars), representing 16.1% and 16.6% of the global total, respectively (CGIAR Science Council, 2005). The US is also increasing funding for more basic agricultural research (Danford, 2006).

Privatization of R&D

In most NAE countries, the private sector has had a longstanding triple role in the public agricultural R&D: firstly through involvement with the management of the publicly provided funds as the primary user, secondly through funding publicly performed research in public sector organizations and universities and thirdly by performing research using public funds. There was also a net flow of public funds to private research (Alston et al., 1998).

Investments of the private sector in agricultural R&D have generally increased since the early 1980s. Growth of the private sector spending slowed at the end of 1990s but the balance continued to shift towards private sector funding. Privately performed R&D has become a prominent feature of agricultural R&D in rich countries including most countries in NAE (Alston et al., 1998; Rubenstein and Heisey, 2005) and constituted by 2000 around 55% of all agricultural R&D in developed countries (Table 4-5; Pardey et al., 2006).

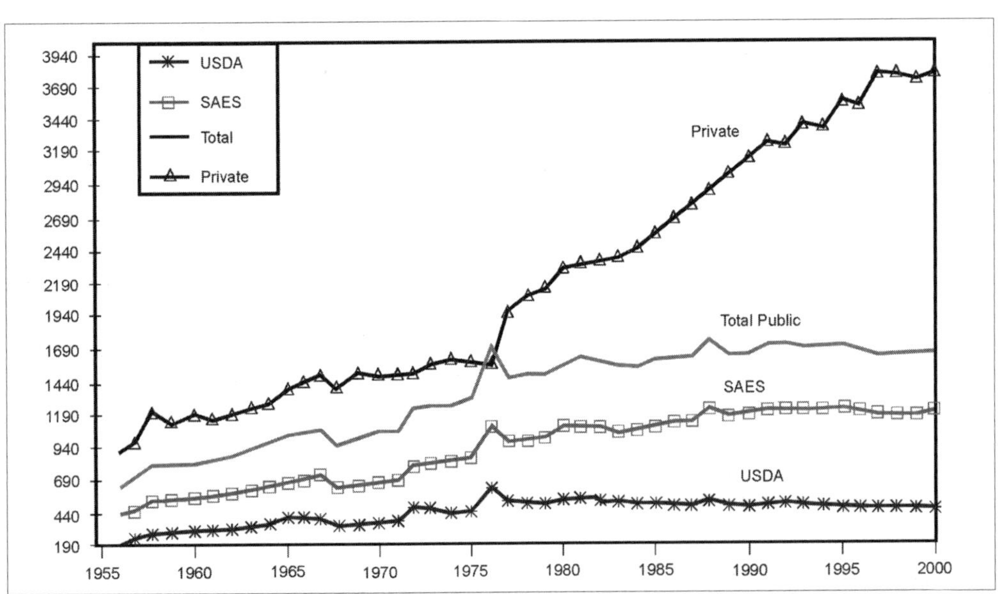

Figure 4-5. *Funding for agricultural research in the US (thousands of constant 1984 dollars; SAES = State Agric. Expt. Stations).* Source: Huffman and Everson, 2006.

The relative importance of private agricultural R&D in total agricultural R&D varies, however, between countries. For example, the private sector performed over 60% of all agricultural research in the UK and more than 50% in the US and the Netherlands by the late 1990s (Alston et al., 1998) (Figure 4-5, Table 4-4).

It has been suggested that the application-orientation of the private sector to some extent fills the gap between technology generation and extension that existed in the public research model. However, there is concern that the shift towards a higher proportion of privately funded agricultural R&D moves the focus too much away from public goods, equity and distributional issues (BANR, 2002). As the private sector can retain few financial returns in the short term from innovations that improve environmental benefits and food safety, the public sector remains the primary source for new technologies with these characteristics (Rubenstein and Heisey, 2005). In recent years, as environmental, food quality and income pressures in agriculture increased, the private sector has started to take a more long-term view and fund R&D into more sustainable farming practices (Morris and Winter, 1999; Walker, 2001; Voluntary Initiative, 2007).

Shifts in R&D agendas

Public research and private sector research inevitably tend to focus on different areas of R&D. For example, approximately 12% of private R&D focused on farm-level technologies compared to around 80% of public R&D in 1993 (Alston et al., 1998). Chemical research accounted for more than 40% of private agricultural research in the US and the UK and for nearly three quarters of privately funded agricultural research in Germany, while 58% of the private research in the Netherlands focused on food products. Particular areas of private agricultural R&D tend to be concentrated in particular countries. For example, Japan, the US and France account for 33, 27 and 8%, respectively, of all food processing research carried out by the private sector in OECD countries. Chemical research related to agriculture is even more concentrated with the US, Japan and Germany representing 41, 20 and 10% of all reported private-sector research (Alston et al., 1998). Data available for the US and the UK show a dramatic shift in private sector expenditures from farm machinery and post-harvest processing in the 1960s to agricultural chemicals, plant breeding, veterinary and pharmaceutical research by the end of the 1990s (USDA, 1995; Thirtle et al., 1997).

Since the Second World War, the scope of agricultural R&D in NAE broadened considerably and increasingly included issues relating to post-harvest, food chain, nutrition, rural development, environment and sustainability (Huffman and Evenson, 1993; OECD, 1999). Funding initiatives to increase integration of social and life sciences and economics have increased in NAE in recent years. Examples include the 6th framework program of the European Commission[10] and the Rural Economy and Land Use Programme (RELU)[11] in the UK. On the other hand, it has been suggested that AKST has made only a limited contribution to national policy making, that this has often been primarily by economic research and that contributions to public debate have been sporadic (OECD, 1999).

Funding and scope of extension

There has also been an increasing involvement of the private sector in agricultural extension (Umali and Schwartz, 1994). The last decade has seen increased demands on the expertise of agricultural advisors, particularly in respect to agri-environmental issues. Yet at the same time public funding for extension services has been reduced throughout much of NAE, which has weakened the links between science and application (Ingram and Morris, 2007; Lambert et al., 2007). Public extension systems have been substantially downsized or phased out altogether in some European countries (Read et al., 1988; OSI, 2006). In North America and Western Europe, technical support to farmers is now to a large extent provided by agricultural specialists who work for private sector firms, especially input supply companies. Some Eastern European countries, such as Poland and Hungary, still have large public agricultural extension systems.

The focus of public sector extension services in parts of NAE has gradually changed from an agricultural production-centered advisory regime to an environmental regime (Winter et al., 2000). There has also been a switch of funding that support farmers' activities to control farming business and to address issues of negative externalities. The emphasis on control is to a large part a result of concerns about issues such as consumers' freedom of choice and crises like BSE, foot and mouth disease as well as avian influenza. Advisors remain an essential component of the agricultural knowledge system despite increased use of other mechanisms that increase farmers' learning, such as demonstration farms, farmer-farmer interaction and group learning. Farm visits by advisors still are the most effective of all methods of communication and the most valued by farmers (Ingram and Morris, 2007). In fact, advisors have become more important as farming, markets and regulations become ever more complex. Their role is further amplified by farmers' increasing reluctance to share knowledge with their peers in order to retain a competitive advantage. However, the role played by different types of agricultural advisors in the transition to more sustainable farming systems is still only partly understood (Ingram and Morris, 2007). Extension services seem also to face problems serving the increasing numbers of part-time farmers (Suvedi et al., 2000).

Recent developments

Governments of the OECD countries have in the 1990s been prepared to fund all or most higher education costs, depending on their general policy on tuition fees (OECD, 1999). However declining student numbers have increased the pressure to reduce public funding. They are prepared to fund also "basic" and "pre-competitive" sectoral research but economic sectors are increasingly encouraged to fund sector specific research. Responsibility for extension/development work has been increasingly shifted towards the clients. A number of countries have a strong commitment to fund public-good type extension, while most extension workers are nowadays involved in monitoring and implementing public regulatory schemes (OECD, 1999). There

[10] http://ec.europa.eu/research/fp6/pdf/fp6-in-brief_en.pdf.
[11] http://www.relu.ac.uk/ (accessed 27 Feb 2007).

Table 4-5. **Public and private agricultural R&D expenditure, circa 2000.**

	Expenditure					
	Million 2000 international dollars			Share (%)		
	Public	Private	Total	Public	Private	Total
Developing countries	93.7	6.3	13,688	12,819	869	100
Developed countries	44.8	55.2	22,767	10,191	12,577	100
Total	63.1	36.9	36,456	23,010	13,446	100

Source: Pardey et al., 2006.

has been a recent trend for governments to fund programs rather than institutions and an effort towards addressing competitive grants to longer-term programs or themes rather than to individual projects (OECD, 1999).

4.5.3.2 NAE in the global context

Total spending on science in the world is highly spatially concentrated. The US, Japan, Germany, France and the UK accounted in 2000 for 68% of the world's total science spending (CGIAR, 2005). These five countries together with Italy, Canada, China, India and South Korea contribute 81.4% of the world total scientific investment. In contrast, the share of the 80 countries that spent least on science had slipped further from only 0.36% from the world's total spending in 1995 to 0.33% in 2000, which represents a decrease by almost 10%. These 80 countries account for 7% of the world's population and 1.7% of the world's GDP (CGIAR, 2005; Figure 4-3).

Concerning agricultural research, over the past two decades, worldwide public investments have increased by 51% in inflation-adjusted terms (from an estimated $15.2 billion in 1981 to $23 billion in 2000) (CGIAR, 2005). However, also agricultural R&D has concentrated in a handful of countries. The US, Japan, France and Germany continue to provide two-thirds of the public research done by rich countries in 2000 with little change compared with two decades before (CGIAR, 2005). Similarly, five transition economies (China, India, Brazil, Thailand and South Africa) accounted for 53% of the developing world's agricultural research, up from 40% in 1981. In particular, China, Brazil and India have expanded their basic research capacity, reducing their dependence on adaptive R&D and becoming potential sources for the poorest countries relying on adaptive research (CGIAR Science Council, 2005).

Spillovers of science and technology are increasingly recognized as an important feature of the history of agricultural development (CGIAR, 2005). Half or more of the local productivity gains in agriculture during the past decades can be attributed to *"spill-in" technologies* developed elsewhere, even if spillovers have turned out difficult to plan for. Unfortunately, spillovers can sharpen the gap between rich and poor countries due to different facilities for utilization (Alston, 2002). For example, research conducted in CIMMYT and IRRI in developing countries provided large economic benefits for the US, due to technology spillover (Pardey et al., 1996). Also, rich countries are increasingly moving away from technologies appropriate for poor farming communities. In addition, regulatory policies like IPR, biosafety protocols, trading regimes and specific restrictions for moving genetic material are increasingly influencing the extent to which spillovers of R&D in NAE are feasible or economically viable (CGIAR Science Council, 2005).

A central element for research and donor organizations in NAE has been the provision of advanced training to help capacity building, so that individuals and institutions in developing countries become more self-reliant in identifying and executing AKST. Capacity building is generally targeted to individuals, e.g., scholarships and fellowships. Examples include IARC Fellowships (CIMMYT, Vavilov-Frankel/IPGRI), Generation Challenge (CGIAR), UN, TWAS, Commonwealth Scholarship and Fellowship Plan (CSFP) and fellowships through research organizations (e.g., Rothamsted International) and universities (CSFP, 2007; Generation Challenge Programme, 2007; IFAR, 2007; Rothamsted International, 2007). However, although money is provided for training, there is usually no funding to help scientists to continue the work and training received when returning to their home institutes.

Changes in funding priorities brought about by government policies in European countries lead to a gradual erosion of scientists qualified to work in agricultural research for developing countries. There has also been a fragmentation of the researcher skills base, so that experts are spread amongst a large variety of research institutes, universities and non-governmental organizations, rather than a small number of specialized departments. The decline in expertise has been exacerbated by the closure of undergraduate courses in agriculture (NRI, 2002; Science and Technology Committee, 2004; Delgado and Ramos, 2006).

4.5.3.3 International AKST

NAE countries play a major role in funding and shaping agendas for international AKST. This subchapter can only provide a short outline of the changes in funding of international AKST in the last decades. A more detailed analysis is provided in the global report of the IAASTD.

CGIAR is funded mainly through the development aid funds of developed countries (many of which are based in NAE), either directly to the international research centers or through contributions to agencies such as the World Bank, the Asian Development Bank and the European Union. The total financial contributions (in US dollars) to CGIAR up to

2005 were 2517 million from European countries, 1536 million from North America, 1488 million from international and regional organizations (including the World Bank), 731 million from Pacific Rim countries, 199 million from foundations and 159 million from developing countries. During the current century, the top three contributors have been (depending on the year) the World Bank, US, Japan, UK and Commission of the European Community (CEC). There has also been a notable increase of 34% in the contribution from developing countries during the last reported year.

The funding of international agricultural R&D has followed a pattern similar to the funding of national public agricultural R&D in the contributing countries and the aid to agriculture (Table 4-6 [in Annex H]), although its expansion phase occurred later. During the first development period, from 1971 to 1982, real spending of CGIAR grew by 14.3% per year and further research centers covering more commodity crops were established (Alston et al., 1998). In the second phase (mid-1980s to 2001) real spending started to stagnate and finally decline, although the scope continued to broaden to cover more commodities, farming systems and environmental R&D. Spending grew only by 1.4% per year from 1985 to 1991 and only 0.7% (corresponding to a decline of 1.8% in real terms) from 1992 to 2001. Simultaneously, the share of restricted funding increased from 36% to 57% from 1992 to 2001. The budget of CGIAR has started to increase again in the present century with an average annual growth rate of 6.1% (CGIAR, 2005). CGIAR currently supports 8500 scientists and staff in 16 centers and more than 100 countries. However, in 2000 CGIAR only represented 1.5% of the 23 billion US dollars' global public sector investments in agricultural R&D and 0.9% of all public and private agricultural R&D spending (CGIAR Science Council, 2005).

The initial objective of international R&D was to increase the amount of food in tropical countries which faced serious scarcity. It therefore gave highest priority to research on cereals. Soon, however, the research portfolio was broadened to include not only wheat, maize and rice but also sorghum, millet, cassava, chickpea, potato, other food crops and pasture plants. Towards the end of the 1970s CGIAR branched out into several other new areas of activity such as livestock research, farming systems, conservation of genetic resources, plant nutrition, water management, policy research and services to national agricultural research centers in developing countries (CGIAR, 2006). During the 1980s, the environmental, multidisciplinary and systems-oriented, as well as cooperative approaches were strengthened, yet were not mainstreamed. At the end of the decade, forestry and agroforestry were also included and during the 1990s fishery and water management (CGIAR, 2006). In the 1990s the mission developed to emphasize sustainability and sustainable agriculture, nutrition and well-being, the interests of low-income people and food security. The productivity-enhancing agricultural research was reduced, while the expenditures on environmental protection and policy improvement increased (World Bank, 2003b). In the 2000s, the World Bank started to emphasize again the importance of raising agricultural productivity but stressed that a global rather than just a national or local view is crucial (World Bank, 2003b; CGIAR Science Council, 2006).

Globally, the real value of total development aid to agricultural R&D in the late 1990s was only 35% of that of the late 1980s (Falcon and Naylor, 2005). Agriculture's share of the total World Bank's lending fell from 25% in the mid-1980s to 10% in 2000 (World Bank, 2003b). This trend was a more general one (Table 4-6). In 2000, 37% of the world agricultural R&D was performed by private firms, but 94% of that in developed countries; while in many developing countries the share of the private sector in agricultural research continues to remain insignificant (Table 4-5).

There has been a widespread scaling back in investments in public R&D in agriculture among NAE countries although this been balanced to some extent through funding of agricultural R&D through non-traditional sources. There has been a shift from public to private agricultural R&D and a shift in governmental spending priorities (Pardey et al., 2006). These developments are likely to affect productivity prospects in NAE and spillover of ideas and technologies to poor countries. The current trend in NAE agricultural R&D away from staple foods to food quality and medical (including functional foods and gene-tailored diets) and other industrial applications of food commodities may contribute to a slowdown in sustainable productivity gains applicable to poor countries.

4.5.4 Changes in structures and management

There has been a general trend in OECD countries from the traditional model, where an agricultural ministry had sole responsibility for agricultural higher education, research and extension, towards a model with a ministry coordinating overall policies of KST. Especially agricultural higher education has moved to ministries overseeing higher education more generally, with some exceptions (such as Sweden where maintaining integration within AKST was considered most advantageous) (OECD, 1999). In the latter group, special coordination mechanisms between AKST and KST have often been developed.

Universities and research organizations in NAE have to a large extent retained their disciplinary structure and indeed new disciplines have emerged. In the CEE, since the break-up of the Soviet Union, more demand for extension services has emerged to compensate for the disappearance of the centralized chain-of-command system (Miller et al., 2000). The disciplinary structure of NAE universities and research organizations has been complemented by separate, issue-centered research institutes and the functions by cooperative, integrated educational and research programs. It has been predicted that the traditional, administration-oriented system of faculties based on basic sciences may disappear (Väyrynen, 2006).

Education

The number of students in agricultural sciences have decreased in North America and Western Europe during the 1990s, a process that has continued into the present century. In contrast, student enrolment in food sciences and engineering as well as nutrition and dietetics has increased. In Canada, for example, student enrolment in food science and engineering has increased by 62% since 1996 and student enrolment in nutrition and dietetics by 53% while student enrolment in agricultural sciences has dropped by 21% over the same period. Agricultural science disciplines are under increasing budget pressures at universities as well as at other research organizations (van der Meer, 1999; Delgado and Ramos, 2006). Also, agriculture has lost its important role in development studies at least in US universities (Falcon and Naylor, 2005). The situation in the CEE is different. At least in Russia the number of agricultural students increased by 50% from 1995 to 2000 (Miller et al., 2000)

Changes in paradigms, implications of increasing globalization and complexity of the rural world, the decline of employment and incomes in the primary sector, complex relationships between production and sustainability, cultural resistance to change of traditional societies and a decline in political influence of rural areas all increasingly challenged traditional higher education in agriculture (Delgado and Ramos, 2006). The syllabus in agriculture lagged behind society demands, student numbers decreased and university reorganizations led to the close of more and more agriculture related faculties. Initiatives have been started to increase internationalization and cooperation as one component of the drive to help higher education organizations meet these challenges (Delgado and Ramos, 2006).

Changes in research structures and management

Public agricultural research systems in NAE vary in terms of who funds, manages and performs research. Changes in scientific, economic and political factors have caused managers of national research organizations serious problems about how the organizations should be restructured over time, especially in face of policy inertia and increased costs (Read et al., 1988; Alston et al., 1998). In the UK and the Netherlands, for example, the public agencies involved with carrying out research have been consolidated and for some important parts commercialized. In the Netherlands, the share of private funding of Wageningen University and Research Centre rose from 25% in the 1970s to 40% in the mid-1990s and the research was rationalized and oversight streamlined. In the UK, the number of publicly funded research institutes fell by more than half during the same period. The agricultural extension services were increasingly commercialized or privatized in several countries in NAE, e.g., in the UK, France and the Netherlands (Read et al, 1988; van der Meer, 1999; Labarthe, 2006; OSI, 2006). The changes were usually temporarily linked with the change to more market-oriented "laissez-faire" governmental policy philosophy.

Comparatively little structural change has taken place in the public research system in the US until recent years. Historically, the US agricultural research system has been characterized by a decentralized, state-led structure, which fosters geographically specific applied research (Schultz, 1971; Huffman and Evenson, 1993). While the Federal Government provided about the half of all the funds during the last 50 years, state institutions have played an increasingly important role in funding and conducting state-level research. Since 1948 the State Agricultural Experiment Stations (SAES) system has been a considerably larger research enterprise than the USDA. In recent years, the proportion of the public agricultural funds spent on federal in-house research has declined to less than 30% (Rubenstein et al., 2003). The major force behind increasing the state share was matching federal funding with other (including state) funding. Farmer support for the US public system of research and extension is high although research suggests that the goals of some programs may be at odds with many farmers' needs and that there is a bias in the types of farms benefiting from land grant university resources, with smaller and diversified farms being largely underserved (Ostrom and Jackson-Smith, 2005). Fears of bioterrorism in the US led a few years ago to the creation of a National Institute for Agricultural Security (NIAS) to facilitate communication between the federal research system and the state-based agricultural research system (Nipp, 2004).

The changes towards more managed competition in agricultural research and from formula funds to competitive grants have been uneven and the institutions formed are country-specific. In the US the trend towards competitive grants in public agricultural R&D was slower than in other OECD countries, representing only 3% of the public agricultural R&D funds in 1995 (Alston et al., 1998) and 15% of USDA-funded state-level research at the end of the 1990s (Rubenstein et al., 2003). Usually allocation is based on *ex ante* claims (proposals) rather than *ex post* assessments about what was achieved. Allocation of funds to competing programs or institutions is at present based on frequent program proposals and reviews. The role of industry has increased in both funding and setting criteria for public funding and notable shifts towards environmental and food safety issues have taken place.

4.5.5 Influence of beneficiaries

There have always been different views of reality and behind them different normative visions of the desirable characteristics of a target food system and a target world to be promoted and sustained (Thompson, 1992). The values and meanings that are given priority depend on the economic, social and cultural circumstances and the political contexts of individuals and groups (Visser, 2001). The size and power of different interest groups can have a major impact on the funding for and direction of AKST (see IAASTD Global Report). Already in the early 1970s different views existed amongst decision makers about whether either high-tech agriculture or increasing the productivity of small-scale subsistence agriculture was the most appropriate strategy to achieve food security (Falcon and Naylor, 2005). Different approaches are likely to be appropriate for different situations and regions. An important factor in making research relevant to the target group and for successful adoption of R&D is to have strong links between research organizations and the people who are meant to use the results. This is especially important in international AKST where differences between economic, social, cultural and political circumstances are more pronounced (Buhler et al., 2002). Barriers and risks that integrative and participatory approaches have encountered have been described above (4.4.3 and 4.4.4).

The establishment of the agricultural research stations and similar institutes in NAE in the first half of the 20th century indicates that research was conducted on the basis of farmer participation. The same is true of the early commodity-based stations run by private enterprises or by the government. This linkage was strengthened by the fact that many of the earlier agricultural scientists came from the farming community. During the Second World War and thereafter, the top-down emphasis and governmental intervention in R&D increased to ensure food security. Even during this time farmers' interest in guiding R&D was strong and they had a major influence in policy (Buhler et al., 2002). In the latter part of the 20th century, the influence of farmers in public R&D diminished while that of larger companies increased. Levy boards remain one avenue through which farmers exert influence on agricultural research agendas (Accenture, 2007). In recent years farmer participation in the development of AKST has increased again in NAE (Romig et al., 1995, 1996; Walter et al. 1997; Wander and Drinkwater, 2000; Dik, 2004; Groot et al., 2004; Morris, 2006; Ingram and Morris, 2007; Timmer et al., 2007). Public consultation processes have been extended to include a wider range of voices in the setting of agendas for publicly funded agricultural research (OSI, 2006).

Concerns have been expressed that the increased influence of some sections of the private sector in the setting of public research agendas have the potential for biased benefits (Ulrich et al., 1986; Constantine et al., 1994). For example, in the US the agricultural research agenda is today heavily influenced by the private input sector and, to a lesser extent, by processing industries. There is also concern that less research is made available in the public domain due to the increased extent of research being conducted and funded by industry, which needs confidentiality to protect investments and stay ahead of competitors (Buhler et al., 2002). The central role of AKST as a driver of industrialization and structural change, especially but not solely of agriculture, has also raised debate about whether even publicly funded agricultural research is targeted to the full range of user and citizens' groups (BANR, 2002).

The number of civil society groups (or non-governmental organizations, NGOs) in Western Europe and North America has increased dramatically since the end of the Second World War, with most of this increase post 1970. In Central and Eastern Europe the number and influence of policy of civil society groups increased substantially after 1989. Civil society groups include e.g., community groups, women's groups, consumer groups, environmental organizations, labor unions, indigenous peoples' organizations, charitable organizations, faith-based organizations, professional associations and foundations. At a national level, civil society groups are still more influential in Western Europe and North America than they are in Eastern Europe. However, this may change in the future as the general tendency towards liberalization continues. Civil society organizations are now included in consultations on national (and also EU) agricultural policy as stakeholders. At an international level, there has been a policy to invite civil society groups to meetings of UN agencies as observers (UNEP, 2002). Consultations are held with civil society groups at a regional level. However, many civil society organizations doubt the extent of civil society influence on agricultural policy, compared with that of agricultural business interests. Others are concerned that the pressure applied by single issue NGOs on agricultural policy is not always evidence-based and often only represents small segments of society.

The current research climate has been criticized as being characterized by short-term perspective and responsive science and as being dominated by industrial and political influences with only a small role for farmers and consumers in setting of agendas (Buhler et al., 2002). Others see the increasing influence of consumers and NGOs on the setting of agendas as one of the main changes in influencing the evolution of AKST in recent years. There is also mistrust amongst consumers and some NGOs that farmers and farmer organizations have too much influence on the setting of agricultural research agendas.

In the international research, the colonial period was characterized by a top-down approach and a focus on cash crops (see 4.3.2). Then few people with influence in agenda setting came from developing countries. After the end of the colonial period, the national R&D structure, methods and even personnel changed only slowly and thus linkage of agricultural R&D to clients was weak. Indigenous agricultural systems received negative rather than positive attention (Boserup, 1965). Since the late 1970s, participatory approaches involving farmers have become the mainstream. The international donor organizations and contributing governments are influential beneficiaries and clients. Their importance has increased further during the last decade, due to the increasing constraints set by donors in respect of the use of funding (see 4.5.3).

4.5.6 Consequences of the changes in structures and funding

The consequences of the changes described have been critically studied and discussed. Questions posed from an economic point of view include: Have the changes improved the economic efficiency of R&D? Has the emphasis on topics changed, such as farming and environment or processing, or between basic and applied research and extension, or among programs and institutions? Are administrational and transaction costs lower? Other questions that need to be posed include: Have there been changes in who now benefits?

At least since the 1950s, studies have shown unusually high productivity gains stemming from public agricultural research (e.g., Schultz, 1953; Griliches, 1958; Ruttan, 1982; Huffman and Evenson, 1993; Fuglie et al., 1996; Alston et al., 1998) with no evidence of a decline (Alston et al., 2000). This would have justified an even higher share of funds allocated to public agricultural research. However, budget pressures have induced administrators and public decision makers to reduce budgets while striving to avoid a significant loss of productivity.

Competitive grants and short-term contracts

To improve productivity the share of funding given out as competitive grants has been increased since the 1970s (Huffman and Just, 2000; Rubenstein et al., 2003). Also, the increasing role of the private sector in management of the public agricultural R&D has caused concern. In response, debates about how to foster, organize and manage agricultural research (as well as of research in general) have intensified during the 1990s (e.g., Buttel, 1986; Just and Huffman, 1992; Alston et al., 1995, 1998; Huffman and Just, 1994, 1999, 2000). This debate builds on earlier discussions surrounding controversial topics such as national priority setting, central planning of agricultural research, over-organization of institutional research, top-down approaches, requirements for elaborate documentation and justification of research (Schultz, 1980, 1982, 1983, 1985; Huffman and Just, 2000). An asymmetry exists in the sharing of transactions costs associated with external peer-reviewed competitive grant programs, especially when the average grant size is small and the average award rate is low (Huffman, 2005).

Other topics discussed as a response to the dominant developments included the character of agricultural research as innovation and the difference between setting efficient incentives and organizational structures for industrial production/marketing and for innovation processes (Schultz, 1980; Anderson and Hardaker, 1992; Huffman and Just, 2000).

Competitive grants are by many scientists seen as leading to an increase in scientific quality. They have in some cases also been successfully used to lever a change or paradigm shift in organizational behavior (Sutherland et al., 2004). The main intentions of the shift towards more competition were to ensure high quality science, high overall productivity and transparency. However, the shift has also had other fundamental consequences. The increase of managed competition in public funding has substantially contributed to prioritization according to the interests of funding agencies which may reflect interests of governmental policies, commercial interests (farming community, large companies, etc.), NGOs and other stakeholders who are represented on the boards responsible for project evaluation and resource allocation (see also 4.4.5). Because of the changing objectives and priority fields of the financiers and varied sources, the opportunity for specialization and competence building for experts and facilities has been reduced in areas of agricultural R&D where there is no sustained funding, even if there are high pay-off potentials (not necessarily economic profits).

The trend towards more short-term contracts (usually limited to three years or less) has improved accountability (Nickel, 1997) but has had a number of negative impacts for AKST. Research has been increasingly directed towards laboratory work rather than the field. There has been less opportunity for empirical studies on sustainable agrifood systems with their inherent long-term perspective. It has been hypothesized that this may partly explain the shift in the focus of life sciences towards research into biotechnology (Buhler et al., 2002). The drive to short-term funding has also resulted in a reduction in NAE scientists with overseas experience in agriculture.

Based on principal-agent theory, the move from formula/program funding to research grant funding may be partly counterproductive for agricultural research due to too much of the best scientists' time being used for proposal writing/evaluation and signaling activities, the risks of conducting research being imposed unduly on scientists, and review committees not sufficiently sampling diversity (Huffman and Just, 2000). There is also concern about the associated increase in bureaucratization of science. An asymmetry exists in the sharing of transactions costs associated with external peer-reviewed competitive grant programs, especially when the average grant size is small and the average award rate is low (Huffman, 2005). Others note that, in the United States, competitive grants have never reached more than about 15% of total USDA research funding to States, nor more than 17% of total public agricultural research expenditures (Rubenstein et al., 2003; Rubenstein et al., 2007). Rubenstein et al. (2003) showed empirically that the US competitive grants focused more on basic research and were distributed among fewer states than other instruments.

Along with the declining program/formula funding of research institutions, recent trends foster more competition for budget funding, application of the short-term project formula, reduction of funds for technical research staff and more direct management of expenditures. In the principal-agent model for agricultural research incentives, these policy changes resulted in an immediate increase in the institutional risks of research (Huffman and Just, 2000). The short-term benefits of these shifts may not outweigh the longer-term costs and agricultural research organizations may not be able to retain important expertise (Alston et al., 1998). Block allocations on the basis of reviews conducted at longer time intervals may be a way of reducing the transaction costs while still preserving a certain level of competition.

Education as well as managed competition, peer-review in project evaluation and priority setting by scientific journals have all played a significant role in strengthening the disciplinary paradigms and increasing method-orientation in science. Use of the most advanced, disciplinarily appreciated methods has become a crucial precondition for funding, journal publications and career development, often overruling the strategic objectives and practical relevance of the work. These changes had significant consequences for international AKST. For example, development as a field within economics may be disappearing due to "the path-dependent and disequilibrium nature" being at odds with the mathematical directions of the present-day economic theory (Falcon and Naylor, 2005).

Privatization

Already in the early 1970s the public agricultural research system in the US was criticized (by J. Hightower and colleagues) for benefiting the large farmers more than small farmers and for providing particular benefits to agribusinesses (Buttel, 2005). The rise in the role of the private sector (including the farming industry) in public R&D management in the last 15 years, which occurred through increased linking of private and public funds through levy schemes, joined funds and by inviting representatives of industry to join prioritizing committees and the increase in the share of private funding in the overall funding of agricultural R&D has aggravated these concerns (see 4.6). The share of private sector expenditure in total agricultural R&D has increased to the extent that it exceeds public sector expenditures (4.5.3) (Fuglie et al., 1996; Huffman and Evenson, 1993; Huffman and Just, 1998). This trend is seen by many as not benefiting society as it is seen as shifting the focus further away from R&D that could benefit resource-poor communities and small rural enterprises, reduce hunger and poverty and improve equity and social sustainability (BANR, 2002; Buhler et al., 2002). The increased privatization of agricultural research has generated a new stream of agricultural research activism, including the anti-biotechnology movement which in parts contests corporate R&D on genetically modified crops (Buttel, 2005). The legislation introduced in the 1980s enabled universities to patent technologies developed with public funding, which resulted in more involvement in technology transfer that yielded royalty income over gratis technology transfer. This change is seen by some as being to the detriment particularly of smaller farmers (Buttel, 2005).

Alternatively, it has been argued that it is a benefit that competitive funding helps to change the direction of public towards more necessary basic research (NAS, 1972; Rockefeller Foundation, 1982; NRC, 1994, 2003) and that more basic research is necessary to maintain historical rates of agricultural productivity growth. In this view, if basic research were reduced, applied research would eventually become unproductive. There are several potential advantages of competitive grants: responsiveness and flexibility; potential to attract the best talent through open competition; potential, through professional and peer review, to ensure that research resources flow in those directions with the greatest expected payoffs; and capacity to balance and complement other research resources and programs (Alston and Pardey, 1996). Hence, it can be argued that finding an optimal balance between competitive and programmatic funding mechanisms may be a key.

The view has further been expressed (Alston et al., 1995; Alston and Pardey, 1996) that agricultural research policy is "a blunt and ineffective instrument for objectives other than economic efficiency" and that attempts to meet other objectives through public agricultural research policy often incur "transactions costs that are not borne equally." This is particularly the case when there are other policy instruments (e.g., tax and income transfer policies) available to also address equity objectives through public policy. The way the "national economic pie" is sliced among varying groups will be affected by the choice of research priorities and in some cases (particularly in countries with weaker institutional structures) the use of other policy instruments may be relatively unavailable; yet the trade-off between efficiency and other objectives "should be limited" (Alston and Pardey, 1996).

Indeed, a number of arguments have been advanced (Ingram and Rubenstein, 1999; Fuglie and Schimmelpfennig, 2000; Brennan and Mullen, 2002; van der Meer, 2002) for the promotion of public-private cooperation in agricultural research: (1) providing a natural response in the provision of "mixed" or "hybrid" goods that have both public and private characteristics, (2) enhancing research efficiency by enabling the public sector focus resources on areas where private incentives are relatively weak, (3) providing different alternatives for maintaining adequate levels of basic research (e.g., by enabling the public sector to concentrate more on basic research while the private sector focuses on nearer-market research), (4) encouraging more innovative efforts and investments by the private sector, (5) increasing business activity that promotes competition and as a result leads to the supply of better or cheaper products and services, and (6) improving the public reputations of companies and public research managers. Therefore these public policy choices and trade-offs are not simple "either-or" propositions.

Rationalization of structures

The trend towards making public agricultural research facilities more cost effective had a positive economic impact where such streamlining took place in response to changes in scientific methods and to take advantage of new economies of size and scope. However, where this "rationalization" was used merely as a justification for reductions in public R&D investments, the impact could be negative or positive depending on whether the rates of return on the investments were higher than the marginal social opportunity cost of funds (Alston et al., 1998). There are concerns that rationalization has in some European countries contributed to a serious fragmentation and weakening of the disciplinary research base and that the strategic planning of public sector funding organizations sometimes has not been joined up enough at a national level to help maintain crucial scientific expertise and facilities. The costs and time needed for re-building expertise have not always been sufficiently included in the evaluation of areas of research considered for closure (OSI, 2006).

Reallocation of research resources

Reallocation of public research resources away from near-market research programs to environmental and food safety issues is seen by many as having provided social gains but there is so far no formal evidence available on the payoff to public R&D into environmental or food safety issues and incentives to adopt results that yield social benefits are usually required to achieve a payoff at all (Alston et al., 1998).

Diversion of public resources towards agribusiness and food processing research (as happened e.g., in the UK) represents another potentially negative consequence of the recent changes in agricultural research policy in NAE. It is not yet clear whether projects funded in these areas approximate public good projects more closely than those they have displaced in the area of farm productivity and this shift of resources may have reduced the rate of return to public research investments (given that near-market agribusiness and food processing are characterized by relatively few firms with no evidence of market failures) (Alston et al., 1998).

One conclusion of the latest review of the CGIAR system (World Bank, 2003ab) (see 4.5.3) was that changes in the funding processes of CGIAR since the mid-1990s resulted in changing CGIAR's authorizing environment from being science-driven to being donor-driven and a general shift from producing global and regional public goods toward providing national and local services. CGIAR management was streamlined in recent years and, rather than increasing participation, the World Bank claimed a more strategic leading role for itself in CGIAR with creation of a legal entity covering CGIAR's central oversight and fund allocation functions (World Bank, 2003b).

4.6 Development of Public Control of Agrifood Systems

The rise of different forms of control of agriculture has had profound effects on agriculture in NAE over the past 50 years. Standards from both private and public sectors shape innovation and technology in agriculture in multiple ways (Bingen and Busch, 2006). Although in recent years de-regulation is often held up as a policy goal and ambition, in fact in relation to product quality, risk, environmental standards, animal welfare and intellectual property standard setting by both private and public sectors determine the space in which producers and companies compete. Standard setting is done by government regulatory agencies, firms, international organizations such as Food and Agriculture Organisation (FAO) and the World Trade Organisation (WTO) and private voluntary organizations such as business associations.

The section that follows looks at different forms of risk regulation and intellectual property regulation in NAE. These two forms of regulation and changes in the way they are implemented and conceived of are particularly important in relation to agricultural inputs and major new technologies in agriculture such as for example biotechnology.

4.6.1 Development of risk regulation

In developing technology for agriculture, as in other areas of innovation, the products that eventually reach the market place, their public benefits and their commercial profitability depend on a complex set of interactions between scientific developments and industry strategies, policies to promote and to regulate innovation and market opportunities, public and stakeholder attitudes and desires.

This subchapter illustrates interactions between public risk regulation and innovation, although national regulatory systems and international protocols are inevitably influenced by public and stakeholder pressures. From the broad range of public regulatory actions applied on agriculture and food systems, this subchapter takes two examples: pesticide regulation and regulation of genetically modified (GM) crops including intellectual property (IP) rights protection. The examples consider the links between these regulations the similarities and discontinuities in the regulatory systems as they evolved in Europe and the US and the outcomes for the international competitiveness of agriculture on these two continents.

Example 1: Pesticide regulation in Europe and the US

Pesticides are presumptively dangerous under US and also EU laws. Accordingly, each regulatory system establishes conditions under which they can be used without evidence of unreasonable harm to humans or the environment and these become mandatory for users. Scientific analysis of pesticide safety has advanced considerably since the 1960s and thus factors that were unknown 40 or 50 years ago are now considered in evaluating pesticide safety.

More skeptical observers have argued that the regulatory systems that have developed since the 1960s for pesticides have been "reactive" in that the industry and its products are controlled by a system set up in response to evidence of adverse, sometimes unexpected, impacts that have been found in products. Once a hazard to health or the environment has been demonstrated, new products in development are screened to ensure that they do not give rise to similar hazards. The regulatory system is thus built up slowly as new products exhibit different, sometimes unexpected, hazards. Decisions about the need for and form of, regulation are taken on the basis of the best available scientific evidence and in relation to the relevant costs and benefits (Tait and Levidow, 1992).

An example of this process is the evidence that accumulated in the 1960s and 70s that commonly used organochlorine insecticides were harming wildlife (Moore, 1987). Thereafter, regulations were introduced to ensure that chemicals which were highly persistent in the natural environment (previously seen as a desirable attribute) would not be approved for use. Potential persistence in the environment then became a reason to reject a new pesticide from the research and development pipeline at a very early stage. A more recent example was the appearance of pesticide residues in drinking water in the EU. Consequently, the Drinking Water Directive (Council Directive on the Quality of Water intended for Human Consumption, 80/778/EEC) prohibited the use of any pesticide, residues of which appeared in drinking water at a concentration of greater than 0.1µg per liter. High mobility in soils, seen as an indicator of the potential of a chemical to reach drinking water supplies, became a reason for early rejection of a chemical from the product development pipeline.

This intensification of pesticide regulation has continued to the present day, although many other regulatory and policy areas have been subjected to de-regulation initiatives with a view to encouraging industry competitiveness. This has created a barrier to entry for small companies on the pesticide sector. Some interesting contrasts in impact on industry strategies can be found, however, between Europe and the US. The US Food Quality Protection Act (FQPA) 1996 had, according to interviews with agrochemical industry managers, fundamentally changed the way companies respond to regulatory signals from the US Environmental Protection Agency (EPA) in the regulation of pesticides (Yogendra, 2004; Tait et al., 2006). The new safety standard—reasonable certainty of no harm—that is required to be applied to all pesticides used on food crops is linked to a system which expedites the approval of safer pesticides (www.epa.gov/oppfead1/fqpa) on a "fast track" basis creating a new competitive advantage as an incentive for development. Such instruments selectively enable some companies (those that have such products in their development pipelines) to gain a competitive advantage over others and can in a very short space of time alter the behavior of a whole industry sector in a positive direction.

In contrast, the European Drinking Water Directive (80/778/EEC) regarded all new chemical entities as equally hazardous. For an example, while one member of the strobilurin fungicides group with a favorable environmental and health related profile was the first product to be registered under the FQPA fast track system, this group narrowly escaped rejection at an early stage of product development because of the mobility in soils and hence the danger of falling foul of the EC Drinking Water Directive. The regulatory systems currently in operation reflect accumulated evidence over decades as we have learned more about the hazards of different classes of chemicals and removed some chemicals from approved lists, opening up opportunities for companies to develop new products to fill particular market niches.

In considering the interactions between regulatory systems and agrochemical company innovation strategies, the highly onerous regulatory demands on companies developing new pesticides have created a barrier to entry for small companies that might attempt to compete with the incumbent multinationals which has been increasing steadily since the 1970s. This means that, in the pesticide sector, there have been no innovative small companies developing products which could compete with the strategies of multinationals in pesticide development. Unlike the situation in the information and communication technology sector, one group of companies with a consistent set of innovation strategies and the ability to sustain investment without any commercial returns over very long lead times has been able to retain a dominant position in technological innovation for agriculture for the last fifty years. This dominance of the agrochemical industry over innovation in technology for agriculture had an important influence on public attitudes to GM technology (see below). This is particularly the case in Europe, where public concerns about the conventional farming systems, which formed the main market for products from the agrochemical industry, had been increasing steadily (Bauer and Gaskell, 2002).

Example 2: Regulation of genetically modified crops

Considering the second example of evolution of public control systems in AKST, even more fundamental differences than concerning pesticides, emerged between EU and US approaches to the regulation of genetically modified (GM) crops in the 1980s. This debate was one of "product vs. process" (Tait and Levidow, 1992) with the US considering GM crops as inherently similar to existing products subject to existing regulatory systems, while the EU viewed the process of genetic modification as potentially leading to novel unpredictable properties requiring a new approach to regulation. The analogy most frequently used in the EU was the introduction of alien species with the attendant risks of uncontrollable spread in the natural environment (RCEP, 1989). This distinction has been a major contributor to understanding trade difficulties the US as with the EU.

In the early stages of development of GM crop technology, the difficulty for international harmonization of European and US regulatory systems arose at least in part from the fact that the two regions chose different and largely incompatible analogies on which to base their regulatory systems for GM crops. The European process-based approach to GM crop regulation, embodied in the Directive 90/220, was initially intended to be more precautionary than the US approach (although this notion is debated by US regulators) and also to be temporary, pending the generation of evidence on the safety of GM crops in use. However, the emergence in Europe of an advocacy coalition (Sabatier and Jenkins Smith, 1993) campaigning very successfully against GM crops has resulted instead in a regulatory environment based on a new revised Directive 2001/18 and subsequent regulations, which are extremely restrictive and are unlikely to be compatible with a profitable European industry sector producing both GM crops and pesticides. Genetically engineered products under development include additional crop species and a more diverse set of traits. They will present challenges for environmental safety evaluation.

In future development and production of GM crops for global markets is likely to be based outside Europe, particularly in the US and potentially also in India and China. If the co-production of GM crops and pesticides, including strategies for using a combination of GM crops and pesticides to give effective insect pest and disease control, becomes the dominant industry strategy, as currently seems likely, then the multinational companies that currently have a strong research base in Europe are likely to move their headquarters to other parts of the world (Chataway et al., 2004; Tait and Chataway, 2006).

In Canada, the regulatory system requires crops with novel traits to be assessed for their environmental safety irrespective of whether they have been produced by genetic modification or conventional breeding methods (Morris, 2007). This applies for example to herbicide tolerant crops, which have been produced using either genetic modification techniques or conventional breeding. Environmental risks associated with the growing of conventionally-bred herbicide tolerant crops and herbicide tolerant GM crops are considered to be very similar if not identical (ACRE, 2006; Morris, 2007). In the EU conventionally bred herbicide tolerant crops can be introduced without prior environmental risk assessment. In contrast, the EU GM directive requires that herbicide tolerant GM crops are not only assessed for potential direct risks but also indirect and management-related risks. Some EU governments currently oppose certain herbicide tolerant GM crops solely because of their management-related impacts on broad-leaved weeds and associated wildlife (Hawes et al., 2003; Heard et al., 2003ab; Roy et al., 2003; Beckett, 2004; Bohan et al., 2005).

Regulatory systems could be managed to give appropriate signals to companies developing the technology, to improve on the potential benefits for sustainable farming systems. The earliest products of innovative technologies have usually given only a hint of potential future benefits and innovation progress relies as much on social learning as it does on scientific knowledge (Williams, 2000).

4.6.2 Intellectual property rights

Intellectual property rights (IPR) are rights awarded to individuals or organizations over creative works. They give the owner the right to prevent others from making unauthorized use of their property for a limited period. Intellectual property is categorized as Industrial Property (functional commercial innovations) and Artistic and Literary Property (cultural creations). Development of forms of protection of agricultural IPR includes patents, gradually expanded to protect the outputs of agricultural research and innovation, plant breeders rights (PBR) and copyright (see Chapter 2). A unique hybrid system of PBRs has evolved that provides a specialized form of IP protection and offers an alternative to the patent system (CIPR, 2002). The International Convention for the Protection of New Varieties of Plants (the UPOV Convention), which was adopted in Paris in 1961 and entered into force in 1968, has provided the basis for international harmonization in this regard (Box 4-4).

There are, however, unresolved issues associated with the development of IPR frameworks at international and national scales, as none of the systems (patents, trademarks, contracts, GI, varieties) offer much protection of rights of farmers and local communities, especially in developing countries. Many NGOs and farmers' organizations are currently active to develop effective protection mechanisms based on traceability and transparency (Bazile, 2006). For a thorough analysis and assessment of roles, impacts and challenges of IPR protection, see IAASTD Global report, Chapters 2 (2.3.1 Genetic resources management) and 3 (3.2.4 Relationships between AKST and coordination and regulatory processes among multiple stakeholders).

4.6.3 Changes in policy goals

Supply driven policies

The recovery from the Second World War of the agricultural sector in Europe, the changes in the share of agriculture in national GDP and in the share of the workforce employed in agriculture in NAE have been described in Chapter 2 of this assessment. Initially the principal policy instrument used to stimulate production was price. Not only were price fixed at levels that would enable farmers to operate profitably but state support systems absorbed much of the risks of markets. When food production started to exceed national consumption, the memories of shortage and of the widespread rural distress of the 1930s meant that governments were unwilling to allow prices to collapse and the emphasis on retaining production capacity was retained and farmers were further helped by a variety of subsidies provided on inputs (OECD, 1967) (see Chapter 2).

In both North America and Europe extension services played a major role in disseminating new technology and in moving farmers towards a more business orientated approach to their activities. In the U.S. the Land Grant Colleges played a major role. In Europe the emphasis was on services provided by the state or regional authorities, often operating in conjunction with farmer cooperatives. In the centrally planned economies of Central and Eastern Europe shortages remained a problem longer than in the west and production was encouraged through targets for delivery and the provision from regional centers of services such as machinery.

In the development and uptake of new technologies the private sector played a major role (see Chapter 2) for the development of new inputs by seed companies, the agrichemical industry as well as livestock breeders. The underlying science was global, often emerging from publicly funded research. Major international companies played an important role both in fundamental research and especially in turning new understanding into profitable products (ICI, 1978).

Market driven policies

The transition from concerns about shortages to problems relating to surplus was a gradual process and to a substantial extent the mechanisms that had been established to develop and apply new technologies in farming remained in place. Within the European Community the issue of surpluses increasingly dominated policy thinking from the late 1970s (see Chapter 2). The emphasis of policy swung from production to supply control and the use of devices such as quotas and set asides to limit the volume of output from EC farms.

The impact of this on AKST was gradual. Substantial funds continued to be allocated to agricultural research and to extension. However, in several countries there was an increasing view that extension and research should be funded by the industry as was the case in other major sectors. Charges were made to farmers for extension services that related to increased profitability on the farm. In Europe national extension services tended to be privatized. Research funding continued to come from the state but an increasing share was expected to be derived from levies on the industry.

Box 4-4. The international Intellectual Property (IP) architecture: Multilateral, regional and bilateral rules.

The architecture of the global IPR regime has become increasingly complex, and includes a diversity of multilateral agreements, international organizations, regional conventions and bilateral arrangements.

Multilateral treaties. Most of these agreements are administered by WIPO, and are of three types:

1. Standard setting treaties, which define agreed basic standards of protection. These include the Paris Convention, the Berne Convention and the Rome Convention. Important non-WIPO treaties of this kind include the International Convention for the Protection of New Varieties of Plants (UPOV) and TRIPS.
2. Global protection system treaties, which facilitate filing or registering of IPRs in more than one country. These include the Patent Cooperation Treaty (PCT), and the Madrid Agreement Concerning the International Registration of Marks.
3. Classification treaties, which organize information concerning inventions, trademarks and industrial designs into indexed, manageable structures for ease of retrieval. One example is the Strasbourg Agreement Concerning International Patent Classification.

Other non-WIPO international agreements with an IPR content include the International Treaty on Plant Genetic Resources for Food and Agriculture and the Convention on Biological Diversity.

Regional treaties or instruments. Examples of these kinds of agreement include the European Patent Convention, the Harare Protocol on Patent and Industrial Designs within the Framework of ARIPO, and the Andean Community Common Regime on Industrial Property.

Regional trade agreements. Regional trade agreements normally have subchapters governing IP standards. For example, the North American Free Trade Association, the proposed Free Trade Area of the Americas, the EU/ACP Cotonou Agreement.

Bilateral agreements. Specifically, these include those bilateral agreements that deal with IPRs as perhaps one of several issues covered. A recent example is the 2000 Free Trade Agreement between the US and Jordan, but there are many others.

Source: UNCTAD/ICTSD, 2001; CIPR, 2002.

References

Accenture. 2007. UK levy boards: Research to underpin the fresh start review. Available at http://www.defra.gov.uk/farm/policy/levy-bodies/freshstart/pdf/research-report.pdf.

Ackoff, R.L. 1983. The future of operational research is past. J. Oper. Res. Soc. 30:93-104.

ACRE. 2006. Managing the footprint of agriculture: Towards a comparative assessment of risks and benefits for novel agricultural systems: Report of the ACRE sub-group on wider issues raised by the farm-scale evaluations of herbicide tolerant GM crops (Consultation Draft). Available at http://www.defra.gov.uk/environment/acre/fsewiderissues/acre-fse-060317draft.pdf. Advisory Comm. Releases Environ., London.

Alrøe, H.F., and E.S. Kristensen. 1998. Sustainability and organic farming. *In* R. Zanoli and R. Krell (ed), Research methodologies in organic farming. No. 58 FAO, Reg. Off. Europe.

Alston, J.M. 2002. Spillovers. Aust. J. Agric. Resour. Econ. 46:315-346.

Alston, J.M., M.C. Marra, P.G. Pardey and T.J. Wyatt. 2000. Research returns redux: A meta-analysis of the returns to agricultural R&D. Aust. J. Agric. Resour. Econ. 44(2):185.

Alston, J.M., G.W. Norton, and P.G. Pardey. 1995. Science under scarcity: principles and practice of agricultural research evaluation and priority setting. Cornell Univ. Press, Ithaca NY

Alston, J.M., and P.G. Pardey. 1996. Making science pay: The economics of agricultural R&D policy. The AEI Press, Washington, DC.

Alston, J.M., P.G. Pardey, and V.H. Smith. 1998. Financing agricultural R&D in rich countries: What's happening and why? Aust. J. Agric. Res. Econ. 42(1):51-82.

Anderson, J.R., and J.B. Hardaker. 1992. Efficacy and efficiency in agricultural research: A systems view. Agric. Syst. 40:105-123.

Apostel, L., G. Berger, A. Briigs, and G. Michaud (ed). 1972. Interdisciplinarity: problems of teaching and research in universities. OECD, Paris.

Ashley, C., and S. Maxwell. 2001. Rethinking rural development. Dev. Policy Rev. 19(4):395-425.

Balmford, A., R.E. Green, and J.P.W. Scharlemann. 2005. Sparing land for nature: exploring the potential impact of changes in agricultural yield on the area needed for crop production. Global Change Biol. 11:1594-1605.

BANR. 2002. Publicly funded agricultural research and the changing structure of US agriculture board on agriculture and natural resources (BANR). Available at http://darwin.nap.edu/execsumm/0309076161.html. Board on Agric. Natural Resources, National Res. Council, Washington, DC.

Bärmark, J., and G. Wallen. 1980. The development of an interdisciplinary project. p. 221-235. *In* K. Kinorr et al. (ed), The social process of scientific investigation. Sociol. Sci. Yearbook. Dordrecht, the Netherlands.

Barney, G.O. 1981. The Global 2000 report to the President: Entering the twenty-first century. Council Environ. Quality and US Dep. State, Washington, DC.

Bauer, H.H. 1990. Barriers against interdisciplinarity. Implications for studies on science, technology and society. Sci. Tech. Human Values 15:105-119.

Bauer, M.W., and G. Gaskell (ed). 2002. Biotechnology: The making of a global controversy. Cambridge Univ. Press, Cambridge.

Bawden, R.J. 1991. Systems thinking and practice in agriculture. J. Dairy Sci. 74:2362-2373.

Bawden, R.J., R.L. Ison, R.D. Macadam, R.C. Packham and I. Valentine. 1985. A research paradigm for systems agriculture. p. 31-42. *In* J.V. Remenye (ed), Agricultural systems research for developing countries. Canberra.

Bazile, D. 2006. State-farmer partnerships for seed diversity in Mali. Gatekeeper Ser. 127. IIED, London.

Beckett, M. 2004. Secretary of State Margaret Beckett's statement on GM policy. 9 Mar 2004. Available at http://www.defra.gov.uk/corporate/ministers/statements/mb040309.htm. Defra, London.

Bentley, J. 1994. Facts, fantasies and failures of farmer participatory research. Agric. Human Values 11(243):140-150.

Bingen, J., and L. Busch (ed). 2006. Agricultural standards—The shape of the global food and fiber system. Springer Verlag, NY.

Bohan, D., C. Boffey, et al. 2005. Effects on weed and invertebrate abundance and diversity of herbicide management in genetically modified herbicide-tolerant winter-sown oilseed rape. Phil. Trans. R. Soc. London B 272(1562):463-474.

Boserup, E. 1965. The conditions of agricultural growth: The Economics of agrarian change under population pressure. Aldine, Chicago.

Bouma, J., J. Stoorvogel, R. Quiroz, S. Staal, M. Herrero, W. Immerzeel, et al. 2007. Ecoregional research for development. Adv. Agron. 93:257-311.

Brennan, J.P., and J.D. Mullen. 2002. Joint funding of agricultural research by producers and government in Australia. *In* D. Byerlee and Echeverría (ed), Agricultural research policy in an era of privatization. CABI Publ., UK.

Britz, J.J., P.J. Lor, I.E.M. Coetzee, and B.C. Bester. 2006. Africa as a knowledge society: A reality check. Int. Inform. Library Rev. 38:25-40.

Brown, A., J. Chapman, and R. Castro. 1988. Possible future directions of farming systems research and extension: A concept paper. A paper prepared for AID/S&T/AGR. Chemonics Bureau Sci. Tech., Washington, DC.

Bruun, H., J. Hukkinen, K. Huutoniemi, and J.T. Klein. 2005. Promoting interdisciplinary research. The case of the Academy of Finland. Available at http://www.aka.fi/modules/upndown/download_file.asp?id=739D547AF6994623AC202B859A1A94DC&tabletarget=data_1&pid=E12080C439CB48A191E80295F382194F&layout=aka_fi_sisa. Publ. Acad. Finland 8/05.

Buhler, W., S. Morse, E. Arthur, S. Bolton, and J. Mann. 2002. Science, agriculture and research, a compromised participation? Earthscan, London.

Buller, H., and C. Morris 2003. Farm animal welfare: A new repertoire of nature-society relations or modernism re-embedded? Sociol. Ruralis 43(3):216-237.

Bunge, M. 1985. Treatise on basic philosophy. Vol. 7. Epistemology & methodology III: Philosophy of science and technology Part II. Reidel, Dordrecht.

Busch, L., J.L. Silver, W.B. Lacy, C.S. Perry, M. Lancelle, and S. Deo. 1984. The relationship of public agricultural R&D to selected changes in the farm sector. Rep. to National Science Foundation. Dep. Sociology, Agric. Exp. Station, College Agric., Univ. Kentucky, Lexington.

Busch, L. 2005. Commentary on "Ever since Hightower": The politics of agricultural research activism in the molecular age. Agric. Human Values 22:285-288.

Buttel, F. 1986. Biotechnology and agricultural research policy: Emergent research issues. p. 312-347. *In* K. Dahlberg (ed), New directions for agriculture and agricultural research. Rowmand & Allanheld, Totowa, NJ.

Buttel, F.H. 2005. Ever since Hightower: The politics of agricultural research activism in the molecular age. Agric. Human Values 22:275-283.

Byerlee, D., and G.E. Alex. 1998. Strengthening national agricultural research systems. Policy issues and good practice. World Bank, Washington, DC.

Carney, D. (ed). 1998. Sustainable rural livelihoods, what contribution can we make? DFID, London.

Carson, R.L. 1962. Silent spring. Houghton Mifflin, Boston.

CGIAR. 2005. CGIAR annual report 2004—Innovations in agricultural research. CGIAR Secretariat, Washington, DC.

CGIAR. 2006. History of the CGIAR. Available at http://www.cgiar.org/who/history/index.html. CGIAR, Washington, DC.

CGIAR Science Council. 2005. Science for agricultural development: Changing contexts, new opportunities. Science Council Sec., Rome.

CGIAR Science Council. 2006. Summary report on systems priorities for CGIAR research 2005-2015. Available at http://www.sciencecouncil.cgiar.org/activities/spps/pubs/SCBrief%20SystPrior.pdf. Science Council Sec., Rome.

Chambers, R. 1983. Rural development: Putting the last first. Longman, London.

Chataway, J., J. Tait, and D. Wield. 2004. Understanding company R&D strategies in agro-biotechnology: Trajectories and blindspots. Res. Policy 33(6-7):1041-1057.

Checkland, P.B. 1981 Systems thinking, systems practice. John Wiley, NY.

Checkland, P.B. 1988 Images of systems and the systems image: presidential address to ISGSR. J. Appl. Syst. Anal. 15:37.

Chesbrough, H.W. 2003. Open innovation. The new imperative for creating and profiting from technology. Harvard Business Sch. Press, Boston.

Chetty, M., E. Blake, and E. McPhie. 2006. VoIP deregulation in South Africa: Implications for underserviced areas. Telecomm. Policy 30(5):332-344.

CIPR. 2002. Commission on Intellectual Property Rights. London.

Clayton, K. 1985. The University of East Anglia. p. 189-196. *In* L. Levin and L. Lind (ed), Interdisciplinarity revisited: Reassessing the concept in light of institutional experience. Linköping Univ. Sweden.

Constantine, J., J.M. Alston, and V.H. Smith. 1994. Economic impacts of California's one variety cotton law. J. Polit. Econ. 102(5): 66-89.

Conway, G. 1997. The doubly green revolution. Penguin Books, London.

Cook, S.M., Z.R. Khan, and J.A. Pickett. 2007. The use of push-pull strategies in integrated pest management. Ann. Rev. Entomol. 52:375-400.

Cooke, P., and K. Morgan (ed). 1998. The associational economy: Firms, regions, and innovation. Oxford Univ. Press, Oxford.

CSFP. 2007. Commonwealth scholarship and fellowship plan. Available at http://www.csfp-online.org. Commonwealth Scholarship and Fellowship Plan, UK.

Dahlberg, K. 1993. Regenerative food systems: Broadening the scope and agenda of sustainability. In P. Allen (ed), Food for the future: Conditions and contradictions of sustainability. John Wiley, NY.

Daily, G. (ed). 1997. Nature's services. Societal dependence on natural ecosystems. Island Press, Washington, DC.

Dalgaard, T., N.J. Hutchings, and J.R. Porter. 2003. Agroecology, scaling and interdisciplinarity. Agric. Ecosyst. Environ. 100:39-51.

Danford, W.H. 2006. Funding basic agricultural research. Science 314:223.

DeFries, R., and L. Bounoua. 2004. Consequences of land use change for ecosystem services: A future unlike the past. GeoJournal 61:345-351.

Delgado, M., and E. Ramos. 2006. Cooperation in higher education in agriculture and rural development at the ETSIAM. Int. Seminar on Higher Agric. Educ. Int. Co-Operation: Role and strategies of universities. Montpellier, 27-29 Sept 2006. Available at http://www.agropolis.fra/pdfiipe/Delgado_texte.pdf.

Deutsch, L. 2004. Global trade, food production and ecosystem support: Making the interactions visible. PhD thesis, Dep. Systems Ecol., Stockholm Univ.

De Wit, C.T. 1986. Introduction. p. 3-10. *In* H. Van Keulen, and J. Wolf (ed) Modelling of agricultural production: Weather, soils and crops. Simulation Mono., Pudoc, Wageningen.

Dik, A.J. 2004. Bulletin OILB/SROP 27 (8) 421-423. Proc. Meeting IOBC/WPRS working groups management of plant diseases and arthropod pests by BCAs and their integration in agricultural systems, Trentino, Italy, 9-13 June 2004.

Douglass, G.K. 1984. The meanings of agricultural sustainability. p.1-29 *In* G.K. Douglass (ed) Agricultural sustainability in a changing world order. Westview Press, Boulder, CO.

Duncker, E. 2001. Symbolic communication in multidisciplinary cooperations. Sci. Tech. Human Values 26(3):349-386.

DuPuis, E.M., and D. Goodman. 2005. Should we go "home" to eat? Toward a reflexive politics of localism. J. Rural Studies 21(3):359-371.

EC (European Commission). 2006. She figures 2006. Women and science. Statistics and indicators. Directorate General for Research, Brussels. http://ec.europa.eu/research/science-society/women/wssi/publications_en.html.

Ellis, F., and S. Biggs. 2001. Evolving themes in rural development in 1950s-2000s. Dev. Policy Rev. 19:437-448.

El Titi, A. 1992. Integrated farming: An ecological farming approach in European agriculture. Outlook Agric. 21(1):33-39.

ETAN. 2000. Science policies in the European Union: Promoting excellence through mainstreaming gender equality. Available at http://cordis.europa.eu/improving/women/documents.htm. Eur. Tech. Assessment Network, Eur. Commission, Brussels.

EURAGRI. 2005. Anticipating the future: Knowledge based policy for European agriculture. Available at http://euragri.csl.gov.uk/aims.cfm. XIX Members Conf. European Agric. Res. Initiative, York, UK.

Evenson, R.E. 1983. Intellectual property rights and agribusiness research and development: Implications for the public agricultural research system. Am. J. Agric. Econ. 65:967-975.

Evenson, R.E., and D. Gollin, 2003. Assessing the impact of the green revolution, 1960 to 2000. Science 300:757-762

Falcon, W.P., and R.L. Naylor. 2005. Rethinking food security for the twenty-first century. Am. J. Agr. Econ. 87(5):1113-1127.

Feller, I. 2005. Whither interdisciplinarity (in an era of strategic planning)? Proc. Ann. AAAS meeting, 2004. AAAS, Washington, DC.

FEMtech. 2006. Facts and figures. Available at http://www.femtech.at/index.php?id=467. Austrian Res. Promotion Agency.

Francis, C., G. Lieblein, S. Gliessman, T.A. Breland, N. Creamer, R. Harwood, et al. 2003. Agroecology: The ecology of food systems. J. Sustainable Agric. 22(3):99-118.

Fuglie, K., N. Ballenger, K. Day, C. Klotz, M. Ollinger, J. Reilly, et al. 1996. Research and development: Public and private investments under alternative markets and institutions. ERS AER735. USDA, Washington, DC.

Fuglie, K.O., and D.E. Schimmelpfennig (ed). 2000. Public-private collaboration in agricultural research: New institutional arrangements and economic implications. Iowa State Univ. Press, Ames.

Galison, P., and B. Hevly.1992. Big science. The growth of large-scale research. Stanford Univ. Press, Stanford.

Generation Challenge Programme. 2007. The generation challenge programme. Cultivating plant diversity for the resource poor. Available at http://www.generationcp.org.

Gibbons, M., C. Limoges, H. Nowotny, S. Schwartzman, P. Scott, and M. Trow. 1994. The new production of knowledge. The dynamics of science and research in contemporary societies. Sage, London.

Grantham, G. 1984. The shifting locus of agricultural innovation in nineteenth-century Europe: the case of the agricultural experiment stations. Res. Econ. Hist. 3 (suppl):191-214.

Green, R.E., S.J. Cornell, J.P.W. Scharlemann, and A. Balmford. 2005. Farming and the fate of wild nature. Science 307:550-555

Griliches, Z. 1958. Research costs and social returns: Hybrid corn and related innovations. J. Polit. Econ. 65: 419-465.

Griliches, Z. 1979. Issues in assessing the contribution of research and development to productivity growth. Bell J. Econ. 10:92-116

Groot, J.C.J., M. Stuiver, and L. Brussaard. 2004. Land use systems in grassland dominated regions. *In* A. Luscher, et al. (ed), Proc. 20th Gen. Meeting Eur. Grassland Fed., Luzern, Switzerland, 21-24 June 2004.

Grove-White, R., P. Macnaghten, and B. Wynne. 2000. Wising up. The public and new technologies. http://csec.lancs.ac.uk/docs/wising_upmacnaghten.pdf. Lancaster Univ., UK.

Hawes, C., A.J. Haughton, et al. 2003. Responses of plants and invertebrate trophic groups to contrasting herbicide regimes in the farm scale evaluations of genetically modified herbicide-tolerant crops. Phil. Trans. R. Soc. London B 358:1899-1913.

Hayami, Y., and V.W. Ruttan. 1971. Agricultural development: An international perspective. Johns Hopkins Univ. Press, Baltimore.

Heard, M.S., C. Hawes, G.T. Champion, S.J. Clark, L.G. Firbank, A.J. Haughton, et al. 2003a. Weeds in fields with contrasting conventional and genetically modified herbicide-tolerant crops. I. Effects on abundance and diversity. Phil. Trans. R. Soc. London B 358:1819-1832.

Heard, M.S., C. Hawes, et al. 2003b. Weeds in fields with contrasting conventional and genetically modified herbicide-tolerant crops. II. Effects on individual species. Phil. Trans. R. Soc. London B 358:1833-1846.

Heemskerk, M., K. Wilson, and M. Pavao-Zuckerman. 2003. Conceptual models as tools for communication across disciplines. Conserv. Ecol. 7(3):8.

Helenius, J. 2006. Food systems research and interdisciplinarity. Proc. Seminar Developing Organics, Session on Interdisciplinary Interaction in Research. (In Finnish). Helsinki, 14 Feb 2006.

Huffman, W.E. 1996. Labor markets, human capital, and the human agent's share of production. p. 55-79. *In* J.A. Antle and D.A. Sumner (ed), The economics of agriculture: Papers in honor of D. Gale Johnson. Vol. 2. The Univ. of Chicago Press, Chicago.

Huffman, W.E. 1998a. Modernizing agriculture: A continuing process, Daedalus (fall issue)

Huffman, W.E. 1998b. Education and agriculture. *In* G. Rausser and B. Gardner (ed), Handbook of agricultural economics. North Holland, Amsterdam.

Huffman, W.E. 2005. Developments in the organization and finance of the public agricultural research in the U. S. 1988-1999. Dep. Econ. Working Pap. Ser. Iowa State Univ., Ames.

Huffman, W.E., and R.E. Evenson. 1993. Science for agriculture: A long term perspective. Iowa State Univ. Press, Ames.

Huffman, W.E., and R.E. Evenson. 2001. Structural and productivity change in US agriculture, 1950-1982. Agric. Econ. 24(2):127-147.

Huffman, W.E., and R.E. Evenson. 2006. Science for agriculture: A long-term perspective. Blackwell, New York.

Huffman, W.E., and R.E. Just. 1994. Funding, structure and management of public agricultural research in U.S. Am. J. Agric. Econ. 76:744-759.

Huffman, W.E., and R.E. Just. 1998. The organization of agricultural research in western developed countries. Staff Pap. 304. Dep. Econ., Iowa State Univ., Ames.

Huffman, W.E., and R.E. Just. 1999. Benefits and beneficiaries of alternative funding mechanisms. Rev. Agric. Econ. 19:2-18.

Huffman, W.E., and R.E. Just. 2000. Setting efficient incentives for agricultural research: Lessons from principal-agent theory. Am. J. Agr. Econ. 82:828-841.

Humphries, S., J. Gonzalez, J. Jimenez, and F. Sierra. 2000. Searching for sustainable land use practice in Honduras: Lessons from a program of participatory research with hillside farmers. Agric. Res. Extension Network Pap. 104. July. ODI, London.

Hunger Project. 2005. The Hunger Project website. Available at http://www.thp.org.

ICI. 1978. Jealott's Hill fifty years of agricultural research 1928-1978. Imperial Chemical Industries Plant Protection Division.

IFAR. 2007. Programs and fellowships. Available at http://www.ifar4dev.org/fellowships.

IFOAM, 2005. Principles of organic agriculture. Available at http://www.ifoam.org/press/press/Principles_Organic_Agriculture.html. IFOAM, Bonn.

IOBC. 1999. Integrated production. Principles and technical guidelines. 2nd edition. E.F. Boller et al. (ed) IOBC/WPRS Bulletin 22(4).

Ingram, J., and C. Morris. 2007. The knowledge challenge within the transition towards sustainable soil management: An analysis of agricultural advisors in England. Land Use Policy 24(1):100-117

Ingram, C.K., and K.D. Rubenstein. 1999. The changing agricultural research environment: What does it mean for public-private innovation? AgBioForum 2(1):24-32.

Institute Rural Sciences. 2007. Europe: The development of organic farming between 1999 and 2006. Available at http://www.organic.aber.ac.uk/statistics/europe05.shtml. Statistics Inst. Rural Sci., Univ. Wales.

James, P.G. 1971. Agricultural policy in wealthy countries. Angus and Robertson, Sydney.

Jantsch, E. 1975. Design for evolution. George Braziller, New York.

Johansson, S. 2005. The Swedish foodprint—An agroecological study of food consumption. PhD dissertation. Acta Univ. Agric. Sueciae 2005:56. SLU Service/Repro, Uppsala.

Jorgenson, D.W., and F.M. Gollop. 1992. Productivity growth in US Agriculture: A postwar perspective. Am. J. Agric. Econ. 74:745-750.

Just, R.E., and W.E. Huffman. 1992. Economic principles and incentives: Structure, management, and funding of agricultural research in U.S. Am. J. Agric. Econ. 74:1101-1108.

Kahiluoto, H., P. Berg, A. Granstedt, H. Fischer, and O. Thomsson (ed). 2006. Localisation and recycling in Baltic rural food systems—interdisciplinary synthesis. Baltic Ecol. Recycl. Agric. Soc. (BERAS) No. 7. Centre Sustain. Agric. (CUL), Swedish Univ. Agric. Sci., Uppsala.

Klein, J.T. 1996. Crossing boundaries. knowledge, disciplinarities, and interdisciplinarities. Univ. Press Virginia, Charlottesville.

Klein, J.T. 1990. Interdisciplinarity. History, theory, and practice. Wayne State Univ. Press, Detroit.

Klein, J.T. 2004. Prospects for transdisciplinarity. Futures 36:512-526.

Klevorick, A.K., R.C. Levin, R.R. Nelson, and S.G. Winter. 1995. On the sources and significance of interindustry differences in technological opportunities. Res. Policy 24:185-205.

Labarthe, P. 2006. Changing European farming systems for a better future: New visions for rural areas. p. 330-334. *In* Proc. 7th Eur. IFSA Symp. New Visions for Rural Areas, 7-11 May 2006, Wageningen.

Lambert, D.M., P. Sullivan, R. Claassen, and L. Foreman 2007. Profiles of US farm households adopting conservation-compatible practices. Land Use Policy 24:72-88.

Lappé, F.M., and J. Collins. 1988. World Hunger: 12 myths. Earthscan, London.

Laszlo, E., and A. Laszlo. 1997. The contribution of the systems sciences to the humanities. Syst. Res. Behav. Sci. 14:5-19.

Leeuwis, C., 2004. Rethinking innovation and agricultural extension. *In* H.A.J. Moll et al. (ed), Agrarian institutions between policies and local action: Experiences from Zimbabwe. Weaver Press, Harare.

Lele, S., and R.B. Norgaard. 2005. Practicing interdisciplinarity. BioScience 55:967-975.

Lipton, M. 2005. The family farm in a globalizing world: The role of crop science in alleviating poverty. 2020 Disc. Pap. No. 40. IFPRI, Washington, DC.

Lockeretz, W., and M.D. Anderson. 1993. Agricultural research alternatives. Univ. Nebraska Press.

MA. 2005. Millennium Ecosystem Assessment Synthesis Report. Island Press, Washington, DC.

MacDonald, D., J.R. Crabtree, G. Wiesinger, T. Dax, N. Stamou, P. Fleury, et al. 2000. Agricultural abandonment in mountain areas of Europe: Environmental consequences and policy response. J. Environ. Manage. 59:47-69.

Mäkelä, J. 2006. Possibilities of triangulation: qualitative and quantitative interpretations of local and organic food. Proc. Seminar Developing Organics, Session on Interdisciplinary Interaction in Research. (In Finnish). Helsinki, 14 Feb 2006.

Mann, C. 1997. Reseeding the green revolution. Science 277:1038-1043.

Mansilla, V.B., and H. Gardner. 2003. Assessing interdisciplinary work at the frontier. An empirical exploration of "symptoms of quality". Rethinking interdisciplinarity. Available at http://www.interdisciplines.org/interdisciplinarity/papers/6/7. Interdisciplines.

Masefield, G.B. 1972. A history of the colonial agricultural service. Clarendon Press, Oxford.

Miettinen, R. 2002. National innovation systems. Scientific concept or political rhetoric. Edita, Helsinki.

Miller, R.J., P. Sorokin, Y.F. Lachuga, S. Chernakov, and A.D. Goecker. 2000. A comparison of agricultural higher education in Russia and the United States. J. Nat. Resour. Life Sci. Educ. 29:68-77.

Moore, N.W. 1987. The bird of time. Cambridge Univ. Press, Cambridge.

Morris, C. 2006. Negotiating the boundary between state-led and farmer approaches to knowing nature: An analysis of UK agri-environment schemes. Geoforum 37(1):113-127.

Morris, C., and M. Winter. 1999. Integrated farming systems: The third way for European agriculture. Land Use Policy 16:193-205.

Morris, S.H. 2007. EU biotech crop regulations and environmental risk: a case of the emperor's new clothes? Trends Biotech. 25(1):2-6.

NAS. 1972. Report of the committee on research advisory to the U.S. Department of Agriculture. Nat. Acad. Press, Washington, DC.

NRC. 1994. Investing in the national research initiative: An update of the competitive grants program in the U.S. Department of Agriculture. Nat. Acad. Press, Washington, DC.

NRC. 2003. Frontiers in agricultural research: Food, health, environment, and communities. Nat. Acad. Press, Washington, DC.

Nickel, J.L. 1997. A global agricultural research system for the 21st century. p. 139-155. *In* C. Bontefriedheim and K. Sheridan (ed), The globalization of science: The place of agricultural research. ISNAR, The Hague.

Nipp, T. 2004. Agrosecurity: the role of the agricultural experiment stations. J. Food Sci. 69: CRH50-CRH54.

NRI. 2002. http://www.nri.org/about/history.htm. Univ. of Greenwich, UK.

Odum, H.T. 1983. Systems ecology. Wiley, New York.

Odum, H.T. 1988. Self organization, transformity, and information. Science 242:1132-1139.

OECD. 1967. Agricultural policies in 1966. OCED, Paris.

OECD. 1999. Summary and evaluation of main developments and changes in organizational forms of and approaches by the AKS in OECD member countries. AGR/CA/AKS(00), Direct. Food, Agric. Fisheries, Committee for Agriculture, OECD, Paris.

OECD. 2002. Dynamizing national innovation systems. OCED, Paris.

OECD. 2005a. Governance of innovation systems. Vol 1. Synthesis report. Available at http://www.oecd.org/document/25/0,2340,en_2649_37417_35175257_1_1_1_37417,00.html. OECD, Paris.

OECD. 2005b. Governance of innovation systems. Vol. 2: Case studies in innovation policy. OECD, Paris.

OECD. 2005c. Governance of innovation systems. Vol. 3. Case studies in cross-sectoral policy. Available at http://www.oecd.org/document/12/0,2340,en_2649_37417_35791756_1_1_1_37417,00.html. OECD, Paris.

OECD. 2006a. Factbook 2006. OECD, Paris.

OECD. 2006b. Main science and technology indicators, 2006-1. OECD, Paris.

OTA. 1992. New technological era for American agriculture. US Gov. Printing Off., Washington, DC.

Olmstead, A.L., and P. Rhodes. 1994. The agricultural mechanization controversy of the interwar years. Agric. Hist. 68:3.

OSI. 2006. Science review of the Department for Environment, Food and Rural Affairs. Available at http://www.dti.gov.uk/science/science-in-govt/works/science-reviews/review/defra/page24808.html. UK Off. Sci. Innov., London.

Ostrom, M., and D. Jackson-Smith. 2005. Defining a purpose: Diverse farm constituencies and publicly funded agricultural research and extension. J. Sustain. Agric. 27:57-76.

Pardey, P.G., J.M. Alston, J.E. Christian, and S. Fan. 1996. Hidden harvest: U.S. benefits from international research aid. Food Policy Report. IFPRI, Washington, DC.

Pardey, P.G., and N.M. Beintema. 2001. Slow magic: Agricultural R&D a century after Mendel. IFPRI, Washington, DC.

Pardey, P.G., N. Beintema, S. Dehmer, and S. Wood. 2006. Agricultural research. A growing global divide? IFPRI, Washington, DC.

Pawson, E., and S. Dovers. 2003. Environmental history and the challenges of interdisciplinarity: An antipodean perspective. Environ. Hist. 9:53-75.

Pidgeon, N.F., W. Poortinga, G. Rowe, T. Horlick-Jones, J. Walls, and T. O'Riordan. 2005. Using surveys in public participation processes for risk decision making: The case of the 2003 British GM Nation? public debate. Risk Analy. 25(2):467-479.

Porter, A., and F.A. Rossini. 1984. Interdisciplinary research redefined. Multiskill, problem focused research in the STRAP framework. R&D Manage. 14(2):105-111.

Pray, C.E., J. Huang, R. Hu, and S. Rozelle. 2002. Five years of Bt cotton in China—the benefits continue. Plant J. 31:423-430.

Pretty, J. 1995. Regenerating agriculture: Policies and practice for self-reliance. Earthscan, London.

Pretty, J.N., A.D. Noble, D. Bossio, J. Dixon, R.E. Hine, F.W.T. Penning de Vries, et al. 2006. Resource-conserving agriculture increases yields in developing countries. Environ. Sci. Tech. 40:1114-1119.

Qin, J., F.W. Lancaster, and B. Allen. 1997. Types and levels of collaboration in interdiscpilinary research in the sciences. J. Am. Soc. Inform. Sci. 48(19):893-916.

Rao, S.S. 2005. Bridging digital divide: Efforts in India. Telemat. Inform. 22(4):361-375.

RCEP. 1989. Thirteenth report: The release of genetically engineered organisms to the environment. R. Comm. Environ. Pollution, HMSO, London.

Read, N., J.J. Quinn, and A. Webster. 1988. Commercialisation as a policy mechanism in UK agricultural research, development and extension. Agric. Syst. 26:77-87.

Rhodes, F.H.T. 2001. The creation of the future. The role of the American university. Cornell Univ. Press, Ithaca.

Robb, C. 1998. PPA's: A review of the World Bank's experience. p. 131-142. *In* J. Holland and J. Blackburn (ed), Who's voice? Participatory research and policy change. IT Publ., London.

Rockefeller Foundation. 1982. Science for agriculture. Rockefeller Foundation, New York.

Romig, D.E., M.J. Garlynd, R.F. Harris, and K. McSweeney. 1995. How farmers assess soil health and soil quality. J. Soil Water Conserv. 50:229-236.

Romig, D.E., M.J. Garlynd, and R.F. Harris. 1996. Farmer-based assessment of soil quality: A soil health scorecard. *In* Methods for assessing soil quality. Spec. Publ. 49. Available at http://www.cias.wisc.edu/wicst/pubs/farmer_based.htm. SSSA, Madison, WI.

Rothamsted International 2007. http://www.rothamsted-international.org.

Roy, D.B., D.A. Bohan, A.J. Haughton, M.O. Hill, J.L. Osborne, S.J. Clark, et al. 2003. Invertebrates and vegetation of field margins adjacent to crops subject to contrasting herbicide regimes in the farm scale evaluations of genetically modified herbicide-tolerant crops. Phil. Trans. R. Soc. London B 358:1879-1898.

Roy, S. 1990. Agriculture and technology in developing countries: India and Nigeria. Sage, New Delhi.

Rubenstein, K.D., P.W. Heisey, C. Klotz-Ingram, and G.B. Frisvold. 2003. Competitive grants and the funding of agricultural research in the U.S. Rev. Agric. Econ. 25(2):352-368.

Rubenstein, K.D., and P.W. Heisey. 2005. Can technology transfer help public-sector researchers do more with less? The case of the USDA's Agricultural Research Service. AgBioForum 8(2/3):134-142.

Rubenstein, K.D., P. Heisey, and D. Schimmelpfennig. 2007. Sources of public agricultural R&D changing. Amber Waves 5(3):7.

Ruttan, V.W. 1982. Agricultural research policy. Univ. Minneapolis Press, Minnesota.

Ryan, J.G. 2001. Synthesis report on workshop assessing the impact of policy-oriented social science research. Scheveningen, the Netherlands, 12-13 Nov. IFPRI, Washington, DC.

Sabatier, P.A., and H.C. Jenkins Smith. 1993. Policy change and learning: An advocacy coalition approach. Westview Press, Colorado.

Schienstock, G. (ed). 2004. Embracing the knowledge economy. The dynamic transformation of the Finnish innovation system. New horizons in the economics of innovation. Edward Elgar, London.

Scholz, R., and D. Marks. 2001. Learning about transdisciplinarity: Where are we? Where have we been? Where should we go? p. 236-252. *In* Transdisciplinarity: Joint problem solving among science, technology and society. Birkhäuser, Basel.

Schultz, T.W. 1953. The economic organization of agriculture. McGraw Hill, New York.

Schultz, T.W. 1971. The allocation of resources to research. p. 90-120 *In* W.L. Fischel (ed), Resource allocation in agricultural research. Univ. Minnesota, Minneapolis.

Schultz, T.W. 1980. The entrepreneurial function in agricultural research. Dept. Econ., Agric. Econ. Pap. No. 80:29. Univ. Chicago, Illinois.

Schultz, T.W. 1982. Knowledge activities of universities. A critical view. Univ. Chicago, Dept. Econ., Human Capital Pap. No. 82:1. Univ. Chicago, Illinois.

Schultz, T.W. 1983. An economist's assessment of agricultural research. Univ. Chicago, Dept. Econ., Agric. Econ. Pap. No. 83:28. Univ. Chicago, Illinois.

Schultz, T.W. 1985. Agricultural research, Canada and beyond. p. 11-20. *In* K.K. Klein and W.H. Furtan (ed) Econ. Agric. Res. in Canada. Univ. Calgary Press, Alberta.

Science and Technology Committee. 2004. The use of science in UK international development policy; thirteenth report of session 2003-04. Vol. 1. House of Commons. The Stationery Office, London.

Sillitoe, P., A. Bicker, and J. Pottier (ed). 2002. Participating in development. Approaches to indigenous knowledge. Routledge, London.

Slaughter, S., and L.L. Leslie. 1997. Academic capitalism. Politics, policies, and the entrepreneurial University. Johns Hopkins Univ. Press, Baltimore.

Song, C.H. 2003. Interdisciplinarity and knowledge inflow/outflow structure among science and engineering in Korea. Scientometrics 58(1):129-141.

Spedding, C.R.W. 1979. An introduction to agricultural systems. Elsevier, London.

Spielman, D.J. 2005. Innovation systems perspectives on developing country agriculture. A critical review. Disc. Pap. 2. ISNAR, Washington, DC.

Stephens, W., and T. Hess. 1999. Systems approaches to water management research. Agric. Water Manage. 40:3-13.

Stokes, D. 1997. Pasteur's quadrant. Basic science and technical innovation. Brookings Inst. Press, Washington, DC.

Suarez-Seone, S., P.E. Osborne, and J. Baudry. 2002. Responses of birds of different biogeographic origins and habitat requirements to agricultural land abandonment in northern Spain. Biol. Conserv. 105:333-344.

Sumberg, J., and C. Okali. 1997. Farmer's experiments: Creating local knowledge. Lynne Rienner, London.

Sumberg, J., C. Okali, and D. Reece. 2003. Agricultural research in the face of diversity, local knowledge and the participation imperative: Theoretical considerations. Agric. Syst. 76:739-753.

Sutherland, J.A., S.B. Mathema, T.B. Thapa, and S.B. Pandey. 2004. Encouragement of effective research and development partnerships through a process of competitive funding in Nepal. Uganda J. Agric. Sci. 9:150-156.

Suvedi, M., M. Knight-Lapinski, and S. Campo. 2000. Farmers perspectives of Michigan State Univ. extension: Trends and lessons from 1996 and 1999. J. Extens. 38(1) Available at http://www.joe.org/joe/2000february/a4.html.

Tait, J., and C. Chataway. 2007. Risk and uncertainty in genetically modified crop development: The industry perspective. Environ. Planning 2007.

Tait, J., J. Chataway, and D. Wield. 2000. Final report, PITA project. Available at www.technology.open.ac.uk/cts/projects.htm#biotechnology.

Tait, J., J. Chataway, and D. Wield. 2006. Governance, policy and industry strategies: Agro-biotechnology and pharmaceuticals. p. 378-401. *In* G. Dosi and M. Mazzucato (ed), Knowledge accumulation and industry evolution. Cambridge Univ. Press, Cambridge.

Tait, J., and L. Levidow. 1992. Proactive and reactive approaches to risk regulation: The case of biotechnology. Futures 24(3): 219-231.

Tansey, G., and T. Worsley. 1995. The food system. A guide. Earthscan, London.

Thirtle, C., J. Piesse, and V.H. Smith. 1997. Agricultural R&D policy in the United Kingdom. *In* J.M. Alton, et al. (ed), Paying for agricultural productivity. IFPRI, Washington, DC.

Thompson, P.B. 1992. The varieties of sustainability. Agric. Human Values 9:11-19.

Timmer, D.K., R.C.D. Loë, and R.D. Kreutzwiser. 2007. Source water protection in the Annapolis Valley, Nova Scotia: Lessons for building local capacity. Land Use Policy 24:187-198.

Ulrich, A., W.H. Furtan, and A. Schmitz. 1986. Public and private returns from joint research: An example from agriculture. Q. J. Econ. 101(10):101-129.

Umali, D.L., and L. Schwartz. 1994. Public and private agricultural extension: Beyond traditional frontiers. Disc. Pap. 236. World Bank, Washington, DC.

UN. 2002. Report of the World Summit on Sustainable Development. Johannesburg, 2002. Available at http://daccessdds.un.org/doc/UNDOC/GEN/N02/636/93/PDF/N0263693.pdf?OpenElement. United Nations, New York.

UN. 2004. A more secure world: Our shared responsibility. United Nations, New York.

UNCTAD/ICTSD. 2001. Intellectual property rights and development. UNCTAD, Geneva.

UNEP. 2002. Enhancing civil society engagement in the work of the United Nations environment programme: Strategy paper. UNEP/GC.22/INF/13. UNEP, Nairobi.

USDA. 1995. Agricultural research. AERI Update No. 5, revised. USDA, Washington, DC.

van der Meer, K. 1999. Decentralization and privatization of agricultural research and extension in the Netherlands. Available at http://www.worldbank/html/aftsr/con31.htm#decentralization. Special Program for African Agricultural Research (SPAAR).

Van der Meer, K. 2002. Public-private cooperation in agricultural research: Examples from the Netherlands. *In* D. Byerlee and R.G. Echeverría (ed), Agricultural research policy in an era of privatization. CABI Publ., UK.

Van Keulen, H. 2008. Historical context of agricultural development. *In* R. Roetter et al. (ed), Science for agriculture and rural development in low-income countries. Springer, NY.

Vandermeer, J., and I. Perfecto. 2005. The future of farming and conservation. Science 308:1257-1258.

Väyrynen, R. 2006. The speech of the director of the Academy of Finland. The 60th Anniv. Faculty Social Sci., Helsinki Univ., Finland.

Venkula, J. 2006. From the illusion of scientific certainty towards facing uncertainty. Proc. Seminar developing organics, session on interdisciplinary interaction in research. (In Finnish). 14 Feb 2006. Helsinki.

Visser, L.E. 2001. Development sociology and the interaction between the social and natural sciences. Asian J. Agric. Dev. 1(2):69-76.

Voluntary Initiative. 2007. Available at http://www.voluntaryinitiative.org.uk/Content/Default.asp.

Von Braun, J. 2000. The new strategic directions for the international agricultural research. Q. J. Int. Agric. 39:233-236.

Walker, J. 2001. Cultivating sustainable agriculture. Unilever Mag. 119(1):1-12. Available at http://www.unilever.com/Images/2001%20Cultivating%20Sustainable%20Agriculture_tcm13-5312.pdf.

Walter, G., M. Wander, and G. Bollero. 1997. A farmer centered approach to developing information for soil resource management: the Illinois soil quality initiative. Am. J. Alternative Agric. 12(2):64-72.

Wander, M.M., and L.E. Drinkwater. 2000. Fostering soil stewardship through soil quality assessment. Appl. Soil Ecol. 15: 61-73.

Way, M.J., and H.F. v. Emden. 2000. Integrated pest management in practice—pathways towards successful application. Crop Prot. 19:81-103.

WCED. 1987. Our common future. World Commission Environ. Dev., Oxford Univ. Press, Oxford.

White, L., Jr. 1967. Historical roots of our ecological crisis. Science 155:1203-1207.

Wibberley, J. 1995. Cropping intensity and farming systems: Integrity and intensity in international perspective. J. R. Agric. Soc. England 156:43-55.

Williams, R. 2000. Public choices and social learning: The new multimedia technologies in Europe. Inform. Soc. 16(4):251-62.

Winter, M., J. Mills, M. Lobley, and H. Winter. 2000. Knowledge for sustainable agriculture. Available at http://www.wwf.org.uk/filelibrary/pdf/sust_agriculture.pdf. WWF, UK.

Wolfe, A.K., D.J. Bjornstad, M. Russell, and N.C. Kerchner. 2002. A framework for analyzing dialogues over the acceptability of controversial technologies. Sci. Tech. Human Values 27(1):134-159.

Wood, D., and J.M. Lenne. 2005. "Received wisdom" in agricultural land use policy: Ten years from Rio. Land Use Policy 22:75-93.

World Bank. 2003a. The CGIAR at 31: An independent meta-evaluation of the Consultative Group on International Agricultural Research. Vol. 1: Overview Rep. World Bank, Washington, DC.

World Bank. 2003b. The CGIAR at 31: Celebrating its achievements, facing its challenges. Precis. Available at http://lnweb18.worldbank.org/oed/oeddoclib.nsf/DocUNIDViewForJavaSearch/D01D16C0D67B0F3085256D56005AE040/$file/Precis_232_CGIAR.pdf. Oper. Eval. Dep. Spring 2003. No. 232. World Bank, Washington, DC.

World Bank. 2006. World development indicators. World Bank, Washington, DC.

Yogendra, S. 2004. The US Food Quality Protection Act—Review of the dynamics of pesticide regulation and firm responses. Innogen Working Pap. 11. Available at http://www.innogen.ac.uk/Publications

Zeddiesa, J., R.P. Schaaba, P. Neuenschwander, and H.R. Herren. 2000. Economics of biological control of cassava mealybug in Africa. Agric. Econ. 24:209-214.

5

Looking into the Future for Knowledge, Science and Technology and AKST

Coordinating Lead Author:
Marie de Lattre-Gasquet (France)

Lead Authors:
Helena Kahiluoto (Finland), Reimund Roetter (Germany)

Contributing Authors:
Tilly Gaillard (USA), Paul Guillebeau (USA), Hugues de Jouvenel (France), Véronique Lamblin (France), Dorota Metera (Poland), W.A. Rienks (The Netherlands)

Review Editors:
John Stone (Canada), Pai-Yei Whung (USA)

Key Messages
5.1 Context 154
5.1.1 Problem statement 154
5.1.2 Review of related studies 155
5.1.2.1 At global level 155
5.1.2.2 At European level 156
5.1.2.3 At North American level 160
5.1.2.4 Relationship of scenarios in different exercises 161
5.2 Indirect Drivers for AKST 161
5.2.1 Demographic drivers 161
5.2.2 Economics and international trade 164
5.2.3 Sociopolitical drivers 164
5.3 Key Direct Drivers for KST: Uncertainties and Consequences for AKST 165
5.3.1 Transformation in models of knowledge production: trends and uncertainties 165
5.3.1.1 Trends 165
5.3.1.2 Uncertainties of the future 166
5.3.2 Transformation in models of innovation: trends and uncertainties 166
5.3.2.1 Number of researchers: trends 166
5.3.2.2 Research and technology organizations: trends 167
5.3.2.3 Universities: trends 167
5.3.2.4 Multinational enterprises and small and medium enterprises: trends 168
5.3.2.5 International, national and regional governments: trends 168
5.3.2.6 Uncertainties of the future 168
5.3.3 Information technology: trends and uncertainties 169
5.3.3.1 Trends 169
5.3.3.2 Uncertainties of the future 169
5.3.4 Evolution of KST with potential impact on AKST 169
5.3.5 Financial resources devoted to science and technology: trends and uncertainties 170
5.3.5.1 Trends 170
5.3.5.2 Uncertainties of the future 171
5.3.6 Attitudes towards science and technology: trends and uncertainties 171
5.3.7 Education in science: trends and uncertainties 172
5.4 Key Direct Drivers for Agriculture: Uncertainties and Consequences for AKST 172
5.4.1 Food consumption and distribution: Trends and uncertainties 172
5.4.1.1 Ongoing trends 172
5.4.1.2 Uncertainties of the future 174
5.4.1.3 Consequences for AKST 174
5.4.2 Policies, trade and markets 175
5.4.2.1 Ongoing trends 175
5.4.2.2 Uncertainties of the future 176
5.4.2.3 Consequences for AKST 177
5.4.3 Farming systems and farm structures 177
5.4.3.1 Ongoing trends 177
5.4.3.2 Uncertainties of the future 178
5.4.3.3 Consequences for AKST 178
5.4.4 Agricultural labor and organizations 178
5.4.4.1 Labor and gender dynamics: ongoing trend 178
5.4.4.2 Organizations: ongoing trends 179
5.4.4.3 Uncertainties of the future 180
5.4.4.4 Consequences for AKST 180
5.4.5 Natural resources availability and management 180
5.4.5.1 Ongoing trends 180
5.4.5.2 Uncertainties of the future 185
5.4.5.3 Consequences for AKST 185
5.4.6 Climate change and variability 186
5.4.6.1 Ongoing trends 186
5.4.6.2 Uncertainties for the future 188
5.4.6.3 Consequences for AKST 190
5.4.7 Energy and bioenergy 191
5.4.7.1 Ongoing trends 191
5.4.7.2 Uncertainties of the future 191
5.4.7.3 Consequences for AKST 192
5.5 Key Drivers for AKST, Agricultural Research and Innovation Systems and their Uncertainties 192
5.5.1 Organizations and funding of AKST 192
5.5.1.1 Ongoing trends 193
5.5.1.2 Uncertainties of the future 194
5.5.1.3 Consequences for AKST 195
5.5.2 Proprietary regimes 195
5.5.2.1 Ongoing trends 195
5.5.2.2 Uncertainties of the future 196
5.5.3 Access, control and distribution of AKST 196
5.5.3.1 Ongoing trends 196
5.5.3.2 Uncertainties of the future 197
5.5.3.3 Consequences for AKST 197
5.6 Future AKST Systems and Their Potential Contributions to Sustainable Development Goals 198
5.6.1 Four normative agricultural innovation systems 198
5.6.1.1 Market-led AKST 198
5.6.1.2 Ecosystem-oriented AKST 198
5.6.1.3 Local food-supply led AKST 199
5.6.1.4 Local-learning AKST 200
5.6.2 Towards options for action 201

Key Messages

1. Choices about agricultural knowledge, science and technology (AKST) relate to paradigms, investment, governance, policy and other ways to influence the behavior of producers, consumers and food chain actors. They will have powerful impacts on which development and sustainability goals are achieved and where, both globally and within NAE. There are many uncertainties of the future, and therefore a number of alternative AKST futures can be identified. It is unlikely that all development goals can be achieved in any of these futures.

As Seneca wrote: "There is no favorable wind for the person who does not know where he wants to go." Depending on which development direction society chooses and how funds are allocated, different drivers will be emphasized. This will affect agricultural systems and related AKST. When making decisions, policy makers will need to consider the opinions of the local population and organizations, and the increasing number of NGOs involved in AKST. Interventions on some trends or in response to some uncertainties can be more quickly implemented and be more effective than on others.

2. The conclusion of a number of recent global and regional foresight exercises on agriculture, rural development and environment is that business as usual will not be good enough. Consumers, producers and information providers will have to rapidly recognize and respect the physical limits of the planet and the biological equilibriums needed to ensure long-term survival. New responses must be found. Different kinds of approaches have been used to address future changes in agriculture. Some have employed projections accompanied by limited policy simulations. Others have proposed scenarios and considered a wide range of uncertainties in an integrated manner. They all explore key linkages between different drivers and resulting changes.

3. Science and technology studies stress the consequences of major technological developments in fields not directly related to agriculture but that could have important potential impact on AKST in the future. These relate, for example, to information and communication technologies e.g., imaging and Radio Frequency Identification, as well as to nanotechnologies, genomics, biotechnologies and physics.

4. NAE agrifood systems will continue to face longstanding problems to increase the output level of agricultural products and services without jeopardizing (1) the natural resource base, (2) food security through equitable access to food and stable food supplies for an aging NAE population and a growing global population, and (3) food safety. The second challenge does not mean producing food to sell or to donate to other countries, but rather cooperating with other countries in developing and sharing AKST that meets this goal.

5. Emerging trends in agriculture are leading AKST to tackle problems that are interacting in a dynamic, complex and mutually reinforcing way, generating long-term impacts, cross-impacts and feedback loops. They are thus requiring new forms of AKST. The main trends are the following:

- Human as well as plant and animal health considerations are becoming more important. Populations in North America and Western Europe, especially the poor, face alarming increases in illnesses associated with inadequate diets and over-processed food. Central and Eastern Europe are likely to face the same problems in the near future. Increased plant and animal diseases, as well as weed and insect problems, both evolving and invasive, are threatening production in certain areas and are leading to overuse of agricultural chemicals and antibiotics, whose lingering residual effects in the environment is threatening human health. This trend could be addressed through new AKST, more information and appropriate regulations, as well as encouragement for individuals and companies to market and consume organic foods.
- Agricultural trade policies and subsidies in NAE tend to undermine the achievement of development goals in other parts of the world. There is uncertainty about whether the World Trade Organization will be effective in harmonizing approaches to internal subsidies, and about whom is likely to benefit, how much and for how long if NAE subsidies are removed. Applying AKST to agricultural policies and property regimes might help balance the needs of vulnerable people in other regions of the world.
- Farms tend to specialize, as they grow in size and decline in numbers. Alternative agrosystems coexist with mainstream agriculture. Farmers are working in larger enterprises, operating through cooperative arrangements and contracts with large businesses. This could lead to greater complexity and monopolies that reduce resilience and choices. AKST is needed to devise alternative agrosystems.
- Businesses in every sector of the food system are concentrating into integrated networks and exerting power by imposing standards on suppliers that challenges their ability to remain viable. Such standards gradually exclude small-scale producers, processors or other enterprises from participation in markets. The rate at which this integration is proceeding and the specific geographic areas and sectors that businesses will choose to enter are uncertain, in part because most business decisions are not transparent.
- Rural populations are dwindling and agro-urban areas are growing. Multiple expectations on farming systems are leading to the development of new enterprises such as agrotourism, and are pushing the farming systems to deliver new services, such as watershed development and landscape protection. But the high demands on agriculture to provide energy could change this trend. AKST is needed to improve the sustainability of food and farming systems, regardless of what is demanded of them.
- Migrant labor represents a growing proportion of the workers in the agrifood sector, especially in parts of the United States and the southern countries of Europe. An increasing number of them are illegal immigrants.

Enforcement of immigration law would force undocumented workers to leave the countries. The impact on the labor force could be solved by policy research and technological advances, but must be accompanied by political and social measures.

- Rising prices of energy, water, minerals and other natural resources could affect outputs, costs and practices in all sectors of the food system. NAE agriculture uses large quantities of natural resources such as oil, water and phosphates, although there are regional differences. Decreasing availability and increasing competition for these resources boosts costs to heights that can have very negative impacts on agricultural production, processing, distribution, retail and purchasing. These effects could be averted by a substantial reduction in the use of these resources thanks to improved management and new technological developments that increase use efficiency, and hence could mitigate the consequences of the current trend.
- Climate change increasingly affects agriculture, which will require a wider and stronger spectrum of adaptation responses as well as efforts to reduce energy needs and emissions. Higher temperatures, more erratic precipitation patterns and increased risks of droughts, particularly in the southwestern parts of the USA and in Europe, coupled with a northern shift of cropping zones, will lead to shifts in agricultural systems and production regions. Extreme events will severely challenge adaptive capacity. Existing AKST needs to be applied and new AKST developed.
- Increased demands are being laid on agriculture for providing energy and biomaterials. Bioenergy that includes the production of liquid fuels from biomass could meet some of the world's growing energy needs. It is unclear to what extent agriculture in NAE will become an energy producer, and how much can be achieved from other renewable energy sources and conservation. The development of bioenergy will increase competition for land and water resources and could lead to higher food prices. Significant technological challenges still need to be overcome for the second-generation technologies to become commercially viable.

6. Emerging and ongoing trends as well as uncertainties in Knowledge, Science and Technology (KST) can be identified and are going to influence the way Agricultural Knowledge Science and Technology (AKST) will be developed:

- Innovation is a strategic element in economic competition, but companies' investments depend on the expected return; the level of private R&D varies from one country to the other. Large multinational companies are increasingly influencing priorities and investments in agricultural science and technology and are highly involved in agricultural extension. Some consider this trend as positive, others as negative.
- The public funding of science and technology is starting to be insufficient to adequately address agricultural problems including satisfaction of consumer demands and the need for more sustainable natural resource management. The decreasing proportion of publicly funded AKST means that less AKST is available in the public domain thus limiting farmers' choices and restricting research on issues such as food security and safety, sustainability, climate change. This also has a negative impact on partnerships with other regions of the world. Halting and reversing this negative trend depends on the will of governments. A reshaping of intellectual property rights and other regulatory frameworks could also modify this trend.
- The interest for science and the number of students in science and technology in most of NAE is declining. The population of European researchers is aging, and students tend to turn away from science and technology, especially when it is research oriented. Measures are needed to bolster school education programs and public awareness in order to draw public recognition to the benefits of S&T. In North America, the number of students in "sustainability programs" is increasing, but few have agricultural backgrounds.
- The present domination of NAE in generating formal knowledge could be challenged. Bigger R&D budgets and better R&D results in Asia are changing the relationship of NAE research with that of the rest of the world. This could lead to more networking and increased competition among agriculture, industry and services.
- The involvement of users in research definition and execution is challenging the traditional research approach. Innovation is a process that integrates various forms of research and the knowledge it creates in a wide range of patterns. Users are increasingly expressing needs which challenge the traditional disciplinary research approaches but may pave the way to a more integrated approach that some researchers find difficult and that, potentially, could be an obstacle to the required innovation.
- The capacity of universities and public research organizations, the private sector and the governments to make their economies competitive by defining research priorities jointly and funding R&D is uncertain. Collaborative research is gaining in importance and measures could be taken to further promote it and improve the general R&D effort.

7. There are several plausible pathways and major differences in the AKST sets of drivers; much depends on the society's choices. The differences lead to alternative pathways for AKST development:

- Economic considerations and drivers could shape a globally-oriented "market-led AKST" wherein multinational corporations and other private sector actors play a major role. In that case, public policies would tend to be reactive, and consumer protection would amount to measures taken after serious problems have occurred. Policies would mainly focus on trade liberalization and assurances of a favorable platform for free competition. The common interests of transnational corporations and wealthier consumers would determine industrial, KST- and capital-intensive solutions marketed under private labels. "Market-led AKST" could effectively decrease hunger and poverty and improve nutrition and human health in NAE and at international levels. However, it would probably contribute little to equity and sustainable economic development.

- Increased government intervention could lead to "ecosystem-oriented AKST" with strong public sector input and interventions to internalize environmental externalities through regulations, taxation, subsidies and international standards. In that scenario, the public sector would invest in centralized, coordinated innovation systems, with few centers of excellence. Education would be a priority and solutions would probably be knowledge-intensive, high-tech and precision oriented. "Ecosystem-oriented AKST" could make a major contribution to improving environmental sustainability through knowledge-intensive technologies that use resources efficiently and to sustaining economic development by investing human and financial capital in the development of green technologies. It could have the potential to level off global imparities. However, little emphasis on social viewpoints might lead to shortcomings regarding issues such as equity and enhanced livelihoods.
- Cross-sectoral public-private governance platforms with emphasis on regional and local decision making along with the subsidiary principle and bottom-up approaches could lead to "local-learning AKST" system. Food system actors and both rural and urban regions would participate in interactive knowledge networks that are decentralized and regionally diversified. Externalities would be internalized through direct response and locally visible impacts of AKST, but local standards would also be developed. "Local-learning AKST" could successfully contribute to the goals of enhancing livelihoods, equity and social capital and to environmental sustainability, especially within the regions. Nutrition and human health would be improved through knowledge-based, safe local diets and a reduction in meat consumption. Balanced urban-rural regional economic development would be promoted by keeping up the added value in the region. Hunger and poverty in other regions would not be a high priority.
- A "local food-supply led AKST" system could arise if research efforts were not coordinated and if budget cuts were took place. "Local food-supply led AKST" is a plausible future which would not contribute to development and sustainability goals.

5.1 Context

Agricultural systems and land use are changing as a consequence of changes in demography, world trade, climate, diets, political unions (e.g., enlargement of the European Union) and technology. The degree and impact of these variations are largely unknown. Although the future is unpredictable, some developments can be foreseen and alternatives explored. This chapter focuses on trends and uncertainties related to the futures of the main drivers of agricultural research and innovation systems and agricultural knowledge, science and technology (AKST).

5.1.1 Problem statement

The future of the agricultural research and innovation systems in North America and Europe is not certain, and current systems may be revised or new ones built. There are several plausible futures, some more desirable than others. Each of them depends on the decisions and actions of today's leaders. Some of the appealing futures appear plausible and feasible and may help decision makers choose strategies to reach those futures. Other futures, although desirable, are utopic and may be of less value for planning the future.

Forecasting and foresight are methods to think about options for the future. They can have a national, a regional or a sectoral focus. They can be based on scientific panels, the Delphi method, scenario development, investigative surveys, working groups or scientific seminars. Foresight activities can focus on the result (e.g., projections or scenarios) or on the process (Godet, 1977; Irvine and Martin, 1984, 1989; Hatem, 1993; Martin, 1995; de Jouvenel, 2004; de Lattre-Gasquet, 2006). Emphasizing the process can help to build strategic capabilities and to inform research and innovation policies ("embedded foresight") (Kulhmann et al., 1999).

Identifying appropriate drivers is the first step in forecast/foresight activities. As defined in chapter 1, a "driver" is any natural or human-induced factor that directly or indirectly causes a change in an ecosystem. Drivers are linked to decision making, as many of the drivers can be influenced by policy choices. A "direct driver" unequivocally influences agricultural production and services and can therefore be identified and measured with differing degrees of accuracy. An "indirect driver" operates more diffusely, often by altering one or more direct drivers, and its influence is established by understanding its effect on a direct driver. The tendential development of each driver must be presented, and curves and potential breaks that could block the tendential development should be explored (de Jouvenel, 2004). In this chapter, uncertainties about the futures have been raised in the form of questions, and no hypotheses about future development have been made.

As described in chapter 1, AKST is knowledge, science and technology pertaining to agriculture. It is a subset of science and technology, located at the intersection of the agricultural system and the knowledge, science and technology system. The futures of AKST depend on the futures of agriculture, the futures of KST, and have their own dynamic. This chapter is built around four questions:

- What are the key drivers for knowledge, science and technology (KST), their major uncertainties and consequences for AKST? (see 5.3)
- What are the key drivers for agriculture, their major uncertainties and consequences for AKST? (see 5.4)
- What are the key drivers for agricultural knowledge, science and technology (AKST) and agricultural research and innovation systems and their major uncertainties? (see 5.5)
- What are some future normative AKST systems and their potential contributions to sustainable development goals? (see 5.6)

For each driver, the questions show that the future is uncertain. Each driver also points to fields where AKST needs to be developed or expanded.

The plausible futures comprise a number of goals for

an agricultural research and innovation system, including promotion of sustainable agriculture and enhancement of nutritional security, human health and rural livelihoods, and AKST depends on the priorities. At the same time, an agricultural research and innovation system and certain AKST could help mitigate environmental degradation and social inequities. Reaching all of these goals will be difficult; various agricultural research and innovation systems favor particular goals at the expense of others. These alternative futures expand the spectrum of possibilities and will facilitate discussions among decision makers about strategic choices.

5.1.2 Review of related studies

A number of recent foresight exercises focusing on agriculture, rural development, environment, science and technology have been undertaken at global and regional levels. Different kinds of approaches have been used to address future changes pertaining to agriculture. Some have employed projections accompanied by limited policy simulations. Others have proposed scenarios and considered a wide range of uncertainties in an integrated manner. They all explore key linkages between different drivers and resulting changes. They all conclude that business as usual will not suffice. However, no assessment has explicitly focused on the future role of AKST.

5.1.2.1 At global level

A number of quantitative models have been developed by such organizations as IFPRI, the Food and Agricultural Policy Research Institute (FAPRI), FAO, OECD, and the Netherlands Environmental Assessment Agency.

Partial equilibrium models (PE) treat international markets for a selected set of traded goods, e.g., agricultural goods in the case of partial equilibrium agricultural sector models. These models consider the agricultural system as a closed system without linkages with the rest of the economy, apart from exogenous assumptions on the rest of the domestic and world economy. The strength of these partial equilibrium models is their great detail of the agricultural sector. The "food" side of these models generally uses a system of supply and demand elasticities incorporated into a series of linear and nonlinear equations, to approximate the underlying production and demand functions. World agricultural commodity prices are determined annually at levels that clear international markets. Demand is a function of price, income and population growth. Regional biophysical information (for land or water availability, for example) is constraining the supply side of the model (IAASTD Global Chapter 5).

Computable general equilibrium (CGE) models are widely used as an analytical framework to study economic issues of national, regional and global dimension. CGE models provide a representation of national economies, next to a specification of trade relations between economies. CGE models are specifically concerned with resource allocation issues, that is, where the allocation of production factors over alternative uses is affected by certain policies or exogenous developments. International trade is typically an area where such induced effects are important consequences of policy choices. CGE models have sometimes been used to provide a scientific guarantee in support of full trade liberalization (Boussard et al., 2006).

Beyond IAASTD, major global environmental assessments include:

- The Millennium Ecosystem Assessment (MA, 2005).
- The Intergovernmental Panel on Climate Change (IPCC) assesses scientific, technical and socioeconomic information needed to understand climate change, its potential impacts and options for adaptation and mitigation. In 2007, IPCC finalized its Fourth Assessment Report.
- The UNEP-led Global Environment Outlook (GEO) project focuses on the role and impact of the environment for human well-being and the use of environmental valuation as a decision tool.
- The OECD environmental outlook to 2030 focuses on environment-economic linkages to 2030. The projections are complemented by qualitative discussions based on extensive OECD analytical work.
- The Comprehensive Assessment of Water Management in Agriculture led by the International Water Management Institute (IWMI) critically evaluated benefits, costs and the impacts of the past 50 years of water development and looks at current challenges to water management.
- The Global Scenario Group (GSG) was convened in 1995 by the Stockholm Environment Institute to examine the prospects for world development in the twenty-first century. Numerous studies at global, regional and national levels have relied on the Group's scenario framework and quantitative analysis (Kemp-Benedict et al., 2002).

Chapters 4 and 5 of the global IAASTD report have reviewed a number of quantitative models extensively (see Table 5-1 in the Global Report):

- IMPACT-WATER. A partial equilibrium agricultural sector model with a water simulation module developed by the International Food Policy Research Institute (IFPRI) (Rosegrant et al., 2002). Using this model, IFPRI has made a number of studies, e.g., Global Food Projections to 2020 (Rosegrant et al., 2001), Global water outlook to 2025 (Rosegrant et al., 2004), Fish to 2020: supply and demand in changing global markets (Delgado et al., 2003), Food security (Von Braun et al., 2005),
- IMAGE. Integrated model to assess the global environment developed under the auspices of the Netherlands Environmental Assessment Agency (MNP) (Bouwman et al., 2006),
- GTEM. Global trade and environment model, a computable general equilibrium model developed by the Australian Bureau of Agricultural and Resource Economics (ABARE) (Pant, 2002),
- WATERSIM. Water, Agriculture, Technology, Environment and Resources Simulation Model developed by the International Water Management Institute (IWMI) and IFPRI (de Fraiture et al., 2006),
- GLOBIO3. Global methodology for mapping human impacts on the biosphere, a consortium that seeks to develop a global model for exploring the impact of en-

Table 5-1. **Overview of quantitative modeling tools used in IAASTD Global Chapter 5.**

Global foresight model	Main focus	Timeline	Approach
Global Scenario Group (GSG)	Sustainable development		Strong focus on storyline, supported by quantitative accounting system
IPCC—Third and Fourth Assessment Reports (TAR3 and TAR4)	Climate change, causes and impact	2100	Storylines supported by modeling.
IPCC—SRES	Greenhouse gas emissions	2100	Modeling supported by storylines.
UNEP: GEO3 & GEO4 RIVM 2004	Environment		Storylines and modeling. Modeling on the basis of model chains/interlinked models
Millenium Ecosystem Assessment —MA	Ecosystems	2050	Storylines and modeling. Modeling on the basis of linked models
OECD—FAO Food outlook	Food Systems	2015	
OECD—FAO Food	Food Systems	2030/2050	
FAO at 2020	Agriculture	2020	Single projection, mostly based on expert judgment
IFPRI World Food Outlook	Agriculture	2020	Model-based projections. Global and regional scenarios.
OECD Environment Outlook	Ecosystems		

vironmental change on biodiversity and was designed to support UNEP's activities (GLOBIO, 2001),

- EcoOcean. A marine biomass balance model of the University of British Columbia,
- GEN-CGE. A computable general equilibrium model for India,
- CAPSiM. A partial equilibrium agricultural sector model for China.

Since 1995, FAO has been using a World Food Model, which is a partial equilibrium model capable of making projections on food demand and supply at the 2030 horizon and 140 countries and 32 products. FAO has published the work of Collomb (1999) and more recently two reports on world agriculture towards 2015-2030 and towards 2030-2050 (Bruinsma, 2003; FAO, 2006). OECD and FAO publish the Agricultural Outlook periodically. The most recent is for 2007-2016 (OECD/FAO, 2007).

Quantitative projections indicate a tightening of world food markets, with increasing resource scarcity, pushing prices up which especially penalizes the poor consumers. Real world prices for most cereals and meats are projected to increase in the coming decades, dramatically reversing trends from the past several decades. Price increases are driven by both demand and supply factors. Population and economic growth in sub-Saharan Africa, together with already high growth in Asia and moderate growth in Latin America drive increased growth in demand for food. Rapid growths in meat and milk demand are projected to put pressure on prices for maize and other coarse grains and meals. Bioenergy demand is projected to compete with land and water resources. Overall growing water demands and land scarcity are projected to increasingly constrain food production growth and have an adverse impact on food security and human well-being goals. Higher prices can benefit surplus agricultural producers, but can also reduce access to food for a larger number of poor consumers, including farmers who do not produce net surplus for the market. As a result, progress in reducing malnutrition is projected to be slow (IAASTD global report, chapter 5).

Although none are identical to the IAASTD exercise in scope and timeframe, many meetings and reports have addressed one or more of the components included in the IAASTD narrative. We have collected and reviewed a number of them focusing on Europe and North America which include elements of the IAASTD exercise.

5.1.2.2 At European level

There are too many foresight activities in Europe to describe them all. We will describe a few exercises and give the references for networks and places where information can be found.

In the European Commission, foresight activities are launched and carried out in several places:

- The European Technology Platforms (ETPs) which provide a framework for stakeholders, led by industry, to define research and development priorities, timeframes and action plans on a number of strategically important issues where achieving Europe's future growth, competitiveness and sustainability objectives is dependent upon major research and technological advances in the medium to long term. More than thirty platforms exist, for example "Food for life," "Plants for the future," "Global animal health," "Forest-based sector technology."

- The Joint Research Centre's (JRC's)/Institute for Prospective Technological Studies (IPTS). The mission of IPTS is to provide technico-economic analyses to support European decision markers. It monitors and analyses S&T related developments, their cross-sectoral impacts, interrelationships and implications for future policy development.
- The European Science Foundation (ESF) which has introduced "Forward Looks" to enable Europe's scientific community, in interaction with policy makers, to develop medium to long-term views and analyses of future research developments with the aim of defining research agendas at national and European level.
- The ForSociety which is a network where national foresight program managers coordinate their activities.
- The Science and Technology Foresight Unit of the DG research whose missions are to promote cooperation in European foresight, to monitor and exploit foresight, informing European research policy developments and contributing to policy thinking in DG research, to implement S&T foresight activities, to promote foresight dissemination and experience sharing, and to prepare a foresight report. Studies are commissioned and expert groups meet. The Science and Technology Foresight Unit has commissioned studies such as "Converging Technologies. Shaping the Future of European Societies" (Nordmann, 2004), the future of Key Research Actors in the European Research Area (Akrich and Miller, 2007; http://cordis.europa.eu/foresight/home.html).
- Different directions can launch foresight activities. For example, in 2007 the Standing Committee on Agricultural Research (SCAR) commissioned a Foresight food, rural and agrifutures (FFRAF) study which is presented below.

The European Parliamentary Technology Assessment (EPTA), the European Organization for Nuclear Research (CERN) and the European Molecular Biology Laboratory (EMBL) all have foresight activities.

The European Foresight Monitoring Network (EFMN) monitors ongoing and emerging foresight activities and disseminates information about these activities to a network of policy researchers and foresight practitioners. It supports the work of policy professionals at regional and national level. The EFMN is part of the European Foresight Knowledge Sharing Platform. It monitors and maps Foresight activities all over the world.

The European Futures Observatory (EUFO) is a UK based not-for-profit company limited by guarantee, formed in October 2004, which aims to foster the development of a European School of Futures Studies. It is starting to carry out studies and has looked at the strategic futures that the US may encounter out to the year 2025.

In Europe, a number of modeling exercises have been designed. The global economy-wide dimension is covered by the economic LEITAP model (a modified version of the global general equilibrium Global Trade Analysis Project, GTAP, model) and the biophysical IMAGE model (developed by MNP).

ESIM (European Simulation Model) is providing more agricultural detail for the EU-25 countries. CAPRI has been developed by the University of Bonn and is a static partial equilibrium model with a dynamic recursive version to simulate policies.

WEMAC, developed by the Institut National de la Recherche Agronomique (INRA), in France, is a partial equilibrium model on crops that can make projections and simulations for cereals and oil crops in Europe.

MEGAAF (modèle d'équilibre général de l'agriculture et de l'agroalimentaire français) is a general equilibrium model to simulate commercial policies for France and the rest of Europe.

Three recent European foresight exercises represent different approaches: Eururalis, Scenar 2020 and FFRAF (Foresight food, rural and agrifutures). Eururalis was launched with the aim to explore alternative future rural development options for EU-25 (Klijn and Vullings, 2005). This Dutch project is developing and analyzing a set of four long-term alternative scenarios to capture major uncertainties. Based on its success in providing sound information on future rural development options during the 2004 Dutch EU Presidency, an extended version of the Eururalis toolbox (no. 2.0) is under development. The new version will be used to analyze a number of specific rural policy questions for EU-25, including issues related to bioenergy and strategic options for the Common Agricultural Policy (CAP) after 2013 and the consequences on sustainability indicators. Such policy questions can be posed for each of the four different world views, as developed in Eururalis 1.0, with regional differentiation and different time horizons: 2010, 2020 and 2030. The aim of the Eururalis toolbox is to help policy makers formulate long-term development strategies for rural areas in Europe (EU-25) (Box 5-1).

Alternatively, Scenar 2020, a recent initiative of European Commission, Directorate General for Agriculture, uses a baseline approach with varying policy options and particular focus on the impact of technological change (especially information communication technology) and food chains on agriculture and rural areas (EC, 2007).This study aims to identify future trends and driving forces shaping the European agricultural and rural economy (EU-27 +) on a time horizon up to 2020. Analyses of trends from 1990 to 2005 provide the basis for developing a reference scenario (baseline) that represents a trend projection up to 2020. Three variants are constructed around the baseline: the baseline with modifications of current policies that are reasonably certain to happen, a "liberalization" scenario and a "regionalization" scenario. The latter two represent alternative policy frameworks with differing degrees of support to the agricultural sector. Drivers of change are grouped into those that are independent of policy influence (at least for the time horizon up to 2020) and those associated with agricultural and environmental policies (EC, 2007). In Scenar 2020, the spatially explicit land use model CLUE-s (Conversion of Land Use and its Effects) (Verburg et al., 2002) is used. The CLUE-s model disaggregates the outcomes of the ESIM-CAPRI-LEITAP/IMAGE suite of models to a temporal reso-

Box 5-1. EURURALIS. Scenario "Competing claims for scarce resources—EU biofuel policy option" Source: W.A. Rienks.

The results of Eururalis outline what could happen in rural Europe towards 2030, based on conditions that differ in nature, course, duration or place. In Eururalis four contrasting scenarios are evaluated. The impact on various people, planet and profit indicators is calculated. One of the scenarios is the Global Economy scenario. This scenario depicts a world with fewer borders and regulation compared with today. Trade barriers are removed and there is an open flow of capital, people and goods, leading to a rapid economic growth, of which many (but not all) individuals and countries benefit. Within this scenario three alternative policy options for biomass production for biofuels have been elaborated (only 1st generation biomass technology being taken into account):

1. no blending obligation for the EU (No BF)
2. 5.75% blending obligation of biomass in transport fuel within the EU (BF 5.75%)
3. 11% blending obligation of biomass in transport fuel within the EU(BF 11.5%)

Results: The figure (see Annex H) shows the impact on agricultural land use (crop area) in EU15 and Brazil in the Global Economy scenario with 3 different policy options regarding the blending of biomass in transport fuel. The graph shows opposite trends for both regions. In the EU15 towards 2030 there is land to spare. Consequently, marginal agricultural regions will face land abandonment. This is driven by higher yields per hectare and low growth of the EU population and its demand for food. In EU15, the abandonment of extensive agricultural land sometimes leads to loss of high nature value farmlands. In Brazil, on the contrary, growing regional and global population and an increased demand for food crops worldwide drive the increase of agricultural land. This will put extra pressure on nature and forest areas. For both EU15 and Brazil there are clear impacts of the EU biofuels policy. The blending obligation for transport fuel increases the needed crop area in both regions. In South America this is putting an extra pressure (of about 20 million ha) on land used currently as nature or pasture land. In Europe the extra demand for biomass is slowing down the trend of agricultural abandonment but it does not stop it. These results clearly show that EU strategic policy has not only impact on land-use within Europe but also a very significant impact elsewhere in the world.

References

Wageningen UR and Netherlands Environmental Assessment Agency. 2007. Eururalis 2.0: A scenario study on Europe's rural Areas to support policy discussion. Eururalis 2.0 CDrom. Alterra, Wageningen Univ., Wageningen The Netherlands.

Verburg, P.H., B. Eickhout, H. Van Meijl. 2007. A multi-scale, multi-model approach for analyzing the future dynamics of European land use. Annals of Regional Science.

Klijn J.A., L.A.E. Vullings, M. Van de Berg, H. Van Meijl, R. Van Lammeren, T. Van Rheenen, et al. 2005. The EURURALIS study: Technical document. Alterrarapport 1196. Alterra, Wageningen

lution of two years and a spatial resolution of 1 km. CLUE-s provides a cross-sectoral approach that includes all land use relevant sectors, while the ESIM, CAPRI and LEITAP/IMAGE models mainly address the land use of agricultural sectors. The results indicate that the structural changes, i.e., decline of agricultural contribution to total income and employment, will continue at national level. Regions with high shares of agriculture and industries may be vulnerable to this process with regard to employment and income growth, as the structural change process is often characterized by adjustment processes and related costs. The impacts of each scenario on production, employment, land use, etc. are detailed in the Scenar 2020 report.

EURURALIS mainly sketches different alternative future directions and their consequences while Scenar 2020 performs a sensitivity analysis with regard to very precise policy modifications. Each has its advantages. SCENAR 2020 identifies demographic dynamics as the strongest driver, now and probable also for the future rural world. In general, the SCENAR study concludes that the economic importance of agriculture will continue to decline although agriculture will remain a significant land use with an increasing role in managing externalities such as landscape and biodiversity. In 2020, there will be fewer farms but they will be more competitive at global scale, and they will enjoy higher average income and higher productivity.

FFRAF (Foresight food, rural and agrifutures) was launched by the Standing Committee on Agricultural Research (SCAR) of the European Commission to identify possible scenarios for European agriculture in a 20-year perspective and priority research needs for the medium and long term. FFRAF shows that the European Union is at the beginning of a major disruption period in terms of international competitiveness, climate change, energy supply, food security and societal problems of health and unemployment. It points to the need for a new strategic framework for research planning and delivery. The framework needs to cater for four broad lines of action and a fifth cross-cutting theme, respectively: sustainability challenge, security challenge, knowledge challenge, competitiveness challenge and policy and institutional challenge (FFRAF, 2007).

A number of exercises have also been conducted for the EU's East European countries, such as Czech Republic, Hungary, Poland, etc. For example, the ForeTech project looked at technology and innovation related to agriculture, food and drinks for Bulgaria and Romania. Another study analyzed the potential evolution of agricultural income and the viability of selected farming systems in the Czech Republic, Hungary, Latvia, Poland and Romania under different Common Agricultural Policy implementation scenarios (Cristoiu et al., 2006).

The UK, Finland, Germany, The Netherlands, Ireland, Norway, Sweden, Romania, France, etc. have all conducted foresight studies on the future of the agricultural sector and/or the future of science and technology in their countries (Table 5-2).

Table 5-2. **References to European foresight exercises related to Agriculture, Food, Science and Technology since 2003.**

Scenarios for the Future of European Research and Innovation Policy	Proceedings of a STRATA/Foresight Workshop. 9-10 December 2003. EUR 21251.	EU Commission, 2003
Prospects for Agricultural Markets in the European Union 2003-2010.	Brussels, June 2003.	EU Commission, Directorate General for Agriculture, 2003
THE AGRIBLUE BLUEPRINT	Sustainable Territorial Development of the Rural Areas of Europe	EU Commission, Directorate General for Research, 2004
Foresighting the New Technology Wave	Expert Group: http://cordis.europa.eu/foresight/ntw_expert_group.htm Dissemination conference: http://cordis.europa.eu/foresight/ntw_conf2004.htm	EU Commission, 2004
Prospective Analysis of Agricultural Systems	European Commission, Technical Report EUR 21311 EN. ftp://ftp.jrc.es/pub/EURdoc/eur21311en.pdf	EU Commission, IPTS, 2004
Key Technologies for Europe	http://cordis.europa.eu/foresight/kte_expert_group_2005.htm The "Key Technologies" Expert Group has approached the future of several key technologies all crucial for Europe's future: biotechnology, nanotechnology, information technologies, communication technologies, transport technologies, energy technologies, environmental research, social sciences and humanities, manufacturing and materials technologies, health research, agricultural research, cognitive sciences, safety technologies, complexity research and systemic, research in the services sector.	EU Commission, 2005.
Emerging Science and Technology priorities in public research policies in the EU, the US and Japan.	EUR 21960 http://ec.europa.eu/research/foresight/pdf/21960.pdf	EU Commission, 2006
Using foresight to improve the science—policy relationship.	EUR 21967 http://ec.europa.eu/research/foresight/pdf/21967.pdf	EU Commission, Directorate General for Research, 2006
The future of key research actors in the ERA.	Synthesis paper. (Madeleine Akrich and Riel Miller).	EU Commission, Directorate General for Research, 2006
Emerging Science and Technology priorities in public research policies in the EU, the US and Japan	EUR 21960. http://ec.europa.eu/research/foresight/pdf/21960.pdf	European Commission, 2006
Prospects for the Agricultural Income of European Farming Systems	Technical Report EUR 22506 EN	EU Commission, IPTS, 2006
EURURALIS	www.eururalis.nl	
Foresighting food, rural and agri-futures	http://ec.europa.eu/research/agriculture/scar/pdf/foresighting_food_rural_and_agri_futures.pdf	FFRAF report
SCENAR 2020	http://ec.europa.eu/agriculture/publi/reports/scenar2020/index_en.htm	A scenario study on agriculture and the rural world.
Plants for the future	Stakeholders Proposal for a Strategic Research Agenda 2025. Including Draft Action Plan 2010. http://www.epsoweb.org/catalog/tp/tpcom_home.htm	
Agri-Food Industries & Rural Economies: Competitiveness & Sustainability: The Key Role of Knowledge		Downey, L. June, 2005
Green Technological Foresight on Environmental Friendly Agriculture	http://www.risoe.dk/rispubl/SYS/ris-r-1512.htm	
Prospective environmental analysis of land use development in Europe	http://www.eea.europa.eu/multimedia/interactive/prelude-scenarios/prelude	

5.1.2.3 At North American level

North America (NA) has a large number of studies on the future of agriculture and/or AKST, but there is no coordination or networking among organizations, hence the studies are difficult to collect. More prominently than in Europe, the role of technology is a commonly addressed element in foresight exercises.

Beyond the Central Intelligence Agency (CIA) which does not reveal the results of studies, the National Intelligence Council (NIC) is a centre for midterm and long-term strategic thinking. The "Mapping the global future" report looks at the world in 2020 (NIC, 2004).

As far as agriculture is concerned, the Economic Research Service (ERS) of the US Department of Agriculture (USDA) conducts a research program to inform public and private decision making on economic and policy issues involving food, farming, natural resources, and rural development. ERS specialists, for example, provide wide-ranging research and analysis on production, consumption, and trade of key agricultural commodities and on agricultural policies of countries and regions important to U.S. agriculture, as well as on international trade agreements and food security issues. The Economic Research Service (USDA/ERS) has developed the SWOPSIM model (Static World Policy Simulation Model) to study the interaction of US policies with those of the rest of world. (See http://www.ers.usda.gov/).

Universities are also very active in trade modeling. The University of Purdue, for example, has developed GTAP (Global Trade Analysis Project), a data base and a model on production, consumption and trade.

The World Technology Evaluation Center, Inc. (WTEC) is a US organization conducting international technology assessments via expert review. For example, report on converging technologies (nanotechnology, biotechnology, information technology and cognitive science) have been written for the National Science Foundation (NSF) (Roco and Bainbridge, 2002; Bainbridge and Roco, 2006).

The Department of Interior (DOI) has conducted a study "Water 2025" which sets a framework to focus on meeting water supply challenges in the future (US DOI, 2005).

The application of nanotechnology in precision agriculture is a recurring theme. Producers could have near real-time data from every plant or animal (Fletcher, 2007; Western Farm Press, 2007); computers would automatically collect and analyze the information. These data would allow producers to detect and correct disease infections, pest infestations, nutrient/water deficiencies, etc. before there is any significant effect on the plant/animal. This type of system would allow precise targeting (and tremendous reductions) of medicines, pesticides, nutrients and water. Much of the process would be completely automated; problems could be addressed or prevented (Catlett, 2003). Combinations of detection technology and global positioning technology would allow detection and precise location information. Pesticides, nutrients and water could be used more efficiently and with fewer environmental effects.

The application of technology will also be a response to demographic changes in North America (NA). Slow population growth, combined with an aging population, will reduce the labor pool available for agriculture. However, increased mechanization of North America agriculture will reduce the number of workers needed for an agricultural operation (McCalla, 2000). Although the workers will have to assess and apply much more information, computer assistance and automated responses will minimize the manpower requirements.

Consumer demands are also a common element in many of the foresight reports. In part, the application of technology will be driven by consumer demands. The North American demand for food quantity is expected to be mostly static, but greater affluence and consumer knowledge will create a demand for product differentiation. An aging, health conscious NA population will ask for greater health benefits and fewer risks from food. Biotechnology can be used to manipulate nutritional qualities of foods and reduce chemical inputs remaining on foods.

Additionally, affluent consumers are more knowledgeable about environmental issues and more likely to pay a premium for products that have been produced/processed with attention to environmental or social issues (Univ. Georgia, 2000). Technology can provide the means to track individual food items or food components from the field to the table (Western Farm Press, 2007). Consumers will be able to make buying decisions based a wide range of nutritional, environmental and social factors.

Greater affluence is also associated with an increased demand for meat in the diet. Because the typical diet in NA is already based on meat, the demand in NA is unlikely to change significantly. However, increasing affluence in other countries will most likely strengthen the export market for meat produced in NA. Additionally, there will be greater demand for grains to produce meat animals.

Aging and affluence will also generate greater demand for additional processing of food products (Western Farm Press, 2007). Aging consumers, in particular, are willing to pay more for convenience. Consequently, there will be a greater demand in NA for prepared foods or products that can be prepared quickly and easily.

All of these consumer factors will combine to create a broad, varied market for differentiated products. Some groups of people will be most interested in food properties (e.g., nutrition, flavor, or convenience); others will choose agricultural products based on concomitant environmental impacts of production. Technology and rapid global communication will allow consumers to evaluate a wide range of factors and to identify/track agricultural products from the field to their home.

There are reports that discuss the importance of multifunctional agricultural systems and underline the need for greater public awareness and support of multifunctionality (McCalla, 2000; Tilman et al., 2002). Affluent consumers are not concerned about food supply and have greater knowledge of the environment. They are more likely to pay for environmental services (e.g., wildlife habitat or watersheds) associated with agricultural production.

Agriculture will provide new products and services. Genetically modified plants and animals will produce many different pharmaceuticals and raw materials for industry. In NA, agriculture will become a major source of energy (Ugarte et al., 2006). Modified plants and agricultural waste products will be converted to fuel. This industry will expand into a leading market for agriculture, providing a major additional revenue stream but possibly creating resource competition between the production of food and fuel. Agriculture will become a more important source of fuel as China and India become key competitors for energy (Vanacht, 2006). It would be particularly important if feed grains (e.g., corn) were massively used for energy, as is done currently, or lose in importance. In the former case, it will become more difficult and expensive to meet a rising demand for meat (Ugarte et al., 2006).

Carbon sequestration may be a new role for NAE agriculture (Skaggs, 2001, US EPA, 2005). As China, India and other countries become more industrialized, it will become more critical to mediate levels of greenhouse gases. Plants can remove carbon dioxide from the atmosphere, a service that agriculture could provide. If carbon sequestration is combined with fuel production, agriculture could provide energy with little or no net gain in greenhouse gases.

The scale and impetus for multifunctional agriculture will depend on locality and the services desired. Many services (e.g., watershed protection) are primarily beneficial to the local area; demand and support for these services will occur at state and local levels (Skaggs, 2001). The federal government will be involved with other services, such as carbon sequestration, that benefit a much larger population and area (Skaggs, 2001; US EPA, 2005).

A number of reports discuss the implications of dualism in NA agriculture. Agriculture will consist almost entirely of very large and small farms. A relatively small number of large farms will produce most agricultural products. Small farms will survive, but operators will also depend on off-farm income; it will be important to provide the related opportunities (Skaggs, 2001). There will be an increased trend for more public-private partnerships (Skaggs, 2001; Univ. Georgia, 2000). A more affluent society will focus private research on convenience/appeal of agricultural products and public research on product safety and environmental impacts.

As knowledge increases, more companies, institutions and individuals will have intellectual property (IP) rights for components that are necessary to further AKST (Atkinson et al., 2003). It is important to revise the current system of IP protection and to harmonize IP security internationally. A new system is needed that will facilitate the sharing of information without eliminating the financial incentive that drives much agricultural research (Table 5-3).

5.1.2.4 Relationship of scenarios in different exercises

All the exercises reviewed have developed assumptions about a number of underlying uncertainties and future development of key driving forces and arrived at different logics regarding the construction of alternative futures. Nevertheless, many scenarios display some similarities, and it has been argued that the enrichment of global scenarios, often through participatory processes, will define an important agenda for policy analysis, scientific research and education. This will require the enhancement of the role of ecosystems in both scenario narrative and quantification. Narratives will need to more richly reflect ecosystem descriptors, impacts, and feedbacks. Models will need to simulate ecosystem services within global assessment frameworks. (Raskin et al., 2005).

5.2 Indirect Drivers for AKST

As indicated in the conceptual framework, the AKST system does not exist in isolation. It interacts with other societal parameters of development: demography, economy, international trade, sociopolitics, science, technology, education and culture. Only some elements will be highlighted here as these indirect drivers are reviewed in detail in chapter 4 of the global report, and some of them pertaining to North America and Europe have been reviewed in chapters 1, 2, 3 and 4 of this report. The predominant drivers of AKST futures are in the KST system and in agriculture.

5.2.1 Demographic drivers

Population growth is an important driver of demand for agricultural products and AKST, but the influence of AKST on population growth is very slow. Food demand is increasing as the world's population grows and migrates. People's requirements for food are related to three factors: quantity, quality (nutrition and safety) and cost. Since climate change, water shortages and soil degradation are rapidly changing the conditions of agricultural production, the Malthusian fears of a widening gap between people's needs and food production are once again coming to the forefront in discussions on the future of the planet. The problem is most acute in the developing countries (Smil, 2000; Raoult-Wack and Bricas, 2001; Gilland, 2002; Von Braun et al., 2005). The global composition of the food demand (e.g., cereals, sugar crops, oil crops, produce, livestock and fish) will be shaped by population growth rates, economic growth, income levels, food safety scares and rapid urbanization in the developing economies, particularly in Asia (Cranfield et al., 1998; Collomb, 1999; Rosegrant et al., 2002; Schmidhuber, 2003; Schmidhuber and Shetty, 2005; Smil, 2005; Griffon, 2006).

Population size and structure are determined by three fundamental demographic processes: fertility, mortality and migration. The common understanding of projections in world demography is that the growth in world population will continue up to a maximum of 7.5 to 9 billion during the second half of the 21st century, followed by a slow decrease (UN Projections).

Between 2007 and 2050, the population of the more developed regions (Europe and North America) will remain largely unchanged at 1.2 billion inhabitants, but the population of the less developed regions is projected to rise from 5.4 billion in 2007 to 7.9 billion in 2050 and the population of the least developed countries is projected to rise from 804 million people in 2007 to 1.7 billion in 2050. Consequently, by 2050, 67% of the world population is expected to live in the less developed regions, 19% in the least developed countries, and only 14% in the more developed regions (UN, 2006).

The European Union no longer has a "demographic motor." Member States whose population is not set to fall before 2050 represent only a small share of Europe's total population. Of the five largest Member States, only Britain and France will grow between 2005 and 2050 (+8% and +9.6% population growth respectively). In some countries population figures will take a downturn before 2015, with a percentage drop of more than 10-15% by 2050 (CEC, 2005).

Table 5-3. **References to Foresight exercises related to Agriculture, Food, Science and Technology at North American level since 2003.**

Welcome to the New World of Agriculture	http://www.extension.iastate.edu/AgDM/articles/others/SetMay00.htm	AgDM. 2000
Critical Dimensions of Structural Change	2nd Annual National Symposium on the Future of American Agriculture, 2000. University of Georgia. http://www.agecon.uga.edu/archive/agsym00.html	University of Georgia. 2000
Agriculture in the 21st Century	CIMMYT—International Maize and Wheat Improvement Center. http://www.cimmyt.org/Research/Economics/map/research_results/	McCalla, A.F. 2000
The Future of Agriculture: Frequently Asked Questions	New Mexico State University (NMSU), College of Agriculture and Home Economics. Technical Report 37. http://cahe.nmsu.edu/pubs/research/economics/TR37.pdf	Skaggs, R. 2001
Agricultural sustainability and intensive production practices	Nature. 418: 671-677	Tilman, D. K.G. Cassman, P.A. Matson, R. Naylor, and S. Polasky. 2002
Intellectual Property Rights: Public Sector Collaboration for Agricultural IP Management	Science, 301: 174-175	Atkinson, R.C., R.N. Beachy, G. Conway, F.A. Cordova, M.A. Fox, K.A. Holbrook, et al. 2003
A Food Foresight Analysis of Ag. Biotechnology		Calif. Dept. of Food and Agriculture Food Biotechnology Task Force. 2003
Futurist View of American Agriculture	New Mexico State University. http://hubbardfeeds.com/swine/MSF03_futurist.shtml/distinguished_economist/4disting_econ_lec/4distecon_contents.htm.	Catlett, L. 2003
21st Century Agriculture: A Critical Role for Science and Technology	http://www.usda.gov/news/pdf/agst21stcentury.pdf	USDA. 2003
Building Science and Technology Capacity for Agriculture: Implications for Evaluating R&D	4th International Crop Science Congress. http://www.cropscience.org.au/icsc2004/symposia/4/5/2107_pardeyp.htm	Pardey, P.G. and J.M. Alston. 2004
U.S. and World Agricultural Outlook	Staff Report 1-05. Iowa State University et University of Missouri-Columbia, Ames, Iowa U.S.A. janvier 2005 http://www.fapri.iastate.edu/ outlook2005/text/FAPRI_OutlookPub2005.pdf	Food and Agricultural Policy Research Institute, 2005
Agriculture's Future: Reading the Tea Leaves	Maple Leaf Bioconcepts. Napanee, Ontario. http://nabc.cals.cornell.edu/pubs/nabc_16/talks/Oliver.pdf	Oliver, J.P. 2005
	Written Testimony before the United States Senate Committee on Agriculture, Nutrition, and Forestry Regarding Benefits and Future Developments in Agriculture and Food Biotechnology	Greenwood, J. 2005
Food Security Assessment	GFA-16, May 2005, USDA/ERS. http://www.ers.usda.gov/Publications/GFA16/	United State Department of Agriculture/Economic Research Service (USDA/ERS, 2005b)
USDA Agricultural Baseline Projections to 2014	Office of the Chief Economist, World Agricultural Outlook Board, U.S. Department of Agriculture. Prepared by the Interagency Agricultural Projections Committee. Baseline Report OCE-2005-1, 116 pp. http://www.ers.usda.gov/Publications/oce051/	Economic Research Service United State Department of Agriculture (USDA/ERS, 2005)
Greenhouse Gas Mitigation Potential in U.S. Forestry and Agriculture	http://www.epa.gov/sequestration/greenhouse_gas.html	U.S. EPA Office of Atmospheric Programs. 2005

Table 5-3. **Continued**

Economic and Agricultural Impacts of Ethanol and Biodiesel Expansion	University of Tennessee Agricultural Economics. http://www.21stcenturyag.org/	Ugarte, D., B. English, K. Jensen, C. Hellwinckel, J. Menard, and B. Wilson. 2006
Six Megatrends in Agriculture	The John M. Airy Symposium: Visions for Animal Agriculture and the Environment, January. http://www.iowabeefcenter.org/content/Airy/VANACHT%20Abstract.pdf	Vanacht, M. 2006
Agriculture Megatrends: Ten Trends Redefining the Practice of Agriculture in the World	http://www.newmediaexplorer.org/steve_bosserman/2007/02/01/agriculture_megatrends_ten_trends_redefining_the_practice_of_agriculture_in_the_ world.htm	Bosserman, S. 2007
Maximizing Productivity of Agriculture: The Food Industry and Nanotechnology	http://www.foresight.org/challenges/agriculture002.html	Fletcher, Anthony. 2007
	http://westernfarmpress.com/news/farming_ags_future_pepper_3/	Western Farm Press. 2007
Harvest on the Horizon: Future Uses of Agricultural Biotechnology	http://pewagbiotech.org/research/harvest	Pew Initiative on Food and Biotechnology. 2007

The average number of persons per household in EU-15 declined from 2.8 in 1981 to 2.4 in 2002 (UN, 2006). Most of the single person or single parent households are located in urban areas. Families with children tend to move out or are pushed out of highly urbanized areas and into new suburban areas (exurbia), but this does not change their need for services such as schools, sports facilities, etc. Rural areas, with shrinking populations cannot readily sustain such services. The general phenomenon of smaller household sizes has a number of direct implications in the structure of the markets that are being served by the food industries: packaged food needs to come in smaller quantities, demand for convenience food grows because singles usually spend little time preparing food, the number of food-catering services tends to go up (Leijten, 2006).

In Europe and North America, 20% of the population is already aged 60 years or over. That figure, with regional differences, is projected to reach 33% in 2050. In 2025, the fertility rate per woman is projected to be higher in the USA (2.18) than in Western Europe (1.62) and Eastern Europe (1.51) (Eberstadt, 2007; US Census Bureau, 2007) (Table 5-4).

In developed countries as a whole, the number of older persons (persons aged 60 or over) has already surpassed the number of children (persons under age 15) and by 2050 the number of older persons is expected to be more than double the number of children in developed countries (UN, 2006). The populations of 46 countries or areas, including Germany, Italy, most of the successor States of the former USSR and several small island States are expected to be smaller in 2050 than in 2005 (UN, 2006).

The contribution of international migration to population growth in the more developed regions has increased in significance as fertility declines. During 2005-2050, the net number of international migrants to more developed regions is projected to be 103 million, a figure that counterbalances the excess of deaths over births (74 million) projected over the period. In 2005-2010, the net migration more than doubled the contribution of natural increase (births minus deaths) to population growth in eight countries or areas, namely, Belgium, Canada, Hong Kong (China SAR), Luxembourg, Singapore, Spain, Sweden and Switzerland. Net migration counterbalanced the excess of deaths over births in eight other countries *viz.* Austria, Bosnia and Herzegovina, the Channel Islands, Greece, Italy, Portugal, Slovakia and Slovenia. In terms of annual averages for 2005-2050, the major net receivers of international migrants are projected to be the United States (1.1 million), Canada (200,000), Germany (150,000), Italy (139,000), the United Kingdom (130,000), Spain (123,000) and Australia (100,000). The countries with the highest levels of net emigration (annual averages) are projected to be China (-329,000), Mexico (-306,000), India (-241,000), Philippines (-180,000), Pakistan (-167,000) and Indonesia (-164,000) (UN, 2006).

In the future, the NAE region will be concerned with food demand from its own population (Tables 5-5 and 5-6) and the needs of the rest of the world, especially the less developed countries. It remains to be seen how NAE will respond to the need to feed the growing populations of Africa and Asia and the need to ensure environmental sustainability in these regions.

Table 5-4. **Projected total fertility rates per woman in 2015.**

Region	Projected fertility rate in 2025
Northern America	2.13
USA	2.18
Western Europe	1.62
Eastern Europe	1.51
Commonwealth of Independent States	1.73

Source: US Census Bureau, 2007.

Table 5-5. UN Population prospects for Europe.

Year	Medium variant	High variant	Low variant
(thousands)			
2005	731,087	731,087	731,087
2015	727,227	743,202	711,151
2025	715,220	752,266	677,662
2035	697,507	757,482	639,351
2050	664,183	777,168	566,034

Source: UN, 2006.

Table 5-6. UN Population projections for North America.

Year	Medium variant	High variant	Low variant
(thousands)			
2005	332,245	332,245	332,245
2015	364,334	372,011	356,656
2025	392,978	413,338	372,678
2035	416,777	452,730	382,037
2050	445,303	517,137	381,551

Source: UN, 2006.

5.2.2 Economics and international trade

Increases in demographic and socioeconomic pressure (increases in average income and labor productivity) in society are the main driving forces of technological development in agriculture (Giampietro et al., 1999).

In 2005, North America represented 15% of merchandise and 17% of commercial services exports. Europe represented 44% and 52%, and the Commonwealth of Independent States (CIS) represented 3% and 2% (World Trade Report, 2006).

The global state of the economy, including gross domestic product (GDP), trade related issues and employment has influenced agriculture and AKST. In the next fifty years, the NAE economy will be mostly challenged by the prices of energy and other natural resources and the competition of products from developing countries. NAE's aging population will generate high expenses and might lead to a shortage of human resources. Currently, the sluggishness of the European economy constitutes a drag on world trade and output growth. The Commonwealth of Independent States (CIS) has strong economic growth thanks to the expansion of the energy sector. For the US, the current account deficit is a major question (World Trade Report, 2006).

The annual World GDP growth rate was 2.8% during 1990-2003 broken down as follows: high-income countries 2.6%, middle-income countries 3.5%, low-income countries 4.7% (World Bank, 2005). This indicates a "catching up" process: the income growth rate is higher for countries with a lower initial GDP level. For the same period, GDP growth in the EU-27 is about 2% per year; this is lower than the growth in other high income countries (EC, 2007).

Future world income growth will be determined by the growth in production factors (labor, capital, land) and the productivity growth of these factors. Continued economic growth is expected over the coming period in almost all regions of the world. This growth will be considerably higher for most of the transitional and developing countries than for the EU-15, the United States and Japan, in particular for Brazil, China, India and the new EU Member States (EC, 2007). In the United States, public debt levels are expected to increase over the next twenty years due to significant increases in public expenditures on health care (OECD, 1995). In the reference case projections, the U.S. economy stabilizes at its long-term growth path by 2010. GDP is projected to grow by an average of 2.9% per year from 2004 to 2030, slower than the 3.1% annual average over the 1980 to 2004 period, because of the retirement of the baby boom generation and the resultant slowing of labor force growth. Canada's labor force growth is projected to slow in the medium to long term, however, as baby boomers retire. The country's overall economic growth is projected to fall from the current average of 2.9% per year to averages of 2.6% per year from 2007 to 2015 and 2.1% per year from 2015 to 2030 (IEO, 2007).

Over the long term, OECD Europe's GDP is projected to grow by 2.3% per year from 2004 to 2030 in the reference case, in line with what OECD considers to be potential output growth in the region's economies. According to the International Monetary Fund, (IMF) structural impediments to economic growth still remain in many countries of OECD Europe, related to the region's labor markets, product markets, and costly social welfare systems. Reforms to improve the competitiveness of European labor and product markets could yield significant dividends in terms of increases in regional output (IEO, 2007).

5.2.3 Sociopolitical drivers

The term "political" refers to factors that are related to politics, that is, to the processes of decision making on public policies at the subnational, national and international level and to the processes of implementing these policies. The term "social" is used here broadly to refer to human society. Political stability is an important factor that influences the direct and indirect drivers of agricultural development. Civil strife and internal and cross-border conflicts and wars can have a considerable negative impact on agricultural production.

It is very difficult to assess potential changes in sociopolitical drivers. In North America and Europe, the main uncertainties are the integration of Eastern European countries in the EU and the situation in the CIS. How will the political regime evolve? What will be the relationships among the states? One of the main problems in relations between Russia and the European Union (EU) is the absence of strategic goals. Russia, having played a critical role in ending the Cold War, has neither found its place in the strategy of EU expansion nor in that of NATO. In 2007, the active Partnership and Cooperation Agreement (PCA) between Russia and the EU, which both sides agree has become outdated and is no longer able to meet today's challenges, is due to expire. The form that any new legal, contractual basis for relations between Russia and the EU may take will have implications not only for stability within Europe, but also for Russia's democratic future (Arbatova, 2007). The future relationship of Russia with the USA is also an important uncertainty.

5.3 Key Direct Drivers for Knowledge, Science and Technology (KST): Uncertainties and Consequences for AKST

5.3.1 Transformation in models of knowledge production: trends and uncertainties

5.3.1.1 Trends

Knowledge is defined today as a learning and cognitive capacity. Most importantly, it has to be apprehended in action. This implies a fundamental distinction between information and knowledge. Traditionally a distinction is made between implicit knowledge (e.g., daily life or common sense knowledge, experience knowledge, local or indigenous knowledge, action knowledge) and explicit knowledge (practical, theoretical or creative knowledge). Other typologies emphasize the context in which knowledge is used, as defined by the knowledge itself (normative and descriptive knowledge, strategic and operative knowledge, scientific and empirical knowledge, past- and future-oriented knowledge). Finally, certain authors focus more on the modes of inscription of knowledge, and thus distinguish between: "embrained" knowledge (based on certain conceptual and cognitive skills), embodied knowledge, "encultured" knowledge (built up in the processes of socialization that lead to shared forms of understanding), embedded knowledge (in systemic routines) and encoded knowledge (which can be considered as equivalent to information) (Amin and Cohendet, 2004).

New forms of knowledge production and new concepts are appearing. We will briefly mention them as they are often used in discussions of future research systems:

Mode 1 and Mode 2. "Mode 1 refers to a form of knowledge production, a complex of ideas, methods, values, and norms that has grown to control the diffusion of the Newtonian (empirical and mathematical physics) model to more and more fields of enquiry and ensure its compliance with what is considered sound scientific practice. Mode 1 is . . . the cognitive and social norm which must be followed in the production, legitimation and diffusion of knowledge." "In Mode 1 problems are set and solved in a context governed by the, largely academic, interests of a specific community. By contrast, Mode 2 knowledge is carried out in a context of application. Mode 1 is disciplinary while Mode 2 is transdisciplinary. Mode 1 is characterized by homogeneity, Mode 2 by heterogeneity. Organizationally, Mode 1 is hierarchical and tends to preserve its form, while Mode 2 is more heterarchical and transient. Each employs a different type of quality control. In comparison with Mode 1, Mode 2 is more socially accountable and reflexive. It includes a wider, more temporary and heterogeneous set of practitioners, collaborating on a problem defined in a specific and localized context." (Gibbons et al., 1994).

Collective intelligence (or Mode 3). This concept is the subject of a lively ongoing discussion, but a working definition is that "collective intelligence is the capacity of human communities to cooperate intellectually in creation, innovation and invention" (Lévy, 2000). This type of general definition only helps to specify the distinctiveness of how "collective intelligence" produces knowledge by stressing how it differs from the lone researcher in Mode 1 or the purposeful process in Mode 2 (cited by Akrich and Miller, 2007).

Triple Helix. The "Triple Helix" model (Leydesdorff and Etzkowitz, 1998) implies university-industry-government relations. It is developing, though at unequal speed depending on the country.

Platform model. The notion of platform devised by Keating and Cambrosio (Keating and Cambrosio, 2003) attempts to formalize the attributes of a network insofar as it connects a set of devices, tools, instruments, technologies and discourses which are used by a heterogeneous group of people, ranging from basic scientists to engineers and users, to pursue a specific goal. The heterogeneity of this grouping may lead to the production of new research "entities", new technologies and new practices, in short, transdisciplinary built-in innovation.

Frontier research. This concept has been devised by experts of the European Commission to characterize the fast-growing space which is at the intersection between basic and applied research. Its position at the forefront of knowledge creation makes frontier research an intrinsically risky endeavor that involves the pursuit of questions without regard for established disciplinary boundaries or national borders.

Questions of intellectual property are linked to the transformation of knowledge production and are equally important. The development of the Web and electronic communication tools facilitates the circulation and also the production of knowledge. This process can be far more flexible than it used to be in traditional research settings and can involve non-professional researchers thus leading to new forms of collective innovation. Yet the way in which intellectual property rights (including contracts and transaction/payment systems) are defined and managed is going to play a crucial part in these developments.

5.3.1.2 Uncertainties of the future

The evolution of KST could create more cooperation in AKST among NAE countries. The Lisbon Strategy recognizes that Europe is lagging behind the United States in terms of science and technology. A number of studies are being carried out in Europe to find ways to catch up. The United States and Europe are often seen more as competitors than as partners.

The involvement of users in research definition and execution is challenging the traditional research approach. Innovation is a process that integrates various forms of research, and the knowledge it creates, in a wide range of patterns. Users are increasingly expressing their needs, thus challenging traditional disciplinary research approaches and creating the need for a more integrated approach, which some researchers find difficult and which could become an obstacle to required innovation.

As far as models of knowledge production, there are a number of uncertainties concerning the future which can be formulated with questions:

- Will the "triple-helix" model that implies university-industry-government relations develop quickly?
- Will knowledge production and innovation become more user-centered? How diverse will the forms of knowledge be? Should knowledge be yoked strictly to industrial research imperatives? Will knowledge production remain highly conventional, with a strong hierarchical and disciplinary structure?
- Will research be harnessed to solving specific problems like health and environmental conditions? Will knowledge production become highly "socialized" with many institutions being involved?
- Will universities remain the arbiters of what is and is not legitimate scientific knowledge?
- Will intellectual property issues evolve as quickly as production modes and new modes of cooperation?
- How will the governance of the whole research and innovation chain adapt to a systemic approach? Will policies take into account the new forms and producers (including individual researchers) of knowledge looking at quality, trust and transparency?

The way these questions will be answered in the different regions of NAE will affect the AKST systems.

5.3.2 Transformation in models of innovation: trends and uncertainties

The innovation systems concept emerged through policy debates in developed countries in the 1970s and 1980s. The concept of national innovation systems rests on the premise that understanding the linkages among the actors involved in innovation is key to improving technology performance. Innovation and technical progress are the result of a complex set of relationships among actors producing, distributing and applying various kinds of knowledge. The innovative performance of a country depends to a large extent on how these actors relate to each other as elements of a collective system of knowledge creation and use as well as the related technologies. These actors are primarily private enterprises, universities and public research institutes and the people within them (OECD, 1997). These systems developed in an institutional (often network-based) setting which fostered interaction and learning among scientific and entrepreneurial actors in the public and private sector in response to changing economic and technical conditions. Over time, the innovation concept has gained wide support among the member countries of the Organization for Economic Cooperation and Development (OECD) and the European Union (World Bank, 2006).

The innovation system perspective brings actors together in their desire to introduce or create novelty or innovation in the value chain, allowing it to respond in a dynamic way to an array of market, policy and other signals. Innovation capacity is sustainable only when a much wider set of attitudes and practices comes together to create a culture of innovation, including a wide appreciation of the importance of science and technology in competitiveness, business models that embrace social and environmental sustainability, attitudes that embrace a diversity of cultures and knowledge systems and pursue inclusive problem solving and coordination capacity, institutional learning as a common routine, and a forward-looking rather than a reactive perspective (World Bank, 2006).

The main sources of information on innovation systems are UNESCO, OECD, OST (Observatoire des Sciences et Technologies) and ISNAR (International Service for National Agricultural Research). For North America, the National Science Foundation is a source of information. For Europe, Cordis provides a lot of information. The Institut Français des Relations Internationales (IFRI) has a research program on the Russian innovation system. These sources show that innovation systems vary in different regions of North America and Europe.

5.3.2.1 Number of researchers: trends

There were about 4.9 million researchers in the world in 2001. In Europe there were about 1.67 million (952,000 in the EU 15 and 503,000 in Russia) and 1.361 million in North America (1.271 million in the USA and 90,000 in Canada) (OST, 2006a). Between 1996 and 2001, the number of researchers decreased substantially in Canada and Russia. In Russia, the most worrying problem seems to be that the average age of researchers is going up. There seems to be an increase in the number of doctoral students, but this does not necessarily mean increased interest in science as a career. Doctoral studies in Russia fulfill several functions e.g., dodging military service and obtaining a scientific title that can also be useful in the business sector (Dezhina, 2005).

The situation has been summarized as: "the population of European researchers is currently facing a demographic problem. As in most sectors, this population is aging, in line with the general trend over the past sixty years. Consequently, huge numbers of researchers are expected to retire over the next few years. It will be necessary to rapidly recruit new researchers, whose numbers will obviously depend on the resources allocated to R&D, which are in part contingent on public policies. This recruitment challenge poses a number of problems. First, students in Europe tend to be turning away from science and technology, especially when it is research oriented. Some see this as a consequence of the more critical attitude that has developed towards technical 'progress,' which is perceived as bringing as many threats as it does hopes. Others stress the lack of attractiveness of careers in these fields in terms of workload, status and pay. In Europe researchers' salaries are relatively low when compared to industry or the service sector" (Akrich and Miller, 2007).

In the context of internationalization of higher education and research, the question of remuneration is crucial. In the absence of European policies that take into account stiff competition to recruit the best PhDs and post-docs, many young European researchers are attracted abroad, especially to the US. For the same reasons, this outward migration is not compensated for by sufficient inward migration, both quantitatively and qualitatively. The research job market in Europe is fragmented, organized on a national or even local scale, with a low level of competition. Selection takes place in a relatively opaque way that often favors local candidates. This mode of functioning does not promote international openness and leads to unequal levels of quality. Many authors agree that the broader a market is, the greater its specialization and the higher the overall level of quality. The low level of internationalization of the European research job market is not offset by intra-European mobility. It remains limited due to the rigidity of statuses and organizations and the absence of systems for managing scientific careers on a European scale, even if young researchers are becoming more mobile thanks to a strong European policy. Scientific dynamics and the capacity to innovate, strongly based on the possibility of establishing original links between separate research currents, would undoubtedly be enhanced by active policies to promote mobility (Akrich and Miller, 2007).

In the USA, according to a report of the National Science Foundation (NSF, 2003), the future strength of the US science and engineering workforce is imperiled by two long-term trends: (1) global competition for science and engineering (S&E) talent is intensifying, such that the United States may not be able to rely on the international science and engineering labor market to fill unmet skill needs; (2) the number of native-born S&E graduates entering the workforce is likely to decline unless the Nation intervenes to improve success in educating S&E students from all demographic groups, especially those that have been underrepresented in S&E careers (NSF, 2003). Indeed, foreign students account for about one-third of the total number of doctoral degrees in the natural sciences and engineering in the United States. Many foreigners stay in the United States after completion of their degrees to work in industry or as postdocs at American universities (Eliasson, 2004). The composition of the American population and the American workforce is changing. The minority populations, African-Americans, Hispanics, Asians and Native Americans, will increase. More of these people will be entering college and subsequently the labor force in the next decade. Today minority groups represent 24% of the American population and only seven percent of the total labor force in science and engineering (Eliasson, 2004). According to the Third International Math and Science Study, American fourth graders did relatively well in both subjects, but by the time they reached their senior year in high school, U.S. students ranked very low compared to students in other countries (NSF 2003). There is a great need for mathematics and natural sciences teachers in U.S. secondary schools.

5.3.2.2 Research and technology organizations: trends

"Research and technology organizations (RTOs) are generally non-profit organizations that provide innovation, technology and R&D services to a variety of clients (firms, public services, administrations). This makes them 'in-between' organizations: their financing includes both private resources (via contracts, patents and licenses) and public funds; they increasingly straddle applied and basic research, and are thereby engaged in 'frontier research', and their work has a distinct multidisciplinary dimension that includes the economic and social sciences. This particular positioning is a source of tension, so that the specificity of RTOs depends on a balance being maintained between their diverse components." (Akrich and Miller, 2007)

Historically and by construction RTOs have tended to encourage multidisciplinarity projects and have been less constrained by the boundaries between basic and applied research. Consequently, they have many assets conducive to playing a strategically important role in the current context. With links to fundamental research, RTOs have expertise in the development of tools and concepts (mathematical modeling, complex systems theory, etc.) that allow them to articulate and blend the sets of heterogeneous knowledge and technology that are major sources of innovation. RTOs are also well configured to take advantage of the increasing number of actors involved in research and the intensified relations between the scientific community and its environment.

There is comparatively little information about R&D laboratories in the United States. Government laboratories or federal laboratories have typically been established to serve a mission of a particular government agency. They include government-owned but contractor-operated labs and federally-funded R&D Centers. In 2002, government laboratories received about 25 of a total of 81 billion dollars of total federal investments in R&D (31%), which can be compared to approximately 10 billion dollars for the academic sector. The biggest recipients are those under the Department of Defense (Eliasson, 2004).

5.3.2.3 Universities: trends

Universities across Europe reflect a multitude of realities. In certain countries they are the main source of research and higher education. In other countries they coexist with large research organizations and even, as in France, with other types of higher education institutions (*Grandes Ecoles*) that are increasingly engaged in research. On the whole, there is less investment in higher education in Europe than in other countries such as the US. Funding is primarily from the public sector, and students pay a relatively low share of the education costs. However, funding for university-based research has increased substantially over the last 15 years. There has also been a diversification of the sources of funding for research institutions that now include national governments, supranational bodies (e.g., the European Commission), regional governments, business enterprises and civil society. The respective weight of teaching and research and the mechanisms through which research activities can be financed and encouraged vary considerably among countries and universities. In general, universities in Europe currently face similar challenges: offering courses to young adults, meeting the demand for ongoing education and training, and participating in knowledge production in increasingly diverse contexts and with an ever-greater variety of partners. The juxtaposition of these different tasks generates strong tension within universities, in part due to limited resources. The situation is exacerbated by the fact that the main missions of universities are often ambiguous; additionally, key stakeholders and managers may not agree on priorities. (Akrich and Miller, 2007).

There are about 4200 universities and colleges in the U.S., and most of the research is carried out at about 263 doctoral/research universities. Universities perform about 13% of total R&D, and 82% of federal support goes to 100 universities. Twenty of them receive about 34% of the government support (Eliasson, 2004).

In 2004, Russia had 1071 higher education establishments (40% more than in 1993). They are starting to be involved in research (OST, 2006a).

5.3.2.4 Multinational enterprises and small and medium enterprises: trends

Today's multinational corporations (MNCs) see innovation as a strategic element in economic competition. The life cycles of products are increasingly short, and firms are encouraged to produce returns on investments more and more quickly. Consequently, an R&D race has developed among multinationals. R&D activities enable firms to build up knowledge about technologies to support their key activities. R&D is also critical to the firm's long-term competitiveness, by enabling them to identify, acquire and apply knowledge that has been developed by others.

MNCs have been expanding R&D outside their home countries in recent decades. R&D investments by MNCs, within their affiliates or with external partners in joint ventures and alliances, support the development of new products, services and technological capabilities. These investments also serve as channels of knowledge spillovers and technology transfer that can contribute to economic growth and enhance competitiveness abroad. International R&D links are particularly strong between USA and European companies, especially in pharmaceutical, computer and transportation equipment manufacturing. More recently, certain developing and newly industrialized economies are emerging as hosts of US-owned R&D, e.g., China, Israel and Singapore (NSF, Science and Engineering Indicators, 2006).

Small and Medium Enterprises (SMEs) are extremely heterogeneous, ranging from high-tech start-ups to small building contractors to the local companies. However the sectoral coverage narrows considerably when the focus is on research related issues. Technology based SMEs account for around 10% of all SMEs (NSF, Science and Engineering Indicators, 2006).

5.3.2.5 International, national and regional governments: trends

A variety of actors, including advisory bodies, national agencies, ministries and specialized institutes are involved in making and implementing national science, technology and innovation policies. These actors engage in a wide range of activities, including planning, forecasting, strategic intelligence and consultation with stakeholders. The national level actors are involved throughout the process, which covers needs identification, agenda-setting, policy implementation, policy evaluation and benchmarking results. The forms of intervention of regional powers in research and technology policies vary.

The defining characteristics of the US public R&D policy are an even stronger impact of the economic factors than in other geographical areas, the enormous influence of defense-related research activities, and the importance given to the high potential areas made up of converging technologies (EU Commission, 2006). North American policies emphasize research support for regional and local universities. Regional authorities have policies for attracting and developing a qualified local workforce; these policies spurred the creation of technology clusters and parks. In the USA, 60% of all R&D is concentrated in six states, with California alone accounting for 20% (UNESCO, 2006b).

In Europe, national authorities generally retain the leading role in policy formulation and implementation, but there are very wide differences among countries in the extent and nature of this leadership (Akrich and Miller, 2007). Europe is much more influenced by societal, i.e., social and environmental, factors than the U.S. as far as R&D policy setting is concerned. Ecological and quality of life issues generally provide a unifying and defining element of European public R&D support policy. Nevertheless, the European landscape is characterized by important inter-country differences. A number of factors account for this, such as GDP, political environment and scientific position. Europe is also faced with policy rigidities that strongly affect the efficiency of public support, influencing both the form in which support is being administered and the research organization itself (EU Commission, 2006). The distribution of prerogatives between regional, national and European government varies from country to country, e.g., the länders are very influential in Germany, and regionalization is being introduced in Spain and the United Kingdom.

5.3.2.6 Uncertainties of the future

There are a number of uncertainties related to the future and the way these questions will be answered in the different regions of NAE will affect the AKST systems. These questions are:

- The capacity of universities and public research organizations, the private sector and the government to jointly define research priorities and fund R&D in order to make their country's economies competitive is uncertain. Collaborative research is gaining ground, and measures could be taken to further promote it and improve the general R&D effort. Will governments be able to develop "innovation plans" that favor interactions between universities, industries and governments? Will the public and the private sector reach a consensus on priorities? Since the KST system is composed of both the public and the private sector, working with the whole system could lead to a consensus on priorities. This would allow the public sector to take better account of the private sector and consumer needs and concentrate on the development of public goods.

- Innovation is a strategic element in economic competition, but companies make investment decisions according to their expected returns. The level of private sector contributions to R&D varies among countries. Large multinationals are increasingly influencing priorities and investments in agricultural science and technology and are heavily involved in agricultural extension. Some see this trend as positive, others as negative. Will policies that enable firms to pursue the "best quality according to international standards" clash with policies aimed at ensuring that "research is a means for local economic development?" Will enterprises be able to earn money from research, invest massively in research and produce significant industrial innovation? How does the internationalization of science interact with the internationalization of industrial R&D? How do innovation systems adapt to maximize benefits and lower costs of internationalization? How will the potential contradictions between local development and internationalization be addressed?
- How far will the current regionalization trend go in Europe? Will excessive competition between regions, in the absence of coordination at the European level, lead to a fragmentation of efforts and the absence of a coherent strategic vision? Will Europe be able to reinforce excellence, especially in new, fast-growing research areas and areas where science and technology are closely interlinked? Will the strengthening of large-scale pan-European projects concentrate and integrate research without accommodating local concerns and context? Will European universities serve the industrial economy, or simply become more closely linked to "external" research? Will there be a more open and dynamic European market for funding post-doctoral researchers, including opening access to non-academic research? Will greater importance be given to service sector activities and SMEs? Will the Russian Federation manage to transform its R&D system and attract young people to R&D?

5.3.3 Information technology: trends and uncertainties

5.3.3.1 Trends

The information technology boom started over thirty years ago. Information technology is the most important among the key technologies because of its dominant role in all other areas and in the convergence of technologies. It deserves continued special attention due to its economic and societal relevance not least for innovation. Information and Communication Technologies, especially Artificial Intelligence and Cognitive Science can help breaking up rigid organizational structures hindering innovation, and do so in harmony with cultural, social and natural heritage. There is a trend towards modeling more and more of reality in computational systems. There is literally no part of reality which might not be subject to such modeling, including intelligent human beings as the most challenging goal. Information Technology is a cross-sectoral discipline par excellence. Its applications virtually cover any sector and any discipline (Bibel, 2005).

New forms of expertise are emerging, facilitated by the development of information and communication technologies (ICT) that allows both access to content and contact amongst actors. ICT will play a part in all fields of science and technology and in agriculture, especially by providing images, real-time data wherever needed (Cuhls, 2006). Imaging will be available very soon (NISTEP, 2005) and will contribute to precision farming and to making agriculture, especially the related resource and land management, more efficient. Radio Frequency Identification (RFID) could replace common barcodes and have a huge impact on agriculture and the marketing of products (Cuhls, 2006). Models and simulations will improve and support crop management, weather forecasts, etc.

Currently, IT availability and use in NAE is uneven among countries and sectors. Europe, in general, is behind North America. Within Europe, there are major differences. Some countries of Eastern Europe and to a lesser extent, Central Europe, have relatively low access to information technologies.

5.3.3.2 Uncertainties of the future

There are a number of uncertainties related to the future and the way these questions will be answered in the different regions of NAE will affect the AKST systems. These questions are:

- Will drastic cost reduction in ICT-based Microsystems and artificial intelligence and knowledge management software lead to widespread self education, training and research generation tools?
- Will Eastern Europe be able to reduce the digital divide with the rest of Europe?
- As far as Information Technology is concerned, will Europe manage to catch up and keep pace with North America?

5.3.4 Evolution of KST with potential impact on AKST

Beyond what is happening in the ICT sector, other developments in the knowledge, science and technology systems could have important consequences for AKST. Technology forecasting and foresighting activities have been carried out at the European (EC, 2006) and national levels (Technologies Clés in France; Futur in Germany; National Intelligence Council's 2020 project in the USA, etc.) to identify emerging priority technologies that will be of paramount importance for Europe in the future. At the European level, forty technologies have been grouped within four main scientific fields (EC, 2006):

- Nanotechnologies, knowledge-based multifunctional materials, new production processes,

- Information society technologies,
- Life sciences, genomics and biotechnology for health,
- Sustainable development, global change and ecosystem.

Two different rationales support the selection of these technologies. The first one is that they are emerging and have been identified through a questionnaire sent to a panel of about 1300 experts in all the countries of the enlarged Europe. The second one is the results of the foresight literature review both in the European and the main competitor countries (EU Commission, 2006). However, if Gross Expenditures for R&D (GERD) stay at the present level and if there is no coherent European or NAE policy, it is unlikely that all of the research can be done. The AKST investments will not be the same if the main drivers are life sciences, sustainable development and economic factors or if they are societal motives.

In the USA, a technical foresight study (Global Technology Revolution 2020) undertaken by RAND Corporation (RAND Corporation, 2006) has identified applications:

- *Cheap solar energy*. Solar energy systems inexpensive enough to be widely available to developing and undeveloped countries as well as to economically disadvantaged populations that are not on existing power grids.
- *Rural wireless communications*. Widely available telephone and Internet connectivity without a wired network infrastructure.
- *Communication devices for ubiquitous information access*. Communication and storage devices—both wired and wireless—that provide agile access to information sources anywhere, anytime. Operating seamlessly across communication and data storage protocols, these devices will have growing capabilities to store not only text but also meta-text with layered contextual information, images, voice, music, video and movies.
- *Genetically modified (GM) crops*. Genetically engineered foods with improved nutritional value—e.g., through added vitamins and micronutrients, increased production; by tailoring crops to local conditions and reduced pesticide use; by increasing resistance to pests.
- *Rapid bioassays*. Simple, multiple tests that can be performed quickly and simultaneously to verify the presence or absence of specific biological substances.
- *Filters and catalysts*. Techniques and devices to effectively and reliably filter, purify and decontaminate water locally using unskilled labor.
- *Targeted drug delivery*. Drug therapies that preferentially attack specific tumors or pathogens without harming healthy tissues and cells.
- *Cheap autonomous housing*. Self-sufficient and affordable housing that provides shelter adaptable to local conditions as well as energy for heating, cooling and cooking.
- *Green manufacturing*. Redesigned manufacturing processes that either eliminate or greatly reduce waste streams and the need to use toxic materials.
- *Ubiquitous radio frequency identification (RFID) tagging of commercial products and individuals*. Widespread use of RFID tags to track retail products from manufacture through sale and beyond, as well as track individuals and their movements.
- *Hybrid vehicles*. Automobiles available to the mass market with power systems that combine internal combustion and other power sources.
- *Pervasive sensors*. Presence of sensors in most public areas and networks of sensor data to accomplish widespread real-time surveillance.
- *Tissue engineering*. The design and engineering of living tissue for implantation and replacement.

Biotechnologies and nanotechnologies are two technologies that are quite controversial in some countries, especially in Europe. They both elicit fear, and their costs and benefits depend on how they are incorporated into societies and ecosystems and whether there is the will to fairly share benefits as well as costs. They may have important potential impacts on agriculture and food systems (Scott and Chen, 2003).

5.3.5 Financial resources devoted to science and technology: trends and uncertainties

5.3.5.1 Trends

The world devoted 1.7% of gross domestic product (GDP) to R&D in 2002. In 2001, this proportion was 2.74 for the United States, 1.91 for EU-15, 1.9 for Canada and 1.29 for Russia (OST, 2006a) (Table 5-7).

North America, Europe and Japan dominate the production of knowledge, but there has been a remarkable growth of gross expenditure on R&D in Asia (27.9% of world share in 1997 and 31.5% in 2002).

In the United States, industry contributes about 64% of gross expenditures on R&D, in Canada, 48%, in EU-25 54% and in Russia 31% (UNESCO, 2006b).

With 25 Members, since the accession of ten new countries from Central, Eastern and Southern Europe in May 2004, the European Union now accounts for 90% of European gross domestic expenditure on R&D. There is no true European R&D market since there are great discrepancies in R&D capacities between the EU Member States. Even if the new Member States will attract R&D investments, the R&D budget of the European Commission represents just five percent of public expenditure on R&D by Member States. In 2001, Europe accounted for 46.1% of the world's R&D publications (OST, 2006b).

Since the disintegration of the USSR more than a decade ago, the R&D systems of all these states have been seriously reduced, yet they remain important. The proportion of GDP spent on R&D by the Federation of Russia, for example, was 1.17% in 2004 (OST, 2006a). Moreover, the number of researchers in Russia, 3,400 per million inhabitants, is the third highest in the world after Japan (5,100) and the USA (4,400) (UNESCO, 2006b). Almost 3,650 organizations represent science and research in today's Russia (OST, 2006a).

The evolution of science and technology is increasingly expensive. Each answer gives rise to new questions. Although nations are very much aware of the importance of science and technology for their economy, there are limits to the amounts of money they are willing to spend on it. Consequently, nations and businesses must choose which areas of science and technology they will support. As a result of competition for resources, researchers must account for the

Table 5-7. **Percentage of resources devoted to R&D, share of world scientific publications and ratio of researchers in three NAE sub-regions.**

	North America	EU 25	Federation of Russia
Percent of GDP devoted to R&D (2003)	2.4	1.8	1.29
Share of gross expenditures on R&D (2003)	36.1	24.3	1.9
Share of gross expenditures on R&D coming from private sector (2003)	62.8	53.7	30.8
Share of world scientific publications (2004)	36.2	34.2	2.4
Ratio of researchers to total population	4.4	2.6	3.4

Source: OST, 2006b.

activities much more than in the past. Research must increasingly justify the resources that support their programs; additional funding is often linked to applied solutions for societal problems.

5.3.5.2 Uncertainties of the future

The present domination of NAE in generating formal new knowledge could be challenged. The growth of gross expenditure on R&D and R&D results in Asia is changing the relationship of NAE research with the rest of the world. This could result in new networks and increased competition among agriculture, industry and services.

Public funding of science and technology is insufficient to adequately address and provide solutions for agriculture that better fulfill the needs of consumers and better respond to the requirements of more sustainable natural resource management. Less AKST is available in the public domain, limiting farmers' choices and the achievement of sustainable agriculture and rural development. This also has a negative impact on partnerships with other regions of the world. Halting and reversing this negative trend depends on the will of governments. Reshaping intellectual property rights and other regulatory frameworks could also modify this trend. Questions concerning options for the future are:

- Will financial efforts and administrative measures break down barriers between the public and the private sectors where such barriers still exist?
- How will the increased productivity of industrial systems affect resources devoted to KST?
- Since budgetary resources are limited, should the public sector support technologies in areas of strength or, on the contrary, areas of specific weaknesses? Should the public sector leave the market and support targeted R&D firms through tax incentives, mobility, etc.? Should it fund most of the research and leave only accompanying measures for the private companies?
- Will Europe be able to mobilize extra financial and human resources for KST to keep pace with the United States and Japan or be taken over by fast-developing Asian countries? Will Europe become attractive for young researchers, irrespective of their country of origin, providing them with the resources needed to develop their full research potential and retain them in Europe? Will a pan-European approach for investing in high-quality frontier research be established?
- What kinds of relationships will North American and European science and technology systems have with Asia? And with the less developed countries?

5.3.6 Attitudes towards science and technology: trends and uncertainties

The NSF Science and Engineering Indicators 2006 reports that although Americans express strong support for science and technology, most people are not very well informed about these subjects. The public's lack of knowledge about basic scientific facts and the scientific process may discourage government support for research, the number of young people choosing S&T careers and the public's resistance to miracle cures, get-rich schemes and other scams.

Americans have more positive attitudes about the benefits of S&T than Europeans and Russians. In recent surveys, 84% of Americans compared with 52% of Europeans (EU-25) and 59% of Russians, agreed that the benefits of scientific research outweighed any harmful results. Most Americans and Europeans know little about genetically modified (GM) foods and related issues. Although attitudes were divided, opposition to introducing GM food into the US food supply declined between 2001 and 2004. This was not the case in Europe. However, the majority of Americans believe that GM food should be labeled (NSF, 2006).

Relations between researchers and society have become stronger during the past few years. The development of a number of controversies in the public sphere has undermined the illusion, harbored by many, that science is able to eliminate all uncertainties.

Researchers can no longer be treated as a population subject to homogeneous organization, structured according to disciplinary divisions, with ties to the social world mediated by administrative and political authorities. On the contrary, they are now a multitude of groups that interact in varied ways, re-arranging or even partially erasing boundaries between disciplines and different forms of knowledge, science being only one of these forms (Akrich and Miller, 2007).

Future uncertainties:

- How will the "precautionary principle" affect scientific advances?
- How will religious fundamentalist groups affect the development of research and technology? How will social values influence interventions on nature?
- What role will civil society organizations play in the determination of research agendas?

- Will there be greater investments in anticipatory processes (e.g., foresight activities, citizen's summit, etc.)?

5.3.7 Education in science: trends and uncertainties

5.3.7.1 Trends. Over the past 15 years, most OECD economies have experienced a large increase in the number of students in higher education. The absolute number of students in science and technology has risen too, but the proportion of university students in S&T has steadily decreased during the same period. Some disciplines, such as mathematics and physical sciences, show particularly worrying trends. Nevertheless, higher education with professional objectives (engineers, technicians, etc.) remains attractive.

Image and motivation surveys show that young people continue to have a largely positive perception of science and technology. S&T are considered important for society and its evolution despite concerns in specific areas often linked to their negative environmental and societal consequences. Scientists are among the professionals the public trusts most, even though their prestige has declined (senior management or government positions are rarely held by scientists or engineers, and media reports on S&T events do not focus on the researchers themselves, who are thus very rarely known by name). Yet parents encourage careers in S&T for their children. There is a sharp difference between the positive opinion of young people towards S&T and their actual wish to pursue S&T careers. S&T professions continue to generate great interest among youth in developing countries, but not in industrialized countries, where especially girls find it unattractive. Many young people have a negative perception of these careers and lifestyles. Incomes are expected to be low relative to the amount of work involved and the difficulty of the required studies.

Students often lack knowledge about what S&T professionals really do and many are unaware of the range of career opportunities stemming from S&T studies. What they do know often comes from personal interactions (mostly S&T teachers, or someone in the family), or through the media. Scientists are usually portrayed as white men in white coats and engineers as performing dirty or dull jobs. As S&T professions evolve quickly, S&T teachers and career advisors often lack up-to-date information to convey to their students. Young people therefore have few opportunities to learn about the lives of S&T professionals. The careers of S&T professionals as a whole have suffered from media reports of poor prospects and funding and increased job insecurity, despite the fact that this applies primarily to researchers. Furthermore, the possibility of reaching a proper balance between a successful career and a fulfilling family life, which is important to young people, is perceived as difficult in S&T professions.

Many initiatives have been launched at different levels to promote S&T careers and studies. Government actions have often been designed to improve the image of science and scientists in society (science weeks, science days, etc.) and more has been done by the professional scientific organizations. Communication tends to focus on science itself, not on the reality of S&T professions. The actual impact of the various actions on both young people's attitudes and their choices of studies or careers is poorly evaluated, however. Furthermore, communication between the various stakeholders is often inadequate.

5.3.7.2 Uncertainties of the future. The interest for science and the number of students in science and technology in most of NAE is declining. The population of European researchers is aging, and students tend to turn away from science and technology, especially when it is research oriented. Measures relate to school education programs and public information to change the public's attitudes about the benefits of S&T. In North America, the number of students in "sustainability programs" is increasing, but fewer have agricultural backgrounds. What will be done in primary and secondary schools and in universities to interest students in scientific research? What will be done in terms of remuneration to attract and keep researchers? How will universities deal with their missions to educate a diverse student body and to carry out research with local industrial communities? Will universities turn to problem solving? Will education become concentrated in a global knowledge oligopoly comprising a small number of giant US, European and Asian firms? Will North America and Europe continue to play an important role in training scientists from developing countries?

5.4 Key Direct Drivers for Agriculture, Uncertainties and Consequences for AKST

AKST will be greatly influenced by changes in agriculture, and can also influence changes in agriculture. At present and for at least the next twenty years, North American and European policies, trade and markets will greatly influence the world agriculture. In this chapter, policies, trade and markets have been considered a key driver of agriculture. Land use change and natural resources have been dealt with together.

5.4.1 Food consumption and distribution: trends and uncertainties

Human as well as plant and animal health considerations are becoming more important. Populations in North America and Western Europe, especially the poor, face alarming increases in illnesses associated with inadequate diets and over-processed food. Central and Eastern Europe are likely to face the same problems. Increased plant and animal diseases, as well as weed and insect problems, both evolving and invasive, are threatening production in certain areas, and lead to overuse of agricultural chemical and antibiotics, whose lingering residual effects in the environment are threatening human health. This could lead to changes in food production and processing. The growing organic food market could counter this trend. The problem could be addressed through well-target information and appropriate regulations, as well as changes in the behavior of individuals and companies.

5.4.1.1 Ongoing trends

Consumers' food preferences cannot be understood or predicted by simple models: food preferences arise from a combination of different factors and drivers; e.g., income, household size, age, ethics such as on animal welfare, influence of policies or media (EC, 2007). Changes in food consumption can be assessed over the years using indicators such as food budget, calorie intake, categories of foodstuffs, home or out-of-home consumption, homemade or precooked meals, quality of food products. Changes in dietary patterns influence food systems, agricultural products and services, (both food and non-food products) and other ecosystem services. While changes in food demand directly affect the types and quantity of food being produced, and thus affect the AKST used in producing this food, changes in AKST driving food supply can also influence food consumption patterns indirectly.

Growing incomes, reduction in household size, increasing number of women in the workforce, changes in the lifestyle with more time constraints, food scares, growing concerns for health and well-being and ethics have influenced food consumption in recent years (EC, 2007)

For the future, the most important trends that can be influenced by AKST seem to be dietary patterns, increased illnesses associated with inadequate diets and over-processed food, consumer attitudes with increased consumption of processed and convenience food, and the effects of mass distribution on food consumption.

The nutritional transformation reached many industrialized countries in the 19th century, and advanced to many developing countries in the last 50 years or so. In the United States, the fraction of expenditure on food was 25% in 1930, less than 14% in 1970, and around 10% in 1995. In the European Union (EU-27), the fraction of expenditure on food decreased from 14.5% in 1995 to 12.8% in 2006 (Eurostat). North America and Europe are in a situation of "food satiety," with an overabundance of food products on the market but a growing health divide between rich and poor. In countries of NAE, more than 80 kg of meat are consumed per capita every year. This high meat consumption entails a huge cereal and water demand and exacerbates some health problems (e.g., heart disease). All meats do not require the same quantity of vegetal calories for production; eleven vegetal calories produce one calorie of beef or mutton; eight calories produce one calorie of milk; four calories are needed for one calorie of pork, poultry or egg (Collomb, 1999 cited by Griffon, 2006). At present, the fish/seafood food group is relatively unimportant as a source of daily protein in Europe (7.2 g^{-1} day^{-1} $person^{-1}$) although its contribution almost matches the average share of beef and veal (7.6 g^{-1} day^{-1} $person^{-1}$). However, the fish/seafood group registers large variations between countries (de Boer et al., 2005). Many foods have excessive fat and sugar, and too much red meat is consumed, partially as a consequence of subsidies given to some agricultural products (Fields, 2004; Birt, 2007).

Growing concerns for health and well-being are influencing consumers' food choices. Consumers are increasingly looking for health foods and "natural" products, which are often associated with organic production. They are looking for food that provides benefits other than just basic needs (functional food), Consumer concern for obesity has created a market for fat-reduced or sugar-reduced products. Consumers are increasingly buying fresh food all year-round from all over the world (EEA, 2005), and are switching to chill-cooked meals made from fresh ingredients. These trends are strongly influenced by the double-income households, the decreasing household size and the aging population.

Food demand is also influenced by the cultural settings. Shapes, textures, flavors and colors of foods help define different cultures. Consumption patterns (e.g., cooking styles, meal organization and eating utensils) are a powerful medium for the construction of cultural identity, but globalization is flattening differences. Moreover, food is different from other consumer products in that it passes through the body. Man is transformed by it to a greater extent than by any other product, and it affects his well-being more directly. Overall food contributes to both sensory and social pleasure and also has considerable effect on Man's sense of individual and collective identity (Fischler, 1990; Raoult-Wack and Bricas, 2001).

The populations of both North America and Europe exhibit alarming increases in diet-related illnesses (e.g., obesity, diabetes and arteriosclerosis). For example, the UK has included studies on "tackling obesities: future choices" in its foresight program. A number of recent crises (e.g., mad cow disease, listeria and foot and mouth disease) have exacerbated consumer concerns about food safety.

Distribution affects food demand. In the agroindustrial age (Malassis, 1997), the food sector consists of Small or Medium-sized Enterprises (SMEs) and large groups. Mass distribution (hypermarket-type food outlets) plays a growing role and influences both food production and food consumption. Supermarkets are playing a major role in determining food consumption patterns and have shaped North American and EU tastes. In Central and Eastern Europe, massive inflows of foreign direct investment and domestic investments are changing the consumption patterns.

The following trends have recently been observed with respect to food distribution (Anania, 2006; Fulponi, 2006; Henson and Reardon, 2006):

- An increasing share of food sold to consumers in large stores everywhere in the world, i.e., in cities in the industrial countries and in rural areas in the developing countries (Dries, Reardon, Swimmen, 2004);
- A rapid increase in the (already extremely high) rate of concentration of the food retail sector;
- The setting, by the retail sector, of more private food safety and quality standards implying more stringent minimum standard requirements than those defined by existing public regulations (such as EurepGap, enforced today for fresh products);
- The "decentralization," by the retail sector to its suppliers of food products, of an increasing number of functions (such as packaging, pricing and logistic tasks needed to guarantee *just-in-time* deliveries);
- The imposition of increasingly more restrictive requirements as a necessary condition for suppliers to be considered as potential sources, such as the capacity to deliver a "basket" of goods (rather than a single one) or to provide large volumes and do so over extended periods of time throughout the year, all aimed at reducing the number of suppliers and, hence, transaction costs;
- An increase in the imbalance in the distribution of market power along the food chain, with the highly con-

centrated retail sector holding significant and increasing market power *vis-à-vis* its suppliers.

These trends could be undermined if consumers in North America and Europe adopted a "sustainable development" perspective, for example by reducing their demand for non-seasonal and non-local crops, meat and fish and adjusting food portions to human needs. A number of NGOs and local organizations are pushing in this direction, and some supermarkets in the EU are also active in that direction. National and international regulations could also have an effect on food demand. Food processing companies are increasingly encouraged to reduce the portion of sugar and starch in their products. There is an increasing demand from consumers for labeling, traceability and other information. Media publish messages on diets. The Codex Alimentarius develops quality food standards, consumer health guidelines, fair trade practices and internationally harmonious food standards. Furthermore, society has become increasingly aware of environmental impacts and animal welfare associated with agriculture. This appears to be causing some changes in buying and consumption habits that may decisively influence consumers' willingness to pay a premium on a product they may perceive as safer, produced in ethical conditions, or more beneficial.

5.4.1.2 Uncertainties of the future

Many uncertainties could greatly affect the food marketplace of the future. This section provides a list for further discussion:

- Food demand at global and NAE levels.
  - Will food continue to be an instrument of cultural identity in many countries? Will food become completely standardized?
  - Will NAE have to contribute to the changes in meat and cereal consumption that will take place in the other regions of the world?
  - How will the consumption of off-season crops evolve? How will the consumption of meat and fish evolve? If there are increases in meat and fish consumption, will the increased demand be met through increased local production or imports? Can increased demand be met through "meat/fish" produced without animals?
  - What will be the changes in the consumption of processed (convenience) food? What will be the consumer attitudes towards preparing food at home?
  - In the past fifty years, there has been a decrease in the real prices of food. Will consumers be ready to pay a premium for "quality" products or will they continue to see the share of food decrease in the share of household expenditures?
  - In Eastern Europe, how fast will food diversification take place?
- Health.
  - At the global, NAE and European level, will there be coordination and harmonization of international food standards? How strict will consumer protection be? Will human health be adequately protected?
  - In NAE, will governmental measures be sufficient to make consumers aware of links between food and health? Will consumers demand foods tailored to specific health needs? Will consumers recognize and demand functional foods? What food safety measures will consumers demand? Will improved analytical methods increase the demand for organic foods or foods free of chemical residues? Will consumers pay a premium for these food services? Will increasingly aseptic foods reduce human immunity? Does increased hygiene increase the risk of resistant pathogens? How to strike a balance between necessary hygiene and excessive hygiene?
- Food manufacturing, processing and distribution.
  - In NAE, will horizontal and vertical integration of the whole food industry continue? Will the development of niche markets influence the ongoing trend of integration? Will farmers be able to choose their production or will they become even more dependent on the food distribution and manufacturing industries? Will home delivery replace conventional food shopping? Will local distribution points be created for food ordered through the internet? Can the relationship between farmers and consumers be strengthened?

5.4.1.3 Consequences for AKST

To achieve nutritional security strategic choices have to be made in the economic and social domains (lifestyles) and in the domains of international and national food regulations and modes of distribution. As far as food consumption is concerned, as in other topics, AKST choices will not only be technical but will also be influenced by actors and their ideologies. The following illustrates the choices that will have to be made:

- To produce safe high quality food, animal and plant genetic resources will need to be evaluated and preserved. Factors determining the shelf life of both fresh produce and processed food, or the stability of plant raw materials after harvest will also be important (ETP, 2005).
- If functional food is developed, then there will be a need for analysis, measure and control, biotechnologies, biochemistry, biology, medicine.
- To create food targeted at specific consumer groups or needs, the identification and characterization of the molecular structure of plant polymers, as well as the characterization of plant metabolites will be very useful, together with molecular breeding and transgenic approaches. This will need an interdisciplinary approach that brings together plant scientists, physicians and nutritionists (ETP, 2005).
- If the emphasis is on food quantity rather than food quality, genomics will be very important.
- If the emphasis is on food quality, functional genomics and systems biology will need to be developed.
- The rapid development of allergies will require the development of special research.
- If transformation is a priority, microbiology will be useful to look at the nutritive qualities of food.
- If a market-led, globalized world develops, food traceability, prevention of bioterrorism and agroterrorism and identification of sabotage will be very important.

There will be a need for nanoscale systems, microsystems technologies, sensors, etc.

- To produce more meat, a major effort will have to be made to produce high quality, sufficient and sustainable feed using biochemical tools and biological assays, molecular mechanisms to decipher the plant-pathogen interaction, the assessment of macro- and micronutrient characteristics, germplasm, etc. (ETP, 2005).
- To produce bioplastics and biomaterials and use renewables, biotechnologies should be very useful.

5.4.2 Policies, trade and markets

Agricultural trade policies and subsidies in NAE tend to undermine the fulfillment of development goals in other parts of the world. There is uncertainty about whether the World Trade Organization will be effective in harmonizing approaches to internal subsidies, and additional uncertainty about whom is likely to benefit, how much and for how long if NAE subsidies are removed. Applying AKST could potentially help to balance the needs of vulnerable people in other regions of the world.

5.4.2.1 Ongoing trends

Agricultural policies

The following agricultural policy/trade developments will be paramount in determining the international competitiveness of NAE agriculture/food industries and the sustainability of rural areas:

- Reform of the EU Common Agricultural Policy;
- NAFTA, CAFTA and other similar trade policies;
- Negotiations under the World Trade Organization (WTO);
- Convention on Biological Diversity (CBD) and International Treaty on Plant Genetic Resources for Food and Agriculture (PGRFA);
- Projected population growth, combined with the greater prosperity of some social groupings;
- Relationships between economic growth and environmental degradation, and the compliance with international, regional and national environmental directives (Kyoto Protocol; EU policies, etc.).

There are three levels of policy framework: international (i.e., WTO, Kyoto agreement, CBD), regional (i.e., EU-CAP, NAFTA), and national/governmental. At all levels, a broad range of agricultural policies relate to different types of institutional support that farmers may be eligible for by complying with specific agreements. Aid, subsidies, tax reductions, special tariffs, etc. could be given to compensate farmers for loss of income opportunities or price gaps they suffer if they produce certain types of crops, tend to the landscape, rest certain areas and/or use new agricultural techniques or practices that authorities deem socially or environmentally preferable. Agricultural policies also relate to natural resources conservation, rural development, agricultural credit, nutrition and international trade.

For Europe, in the next 20 years, there could be a number of trade policy developments, such as the reduction of border barriers to trade, both within the European Union and elsewhere, the enlargement of the European Union, the liberalization of trade in agricultural and food products within the Euro-Mediterranean Association Agreement framework, the liberalization of trade for agricultural and food products resulting from the EPAs (Economic Partnership Agreements) between the EU and the ACP (African, Caribbean and Pacific) countries, etc.

In the EU, the general scheme of the Common Agricultural Policy aid includes market supporting policies and structural policy aid. Examples of market policies include area-based subsidies, production/processing subsidies, consumption subsidies, and agri-environmental aid. Some market policies are directly related to specific alternative agrosystems or their practices. Structural policy aid focuses on elements like modernizing/improving farms and facilitating young people's access to farming. The Common Agricultural Policy (CAP) reform proposed by the Commission in 2002 introduced a major change in the income support regime: the decoupling of direct payments from production with potentially marked effects on land use. Other important reform measures have been the introduction of obligatory, modulated payments to generate funds for agri-environmental and rural development programs, and reduced price support for dairy (partly compensated by direct payments). The intention behind these reforms has been to increase the market orientation of EU agriculture (through decoupling). Concern for less favored agricultural regions, has led to a complex "policy cocktail" (Britz et al., 2006). Several studies conclude that the effect of decoupling will most likely be a decline in cereal and silage maize acreage and in ruminant production in EU-15. A further change can be expected in the economic resources devoted by the EU to rural development, food safety and environmental protection.

Although the IAASTD report does not include Mexico in the NAE assessment, Mexico's trade policies are closely tied to policies in the United States and Canada. All three countries have institutionalized income supports that provide additional assistance to producers when commodity prices (or net farm revenues, in the case of Canada) decline. Additionally, Canada has crafted new approaches to food safety/quality, protection of the environment, the role of science in agriculture, and the overall reinvigoration of the agricultural sector. The United States is proceeding with a comprehensive buyout of tobacco quotas while expanding its efforts in conservation, placing greater emphasis on the continued use of land for production rather than land retirement. However, in all three countries, ample fiscal resources allow agricultural policy to proceed in a direction that is not altogether different from its previous course. However, fiscal constraints could affect the size and content of future agricultural policies in each country (Zahniser et al., 2005).

Interactions between ministries or states often define the policy framework at the national level. At one extreme, regulation is fragmented with little interaction between different ministries. One agency is responsible for health and food safety; another deals primarily with the environment. Other agencies focus on agriculture and transportation/distribution. Interagency issues are often given low priority; consequently, each ministry has limited knowledge of the systemic needs of a regionally based agri-commodity value-chain. At the other extreme, different agencies synchronize public programs. Regional authorities bring independent policy interventions together in one region so as to have the greatest impact on the regional economy; nature is planned.

Agricultural trade and markets

Globalization means changes in the world economy that tend to create a world market for work, capital, goods and services. It is not a new phenomenon but has increased over the last thirty years, largely because of lower transportation and communications costs. Globalization has changed production areas, markets, trade and travel with concomitant effects on food consumption. In many countries, global imports mean that seasonal agricultural products can be eaten all year-round.

Globalization has also increased competition. Some crops, such as cotton, are produced in both industrial and developing countries, but American cotton producers receive much higher subsidies than cotton producers elsewhere in the world. Competition is strong, and countries try to develop policies that favor their growers.

The share of agricultural products (including processed products) in world merchandise exports has decreased steadily over the last six decades, from over 40% in the early 1950s to 10% in the late 1990s, as both volume and price trends have been less favorable than for other merchandise products. Among manufactured goods, it is estimated that the largest value increases were for iron and steel products and for chemicals (WTO, 2006). There are three explanations for this trend: the increase of manufactured products in trade coming from developing countries, the decrease of agricultural prices and the late opening of the agricultural sector to world markets (IFRI, 2002).

Nevertheless in 2005 agricultural products represented an important share of exports of primary products for North America and Europe, less for CIS (Table 5-8). It represented an important share of imports of primary products for Europe and CIS. Significant market changes would have important implications for agriculture and AKST in these regions.

Exports of agricultural products and agroindustrial products are extremely concentrated in North America and Europe (IFRI, 2002). Over the last few years, new actors have entered the game and changed the rules. For example, in the wheat market, there is increasing competition between traditional world leaders (USA, Canada, EU, Australia) and the Black Sea region countries (Ukraine and Kazakhstan). Volumes of world wheat imports are expected to increase further due to ever-growing demand for wheat in Third World countries (Egypt and Nigeria), Brazil and Mexico (Garnier, 2004; FAO, 2006).

5.4.2.2 Uncertainties of the future

A number of uncertainties and questions for the future can be raised relating to trade and policies:

- What will be the impact of the increase of commodity prices on the rural poor and developing countries' farmers, and how will it affect their capacity to take advantage of AKST?
- If there is further liberalization of agriculture, how can the effects of subsidies in NAE be offset for the small producers of the rest of the world?
- What role will some NAE countries play to improve the governance of trade and markets, to make negotiations more transparent and participatory, to strengthen the negotiating capacity of developing countries, to promote regional integration and negotiation from shared platforms?
- What will be the consequence of the new use of agricultural products on agricultural trade?
- How much will the countries of the Black Sea region change NAE's agricultural market?
- How will the EU develop? Will it continue to expand with new member states (EU-30, EU-40) or will it divide? What will be the consequences of changed development policies and stronger collaboration with the Southern Mediterranean countries and Russia on policies, trade and agricultural systems of NAE? What political and economic coalitions will develop outside NAE, and how will that affect agricultural markets and trade?
- How will increased international coordination in areas such as trade, commercial and consumer protection law, and defense and security develop and affect policies and trade?
- What effects will demographic trends have on future policies? Will current trends of stagnating and declining populations in large parts of NAE continue? Can out-migration from more remote rural areas to urban centers be halted? Will there be sufficient incentives to attract investments in rural areas? In which sub-regions within NAE will agriculture vanish?
- Will migration of skilled labor within NAE be permitted? Where will the main migrations take place, and will they help to increase the economic viability of rural areas? To what extent will urban commuters and new well-to-do residents be able to contribute to sustainable rural development?
- Will agriculture and rural areas in NAE develop sufficient adaptive capacity to overcome threats and risks imposed by future environmental change (including climate change)? Will more stringent environmental regulations be agreed upon, together with stronger internalization of externalities? How will that affect agricultural production and production orientation in NAE? How will the impacts of climate change in other world regions affect changes in NAE policies and trade?
- How will a WTO extension of the scope for the exchange of goods, services, labor and capital between countries affect agricultural systems? What will happen if almost all trade barriers for agricultural products and subventions are eliminated? To what extent will that increase environmental risks?
- To what extent will producer subsidies further decline—and how fast? And, how will the money saved in that manner be spent? Will it be invested to alleviate poverty and (thereby) reduce environmental degradation, or for other challenges?
- How will the demand for the major agricultural products of the region evolve?
- How will the share of agricultural products (food and raw materials) in the NAE region develop—will it drop further? How will intra- and inter-regional trade evolve?

Table 5-8. **Share of agricultural products in trade in total merchandise and in primary products in NAE regions, 2002.**

	Share of agricultural products in trade in total merchandise (2005)		Share of agricultural products in trade in primary products (2005)	
	Exports	Imports	Exports	Imports
World	8.4	8.4	32.8	32.8
North America	9.2	6.0	43.3	26.8
South and Central America	26.4	8.9	41.6	31.0
Europe	9.1	9.4	49.4	39.0
CIS	7.8	13.2	11.5	53.1
Africa	10.9	13.9	14.3	50.2
Middle East	2.3	10.0	3.1	56.5
Asia	5.6	7.5	37.9	24.6

Source: WTO, 2005.

5.4.2.3 Consequences for AKST

There are thus a large number of possible future pathways for agricultural policy and trade at national and supranational level within NAE and outside, which in turn will generate different types of farming and agricultural systems.

If the future is more ecosystem oriented, with externalities increasingly internalized, e.g., by progressively decoupling subsidies from production, more stringent environmental regulations, the introduction of special taxes and different product pricing methods, then AKST should be organized to better support the development of more environmentally-friendly and resource-use efficient technologies and production systems, including all kinds of "green technologies" and supportive policies that contribute to the adoption of such technologies to reduce resource use and farm emissions. Such direction would certainly lead to more integration of agricultural and environmental sciences and more cooperation with the various interest groups involved in natural resources management at different levels. AKST in this setting would still be strongly oriented towards feasible technical solutions and require longer term planning and investments.

If we live, however, in a market-led future, the influence of consumers and their preferences on demand for research would become stronger: issues like food safety (labeling, traceability, etc.) would be in the center and require more comprehensive attention by AKST than currently. Such AKST would be organized differently, and multinational companies might have the lead. In a future that would favor regionalization and local approaches, social equity, reduction of income disparities between urban and rural areas, and more power and political influence to local people, the requirements for AKST would again be very different (Kahiluoto et al., 2006). Such a future would also very likely imply changes in attitudes towards consumption and diets, e.g., less meat. Though objectives, organization and funding of AKST have already drastically changed over the last 10 to 20 years (Van Keulen, 2007), further policy adjustments would be required to support the development of mechanisms for increased involvement of stakeholders, and a more demand-driven AKST that is increasingly built on interactive knowledge networks (OECD, 1999), and serves the multiple development goals of rural areas, e.g., through supporting the development of multifunctional agricultural systems. Some recent trends, like special payments for rural development would need to be intensified. The AKST required in such a future, would also need to support the realization of full participation of stakeholders in decisions concerning the design and implementation of agricultural and environmental policies.

This might be realized by harnessing the power of ICT and appropriate databases with new tools for interactive analysis of alternative land use and policy options for sustainable regional development (Van Ittersum et al., 2004). AKST would seek solutions through behavioral changes. It would also need to generate the information required to compare the environmental and social effects of integrated, local versus more specialized, world-market oriented farming systems. The type of AKST required would be fairly interdisciplinary and oriented towards locally tailored solutions and their implementation.

5.4.3 Farming systems and farm structures

Farmers are increasingly operating in larger enterprises and within cooperative arrangements as well as through contracts with large businesses. This could lead to greater complexity and monopolies which could reduce resilience and choices. There is uncertainty about how long this trend will last. It could be altered, for example, by changes in organizational practices and consumer demand and socioeconomic research.

Population figures in rural areas are declining and agro-urban areas are growing. Multiple expectations on farming systems are leading to the development of new enterprises such as agrotourism and are placing emphasis on farming systems that can deliver new services, such as watershed and landscape protection. High demands on agriculture for providing energy could change this trend.

5.4.3.1 Ongoing trends

The term agricultural system (or agrosystem) is a concept that has been in continuous evolution over the last few decades. The great number of elements involved in its definition and their interrelations are partially responsible for this evolution. An extended definition is "the system of production used by a farmer as specified by the technology used, resources available, preferences held and goals pursued within a given agroecological and socioeconomic environment" (Dillon and Hardaker, 1993).

In the arena of discussion about the agricultural systems in Europe, references to the dichotomy between traditional or mainstream systems, on the one side, and emerging or alternative systems, on the other side, are frequent. However, there is no clear consensus about the scope of these concepts. As a first approach (Grudens-Shuck et al., 1998), alternative agricultural systems could be systems that include non-traditional crops, livestock and other farm products; services, recreation, tourism, food processing, forestry and other enterprises based on farm and natural resources; unconventional production systems such as organic farming; or direct marketing and other entrepreneurial marketing strategies. A European prospective analysis of agricultural systems (Libeau-Dulos and Cerzo, 2004) shows that the principal alternative agrosystems coexisting with mainstream agriculture are organic farming, integrated production, conservation agriculture and agriculture under guaranteed quality. Other, less widely used agrosystems in the EU, include precision agriculture, short-chain agriculture, urban agriculture, *agriculture paysanne* and permaculture.

Farms are becoming specialized, increasing in size and declining in number. In Eastern Europe, farms were first industrialized after WWII, although private small-scale farming continued to exist. Food chain organizations developed towards global, linear and centralized structures with regional specialization (McFetridge, 1994; Royer, 1998; Cook and Chaddad, 2000; Reardon and Barrett, 2000; Hendrickson et al., 2001; Harwood, 2001).

5.4.3.2 Uncertainties of the future

Examples of questions about the future are:

- What is the economic viability of family farm systems? Will the trend toward larger, capital intensive farms continue? Will the marketplace support farms that produce specialty products for niche markets?
- Will prices and subsidies lead to the broadening of agricultural systems, or on the contrary to their reduction? What role will the transfer of existing technologies and the development of new ones play? How will improved analytical methods, increased traceability and reduced risks of fraud in the agricultural industry develop? Will the dissemination of biotechnology facilitate the emergence of new alternative systems? What could be its impact on precision agriculture, for example?

5.4.3.3 Consequences for AKST

The adoption of a new agricultural production system involves changes in the way holdings are managed; this makes the presence of a science and technology transfer system capable of meeting the new requirements of farmers especially important. The availability of such a system therefore strongly influences the choice of production systems that involve substantial changes, as is the case with organic farming, (which recovers traditional practices) and conservation agriculture (which experiments with new practices). The influence of this factor on the adoption of agriculture of certified quality is dictated by marketing and distribution criteria; in fact, this agrosystem facilitates acquisition of better knowledge, and ergo fulfillment of consumers demands.

Farmers' willingness to make the transition from mainstream agricultural practices is not enough if they do not have access to the technology required. Hence this factor strongly affects the selection of agrosystems whose practices require the use of new technologies (e.g., integrated farming and conservation agriculture). The choice of organic farming involves the use of natural resources, thus requires good knowledge about soils, biological pest and disease control, organic fertilizers. If conservation agriculture develops not only in large farms, for specific production types (cereals, wood crops), but also in smaller farms, substantial investments in special machinery will be necessary. Production and distribution of AKST must be carefully examined if alternative agricultural systems are to be developed.

5.4.4 Agricultural labor and organizations

Migrant labor represents a high proportion of the workers in the agrifood sector, especially in parts of the United States and the southern countries of Europe. An increasing number of these laborers have come illegally. Enforcement of immigration law would force undocumented workers to leave the countries. The loss of labor force cannot be offset by mechanization and technological advancements alone. Changes in migrant labor could lead to higher wages, and thus higher prices, going out of business or moving production overseas.

5.4.4.1 Labor and gender dynamics: ongoing trends

In 2003, in the European Union, agriculture provided jobs for 13.3 million people, representing 6.6% of total employment. The national distribution of employment in agriculture was extremely uneven. There were 5.8 million people employed in agriculture in the 13 "old" Member States, where employment in agriculture made up only 3.6% of total employment. In the Eastern European countries of the EU, there was an average of 12.4% of total employment in agriculture (EIROnline, 2005).

The composition of labor in agriculture has changed over time, particularly with the sector being affected by different stages of economic development (Hayami and Ruttan, 1985). Four major trends affect the labor situation: important use of migrant labor in agriculture, growing unemployment in rural areas, aging farmers and enlargement of skills needed to be a farmer. There are no major territorial discrepancies in these trends (Brouwer, 2006).

In North America and Europe, an important proportion of workers in the three agrifood sectors (farming, fishing and forestry; meat and fish processing; food service) are migrants. They are especially important for crop agriculture. In the United States, a significant majority of farmworkers lack proper work authorization and immigration status (Kandel and Mishra, 2007). Two major proposals for immigration reform could lead to reduction in the farm labor supply. Enforcement would force undocumented workers to leave the countries. Legalization would give workers greater flexibility to seek other jobs and wages would probably rise. Possible responses to wage increases by firms would be to increase prices, to produce other crops/products, to adopt labor-saving technology, or to go out of business or move production overseas. In crop agriculture, fruit, vegetable and horticultural producers have high farm costs and would be most affected by immigration reforms. In the United States, Hispanics were the principal operators of 51% of the farms and ranches in the 1997-2002 period (Dohm, 2005). This trend might become even stronger in the future.

Unemployment in rural areas is exacerbated by a trend for farms to be abandoned or sold for other purposes (EC, 2004, 2007). To realize an adequate income, farmers leave their farms or combine farming with another job. Women have a higher tendency than men to leave rural areas. Conversely, larger farms have difficulty finding enough qualified personnel. Better-educated and skilled persons seek other opportunities because the hard and dirty work of agriculture is unattractive. In the future, without sufficient labor, many farms will be forced out of business.

More than half of all farm holdings in EU-15 are owned by farmers above 55 years of age, and one out of three farms, by farmers above the age of 65. Less than one out of twelve farm holdings in EU-15 is owned by farmers under the age of 35 years. The economic transformation in countries of Central/Eastern Europe and Asia caused significant changes in agricultural labor use. Estonia, the Czech Republic and Slovakia all have an aging agricultural population. For the other countries, the relative importance of the oldest age group fell in the period up to 2000 (IAMO, 2003). The average agricultural labor force migration rate varies between approx. 8% in Estonia and 10% in Georgia (Herzfeld and Glauben, 2006).

Success in agriculture has been based on production skills for at least 10,000 years. Producers learned about crops and animals and understood seasonal cycles and the need to adapt to climate and pest unpredictability. Knowledge was transferred from parent to child and from neighbor to neighbor. Today's farmers need a larger range of skills. They need relationship skills to effectively cooperate with input and information suppliers. Farmers need knowledge and market skills, particularly to reach emerging markets. They frequently enter collaborative agreements with fellow producers in new models of cooperation. In addition to production skills, today's growers need mechanical/technical skills and financial management skills (Butler-Flora, 1998). But there are still many poorly educated farmers in North America and Europe. Most of the people living in rural Poland (aged 13 years and more) have no more than a secondary education (Central Statistical Office of Poland, 2007). On the other hand, in Estonia and Hungary, almost 10% of those active in agriculture have a university qualification or the equivalent (IAMO, 2003).

5.4.4.2 Organizations: ongoing trends

Farmer associations. Today in North America and Europe, most farms and ranches are still small (Dohm, 2005), but they are getting larger and more concentrated. Many farmers sign contracts with large businesses to secure outlets for their products. Others sell their products themselves elsewhere, such as on commodity exchanges but they have greater exposure to the risks and vagaries of the open market. There is great variation in the level of influence of farmers' organizations. In North America and most of Western Europe, some groups (e.g., cotton or wheat in the USA) are well organized politically and have a platform to directly influence resources that support their commodity.

Inputs enterprises. These companies supply seed, fertilizers, pesticides and other components needed to produce crops. Within the last fifteen years, agricultural inputs have become highly concentrated within a small number of companies. Less than ten multinational companies control the lion's share of the global pesticide and the global seed market. These companies also control nearly all of the private sector agricultural research.

Processing/marketing enterprises. These companies buy agricultural products and process them for the marketplace or make them available to consumers without further processing. The largest of these companies are multinational in scope and wield tremendous influence on agriculture and AKST. For example, Frito-Lay which controls about 40% of the snack food market worldwide and is the largest snack food company in more than thirty countries. If the company needs a certain type of agriculture product or refuses a certain type of commodity, agriculture and AKST will be revised to accommodate them. Even though the genetically engineered NewLeaf potato was a valuable tool for pest management, potato farmers in the United States quit growing them largely because McDonald's corporation told their suppliers not to use NewLeaf potatoes in their french fries.

Media. The media has a powerful influence on consumer preferences; consumers reflect their desires in the marketplace and the polling booth. The tremendous growth of the organic market, for example, is largely driven by the media depiction of pesticide risks; whether or not the risks are accurately depicted is largely irrelevant. The marketplace determines what agricultural products will be produced and how they will be distributed. Elected officials determine resource allocation and a broad range of policies and regulations affecting agriculture and AKST.

Agricultural universities/colleges. Universities and colleges conduct most of the public-sector research. Researchers typically have a long career with a single institution. Hiring decisions by the university or college can have substantial implications for the direction and progress of AKST.

Although these actors have been presented individually, their influence is a much more complicated interaction. For example, a processing company may use the media to promote cotton as a clothing material. As consumer demand for cotton increases, cotton producers need to increase productivity. The university recognizes a need for a cotton AKST position to help cotton growers achieve production goals. The companies that provide inputs for cotton production introduce new plant varieties and chemicals that the cotton researcher incorporates into a more efficient production system. The cycle repeats as the media report that cotton production degrades the environment; the processing company demands more environmentally-friendly cotton; the university turns its attention to more sustainable production methods; the input companies produce less dangerous chemicals and so on.

The increasingly integrated global trade environment leads to convergence in dietary preferences and patterns across countries and this, in turn, is stimulating the ongoing structural changes in food processing and retailing. Thus, to a large degree, multinational food companies are the cause *and* the consequence of the evolving global food system. By their nature, these multinational food companies transcend national borders and give rise to greater interdependence of economies and larger trade flows. To manage and harmonize product flows along the food chain, they also are at the basis of vertically cocoordinated marketing systems. The purpose of these systems is to ensure that product and process requirements for food products are met at all stages of the supply chain, thereby reducing transaction costs. Thus, evolving globalized systems of food production and retailing are becoming an element of increasing importance with respect to the integration of developing countries into global food markets (OECD/FAO, 2005).

5.4.4.3 Uncertainties of the future

There are generic and specific uncertainties related to labor and organizations. Here are some of them.

- *Farmers' age and gender.* Will measures be taken to formalize women's status in the farm enterprise? Will women manage an increasing number of farms? In the EU, will there be enough young people interested in farming and capable of managing sustainable production methods that meet environmental and societal goals while providing an adequate income?
- *Employment.* How can unemployment/underemployment in rural areas be solved? Will farmer education and the creation of non-farm jobs in rural areas be addressed simultaneously? How will the pluriactivity of men and women in rural areas be taken into account? How will pluriactivity influence benefits and resources available to farmers? How will structural unemployment in agriculture be tackled, especially in the Eastern European countries?
- *Migration.* How will NAE political leaders address the problems associated with illegal migrants coming to rural areas for permanent or seasonal agricultural work?
- *Education, skills.* Will there be training courses to help farmers become entrepreneurs who can compete in global agricultural markets while achieving the goals of sustainability and multifunctionality? Will there be administrative and financial measures to facilitate young farmers' training and installation?

5.4.4.4 Consequences for AKST

Decisions related to labor will have consequences on AKST. For example, if migration is permitted and people from outside NAE move to rural areas for seasonal work, the need for research on crop harvesting, etc. will not be great. On the other hand, strict migration policies will lead to research on productivity improvement. Another example: the demand for mechanization, computer assistance and automated responses will also not be the same if NAE is able to attract young, well-trained, entrepreneurial farmers, or if the rural population continues to age, is not very well trained, and labor is not available.

5.4.5 Natural resources availability and management

Increasing prices of energy, water, minerals and other natural resources could affect outputs, costs and practices in all sectors of the food system. Decreasing availability of natural resources, for example oil, water and phosphate, and increasing competition for the use of these resources are leading to rising costs which could have very negative impacts on agricultural production, processing, distribution, retail and purchasing. A substantial reduction of the use of these resources in agricultural production through savings, improved management and new technological developments that increase use efficiency, etc., could alleviate the consequences of this trend.

5.4.5.1 Ongoing trends

Agriculture has a complex relationship with natural resources and the environment. It is a major user of land and water resources yet needs to maintain the quantity and quality of these resources in order to remain viable.

Natural resources, including raw materials, comprise minerals, biomass and biological resources such as forest, soil, water, air, energy resources such as fossil fuels, wind, geothermal, tidal and solar energy and land areas. Whether these resources are utilized as materials/inputs for production, or as environmental buffers or sinks, most of them are essential for the functioning of agroecosystems and socioecological systems at large. The way and speed in which renewable and non-renewable natural resources are being used strongly determines the basis for sustainable development (Millennium Ecosystem Assessment, 2006). The climate system is an important issue since it is an important natural resource (see 5.4.6: Climate change and variability); energy and bioenergy issues are also important (see 5.4.7).

The linkages between natural resource availability and agricultural management practices are considerable. For example, the need for irrigation will not be the same if and where climate becomes drier and water gets more polluted and the frequency of major floods increases, etc.

Agriculture utilizes natural processes to produce the goods (food and non-food) that we need to support the demand of an ever growing population (Verhagen et al., 2007). While acknowledging that population trends and projections for NAE show stagnation and decline, the region will most likely continue to produce for and export to other regions of the world to help satisfy their needs and requirements. Both, renewable resources like agricultural soils, and non-renewable resources like the world's fossil fuels, have their limits. The most limiting resources to food production and other goods provided by agroecosystems in NAE are land and water. Agricultural systems are typically managed to maximize provisioning services to provide food, but they require several other supporting and regulating services to support production. Agriculture both depends on and generates ecosystem services. Agricultural ecosystem services have been grouped into three categories: services that directly support agricultural production (such as maintaining fertile soils, nutrient cycling, pollination), services that contribute directly to the quality of human life (such as cultural and aesthetic values of the landscape) and services that contribute towards global life-supporting functions (such as carbon sequestering, maintenance of biogeochemical cycles, supply of fresh water, provision of wildlife habitats) (Björklund, 2004). Growing populations and activities put increasing pressure on land, soil and water resources. Current estimates suggest that 10-20% of the global terrestrial area has degraded soils, and that that area is extending. Pressure on land and water will be further exacerbated by climatic change. Lack of access to natural resources is a major reason for many local, regional and (trans-) national conflicts. This applies, currently, to low-income countries, where food, forests, wildlife, fisheries and energy sources, which are bound to land and water, form the basis for the livelihood of a large share of the population.

Resource use in the NAE region has been and remains very high. At the same time, resource used by growing economies such as China, India and Brazil increases at an accelerated pace. If the world as a whole would follow the patterns of consumption experienced in NAE, global resource use is estimated to double within the next 10-15 years. However, there is still an enormous slack in resource use efficiency, namely water and nutrient use efficiency, leaving much scope for improvement (Smil, 2000). Inefficient use of resources and overexploitation of non-renewable resources are obstacles, whereas sustainable production and consumption are key to sustainable development (within NAE and globally).

Agriculture generates waste and pollution, yet it also conserves and recycles natural resources, and can significantly contribute to the enrichment of landscapes and creation of habitats for wildlife.

Agriculture both causes and is affected by changes in natural resource availability and quality.

In the following paragraphs we describe major trends and uncertainties related to changes in and threats to agriculture resulting from changes in natural resources and vice versa, agriculture's impact on natural resource availability and quality.

Among the major threats affecting agriculture in the NAE region are climatic change, water scarcity, soil erosion and biodiversity loss (see http://ec.europa.eu/environment/agriculture/index.htm).

On the other hand, NAE agriculture affects natural resource availability and quality mainly through its demands on land, soil, water and energy for producing biomass (food, feed, fiber and fuel), its impacts on the environment from inappropriate management practices such as soil, water and air pollution through excessive use of agrochemicals, soil degradation (erosion, organic matter decline and compaction) and biodiversity loss (see http://ec.europa.eu/environment/soil/pdf/soillight.pdf;). However, there is also a range of environmental benefits created by agriculture such as maintenance of semi-natural habitats for wildlife and of agricultural landscapes thanks to its important environmental services (see http://ec.europa.eu/agriculture/publi/fact/envir/2003_en.pdf).

Effects on agriculture

Favorable climatic and soil conditions are the basis of fertile, diversified and rich agricultural landscapes in the NAE region. The impacts of natural resources are often concentrated locally and regionally, although some are of national and international significance. Land, water and other natural resources are limited. Resource scarcity and competing claims for scarce natural resources, among different agricultural land use types and with other land uses are increasing. That competition is currently very alarming in the very densely populated agricultural lowlands of Asia where fertile arable land is reduced by its conversion for other than agricultural uses (Van Ittersum et al., 2004). In the NAE region, under current climatic conditions, water is at times scarce in parts of NAE such as in the Mediterranean region. That water scarcity will become more severe with anticipated climate change. More extreme weather conditions will lead to more frequent drought and heat stress, more intensive precipitation, frequent flooding, erosion and poor trafficability of agricultural land. Despite many efforts in the NAE region to reduce environmental degradation and improve the quality and availability of the natural resource base, policies and new technologies have not been sufficient to reverse unsustainable trends (Van Camp et al., 2004).

Agricultural impacts on natural resource availability and quality

Agriculture has a significant effect on the environment in the NAE region. In the European Union, for instance, about 50% of the lands are farmed. Many of the environmental effects of agricultural activities are confined to the sector itself, but off-farm effects are also important. In its study "The Limits to Growth" more than 30 years ago, the Club of Rome showed how population growth and natural resources interact and impose limits on industrial and economic growth. As an example, the first global assessment of soil degradation found that 38% of currently used agricultural land has been degraded. Such phenomena are signs of an "overshoot"[12] or, an imbalance between availability, quality and claims on the earth's natural resources, beyond what can be sustained over time. A core question of the various "limits to growth" scenarios was: How may the expanding global population and economy interact with and adapt to the earth's limited carrying capacity over the next 100 years? The simulation model applied to that end has been criticized for underestimating the power of technology and for not adequately representing the adaptive capacity of the free market. Its "30 years update" (Meadows et al., 2004) concludes that: "We are still drawing on the world's resources faster than they can be restored, and we are releasing wastes and pollutants faster than the Earth can absorb them to render them harmless." This is in line with analyses by European research agencies that led to, among others, the recent EU strategy on soil protection (e.g., Van Camp et al., 2004), and the EU Thematic Strategy on the sustainable use of natural resources. Human demand started to exceed

[12] To go too far, to grow so large so quickly that limits are exceeded (after Meadows et al., 2004)

nature's supply as of the early 1980s and has exceeded it by about 20% since 1999 (Wackernagel et al., 2001). This kind of "footprinting" is a way to translate human activities into appropriate areas. There are different approaches to this exercise (e.g., Johansson, 2005). Although the method of calculating the ecological footprint just using one single measure has its limits and may be criticized, the basic message has been confirmed by the Millennium Assessment (2006) and other recent studies (e.g., www.RedefiningProgress.org).

To use a concrete example, in Sweden, thanks to its large forest resources, the total ecological footprint per citizen is 8.17 global ha per capita with no deficit (Wackernagel et al., 2001). However, even Sweden is extremely dependent on areas outside its borders for its food consumption (Deutsch, 2004; Johansson, 2005). There has been a decrease in agricultural land in Sweden after WWII. Between 1951 and 1992 about 20% of Swedish agricultural land has been reallocated; most of it has been afforested or urbanized (Björklund et al., 1999). Furthermore, the direct foodprint has decreased in size due to agricultural intensification with increased use of external inputs.

The total land area of Sweden is 41.1 million ha, of which a major proportion is mountain and forest area, not suited for cultivation. In 1997-2000 Sweden had an average agricultural area of 3.2 million ha, with 2.8 million ha being arable land and more than 0.4 million ha permanent pasture land. This corresponds to 0.31 ha of arable land per capita in Sweden, (compared to the world average of 0.23 ha per capita), and 0.05 ha of pasture land not suited for cultivation, (compared to the world average of 0.58 ha per capita) (FAOSTAT, 2003). During that same period, one-third of the area, which Sweden required for food consumption, was outside Swedish borders (Johansson 2005). In 1999, almost 80% of the agricultural area needed to produce manufactured feed for Swedish animals was outside Swedish borders and 60% of all imports were for animal feed (Deutsch, 2004). The total agricultural area, in Sweden and worldwide, supporting Sweden's annual food consumption in 1997-2000 was, on average, approximately four million ha, or 0.44 ha per capita (Johansson, 2005).

As in any economic activity, in the farm, various production factors are combined in different proportions with the aim of producing foods and raw materials. This process varies between the different existing systems and is based on specific techniques or production practices which could be defined as an ensemble of knowledge, resources and proceedings used by a system to obtain a particular product.

In many of the densely populated parts of northwestern Europe and since the late 1980s also in the new member states, fertile land is lost and soil is sealed by urbanization, with increasing demand for built-up area per capita, roads, industrial terrain, etc. In the Netherlands, the land covered by built-up areas is already around 10% (Klijn and Vullings, 2005). In its communication on soil protection the Commission of the European Communities states that there is evidence that soil may be increasingly threatened by a range of human activities, which may degrade it and its functions, so vital for life, thus undermining sustainability (CEC, 2002). In the EU, an estimated 52 million hectares, representing more than 16% of the total land area, are affected by some kind of degradation process. In the new member states this figure rises to 35%. Soil degradation in dry areas is also known as desertification. Areas that risk desertification include central and southeast Spain, central and southern Italy, southern France and Portugal and large parts of Greece. The major threats to soil functions in Europe are erosion, a decline in organic matter, local and diffuse contamination, sealing, compaction, a decline in biodiversity and salinization (Van Lynden, 2000; CEC, 2002). These threats are complex and interlinked and although unevenly spread across Europe, their dimension is continental. The biggest threat is soil erosion by water. Within EU-25 it is most serious in central Europe and the Mediterranean region, where 50-70% of agricultural land is at moderate to high risk.

Water

In addition to domestic supplies, water is also provided for (Ashley and Cashman, 2006):

- Agriculture: irrigation of crops, livestock, horticulture, very dependent on activities, local soils and resources and climate;
- Trade and industry: factories, shops and institutions such as hospitals, also for power generation and cooling. Consumption is very specific to the nature of the activity, but in a number of developed countries industrial demand has fallen due to a general decline in heavy industry in favor of service industries; better use of recycling and reuse/recovery of water locally; and better water accounting and auditing, reducing wastage and unnecessary use. Overall, demand in this sector is expected to rise by a small percentage worldwide from current levels of about 20% of global water use.
- Public amenities: parks, street washing, firefighting, flushing mains and sewers. This may be water provided free of charge (and unmeasured) where the water service provider (WSP) is a municipality. Firefighting is a major reason for ensuring that water main pressures are maintained and for supplying high-rise buildings.
- Losses: in distribution systems, domestic leaks and dripping taps, where "unaccounted for" water is due to metering errors, unauthorized use and general unrecorded consumption (Alegre et al., 2000). Unaccounted for water (including all losses) may comprise from 6% up to 55% of the total water supplied in areas with aging mains and service pipes.

Agriculture consumes about 70% of all freshwater withdrawn from lakes, waterways and aquifers around the world (FAO, 2007). The same figure holds true for NAE (Shiklomanov, 1999). It takes 1,000 to 2,000 liters of water to produce one kilogram of wheat and 13,000 to 15,000 liters to produce the same quantity of grain-fed beef (FAO, 2007).

Water use by agriculture is primarily determined by the development of irrigated land use, but also by cattle-rearing and people's domestic needs (Figures 5-1 and 5-2. The EU has 9% of its agricultural production under irrigation (13 million ha), over 75% of this is in Spain, Italy, France and Greece. More than 22 million ha (18% of total cropland) are irrigated in the US, over 80% of which is in the West (Gollenhon et al., 2006). In agriculture the efficiency of wa-

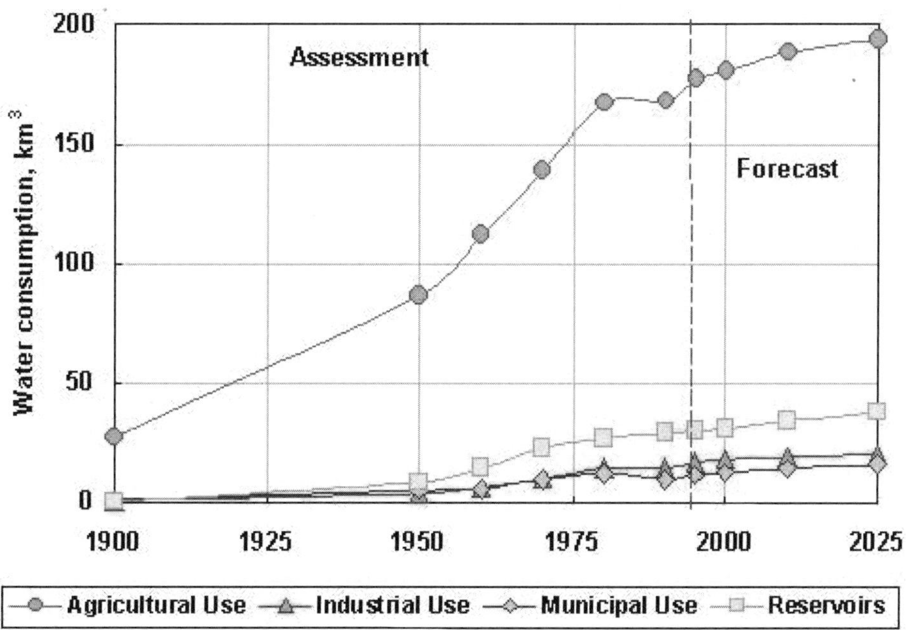

Figure 5-1. *Dynamics of water consumption in North America by type of economic activities from 1900 to 2025.* Source: Shiklomanov, 1999

ter use, per unit, would increase substantially through the ability to target and tailor the application of water coupled with an improvement in crop strains. The greatest impact could be felt in the area of biotechnology, with the possibility of engineering more water-efficient cultures, and ICT, which would bring about more effective water use in agriculture. Improvements could also come from a greater acknowledgement of the need to better manage the role of "virtual water" (water used to produce products) and changes in crop production (in developing countries) and import patterns (in developed countries).

Water use efficiency depends upon agricultural practices and water management techniques. In agriculture the amount of fertilizers and animal manure applied often far exceed crop demands (Wolf et al., 2005). Nutrient surpluses cause problems for human beings, plants and animals. Excesses or nutrient emissions to the environment are being reduced very slowly, *inter alia* through the implementation of the EU nitrate directive. In North America, in an increasing number of watersheds, water supply limits have already been exceeded. In the Midwest of the US, the Ogalallah aquifer in Kansas is overdrawn by 12 km^3 each year. So far its depletion has caused 1.01 million ha of farmland to be taken out of cultivation.

Forestry

Forests: the services, goods and products they provide affect the daily lives of most, if not all citizens. Within EU-25, forests cover 140 millions ha, or about 36% of the land area. Europe's forests are extending in area, increasing in growth rate and expanding in standing volume due to under-exploitation. In EU-25, there are over 4 million people directly or indirectly employed in forestry and forest-based industries, mainly in rural areas. Europe produces 28% of the world's paper supply and is a major operator in wood-based panels and engineered wood products; the contribution of the forest sector accounts for 8% of Europe's added value (i.e., 600 billions euros). With five percent only of the world forest area, Europe produces 25-30% of the world production of forest-based products. The forestry sector's main asset is based on the renewable natural resources and the use, to a large extent, of environmentally-friendly processes. Forest-based industries are very efficient in recovering, reusing and recycling their materials and products, for the manufacturing of new products as well as for energy production. Rigorous life cycle assessments of forest products have shown that they have a strong comparative advantage vis-à-vis other materials. More utilization of forest biomass as a source for energy will be of high importance for a more environmentally-friendly energy secure, sustainable Europe.

Fisheries and aquaculture

In a little more than half a century, the situation of the world fisheries has undergone dramatic change. After the Second World War, fishery landings quadrupled from 20 to 80Mt. This progression was due to the successive opening of new resources to exploitation and greater fishing capacities. In the 1970s and 1980s, the pace slowed down, and for the last two decades, world production has stagnated. Fleets are at over-capacity, and the states of many stocks are degraded. Since the 1970s, the proportion of overexploited stocks has been increasing, that of the under or moderately exploited stocks decreasing, and that of fully exploited stocks, largely

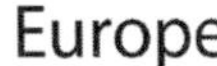

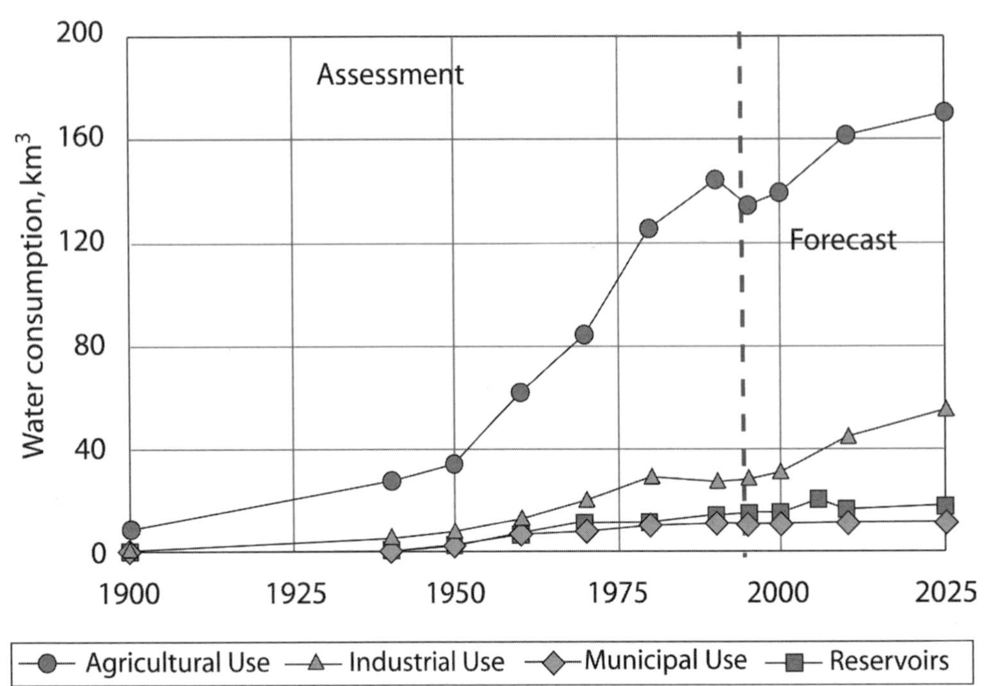

Central and Western Europe

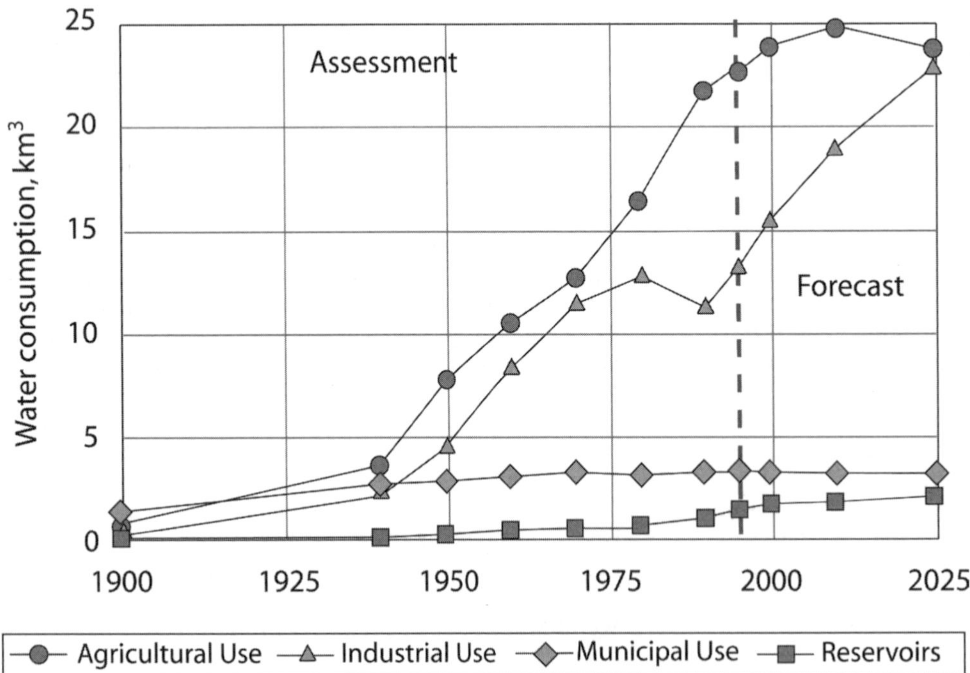

Figure 5-2. *Dynamics of water consumption in Europe and Central and Western Europe by kind of economic activities from 1900 to 2025.* Source: Shiklomanov, 1999.

stable (50%). There is probably no new stock resource, underexploited or unexploited, anymore. Overexploitation has been controlled more quickly in zones exploited by developed countries (Northern Atlantic, Northern Pacific) but now, in varying degrees, affects all the oceans. The Northwestern Atlantic fisheries have experienced one of the most spectacular collapses because of cod stocks, which had been fished for five centuries. Since the moratorium on cod fishing in 1993 in Canada, stocks have not been replenished. The commercial fisheries of the Northeast Atlantic are fully exploited, overexploited or depleted. If the total captures are seemingly stable, it is because of the transfer of fishing from the traditional and high trophic species (cod, haddock) towards species of lesser value (blue whiting, sandeel) or temporarily productive stocks threatened with depletion in the short term (deep-sea species).

5.4.5.2 Uncertainties of the future

While progress has been made in developing new technologies and new institutions and in creating awareness of environmental problems, the outlook today on natural resources is no better than in the early 1970s. There are a number of uncertainties involved concerning the future availability and quality of natural resources, land use and environment in NAE, some of them arising from or being aggravated by global trends such as trade liberalization and climatic change. Among the major factors influencing natural resource availability and land management in NAE is the rise in the consumption for food, feed, fiber and fuel in and outside the region. How will demand for these goods develop in the next decades, and what can and will the NAE supply in order to meet these demands? Will growth in production continue as in the past?

- How will the demographic and economic development within the different regions of NAE affect the severity of the different claims on land, water and other natural resources and the competition between agriculture and other land uses?
- More specifically, related to the supply of food and non-food by agriculture, is the question of the future availability of water, especially in the face of climatic change. How will water availability develop, and to what extent will it restrict agricultural production and/or contribute to environmental degradation? How polluted will water be and what kind of efforts will be made to depollute, desalinize and reuse such water?
- How much suitable agricultural land will be shifted to other land uses? Will less suitable lands be cultivated? What effects will that have on the use of agrochemicals, biodiversity and environmental risks?
- Within agriculture, what will be the share of biofuel crop cultivation in the future, and what implications will the expansion of biofuel crops have on the supply of other agricultural products and on natural resource quality in the different sub-regions of NAE?
- How will the required goods be produced, and how will that affect the quality of water, soil, air and land use?
- What gains in efficiency and increases in water, land, energy and labor for agriculture would be needed to avoid jeopardizing future environmental sustainability? What gains could be achieved by new, improved production technologies and better water resources management? Can such knowledge be generated and be adequately disseminated and implemented in a timely manner? Will policy interventions be sufficient to overcome expected shortages?
- Will there be crops that require fewer fertilizers and other agrochemicals and that also require less water resources, obtained as a result of a fuller understanding of factors regulating nitrate and phosphate utilization, water use efficiency and their impact on natural resources?
- What will happen to natural resource quality if the viability of rural areas in NAE declines?
- Will current trends towards more consumer concern for environment and health, greater demand for food safety (labeling and traceability), organic products, less meat and more convenient foods continue? What will be the implications for natural resource use, land use practices and environmental quality?
- In order to improve the sustainability of coastal capture fisheries and increase their productivity, will research be carried out on efficient management systems, taking into account the ecosystem and improved fishing technologies?
- Will NAE develop its aquaculture production? Will there be more research on the aquatic environment for aquaculture?

Agricultural land use has the potential to damage or destroy the natural resource base and in so doing undermine future needs and development. It also has the potential to conserve agricultural landscapes. Most often, it focuses on short-term economic gains, disregarding long-term impacts and needs and thus contributing to environmental degradation. Clearly part of the solution lies in a change in demands from society, e.g., via changes in dietary preferences and lifestyle, but it also devolves to the agricultural sector to assume responsibility and find ways to reduce the negative environmental impacts by developing appropriate AKST.

5.4.5.3 Consequences for AKST

Agriculture is a major user of land and water resources and is in competition with other users for these limited resources. The sustainable development challenges for agriculture are strongly related to this competition and the role agriculture has in rural development. The pleas made 15 years ago and expressed in Agenda 21 are also valid for today: "Major adjustments are needed in agricultural, environmental and macroeconomic policy, at both national and international levels, in developed as well as developing countries, to create the conditions for sustainable agriculture and rural development" (UN, 1993).

The concepts of production ecology are very helpful in structuring the interrelationships between agriculture, natural resources and environmental quality (Van Ittersum and Rabbinge, 1997). Cropping activities, for instance, are defined by the mix of inputs to produce given target yields. The level of undesired outputs (i.e., nitrate leaching, pesticide leaching, or unproductive evaporation) associated with a given target yield will critically depend on the production technology (i.e., the various resource management practices and their use efficiencies) applied. Nutrients, pesticide and water loss will critically depend on the timing and splits of fertilizer application, type of crop protection and tillage. Policies need to support the diffusion of improved or "best practices" by environmental regulations that aim at reducing nitrate and pesticide leaching. The rigorist approach of such regulations depends on societal choices, which in turn also co-determine the preferred production orientation and farming systems.

Striving for food security and responding to the consequences of globalization of markets and global environmental change (including climate change) are some of the major challenges of our time (CGIAR Science Council, 2005; Roetter et al., 2007). In the future, particular attention needs to be given to climate change and possible (mitigative) adaptation options, as it is superimposed on and will influence other major challenges for agriculture such as the production of sufficient, affordable, high-quality, safe food, as well as feed, fiber and biobased fuel. So far, climate-induced risks and opportunities for agricultural systems have not been sufficiently addressed by AKST.

One of the challenges for AKST is to improve its adaptive capacity. This will be required and beneficial for the sector irrespective of the precise impact of global environmental change.

Closely related to this is the development of modern, resource-use efficient and low emission farming systems and agricultural practices. For the design and *ex ante* evaluation of such systems, the development of better tools like crop models, farm household models and regional land use (optimization) models—linked to GIS—can be very helpful. Such tools will be crucial for analyzing the consequences of possible alternative development pathways on agricultural production and natural resource use. Improved methods and tools together with appropriate stakeholder participation have a high potential to support and promote well-informed policy designs and the implementation of effective policies.

Directly related to this is the challenge for AKST to generate the means that can contribute to conflict resolution regarding competition for scarce natural resources. During the 1990s, some public AKST systems (CGIAR and NARS partners world-wide) have tried to respond to that challenge seriously, e.g., by developing ecoregional research methodologies (Bouma et al., 2007). Both, top down and bottom up approaches to Natural Resource Management (NRM) have been developed (Van Ittersum et al., 2004), with the top down approaches directed more towards policy makers and regional resource managers and the bottom up approaches more towards participatory technology development and support for decision making on optimizing resource use at the local level. Both approaches are required and need to be interlinked in the future to effectively support NRM by improving decision making on land/resource use issues. If the future world opts to achieve sustainability goals mainly through technological solutions and refuses to change its attitude towards consumption and dietary issues, AKST will have to be organized differently than in a world that considers solutions only sustainable if they increase equity, are owned and accepted by local resource managers and contribute to environmental sustainability. In the first case, AKST should be organized to seek local solutions by linking local knowledge networks tightly to global networks of excellence. Whereas, in the latter case, a local learning approach should be promoted to better integrate the different local knowledge centers and link them to global centers of excellence for tapping the relevant disciplinary knowledge. Likewise, in a world that favors technological solutions above behavioral change, AKST will have to focus more on technological improvements in precision agriculture and conventional, specialized agriculture to restrict negative environmental effects than on integrated systems of organic agriculture that minimize emissions through recycling and avoid the use of agrochemicals. The focus of AKST will also depend heavily on whether choices clearly support a biobased economy in which biofuels play a big role. Given the threats of global environmental change, a AKST that directs its efforts towards the development of sustainable, (energy, water, nutrient, and labor use efficient), economically viable farming and land use systems that serve the multiple development objectives of rural areas will be beneficial for natural resources quality and the environment under different plausible futures. Finally, if society decides to make a serious effort to overcome environmental degradation and resource depletion, well designed technologies will be effective tools in supporting sustainable development.

To enhance the aesthetic value and sustainability of the landscape, research will be needed on ornamental plants, genetic exchanges with wild species and improved management strategies to preserve the natural biodiversity of local crops as well as wild species and to contribute to sustainability issues, such as recycling strategies, energy production and fire prevention (ETP, 2005).

Last, little research has been carried out on the sustainability of coastal fishing production systems which are still intensive, while aquaculture production systems, on the contrary need to be intensified and new species introduced. The priority given to fisheries and aquaculture will differ according to the type of agricultural research and innovation system. Ecosystem-oriented AKST will favor the sustainability of coastal fishing while AKST directed to local food supply should favor aquaculture. Market-led AKST will probably put little priority on these themes in their present condition.

5.4.6 Climate change and variability

To counter the increasing effects of climate change on agriculture will require a wider and stronger spectrum of adaptation responses as well as efforts to reduce energy needs and emissions. Increasing temperatures, more erratic precipitation patterns and increased risks of droughts, particularly in the southwestern parts of USA and Europe, coupled with a northern shift of cropping zones, will lead to changes in agricultural systems and production regions. Extreme events will severely challenge adaptive capacity. AKST could be developed to provide better adaptation and mitigation responses.

5.4.6.1 Ongoing trends

Agricultural systems, forestry and fisheries are quite sensitive to climate change and variability and can be strongly affected by them. Concurrently, land use and land use change, particularly through agricultural and forestry activities, can strongly influence climate. There is now unequivocal evidence that the Earth's climate has demonstrably warmed since the pre-industrial era and that most of the warming over the last 50 years is very likely to have been due to increases in greenhouse gas[13] concentrations in the atmosphere. Atmospheric concentrations of these gases are at their highest recorded levels and continue to go up, mainly due to combustion of fossil fuels, agriculture and land-use change (Figure 5-3). It is generally not the changes in the means of weather variables that impose the greatest risks, but the increase in frequency or intensity of extreme events that pose challenges to agricultural systems. The full appearance of many of the impacts of these changes is delayed by inertia in the climate system and in the behavior of ecosystems (IPCC, 2007ab).

Agricultural climate change response options are often taken in the context of other stresses and objectives through a range of technological, behavioral and policy changes. While the impacts of a changing climate are complex, farmers have shown a considerable capacity to reduce emissions from agriculture and adapt to climate change by adopting appropriate agricultural practices and systems. To manage current climatic risks and increase resilience to likely future changes, mitigation measures such as cultivation practices that increase soil carbon sequestration, manure management and reforestation need to be continued. The earlier and stronger the cuts in emissions, the quicker concentrations will approach stabilization (although the effects of such measures on the climate will only emerge several decades after their implementation). Regardless of these mitigation measures, global warming will continue and the associated climate changes during the 21st century are expected to exceed any experienced in the past thousands of years over which agriculture has been practiced in the NAE region. While mitigation measures clearly need to be pursued to reduce emissions from agriculture, some changes are now inevitable and will require adaptation responses.

Large parts of North America and Europe are located in the temperate climatic zone characterized by favorable agroclimatic conditions, i.e., neither too dry nor too hot—with ample, well-distributed rainfall and relatively mild winters. The NAE region also includes areas in which current climatic risks such as drought, frost and flood play a considerable role, but the risk-prone areas are proportionately smaller than in other regions. Drought-prone regions include large parts of southwestern US, the Canadian Prairies and the Mediterranean, while frost risk and low temperatures limit agricultural activities in large parts of Canada, the Nordic countries and Russia. The highest emissions of greenhouse gases from agriculture are generally associated with the most intensive farming systems whereas some of the low input farming systems currently located in marginal areas may be the ones that are the most severely affected by climate change (IPCC, 2007b).

Agriculture contributes significantly to methane and nitrous oxide emissions. Land-use change can also provide a significant contribution to carbon dioxide emissions, but emissions connected to the use of fossil fuel for machinery and heating are considerably worse (Figure 5-4) (Rosenzweig and Hillel, 2000; Stern et al., 2006; UNESCO, 2006a). In the NAE region, greenhouse gas (GHG) emissions from agriculture are in the range of 7-20% of total country emission inventories (in terms of radiative forcing). Latest estimates suggest that agriculture accounts for 48% of CH_4

[13] Greenhouse gases and clouds in the atmosphere absorb the majority of the long-wave radiation emitted by the Earth's surface, modifying the radiation balance and, hence, the climate of the Earth. The primary greenhouse gases are of both, natural and anthropogenic origin, including water vapor, carbon dioxide (CO_2), methane (CH_4) nitrous oxide (N_2O) and ozone (O_3), while halocarbons and other chlorine- and bromine-containing substances are entirely anthropogenic.

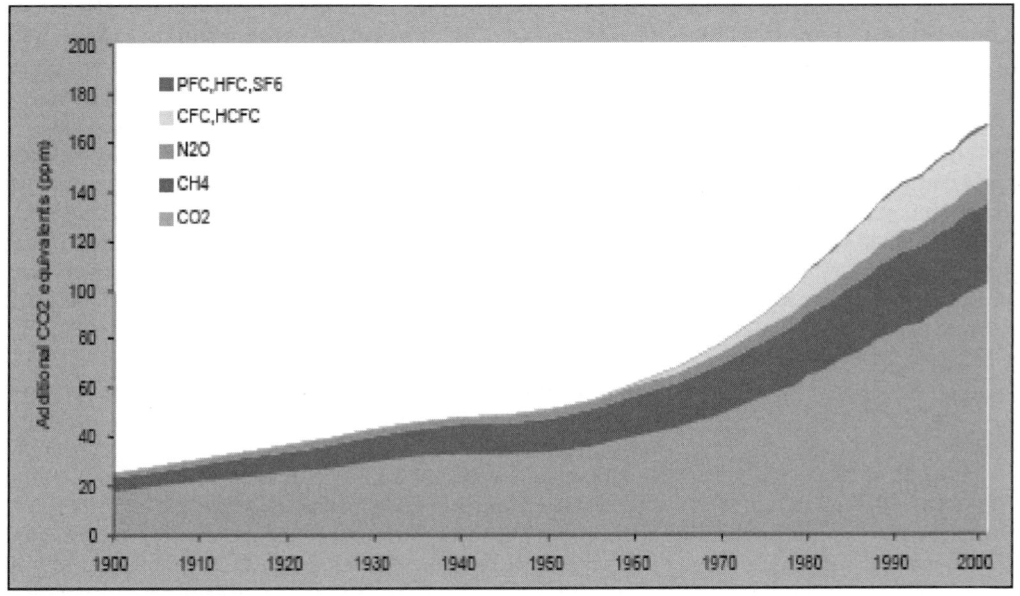

Figure 5-3. *Rise of greenhouse gases (CO2, methane and nitrous oxide and others) 1900-2000 as compared to reference year 1750.* Source: European Environment Agency, 2004.

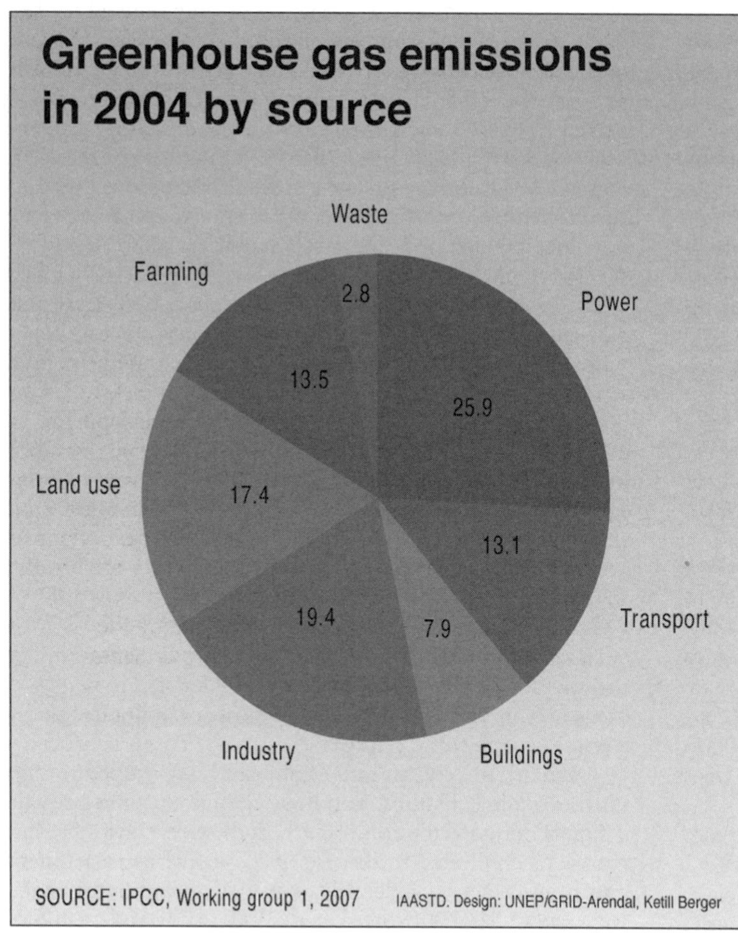

Figure 5-4. *Sources of GHG emissions.* Source: IPCC, 2007.

emissions and 52% of N_2O emissions in the EU. The role of agriculture both as a source of and as a sink for GHGs varies significantly across the NAE region because of the different agricultural policies and practices. Emissions also come from changes in forests and other woody biomass stocks, forest and grassland conversions and from the soil (IPCC, 2000b; UNESCO, 2006a). There is a clear trend across the whole NAE region to boost efforts to decrease emissions by replacing fossil fuels with liquid biofuels (IEA, 2006).

The effects of climate change on agriculture (including forestry and fisheries) are already visible in different parts of NAE (IPCC, 2007b). During the 20th century, for instance, as a result of spring and summer warming and a shorter period of snow, the thermal growing season (with daily mean temperatures above 5°C) was lengthened by about ten days in southern Finland (Carter, 2007).

5.4.6.2 Uncertainties for the future

How might GHG emissions develop in the future?

There are a number of uncertainties involved in predicting the future development of GHG emissions (IPCC, 2000a). Some of the uncertainties relate to economic development, energy supply and use as well as consumer behavior around the world (Sachs, 2006; EC, 2007). Other uncertainties relate to the operation of the carbon cycle which is crucial in translating emissions into concentrations as well as the magnitude and behavior of vulnerable carbon pools (UNESCO, 2006a; IPCC, 2007a): Natural carbon pools could well turn into sources as global warming and deforestation continue. Some of the most vulnerable pools are (1) carbon in frozen soils, (2) carbon in cold and tropical peatlands, and (3) biomass-carbon in forests vulnerable to fire and insect infestations. Within the time span of our assessment (up to 2050) most of the IPCC emission scenarios are indistinguishable because of the inertia in our economic and technological systems. Furthermore, and as a result of this and the inertia also in the climate system, climate projections in the NAE region until 2050 are quite similar.

Possible evolution of NAE climate and possible consequences for agriculture

Climate projections indicate that annual temperatures over Europe will continue to warm at a rate of between 0.1 and 0.4° C per decade. The greatest increases are expected over southern Europe and northeast Europe (Parry, 2000). Higher temperatures will increase evaporation from plants and soil, worsening the water problems that already afflict the hotter (southern) regions of NAE. Annual precipitation is expected to increase by 1-2% per decade^{-1} in northern Europe. There will be little decrease (at maximum -1% decade^{-1}) in southern Europe, and hardly any change over central Europe. In North America trends towards increased temperatures and changes in the frequency of heavy precipitation over most land areas are expected to continue. Furthermore, extreme events are likely to increase in frequency and severity (IPCC, 2007a).

Warming in NAE will generally lead to a northward expansion of suitable cropping areas, and an increase in the length of the growing season for indeterminate crops (whose growth is determined primarily by environmental conditions e.g., root crops) but a reduction for determinate crops (that develop through a pre-determined set of stages, from germination to ripening e.g., cereals). It is assumed that about 10-20% of the increased crop productivity, which has doubled over the last 100 years, may be due to the growth-enhancing effect of CO_2. It is unclear whether this will continue and to what extent this fertilization effect will be reduced by combinations of multiple biotic (pests, diseases) and abiotic (drought, heat) stresses. The increase of atmospheric CO_2 concentrations may increase water use efficiencies (Roetter and van de Geijn, 1999; IPCC, 2007a). However, the expected frequency of extreme weather (flooding and droughts) will possibly offset the potential benefits to Europe (Olesen and Bindi, 2002) as well as to Canada and the United States (Reilly et al., 2003; Easterling et al., 2004; Lemmen and Warren, 2004). Northern Hemisphere snow cover, permafrost and sea-ice extent are projected to decrease further. In some areas, the timing of water availability is expected to change—more precipitation falling as rain in winter, earlier snow-melt and more frequent dry spells in summer (IPCC, 2007a). In regions where crop production is affected by water shortages, such as in southern Europe, increases in the year-to-year variability of yields in addition to lower mean yields are predicted. Extreme high or low temperatures during crucial stages of plant growth can lead to considerable yield loss. Sea level rise could lead to larger areas being susceptible to flooding and saltwater intrusions with potentially disastrous effects on harvests.

In NW Europe, climate change may lead to positive effects for agriculture by triggering the introduction of new crop varieties and species, higher crop production and expansion of suitable agricultural land area. However, climate change may have negative effects on infectious diseases of plants (Chancellor and Kubiriba, 2006) and may motivate a demand for different pest management practices and for measures to reduce nitrate leaching and the turnover of soil organic matter (Olesen and Bindi, 2002). Estimated increases in water shortages and extreme weather events may result in lower yields (and harvest indices), greater yield variability and a reduction of suitable areas for traditional and region-specific crops. Such effects will most likely aggravate the current trends of agriculture intensification in NW Europe and extensification in the Mediterranean and SE parts of Europe.

In the US and Canada, future climate change is likely to result in agricultural shifts toward higher latitudes and elevations. Moderate increases in temperature (1-3° C) along with elevated CO_2 and changes in precipitation will have small beneficial impacts on crops such as wheat, maize and cotton. Further warming, however, will probably have increasingly negative effects (Lemmen and Warren, 2004; Easterling et al., 2004; Stern et al., 2006). Some authors have reported positive crop yield responses to temperature increases of about 2°C, but negative yield responses at increases over 4°C. Higher temperatures and warmer winters could reduce winterkill of insects and broaden the range of other temperature-sensitive pathogens (Rosenzweig et al., 2000). It is still not clear whether North American agriculture as a whole will be affected negatively or positively by climate change. Part of the reason for this is the difference in assumptions regarding agriculture's adaptation potential. The growth enhancing effects of increasing CO_2 concentrations (currently around 380 ppm and increasing at an annual growth rate of 2 ppm) on crops may mask much of the negative effects of changed temperature and precipitation patterns. Agriculture will likely be vulnerable to higher frequency and severity of extreme events—as was demonstrated during the summer 2003 European heat wave that was accompanied by drought and maize yield reductions of 20%, representing the largest yield decline since the 1960s.

How could technological innovations influence the ability of agriculture to mitigate and adapt to climate change?

Although unable to erase uncertainties, technological innovations may greatly influence the ability of agriculture to mitigate and adapt to climate change. For Europe, mitigation and adaptation are necessary and complementary for a comprehensive and coordinated strategy (Olesen and Bindi, 2002; Metzger et al., 2006). Adaptation is an important complement to greenhouse gas mitigation measures and policies. Adaptation to climate variability and change is not a new concept. Managed systems are likely to be more amenable than natural systems, and some regions will face greater obstacles than others. Throughout human history, societies have shown a capacity for adapting—though not always successfully (Lamb, 1995; Diamond, 2005). However, adapting to climate change will not be an easy, cost-free task, and adaptation decisions in one sector (e.g., water resources) might have implications for other sectors. Many of the existing adaptation strategies may be strained by the expected changes in climate, particularly extreme events. Adaptation technologies include changing varieties/species to fit in better with changed thermal and/or hydrological conditions, changing irrigation schedules and adjusting nutrient management, applying water-conservation technologies (such as conservation tillage), altering timing or location of cropping activities, etc. Some of those adaptation measures also have mitigative effects—such as applying "zero tillage" practices or using cover/catch crops in spring to reduce leaching and erosion. The provision of appropriate enabling environments and policies such as technology and knowledge generation and dissemination mechanisms will also be important considerations (Easterling et al., 2004; Kabat et al., 2005; Carter, 2007).

Adaptive capacity and sustainability

The essence of sustainable development as defined by the Brundtland Commission (WCED, 1987) is meeting fundamental human needs while preserving the life support systems of the earth (Kates et al., 2000). Actions directed at coping with the impacts of climate change and efforts to promote sustainable development share some important common goals and determinants such as access to resources, equity in the distribution of resources, and abilities of decision-support mechanisms to cope with risks. Sustainable development can result in improved adaptation to climate change and enhance adaptive capacity (IPCC, 2007b; Verhagen et al., 2007). Climate change adds an extra challenge or constraint to existing obstacles to achieving the various social, ecological and economic objectives defining sustainable development. For agrosystems, any changes in technologies and institutional arrangements that increase flexibility and resilience regarding the different sustainability dimensions, will, in turn, increase their adaptive capacity/capability to cope with climate change.

Impact of climatic change (a function of exposure and sensitivity of a system) and adaptive capacity determine the vulnerability of socioecological systems to climate change (Yohe and Tol, 2001).

For Europe, the ATEAM (Advanced Terrestrial Ecosystem Analysis and Modeling) project constructed scenarios for a range of possible changes in socioeconomic conditions, land use patterns and climate to assess the vulnerability of the human-environment system to global change (Ewert et al., 2005; Schröter et al., 2005). Results from that assessment show that global change will have a large influence on ecosystem service provision in Europe. There is, however, a large heterogeneity in the projected vulnerability between regions. The Mediterranean region is projected to be most vulnerable, while northwestern European countries face the lowest impacts and show the greatest adaptive capacity (Metzger et al., 2006).

For the United States, US agriculture on the whole can adapt (with either some net gains or some costs) if warming occurs at the lower end of the projected scale of magnitude (i.e., 2 to 3° C by the end of the century) and the variability level stays constant (Easterling et al., 2004). However, with a much larger magnitude of warming, even under optimistic assumptions about adaptation capabilities, many sectors would experience higher losses and costs (Easterling et al., 2004). Canada will likely experience similar effects (Lemmen and Warren, 2004). In this context, another feature that clearly distinguishes NAE agriculture from other regions is the significant high level of its current adaptive capacity. This is mainly due to the region's access to important economic, technological and other resources which is better than that of other regions (Adger et al., 2005). It is also co-determined by the fact that relatively large areas have a relatively low exposure to climate change, compared to other regions.

5.4.6.3 Consequences for AKST

Options for dealing with the threats of climate change require examination at regional and local scales. Questions include: how can emissions from agriculture and forestry be effectively reduced, how can agriculture and forestry best adapt under given local conditions, what role can biofuels play and, finally, what are the implications for AKST?

There will be different requirements for AKST, depending on future policy and societal choices, such as the degree of emission reduction, energy price increases, reduced consumption, proactive adaptation and enhanced adaptive capacity.

Some of the obvious consequences for AKST are given below. Furthermore, some suggestions are given on the efficacy of different measures in reducing the vulnerability of agriculture and rural areas to climate change:

1. AKST needs to generate the information required to improve climate modeling and scenario development. This includes developing improved methods for determining GHG emissions from agricultural activities and improving our understanding of the carbon cycle.
2. Another area that requires attention is the effectiveness of adaptation to today's climate variability (Adger et al., 2005); such lessons are important for better understanding of vulnerabilities and measures needed for different climatic risks.
3. Improvement is also required in the area of climate change impact assessment methodologies—this refers to the modeling of multiple stresses as well as to the quantification of climate change scenarios on the whole range of ecosystems goods and services (Carter, 2007) and the effects of climate change on the quality of crop and animal production.
4. More effort is required to develop knowledge and tools needed to support the design and evaluation of mitigation and adaptation options for agriculture; this also includes more comprehensive cost-benefit analysis than now available (Stern et al., 2006; Carter, 2007). Comprehensive energy-efficient agricultural systems need to receive particular attention.
5. Likewise, more consideration needs to be given to the establishment of AKST multistakeholder approaches for designing and implementing feasible strategies at the farm and subnational scale. All actors need to be involved in a participatory planning process.
6. There needs to be more focus on regional studies of impacts and mitigation/adaptation of climate change in agriculture, including assessments of the consequences on current efforts in agricultural policies for sustainable agriculture that also preserve environmental and social values in rural communities.
7. The development of strategies to enhance the adaptive capacity of agroecosystems is a related issue that dwells on the generation of interdisciplinary knowledge and a willingness to better integrate different AKST activities across sectors and among stakeholders so that they become less vulnerable and risks are better managed.
8. Finally, research should focus on creating productive and multifunctional land use systems in rural areas that aim to provide sustainable ecosystem services and employment. This should include, where necessary, restoration of degraded lands and the integrated management of natural resources.

Where governments and citizens assume more responsibility for the environment and are proactive in terms of alleviating the threats of climate change, AKST activities will be more far-reaching and will require the provision of better information, appropriate technologies and multifunctional agricultural landscapes. However, where decisions on natural resources and the environment (including climate system) are not integrated with economic decisions, AKST will be reduced to contributing to the fulfillment of consumers' requirements regarding food and non-food products.

5.4.7 Energy and bioenergy

Increased demands are being levied on agriculture to provide energy and biomaterials. Bioenergy that includes the production of liquid fuels from biomass could meet some of the world's growing energy needs. It is unclear to what extent agriculture in NAE will become an energy producer, and how much can be achieved from other renewable energy sources and conservation. The development of bioenergy will increase competition for land and water resources and push up food prices. Social, technological and economic studies are badly needed.

5.4.7.1 Ongoing trends

Since World War II, global energy consumption has increased more than six-fold. In the same period, per capita energy demand has more than doubled. The energy demand growth rate is not slowing down in spite of record oil prices and global primary energy demand is expected to grow by more than 50% by 2030 (Fresco, 2006; IEA, 2006). According to the World Energy Outlook (IEA, 2006), in the reference scenario, the average annual percent change is expected to be 1.8 for the world, 3.0 for non-OECD Asia, 1.0 for the USA, 1.2 for Canada, 0.4 for OECD Europe and 1.3 for Russia. In the case of a low growth rate, the average annual percentage change is expected to be 1.4 for the world, 0.6 for the USA, 0.8 for Canada, 0.1 for OECD Europe and 0.8 for Russia.

Energy is a key driver in agriculture through the consumption of fossil fuels and fertilizer production. Agriculture also can be a source of energy. Energy consumption in agriculture depends on the type of crop, the production system and agroclimatic conditions and the farm size. Irrigation accounts for the largest share and is thus especially vulnerable to changes in energy prices. It has also been observed that the application of farmyard manure, another source of energy, has been decreasing over time. Application of mineral fertilizers, for improving yield and productivity has been on the increase but a stringent EU policy framework and related national policies have led to a decline in recent years (Wolf et al., 2005; EC, 2007). At present in the USA (Konyar, 2001), average direct and indirect energy account for 19% of the total variable costs, ranging from ten percent for soybeans and up to 27% for cotton. For irrigated crops, energy constitutes an average of 33% of the total variable cost, and ranges from 26% for hay to 51% for sorghum. These proportions could change with the use of biobased fuels. The availability and price of energy also influences the transport of agricultural products and hence global trade.

Biofuels that can be used for transport include bioethanol, biomethanol, biodiesel, biogas, biohydrogen and pure vegetable oil as well as solids such as agriculture and forestry wastes (Schröder and Weiske, 2006). The two primary biofuels in use today are ethanol and biodiesel, both of which can be used in existing vehicles. Ethanol is currently blended with gasoline, and biodiesel with petroleum-based diesel for use in conventional vehicles. Globally, ethanol accounts for about 90% of total biofuel production, with biodiesel making up the rest (Marris, 2006; Sanderson, 2006). Global fuel ethanol production more than doubled between 2000 and 2005, while production of biodiesel, starting from a much smaller base, expanded nearly fourfold. By contrast, world oil production increased by only seven percent during the same period.

Petroleum refining is being developed on a very large scale; biofuels are produced in lower volumes and, currently, much more decentralized. According to the World Energy Outlook (IEA, 2006), significant technological challenges still need to be overcome for the second-generation technologies to become commercially viable. In the case of biodiesel in particular, where a wide range of plant and animal feedstock can be used, production facilities tend to be rather dispersed. Ethanol fuel production has tended to be more geographically concentrated than biodiesel e.g., in the United States, predominantly in the Midwestern states that have abundant corn supplies (Worldwatch Institute, 2006).

The various biomass feedstock used for producing biofuels can be grouped into two basic categories. The first is the currently available "first-generation" feedstock, composed of various grain and vegetable crops that are harvested for their sugar, starch, or oil content and can be converted into liquid fuels using conventional technology. The yields from the feedstock vary considerably, with sugar cane and palm oil currently producing the largest volumes per hectare (Marris, 2006). By contrast, the "next-generation" biofuel feedstock comprising cellulose-rich organic material will be harvested for its total biomass (Fresco, 2006). To convert these fibers into liquid biofuels requires advanced technical processes, many of which are still under development. Advanced biofuel technologies could allow biofuels to replace 37% of US gasoline within the next 25 years, with the figure rising to 75% if vehicle fuel efficiency were doubled during that same period. The biofuel potential of EU countries is in the range of 20-25% (EEA, 2006) if strong sustainability criteria for land use and crop choice are applied and bioenergy use in non-transport sectors grows in parallel.

5.4.7.2 Uncertainties of the future

As far as energy and bioenergy are concerned, there are three major uncertainties for the future:

- To what extent will bioenergy supply develop globally?
- Which considerations will determine future bioenergy use in NAE?
- Will agriculture be able to substantially reduce energy required for production?
- What will be the consequences of bioenergy production on food prices and water usage?

Among the major considerations in NAE that will influence the energy market will be the energy security aspect. Second, there will be the increasing awareness of the need to protect the Earth's climate system through the reduction of greenhouse gas emissions (GHG). The recent (March 2007) agreement of EU leaders on greenhouse reduction targets and renewable energy use is a milestone that may well trigger changes in energy policy elsewhere (i.e., the US, Russia, China). There continues to be a lively debate regarding the trade-offs between economic growth and energy. Some experts claim that energy costs will rise sharply if we increase the share of biofuels in the energy supply mix. This does not consider the many opportunities for reducing the use of fossil fuel e.g., by applying energy-saving technologies and choosing low-emission activities (as has already been demonstrated by many NAE multinational companies such as BP, Shell, Bayer, General Electric) (Fresco, 2006).

The "next-generation" biofuels are based on cellulose biomass such as tall grasses as well as wood and crop residues that are generally abundant and can be harvested with less interference with the food system and potentially will put less strain on land, air and water resources. Another potential "next-generation" feedstock is the organic portion of municipal solid waste. The use of "next-generation" cellulose biomass feedstock has the potential to dramatically expand the resource base for producing biofuels in the future (Fresco, 2006; Marris, 2006). Over the next 10-15 years, lower-cost sources of cellulose biomass, such as the organic fraction of municipal waste and the residues from the processing of crops and forestry products, are expected to provide the initial feedstock. Many questions arise in this context. One is, to what extent can these technological developments be accelerated by further supporting policy interventions, better public–private research cooperation and increased investment?

Research and development efforts to date have demonstrated the feasibility of producing a variety of liquid fuels from cellulose biomass for use in existing vehicles. As of mid-2006, however, the costs of producing such liquid fuels were not competitive with either petroleum-derived fuels or more conventional biofuels. The diffusion of "Flex Fuel Cars" (currently about 50% of the cars in Brazil) introduces flexibility to respond to fuel price fluctuations. Various government and industry-sponsored efforts are under way to lower the costs of making liquid fuel from cellulose biomass by improving the conversion technologies. (Worldwatch Institute, 2006). How fast these developments will proceed is still unclear. Unambiguous cost signals as well as information regarding the availability of new technologies will influence consumer preferences and behavior. These developments will depend on economic growth and sustainable development outside the NAE region. According to recent projections China and India are expected to account for 30 to 40% of energy demand by 2030 (IEA, 2006).

The dual challenge is to secure adequate energy at affordable prices and, at the same time, limit consumption such that it does less environmental harm. It is unclear to what extent agriculture in NAE will become an energy producer, and how much its energy-efficiency can be increased. This depends on AKST as well as on other KST efforts. More centralized and technology-intensive renewable forms of energy may well outweigh agriculture as an energy-producer.

5.4.7.3 Consequences for AKST

Actors in AKST need to pay more attention to the following energy-related issues:

- Research into new farming systems that are able to satisfy their own energy needs and defray their own costs by producing biofuels, as well as installing other renewable sources of energy such as wind and solar power.
- Generation of knowledge that allows sustainable production of biofuels, i.e., in an economically-viable, environmentally-friendly and socially-acceptable manner.
- Proper accounting for the full energy demand of the agricultural sector in environmental impact assessments.
- Biochemistry and ecosystem studies to eliminate agricultural and forestry residues or use it to produce bioenergy.

Furthermore, the following general issues need to be considered:

- Evaluation of investment options in the short, medium and long term for energy exploration and production infrastructure.
- Increasing energy efficiency, identifying measures to reduce the demand from the transport sector, promoting the development and deployment of technology.
- Assessing options for next 50-100 years, e.g., potential for biofuels and other renewable sources like wind, solar, tidal, etc.
- Making use of new technologies to combine energy sources in an efficient way (photovoltaic with fuel cells or new large accumulators), especially in decentralized systems.

5.5 Key Drivers for Agricultural Knowledge, Science and Technology (AKST) and Agricultural Research and Innovation Systems and Their Uncertainties

Agricultural R&D is not conducted in isolation; it is strongly influenced by the rest of science. In 2000, the world invested 725 billion dollars in all the sciences carried out by both public agencies and private firms—that is about one third more than in 1995—with the biggest increases in the Asia and Pacific region. However, there is evidence of a huge, and partly growing, divide between the "scientific haves and have nots." The total amount spent on sciences is approximately 1.7% of the world's GDP worldwide. Public agricultural R&D funds amounted to 23 billion dollars in 2000, about 3% of the total science spending (CGIAR Science Council, 2005).

Today's agricultural research systems are increasingly being asked to tackle problems that are, strictly speaking, external to agriculture. The emphasis is shifting away from the development of productivity and increasing technologies towards that of new approaches to social and environmental issues, such as the protection of natural resources, food safety and animal welfare. The challenge is to promote development that balances equity and environmental interests with those of economic growth, while limiting the negative external effects of agriculture (ISNAR, 2003).

5.5.1 Organizations and funding of AKST

The futures of organizations for AKST are going to be influenced by changes in the Agricultural System and in the KST systems. In this subchapter, we will briefly describe these organizations in the different regions of North America and Europe, and shed light on a number of uncertainties for the future. Funding will also be considered.

AKST organizations in North America and Europe include all the formal and informal organizations controlling, generating, distributing and utilizing agricultural knowledge, science, technology, inputs, markets, credits, capital and assets. This implies primarily research, education and extension organizations, but also government agencies, administrative and political decision-making bodies, NGOs and associations, and private enterprises, acting within the food chain and interacting with it e.g., in regulation, input production, waste management, markets and financing. The North American and European agricultural innovation systems have had a major impact on shaping a broad range of AKST organizations outside the region, for example, transnational private companies and NAE-based and -dominated international organizations, and the CGIAR, as well as many national organizations of Africa, Asia and Latin America.

5.5.1.1 Ongoing trends

Formal AKST structures started to take shape in the late 1800s. In the USA, contrary to most of Europe, education, research and extension were integrated among each others (Huffman and Evenson, 1993), while in Russia they were separate with no public extension service (Miller et al., 2000). In USA, decisions on AKST were taken at state level which fostered innovation and diversity, while in Eastern Europe there was a strictly centralized top-down model (Miller et al., 2000). The governmental responsibility for AKST in NAE rested traditionally with an agricultural ministry, but now is increasingly been brought into closer connection with the general public KST and innovation policy (OECD, 1999, 2005abc). To counteract consequent disintegration of components of AKST, cooperation between them and across institutes (especially between research institutes and universities), disciplines and territories, is increasingly being encouraged, also by specific funds. This has been more successful for research and extension entities that for universities. The organizational structure chosen for AKST components seems to have profound influences, and effective cooperation across ministry boundaries seems to be very challenging (OECD, 1999).

AKST grew during the first half of the 1900s, and the pace accelerated after World War II. The share of public AKST funds to universities increased from the 1970s onwards. Since the 1980s the number of facilities has declined, and they have been privatized and rationalized (Alston et al., 1998). Although the share of agriculture in total R&D funding has declined, the agricultural research intensity ratio (agricultural public R&D relative to agricultural GDP) has risen more than the average science and technology research intensity ratio (Alston et al., 1998). In general, the level of support reflects the size of the country's agricultural sector. The largest budgets for agricultural research are found in the US, Japan, France, Canada, the UK, Italy, Germany and the Netherlands (Pardey et al., 1999). Spending on private-sector agricultural research is greatest in the US, Japan, the UK, France and Germany, mainly thanks to the various multinational conglomerates that have their headquarters in these countries. The figures suggest that private- and public-sector research complement rather than substitute each other. Countries that have traditionally provided substantial support for public-sector research have created an enabling environment for research and technology development, which motivates the private sector to advance its own research. Between 1981 and 1993 private-sector research expenditure grew by 5.1% per year while public-sector research expenditure grew by only 1.8% (Alston et al., 1998). By 2000, private sector investments accounted for around 55% of all agricultural R&D in developed countries, but in low-income countries, it was negligible (CGIAR Science Council, 2005). The growth in aggregate agricultural research (public and private sector) continues at a rate of approximately 3.4% per year, slightly lower than the 4% growth rate in total research (ISNAR, 2003).

NAE governments are funding higher education with an increasing tendency towards tuition fees, and also "basic" and "pre-competitive" sectoral research, but economic sectors are increasingly encouraged to fund sectoral research, and extension/development costs are addressed to clients (OECD, 1999).

The involvement of the private companies in agricultural extension has also gone up (Umali and Schwartz, 1994), while public extension services have become increasingly chargeable and have been down-sized (Read et al., 1988; OSI, 2006) except in some European countries with small farm-dominated agriculture or a conscious choice for independence of commercial interests (OECD, 1999). However, this proportion of private funding is about the same as the general repartition of private funding of R&D.

The model for international research centers was introduced after World War II, and in the 1970s they were united to form CGIAR, whose centers grew in size and numbers but whose budget in the 1990s stagnated and then took a downturn, until the year 2000 when it started recovering. In 2000, CGIAR represented 1.5% of the global public sector investments in agricultural R&D and 0.9% of all public and private agricultural R&D spending (CGIAR Science Council, 2005).

In NAE agricultural research organizations, there appears to be a decrease in the importance of traditional productivity-oriented agricultural research and an increase in research on socially relevant themes such as environment and food safety. A similar (although less pronounced) change is also apparently occurring in many developing countries (ISNAR, 2003).

Agricultural research policy is now less frequently coordinated and formulated in agricultural research institutions and is increasingly becoming the responsibility of government ministries or science and technology councils. In addition, agricultural research policy is increasingly being integrated into general science policy. When agricultural research institutions operate as commercial suppliers of research, for example under contract, they are likely to develop a strong client focus, moving close to the goals defined by their clients. Indeed, some institutions are implementing active commercial strategies in order to attain these goals. If the institutions' legal frameworks permit, their client base may very varied and include government ministries, regional and local government entities, industries and farmers' associations. This development is not welcomed in all circles (ISNAR, 2003). In the USA, for instance, researchers' commercial activities tend to reduce their creativity and their willingness to undertake basic research (Huffman and Just, 1999).

Over the past decade, the structure and organization of agricultural research have been subject to accelerated change. This reflects new ideas about interactions between the public and private sectors, such as the client focus (Persley, 1998). Some of the most rapid changes have occurred in the UK, where the government has sharply reduced its support for agricultural research. In other countries, such as the Netherlands, the government is abandoning institutional financing but still finances a substantial research program through contracts with long-standing research organizations that now function almost as private sector entities. Privatization is not the only means of improving control over agricultural research and client responsiveness. Innovative research methods are being established that combine public and private sector research. And scientific capacity is well maintained in the majority of EU-15 and North America thanks to the important role played by their universities.

Drivers

Major drivers for expansion of formal AKST organizations were industrialization, advances in technology and knowledge, and an optimistic view of societal benefits, affected by demand and mediated through policy (Alston et al., 1998; Van Keulen, 2007). Privatization was fostered by the introduction of Intellectual Property Rights, advances in genetics and new research policies (Alston et al., 1998). Public funding has taken a downturn since the mid 1970s mainly for the following two reasons: first, a general paradigm shift in the society towards a smaller role for public policy and a larger role for the marketplace and second, lesser societal benefits, eradication of food insufficiency, and a smaller share of GNP in NAE. This is true, although there is evidence of continued high returns to investments in public AKST (Alston et al., 2000). Many governments are giving AKST another opportunity to show its comparative advantage in contributing to emerging wider societal interests through innovative, interactive AKST, even if rewarding mechanisms still need further development (OECD, 1999). The limited contribution of AKST to public debate and policies during the recent decade is seen as a major challenge (OECD, 1999).

Growth in size, specialization, consolidation of food chain organizations and increasing domination by multinational corporations was driven first by industrialization and later by liberalization of international trade, mobility of capital and people, new technologies (Galizzi and Pieri, 1998) and by regulatory barriers discriminating small enterprises. Public AKST had at least as much importance as private R&D and market forces in bringing about changes in livestock specialization (but not in crop specialization), farm size and farmers' off-farm activities (Busch et al., 1984; Huffman and Evenson, 2001), supported by well-targeted agricultural policies. (Van Keulen, 2007). Differences among NAE regions have been mainly due to differences in political-economic history.

5.5.1.2 Uncertainties of the future

Public funding to develop AKST organizations

Success in meeting the challenge of changing societal demand, whether public or private, will crucially affect public and societal support for development of AKST in the future. Questions about the future concern the following: Will research questions be shared and will public sector research be increasingly oriented toward the generation of knowledge? Will the view of the societal potential for AKST widen to emphasize the notion of multifunctionality and ethical consumption in order to attract public acceptance for funding AKST? Will the share of agriculture in the GDP decline? Will food insecurity and the central role of NAE AKST beyond its borders turn the view of the societal potential of NAE AKST positive? Will AKST adjust its paradigms and image by adopting a wider, more diverse and flexible agenda to realize its comparative advantages in meeting the changing societal demand?

Will organization structures become flexible enough to promote changes in scopes and targets?

Role of private AKST organizations

Technological developments (such as functional foods, gene-tailored diets, photosynthesizing microbes for energy, GMOs, nanotechnologies, information technologies) tend to increase the role of private companies in science and technology, thus compensating the decline in public funding. However, the demand for public goods, including food security, will continue to grow. Policies determine whether the internalization of externalities make public goods economically rewarding to provide through private AKST. If not, will the companies cream off or manage to segment supply for different markets, thus better contributing to meeting the development and sustainability goals of this assessment? Will public and private AKST organizations manage to increase synergy and intermediate spaces? Or will public AKST develop the public goods and set regulations that constrain the private sector?

Dis/integration of organizations at global and national level and within AKST

On the one hand, the integration of KST and AKST is increasingly being sought (OECD, 1999). On the other hand, we have learned from the US success story of integrating of research, education and extension. These two targets may, in some cases, be contradictory. There is overall agreement on the need for integration within AKST to increase the multifunctionality of food systems and agriculture. This has to proceed on and among different levels, starting from policy coherence at the level of ministries and administrative bodies, to increased communication among food system actors and among disciplines within the formal knowledge systems. Interdisciplinarity is getting wide acceptance as the preferential strategy in the latter. This avoids the endless emergence of new sciences and borders through the unification of existing ones. However, there are multiple barriers to this kind of development, such as risks related to integration, especially for the necessary advancement of the disciplinary bases.

Views opposing integration are also being considered (Sumberg et al., 2003; OSI, 2006). Will the barriers and risks be avoided, and will integrative approaches in structural development of AKST organizations take over as predicted for universities (Väyrynen, 2006), possibly based on flexible models of interacting scientific communities (Lele and Norgaard, 2005)? Will incentives and tools be created for public NAE science and technology organizations to intensify links with CGIAR and NARS organizations, to ensure appropriateness and adoption of technologies?

De/centralization and consolidation

Physical distancing in food chain and regional specialization has economic benefits but has often had negative environmental and social impacts. Food and product chains and marketing channels seem to be diversifying, due to more varied product combinations and demand segments. Will these developments and the notion of multifunctionality cause a paradigm shift towards the development of diversification and integration within regions, or will farm and regional specialization continue, using new tools to meet the environmental and social challenges?

And do policies, demand and formal AKST lead to diversification of on-farm supply, or only operate at the regional or even national or international level, through comparative economic advantages? Where in this scenario is the lower limit of the economic scale set?

Will paradigms and their operationalization in policies and demand lead to centralized transnational organizations, possibly to the existence of a small number of discipline-based international centers of excellence for the whole world, generating knowledge, technologies and products and segmenting their activity for diverse markets? Or does contextually and local adaptation proceed through decentralization and regionalization of AKST?

Contextualization of AKST organizations

There are means to adjust the societal and organizational situation to the requirements of capital-intensive agricultural technology, a technology which is less appropriate for resource-poor farming communities as such. One example is the Grameen Bank, founded by Muhammad Yunus, 2006 Nobel Prize winner. The Grameen Bank provides microloans for the poor, with an adjusted guarantee system. Will such models for more diverse and contextualized organizational structures, e.g., banking systems, be developed and popularized?

5.5.1.3 Consequences for AKST

To achieve the development and sustainability (D&S) goals requires reconsideration of appropriate organizational structures for AKST. Societal support for the development of AKST and relevance for the crucial challenges demand broad dialogue and a broad range of perspectives implied in flexible, diverse, integrative organizational structures. The share of public and private sectors in AKST is decisive for the kind of public regulatory arrangements that best meets the public goals. Regulatory regimes can be limited to covering cover transparency and communication, or can set economic incentives for the mainly private organizations to promote the goals, or directly regulate their activity through rules and legislation or through a public organizational structure. Public economic incentives may increase feasibility and result in higher equity than full reliance on price premia paid by consumers. In any case, proactive policies are required to shape AKST organizations and their activity.

The integration of organizations of knowledge generation and dissemination can promote goals. However, if focus is on globally coherent and centralized policies and AKST organizations, the strengths and weaknesses of society will be very different than if focus is on locally coherent and decentralized policies and AKST organizations. Global models with few centers of excellence and top-down approaches in science might be better in meeting global environmental problems, while local horizontally-integrated models and bottom-up approaches might have greater social and cultural benefits. Integration among organizations representing AKST components may produce more traditional solutions that are still highly relevant for present actors, while linkages to KST components foster more substantial changes and innovations with higher risks and opportunities for meeting the D&S goals. Relevance and contextuality of the latter might depend on importance given to the social sciences.

5.5.2 Proprietary regimes

The private sector invests in agricultural research purely to make a profit. A legal framework that adequately protects intellectual property rights is therefore very important. Interacting factors determine the effectiveness of patents awarded in any country: (1) the scientific fields in which a patent can be obtained; (2) international treaties that guarantee the respect for patents awarded in other countries and vice versa; (3) the ability to maintain an obsolete patent; (4) the ability to sanction patent violations; and (5) the duration of patent protection (Ginarte and Park, 1997; ISNAR, 2003).

5.5.2.1 Ongoing trends

The assignment of intellectual property rights to living things is of relatively recent origin in developed countries. Vegetative propagated plants were first made patentable in the US in 1930. And the protection of plant varieties (or plant breeder's rights—PBRs), a new form of intellectual property, only became widespread in the second half of the 20th Century. Intellectual property laws vary from jurisdiction to jurisdiction, such that the acquisition, registration or enforcement of IP rights must be pursued or obtained separately in each territory of interest. However, these laws are becoming increasingly harmonized through the effects of international treaties such as the 1994 World Trade Organization (WTO) Agreement on Trade-Related Aspects of Intellectual Property Rights (TRIPs), while other treaties may facilitate registration in more than one jurisdiction at a time.

If the aim of plant variety protection is to provide incentives to breeders, one of the questions that arises is how the contribution of farmers to the conservation and development of plant genetic resources should be recognized and preserved. Building on the principles embodied in the CBD, PGRFA seeks to establish principles for facilitating access to plant genetic resources and establishing fair and equitable mechanisms of benefit sharing. The International Union for the Protection of New Varieties of Plants (UPOV) aims to encourage the development of new varieties of plants for the benefit of society by codifying intellectual property for plant breeders. In 2005, 58 countries had joined UPOV. For plant breeders' rights to be granted, the new variety must meet four criteria under the rules established by UPOV. The new plant must be novel, which means that it must not have been previously marketed in the country where rights are applied for. The new plant must be distinct from other available varieties. The plants must display homogeneity. The trait or traits unique to the new variety must be stable so that the plant remains true to type after repeated cycles of propagation. Protection can be obtained for a new plant variety how ever it has been obtained, e.g., through conventional breeding techniques or genetic engineering. (UPOV, 1991)

In 2001, the FAO Conference adopted the International Treaty on Plant Genetic Resources for Food and Agriculture. This legally binding Treaty covers all plant genetic resources relevant to food and agriculture and is in harmony with the Convention on Biological Diversity. The Treaty is vital in ensuring the continued availability of the plant genetic resources that countries will need to feed their people. Through the Treaty, countries agree to establish an efficient, effective and transparent Multilateral System to facilitate access to plant genetic resources for food and agriculture and to share the benefits in a fair and equitable way. The Multilateral System applies to over 64 major crops and forages. The Governing Body of the Treaty, which will be composed of the countries that have ratified it, will set out the conditions for access and benefit-sharing in a "Material Transfer Agreement" (MTA).

There have been several extensions of patenting, especially in the direction of patenting gene sequences, totally or partially. The United States has now issued patents on protein coordinates (i.e., on the result of physical measurements of proteins to define their precise shape). The monopoly that is actually claimed in these patents is the use of the measured coordinates in computer programs to attempt to model the interaction of the protein with other chemicals that might be candidates for therapeutics (Knoppers and Scriver, 2004).

5.5.2.2 Uncertainties of the future

Bits of information or research tools are contributions to product development, but economically, there is little or no independent value in these piecemeal inventions or discoveries. The economic value derives from the final product. Why can't firms or public research rely completely on biotechnology firms to improve their products? What kinds of incentives must be offered to develop new research tools in public research? What will be the impact on industry of products coming off patent? Will industry continue to be interested in high-risk low-payoff products or will it concentrate on blockbusters? Will the public procurement model be developed, especially for products such as vaccines?

How far will the World Intellectual Property Organization (WIPO) go to harmonize international patent law? Will patent law ever be harmonized? Will world patents be created? How far will the collective networks in the field of agricultural biotechnologies manage to achieve co-development and patents for novel technologies?

5.5.3 Access, control and distribution of Agricultural Knowledge, Science and Technology

In this subchapter we look at what kind of major arrangements there are for access, control and distribution of AKST, how they evolved until now and why: which were the main drivers? How do they differ among North American, European Union and non-EU Eastern European countries and Russia and why: what kind of differences were there in the drivers?

Access, control and distribution of AKST covers issues of funding and management of formal AKST structures, participation of different stakeholders and beneficiaries in agenda setting, R&D processes, interpretation and application of results, dissemination, extension and communication processes, relevance of solutions, appropriateness of technologies and options for spillovers for different beneficiaries. The futures of access, control and distribution of AKST is very much influenced by the futures of actors of the KST systems and models of knowledge production (5.3).

5.5.3.1 Ongoing trends

Access of farmers was arranged in USA through decentralized, integrated AKST and in Russia and part of CEE through top-down "chain-of-command" with no public extension service (Miller et al., 2000). Decline in public funding has been linked with even higher decline in public control of AKST since the 1980s. The role of the private sector has increased in the management of public funds and publicly funded and performed R&D, with a decreasing net flow of public funds to private research (Alston et al., 1998). Due to privatization, there is less focus on farm-level technologies and on equity and distributional issues and on public goods (Alston et al., 1998; BANR, 2002) and less AKST is available in the public domain. Again on public support, only £219 million of the annual UK government subsidy of £3102 millions to agriculture (not including the additional subsidies for foot and mouth disease) was used to create positive externalities (Pretty et al., 2005). Farmers' influence and participation since WWII declined but has recently been increasing (Romig et al., 1995; Walter et al., 1997; Wander and Drinkwater, 2000; Groot et al., 2004; Morris, 2006; Ingram and Morris, 2007). However, technologies have sought to increase the scale of food chain actors and the industrialization of the farm sector, and are less appropriate for poor farming communities (Alston et al., 1998; BANR, 2002). The power of the retail end of the food chain has increased, but whether consumers now have more influence is open to debate (Buhler et al., 2002).

Since the 1970s, competition and short-termism have been penetrating in public AKST to broaden its scope and make it more transparent and efficient (Alston et al., 1998; Buhler et al., 2002). We might ask whether economic efficiency has failed to reach its goal. (Buttel, 1986; Huffman and Just, 1999, 2000) According to creativity research, extreme competition and lack of safety are a serious threat to creativity and true innovation. Recently, governments have been shifting towards funding multi-annual programs and long-term thematic areas with a considerable stakeholder involvement in the process and stronger links among AKST components, to increase efficiency and reduce fragmentation of solutions (OECD, 1999). The target is seen in innovative, interactive AKST, and the role of AKST in becoming a partner by contributing to the decision-making processes rather than prescribing optimal solutions (OECD, 1999).

Drivers

A major driver for privatization (see also 3.4.1) was the shift in paradigms towards the recognition of markets as better regulators than policies. The consequent "laissez-faire" role of the government in the management of the national economy (Alston et al., 1998) led to budget cuts and to the protection of space for large private companies to act through regulations, e.g., pesticide regulations and IPR (Bauer and Gaskell, 2000). Trade liberalization contributed to giving more voice to transnational companies. Advances in genetics and intentional research policy (Alston et al., 1998) enhanced control by the private sector. The failure of public AKST to serve all the target groups might have left empty niches for private companies too. These developments and the imposition of more targets for low-income countries as a precondition for support (e.g., the increase of restricted funding for CGIAR; World Bank, 2003) left more room for NAE policies after WWII also beyond NAE. The growing position of NGOs in AKST since the 1970s was a reaction to negative externalities, which over and above the increased role of agri-business, again contributed to short-termism and competitive grants. The re-emergence of longer term and bigger programs was fostered by strife for governmental efficiency. A paradigm of "new public management" increased stakeholder participation in the 1990s: no more, or less, government, but better government, implying more enlightened regulation, improved service delivery, devolution of responsibility, openness, transparency, accountability and partnership (OECD, 1999).

5.5.3.2 Uncertainties of the future

There are a number of uncertainties for the future of AKST access, control and distribution in North America and Europe and thus for the impacts on development and sustainability goals at global level.

Privatization. Public goods, the poor and hungry, and rural livelihoods are target groups with the least voice on the market at present, and the private sector is led by markets. Markets can be directed to work for the social optimum through internalization of externalities, i.e., including the negative and positive externalities, in prices. Instruments include penalties (Jackson, 2005), reallocation of all taxes, subsidies and incentives, and institutional and participatory mechanisms (Pretty et al., 2001). Regulation can be used to set limitations.

Will private sector control in NAE AKST continue to grow, or will the public sector take more control, either through direct funding and control, or by helping the market forces to work for social optimum in terms of sustainability and food security? How will incentives to supply public goods through multifunctional farming be created: regulations, internalizing externalities by reallocating subsidies and taxes, creating new markets, e.g., for GHG emission quotas, or consumer certificates and price premia?

Integration of perspectives. Access, control and distribution of AKST does not only depend on who pays, they also depend on the perspectives and competences represented in AKST processes. There is evidence of reduced efficiency due to excessive introduction of competition and short-term thinking in management of formal public science and development structures (Huffman and Just, 2000). The risks of short-term thinking are especially serious with regard to learning-intensive integrated approaches and sustainability objectives which have an inherent long-term perspective. Will time-consuming and learning-intensive integration win the fight for paradigm shift or become impossible in a context of potentially declining resources and growing competition based on expert values, disciplinary quality and merit criteria?

Control by beneficiaries: The perspectives of solvent, large-scale industry might steadily be given more emphasis in the knowledge networks if public funding declines and if there are no new creative solutions to diversify perspectives. Multifunctionality of agricultural production and diversification of marketing channels and actor networks could decrease dependence on one market and thus give farmers and the supply chain a better position to negotiate with other actors on the market. Locally-oriented AKST might require less public support to achieve influence and outcomes equal to that of globally-oriented AKST.

Do policies, demand and formal AKST lead to diversification of supply and distribution channels and thus increased independence from retail, mainly at the farm, regional or national level?

Will the responsibility be put on consumers and other actors, or will more emphasis be placed on public control as a means to enhance sustainable consumption? Will the competence and viewpoint of beneficiaries with the least voice in low-income countries—the poor and hungry—be integrated in knowledge and technology generation in the worldwide influential NAE AKST, to prevent past failures and to shape future food systems to meet D&S goals?

Dissemination of information. In a situation of increasing transfer of control from political decision-makers to the market, adequate, accessible market information is essential. Well-informed choices by consumers and other food system actors through education of "food competent citizens" is a precondition for promoting D&S goals through consumer choices. Appropriate standards and price premiums create incentives. The option of different consumer segments to influence on the market is not equal, but depends on their purchase power. In addition to economic barriers there are social and psychological barriers for consumption (Jackson, 2005).

Will the dominant trend for down-sized, client-charged information to farmers continue with the increasing niche being filled in by agri-business companies, or will there be a demand for independent extension services? Or will the increasingly integrative approaches and structures extensively incorporate clients in interactive communication networks to generate and utilize knowledge and technologies and thus decrease the significance of separate extension services?

Will the opportunities offered by modern communication and information technologies be successfully utilized to increase communication and enhance access to knowledge, technologies and markets, avoiding further growth of the "digital barrier"?

5.5.3.3 Consequences for AKST

More and more agri-business companies are transnational, thus creating a risk of homogenization of practices and less competitiveness for resource-poor farmers. Yet access to knowledge, technology and resources requires participation in AKST processes through equal dialogue, among the various beneficiaries, with their specific value systems, perspectives and skills. This requires a shift from technology transfer approaches to interactive social learning networks. Such a shift is easier to introduce in the more local than global agricultural innovation systems. A word of caution: equity in the daily environment can hide the consequences of global disparities. Global equity requires effective global communication networks based on modern technologies and inter-regional global regulatory frameworks. Meeting D&S goals more broadly will require the integration of varied perspectives of ecological, social and economic sustainability, different parts of the food and non-food chains, and various stakeholders. More emphasis on policies allows for effective internalization of externalities also in terms of D&S goals, while less regulation requires more emphasis on education, information and standards and tends to lead to lower market equity. Regionally and locally oriented AKST systems enhance transparency and direct feedback from consumers, (citizens and communities), as well as from local ecosystems to the production entities and thus complement regulatory and information systems.

5.6 Future AKST Systems and Their Potential Contributions to Sustainable Development Goals

5.6.1 Four normative agricultural innovation systems

Despite ongoing trends, there are many uncertainties about the futures of indirect and direct drivers of AKST systems in North America and Europe. Sketching four normative futures shows that there is no one best future because the future is a realm of freedom, power and will (de Jouvenel, 2004) and depends on the strength of the actor(s). Not all options are compatible and coherent. Choices will have to be made. Reality will probably be a mix of options.

5.6.1.1 Market-led AKST

The S&T policies of North America, the European Union and Russia and the non-EU Eastern European countries converge and favor the private sector. Market-led AKST decreases hunger and poverty and improves nutrition and human health in NAE and at international levels. However, it contributes little to equity and sustainable economic development.

Multinational Corporations (MNCs) in association with a few universities and small innovative firms develop and fund most AKST. Elite research groups throughout North America and Europe form technology clusters with firms. Research is not location specific. It is done where human resources are the best. MNCs and a few universities control and sell most AKST. Important research investments are made to support two markets: functional food for the high revenue consumers and inexpensive safe food for the low revenue customers. International agricultural knowledge centers conduct most public research. The European Research Space is a great success. Centers of excellence at international level associate R&D public institutes and major firms with the objective of developing new activities or markets through innovation.

Private companies benefit from strong intellectual property rights and the privatization of living organisms. Legislation makes it possible for universities, non-profit organizations and small businesses to keep ownership of intellectual property developed with the support of public funds. Common regulations and standards are designed to facilitate generation and distribution of knowledge. Tax incentives encourage companies to invest and to collaborate with each other and with universities. Large vertically integrated firms own farm enterprises and control access and distribution of inputs and capital.

As far as the generation of knowledge is concerned, production and problem-oriented multidisciplinary work is encouraged. Despite managerial discourses on sustainable development, AKST generated by MNCs does not prevent certain areas, such as marine ecosystems and biofuels, from being left out of research agendas. Therapeutic successes and widespread application of nanotechnologies lead progressively to a global conception of nature and life. The frontiers between the different worlds of human beings, animals and plants are fading.

5.6.1.2 Ecosystem-oriented AKST

In ecosystem-oriented AKST, there is no clear demarcation between university science and industrial science, between basic research, applied research and product development, or even between careers in the academic world and in industry. Ecosystem-oriented AKST can make a major contribution to at least three development and sustainability goals:

1. Environmental sustainability by the development of novel, knowledge-intensive and resource use-efficient technologies,
2. Sustainable economic development, by investing human and financial capital in the development of "green technologies", and
3. Enhanced livelihoods and equity by developing a broad range of technologies (both low and high cost) and by making these widely accessible so that also poor and small farmers can benefit from them.

Many subsidies and most trade barriers have been eliminated. Support payments reward farmers for the provision of services other than food. In the EU and North America, agricultural policies promote the multifunctional nature of agriculture and the improvement of natural resource quality through strict adherence to stricter environmental regulations. In Eastern European countries and Russia, governments and farmers' associations are conscious of the disasters created by excessive usage of agrochemicals combined with poor infrastructures. Drastic reforms are being implemented to improve environmental policies.

Laws facilitate the ownership of knowledge by all those who have contributed to this ecosystem-oriented AKST. Policies support increased scientific cooperation among NAE countries. Special emphasis is on strengthening cooperation within NAE, especially EU and North America with Eastern European countries and Russia. Innovation, public/private interactions and collaboration with the less developed countries is also being encouraged. Researchers collaborate with a broader range of organizations and disciplines. Problem-oriented, demand-driven approaches prevail, and there is a great deal of research integration (multidisciplinary and interdisciplinary work, systems approach). Incentives are being used to attract young students to science and technology, especially the environmental and agricultural sciences. Efforts are being made to promote new scientific fields in universities and to renew interest in important fields that have been ignored.

AKST increasingly serves homogeneous consumer preferences and diets. Lifestyles and social awareness are boosting the demand for convenience and functional foods. Although the demand for organic products is going up, the new technical, convenience-led food solutions (e.g., ready-meals) clearly predominate. Efforts are being made to increase national and international budgets for more research and cooperation world-wide concerning access, control and distribution of inputs. Research investments are concentrated on global and regional centers of excellence conducting both basic and applied research. Emphasis is on investments that support a knowledge- and bio-based economy.

In the field of climate change mitigation and adaptation, policies related to spatial planning stimulate the reduction of greenhouse gas emissions and protect NAE against climate change. Spatial planning has led, for instance, to the diffusion of new technologies such as floating greenhouses (e.g., in the Netherlands in response to the rising sea level), non-animal meats, or low-emission animal farms (to avoid pollution) and roof farming (natural cooling in urban areas). At the same time, conventional agricultural techniques are being further improved with considerable effort to heighten resource use efficiency, especially for water, nutrients and energy (precise provision in time and space). In many regions, farms specialize in either specific livestock or arable farming, depending on their local soil and climatic conditions.

Relatively inefficient cultivation of biofuel crops e.g., rapeseed oil, barley, sunflower, has been replaced by second generation biofuel production. Agriculture is both an energy producer and an efficient energy consumer. However, the energy-producing capacity of agriculture is outweighed by other more centralized and technology-intensive renewable forms of energy such as artificial photosynthesis (combining sub-processes of photosynthesis), a favored source in large-scale energy labs. Many farms are able to cover their energy needs and costs by producing biofuels and installing eolian and solar parks on their fields. New knowledge allows for the sustainable production of biofuels and innovative, environmentally-friendly farming systems.

Results from research into knowledge-intensive technologies supported by information technologies (such as GIS, remote sensing, GPS-controlled robots, detailed soil databases, etc.) allow wide implementation of precision farming. Food processing is taking place in new energy- and/or labor-saving forms, such as intelligent greenhouses (with virtually no labor) and multistory food factories—as developed in the Netherlands (agrometropoles). GMOs are widely accepted (but less in EU-15 than in America and Eastern Europe) and play a significant role in reducing sions from agriculture to the envir

Research is also done to bette and circumstances that influence choices. This information leads to preferences. Advances in researc balance of foods and optimize nutritional teractions in crops and livestock. Better understanding of the system leads to improvements in regulatory frameworks.

5.6.1.3 *Local food-supply led AKST*

Local food-supply AKST is a multi-actor system with little coordination between organizations: the AKST systems in North America, the European Union and the non-EU Eastern European countries are very different from one another. The AKST systems manage to contribute to improved nutrition and human health at national level, but most rural areas are driven by urban economies. The importance of agriculture in rural activity differs between regions. At the international level, AKST systems have little impact on hunger, poverty and environmental sustainability.

No coherent research, innovation and IPR policies are designed in NAE, and the policies there are, are not always consistent at the national level. Each country has its own distinctive educational and cultural features. Efforts are being made to improve secondary education and to put students through the first years of universities, but not many students become science majors. The quality and quantity of research personnel is deteriorating.

In most countries, the access, control and distribution of knowledge, science and technology remains linear. Fundamental research, applied research, extension and education are done in separate organizations. There is little synergy among the many different types of organizations involved. A few large private companies have their own research capacities and are highly integrated. However, as their investments are relatively small, they cannot influence the global research agenda. In the USA and Canada, land grand universities are fading away because of the competition for scarce funding. In the countries of the European Union, governments continue to provide some funding for public research to avoid conflicts with farmers and researchers, but funds given to KST in real terms are below what they were at the beginning of the century. Local universities and public research organizations continue to provide public goods; however they are often in conflict with private companies and accuse them of privatizing knowledge. In Russia and non-EU Eastern European countries, AKST is not a priority; the little research that is done focuses on the large-scale cereal-vegetable farming systems.

The size of holdings varies greatly which explains the great inequalities in access, control and distribution of inputs and capital. Family farms are still the most prevalent, but they have limited access to inputs and capital.

Knowledge generation mainly concerns conventional food production and protection. Except in North America, little is done to investigate or use genetically modified crops and animals. Research tends to ignore growing problems such as water scarcity, soil depletion and socioeconomic viability of agricultural systems.

Local-learning AKST

learning AKST is regionally focused and proactive in eting local development and sustainability goals. It is a well coordinated multi-actor system that successfully integrates the different goals at regional and local levels. It successfully contributes to the goals of enhancing livelihoods, equity and social capital and environmental sustainability. Nutrition and human health are improved through knowledge-based sustainable, fresh and safe local diets and a reduction in meat consumption. Balanced regional economic development and stewardship of natural resources are promoted by keeping the added value and employment of input production, processing, transportation and marketing in the region and through investments in quality growth and welfare services. Due to the local orientation, there is little exportation of products or knowledge outside of NAE, but more resources of low-income countries are left untouched by NAE so they can serve other purposes including the provision of food, fiber and fuel for their own consumption. Nevertheless, many technologies developed for NAE could be appropriate for resource-poor rural communities also in low-income countries.

Policies and governance are based on cooperation among different sectors, utilizing trans-ministry and public-private platforms, i.e., regional food, agriculture, health, environment, rural, trade, and KST policies are fully integrated. Development is knowledge-intensive, and the importance of science policies is widely recognized. Environmental policies are increasingly focused on local and regional issues rather than on global change issues. Agricultural policies allocate subsidies to internalize positive ecological, socio-cultural and economic (widening of spatial and temporal scales) externalities. Diverse and flexible financing and credit systems flourish, and rural capital is primarily addressed to serve local/regional rural needs. Systems to balance regional imparities in capital supply are being created. Global issues are being addressed thereby enhancing understanding through worldwide regional networks and, consequently, learning from and developing local solutions. Intellectual protection is not strict, and therefore many research results are available for less developed countries, and gene resources are owned by local communities. National and international trade is open, but the effects of internalized factors pertaining to climate and energy resources push up transportation prices. Intensive use of modern communication technologies and rural and nature tourism can replace long-distance traveling and, furthermore, broaden the mindset and provide entertainment. Regarding development collaboration, each sub-region of NAE has close links to its neighboring countries to the south. Universities and the private sector are encouraged to pool patents through licensing, moreover licensing is free for the developing world.

The agrifood system actors (producers, traders, processors, waste managers, input producers, financers, institutional kitchens and private consumers), together with citizens, NGOs (representing public goods), municipalities, county agencies and scholars form an interactive, open learning network with different platforms designed for different needs. These networks are connected with the networks of other regions of the world on the basis of interests/needs/goals or to connect actor groups/professions/competences. The regional networks work closely with regional, decentralized university systems to develop local and regional agrifood systems. They utilize the international knowledge networks and carry out disciplinary and interdisciplinary research. The networks are linked with the boards of universities; they provide funds and participate in formulating the agendas, planning and performing knowledge and technology generation, and interpreting and evaluating results. The interactive networks ensure that the generated knowledge and technology are highly relevant, locally adapted and socially contextual. They also ensure that agrifood system actors have full access to the results and get the necessary underlying understanding and technical knowledge from the universities.

Within the universities, disciplinary science communities and cross-cutting interdisciplinary science communities utilizing the developments of disciplinary work, systematically interact with trans-disciplinary stakeholder platforms. Research leads to collaborative, reflexive, democratic processes to develop sustainable, local food systems. Progress provides the capacity to internalize externalities such as food, fiber and fuel that enable consumers to make knowledgeable choices. Purportedly open to all citizens, education emphasizes increased understanding of different values and goals, the multiple impacts of food choices and communicational and team working abilities. In scholarly education, attention is given to systems, interdisciplinary and participatory approaches, a robust scientific-philosophical base and conceptual tools to promote understanding of and communication across different disciplinary paradigms. Advanced communication technologies are used at that level. Universities also interact with actors from low-income countries to integrate their views in knowledge generation and to strengthen their capacities.

AKST serves diverse, locally- and regionally-adapted, sustainable dietary and food, fiber and energy systems. Health and nutrition draw on a knowledge-based understanding of farming systems and of local diets, composed of fresh, seasonal foods rather than gene-tailored, functional food ingredients. Consumers appreciate the cultural heritage. They rely on and ensure protection for the local and regional ecosystems, with their goods and services. Local bioenergy and renewable energy-based, energy-efficient and integrated agrifood systems are being developed and continuously improved. Predominant farming systems are based on biologically fixed nitrogen, recycling materials (nutrient cycling) and energy flows within local agriculture and as returns from the local demand-chain that includes processing, and from watercourses. Thus, bioenergy, food and also wood production are integrated, and their waste is used for energy and fertilizers. Small-scale solar and eolian energy sources are connected in the regional electricity network. New plant and animal varieties are developed; those fit in with the integrated systems and often carry the significant amount of diversity needed to adapt to different locations. Urban agriculture is an inherent part of spatial and city planning. Regional and local food processing and retailing outfits utilize farm- and waste-based energy and have local contract networks to purchase inputs. Life-cycle and sustainability assessments are carried out on the impact of land-use changes and features of production and food systems, but the emphasis is on direct communication and feedback from local communities and ecosystems.

Local and regional markets that are being developed give special attention to energy-efficient logistical arrangements. Different forms of community-supported agriculture, with shared risk and labor between producers and consumers, food circles, farmers' markets and direct sales flourish besides the horizontally integrated production-trade-consumption chains. The use of fossil energy for transportation is reduced accordingly and the added value of the food chain is kept in the region. Externalities are internalized, but that does not only depend on public regulation, taxation and economic incentives with regional variation. An important part of internalization depends on the proximity of different actors, mutual trust and social capital, and thus on direct communication and feedback from the local socioecological context. Local labels embracing the whole chain are being successfully introduced, and regional marketing ensures an adequate sales level.

5.6.2 Towards options for action

Choices about agricultural knowledge, science and technology (AKST) relate to paradigms, investment, governance, policy and other ways to influence the behavior of producers, consumers and the rest of the food chain actors. They will have powerful impacts on which development and sustainability goals are achieved and where, both globally and within NAE. It is unlikely that all development and sustainability goals can be achieved in any of these futures.

Outlining these four normative agricultural innovation systems before proposing options for actions should help decision makers to make coherent choices. As Seneca wrote "There is no favorable wind for the person who does not know where he wants to go." Knowledge about ongoing trends, uncertainties and possible AKST systems should help decision makers to choose among options for actions presented in chapter 6. Appropriate AKST investments and policies will require an appropriate mix of strategies that are in line with the potentials and constraints of different NAE regions and countries, but they must also address the broader changes taking place.

References

Adger, W N., N.W. Arnell, and E. Tompkins. 2005. Successful adaptation to climate change across scales. Global Environ. Change 15(2):77-86.

Akrich, M., and R. Miller. 2007. The future of key actors in the European research area: Synthesis paper. Technology Foresight Group, DG Research, European Commission, EU 22961 EN.

Alegre, H., W. Hirner, J.M. Baptista, and R. Parena. 2000. Performance indicators for water supply services. Manual of Best Practice Ser., IWA Publ., London.

Alston, J.M., P.G. Pardey, and V.H. Smith. 1998. Financing agricultural R&D in rich countries: What's happening and why? Aust. J. Agric. Resour. Econ. 42(1):51-82.

Alston, J.M., M.C. Marra, P.G. Pardey, and T.J. Wyatt. 2000. Research returns redux: A meta-analysis of the returns to agricultural R&D. Aust. J. Agric. Resour. Econ. 44(2):185.

Amin, A., and P. Cohendet. 2004. Architectures of knowledge: Firms, capabilities, and communities. Oxford Univ. Press, Oxford.

Anania, G., 2006. An assessment of the major driving forces in the area of "economy and trade" which will contribute to shape the future of agriculture in Europe. Short Report to the SCAR Expert Working Group/EU Commission. Foresighting in the field of agricultural research in Europe. EU, Brussels.

Aoki, M. 2001. Toward a comparative institutional analysis. MIT Press, Cambridge, MA.

Arbatova, N. 2007. Russia-EU beyond 2007. Russian domestic debates. Russie.Nei.Visions No. 20. IFRI Russia/NIS Center.

Ashley, R., and A. Cashman. 2006. The Impacts of change on the long-term future demand for water sector infrastructure. Infrastructure to 2030: Telecom, land transport, water and electricity. OECD, Paris.

Atkinson, R.C., R.N. Beachy, G. Conway, F.A. Cordova, M.A. Fox, Holbrook, et al. 2003. Public sector collaboration for agricultural IP management. Science 301:174-175.

Bainbridge, W.S., and M. Roco (ed). 2006. Managing nano-bio-info-cogno innovations. Converging technologies in society. NSF. Available at http://www.wtec.org/ConvergingTechnologies/3/NBIC3_report.pdf.

BANR. 2002. Publicly funded agricultural research and the changing structure of US Agriculture Board on Agriculture and Natural Resources. Committee to review the role of publicly funded agricultural research on the structure of US agriculture. National Res. Council, Washington, DC.

Batjes, N.H. 2000. Soil degradation status and vulnerability assessment for Central and Eastern Europe—Preliminary Results of the SOVEUR Project. Proc. Concluding workshop. Busteni. Report 2000/04. FAO, Rome.

Bauer, M.W., and G. Gaskell (ed). 2002. Biotechnology: The making of a global controversy. Cambridge Univ. Press, Cambridge.

Bibel, W. 2005. Information technology. Background paper for the European Commission "Key Technologies" Expert Group. DG Research, EC, Brussels.

Birt, C. 2007. A CAP on health? The impact of the EU Common Agricultural Policy on public health. Faculty of Public Health, London.

Björklund, J., K.E. Limburg, and T. Rydberg. 1999. Impact of production intensity on the ability of the agricultural landscape to generate ecosystem services: an example from Sweden. Ecol. Econ. 29(2):269-291.

Björklund, S. 2004. Ecosystem services in an agricultural context. p. 21-24. *In* G. Agerlid (ed), Ecosystem services in European agriculture—theory and practice. Swedish Acad. Agric. Forestry 143:1.

BMBF. 2003. Futur: Der deutsche Forschungsdialog. Eine erste Bilanz. Bundesministerium für Bildung und Forschung. Bonn, Germany.

Botkin, J.W., M. Elmandjra, and M. Malitza. 1979. No limits to learning: Bridging the human gap. A report to the Club of Rome. A. Wheaton and Co., Exeter.

Bouma, J., J. Stoorvogel, R. Quiroz, S. Staal, M. Herrero, W. Immerzeel, et al. 2007. Ecoregional research for development. Adv. Agron. 93:257-311.

Boussard, J.M., F. Gérard, M.G. Piketty, M. Ayouz, and T. Voituriez. 2006. Endogenous risk and long run effects of liberalization in a global analysis framework. Econ. Modelling 23:457-475.

Bouwman, A.F., T. Kram, and K.K. Goldewijk (ed). 2006. Integrated modelling of global environmental change: An overview of IMAGE 2.4. Rep. 500110002. MNP, Netherlands.

Braun, J., M.W. Rosegrant, R. Pandya-Lorch, M.J. Cohen, S.A. Cline, M.A. Brown, and M.S. Bos. 2005. New risks and opportunities for food security. Scenario analyses for 2015 and 2050. 2020 Disc. Pap. 39. IFPRI, Washington, DC.

Britz, W., T. Heckelei, and I. Perez. 2006. Effects of decoupling on land use: An EU wide, regionally differentiated analysis. Agrarwirtschaft 55(5/6):215-226.

Brouwer, F. 2006. Main trends in agriculture. Policy Brief 1 (D14). LEI Agric. Econ. Res. Inst., Wageingnen.

Brundenius, C., B. Göransson, and J. Agren. 2006. The role of academic institutions in the national system of innovation and the debate in Sweden. Paper presented at Universidad 2006, 5th Int. Cong. Higher Educ. Cuba, 13-17 Feb 2006.

Bruinsma, J. (ed). 2003. World agriculture towards 2015-2030. An FAO perspective. Available at http://www.fao.org/documents/show_cdr.asp?url_file=/docrep/005/y4252e/y4252e00.htm. FAO, Rome.

Bruun, H., J. Hukkinen, K. Huutoniemi, and J.T. Klein. 2005. Promoting interdisciplinary research. The case of the Academy of Finland. August 2005. Academy of Finland, Helsinki.

Buhler, W., S. Morse, E. Arthur, S. Bolton, and J. Mann. 2002. Science, agriculture and research, a compromised participation? Earthscan, London.

Burch, D., and G. Lawrence. 2005. Supermarket own brands, supply chains and the transformation of the agri-food system. Int. J. Soc. Agric. Food, 13(1):1-18.

Busch, L., J.L. Silver, W.B. Lacy, C.S. Perry, M. Lancelle, and S. Deo. 1984. The relationship of public agricultural R&D to selected changes in the farm sector. Rep. Nat. Sci. Foundation. Univ. Kentucky, Lexington.

Butler Flora, C. 1998. Skills for the 21st century relation-building. Rural Dev. News. Summer 1998.

Buttel, F. 1986. Biotechnology and agricultural research policy: Emergent research issues. p. 312-347. *In* K. Dahlberg (ed), New directions for agriculture and agricultural research. Rowmand & Allanheld, Totowa, New Jersey.

Carter, T. (ed). 2007. Assessing the adaptive capacity of the Finnish environment and society under a changing climate: FINADAPT. Summary for Policy Makers. Available at www.environment.fi/publications. Finnish Environ. Inst., Helsinki.

Catlett, L. 2003. Futurist view of American agriculture. New Mexico State Univ.

CEC. 2002. Towards a thematic strategy for soil protection. Communication from the Commission to the Council, The European Parliament, The Economic and Social Committee and the Committee of the Regions. COM (2002) 179 final. Commission of the European Communities, Brussels.

CEC. 2005. Green Paper Confronting demographic change: A new solidarity between the generations. COM (2005) 94 final. Commission of the European Communities, Brussels.

CGIAR Science Council. 2005. Science for agricultural development. Changing contexts, new opportunities. Available at www.sciencecouncil.cgiar.org. CGIAR Science Council, Rome.

Chambers, N., C. Simmons, and M. Wackernagel. 2000. Sharing nature's interest: Ecological footprints as an indicator of sustainability. Available at http://www.ecologicalfootprint.com. Earthscan, London.

Collomb, P. 1999. A narrow road towards food security in 2050. Available at http://www.fao.org/DOCREP/003/X3002F/X3002F00.htm. FAO, Rome.

Cook, M.L., and F.R. Chaddad. 2000. Agroindustrialization of the global agrifood economy: Bridging development economics and agribusiness research. Agric. Econ. 23:207-218.

Cranfield, J.A.L., T.W. Hertel, J.S. Eales, and P.V. Preckel. 1998. Changes in the structure of global food demand. Amer. J. Agric. Econ. 80(5):1042-1050.

Cristoiu A., M. Canali, and S. Gomez y Paloma (ed). 2006. Prospects for the agricultural income of European farming systems: Summary results. EUR Number: 22506 EN. EC and Inst. Prospective Tech. Studies (IPTS), Seville.

Cuhls, K. 2006. Science, technology and innovation drivers. Foresighting in the field of agricultural research in Europe. SCAR Expert Working Group/ EU Commission, Brussels.

Daily, G. (ed). 1997. Nature's services. Societal dependence on natural ecosystems. Island Press, Washington, DC.

De Boer, J., M. Helms, and H. Aiking. 2005. Protein consumption and sustainability: Diet diversity in EU-15. Ecol. Econ. 59(3):267-274.

De Fraiture, C., D. Wichelns, J. Rockstrom, and E. Kemp-Benedict. Scenarios and future outlook. 2006. Chapter 4. *In* D. Molden (ed), Water for food, water for life. A comprehensive assessment of water management in agriculture. Earthscan, New York.

De Jouvenel, H. 2004. An invitation to foresight. Futuribles Perspectives Ser. Éditions Futuribles, Paris.

Delgado, C.L., N. Wada, M.W. Rosegrant, S. Meijer, and M. Ahmed. 2003. Fish to 2020: Supply and demand in changing global markets. Available at http://www.ifpri.org/pubs/books/fish2020/oc44.pdf. IFPRI, Washington, DC.

De Lattre-Gasquet, M. 2006. The use of foresight in agricultural research. *In* L. Box and R. Engelhard (ed), Science and technology policy for development. Dialogues at the interface. Anthem Press, London.

Deutsch, L. 2004. Global trade, food production and ecosystem support: Making the interactions visible. Dep. Systems Ecol. Stockholm Univ., Sweden.

Dezhina, I. 2005. Russian scientists: Where are they? Where are they going? Human resources and research policy in Russia. Russia.Cei.Visions No. 4. IFPRI, Washington, DC.

Diamond, J. 2005. Collapse. How societies choose to fail or survive. Penguin Books, London.

Dixon, J., A. Gulliver, and D. Gibbon. 2001. Global farming systems study: Challenges and priorities to 2030—Synthesis and global overview. Consultation Document. World Bank, Washington, DC, and FAO, Rome.

Dohm, A. 2005. Farming in the 21st century. A modern business in a modern world. Occup. Outlook Quar.. 49(1):18-25. Available at http://www.bls.gov/opub/ooq/2005/spring/art02.pdf.

Dorward, A., J. Kydd, J. Morrison, and I. Urey, 2004. A policy agenda for pro-poor agricultural growth. World Dev. 32(1):73-89.

Dries L., T. Reardon, and J.F. Swinnen. 2004. The rapid rise of supermarkets in Central and Eastern Europe: Implications for the agrifood sector and rural development. Devel. Policy Rev. 22(5):525-556.

DuPuis, E.M., and D. Goodman. 2005. Should we go "home" to eat? Toward a reflexive politics of localism. J. Rural Studies 21(3):359-371.

Easterling, W.E., B.H. Hurd, and J.B. Smith. 2004. Coping with global climate change. The role of adaptation in the United States. Pew Center on Global Climate Change, Washington, DC.

Eberstadt, N., 2007. Etats-Unis: l'exception démographique. Éditions Futuribles (Paris) 333:19-34.

EC. 2004. The agriblue blueprint. Sustainable territorial development of the rural areas of Europe. Dissemination conference. Sept 2004. EC, Brussels.

EC. 2005. Frontier research: The European challenge. High-Level Expert Group Report. EUR 21619. EC, Brussels.

EC. 2006. Emerging science and technology priorities in public research policies in the EU, the US and Japan. Final report EUR 21960. Available at http://ec.europa.eu/research/foresight/pdf/21960.pdf. EC, Brussels.

EC. 2007. Scenario 2020. Scenario study on agriculture and the rural world. No. 30 CE 0040087/00-08. Available at http://ec.europa.eu/agriculture/publi/reports/scenar2020/index_en.htm. EC, Brussels.

EEA. 2004. Greenhouse gas statistics. Available at http://www.eea.eu.int. EEA, Copenhagen.

EEA. 2005. European environment outlook. EEA 410 Report No 2/2005. Eur. Environ. Agency, Copenhagen.

EEA. 2006. How much biomass can Europe use without harming the environment? Available at http://reports.eea.europa.eu/briefing_2005_2/en/briefing_2_2005.pdf. Eur. Environ. Agency, Copenhagen.

EIROnline, 2005. Industrial relations in agriculture. Available at http://www.

eurofound.europa.eu/eiro/2005/09/study/tn0509101s.htm.

Elliasson, K. 2004. American science—The envy of the world? An overview of the Science System and Policies in the United States. A report commissioned by the Swedish Ministry of Education and Science. ITPS, Sweden.

ETP. 2005. The vision for 2020 and beyond. Available at http://etp.ciaa.be/documents/BAT%20Brochure%20ETP.pdf. EU, European Technology Platform on Food for Life.

Ewert, F., M.D.A. Rounsevell, I. Reginster, M.J. Metzger and R. Leemans. 2005. Future scenarios of European agricultural land use: I. Estimating changes in crop productivity. Agric. Ecosyst. Environ. 107:101-116.

FAO. 2006. World agriculture towards 2030-2050. An interim report. Prospects for food, nutrition, agriculture and major commodity groups. FAO, Rome.

FAO. 2007. FAO urges action to cope with increasing water scarcity. Improving agricultural practices key. Newsroom, 22 Mar 2007. FAO, Rome.

FAOSTAT. 2003. Available at http://faostat.fao.org/default.aspx?alias=faostat&lang=fr (accessed 6 November 2007)

FAPRI. 2005. U.S. and world agricultural outlook. Staff Report 1-05. Available at http://www.fapri.iastate.edu/outlook2005/text/FAPRI_OutlookPub2005.pdf. Food Agric. Policy Res. Inst., Iowa State Univ. and Univ. Missouri-Columbia, Ames.

FFRAF. 2007. Foresighting food, rural and agri-futures. Available at http://ec.europa.eu/research/agriculture/scar/pdf/foresighting_food_rural_and_agri_futures.pdf. FFRAF report, EU, Brussels.

Fields, S. 2004. The fat of the land. Do agricultural subsidies foster poor health? Environ. Health Perspect. 112(14):820-823.

Fletcher, A. 2007. Maximizing productivity of agriculture: the food industry and nanotechnology. Available at http://www.foresight.org/challenges/agriculture002.html.

Fischler, C. 1990. L'hominvoire : Le gout, la cuisine et le corps. (Without English abstract). Éditions Odile Jacob, Paris.

Fresco, L.O. 2006. Biomass for food and fuel: Is there a dilemma? The Duisenberg Lecture, Singapore, Sep 2006. Univ. Amsterdam, The Netherlands.

Fulponi, L. 2006. Private voluntary standards in the food system: The perspective of major food retailers in OECD countries. Food Policy 31(1):1-13.

Furobotn, E.G., and R. Richter. 1998. Institutions and economic theory. The contribution of the new institutional economics. Univ. Michigan Press, Ann Arbor.

Galizzi, G., and R. Pieri. 1998. Information technology as a development factor in the food sector and in customer care. Rivista di politica Agraria, Rassenga della Agricoltura Italiana 16(4):3-18.

Gallopin, G., A. Hammond, P. Raskin, and R. Swart. 1997. Branch points: Global scenarios and human choice. Resource Pap. Global Scenario Group. Stockholm Environ. Inst., Sweden.

Gallopin, G., and F. Rijsberman. 2000. Three global water scenarios. Int. J. Water 1(1):16-40.

Garnier, H. 2004. Marché mondial du blé. Grain Mag. Jan 2004.

Giampietro, M., S.G. Bukkends, and D. Pimentel. 1999. General trends of technological changes in agriculture. Crit. Rev. Plant Sci. 18(3): 261-282.

Gibbons, M., C. Limonges, H. Nowotny, S. Schwartzman, P. Scott, and M. Trow. 1994. The new production of knowledge: The dynamics of science and research in contemporary societies. Sage Publ., London.

Gibbons, M. 1999. Science's new social contract with society. Nature 402:C81.

Gilland, B. 2002. World population and food supply. Can food production keep pace with population growth in the next half-century? Food Policy 27:47-63.

Ginarte, J., and W. Park. 1997. Determinants of patent rights: A cross national study. Res. Policy 27:283-301.

GLASOD. 1992. World map of the status of human-induced soil degradation. UNEP and Int. Soil Ref. Inform. Centre. GLASOD project. Winand Staring Centre, Wageningen.

GLOBIO. 2001. Global methodology for mapping human impacts on the biosphere. Available at http://www.globio.info/. UNEP/DEWA, Nairobi.

Godet, M. 1977. Crise de la prévision: Essor de la prospective. Presse Universitaire de France, Paris.

Griffon, M. 2006. Nourrir la planète. (Without English Abstract). Éditions Odile Jacob, Paris.

Groot, J.C.J., M. Stuiver, and L. Brussaard. 2004. Land use systems in grassland dominated regions. p. 1202-1204. *In* A. Luscher, et al. (ed), Proc. 20th General Meeting of the European Grassland Federation. Luzern, Switzerland, 21-24 Jun 2004.

Grudens-Schuck N., W. Allen, T.M. Hargrove, and M. Kilvington. 1998. Renovating dependency and self-reliance for participatory sustainable development. Agric. Human Values 20:53-64.

Hafner, S. 2003. Trends in maize, rice and wheat yields for 188 nations over the past 40 years: a prevalence of linear growth. Agric. Ecosyst. Environ. 97:275-283.

Harwood, R.R. 2001. Sustainability in agricultural systems in transition—at which cost? Proc. Int. Symp.: Sustainability in Agricultural Systems. ASA, CSSA, World Bank and IFAP, Baltimore.

Hatem, F. 1993. La prospective, pratiques et méthodes. (Without English Abstract). Economica, Paris.

Hayami, Y., and V.W. Ruttan. 1985. Agricultural development. Johns Hopkins Univ. Press, Baltimore.

Heemskerk, M., K. Wilson, and M. Pavao-Zuckerman. 2003. Conceptual models as tools for communication across disciplines. Conserv. Ecol. 7(3):8.

Hendrickson, M.K., W. Heffernan, P. Howard, and J. Heffernan. 2001. Consolidation in food retailing and dairy: Implications for farmers and consumers in a global food system. Sustainable Agric. 13(1):10-11.

Henson, S., and T. Reardon. 2006. Private agri-food standards: Implications for food policy and the agri-food system. Food Policy 30:241-253.

Herzfeld, T., and T. Glauben. 2006. Labor mobility in transition countries and the impact of institutions. p. 451-466. *In* J. Curtiss et al. (ed), Agriculture in the face of changing markets, institutions and policies challenges and strategies. IAMO, Studies on the Agric. Food Sector in Central and Eastern Europe, Vol. 33. Leibniz Inst. Agric. Dev. Central and Eastern Europe, Halle, Germany.

Huffman, W.E., and R.E. Evenson. 1993. Science for Agriculture: A Long Term Perspective. Ames, IA, Iowa State Univ. Press.

Huffman, W.E., and R.E. Evenson. 2001. Structural and productivity change in US agriculture, 1950-1982. Agric. Econ. 24(2):127-147.

Huffman, W.E., and R.E. Just. 1999. Benefits and beneficiaries of alternative funding mechanisms. Rev. Agr. Econ. 19:2-18.

Huffman, W.E. and R.E. Just. 2000. Setting efficient incentives for agricultural research: Lessons from principal-agent theory. Amer. J. Agr. Econ. 82:828-841.

IAMO. 2003. Social security systems and demographic developments in agriculture in the CEE candidate countries. Dir. Gen. Agric., Inst. Agric. Dev. Central and Eastern Europe. EC, Brussels.

IEA. 2006. World energy outlook 2006. Int. Energy Agency, Paris.

IEO. 2007. International energy outlook, 2007. Official energy statistics from the US government. Available at http://www.eia.doe.gov/oiaf/ieo/index.html.

IFRI. 2002. Le commerce mondial au XXIème siècle. (Without English Abstract).Institut Français des Relations Internationales, Paris.

Ingram, J., and C. Morris. 2007. The knowledge challenge within the transition towards sustainable soil management: An analysis of agricultural advisors in England. Land Use Policy 24(1):100-117.

IPCC. 2000a. Emissions scenarios. Summary for policymakers. IPCC Special Report, Working Group III. Cambridge Press, Cambridge.

IPCC. 2000b. Land use, land-use change, and forestry. A Special Report of the

Intergovernmental Panel on Climate Change. Geneva.

IPCC. 2007. The physical science basis. Summary for policymakers. Contribution of Working Group I to the Fourth Assessment Report of the Intergovernmental Panel on Climate Change. Cambridge Press, Cambridge.

Irvine, J., and B.R. Martin. 1984. Foresight in science: Picking the winners. Pinter, London.

Irvine, J., and B.R. Martin. 1989. Research foresight: Priority setting in science. Pinter, London.

ISNAR. 2003. Trends in the organization and financing of agricultural research in developed countries. Implications for developing countries. Res. Rep. 22. ISNAR, The Hague.

Jackson, T. 2005. Motivating sustainable consumption. A review of evidence on consumer behaviour and behavioural change. Sustainable Dev. Res. Network, London.

Johansson, S. 2005. The Swedish foodprint—An agroecological study of food consumption. Department of Ecology and Crop Production Science. Swedish Univ. Agricultural Sciences, Uppsala.

Juma, C., K. Fang, D. Honca, J. Huete-Perez, V. Konde, S.H. Lee, et al. 2001. Global governance of technology: Meeting the needs of developing countries. Int. J. Tech. Manage. 22(7/8):629-55.

Juma, C., and L. Yee-Cheong. 2005. Innovation: applying knowledge in development. UN Millenium Project, Task Force on Science, Technology, Innovation, New York.

Kabat, P., W. van Vierssen, J.A. Veraart, P. Vellinga, and J. Aerts. 2005. Climate proofing the Netherlands. Commentary. Nature 438:283-284.

Kahiluoto, H., P. Berg, A. Granstedt, H. Fischer, and O. Thomsson (ed). 2006. Localisation and recycling in Baltic rural food systems. Interdisciplinary Synthesis. BERAS Nr. 7. Centre Sustainable Agric. (CUL). Swedish Univ. Agricultural Sciences, Uppsala.

Kandel, W., and A. Mishra. 2007. Immigration reform and agriculture. Paper presented at the USDA Agricultural Outlook Conf., Mar 2007. ERS, USDA, Washington, DC.

Kates R.W., W.C. Clark, R. Corell, J.M. Hall, C.C. Jaeger, I. Lowe, et al. 2000. Sustainability science. Science 292:641-642.

Keating, P., and A. Cambrosio. 2003. Biomedical platforms. Realigning the normal and the pathological in late twentieth-century medicine. MIT Press, Cambridge, MA.

Kemp-Benedict, E., C. Heaps, and P. Raskin. 2002. Global scenario group futures. Tech. Notes. SEI PoleStar Series Rep. 9. Stockholm Environ. Inst., Sweden.

Keyzer, M.A, M.D. Merbis, I.F. Pavel, and C.F.A. van Wesenbeeck. 2005. Diet shifts towards meat and the effects on cereal use: Can we feed the animals in 2030?. Ecol. Econ. 55(2):187-202.

Klein, J.T. 1996. Crossing boundaries: Knowledge, disciplinarities, and interdisciplinarities. Univ. Virginia Press, Charlottesville.

Klijn, J.A., and L.A.E. Vullings (ed). 2005. The EURuralis study: Technical document. Rep. 1196. Alterra, Wageningen.

Knoppers, B.M., and C. Scriver. 2004 Genomics, health and society. Emerging Issues for Public Policy, Canada.

Konyar, K. 2001. Assessing the role of US agriculture in reducing greenhouse gas emissions and generating additional environmental benefits. Ecol.l Econ. 38(1):85-103.

Kulhmann, S., P. Boekholt, L. Georghiou, T. Lemola, K. Guy, and A. Rip. 1999. Improving distributed intelligence in complex innovation systems. Final report of the Advanced Science and Technology Planning Network (ASTP). Available at http://publica.fraunhofer.de/starweb/servlet .starweb?path=pub.web&search=N-55510.

Lamb, H.H. 1995. Climate history and the modern world. 2nd ed. Routledge, London.

Leijten, J. 2006. Between flat and spiky world forces. An exploration of the consequences of demographic and social changes for rural areas. Short Report to the SCAR Expert Working Group/EU Commission. Foresighting in the field of agricultural research in Europe. EU, Brussels.

Lele, S., and R.B. Norgaard. 2005. Practicing interdisciplinarity. BioScience 55:967-975.

Lemmen, D.S., and F.J. Warren (ed). 2004. Climate change impacts and adaptation. A Canadian perspective. Available at http://www.adaptation.nrcan.gc.ca.

Leeuwis, C. 2004. Rethinking innovation and agricultural extension. *In* H.A.J. Moll et al. (ed), Agrarian institutions between policies and local action: Experiences from Zimbabwe. Weaver Press, Harare.

Lévy, P. 2000. Collective intelligence: Mankind's emerging world in cyberspace. Perseus Books Group, New York.

Leydesdorff, L., and H. Etzkowitz. 1998. The triple helix as a model for innovation studies. Sci. Public Policy 25(3):195-203.

Libeau-Dulos, M., and E. Rodriguez Cerzo (ed). 2004. prospective analysis of agricultural systems. Inst. Prospective Tech. Studies, Tech. Rep. EUR 21311 EN. EC, Brussels.

Liefert, W. 2002. Comparative (dis?)advantage in Russian agriculture. Amer. J. Agric. Econ. 84(3):762-767.

Liefert, W., S. Osborne, O. Liefert, and M. Trueblood. 2003. Can Russia be competitive in agriculture? Eurochoices 2(3):18-23.

Lockeretz, W., and M.D. Anderson. 1993. Agricultural research alternatives. Univ. Nebraska Press. <B>[[CITY?]]<B>

Lynden, G.W., and J. Van. 2000. Soil degradation in Central and Eastern Europe: The assessment of the status of human-induced soil degradation. FAO Rep. 2000/05. FAO and ISRIC, Rome.

MA (Millennium Ecosystem Assessment). 2005. Ecosystems and human well-being: Synthesis. Island Press, Washington, DC.

MA (Millennium Ecosystem Assessment). 2006. Ecosystems and human well-being: Scenarios—Findings of the Scenarios Working Group. Island Press, Washington, DC.

Malassis, L. 1997. Les trois âges de l'alimentaire. (Without English Abstract). Cujas, Paris.

Marris, E. 2006. Drink the best and drive the rest. Nature 444:670-672.

Martin, B.R. 1995. Foresight in science and technology. Tech. Anal. Strateg. Manage. 7(2):139-168.

McCalla, A.F. 2000. Agriculture in the 21st century. CIMMYT, Mexico.

McFetridge, D.G. 1994. The economics of vertical integration. Can. J. Agric. Econ. 42(4):525-531.

Meadows, D.H., J. Randers, and D.L. Meadows. 2004. A synopsis. Limits to growth. The 30-Year update. Earthscan, London.

Metzger, M.J., M.D.A. Rounsevell, L. Acosta-Michlik, R. Leemans, and D. Schröter. 2006. The vulnerability of ecosystem services to land use change. Agric. Ecosyst. Environ. 114:69-85.

Miller, R.J., P. Sorokin, Y.F. Lachuga, S. Chernakov, and A.D. Goecker. 2000. A comparison of agricultural higher education in Russia and the United States. J. Nat. Resour. Life Sci. Educ. 29:68-77.

Mindeli, L. 2002. Overview of S&T and innovation policies in Russia. *In* Dialogue on S&T between the European Union and the Russian Federation. CSRS, Moscow.

Morris, C. 2006. Negotiating the boundary between state-led and farmer approaches to knowing nature: An analysis of UK agri-environment schemes. Geoforum 37(1): 113-127.

Morris, S.H. 2007. EU biotech crop regulations and environmental risk: a case of the emperor's new clothes? Trends Biotech. 25(1):2-6.

NIC. 2004. Mapping the global future. Report of the National Intelligence Council's 2020 Project. Available at http://www.dni.gov/nic/NIC_2020_project.html.

NISTEP. 2005. Science and technology foresight survey, Delphi Analysis. Sci. Tech. Foresight Center, Ministry of Educ., Culture, Sports, Sci. Tech. (MEXT): Report no. 97. Nat. Inst. Sci. Tech. Policy, Tokyo.

Nordmann A., 2004. Converging technologies. shaping the future of European societies. Report of the High Level Expert Group Foresighting the New Technology Wave. EC, Brussels.

NSF. 2003. The science and engineering workforce. Realizing America's potential. Report of the National Science Board. NSB 03-69. Nat. Sci. Foundation, Washington, DC.

NSF. 2006. Science and engineering indicators 2006. Nat. Sci. Foundation, Washington, DC.

North, D.C. 1990. Institutions, institutional change and economic performance. Cambridge Univ. Press, Cambridge.

OECD. 1995. Effects of ageing populations on government budgets. Economic outlook and policies. OECD, Paris.

OECD. 1997. National innovation systems. OECD, Paris.

OECD. 1999. Summary and evaluation of main developments and changes in organizational forms of and approaches by the AKS in OECD member countries. Dir. Food, Agric. Fisheries, Committee for Agriculture. AGR/CA/AKS(00)4. OECD, Paris.

OECD. 2005a. Governance of innovation systems Vol 1. Synthesis report. OECD, Paris. http://www.oecd.org/document/25/0,2340,en_2649_37417_35175257_1_1_1_37417,00.html.

OECD. 2005b. Governance of innovation systems. Vol. 2. Case studies in innovation policy. OECD, Paris.

OECD. 2005c. Governance of innovation systems. Vol.3. Case studies in cross-sectoral policy. OECD, Paris. http://www.oecd.org/document/12/0,2340,en_2649_37417_35791756_1_1_1_37417,00.html.

OECD/FAO. 2005. Agricultural outlook: 2005-2014. OECD, Paris and FAO, Rome.

OECD/FAO. 2007. Agricultural outlook: 2007-2016. OECD, Paris and FAO, Rome.

Olesen, J.E., and M. Bindi. 2002. Consequences of climate change for European agricultural productivity, land use and policy. Eur. J. Agron. 16:239-262.

OSI, 2006. Science review of the Department for Environment, Food and Rural Affairs. http://www.dti.gov.uk/science/science-in-govt/works/science-reviews/review/defra/page24808.html. UK Office Sci. Innovation, London.

OST. 2006a. Les systèmes nationaux de recherche et d'innovation du monde et leurs relations avec la France : la Russie. Observatoire des Sciences et des Techniques, Paris.

OST. 2006b. Indicateurs de sciences et de technologies. Rapport de l'Observatoire des Sciences et des Techniques. Edition 2006. Paris.

Pant, H.M. 2002. Global Trade and Environment Model (GTEM): A computable general equilibrium model of the global economy and environment. Australian Bureau Agric. Resource Econ., Canberra.

Pardey, P.G., J. Roseboom, and B.J. Craig. 1999. Agricultural R&D investments and impact. *In* J.M. Alston et al. (ed), Paying for agricultural productivity. Johns Hopkins Univ. Press, Baltimore.

Parry, M.L. (ed). 2000. Assessment of potential effects and adaptations for climate change in Europe. The Europe ACACIA project. Jackson Environ. Inst., Univ. East Anglia, Norwich.

Persley, G.J. 1998. Investment strategies for agriculture and natural resources: Investing for knowledge and development. World Bank, ACIAR, DFID, GTZ, Washington, DC.

Poux, X., J.B. Narcy, and V. Chenat. 2005. Agriculture et environnement: 4 scénarios à l'horizon 2015. Ministère de l'écologie et du développement durable, Paris.

Pretty, J., C. Brett, D. Gee, R. Hine, C. Mason, J. Morison. 2001. Policy challenges and priorities for internalizing the externalities of modern agriculture. J. Environ. Planning Manage. 44(2):263-284.

Pretty, J., A.S. Ball, T. Lang, and J. Morison. 2005. Farm costs and food miles: An assessment of the full cost of the UK weekly food basket. Food Policy 30:1-19.

RAND Corporation. 2006. The global technology revolution 2020. Executive summary and in-depth analyses: Bio/Nano/Materials/Information trends, drivers, barriers and social implications. MG-475-NIC and TR-303-NIC. Rand Corp., Washington, DC.

Raoult-Wack, A-L., and N. Bricas. 2001. Food sector development: Multifunctionality and ethics. Agric. Engineering Int: CIGR J. Sci. Res. Dev. 3. Available at http://de.scientificcommons.org/29695540.

Raskin, P., F. Monks, T. Riberio, D. van Vuuren, and M. Zurek. 2005. Global scenarios in historical perspective. p. 35-43. *In* S.R. Carpenter et al. (ed), Ecosystems and human well-being: Scenarios. Findings of the scenarios working group of the Millennium Ecosystem Assessment. Island Press, Washington, DC

Read, N., J.J. Quinn, and A. Webster. 1988. Commercialisation as a policy mechanism in UK agricultural research, development and extension. Agric. Systems 26:77-87.

Reardon, T., and C.B. Barrett. 2000. Agroindustrialization, globalization, and international development: An overview of issues, patterns, and determinants. Agric. Econ. 23:195-205.

Reilly, J., B. Tubiello, B. McCarl, and J. Melillo. 2000. Climate change and agriculture in the United States. p. 379-403 *In* Climate change impacts in the United States. The potential consequences of climate variability and change. Report for the U.S. Global Change Research Program. Cambridge Univ. Press, Cambridge.

Reilly, J., F. Tubiello, B. McCarl, D. Abler, R. Darwin, K. Fuglie, et al. 2003. US agriculture and climate change. New Results. Clim. Change 57:43-69.

Roco, M., and W.S. Bainbridge. 2002. Converging technologies for improving human performances. Nanotechnology, biotechnology, information technology and cognitive science. A NSF/DOC-sponsored report. Available at http://www.wtec.org/ConvergingTechnologies/1/NBIC_report.pdf.

Roetter, R.P., and S.C. van de Geijn. 1999. Climate change effects on plant growth, crop yield and livestock. Clim. Change 43:651-681.

Roetter, R.P., H. van Keulen, M. Kuiper, J. Verhagen, and H.H. van Laar (ed), 2007. Science for agriculture and rural development in low-income countries. Springer, Dordrecht.

Romig, D.E., M.J. Garlynd, R.F. Harris, and K. McSweeney. 1995. How farmers assess soil health and soil quality. J. Soil Water Conserv. 50:229-236.

Rosegrant, M.W., X. Cai, and S.A. Cline. 2002. World water and food to 2025: Dealing with scarcity. IFPRI, Washington, DC.

Rosegrant, M.W., X. Cai, and S.A. Cline. 2004. Global water outlook to 2025, averting an impending crisis. IFPRI, Washington, DC.

Rosegrant, M., M.J. Paisner, S. Meijer, and J. Witcover. 2001. Global food projections to 2020: emerging trends and alternative futures. IFPRI, Washington, DC.

Rosenzweig, C., A. Iglesias, X.B. Yang, P.R. Epstein, and E. Chivian. 2000. Climate change and U.S. agriculture: The Impacts of warming and extreme weather events on productivity, plant diseases, and pests. Center Health Global Environ. Harvard Univ., Cambridge, MA.

Rosenzweig, C., and D. Hillel, 2000. Soils and global climate change: Challenges and opportunities. Soil Sci. 165:47-56.

Royer, J.S. 1998. Market structure, vertical integration, and contract coordination. p. 73-98. *In* J.S. Royer and R.T. Rogers (ed), The industrialization of agriculture: Vertical coordination in the US food system. Ashgate, London.

Sachs, J. 2005. The end of poverty. Economic possibilities for our time. Penguin Press, New York.

Sachs, J. 2006. Global warming. Pay for it now, or pay for it later. The Globe and Mail, 19 Jun 2006.

Sanderson, K. 2006. A field in ferment. Nature 444:673-676.

Schmidhuber, J. 2003. The outlook for long-term changes in food consumption patterns: concerns and policy options. Paper prepared for the FAO Scientific Workshop on Globalization of the Food System: Impacts on Food Security and Nutrition, 8-10 Oct 2003. FAO, Rome.

Schmidhuber, J., and P. Shetty. 2005. The nutrition transition to 2030. Why developing countries are likely to bear the major burden. Plenary Paper 97th Sem. Eur. Assoc. Agricultural Economists, Univ. Reading, 21-22 Apr 2005.

Schröder, G., and A. Weiske. 2006. Greenhouse gas emissions and mitigation costs of selected bioenergy production chains. Impact of environmental agreements on the CAP. MECAP Doc. WP3 D15b. IEEP, London.

Schröter, D., W. Cramer, R. Leemans, I.C. Prentice, M.B. Araujo, N.W.Arnell, et al. 2005. Ecosystem service supply and vulnerability to global change in Europe. Science 310:1333-1337.

Scott, N., and H. Chen. 2003. Nanoscale

science and engineering for agriculture and food systems. A report to the Cooperative State Research. USDA, Washington, DC.

Shearer, A.W. 2005. Approaching scenario-based studies: Three perceptions about the future and considerations for landscape planning. Environ. Planning B 32:67-87.

Shiklomanov, I. 1999. World water resources and their use. Available at http://webworld.unesco.org/water/ihp/db/shiklomanov/part'3/HTML/Fi_17'2.html. SHI/UNESCO, St.Petersburg, Russia.

Silberglitt, R., P. Anton., D. Howell, and A. Wong. 2006. The global technology revolution 2020. RAND, Washington, DC.

Skaggs, R. 2001. The future of agriculture: Frequently asked questions. Tech. Rep. 37. New Mexico State Univ. College Agric. Home Economics.

Smil, V. 2000. Feeding the world. The MIT Press, Cambridge, MA.

Smil, V. 2005. Feeding the world: How much more rice do we need? Keynote Lecture. p. 1-3. *In* K. Toriyama, et al. (ed), Rice is life: Scientific perspectives for the 21st century. World Rice Res. Conf., Tsukuba 4-7 Nov 2004, Japan, WRRC 2004. CD-ROM Proc. Tokyo and Tsukuba.

Stern, N. 2006. Stern review on the economics of climate change. Available at www.sternreview.org.uk. HM Treasury, UK.

Sumberg, J., C. Okali, and D. Reece. 2003. Agricultural research in the face of diversity, local knowledge and the participation imperative: theoretical considerations. Agric. Syst. 76:739-753.

Tilman, D., K.G. Cassman, P.A. Matson, R. Naylor, and S. Polasky. 2002. Agricultural sustainability and intensive production practices. Nature 418:671-677.

Ugarte, D., B. English, K. Jensen, C. Hellwinckel, J. Menard, and B. Wilson. 2006. Economic and agricultural impacts of ethanol and biodiesel expansion. 21st century agricultural impacts project, Univ. Tennessee. Available at http://www.21stcenturyag.org/.

Umali, D.L., and L. Schwartz. 1994. Public and private agricultural extension: Beyond traditional frontiers. Disc. Pap. 236. World Bank, Washington, DC.

UN. 1993. Agenda 21: Earth summit. The United Nations Programme of Action from Rio. UN, New York and Geneva.

UN. 2004. World population prospects: The 2004 revision. Highlights. UN, New York.

UN. 2006. World population prospects: The 2006 revision. Highlights. Available at http://www.un.org/esa/population/publications/wpp2006/wpp2006_highlights.pdf. UN, New York.

UNESCO. 2006a. The global carbon cycle. Scope Policy Brief 2. UNESCO, Paris.

UNESCO. 2006b. UNESCO science report 2005. UNESCO, Paris.

Univ. Georgia. 2000. Critical dimensions of structural change. 2nd Ann. Nat. Symp. Future of Am. Agric., 2000. Univ. Georgia. Available at http://www.agecon.uga.edu/archive/agsym00.html.

US Census Bureau. 2007. International database. US Census Bureau, Washington, DC.

USDA. 2005. Food security assessment. GFA-16, May 2005. Available at http://www.ers.usda.gov/Publications/GFA16/. ERS, USDA, Washington, DC.

US DOI. 2005. Water 2025: preventing crises and conflict in the West. A status report. Available at http://www.doi.gov/water2025/. US Dep. Interior, Washington, DC.

US EPA. 2005. Greenhouse gas mitigation potential in U.S. forestry and agriculture. Available at http://www.epa.gov/sequestration/greenhouse_gas.html. USEPA, Washington, DC.

Vanacht, M. 2006. Six megatrends in agriculture. The John M. Airy Symposium: Visions for animal agriculture and the environment, Kansas MO. Available at http://www.iowabeefcenter.org/content/Airy/VANACHT%20Abstract.pdf.

Van-Camp, L., B. Bujarrabal, A.R. Gentile, R.J.A. Jones, L. Montanarella, C. Olazabal, and S.K. Selvaradjou. 2004. Reports of the Technical Working Groups established under the Thematic Strategy for Soil Protection. EUR 21319 EN/1. EC, Luxembourg.

Van Ittersum, M.K., and R. Rabbinge. 1997. Concepts in production ecology for analysis and quantification of agricultural input-output combinations. Field Crops Res. 52:197-208.

Van Ittersum, M.K., R.P. Roetter, H. Van Keulen, N. De Ridder, C.T. Hoanh, A.G. Laborte, et al. 2004. A systems network (SysNet) approach for interactively evaluating strategic land use options at sub-national scale in South and South-east Asia. Land Use Policy 21:101-113.

Van Keulen, H. 2007. Historical context of agricultural development. *In* R.P. Roetter et al. (ed), 2007. Science for agriculture and rural development in low-income countries. Springer, Dordrecht.

Väyrynen, R. 2006. The Speech of the Director of the Academy of Finland. The 60th Anniversary of the Faculty of Social Sciences, Helsinki Univ., Finland.

Verburg, P.H., W. Soepboer, A. Veldkamp, R. Limpiada, V. Espaldon, and S. Mastura. 2002. Modeling the spatial dynamics of regional land use: the CLUE–s model. Environ. Manage. 30(3):391-405.

Verburg, P.H., M.D.A. Rounsewell, and A. Veldkamp (ed). 2006. Editorial. p. 1-6 *In* Scenario-based studies of future land use in Europe. Special Issue, Agric. Ecosyst. Environ. 114(1).

Vereijken, P.H., and C.M.L. Hermans. 2006. How sustainable can agriculture be in an EU-scenario of free market and social policy? p. 23-24. *In* H. Langeveld, and N. Roeling (ed), Changing European farming systems for a better future: New visions for rural areas.

Verhagen, J., H. Wösten, and A. De Jager. 2007. Agriculture and environment. *In* R.P. Roetter et al. (ed), Science for agriculture and rural development in low-income countries. Springer, Dordrecht.

Von Braun, J., M. Rosegrant, R. Pandya-Lorch, M.J. Cohen, S.A. Cline, M.A. Brown, and M.S. Bos. 2005. New risks and opportunities for food security. Scenario Analyses for 2015 and 2050. 2020 Disc. Pap. 39. IFPRI, Washington, DC.

Wackernagel, M., and W.E. Rees. 1996. Our ecological footprint: Reducing human impact on the Earth. Available at www.newsociety.com/oef.html and www.RedefiningProgress.org. New Society Publ., Gabriola Island, BC.

Wackernagel, M., D. Deumling, C. Monfreda, A.C. Linares, I.S.L. Falfan, and M.A.V. Sanchez. 2001. Ecological footprint of nations. Sustainability Issue Brief, Dec 2001. Available at http://www.rprogress.org/publications/2001/1997_efoot.pdf.

Walter, G., M. Wander, and G. Bollero. 1997. A farmer-centered approach to developing information for soil resource management: The Illinois Soil Quality Initiative. Am. J. Alternative Agric. 12(2):64-72.

Wander, M., and L.E. Drinkwater. 2000. Fostering soil stewardship through soil quality assessment. Appl. Soil Ecol. 15: 61-73.

WCED. 1987. Our common future. World Commission Environ. Dev. Oxford Univ. Press, Oxford.

Weber, K.M. 2005. Environmental technologies. Background paper European Commission's High Level Group on Key Technologies. 4 July 2005. EC, Brussels.

Westhoek, H.J., M. Berg, and J.A. van den Bakkes. 2006. Scenario development to explore the future of Europe's rural areas. Special Issue, Agric. Ecosyst. Environ. 114(1):7-20.

Williamson, O. 2000. The new institutional economics: Taking stock, looking ahead. J. Econ. Lit. 38(3):595-613.

Winter, M. 2003. Embeddedness, the new food economy and defensive localism. J. Rural Studies 19(1):23-32.

Wolf, J., R.P. Roetter, and O. Oenema. 2005. Nutrient emission models in environmental policy evaluation at different scales—experience from The Netherlands. Agric. Ecosyst. Environ. 105:291-306.

World Bank. 2003. The CGIAR at 31: An independent meta-evaluation of the Consultative Group on International Agricultural Research. Vol.1. Overview report. World Bank, Washington, DC.

World Bank. 2005. World development indicators. Available at http://devdata.worldbank.org/wdi2005/Table4_1.htm. World Bank, Washington, DC.

World Bank. 2006. Enhancing agricultural innovation: How to go beyond the strengthening of research systems. World Bank, Washington, DC.

WTO. 2005. International trade statistics 2005: Trade by sector. WTO, Geneva.

WTO. 2006. World Trade Report. Available at http://www.wto.org/english/res_e/reser_e/world_trade_report_e.htm. WTO, Geneva.

Worldwatch Institute. 2006. Biofuels for transportation. Global potential and implications for sustainable agriculture and energy in the 21st Century. 7 June 2006. Washington, DC.

Yohe, G., and R. Tol. 2001. Indicators for social and economic coping capacity: Moving toward a working definition of adaptive capacity. Global Environ. Change 12:25-40.

Zahniser, S., E. Young, and J. Wainio. 2005. Recent agricultural policy reforms in North America. WRS-05-03. April 2005. USDA, Washington, DC.

6

Options for Action

Coordinating Lead Authors:
Marianne Lefort (France), Angus Wright (USA)

Lead Authors:
David Andow (USA), Pavol Bielek (Slovakia), Rodney Brown (USA), Costas Gregoriou (Cyprus), Trish Kammili (France), Jacques Loyat (France), Denis Lucey (Ireland), Joe Morris (UK), Isaak Rashal (Latvia)

Contributing Authors:
Loic Antoine (France), Gilles Aumont (France), Yves Birot (France), Gerard Buttoud (France), Bernard Chevassus (France), Jean-François Dhote (France), Georges Eliades (Cyprus), Ghislain Gosse (France), Anil Graves (UK), Jean-Marc Guehl (France), Christoph Metochis (Cyprus), Jean-Luc Peyron (France), Pierre Ricci (France), Alain Ruellan (France), Yves Savidan (France), Bernard Seguin (France), Andrée Sontot (France)

Review Editors:
Michael Appleby (UK), Maria del Mar Delgado (Spain)

Key Messages

6.1 Paradigm Shift and Key Issues for AKST to Meet Development and Sustainability Goals 212

6.1.1 Why recognize a paradigm for research and action? 212

6.1.2 New research approaches and supportive institutional change 213

6.1.3 Achieving the proper balance between the public and the private sectors 213

6.2 Future Needs and Priorities for AKST 214

6.2.1 Responding to climate change 214

6.2.1.1 Mitigate climate change through agriculture 214

6.2.1.2 Reduce agriculture's vulnerability to climate change 215

6.2.2 Facing new and emerging human, livestock and plant diseases 216

6.2.2.1 Human and livestock diseases 216

6.2.2.2 Insect pests, weeds and diseases of plants 217

6.2.3 Contributing to a global strategy for a low carbon economy 218

6.2.3.1 Biofuels 218

6.2.3.2 Improve energy efficiency of supply chains: food miles and life cycle analyses 220

6.2.4 Trade, markets and agricultural policies 221

6.2.5 Promoting food quality and safety in diverse food and farming systems 223

6.2.5.1 Improve food quality 224

6.2.5.2 Develop diversified, fair and equitable food and fiber supply chains 225

6.2.6 Promoting environmental sustainability through ecological management 226

6.2.6.1 Potential contribution of AKST for long term soil preservation 226

6.2.6.2 Contribution of AKST to water management 227

6.2.6.3 Potential contribution of AKST to biodiversity and genetic resource management 228

6.2.6.4 Potential of AKST for developing energy efficient food and farming systems 230

6.2.6.5 Reducing pressure on natural resources through the ecological footprint method 231

6.2.7 Developing innovative crops and livestock food and farming systems 231

6.2.7.1 The potential of genetics and biotechnology for crops and livestock breeding 231

6.2.7.2 The potential of nanotechnologies in the food and fiber supply chains 234

6.2.7.3 Contribution of AKST to the development of improved pest management 235

6.2.7.4 Development of alternative resource management strategies 236

6.2.8 Developing sustainable systems for forestry 236

6.2.9 Developing sustainable systems for fisheries and aquaculture 239
6.2.9.1 Coastal capture fisheries in the NAE region 239
6.2.9.2 Aquaculture 240
6.2.10 Ensuring socioeconomic viability of the systems and improving rural livelihoods 241
6.2.10.1 Social issues 241
6.2.10.2 Economic issues 243
6.2.10.3 Sustainable rural livelihoods 244
6.2.10.4 Understanding farmer attitudes and behavior 245
6.2.10.5 Rural development 246
6.3 Development of Human Capital, Organizations and Institutions 246
6.3.1 Towards interactive knowledge networks 246
6.3.1.1 Promote stakeholder interaction 247
6.3.1.2 Recognize the importance of indigenous and traditional knowledge 247
6.3.2 Towards meaningful interdisciplinarity 248
6.3.2.1 Enlarge the scope of agricultural knowledge systems 248
6.3.2.2 New skills for AKST personnel 249
6.3.2.3 Need for new learning opportunities 250
6.3.2.4 Interactions with policy makers and political leaders 250
6.3.2.5 Public understanding of the multiple roles of agriculture 250
6.3.2.6 Initial education/training for farmers 250
6.3.2.7 Stimulate links between higher education and research and facilitate the harmonization of the different education systems 251
6.3.2.8 Promote lifelong learning and create a learning society 251
6.3.3 Strengthening information and knowledge-based systems 251
6.3.3.1 Reducing the "Digital-divide" 252
6.3.3.2 Reconfiguration of information systems 252
6.3.3.3 From information systems to knowledge based systems 252
6.3.4 Promoting appropriate institutional and organizational arrangements 252
6.3.4.1 Towards new and "engaged" public institutes 252
6.3.4.2 Innovative education and research models 253
6.3.4.3 Setting up institutional and organizational arrangements for knowledge based systems 254
6.3.4.4 AKST interactions between NAE and other regions 254
6.4 Reshaping Policy Environment and Governance Systems 258
6.4.1 Developing a coherent policy framework 258
6.4.2 Developing AKST in response to international agreements 259
6.4.3 Enlarging the range of proprietary regimes 260
6.4.3.1 General issues concerning proprietary regimes and IPR 260
6.4.3.2 Intellectual property rights 261
6.4.3.3 Access to genetic resources for food and agriculture 261
6.4.4 Setting up new modes of governance 262
6.4.4.1 General governance issues in food and farming systems 262
6.4.4.2 Fisheries issues 263
6.4.4.3 Forestry issues 263
6.5 Funding Investments in Research and Development (R&D) for Agriculture 264
6.5.1 Spending on R&D 264
6.5.2 Paying for agricultural R&D 264
6.5.3 Undertaking agricultural R&D 265
6.5.4 Deciding on R&D 265
6.5.5 Institutional arrangements and collaboration 265
6.5.6 Funding mechanisms and enablers of AKST 266

Key Messages

1. Successfully meeting development and sustainability goals and responding to new priorities and changing circumstances will require widespread recognition of a paradigm shift that is already in process. This new paradigm accords increased importance to the multiple functions of agriculture and its adaptability to local environment and social contexts. This multifunctionality* of agriculture can only be understood and managed by developing new conceptual tools to take into account the complexity of agricultural systems and by placing agriculture in its social and ecological context. New institutional and organizational arrangements are essential to support a more integrated approach to the development, dissemination and uptake of AKST, with increased emphasis on interactive knowledge networks between research, education and extension, multidisciplinary research programs, the involvement of stakeholders in defining research agendas and the provision of education, training and advisory programs that enable a wider group of stakeholders to address these new complexities. Working successfully on this new agenda for agriculture suggests a need for ongoing attention to achieving the proper balance between public and private involvement in AKST with respect to funding, property regimes, delivery and overall governance. Successfully meeting development and sustainability goals requires a range of interactions among the various regions of the world.

AKST Options for Addressing Global Issues

2. Develop strategies to counteract detrimental effects of the agrifood system on climate change and reduce vulnerability to such change. Reducing agricultural emissions of greenhouse gases within NAE will require changes to farming systems, land use and practices throughout the agrifood system, such as increasing energy efficiency and carbon sequestration. In addition, AKST can be developed and used to reduce the adverse effects of climate change on agriculture in NAE and other regions, for example through drought, pest, temperature and salinity tolerant plants.

3. Develop interventions that aid in prevention and better management of new and emerging human, plant and livestock diseases as well as weed and insect problems. The epidemiological dynamics of the overall system from both spatial and temporal scales require better understanding and the development of suitable surveillance and response networks. Early detection and new diagnostic and curative tools are important.

4. Develop and evaluate biofuels. Innovations in AKST can contribute to the development of economically feasible biofuels and biomaterials that have a positive energy and environmental balance and that may be ethically justified by not compromising the world food supply. Research could

* The term *multifunctionality* has sometimes been interpreted as having implications for trade and protectionism. It is used here solely to express the inescapable interconnectedness of agriculture's role and functions.

focus on improving the energy content of biofuel crops and other raw materials and the overall energy efficiency of these systems, as well as new systems that do not compete with food production for land and water such as marine algae and cyanobacteria.

5. Understand the processes and consequences of international trade and market liberalization and identify actions to promote fair trade and market reform to achieve development and sustainability goals. NAE has an obligation to facilitate AKST, which can enhance capacity together with other regions regarding:

- Viable production systems to achieve food security and sustainable rural livelihoods,
- Improved access to and further development of global and local markets,
- Policies to promote fair trade and address market failure, including review of practices such as the use of subsidies, dumping and regulatory regimes and
- Mechanisms for interactive knowledge and technology exchange among NAE and other regions, including the participation of international governmental and government organizations and trade and farmer associations.

AKST Options for Improving Food and Farming Systems and their Sustainability

6. Intensify the focus on nutrition, health, food quality, diversity and safety through different agricultural systems ranging from intensive systems providing basic commodities to more extensive and local systems providing differentiated products. Research and technological developments in new food systems can usefully continue in several directions: to obtain a deeper understanding of the relationships between food, diet and health; to improve quality of raw materials; to enhance the ability to trace along the food chain to support quality and safety assurance; to devise better systems to control food safety vis-à-vis microbial contamination, mycotoxins and xenobiotics.

7. Enhance research in ecological and evolutionary sciences as applied to agricultural ecosystems to devise, improve and create management options that contribute to multifunctionality. Such options call for an ecological approach to agroecosystems for better water, soil and biodiversity management at landscape scales and improved preservation of genetic resources in special collections and in natural conditions.

8. Improve standards of soil and water management among farmers, including irrigation, as a critical component of sustainable farming systems. There is continuing need to improve the scientific understanding of soil and water processes, simultaneously drawing on local knowledge, in order to support the wise use of these fundamental agricultural resources through the development and widespread adoption of appropriate farming technologies.

9. Strengthen breeding activities, generate basic and applied knowledge and further develop relevant technologies, including biotechnologies. It is essential that classical breeding be maintained and expanded to a wider diversity of species. The use and development of both functional genomics and systems biology and the establishment of new breeding methods integrating genomics information will be essential. There are varying opinions in NAE on the potential benefits and risks of transgenic organisms as well as the required regulatory framework. Assessment of new breeding products requires evaluation of the social, economic, environmental and health implications of their dissemination and must have a long term and wide scale perspective.

10. Reduce environmental impacts through diversification and selection of inputs and management practices that foster ecological relationships within agroecosystems. These conservation agriculture practices include ecologically based pest management, minimum tillage, protected cropping and precision farming, among others. AKST is required to analyze the environmental footprint of agriculture and determine the environmental limits within which it must operate. New research in AKST can help design management practices and policy measures that improve environmental performance as a critical component of sustainability.

11. Assess impacts of management systems on animal welfare and develop and promote humane practices. Ethical standards of animal handling and slaughter and attention to the environment in which domestic livestock are raised can significantly reduce stress and suffering of domestic livestock and should be included in future management.

12. Explore, promote and manage the multiple roles of forests to conserve soil, maintain water quality and quantity, protect biodiversity and sequester carbon. Assigning value to ecosystem services and forest resources and improving long-term sustainability and resilience to environmental change will enhance forest stewardship and the livelihoods of people dependent on forest resources.

13. Improve the sustainability of coastal capture fisheries and aquaculture. Fisheries and aquaculture management will benefit from ecosystem management and monitoring that reduce the ecological effects of fishing technology, facilitate selective fishing and create markets for by-catch. Aquaculture can be improved by better understanding the relationship between fish immunity and disease and reducing effects of escapes on native fish. Reducing impacts of waste and developing more sustainable alternative sources of fish feed are critical needs.

14. Comprehensively assess new technologies for their impact on the environment, economic returns, health and livelihoods. All new technologies (transgenics, nanotechnology, biofuel production, etc.) will benefit from thorough analysis with tools such as life-cycle impact analysis and social, economic and vulnerability impact assessment. In the past, the rapid application of technology before full assessment has led to unforeseen problems. Analytical tools that allow the examination of effects on different stakeholders, different agrifood sectors and different dimensions (e.g., environmental and social) are essential.

15. Improve the social and economic performance of agricultural systems as a basis for sustainable rural and community livelihoods:

- Improve the understanding of factors affecting social welfare and the vulnerability of farming communities at the local scale including institutions that govern access to and use of natural resources, systems of incentives and rewards and sources of conflict in rural communities.
- Evaluate the range of goods and services deriving from agriculture and design economic instruments that promote an appropriate balance of private and public goods.
- Assess the performance of farming systems at the farm, regional and national scales that accommodate the multifunctional role of agriculture.

16. Determine research and policy changes that lead to improvement in the welfare of migrant and/or temporary farm labor. Appropriate measures could help ensure the availability of qualified labor for agriculture while reducing inequalities. Much agricultural labor is done by immigrants with precarious legal status in NAE. Changes to immigration law may be required to improve the situation of farm labor.

17. Respond to gender related issues in agricultural research and the agricultural economy. These include equity considerations in research and educational institutions; farm ownership and gendered work roles among farm families and hired labor; problems posed by family fragmentation among migrant workers.

AKST Options for Strengthening Human Capital and Organizational Arrangements

18. Strengthen human capital and reconfigure organizational arrangements to facilitate the development, dissemination and wider use of AKST.

- Strengthen interactive knowledge networks involving multiple and more diverse stakeholders among the research, education and extension components of the AKST systems. These cooperative efforts could be encouraged by governments.
- Improve AKST processes for involving, informing and empowering stakeholders, in particular women and others whose interests have previously been inadequately addressed. New stakeholder involvement mechanisms are advisable for developing and using methods to establish standards of legitimacy for inclusion in these processes.
- Develop and utilize new skills and learning opportunities for existing and future AKST personnel and their various clients so that they can understand and function more comfortably in the context of the wider multifunctional vision of agriculture.
- Enhance meaningful interdisciplinary research, educational programs and extension/development work without compromising disciplinary excellence while identifying and surmounting systemic barriers to interdisciplinarity.
- Strengthen links between research and higher education to promote lifelong learning and the creation of a learning society.
- Strengthen information and knowledge-based systems to enable a rapid, bi-directional flow and utilization of information and knowledge between the wider agricultural sector and the AKST system.
- Promote appropriate organizational arrangements that facilitate the development of human capital within the AKST system.

19. Devise, evaluate and institute new patterns of ownership and employment. This would promote effective participation, equity, development of human capital, cultural change and ongoing education and training.

20. Recognize more fully the important role that traditional and indigenous knowledge plays in agriculture and in the culture and welfare of particular people. Respectful interaction with indigenous peoples and traditional practitioners and serious consideration of the value of their knowledge, experience and techniques can contribute broadly to sustainable and equitable agriculture and the development of new AKST.

21. Reinforce partnerships between NAE and other regions that empower poor and disadvantaged people and organizations. Strengthening interactive knowledge networks and integrated trans-disciplinary research and educational programs facilitates the development of working relationships among AKST organizations worldwide.

22. Increase NAE receptivity to innovative proposals from other regions for mutual capacity building. Harness human and organizational capacities, especially focusing on the capacity to build capacity. Regional and global forums can facilitate this networking and promote enhanced contributions to the global knowledge economy by AKST organizations.

AKST Options for Improving Policy and Governance

23. Support coherent policy frameworks for agricultural and rural development and ensure that relevant government departments collaborate with each other and the private sector and NGO actors in their development. Coordination between government functions can facilitate a balance among the goals of feeding an expanding population, using natural resources efficiently and sustainably and promoting economic development and cultural uses at the local, regional and global levels.

24. Strengthen connections among all actors within the food chain and better balance power among all actors in food chain governance. This requires policies to strengthen business and marketing skills among producers, build mutually beneficial relationships among all members of the food supply chain and educate consumers about farming and food products and systems.

25. Develop policy instruments to internalize current environmental and social externalities of agricultural production and reward the provision of agroenvironmental services. Examples include financial instruments to discourage use of environmentally harmful inputs and promotion of agricultural practices with low carbon emissions, watershed and landscape eco-management and carbon sequestration through agroforestry.

26. Develop policy instruments to remove incentives for farm concentration and agribusiness concentration. These include anti-trust measures, improved competition policies, more stringent corporate social reporting and greater transparency in corporate transactions.

27. Implement more fully and further develop those treaties and conventions that promote development and sustainability goals. These include such areas as climate change, biodiversity conservation, genetic resource conservation, toxics control, desertification, sanitary/phytosanitary, intellectual property and biopiracy.

28. Further consider and develop regimes that define rights of use and of property. The development of "common property regimes" for scarce natural resources such as water that go beyond either public or private ownership could be further considered. Significant public policy discussions of the implications and nature of these proprietary regimes for the future are needed to explore the full implications.

29. Reshape intellectual property rights and associated regulatory frameworks where necessary to facilitate the generation, dissemination, access and use of AKST and recognize the need to improve equitability among regions in use of intellectual property rights. To achieve a better balance between public and private interests and between rewards for innovation and accessibility, consideration could be given to patents that would be narrower, cross-licensing that would result in pooling of patents between universities and the private sector, compulsory or obligatory licensing when deemed necessary, broadening of exemptions of patents to facilitate research and open source technology that leads to collaborative invention.

30. Devise modes of governance at the local level that integrate a wider range of stakeholders' perspectives. Examples such as food policy councils in the US and water management groups that implement the European Water Framework Directive (France, UK, Ireland) already exist to a limited extent in NAE and should be promoted.

AKST Options for Funding

31. Multifunctionality calls for new, increased and more diverse funding and delivery mechanisms for agricultural research and development (R&D) and human capital development. Depending on circumstances, these could include:

- Public investment to serve the public good, addressing strategic, "nonmarket" issues that do not attract private funding, such as food security and safety, climate change and sustainability;
- Public investment to strengthen human capital development and education programs, including multidisciplinary research;
- Private investments made by farming businesses and farmer associations as an important and growing source of new AKST;
- Adequate incentives and rewards to encourage private investors to invest in new R&D, including supporting commercial services such as market information and credit;
- Public–private partnerships to provide technical assistance and joint funding of R&D investments, especially where risks are high and where research developments in the private sector can significantly enhance the public good; and
- Nongovernmental organizations to act as an alternate channel for public and private funding of technical assistance, knowledge transfer and applied research at the local scale. Further support will be needed to facilitate this.

32. Establish effective procedures for funding rural and agricultural development by national and international agencies. This recognizes the strategic role of the agricultural and rural sectors in meeting development and sustainability goals within the NAE regions and globally, allocating funds and managing investment programs for these purposes.

6.1 Paradigm Shift and Key Issues for AKST to Meet Development and Sustainability Goals

6.1.1 Why recognize a paradigm for research and action?

Advances in agricultural knowledge, science and technology (AKST) have been critical in making it possible to meet many of the needs for food and fiber in North America, Europe and other parts of the world. Agriculture is now being required to be responsive to new priorities, expectations and changing circumstances. Many of these are stated in development and sustainability goals, namely: reduce hunger and poverty, improve rural livelihoods and health, increase incomes and facilitate equitable, environmentally, socially and economically sustainable development. Meeting these multiple objectives is made more complicated by a variety of foreseeable and unforeseeable changes. These challenges necessitate emphasis on a new way of considering research, technology development, education and knowledge exchange. The new way of thinking requires that those working in the fields affecting AKST:

- Recognize the importance of the multiple functions of agriculture not only in providing food and fiber but also in providing a range of environmental goods and services associated with land, water and living systems.
- Engage the participation of all people concerned in the process of defining needs and solutions.
- Be specific to local environmental, social and economic context.
- Be adaptive to social and environmental change, including climate change.

Although the "farming systems approach" and other research strategies in recent decades extended the boundaries of consideration for AKST, research and development has remained largely focused on the farm economy itself, with production at its center. The externalized costs of unintended and/or unanticipated negative social and environmental consequences of AKST have been dealt with largely through post hoc regulatory and policy approaches. Although such post hoc responses have in some cases encouraged or compelled positive technological innovation, they have often failed to resolve important problems that might have been better addressed at an earlier stage in the creation, design and implementation of AKST. There will be advantages for such broad anticipatory approaches to complement more narrowly focused R&D. Agriculture and AKST development could be re-conceptualized within the entire context of society and environment, introducing new levels of complexity in understanding and responding to future needs. This requires recognition of a paradigm shift in the way AKST is to be produced and delivered. Elements of this shift are already in process and have appeared throughout the NAE.

Shaping a newly recognized paradigm shift and learning to work successfully within it will require continuing work on the integration of knowledge across a wide range of disciplines. Researchers and policy makers will require new conceptual tools to better address complex questions and help in understanding the dynamic and interactive relationships among multiple relevant factors.

Working more effectively within the new paradigm will also likely require new institutional arrangements. These arrangements could be designed to support a more integrated approach to the development and dissemination of AKST. Methods for such integration will include the creation of multidisciplinary research programs, the involvement of stakeholders in defining such agendas and the provision of education, training and advisory programs to support the exchange of knowledge competencies to deal with these new complexities. In addition, it will be necessary to carry on a continual re-evaluation of the proper balance between public and private interests and investments in the development of AKST.

6.1.2 New research approaches and supportive institutional change

Universities, other research organizations, training institutes and extension services may frequently find it advisable to renew and upgrade their capabilities to operate effectively within a new paradigm recognizing the complexity of agricultural AKST. The nature of the new challenges calls on universities and other organizations to greatly increase the emphasis on multidisciplinary and interdisciplinary research. This can be done without sacrificing disciplinary excellence, which is the foundation for successful multidisciplinary work. Much of this work could be focused directly on meeting development and sustainability goals.

Agricultural research and education, in all its forms, is faced with the challenge presented by what has been termed the "disaggregation" or "disintegration" of agricultural science. In recent decades there has been a strong tendency for many of the most important advances in AKST to come from the basic science and social science disciplines outside of the agricultural sciences. This trend creates an imperative for the agricultural sciences to interact more with the other disciplines both to fully capture the advantages that such contributions from outside the agricultural sciences offer and to help guide the research in directions most useful for agriculture. Organizations and institutions have sometimes recognized and should continue to consider that this cannot depend on individual researchers alone, but comes from changes in the ways that research and educational organizations are structured. Changes in the incentive systems for educators and researchers, such that multidisciplinary efforts are properly recognized and rewarded rather than ignored or punished, facilitates multidisciplinarity.

The development and pursuit of research agendas could involve interactive knowledge systems that call on the more active and effective participation of people outside academic disciplines. The multiple functions of agriculture and the imperative to rise to new challenges can only be met if there is active and effective participation by farmers, farm labor, consumers, environmentalists and other interested parties in the development of AKST. Links between research development on the one hand and education, training and extension on the other could be reinforced and where necessary redesigned. Multiple entry points for farmers and other agricultural practitioners into the AKST system can aid in both the identification of new research needs as well as in the implementation and application of new AKST. The role of farmer-to-farmer education and increased interaction of farmers and researchers with consumers, farm workers and environmentalists are some options that could be more seriously incorporated into the AKST system. Such increasingly interactive systems of research, education and extension will be essential in the innovation necessary to achieve development and sustainability goals.

6.1.3 Achieving the proper balance between the public and the private sectors

Working successfully on a new agenda for agriculture will necessitate ongoing attention to achieving the proper balance between public and private involvement in AKST with respect to funding, property regimes, delivery and overall governance. The recent trend toward privatization of agricultural goods and services has contributed to competitiveness, innovation and efficiency in many aspects of AKST development. However, there are compelling reasons for ongoing reconsideration of how to best protect the specifically public interest aspects of AKST development.

Agricultural production has its foundation directly in the biological world. It is also rooted in particular patterns of culture and economic organization that are specific to agriculture but vary in important ways from region to region. For this and other reasons, the balance of public and private interests and investments in agriculture is different from that in other economic activities. In the last century, government agencies, public organizations and publicly-funded universities and research institutions have worked in partnerships with private organizations and firms in a way that both served private interests and protected certain key public interests, such as relatively open access to seed varieties. Shifting the balance with regard to property regimes and governance within that partnership towards stronger private control has special implications for agriculture. For example, the increasing private ownership of intellectual property rights to seed varieties and genetic material has raised profound economic, environmental and cultural issues whose implications for society bear serious examination. The same can be said for property regimes regarding access to water and other resources. This is particularly significant when the multiple functions and changing circumstances of agriculture in the future are properly taken into consideration.

The increasing globalization of property regimes and forms of public/private interaction and the strongly influential role NAE has in shaping these changes have powerful implications for the rest of the world. Upon consideration of changes that may be appropriate for the NAE region, it is proposed that NAE not impose those changes on nations and regions that may have good reasons for choosing other legal and institutional arrangements. Achieving the best balance between the value of internationally uniform arrangements and the value of arrangements adapted to place and context can be a key issue for achieving development and sustainability goals.

6.2 Future Needs and Priorities for AKST

In NAE the evolution of agricultural science and technology during the last decades has been largely driven by academic and disciplinary approaches with the ambition to better understand biological and agronomical mechanisms of simplified and focused systems. Such approaches have led to high-level science that has in some respects ignored organizational impacts, particularly contextual elements (from biological sciences as well as from social sciences) affected by the deployment of that science, in nonlinear and unpredictable ways. These disciplinary approaches are not sufficient to address a complex problem—as a whole—and could be supplemented with more systems and overall approaches such as "complex system"[14] approaches. These new approaches are today more developed in the ecological domain and consider all relevant sub-systems or components and their inter-relations as well as their associated social, economic and policy frameworks. They require a multiple scale approach, both from a spatial (from local to global) and temporal (from short to long term) point of view.

Putting the focus on complexity and trans-disciplinary approaches does not devalue disciplinary efforts that supply basic knowledge for some of the components of the overall complex system. But, it highlights the importance of mobilizing AKST more in this direction that has been underdeveloped until now and is essential for understanding both the operation and the evolution of the whole system. This will be all the more important as the number of variables and their interrelations increase, many of them being uncertain and addressing different scaling systems (Box 6-1).

As far as global phenomena are concerned, one of the major challenges of the next decades is to develop agricultural activities that respond better to climate change: NAE could play a leading role in this domain. NAE could also consider its role in helping to deal with the spread and emergence of disease: the anticipation and management of new and emerging diseases, the occurrence of which is partly due to climate and partly due to rapid globalization. One other area where AKST can contribute to is to reduce the dependence of the NAE region on petroleum based fuels by developing alternative sources of energy and also by developing energy efficient supply chains at the global level. The NAE region has supported the implementation and development of agricultural activities in many other regions to enrich NAE's own food and nonfood systems. Another challenge for the next 50 years will be to contribute to a sustainable economic, social and environmental development in these regions.

As far as local phenomena are concerned, future agricultural research and development must consider broadening its concerns to address explicitly and directly the multiple functions of agriculture (production of food and fiber including land conservation, maintenance of landscape structure, sustainable management of natural resources, biodiversity preservation and contribution to the socioeconomic viability of rural areas (OECD, 2001) both in Europe and North America. Several broad areas of research are required in order to move towards this goal in a deliberate and logical fashion as detailed in the following sections.

6.2.1. Responding to climate change

Greenhouse gas (GHG) emissions from agriculture are in the range of 7-20% of total country emission inventories (by radiative effect) for NAE and are a contributor to climate change. AKST could be mobilized to mitigate this change while helping agriculture adapt to these changes.

6.2.1.1 Mitigate climate change through agriculture

The influence of agriculture on climate is significant but complex. Agriculture could help in reducing the increase in greenhouse gas emissions and in some cases, through expansion of some agricultural practices and land-use changes and development of new ones, can also contribute to a decrease in GHG. Some examples of agricultural practices and their potential benefits are given below:

- increase carbon sequestration in agricultural soils for example through no or minimum tillage, cover crops and green manures leading to an increase in soil carbon levels. Additional research on the enabling conditions and the magnitude of the net effect on GHG emissions could be useful;
- directly sequester carbon from flue gasses in intensively grown crops in closed conditions (Betts et al., 2007);
- increase carbon sequestration via land use change (Bro-

[14] A "complex system" is a network of many components whose aggregate behaviour is both due to, and gives rise to, multiple-scale structural and dynamic patterns that are not inferable from a system description that spans only a narrow window of resolution (Parrot and Kok, 2000). It leads to emerging new features or proprieties that cannot be predicted from the components. Complexity differs from other analytical approaches in that it is based on a conceptual model in which entities exist in a hierarchy of interrelated organisational levels. The main features of complex systems are (1) the non-linearity of relationships, (2) the occurrence of both negative and positive feedback loops, (3) their openness (show pattern of stability, even if usually far from energetic equilibrium), (4) their history, keeping memory of past events, (5) they may be nested, each component of a complex system may itself be a specific "complex system."

Box 6-1. Contribution of new complex systems science to elucidate agricultural systems

The science of complex systems makes four main contributions:

1. A better understanding of the components of the system and their interactions
2. A better control of the development of dynamic complex sociotechnical systems, e.g., new processes and materials, multi-site factory production and supply chain dynamics
3. A better understanding of the complex environment in which engineered systems exist, e.g., ecology, regulation, ethics, markets and
4. A better understanding of the design, engineering and management process that is often itself a creative, multilevel, complex human system, capable of great successes but inherently liable to spectacular failures (Bourgine and Johnson, 2005).

For these reasons a major effort is required in developing complex system science and education applied to agriculture. Specifically AKST needs to be mobilized for:

- Developing a meaningful knowledge representation and modeling of an agricultural system as a whole;
- Identifying and storing relevant information as well as developing methods to aggregate this information through the establishment of meaningful indicators regarding the functioning of the whole agricultural system; and
- Building infrastructure to facilitate the storage of information from complex agricultural systems approach.

vkin et al., 2004; Soussana et al., 2004). The conversion of arable lands to grasslands and afforestation is one of the important local options. Their net effect on GHG emissions in variable environments could be determined through additional research (Dupouey et al., 2006);

- analyze the effect of extreme heat or cold episodes on carbon accumulation. The long-term benefits of these changes and management systems could be further evaluated;
- manipulate livestock diet to reduce nitrogen losses from animals and/or reduce pH of excreta and to reduce methane emissions by ruminants (Lassey, 2005);
- use husbandry methods, management techniques and novel varieties to minimize the inputs of energy, synthetic fertilizers and agrochemicals on which present industrialized farming methods depend;
- reduce energy use via reduced use of fossil fuels in farming and food processing; and
- conduct high quality whole system studies and develop easy to use decision systems to ensure advantages in one area do not have ill effects in other areas (Seguin et al., 2005).

6.2.1.2 Reduce agriculture's vulnerability to climate change

A change in the climate that has been witnessed particularly over the past 50 years is likely to be reinforced in the next five decades. Some of the most prominent consequences of this change have been in the following areas: acceleration of several physiological processes accompanied by a greater demand in water and nitrogen, variations in rainfall (frequency and quantity), change in the radiative balance, increase in the frequency of extreme episodes and changes in biotic stress.

The geographic distribution of agricultural production within and outside NAE is likely to change considerably due to climate changes for the next 50 or 100 years, even if uncertainties remain in the timing and geographic details of these effects. Two strategies that could be pursued to address these uncertainties are (1) improving the ability to predict future effects of climate change and (2) adapting food production system to minimize adverse effects on food supply and avoid exacerbating hunger.

Improving capacity to predict future effects of climate change on the geographic distribution of agriculture and overall food production in NAE and in other regions

One of the major challenges is to better understand better the consequences of climate change where there are still considerable uncertainties. Although there is a consensus regarding an elevation of the temperature or an increase in the concentration of greenhouse gases, there is less certainty regarding other effects including change in the nature and timing of biotic stress due to the phenological shift of the host plant, outbreaks of new parasites and ways of combating them, variation in the rainfall, increased frequency of extreme episodes (e.g., summer of 2003 in Europe). Taking into account these uncertainties (rainfalls, biotic stress, political and economic choices, etc.) as well as short and long-term effects could help in the understanding of such complex questions and dealing with them. In addition, collecting serial data through appropriate long-term observations could facilitate the construction and validation of previous models and shed more light on these unanswered questions. (Seguin et al., 2006)

Reconfiguring NAE production areas to adapt and optimize available space and resources in new "environments"

Geographic shift in crop and forest production. Many studies suggest that rising temperatures could result in a shifting of crops and forests towards the north where temperatures in the future will most probably be equivalent to current temperatures in the south (Olesen and Bindi, 2002). In Europe, for example, cereals in Finland will shift 100-150 km towards the pole for each 1°C rise in temperature (IPCC, 2001). Continental and mountain forests are expected to occupy less surface area in the future compared to their present distribution as they are sensitive to high temperatures and extreme drought conditions.

The NAE region could anticipate some of these profound changes in the geographic organization and utilization of agricultural lands and study (Easterling, 1996; Watson et al., 1997):

- Possibilities of extending crop productive agricultural lands to Siberia and northern Canada;
- Optimal shift of perennial horticultural crops by optimizing the interactions between varieties, new cultivation environment and crop management systems;
- Occupation of the most sensitive regions (irregular rainfall alternated with intense droughts) by plants that are more robust and have a high plasticity; and
- New and changed species composition of forest areas and its consequence on the amount of forest biomass available.

This adaptation and preparation will be more robust if data have a higher degree of certainty than before, as with the data from IPCC used for predictions of climatic changes (IPCC, 2007).

Development of new and adapted agricultural practices and crop varieties. Simultaneous development of new varieties, either new crops or agricultural crops adapted for predicted climatic changes and agronomic practices appropriate for those crops under predicted climatic conditions may be required.

Some of the desirable traits for these new varieties are better suited for high temperatures, with increased or stable growth with less water and and/or transient drought tolerance, longer durations of vegetative growth and grain filling periods, early budding and better frost resistance for orchard varieties and field crops (Seguin et al., 2006).

Some practices include planting earlier so that crop development would be more advanced in the case of a summer drought, using longer-season cultivars, mixing cultivars and planting seeds deeper and harvesting earlier. Early planting might also eliminate the necessity of artificial drying of grain. Soil moisture may be conserved by using conservation tillage methods, modifying the farm microclimate for example by integrating trees as shelterbelts and changing the way irrigation, fertilization and crop-protective sprays are scheduled, so that inputs are applied according to crop needs or field conditions.

A reevaluation and adjustment of these above mentioned options may prove to be useful, for example by taking into account new rhizosphere communities that develop due to climate change and the effect of these communities and their interactions with the surrounding agroecosystem.

Development of new social systems to enable smooth transitions of rural economies and maintenance of world food supplies. Mass human migrations stimulated by scarcity are often highly disruptive and damaging. If global climate change undermines the basis for agricultural production in rural NAE, it may cause dust bowl-like migrations such as those that occurred in the US during the 1930s. New social programs could be designed to face this scenario and to help alleviate rural poverty and facilitate the economic transformation of rural NAE. Ensuring a stable production of agricultural products so that world food supplies are maintained during these transitions could be one of the main goals of these programs.

6.2.2 Facing new and emerging human, livestock and plant diseases

6.2.2.1 Human and livestock diseases

The past few decades have seen an alarming increase in new and emerging diseases such as AIDS, BSE, SARS, avian influenza, foot and mouth disease and others. These diseases are seen as a threat to global animal, plant and human health. One reason for this upsurge is the increased exposure of humans to infectious agents through changes in lifestyle, international travel and industrialization and globalization of the food industry. However, adequate understanding of the root causes of this upsurge is still lacking. Clearly it will not be possible to meet development goals unless the AKST system responds to the challenge of emerging diseases.

AKST could be used to elucidate the following aspects for a better management of these diseases through the following:

- Understanding the origin of new and emerging diseases
  - Differentiate between "new" and "emerging" diseases: some of these diseases may be old diseases with newly recognized etiologies. Others are diseases that did not exist more than 100 years ago. This difference is important to understand to be able to project the future occurrence of new and emerging diseases (Desenclos and de Valk, 2005);
  - Understand the ecological and evolutionary dimensions leading to the development of new and emerging diseases.
- Predicting epidemics and pandemics across both spatial and temporal scales
  - Identify factors that increase the risk of developing infectious diseases: new areas of risk factor research include the relationship between changes in the environment (such as climate change) and the incidence and distribution of diseases; and the influence of crop and livestock genetic makeup on their susceptibility to disease and response to treatment (Desenclos and de Valk, 2005);
  - Develop basic fundamental research about hosts, pathogens and their interactions at different levels (molecular, cellular and superior integrative levels) (Horwitz and Wilcox, 2005):
    - Hosts: physiopathology, immune response;
    - Pathogens: ecology and biology of the pathogens, vectors; and
    - Host-pathogen interactions: cellular and molecular mechanisms, evolutionary potential (develop a better understanding of how pathogens mutate and migrate and how they skip host species barriers) and in particular research on resistance to anti-infectious drugs.
- Construct models for the system as a whole:
  - Multidisciplinary groups of scientists studying the ecology of an emerging infectious disease could help in the building of these models. These models are parameterized with data from field studies and pathological and microbiological investigations. These studies enhance classic epidemiology by involving an array of medical, veterinary, health and

ecologic scientists and others in a dialogue between model building, parameterization and further refinement of models (Daszak et al., 2004);
 - Integrating various disciplines (evolutionary, social, anthropologic, geographic, economic and public health sciences) could help in understanding the determinants of new and emerging diseases (Daszak et al., 2004; Desenclos and de Valk, 2005).

Building surveillance and response networks

Early detection allowing a rapid response to emerging infectious diseases is essential (WHO, 1998). However, this depends upon the application of the latest diagnostic tools along with developing newer tools and appropriate, predictive epidemiological analysis (Thompson, 2000). Technological advances that require interaction between government, policy advisers and scientists could be applied as part of surveillance strategies (Hughes, 2001).

Some of the options that could strengthen the ability of veterinary and human quarantine systems to cope with the growing threats and build comprehensive, strategic and effective surveillance are to:

- Set up observatories at appropriate scales to collect data over long periods as they could help understand the temporal and spatial dimensions of the epidemiology of the different diseases;
- Develop diagnostic tests and systems that are reliable when the disease is rare; and
- Develop new methods of disinfection to avoid propagation: assess new methods for sterilization of food and reduce contamination of water.

Other innovations that may prove essential are developing new types of cures through newer forms of drug discovery and also through immunizations using nanotechnology or biotechnology to quickly vaccinate livestock and wildlife to cure the disease and thus to lower the chances or delay the disease jumping to humans.

Building and strengthening coordination between veterinary and public health KST infrastructure and training

The coordination between veterinary and public health infrastructure is the underlying foundation that supports the planning, delivery and evaluation of public health activities and practices (Salman, 2004). Three of the main areas that could be developed to help build efficient infrastructure are listed below:

- Enhance epidemiologic and laboratory capacity: the "new" tools of molecular epidemiology could be rapidly deployed to counteract the potentially devastating effects of emerging and re-emerging infectious diseases. In particular, accurate and sensitive DNA-based diagnostics and mathematical models can be used to provide optimum surveillance and the ability to predict the occurrence and consequences of disease outbreaks so that the necessary "preparedness to respond" is available and control strategies can be established (Thompson, 2000). This would play an important role to better understand the advances in investigations of outbreaks, assessment of vaccine efficacy and monitoring of disease trends;
- Provide training opportunities in infectious disease epidemiology and diagnosis in the NAE region and throughout the world with the goal to train laboratory scientists to become leaders in public health laboratories, especially at the state and local levels (Hatch and Imam, 1996); and
- Increase funding given that, in the future, the development of prediction and prevention programs to eradicate or minimize these emerging infectious diseases at a global level may require more global resources accompanied by a greater involvement of international non-governmental development and aid organizations. It is vital that a coordinated global civil society rather than an exclusively governmental approach be implemented in the prevention of these diseases (Harrus and Baneth, 2005).

6.2.2.2 Insect pests, weeds and diseases of plants

Similar to new and emerging human and livestock diseases, there has been an upsurge of new insect pests, weeds and pathogens of plants in the past few decades. Recent examples with important economical or social consequences include epidemics of sudden oak death disease caused by *Phytophthora ramorum* (Rizzo et al., 2005); new genotypes of potato late blight in the US (Fry and Goodwin, 1997); the appearance of Phylloxera, a root-feeding aphid, on grapevines in Europe; and increased parasitic weeds, especially in Europe. In each case, society and/or agricultural practices were severely affected.

Currently, weeds are the major biotic constraint on crop production and the farmers' major variable inputs are for weed control. In NAE nearly 70% of pesticide applications are of herbicides for weed control (over 50% worldwide) and much tillage is to control weeds. The success of chemical control of weeds has led to a weakening of AKST in dealing with weeds, both in the public sector in dealing with ecological and physiological relations between weeds and crops and the whole ecosystem and the private sector has come up with only one new target site for herbicides in the past two decades. The result has been deleterious changes in weed spectra in ecologically preferable minimum tillage systems that reduce erosion and chemical run-off to harder to control perennial weeds.

Farmers are troubled in trying to balance the demands of multifunctionality and weed biodiversity with the needs of productivity and supplying the demands of food fiber and fuels, as the weeds that supply food to wildlife in the field are often secondary hosts of disease and insect pests as well as direct competitors with crops for resources. Weed control is the major constraint to organic agriculture, where considerable soil degrading tillage and backbreaking manual labor is required to deal with weed problems and the ensuing ethical dilemmas. Despite the present and future problems posed by weeds, both public and private sector AKST in weeds is disproportionately low compared to the AKST investment in dealing with other biotic stresses.

As far as pests are concerned, although fewer studies exist compared to their counterparts affecting humans and animals, the emergence or re-emergence of these pests is considered to be linked with several concurrent factors among which are (Anderson et al., 2004):

- Increased global travel and global trade of plant materials, including crop plants but also exotic species used as garden or ornamental plants (Mack and Erneberg, 2002); this trade results in increased risks of dispersing pests onto new hosts and/or into new geographical areas;
- Climatic modifications such as global warming that have already resulted in the extension of the range of some insects, including vectors of pathogens; although currently limited, this effect is predicted to see its importance increase during the forthcoming century;
- Modification of farming practices, with a strong trend towards a reduced diversity of crops and an increased contribution of monoculture;
- Increased occurrence of resistance to pesticides in both insects, weeds and pathogens, further reducing our ability to control these pests and resulting, in some situations, in the build-up of large, difficult to control pest populations; and
- Evolution of the pests themselves, expanding host range of weeds, insects and pathogens and increasing occurrence of feral, weedy forms of many crops.

AKST could be developed and used to understand the root causes of these new and emerging pests to shift focus to preempting new pest emergence, rather than just responding to it. Some of the main options for action in this domain are listed below:

Better understand the origin of and the factors responsible for the invasiveness of insect pests, weeds and pathogens of plants

- Study the factors that determine the invasive potential of these pests:
  - Genetic factors, including genetic makeup, gene expression and its influence on the adaptation of these new pests to the new environment;
  - Ecological factors, including the conditions that could either inhibit or stimulate the invasive potential of new pests.
- Understand how these new pests alter ecological community structure, which in turn can facilitate the development and propagation of these pathogens;
- Study weed ecology, to allow maximum biodiversity with minimum impact on productivity and crop health;
- Conduct retrospective studies on biotic invasions to better understand the factors that stimulated the invasive potential; and
- Increase international collaboration to facilitate the exchange of biological and ecological information associated with insect pests, weeds and pathogens with high invasive risk potential.

Build surveillance and detection networks

- Track the changing geographic distribution of potentially dangerous invasive pests with associated ecological data;
- Develop improved techniques/models/strategies/frameworks for Pest Risk Analysis i.e., the capability to predict the potential risk(s) linked to the introduction of pests into region(s) where it is absent; such strategic analyses could help to focus some monitoring efforts on "high risk" agents and to de-emphasize efforts on "low-risk" agents that have nevertheless made their way to quarantine lists;
- Implement surveillance and efficient alert systems:
  - Develop internet databases with taxonomical and biological data on these pests and pathogens and store samples with their respective data at the regional or national level; this calls for more research in taxonomy of pests and pathogens;
  - Train field workers (agricultural cooperatives, entomologists, naturalists, etc.) to detect the presence of pests and pathogens rapidly and not only alert the other actors involved but also to contribute and supply data to regional and national databases; and
  - Develop new molecular detection tools (e.g., gene chips) that could in certain cases be used *in situ* directly on the fields for cost-effective detection of potentially invasive species and rapid assessment of both qualitative (presence or absence) and quantitative (number) changes observed in affected biological communities.

Developing appropriate management and regulatory measures

- Develop a database with control methods for these pests, preferably based on sustainable, ecological pest management methods coupled to surveillance and detection networks;
- Build systems for effective border control to deal with risks from pests and disease-causing pathogens;
- Develop adaptive management systems to be able to adjust rapidly management and regulatory measures;
- Develop newer and safer pesticides as well as breed pest-resistant crops; and
- Develop new biological control agents to suppress pests and replace chemical controls and tillage.

6.2.3 Contributing to a global strategy for a low carbon economy

6.2.3.1 Biofuels

The heavy dependence of the NAE during past century on petroleum is a major challenge. AKST can be deployed to develop agricultural production of biofuels while decreasing net carbon dioxide output.

Some of the sources that could be used for producing biofuels are: cereal grains and oilseeds to produce bioethanol[15] and biodiesel[16] (1st generation biofuels), cellulosic materials (2nd generation biofuels) and, algae and cyanobacteria (3rd

[15] The complex carbohydrates in plant material are hydrolyzed to simple sugars that are fermented into ethanol or butanol, which can be used in internal combustion engines.

[16] Crop oils are increasingly being used for use as biodiesel. The crop oil is de-esterified to release the fatty acids for use as biodiesel, and glycerin is a byproduct.

generation biofuels). The possibility of producing biohydrogen and bioelectricity using nature's photosynthetic mechanisms (4th generation biofuels) might be explored as well.

Ethical considerations—effects on food security

The rapid growth of biofuels industry is likely to keep farm commodity prices high through the next decade as demand rises for grains, oilseeds and sugar from 2007 to at least 2016 (OECD/FAO, 2007). This will substantially increase meat and milk prices in the NAE and decrease the amount of grain available to the poorer parts of the world both as direct imports and food aid. Brazil's earlier diversion of cane sugar to ethanol stabilized high sugar prices, assisting farmers worldwide, but at a higher consumer price. The diversion of grain to fuel can negatively affect the millennium goal of alleviating hunger throughout the world in the short term, but might have a positive long-term effect. Because heavily subsidized NAE grain will no longer be "dumped" on developing world markets below production costs, the subsistence farmers in the developing world could switch from subsistence to production agriculture, increasing yields and self-sufficiency. There are also eco-ethical considerations; putting more ecologically fragile and necessary lands into production of biofuels; whether oil palm production in Southeast Asia at the expense of jungles, or soybean production at the expense of rangeland or rain forest. It may not be morally justifiable to purchase oils for biofuels from areas where the environment is being negatively exploited. Proponents of some biofuel crops state that they will be grown on "marginal" land. Such lands may not be marginal to biodiversity and wildlife, posing another ethical issue. As discussed below, a major NAE biofuel crop, oilseed rape emits a major greenhouse gas and Jatropha, which is promoted by many NAE organizations in Africa and Asia is highly poisonous. There are ethical issues about promoting the cultivation of such crops. Thus alternative feedstocks that do not increase agricultural area are needed for biofuel production, such as waste cellulosic material, algae and cyanobacteria.

First generation biofuels (cereal grains and oilseeds)

The energy, economic and environmental results of the 1st generation liquid biofuels cannot make them a substantial alternative to fossil transport fuels (IEA, 2004; Sourie et al., 2005). Co-production can improve energy and economic balance and biofuel costs will go down as the technologies improve in production efficiency and economies of scale are realized (Farell et al., 2006). In addition, the amounts of land that would be required to obtain self-sufficiency in biodiesel using oilseed crops alone varies from 9-122% of the global cropping area (Table 6-1), which makes it clear that both fuel and food needs cannot be supplied by standard crop agriculture alone.

Second (cellulosic substrates) and third generation (algae and cyanobacteria) biofuels

New developments in biofuel production seem necessary. Two types of cellulosic second generation substrates for biofuel production are being considered: straws and specially cultivated material. The use of cellulosics will have a higher net energy gain than seed grains/oilseeds (Samson et al., 2005; Farrell et al., 2006), but the present technologies are less environmentally friendly than those using grain, as they use dilute sulfuric acid and heat to separate lignin from the carbohydrates. Third generation sources, such as algae, may be even more environment friendly as well as cost effective. Concentrating future R&D options on the following areas can help make second and third generation sources viable:

Research can help define plant ideotypes that fulfill certain criteria and respond to certain needs:

- Assimilation of carbohydrates (starch and sucrose) at the detriment of proteins. This is cost effective as the crop requires a lower quantity of inputs (particularly water and nitrogen) and has less hauling requirements; some examples are leguminous plants, as they require less fertilizer and the cultivation of C4 plants adapted to low temperatures (Heaton et al., 2004);
- Production of fermentable 5- and 6-carbon sugars that can subsequently be converted to ethanol;
- Increasing the amount of cellulose (especially at the expense of lignin), or modifying its structure such that more is available to cellulases could increase the bioethanol yield of straws and specialty grasses, while lowering demands for acid and heat for hydrolysis (Attieh et al., 2002);
- Increasing the lignin content and the digestibility of straws through transgenics. The solution to increasing digestibility without affecting important traits is to transform elite material to contain modified lignin and cellulose contents, e.g., by partial silencing of the pathway enzymes leading to lignin (Gressel and Zilberstein, 2003), or by enhancing cellulose synthesis (Shani et al., 1999);
- Lowering the presence of polluting silicon in both straws and cultivated grasses;
- Lowering emissions of methyl bromide from oilseed rape or canola (Gan et al., 1998);
- Developing lodging-tolerant varieties and when necessary dwarfed varieties;
- Developing insect resistant varieties, as lower lignin can lead to pest infestation;
- Improving the stand establishment of perennial grasses (Schmer et al., 2006);
- Compensating for the reduction of soil carbon and its consequences on soil quality due to straw harvesting; and
- Explore the harvesting of unwanted aquatic weeds such as water hyacinth for biofuel production.

Biotechnologies including genetic engineering could potentially achieve most of the above mentioned targets by modifying certain metabolic pathways. Development of such modified varieties requires taking into account appropriate safety concerns.

Research and technological developments could focus on increasing processing efficiency by:

- Increasing the efficiency of cellulolytic enzymes, e.g., by gene shuffling to increase activity, stability, temperature optima;
- Improving the pretreatment step that disrupts the structure of the biomass and releases 5-carbon sugars from hemicellulose, hydrolysis of the cellulose to form 6-car-

Table 6-1. **Oilseed crop typical yields and land requirements for self-sufficiency.**

	Typical yields	Area necessary to meet demand	Arable land necessary
	Tonnes oil/ ha/yr	million hectares	% of global total
Oil palm	5	141	9
Jatropha	1.6	443	29
Oilseed rape	1	705	46
Peanuts	0.9	792	51
Sunflower	0.8	881	57
Soybean	0.4	1880	122
Algae/ cyanobacteria*	52.8	4.5*	2.5

*Yield data for 30% oil content (low), area necessary to meet 50% USA diesel demand.

Source: Chisti, 2007.

bon sugars and the fermentation of sugars to ethanol. This remains a technical challenge due to the nature of lignocellulosic feedstocks; and
- Improving fractionation technology while reducing the sulfuric acid and heat to hydrolyze lignocellulosics.

Research options for third generation algal/cyanobacterial biofuels could focus on increasing organism survival, growth and lipid content, carbon dioxide enrichment and yields:
- Organism survival: the best laboratory strains become contaminated and taken over by indigenous local organisms under field conditions. Transgenes conferring herbicide resistance might overcome this problem;
- Growth & lipid content: algae either grow *or* alternatively they produce lipid (fat) bodies, but not do both simultaneously. This requires batch culture or separate growing ponds and lipid producing ponds, increasing production costs. Pathways and genes for continuous production of the best lipids for biodiesel are becoming known and could be explored (Ladygina et al., 2006);
- Carbon dioxide enrichment: the algal response to added carbon dioxide is not as good as it could be; molecular research in photosynthesis could potentially increase yield (Ma et al., 2005);
- Seasonal high yields: algal growth is a function of temperature—when it is too cold they grow less and most do not do well at high summer temperatures; recent and future AKST with plants will have much to offer to overcome this problem (Shlyk-Kerner et al., 2006); and
- Poor light penetration to cultures requires shallow ponds and lower yields; further research in trimming photosystem antennae size could greatly increase efficiency and yields (Tetali et al., 2006).

Research is also required to establish the ecobalance and life cycle analysis for each source.

Fourth generation: producing biohydrogen and bioelectricity

Biophysicists have seen it as an intellectual and practical challenge to harvest solar energy to produce hydrogen or electricity by directly using nature's photosynthetic mechanisms, or by embedding parts of the photosynthetic apparatus in artificial membranes, or using algae to produce sugars and yeast or bacterial enzymes to produce electrochemical energy (Tsujimura et al., 2001; Chiao et al., 2006; Logan and Regan, 2006). This will necessitate considerable long term multidisciplinary efforts to become more than a laboratory curiosity. The informational gains, as well as the new fuel gains about basic biophysical processes are bound to be exceedingly important to AKST.

Biofuels and global carbon balance

The grains, oilseeds and specially cultivated grasses (switchgrass and *Miscanthus)* used for biofuels require considerable fuel in their production and processing. The straws by-products require energy only in harvest and processing and will give a much more favorable carbon balance. Algae and cyanobacteria can achieve the highest carbon balance, as they can be directly "fertilized" by industrial flue gasses, directly removing them from the environment (Brown and Zeiler, 1993).

6.2.3.2 Improve energy efficiency of supply chains: food miles and life cycle analyses

Changes in food production and marketing systems, accompanied by changes in transport technologies, have led to increased transportation of agricultural and food commodities, both in raw and processed forms. Modern food supply chains, transforming goods from field to fork, tend to have greater "food miles" per unit of final consumption than in-season, locally procured items. Transportation enables producers to exploit comparative advantage in farming and food by extending the spatial distribution and size of markets for their produce, to the mutual benefit of producers and consumers, thereby enhancing overall economic efficiency. This applies to produce transported within NAE and between NAE and other regions.

High food miles, especially involving heavy road vehicles, can, however, have negative impacts on sustainability associated with energy use, congestion, pollution and accidents (Smith et al., 2005). These transport related impacts can be significant at the local and national scales. Furthermore, failure to fully attribute the costs of these impacts to transport could give unfair advantage to distant compared to local produce. A large share of total food miles, however, is attributable to shopping by car at out-of-town supermarkets, reflecting changes in retailing and in purchasing habits (Pretty et al., 2005; Smith et al., 2005).

Thus, the relationship between food miles and sustainability of food supply is complicated (Smith et al., 2005). Aggregate travel distance is not in itself a useful indicator. Shorter distances are not necessarily more sustainable due to differences in the characteristics of food and supply systems (Smith et al., 2005; Saunders et al., 2006). Much depends on modes of transport, economies of scale in transportation, complementary functions such as refrigeration and differences in the overall cost and energy efficiency of food production, processing and delivery systems as these affect their comparative advantage over time and space.

Total energy use and CO_2 per unit commodity delivered to end users are more meaningful indicators, requiring a whole life cycle perspective, including production stages (Audsley et al., 1997; Smith et al., 2005; Williams et al., 2006). Thus, while increasing transportation efficiency is a valid target, minimizing transport miles and costs in themselves are not and to do so could increase prices and reduce the range and quality of produce available to consumers and compromise the livelihoods of low cost but distant producers, especially those in developing countries. Some of the options cited below could help guide decisions on sustainable transport throughout the supply chain and its organizations:

- Develop crop varieties by breeding and biotechnology that can be shipped by sea instead of air while maintaining quality.
- Further develop treatment technologies (chemicals, storage conditions, irradiation, biologicals) that preserve shelf life of agricultural commodities allowing shipping by sea instead of air.
- Develop and apply databases and routines to assess transport efficiency and total energy/CO_2 emissions within a whole life cycle, field to fork approach, including full environmental accounting for transport functions.
- Develop methods for carbon and energy accounting, reporting and labeling for food commodities, appropriately communicated to consumers.
- Develop methods to enhance consumer understanding and appreciation of sustainable food procurement and transport systems to inform consumer choice.
- Develop increased efficiency in food transport technology and logistic management systems, including promotion of sustainable transport.

6.2.4 Trade, markets and agricultural policies

As a major importer of commodities, labor and resources and an exporter of products, investment and AKST, NAE has influenced food and agriculture systems throughout the world. Regardless of which scenario will play out in the future, NAE's influence on other regions will continue. It is to NAE's advantage to ensure sustainable development of the whole world's food and agriculture system as well as its own. This task includes environmental, economic and social considerations in a context of autonomy for everyone (Box 6-2).

Development of competitive and viable local production systems could be based on measures to ensure food security, improve farmers' livelihoods and assure sustainable development for both NAE and the concerned regions. Since exchanges between NAE and the other countries are presently through the trade system, NAE has the potential to participate in the continued evolution of the world trading system to ensure that it becomes more fair and equitable.

Develop competitive and viable local production systems

The diversity of agricultures throughout the world is a consequence of the heterogeneity of available natural resources and local, social and historical contexts (Mazoyer and Roudart, 2006). A role for AKST could be to analyze this diversity of agricultures, their resources and constraints and their potential in terms of production, environmental services, social contribution, public goods and externalities as well as an analysis of production systems. The development and implementation of AKST could be based on the following principles (CBD, 2005):

- Focus on food security and improvement of farmers' livelihoods;
- Build on previous experience and knowledge, through combining the skills and wisdom of farmers with modern scientific knowledge;
- Focus on integrated holistic solutions and technical adaptation to local contexts within a clear framework that builds on the principles of the agroecosystem approach;
- Promote cross-sectoral approaches to address different perspectives (social, political, environmental) through association and flexibility; and
- Prioritize actions based on country goals and the wants of direct beneficiaries and locally validate such actions through the full participation of all actors.

Continued attention to family farms is important as they were the basis of agricultural development and the forerunner of industrial development in NAE (Danbom, 2006). In many countries today such farms represent a major part of the rural population. If efficient and competitive in production and trade, these small producers could significantly contribute to achieving a higher and more sustainable pace of development, thereby promoting economic growth and social cohesion (IFAD, 2001).

In order to reach development and sustainability goals, NAE's contribution to AKST in other regions could result in alleviating rural poverty by improving access to resources and improving skills and institutional support so that the rural poor can benefit from:

- Improved access to natural resources, especially land and water and improved natural resource management and conservation practices ;
- Improved agricultural technologies and effective production services;
- Broad range of financial services;
- Transparent and competitive markets for agricultural inputs and produce;
- Opportunities for rural off-farm employment and enterprise development and
- Local and national policy and programming processes

The principle of division of labor has been a common feature of AKST (Herman, 2001). Product specialization based on resource endowments and linked with appropriate AKST will increase productivity (Mattson et al., 2006).

Develop a fair and equitable trade system.

Market forces are shaping and will continue to shape the future of the world's agriculture and food system (Brown, 2002). Private enterprise operating through the market is the main engine of sustained economic growth, but it requires

Box 6-2. Contrasting views on agricultural development and markets

From a sustainability point of view, a society must provide for the replacement and growth of its capital, including both human reproductive capital and replenishment of natural resource capital. Capacities can be constrained severely by scarcity of soils, water, and energy, among other factors, when there is growing demand for food and energy. Scarcity of renewable natural resources is contingent with their use.

Markets are necessary, but do not guarantee sustainability of public goods such as food security, conservation of natural resources, or protection and enhancement of the environment. There are incentives to produce goods with negative externalities because producers may not pay for damage caused to public goods (Stiglitz, 2006).

Agricultural production happens within complex agrarian systems whose capacities can be constrained by a lack of resources or lack of autonomy. Sustainable development should be based on the pillars of endogeneity (as opposed to mimetic growth), self-reliance and self-confidence (as opposed to dependence), be need-oriented (as opposed to market-led), in harmony with nature and open to institutional change (Sachs, 2002). Resilience, the capacity to absorb shocks, is necessary to prevent dependency. A compromise is necessary between the two extremes: autarky and completely free trade. It is necessary to imagine an optimum position that ensures viability and resilience, such as is the principle of biological systems (Tabary, 1993).

This perspective contrasts with the view that puts much more emphasis on the role of less regulated markets and does not see agriculture as an activity that is different in character from other economic activities. In this view, the private sector must be the engine of economic growth; inflation must be low to maintain price stability; state bureaucracies must be small; government budgets must be close to balanced; tariffs on imported goods must be lowered or eliminated; restrictions on foreign investment must be removed; industries, and stock and bond markets must be open to foreign ownership and investment; quotas and domestic monopolies must be removed; exports must increase; state-owned industries and utilities must be privatized; capital markets must be deregulated and currencies made convertible; the economy must be deregulated to promote domestic competition; government corruption, subsidies and kickbacks must be eliminated; banking and telecommunications systems must be opened to private ownership and competition; and citizens must be allowed to choose from among competing foreign and domestic pension options and mutual funds (Friedman, 1999).

For a variety of reasons, the previous position has been a point of major conflict in international trade and financial negotiations over the last two decades. A long-standing perspective within the field of agricultural economics contests this position with one that emphasizes the particular nature of agriculture in its social and biological context. This position argues out that in the agricultural sector the general equilibrium model of the economic theory with a unique and social optimal equilibrium price cannot, indeed, a fortiori at a world level, be simply applied, for the following reasons (Loyat, 2006):

- Certain assumptions for a competitive equilibrium are not met (market failures, asymmetry of information, great differences in productivity levels between agricultures), making any optimal equilibrium illusory;
- Public goods, such as food security or protection of biodiversity, are not recognized by the market. Consequently, the market price will not be able to guarantee these public goods;
- General Equilibrium model cannot represent the diversity of agricultural economics. The equilibrium price on the world market is disconnected from the real costs of production because of imperfect competition, dumping practices and the heterogeneity of resource endowments and labor productivity. This situation can be detrimental for most of local farm systems; and
- Agriculture relies on complex short and long-term interactions. Non-consideration of food security, biodiversity and environmental impact impede price signals from being socially efficient.

that states ensure that the investment climate is conducive to growth by equitably upholding property rights and contracts, maintaining political and macroeconomic stability, providing public goods, using regulation and public services to fill gaps left by markets and investing in the education, health and social protection of its people (Wolfensohn and Bourguignon, 2004).

From a market point of view, AKST can contribute to developing a fair and equitable trade system in NAE and in the rest of the world. A better understanding of what a fair and equitable trade system is, including further examination of the potential negative effects of measures such as dumping, may be required. There are contrasting views on this (see Box 6-2). AKST can help to (1) better understand the market mechanisms; and (2) improve the modeling representation of the agricultural systems and their dynamics, including all the players and their inter-linkages through the markets.

(1) To understand better the market mechanisms

In the field of market mechanisms, the questions for research can relate to:

- Institutional analysis of local, regional and international markets and their mechanisms;
- Modes of cooperation and coordination between players of the food system and the distribution of the added value;
- Economic, institutional and social conditions for an access to the markets for all the actors;
- Promotion of fair trade through prevention of monopolistic practices;

- Attributes of the quality of the products (origin, know-how, practices and manufacturing process) and the way in which markets are able to recognize the qualification process of quality;
- Institutional arrangements necessary for the remuneration of the positive externalities and for an adequate level of public goods production leaving transaction costs between potential beneficiaries sufficiently low and
- Public policies able to generate a fair and equitable global trade system.

AKST could also investigate whether and how comparative advantage from specialization coupled with trade really favors smaller economies more then larger economies (Anderson, 2004).

(2) To improve the representation of complex agricultural systems in the models

Agricultural trade has been one of the most contentious issues in multilateral trade negotiations in recent years due to the effects it could have on developed and developing countries. The trend has been towards more open markets, suggesting that worldwide agricultural production is likely to become more competitive. Analyses suggest that the impact of trade on developing countries will be very uneven. Some simulations even go so far as to suggest that the effects of agricultural trade liberalization will be small, overall and are likely to be negative for a significant number of developing countries (Bureau et al., 2005; Polaski, 2006). Policy recommendations derive too often from static, perfect competitive simulation models. More emphasis could be laid on technological change-induced inequalities, missing market effects on inequalities, dynamic adjustments impact assessment (Chabe-Ferret et al., 2005).

In order to improve the representation of the models (Box 6-3) it is important to distinguish between the various groups of developing countries (net food exporters vs. net food importers, least developed countries benefiting from huge trade preferences, least developed countries with main exports severely penalized by tariff peaks). It is essential to take into account the complex effects of the various types of domestic support, trade preferences (which are presently well utilized in the agricultural sector), regional agreements and the effect of trade liberalization on them (Loyat, 2004).

A larger discussion of the specification of the trade models used for the simulations of market liberalization and policy consequences might be required. This discussion could include representation of labor markets; imperfect information; price instability; uncertainty and risk; dynamics; and environmental externalities.

6.2.5 Promoting food quality and safety in diverse food and farming systems

Research and development in the last 50 years focused mainly on a unique model of development based on an increase of productivity. This trend is changing as recent evolution of agriculture shows that more than ever consumers are emphasizing on food quality, food safety and the relationship between diet and health to combat malnutrition and obesity (WHO, 2003; EC, 2005; USHHS and USDA,

Box 6-3. A specific need for agricultural research in economic modeling: The case of CGE models

The computable general equilibrium (CGE) models have become major instruments supporting trade negotiations. These models provide quantitative estimates of benefit, as well as how benefits are shared among stakeholders. Agriculture is not treated differently than any other economic activity. The validity of this approach can be questioned.

There are three main criticisms to CGE models:

- The most liberalized situations depicted through these models are theoretically efficient and Pareto optimal.[1] But they rely on a particular income distribution, resulting from rewards to factors of production such as land, labor and capital which reflect their relatively scarcity (i.e., economic rents)[2] which are themselves not necessarily socially optimal. Other Pareto efficient situations, with different income distributions, could be deemed more socially desirable;
- Only those commodities which are subject to market exchanges are accounted for externalities, such as water pollution, factors that are ignored by the market are also ignored in the CGE benefit/cost balances analyses; and
- A CGE model assumes markets are functioning efficiently, i.e., marginal costs are equal to marginal returns everywhere, producers and consumer adjust their plans immediately in response to observable equilibrium prices (hence the reference to "equilibrium") (Boussard et al., 2005).

The existence of price instability confounds the price signal and renders it economically inefficient. Thus it appears that these models have no connection with reality. Furthermore, agricultural markets often operate imperfectly because of restrictive practices by dominant players and high levels of risks and uncertainty, especially associated with variation in supply (Boussard et al., 2005).

It follows from the previous considerations that there are gaps in research on the ways to manage supply and demand for agricultural products, knowing that, regardless of the scale: prices on agricultural markets are unstable and volatile, the supply of agricultural products is unstable, sometimes chaotic and subject to uncertainty and risks, especially for the poorest decision makers lacking in resources who are more risk averse than others. Such research may lead to specific policy considerations to improve modeling of agricultural markets and correct for market imperfections (Box 6-2).

[1] *Pareto optimality* is an important notion in neoclasical economics. Named after Italian sociologist and economist Vilfredo Pareto (1848-1923), Pareto optimality is a situation which exists when economic resources and output have been allocated in such a way that no-one can be made better off without sacrificing the well-being of at least one person.

[2] David Ricardo's Concept of Economic Rent on land is the value of the difference in productivity between a given piece of land and the poorest (and/or most distant), most costly piece of land producing the same goods under the same conditions (of labour, capital, technology, etc.).

2005). AKST could be used to understand and respond to these new expectations.

In addition, a new profile for agriculture is taking shape, with two major poles (Loyat, 2006; Hubert et al., 2007): agriculture directed by the demand for products of standard quality; and agriculture directed by the supply of specific products, identified by their origin or their manufacturing process (Box 6-4). In both cases AKST could be used to support the food and fiber supply chains that connect producers to markets, provide incentives and just rewards to producers, processors and marketing agents, provide products of value to consumers and society and also support rural livelihoods. The organization and operation of these supply chains vary considerably across NAE and also among different commodity chains. Global supply chains in bulk commodities often run along side local procurement networks of highly differentiated products.

Some of the areas where AKST can intervene and help respond to the above mentioned expectations are presented below.

6.2.5.1 *Improve food quality*

AKST can explore the relationship between food, diet and health by taking into account the cultural diversity of food systems and the diversity of human responses within a given food system.

AKST can understand better the relationship between diet and health by:

- Investigating basic mechanisms by which nutrients or specific food components may act on biological mechanisms (gene expression, cell signaling and cell function, integrated physiology) (Young, 2002);
- Performing high throughput analysis of biological responses with techniques such as transcriptomics, proteomics and metabolomics (Afman and Muller, 2006; Trujillo et al., 2006); and
- Investigating how genetic polymorphism and metabolic imprinting (influence of early nutrition) of individuals result in the variability of physiological responses to diet (Waterland and Garza, 1999; Miles et al., 2005).

AKST can take into account the various determinants of food choices and their influence on health by:

- Investigating the biological, psychological, historical and socioeconomic factors that affect food choices, as well as their interactions (Bellisle, 2003); and
- Identifying the early events taking place in infancy and childhood that are critical for the development of food preferences e.g., predilection for more diversified foods, fat free foods, high protein foods, etc. (Hetherington, 2002).

AKST can improve the nutritional composition of food for health purposes by (Roberfroid, 2002; Richardson et al., 2003; Azais-Braesco et al., 2006):

- developing methods of assessment of "nutritional profile" of foods that allows a comparison between various food products regarding their contribution to the overall balance of the diet;
- developing functional foods and confirming related health claims; and
- developing and applying methods to remove anti-nutrients, allergens and toxins from the food chain.

Box 6-4. Bipolarisation of agricultural demand

A new profile for agriculture is taking shape, with two major poles.

A demand for common products

The first pole corresponds to an agriculture that provides *basic common commodities*. From an economic point of view, the sustainability of this agriculture is guaranteed thanks to a combination of land, capital and labor with competitive production costs on the international markets. On environmental aspects, standard operating procedures provide information on quality and a sanitation and environmental profile of each good. Farming systems tend to be large scale, specialized with high level of division of labor into particular tasks.

In this type of agriculture the majority of the farmers gradually ceased direct marketing and processing and became suppliers of raw material at low prices (Bonny, 2005). The food processing chain is more complex, made up of players whose economic dimension and the number on each level is very variable, *for example*, a significant number of heterogeneous consumers, farmers generally of modest economic size, a central group of players with a lot of influence on the chain (e.g., central purchasing agencies).

The agro-industries and the distribution companies capture a growing part of the added value. However, in the past few years the downstream sector has developed a strategy of differentiation of its supply and has increased contracts for such with the producers.

A demand for identified products

On the other pole, agricultural *products are identified by their origin, with characteristics specific to a particular region or "terroir" with strong value added linked to niche markets.* The use of controlled labels of origin for wine was one of the first applications of niche marketing in Western Europe.

The territorial identity results from several factors, like the identification of places and, the types of products. It is accompanied by the organization of particular supply chains with a guarantee on the origin and the manufacturing processes, through specific qualification procedures: by the origin of products, by the production process (organic farming, certifications of conformity) and by marketing (fair trade, direct sales).

Improve the standard quality of unprocessed agricultural products and their processing

AKST could contribute to the improvement of nutritional, organoleptic and health quality of unprocessed agricultural products. The role the environment (i.e., soil, air, pathogens) and agricultural practices and their various interactions, play in determining the quality and stability of these unprocessed products (e.g., the production of mycotoxins, polluted soils and the transmission of xenobiotics to food plants and animals, pesticide application and the detection of their residues in food, etc.) could be taken into account. Such improvements in unprocessed agricultural products are particularly appropriate for bulk production (standardized quality products) (Box 6-5).

The overall objective of processing is to be able to design and produce food that meets a large set of criteria (safety, nutrition) and is accepted by consumers (Bruin et al., 2003). The "Preference, Acceptance, Need" set of expected properties (PAN) is an important objective and tailor made food is one example that corresponds to these properties (Windhab, 2006). In addition to processing control, the conservation of properties on the shelf life appears to be of much importance too. This includes packaging. An important goal is to control the processes in order to simultaneously reach all the objectives: food quality and energy and environmental considerations (Dochain et al., 2005). Finally, even if the control of specific properties (nutrition mainly) is of great importance, the ability to control food safety and hygiene appears to be equally essential (Napper, 2006).

Box 6-5. AKST options to improve the quality of unprocessed plant and animal products

AKST focused on the following issues could facilitate improving the quality of unprocessed agricultural commodities:

In the plant domain:

- Understanding plant metabolism and developing plants containing higher levels of important macro- and micronutrients (essential fatty acids, oils, vitamins, amino acids, antioxidants, fibers, etc.) and reduced allergen levels;
- Developing the taste and quality of products, particularly fruits and vegetables, while improving the post harvest quality and storage capacity; and
- Selecting plants with low input requirements to reduce the risk of residues in plant-derived food, particularly pesticide residues, nitrates and other potentially toxic elements.

In the animal domain:

- Understanding the functioning of the rumen ecosystem to underpin the development of improved animal nutrition strategies and technologies for the production of healthy milk and meat;
- Improving the nutritional value and human health features (e.g., the fatty acid composition of meat and milk, the nutritional quality of eggs) as well as sensory qualities such as tenderness, flavor, visual appeal, and processing characteristics;
- Improving livestock resistance to spreading zoonotic diseases, for example through improved immune system function, to improve food safety; and
- Selecting animals that are more robust and able to adapt easily to the production environment (e.g., feeding system, climate, housing/grazing system), to reduce the need for medicines and thus the risk of residues in animal-derived food.

In both plant and animal domains:

- The influence of genetic factors, production methods and contamination by mycotoxins and pathogenic microorganisms on the variability of raw materials and on human nutrition; and
- The development and expansion of technologies that preserve foodstuffs germ-free without refrigeration, such as novel packaging technologies, irradiation, etc.

Develop quality specific products distinguished by their place of origin

Options for research on these products could include:

- Study the attributes of quality (Allaire, 2002);
- Develop processes for the qualification of food products according to their origin, methods of production or marketing (Bérard and Marchenay, 2004, 2007); and
- Encourage the normalization of local knowledge and practices (Bérard and Marchenay, 2006).

Reinforce traceability: from raw materials to marketed products

Spurred on by recent food scares around the world, some governments are forcing the adoption of food traceability systems. The ability to trace products and their components throughout the food chain is becoming more important in markets for safety and quality assurance.

Methodological and technological developments required for efficient traceability in standardized production could include among others:

- Development of new generation of analytical methods based on micro and nanotechnology solutions that comply with the requirements for ubiquity, fast response, low cost, simple use, etc. (European Commission, Framework Program VI Information society technology 2005-2006: Good Food project);
- Development of microsystems technology solutions for the rapid detection of toxigenic fungi and mycotoxins by natural bioreceptors, artificial receptors and nanoelectrode devices;
- Development and characterization of different sensors, based on innovative DNA sensing technologies for direct and real time measurement of target DNA sequences of pathogens present in the food matrix. (European Commission, Framework Program VI Information society technology 2005-2006: Good Food project); and
- Promotion of innovations in DNA fingerprinting, nanotechnology for miniature machines and retinal imaging and their increased integration into plant and livestock industries for improving the speed and precision of traceability (Opara, 2003).

6.2.5.2 Develop diversified, fair and equitable food and fiber supply chains

Many NAE food supply chains operate at the cutting edge of marketing technologies and have reaped increasing profits and market-share. Trends in agricultural and food commodity markets, however, show that most NAE farmers have not only become separated from consumers, but food supply chains are dominated by processors and retailers (Fearne, 1994; Lyson and Raymer, 2000; PCFFF, 2002; Vorley, 2003). Vertical integration of successive stages in agricultural and food supply chains under the control of single corporate organizations or clusters of corporations can reduce the competitive power of farmers (Lamont, 1992; OFT, 2006; UNCTAD, 2006) who have become disadvantaged, inadequately rewarded "price takers" facing limited market opportunities for their produce. The gap between farm and retail prices is growing and is wider in countries where transnational corporations (TNC) have concentrated market power. The farm retail price gap is costing commodity-exporting countries more than US$100 billion each year and anticompetitive behavior by agrifood TNCs is said to be a key cause (Morisset, 1997).

There is thus an urgent need to develop policy instruments to remove incentives for farm concentration and agribusiness concentration (Action Aid International, 2006; SOMO 2006; UK Food Group, 2005). These include:

- Improve competition policies within agrifood markets, for instance, by monitoring corporate concentration, mergers and strategic business alliances and their anti-competitive effects across national borders;
- Apply stringent anti-trust measures that dissuade global price-fixing cartels;
- develop strict monitoring and external verification systems to assess and increase the credibility and transparency of corporate social responsibility;
- Develop international organizations to monitor the concentration and behavior of TNCs involved in agricultural trading and food retailing at a global level. These organizations could be given the task of collecting information, researching policy advice and developing standards of corporate behavior.

Another area of equal importance is that of improving the "connectivity" between food producers and consumers and increasing the competitive power of farmers. Some of the measures that could facilitate this are to:

- Improve the market orientation and responsiveness among producers through training and technical assistance in marketing and related business management skills;
- Improve market intelligence and transparency throughout the supply chain;
- Extend existing and develop new supply chains within NAE and externally that distribute profits more equitably among actors through negotiated multistakeholder arrangements;
- Support actions to add value on or near the farm, through on-farm processing and/or product differentiation, including for example organic and fair trade products and products distinguished by geographical origin or appellations;
- Develop collective business and marketing capability among farmers, through for example farmer groups, cooperatives and trade association in order to improve their bargaining position;
- Increase investments in market development and in marketing infrastructure for local and regional marketing such as storage, processing, refrigeration and transport.

6.2.6 Promoting environmental sustainability through ecological management

From an environmental perspective, the sustainability concept calls for an ecological and evolutionary approach. The understanding of specific ecosystems and the ecological principles by which they function are key elements for the design and management of agricultural systems—simultaneously ensuring both productivity and natural resource preservation (Altieri, 1995; Vandermeer, 1995; Gliessman, 1997).

The design of such agroecosystems is based on ecological principles (Reijntjes et al., 1992) that may be applied using a range of techniques and strategies (Altieri, 2005): "(1) enhancing recycling of biomass, optimizing nutrient availability and balancing nutrient flow, (2) securing favorable soil conditions for plant growth, particularly by managing organic matter and enhancing soil biotic activity, (3) minimizing losses due to flows of solar radiation, air and water by way of microclimate management, water harvesting and soil management through increased soil cover, (4) increasing species and genetic diversification of the agroecosystem in time and space and (5) enhancing beneficial biological interactions and synergisms among agrobiodiversity components thus resulting in the promotion of key ecological processes and services."

AKST needs to fully take into account this ecological perspective on agriculture and its dynamic evolution over time and space. In this context, biodiversity—viewed as the multitude of interactions among all living organisms in the soil and water as well as on the ground and in the air—plays a central role in the preservation and the enhancement of the multiple functions of the agroecosystem (Griffon and Weber, 1996; Altieri and Nicholls, 1999; Thies and Tscharntke, 1999) and particularly with respect to productivity (Hector et al., 1999).

6.2.6.1 Potential contribution of AKST for long term soil preservation

In the last few decades there has been an intensification of human activities on soil (industry, agriculture, urbanization, cemeteries, recreation, etc.). This has largely been achieved without considering soil diversity and its suitability to accommodate these different activities. Consequently there has been a pronounced degradation of soil with negative consequences for a range of soil functions, including the regulation of hydrological and atmospheric gas processes and the provision of habitats for flora and fauna.

In view of an ecological management of agroecosystems, there are some areas where AKST can be developed and help remedy the current situation of soils as mentioned below:

Understand soils better: including past, present and current dynamics.

- Soil is a continuous milieu wherein there are vertical as well as lateral organizations and dynamics. We are in a better position today to understand the vertical organization but more research is essential to understand the lateral organization and dynamics of pedological cov-

ers, that can in turn s hed light on the different existing pedological systems[17] and their differentiation process (Ruellan, 2000, 2005). In particular, it is important to understand better the long distance transportation of organic matter, fertilizers, pesticides and pathogens through this milieu.
- Elucidate the relationship that exists between pedological systems and the current or future social systems (Lahmar et al., 2000).
- Study the rate of evolution of the different characteristics and properties of soil. (AFES, 1998).
- Develop the notion of soil as being not just a part of the larger ecosystem but also as an ecosystem in itself (Lal, 2002).

Link soils and human activity (AFES, 1998; Lahmar et al., 2000; Lal, 2002; Van Camp et al., 2004)

There is need of improved understanding of the influence of human activity on rate of evolution of soils, mechanisms of soil formation, modification of biological activities and their consequence on soil formation, modification of the alteration rate of rocks, etc.
- Understand the effect of climate change on soil evolution and the subsequent re-utilization of these soils in a better way.
- Better understand soil degradation and its consequences on the surrounding environment (air, water, life) and human health.
- Identify the interactions between agricultural practices and soil degradation.
- Where it has not been done, develop a portfolio of soils at the national level that would help in classifying soils according to their properties, functions and appropriate utilization. For example, certain soils can be categorized under "soils meant for agriculture" whereas others can be "sealed" (used for construction or other purposes).

Develop appropriate soil related technology and agricultural practices

Develop agricultural practices that take into account the diversity of soils, thereby matching their properties to their use and management.
- Design tools to improve soil productivity while promoting renewal of soils (Van Camp et al., 2004).
- Develop new methods to remediate soils like phytobial remediation, a new process that combines the best of both traditional bio and phytoremediation using microbes. Plants are grown whose roots are colonized by symbiotic microbes that degrade toxicants and assist plants in taking up toxic materials (Lynch and Moffat, 2005). Other novel bioremediation technologies include transgenic technologies where the bacterial genes are inserted directly into the plant (Mackova et al., 2006). These plants could be used to accelerate the decontamination processes to more rapidly remediate sites and bring or return contaminated areas into production or other use. More research on heavy metals sequestered in biomass could be helpful.
- Develop nanosensors for monitoring soil health.
- Develop and implement accessible information systems and extension services including remote sensing technologies for better soil management.
- Decrease soil degradation and/or increase soil fertility using technologies that permit:
  - an increase in porosity that would prevent soil compaction and promote a decrease in the rate of erosion (excluding arid areas where increased water retention is the primary focus);
  - an increase in water retention and act against drought and land desertification; and
  - an improvement in the retention of organic matter present in soils.

In addition to all of the above, a better integration of existing and new knowledge on soils and soil practices into the legal regimes of states or regions as well as national and international policies could be useful.

6.2.6.2 Contribution of AKST to water management

Water is an essential input for agricultural production for which there is no substitute. It is imperative that NAE achieves sustainable use of water resources in the agricultural sector within the region, as well as contributing to sustainable water management in a wider global context.

Agricultural water management is likely to become more challenging in the future due to increased human demand for water, climate change and limits imposed by available water (Evans, 1996; EEA, 1999, 2001a, 2001b; Hamdy et al., 2003; FAO, 2004a; NRC, 2004; Dobrowolski and O'Neill, 2005; OECD, 2006; Morris, 2007; Rosegrant et al., 2007). Keeping in mind the perspective of ecological design and management of agroecosystems, several options could be considered for AKST to contribute to water management (quantity as well as quality) as follows:

Water Quantity

Irrigation water management

Irrigation is critical for some areas within NAE, especially southern Europe and the western United States (Hutson et al., 2005). The EU has 9% of its agricultural production under irrigation (13 million ha), over 75% of this in Spain, Italy, France and Greece (EEA, 1999; Kasnakoglu, 2006). More than 22 million ha (18% of total cropland) are irrigated in the U.S., over 80% of which is in the West (Gollehon et al., 2006). In Europe, agriculture accounts for 30% of total water abstraction and 55% of consumptive (nonreturnable) use. In parts of southern Europe, these figures are typically over 70% and 60% respectively (EEA, 2001a; Berbel et al., 2004).

With regard to irrigation water management, the following are some of the priorities for the future:
- develop new irrigation technologies and practices (Stringham and Walker, 1987) that further increase water use efficiency (that is "crop per drop"), including controlled water placement and use, cultivations to conserve moisture and introduction of low water demanding crops;

[17] A pedological system is a soil cover that, by its constituents, structures and dynamics (vertical and lateral distribution and functioning), constitutes a unity.

- promote irrigation water auditing and scheduling systems, including remote sensing to monitor crop for optimal timing of irrigation;
- increase training and incentives for farmers to adopt improved irrigation practices; and
- improve water harvesting methods including the construction of dams and water distribution systems.

Removal of excess water

In the future, under conditions of climate change, new integrated land drainage technologies will be required. These can include on-farm water treatment and storage, which can help cope with greater variation in precipitation and temperature and periods of excessive rainfall and river flows, help mitigate salinity and improve overall water resource management (O'Connell et al., 2004; Lane et al., 2006; Morris and Wheater, 2006; Thorne et al., 2006).

Genetic developments (using conventional and transgenic technologies) to reduce drought stress

Technologies are now available to alter the metabolism of plants to make them more tolerant to water induced stress. Research could help in determining optimal water requirements of important crops and in developing new stress resistant varieties. There is need to develop crops which are tolerant to low water quality, especially associated with salinity.

Water Quality

There are major concerns in many parts of NAE about water quality and the consequences for ecosystems and human health (Costanza et al., 1989). In the EU, the Water Framework Directive sets the context for this over the next 20 years (Morris, 2007). Diffuse pollution from agriculture is of major concern in many parts of Europe (Pretty et al., 2000; EFTEC and IEEP, 2004; Bowes et al., 2005; Neal et al., 2005) and increasingly the subject of targeted control measures (EA, 2002). In this respect, some of the priorities for AKST include:

- An integrated approach to water resource management, of which agriculture is part, at the catchment scale (English Nature, 2002).
- Improved integrated understanding of pollutant behavior and transport mechanisms within the landscape (nitrates and pesticides in particular).
- Suitable measures to reduce diffuse pollution from farmland.
- Evidence of the link between land management, runoff and flood generation and options for on farm water retention and storage.
- Methods for on-site passive water treatment systems such as reed-beds, industrial or energy crops and active systems such as nano-based filtration and purification techniques using membrane systems to detect and neutralize undesirable chemical, physical and biological properties.
- Improved understanding of the link between environmental water quality and public health (Hallman et al., 1995).

Water policies and water reuse

Ownership and rights to use water are becoming more contentious as aquifers are depleted faster than they are recharged (Engberg, 2005). Water law and entitlements are typically more complex and less well defined than those for land (Sokratous, 2003; Caponera and Nanni, 2007). Water reuse is a rapidly evolving water-management tool for supplementing limited water resources around the globe (Lazarova and Bahri, 2004). In this context, future research and investment could help to:

- Better understand the agricultural use of water and the cost of providing water services.
- Better appreciate the social, economic and environmental value of water as a basis for sustainable water resource management.
- Develop water allocation and distribution schemes to balance food and agriculture with other water needs (Engberg, 2005).
- Provide water managers and policy makers with decision support tools to guide water resource management and policies that lead to behavioral change and a reduction in water conflicts.
- Understand the role of water property rights and laws, the benefits of local management of water and the role of collective action by water user groups.
- Support schemes for water licensing, pricing and, where appropriate, trading to promote water use efficiency.
- Develop integrated programs to address water reuse, conservation and wastewater reuse for agricultural, rural and urbanizing watersheds after having assessed the social and economic feasibility and impacts of water reuse projects.
- Develop technologies for the exploitation of alternative water sources (e.g., sea water after desalination, air humidity after condensation, production of drinking and irrigation water in seawater greenhouses (Pearce and Barbier, 2000).
- Develop education/outreach programs to foster the development of criteria and standards for economic and sustainable solutions that will help protect public health and the environment.

AKST will have a critical role in managing the potential benefits and risks of agricultural water use as the resource becomes more scarce and valuable. AKST is required to achieve a much greater integration of ecological land and water management as a basis for sustainable, multifunctional agriculture.

6.2.6.3 Potential contribution of AKST to biodiversity and genetic resource management

Agriculture as a whole is based on the human utilization of biodiversity: soil biodiversity, aquatic biodiversity, as well as diversity of plants, animals and microorganisms. Historically, agronomy has led to increased uniformity in the whole farming system in order to facilitate the mechanization and industrialization the whole process (from seedbed to harvest and post-harvest periods) to the detriment of biodiversity. Considering recent advances, it is now obvious that diversity and productivity are linked (Hector et al., 2000; Loreau et al., 2001; Reich et al., 2004; Van Ruijven and Berendse, 2005; Tracy and Faulkner, 2006).

AKST could play a significant role in managing and enhancing present biodiversity, which is the foundation of ecosystem services, namely provisioning (e.g., food, wood, fiber, fuel) and regulating (climate and flood regulation, disease control) services (MEA, 2005). There are many options for AKST to play such a role through agricultural practices and land management as well as through genetic resources preservation as long as the political and regulatory context allows it.

Biodiversity enhancement in agricultural activities and land management

To promote an ecological approach of the agroecosystems, AKST could focus on a better understanding of the impacts of different cropping and livestock systems both on the spatial distribution and the evolution of the overall biodiversity at the landscape level and, as well as the effect of biodiversity on the productivity and quality of the systems including soil and water resources. Some of the options among others are to (Jackson, 1980; Soule et al., 1992; Kerr and Currie 1995; Caughley and Gunn, 1996; Johnson et al., 1996; Srivastava et al., 1996; Johnson et al, 1999; McNeely et al., 2002; Graf, 2003):

- Design rural landscapes with biodiversity enhancement in mind. This might include consideration of such critical issues as mixed and strip cropping for annual crops at the farm level, as well as the creation of migration corridors and improvement in habitat quality at the appropriate scale. It could also include enhancing knowledge of the functional role of nonagricultural biodiversity in achieving specific regulating services at the landscape level (pollination, pest and disease regulation, natural hazard protection, etc.).
- Continue research on radical new types of agricultural production that would be based on biodiversity enhancement while increasing productivity and offering other advantages, including reduced reliance on chemical inputs, lower energy costs and reduced soil degradation and erosion for example:
  - Further research and experimentation on pesticide use and pesticide hazard reduction plans (at national and regional level) that could result in yield gains while enhancing biodiversity and safeguarding human health.
  - Changes in fertilization and tillage practices that could enhance beneficial soil flora and fauna as well as alleviate the contamination of waterways that has multiple effects on wildlife. There is wide recognition of the desirability for substantial additional research on both technical and policy options that ensure a wide and consistent implementation of these changes.
  - Improvements in water use efficiency through technical improvements and policy tools to reduce the impact of agricultural water demands on the environment and biodiversity.
- Better understand the role of both forests and grasslands and their management in the preservation of biodiversity and ecological processes.

AKST could also focus on the optimization of spatial and temporal management of crops and livestock biodiversity at the landscape level and as part of a global agroecosystem (Loreau et al., 2003) to contribute to the sustainability of the whole system by:

- Better understanding the effect of spatial and temporal distribution of varieties (for example among crop plants possessing different pest resistance genes) and the associated organizational and technical practices on the evolution of both pathogen and pollinator communities at the landscape level. This requires the understanding of the biological mechanisms of host-pathogen co-evolution and their susceptibility to a fluctuating environment;
- Improving knowledge on the diversification of production and associated practices and its effect on the productivity as well as on the supply of environmental services (provision of public goods, positive externalities);
- Better understanding the spatial organization and relative proportion of cultivated areas on the one hand and grassland and forests on the other as well as their interaction with urban areas, in the study of water and fertilizers transportation within the territory and as parts of the whole landscape living ecosystem;
- Developing GIS tools and Multi-Agent Systems that help farmer communities and associations determine appropriate locations of various food and farming systems (crops, animals, enterprise, grasslands) to improve production efficiency and meet environmental challenges including biodiversity preservation.

Genetic resources preservation

The global distribution of genetic diversity and the interdependency of all countries vis-à-vis genetic resources call for greatly improved cooperation and coordination mechanisms at the global as well as the local level. Much work will be required in order to upgrade, rationalize and coordinate the global design for *ex situ collections* (based on local, national and regional genebanks): the Future Harvest Centers and the Global Conservation Trust could play a major role in the coordination and support of all the components of such design. This effort has to be accompanied by a more systematic characterization, evaluation and documentation of genetic resources to allow their wide use.

Today, there is a strong scientific consensus that the viability of genetic resources in agriculture depends on *in situ* preservation efforts that allow the development of resources' adaptative capacities and act as a complement to *ex situ* stored collections. These must be carried out in a large variety of ecological and cultural circumstances that can only be achieved through international cooperation and funding. So, in addition to the static conservation of genetic resources, more attention could be paid to the dynamic processes that allow potential evolution in changing environment through *in situ* preservation.

Considering livestock, it will be important to:

- Understand better the evolution of genetic diversity in intensive and extensive breeding populations and develop tools to monitor and control the genetic drift within such populations.
- Develop specific and breeding efforts on locally adapted

populations in order to meet the challenge of specific local demand and maintain a large genetic diversity.

Considering crops, important measures include:

- Develop methods and tools to accompany the preservation of genetic diversity on farm.
- Broaden the conservation circles to establish closer collaboration with grassroots conservation movements and community seed banks.

6.2.6.4 *Potential of AKST for developing energy efficient food and farming systems*

Farming and food systems (FFS) in NAE are energy intensive. Even though farming in general accounts for only about 5% of total energy consumption in the most of the NAE region, this share increases to over 20% of total energy use once food processing, packaging and distribution are included (Fluck, 1992; Giampietro and Pimentel, 1993; Pimental and Giampietro, 1994; Heller and Keoleian, 2000; Heller and Keoleian, 2003; Murray, 2005; Williams et al., 2006). At the farm scale, about 85% energy inputs in NAE farming systems are carbon fossil based: other sources are relatively undeveloped. 50% of farm energy relates to agrochemicals, mainly nitrogen fertilizer, 30% to field machinery and transport and 20% to energy services linked to heating, lighting and materials handling (Box 6-6).

Current NAE farming and food systems and related livelihoods are especially vulnerable to increased energy prices and reduced fossil fuel supplies. Although high energy prices will continue to increase the scope for bioenergy crops, energy efficiency will remain a critical component of their feasibility (Stout, 1991).

AKST is a critical factor in understanding and influencing the farming-energy relationship. In the face of rising energy prices, the following possible priorities are identified:

- Enhanced understanding of energy use and efficiency in farming systems, including synergies and trade-offs with other "performance" indicators such as yield, quality, added value and environmental impacts.
- Development of data bases and evaluation methods such as energy auditing, budgeting and life cycle analysis. Improved farmer and operator skills in energy auditing and management for field and farmstead operations.
- Adapting existing and development of new energy saving technologies for crop and livestock production addressing major field and farm operations and processes, including:
  - Improved minimum cultivation systems
  - Combination tillage and crop establishment field operations, including gantry systems
  - Precision application of fertilizers and crop protection chemicals
  - Whole crop harvesting systems
  - Handling, storage and treatment of materials and deriving energy from "wastes"
  - Irrigation application systems
- Technology development in alternative energy sources, including on-farm wind, solar and groundwater heat and use of ambient conditions to provide energy services in drying and storage.
- Genetic development, using conventional and transgenic

Box 6-6. Energy efficiency in NAE food and farming systems

Energy efficiency in farming can be measured in terms of the ratio of the energy content of output to the energy content of inputs, excluding solar energy in crop photosynthesis and measured in joules or equivalent.

Energy ratios vary across the NAE region according to average yield levels (t/ha) that in turn are a function of the environmental factors and the relative scarcity of land and labor (Pimentel and Giampietro, 1994). Where population pressure is relatively high and land is relatively scarce, such as in many parts of western and northern Europe, high yielding agriculture tends to have high energy inputs per ha and per tonne of product. This gives relatively low energy ratios, of about 1 or less. Where land is relatively plentiful and labor is scarce (and relatively expensive), such as in North America, farming systems are more extensive, have lower energy inputs per ha and per ton of product, but higher (almost 5 times more) energy input per farm worker.

In Eastern Europe and Russia, conditions vary considerably, but relatively low energy inputs per ha and per worker are associated with relatively low yields. In some parts of NAE, some small-scale, peasant-type farming systems can display high energy ratios, but low yields and low added-value are often associated with low incomes and poverty.

The enhanced yield performance of crop and livestock systems in the NAE has thus been based on low cost, readily accessible energy supplies. Furthermore, commonly promoted strategies for adding value to farm products and increasing farm incomes, such as quality assurance, product differentiation and on-farm processing, tend to be energy intensive. Although organic production, now finding favor amongst some consumers, uses less agrochemical energy, inputs of labor and mechanization tend to be higher and overall yields lower than conventional methods. This results in similar, if not reduced, energy efficiency compared with conventional methods.

There are also important links between energy use, greenhouse gases and global warming potential (GWP). For the most part in agriculture, they are indirect, given that most energy is associated with the use of fertilizers and machines. Nitrous oxide (N_2O) in particular and methane (CH_4) emissions (from ruminate livestock) have the greatest impact GWP, more so than CO_2 emissions. However the origin of N_2O is linked to high fertility soil so there is little difference between organic and conventional systems (Williams et al., 2006). There are also other important links with other environmental impacts, such as soil erosion and compaction, water pollution and worker and animal welfare. At the same time, however, energy intensification has helped to reduce drudgery in farm work and has improved the health and life-expectancy of farm workers, and enhanced the skill base and rewards for farm workers, factors which are important in the recruitment and retention of people in farming.

technologies, to improve energy conversion in crop and livestock systems and reduce agrochemical dependency as well as to increase the shelf life of agricultural products with reduced refrigeration.

- Improved design for energy efficiency in farm machinery and equipment.
- Development of energy efficient protected cropping buildings and animal housing, including heating and refrigeration systems.
- Improved methods for recovery and reuse of residues and wastes as "resources"—including fertilizer, heat and power from farm wastes and other off farm waste (such as biosolids).
- Re-development of indigenous, energy saving technologies.
- Improved understanding among consumers of excessive energy costs of "out of season" vegetables, in order to modify purchasing behavior.
- Development of whole supply chain energy auditing and reporting systems, including energy labeling, to inform consumers and policy makers.
- Design of suitable policy instruments to promote energy efficiency in food and fiber supply chains.

6.2.6.5 Reducing pressure on natural resources through the ecological footprint method

The ecological footprint is a method for comparing the sustainability of resource use—mainly energy—among different populations (Rees, 1992). The ecological footprint was defined in terms of land area needed to meet the consumption of a population and absorb all their wastes (Wackernagel and Rees, 1995). Although the concept has been subjected to considerable criticism, recent advances include input-output analysis (Bicknell et al., 1998; Hubacek and Giljum, 2003), land condition indicators and land disturbance analysis (Lenzen and Murray, 2001). These advances have enabled calculation and comparison of ecological footprints across widely divergent scales, from countries to families and categorization of the ecological footprint into commodities, production layers and structural paths. Such analyses provide detailed information on which to base policy decisions for reducing pressure on energy consumption of different types of populations (Lenzen and Murray, 2003).

6.2.7 Developing innovative crops and livestock food and farming systems

AKST could be mobilized at the farm level for developing innovative crop and livestock farming systems by breeding plants and animals with high quality performance both from environmental and production perspectives and breeding of underutilized species. AKST could also contribute to the development of innovative modes of production and evaluating of diversity. These new systems could facilitate better interactions among crops or livestock, production methods and the environment.

6.2.7.1 The potential of genetics and biotechnology for crops and livestock breeding

Breeding has the potential to be a key element to contribute to the realization of development and sustainability goals, both in the areas of food security and safety and to contribute to environmental sustainability (FAO, 2004; Plants for the future, 2005; FABRE, 2006). It would be appropriate to tightly bind breeding with crop or animal system management and with the local environment. The potential of AKST to support breeding activities is enormous—due to the recent progress in genetics especially in molecular genetics and genomics whose continuation is important—and offers new possibilities for breeding methods that could be better explored. Also, these future innovations raise new concerns in terms of possible wider effects and unforeseeable consequences (Boxes 6-7 and 6-8), calling for new ways of assessment and follow up.

Considering basic knowledge, a huge effort has been invested in the last 20 years to explore the structure and functions of the genomes of several living organisms. It enhanced knowledge in genome sequencing, of gene structure, expression and function and in genome structures (physical maps, duplications of chromosomes fragments and deletions, mobile element invasiveness; comparative genomics, etc.) through a more systematic and industrialized approach of the cell/tissue products (transcripts, proteins and metabolites).

Much previous research had been based on an understanding of genetics that has assumed "a direct path from gene to protein and to function as well as the presence of preset responses to external perturbations" (Aebersold, 2005). While it led to the accumulation of large amounts of detailed knowledge that constitutes an important data investment, its limitations have also become apparent: little is known about how cells integrate signals generated by different receptors into a physiological response and few biological systems have produced a consistent set of data that allows the generation of mathematical models that simulate the dynamic behavior of the system. Some of the priorities for research to help better understand these processes could be to:

- Maintain the effort in genomics data acquisition to accumulate knowledge in structure and functions of specific genes and particularly those the expression of which may contribute to development and sustainability goals (FAO, 2004; Plants for the Future, 2005; FABRE, 2006);
- Strengthen the efforts of basic physiology through functional genomics and systems biology that continue to break through the major limitations inherent in previous approaches (Minorsky, 2003). This requires enormous sets of data as well as a sophisticated data infrastructure with a high level mathematical framework (Minorsky, 2003; Wiley, 2006). These efforts will also lead to a better understanding of the interactions between the metabolic pathways and of their role in the expression and the regulation of specific traits;
- Explore further the role of epigenetic mechanisms (DNA methylation, histone acetylation, RNA interference) in the regulatory framework of specific gene sets (Grandt-Dowton and Dickinson, 2005, 2006);
- Increase the understanding of mechanisms of reproductive biology and regulation of ontogenesis that allows elaboration of methods of rapid multiplication of appropriate genotypes (cloning, apomixis, etc.) (FAO, 2004b; FABRE, 2006);

Box 6-7. Plant and animal breeding targets to contribute to development and sustainability goals

For crops AKST could contribute to the following:

- Focus on characters and functions involved in plant susceptibility and resistance to pests, diseases, weeds (weed control in one of the largest input costs in agriculture) and environmental stress (expected climate changes may increase the diversity and spread of pathogens and impose additional heat, cold and drought stresses on plants);
- Develop crops that require less fertilizer and other agrochemicals, and that also require fewer water resources, based on a fuller understanding of factors regulating nitrate and phosphate utilization, water-use efficiency and impact on natural resources;
- Develop crops for different types of agriculture: intensive, but also extensive and organic;
- Understand the genetic and physiological determinants of genetic and phenotypic "plasticity" and develop crops that have capabilities to adapt to environmental change;
- Understand plant metabolism in order to develop plants containing higher levels of important macro- and micronutrients (essential fatty acids, oils, vitamins, amino acids, antioxidants, fibers, etc.) and reduced allergen levels, reduced anti-feedants; and better understand plant carbohydrate metabolism, especially control of source-sink relationships. Use this knowledge to breed healthier, better tasting crops, as well as better food, feed, and biofuel crops;
- Enhance breeding efforts enabling the use of a wide range of species, particularly under-utilized species of medicinal and aromatic plants possessing high health and economic potential; and
- Ascertain how to do the above while maintaining yields at levels that will not require putting more land under the plow.

For livestock, to improve the efficiency and sustainability of production in terms of food quality and safety, the environment, zoonoses and animal welfare concerns, AKST should contribute to the following:

- Identify genes and gene networks that control immunoresistance in livestock, including pigs, poultry and fish, leading to improved disease prevention strategies for persistent and costly diseases;
- Revisit gut physiology (for improved efficiency and decreased pollution and disease), understand the functioning of the rumen ecosystem to underpin the development of improved animal nutrition strategies and technologies for the production of health-enhancing milk and meat, and the reduction of gaseous emissions, especially methane production by cattle;
- Identify genes and gene networks relevant for fertility in all species and reduce the growing infertility problem of high-yield dairy cows;
- Adapt animals to less intensive production systems (plant-based feed and saline water for fish, high digestibility cereal grains for nonruminant animals and poultry);
- Improve nutrition and hygiene in intensive productions to reduce pollution and to control diseases;
- Improve animal welfare: upgrade existing minimum standards; promote research and alternative approaches to animal testing; introduce standardized animal welfare indicators; develop new tools enabling breeders to handle welfare traits more objectively than at present (new biological insights into brain function, the genetics of behavior and physiological indicators of stress and wellbeing); develop efficient information management systems for health monitoring, health detection, etc.; inform animal handlers and the general public on animal welfare issues; support international initiatives for the protection of animals (FAO 2004b; FABRE, 2006; Plants for the Future, 2005).

- Develop comparative biology including comparative genomics (Sankoff and Nadeau, 2000) to ensure the dissemination of knowledge on a wide range of food species including under-utilized ones (FAO, 2004b); and
- Invest in metagenomics, the potential of which is considerable considering applications in agriculture, land environmental remediation, bioenergy, etc. (NRC, 2007).

Concerning applied research, AKST could be pursued to accompany breeding activities focused on functions and mechanisms that contribute to the adaptability of crops and animals to extreme stress—both biotic and abiotic—to quality and safety of food as well as to the sustainability of food and farming systems (Box 6-7). It could be useful to develop these activities on a wide range of food species to maintain progress in both industrial and under-utilized species.

AKST could also explore the potential of more diversified and heterogeneous variety types namely to better meet environmental concerns: for example, it would be interesting to generate a variety of wheat that has three different leaf and stem architectures but is otherwise isogenic; such variety, planted with its mixed morphotypes, could be better at capturing sunlight and carbon dioxide and better at competing with weeds; also, a variety of wheat or maize having different types of root systems (a superficial one with a large covering area and a deep one more localized) could better benefit during restricted water availability in the different soil depth. In this case, the "uniformity" paradigm for variety registration procedures will have to change to integrate and favor diversity.

AKST could also be mobilized to develop innovative breeding strategies and technologies (marker/genomics-assisted selection, gene transfer, targeted mutagenesis, etc.) for the efficient introduction of desired traits into high-yielding

Box 6-8. Genetic engineering and development and sustainability goals

Genetic engineering is distinguished from conventional plant breeding by its reliance on molecular methods (i.e., not including sexual reproduction) to introduce genetic variation into the cells of a target population. In agricultural applications, transgenesis is currently the most common kind of genetic engineering. Transgenesis uses a vector to introduce segments of DNA isolated from one or more organisms into the cells of another organism where it is integrated into the genome. Transgenic annual crop plants are used widely in the United States, Canada, Argentina, Brazil, India and China, and many farmers using them have benefited; the number of farmers planting transgenic crops continues to grow in the NAE and elsewhere. Many new transgenic plants and animals are being developed for use in agriculture. In addition to transgenesis, several other molecular methods are being used to introduce significant genetic variability into agriculturally important species directed evolution and site-specific mutagenesis. In the future it is likely that these and other, yet to be developed methods, will become more common.

However, transgenic organisms have engendered controversy as they have been developed and used. The controversies have revolved around three interlinked issues: policy priorities, self-determination and ownership, and risk and consumer acceptance (NAFTA-CEC, 2004; Andow and Zwahlen, 2006). These controversies have themselves affected the organization of AKST in the NAE. It is likely that the many controversies will not be resolved in the next 5-10 years.

The policy divide, recently reflected by the WTO dispute between the United States, Canada and Argentina versus the European Commission, has resulted in policy instability that has delayed the development and implementation of agricultural genetic engineering. This divide not only occurs between countries in the NAE, but between the NAE and other parts of the world. There is a need to stabilize the policy environment, beginning with clarification of the differences.

Genetic engineering has sharpened some tensions between ownership rights and the rights of farmers and individuals in general. Biological patents remain controversial in many parts of the world, but in the NAE they have accelerated the commercialization of biological products in many fields outside of agriculture as well as in agriculture. These patents have helped stimulate the fusion of molecular biology with plant and animal breeding, which has led to new areas of investigation in the plant sciences. At the same time, they have contributed to a weakening of public sector capacity to conduct innovative research in agricultural biotechnology, and have contributed to the concentration of ownership of the seed industry. The rights of peoples to determine how transgenic organisms enter nations has been a subject of much international negotiation (e.g., under the Cartagena Protocol on Biosafety) and the terms under which they enter into individuals' lives is still a matter of much discussion. These controversies have become more complicated as they have entangled with many other issues, including indigenous peoples' rights, biodiversity conservation and food aid. The consequences of these and related changes need to be understood for the NAE and the rest of the world, to better assess the need for mitigation measures, and if needed, what measures would be appropriate.

The development of transgenic crops has focused attention on risk and consumer preference. Risk assessment has focused on human health and environmental risks, which has led to renewed examination of the methods of risk assessment and agricultural technology assessment, particularly concerning benefits, opportunity costs, long term adverse effects, and the distribution of benefits and risks in society (Snow et al. 2005). Consumer preferences increasingly influence the development of nearly all agricultural technologies, including transgenic crops. These preferences have contributed to the stratification of commodity markets (corn is no longer just "corn"), and have thus undercut, not without some tension, the traditional supply-side approach involving undifferentiated commodity streams throughout the supply chain. The increased attention on risk and technology assessment, and the increasing strength of consumers to influence the development of agricultural technology will be important touchstones for NAE AKST in the coming decades.

IAASTD goals include elimination of hunger and malnutrition by 2050. To accomplish this will require making greater quantities and more nutritious food available to the poor (Sen, 1981), which will require improving access to, increasing production of and decreasing losses of global food supplies. Several reports of international bodies suggest that transgenic organisms will help meet this goal (e.g., FAO, 2004b), while others are less sanguine (e.g., UNECA, 2002). Unlike the Green Revolution, genetic engineering is not a single technology package, so its potential to contribute to development and sustainability goals must be assessed on a case-by-case basis. We can conclude with confidence that genetic engineering is positioned to help meet development and sustainability goals, and we can even say that some (future) products of genetic engineering will likely help meet development and sustainability goals. However, each case must be examined on its own merits. This is the challenge for the future. There is no simple path for the use of genetic engineering that will assure that its products will contribute to meeting development and sustainability goals. Likewise, there is nothing about the technology itself that is inimical to the attainment of those goals. Like other agricultural technologies, we will need to understand better how the socioeconomic and environmental context for the use of transgenic organisms enables them to contribute to these goals.

crops and animals, using the vast potential available in genetic resources' collections of both widely used and underutilized species (crops as well as wild relatives). Among other options AKST could contribute and help (Plants for the Future, 2005; FABRE, 2006):

- Develop innovative breeding methodologies based on sexual reproduction to integrate present genetic and genome knowledge (marker-assisted selection, new mathematical models and software for genetic evaluation and selection—taking into account new data on gene regulation, imprinting, silencing, genome dynamics, whole genome sequencing, etc.); and
- Develop technologies that can lead to "breakthrough innovations" through genetic engineering (for example to move needed genes and pathways from species where they exist to species where they are needed, tissue specific promoters with genes or RNAi to turn on or off genes that are needed only in specific tissues), cloned animals and other methods that do not include sexual reproduction.

It is important that the development of such innovations does not negatively affect other desirable traits or basic physiology of crops/animals and is not harmful either for the environment or for human health and that it benefits many people around the world, namely through their contribution to the achievement of development and sustainability goals (Box 6-8 and 6-9).

Animal welfare has an important high priority place in the agenda for the future. Most livestock production (pigs, poultry, dairy cattle, beef cattle) will probably be in large-scale production systems. AKST may be mobilized to ensure that minimum standards for the protection of farm animals are set and respected (Box 6-7).

More generally, the wide development and dissemination of innovations has to be anticipated and assessed to understand how it might or might not contribute to development and sustainability goals, considering all dimensions of sustainability and integrating appropriate spatial and temporal scales. New AKST developments could accompany this new form of innovation process through:

- a change in the evaluation process, that could move towards a more systems and dynamic approach and take into account all the potential impacts (both positive and negative) of the innovation (ACRE, 2006): (1) from environmental, health, social, ethical and economic point of views, (2) both short and long term, (3) at the pertinent spatial scale. The evaluation of these impacts before and after implementation through appropriate means is important. AKST research could contribute to the developments of methods and tools that could help in the renewal of this evaluation at the different steps of innovation process; and
- a renewal of policy design (associated with systems evaluation process), which call for a priori evaluation as well as follow-up designs and a posteriori analysis.

6.2.7.2 The potential of nanotechnologies in the food and fiber supply chains

Nanoscience involves the study of the characteristics and manipulation of materials at the scale of atoms and mol-

Box 6-9. Animal biotechnology developments and development and sustainability goals

There is considerable potential associated with the use of animal biotechnology.

- Future research on animal cell differentiation may open the way to the production of gametes from stem cells. Coupled with predictive biology and statistical techniques such as genome-wide selection, these approaches could make it possible to produce and select multiple generations in the petri dish;
- The use of nuclear transfer ("cloned") animals for breeding could allow the rapid and wide dissemination of important genes contributing to the realization of development and sustainability goals;
- Genetic modification could be powerful, particularly when considering its potential to immunize animals against specific viral diseases. For example, RNA interference technology could be used to make chickens resistant to avian influenza and reduce the risk of a human flu pandemic; and
- There are many foreseen applications in the medical field: animal models, animals as bioreactors, and animals for xenotransplantation.

Although genetic technology is often claimed to be precise in targeting specific genes, possible broader effects may not be easy to predict and unintended consequences need to be better anticipated and assessed (Straughan, 1999). A number of other concerns have been expressed and debated (Rollins, 1995) including (1) the speed with which animal biotechnology can effect changes in animals, (2) the possibility that intensive use of biotechnology might narrow the gene pool and reduce genetic diversity through the wide use of specific transgenes and intensive cloning of elite animals, (3) that the accidental or deliberate release of genetically engineered animals might be akin to the introduction of alien species, which has been known sometimes to cause serious ecological harm.

As is the case for plant genetic engineering (see box 8), animal biotechnology is not a single technology package and its potential to contribute to development and sustainability goals requires detailed analysis on a case-by-case basis weighing possible costs against possible benefits whether environmental, sanitary, social or economic. Trying to decide in any area what level of risk-taking is ethically justifiable is an important societal decision, even if it is rather difficult to assess; with animal biotechnology, however, the issue becomes even more complex and controversial, because the costs and benefits will be experienced by two different groups with different interests—human beings and animals.

FABRE, 2006; Rollin, 1995.

ecules (RS&RAE, 2004; BSI, 2006). Nanotechnology involves the "design, characterization, production and application of structures, devices and systems by controlling shape and size at the nanometer scale" (RS&RAE, 2004). At extremely small dimensions, materials exhibit different properties and behaviors, for example possessing greater strength or tolerance, or reacting differently to physical or chemical stimulation. Nanotechnology is already incorporated into commercial products such as pharmaceuticals, chemicals, transport and energy products, packaging, coating and lubrication and electronic products. Nanotechnology is also used for environmental sensing and remediation (Zang, 2003; Kuzma and Verhage, 2006).

Nanotechnology is potentially applicable at all stages of the food and fiber supply chain (Joseph and Morrison, 2006; Kuzma and Verhage, 2006). On farm nanotechnologies have potential to:

- improve crop fertilization and protection by improving the precision of application and enabling activation of chemicals at agronomically and environmentally appropriate times;
- identify and immediately treat crop and livestock pathogens and signal contamination of food products by microorganisms;
- apply additives such as minerals in crop treatments that can be recovered on harvest; and
- enhance operational properties of farm machinery through reduced mass, improved coatings and reduced maintenance.

Nanotechnologies using "smart devices" can help detect environmental damage and target remediation, such as water pollution and clean up. With respect to food processing and marketing, nanotechnologies have potential to enhance the nutrient and dietary properties of foods, improve packaging, detect contaminants in foods including toxic substances and extend product life.

There are, however, actual and perceived risks associated with nanotechnologies and their application in farming and food (RS&RAE, 2004; DEFRA, 2005a), with respect to occupational health (Aitken et al., 2004; Health and Safety Executive, 2004), public health (Warheit, 2004; SCENIHR, 2005) and environmental risk (Colvin, 2003; Guzman et al., 2006). For example in 2004, the Royal Society calls for a precautionary approach until the uncertainties associated with "potential toxicity and persistence can be ascertained". In this context public agencies could develop proactive approaches to nanotechnology management by reviewing potential benefits and risks, understanding and informing public perceptions of risks and benefits and prioritizing the farming and food as a pioneer sector for the beneficial use of nanotechnology (Kuzma and Verhage, 2006). For others such as Friends of the Earth and the UK Soil Association, the risks of "nanofoods" to human and environmental health outweigh benefits especially compared to organic options.

Although most investments in nanotechnology are commercially driven, there are important social, environmental and ethical implications that justify government participation in nanotechnology research as well as management. Furthermore, the development of nanotechnology in NAE has potential to influence its use in other regions of the world.

In the context of the use of nanotechnology in farming and food and related environmental management, capacity could be developed in:

- Ensuring that public investments in nanotechnology are aimed at meeting critical societal needs;
- Supporting fundamental studies in nanoscience to improve the understanding of nanoparticle interactions with biological materials and organisms;
- Maintaining a registry of nanotechnology applications, linked to product and process information and "labeling";
- Applying appropriate methods for testing, risk assessment and monitoring of impacts, including epidemiological, occupational and environmental aspects;
- Educating the public and consumers on the benefits and risks of nanotechnology, including product information, to enable informed choice;
- Developing suitable regulatory frameworks for new nanotechnology applications, including specifications and commodity and trade descriptions, working with existing standards agencies;
- Promoting beneficial development and use of nanotechnology in the public interest through scientific research and joint government-industry partnerships; and
- Building international partnerships to promote appropriate nanotechnologies to meet the needs of developing countries (Salamanca-Butello et al., 2005).

6.2.7.3 Contribution of AKST to the development of improved pest management

New technologies and production practices have been developed to reduce the environmentally detrimental effects of pest management in agricultural production. Many of these methods will require further research to improve both their productivity and environmental performance. Such research will continue to contribute to innovation in the development of technologies and practices. Broadly speaking, the new approaches fall under the term ecologically-based pest management (EBPM) as based on a working knowledge of the agroecosystem, including natural processes that suppress or reduce pest populations (National Academy of Sciences Board, 2000).

Management techniques reflecting such an ecologically-based approach can include Integrated Pest Management (IPM), conservation biological control integrated plant nutrient systems (IPNS), no-till (or minimum tillage) conservation agriculture, precision, spatial variable farming and livestock breeding and feeding and housing regimes that reduce environmental load (Elliot and Dent, 1995). Significant uncertainties and controversies remain to be resolved by future research regarding the nature and efficacy of many of these techniques and their many variants (NRC, 2001; Ehler et al., 2005).

As elaborated in the cited sources above, AKST can contribute to the further development and dissemination of such practices by:

- Investment in pest ecology, including insect, weed and pathogen ecology, to allow maximum biodiversity with minimum impact on productivity and crop health;

- Broadening and deepening the research on environmental and human health implications of pesticides. Continuing improvements in human epidemiology and environmental assessment will be useful to better identify and measure the adverse effects of pesticides and thus guide further research into their safe use;
- Ensuring that research on pest management is locally appropriate and effective by drawing on the existing locally developed knowledge and developing new knowledge of local conditions;
- Developing better tools for prior evaluation of the unintended effects of pesticide use and for monitoring and evaluation of negative effects after adoption; and
- Developing new approaches to integrated pest management and organic agriculture (Box 6-10) based on integrating advances in ecological sciences. Better ecological understanding of both the field environment itself and the wider ecosystem will be essential in this respect.

6.2.7.4 Development of alternative resource management strategies

A variety of approaches show significant promise in improving overall resource management and environmental performance of agriculture. Many of these have incidental or significant roles in pest management strategies, but they are designed with wider purposes in mind. They include:

- Optimizing integrated plant nutrient systems by maximizing plant nutrient use efficiency through recycling all plant nutrient sources within the agroecosystem and by using nitrogen fixation by legumes, balancing the use of local and external sources of plant nutrients while maintaining soil fertility and minimizing plant nutrient losses);
- Adapting no till and conservation tillage technologies to environmental, social and economic conditions within specific territories, using both validation and demonstration steps in representative farms (FAO, 2002);
- Setting up participatory mechanisms associating scientists, farmers and extension services to further develop the incorporation of the above technologies into location-specific sustainable resources management systems;
- Developing controlled agriculture (greenhouse and hydroponics) in periurban areas to produce food for the ever increasing urban population (Littlefield, 1998; Savvas and Passam, 2002). Further development of new and innovative systems that are less consuming in energy and inputs is required (John, 2001; FAO, 2005a);
- Developing precision farming to use real-time, site-specific information in crop management, for example:
  - Accurate field mapping with information collected from soil samples, pest monitoring and harvest yield data allows farmers to target the use of plant nutrients and crop protection products, leading to an efficient and judicious use of these products (SEENET, 2007);
  - Highly developed systems use computers installed in farm machinery such as harvesters, fertilizer spreaders and crop sprayers, combined with mobile satellite global positioning systems, enabling farmers in some situations to spatially vary the rate of input application and management operations, thereby optimizing the productivity of the crop based on accurate determination of soil and crop needs (SEENET, 2007).

Box 6-10. Organic agriculture

Although it only represents a small percentage of the total utilized agricultural area, organic farming has developed into one of the most dynamic agricultural subsectors. Organic production has been encouraged by policies to promote sustainable food and farming in many NAE countries. Organic production has the potential to reduce environmental risks associated with use of agrochemicals, market advantage for producers. (EU, 2007; OACC, 2007; USDA, 2007).

In view of the growing production and expanding market due to increasing consumer demand for organic foods, the Codex Alimentarius Commission has developed Guidelines for the Production, Processing, Marketing and Labeling of Organically Produced Foods in order to provide a clear description of the "organic" claim and thereby ensure fair trade practices in this area. The Guidelines are a dynamic text that can be amended as new proposals are put forward in view of the experience gained by member countries as the organic sector develops (Padel and Midmore, 2005; FAO/WHO, 2006; Stolze et al., 2000).

The main factors and activities to be considered in order to promote organic agriculture are:

- To ensure that all stages of production, preparation, storage, transport and marketing are subject to inspection and comply with the guidelines;
- To develop and promote AKST for new techniques for the production and processing of organic products, including skills and training to support adoption; and
- To develop consumer awareness and marketing systems for organic produce as part of a strategies for sustainable food and farming.

The contribution of organic agriculture to food security is open to debate and subject to divergent views, especially as information is scattered and sometimes speculative. This is a topic worthy of further research, especially given the potential for organic farming to support livelihoods amongst relatively resource poor farmers and rural communities, as well as "reconnecting" consumers with farming through locally or regionally produced organic foods.

6.2.8 Developing sustainable systems for forestry

Over the past 20 years in some parts of NAE there has been a move away from productivity as a driver of forest management, with more emphasis being put on environmental and social issues. Today awareness has emerged worldwide regarding other forest values or its "multifunctionality."

Forests, especially mixed forests, are recognized as reservoirs of biodiversity, as contributors to improved water quality and availability and as an important component of the carbon economy.

Sustainable forest management cannot be considered without taking into account the many changes that affect the environmental, societal and economic context. Competing land uses can result in forest fragmentation and can contribute to climate change, invasive species, increase of energy prices of fossil fuels and the new cost of carbon emissions. All of these, among other factors, can have implications for social expectations of forests and forest management as well as natural areas in general (Raison et al., 2001; Houllier et al., 2005; Dekker et al., 2007).

Integrated forest management methods addressing both timber production and ecosystem management have appeared in the NAE (Rauscher, 1999). This management can lead to sustainable development of forests with overall economic, social and environmental benefits. More research will assist in better understanding the multifunctional role of forests from an economic, social and environmental perspective and promote it through appropriate sustainable management methods.

Integrating the multifunctional role of forests. The definition of forest multifunctionality has changed over the decades from a simple three-fold categorization of production of wood, protection and restoration of the environment and social functions into something more complex with multiple functions added under each category. For example, the "production" function now comprises a larger range of forest products such as wood products, bioenergy, green specialty chemicals, novel composites (Table 6-2).

In order to integrate the multifunctional role of forests better it is important to understand how forests can contribute to these functions and the extent to which these functions could be simultaneously realized through definition of optimal management methods that guarantee the provision of these functions as explained below.

Table 6-2. **Functions and objectives of multifunctional forest management.**

Functions	Sub-categories	Specific objectives
Production	Timber products	Sawtimber, veneer, pulp and paper, panels, bark
	Bioenergy	Firewood, charcoal, biofuels
	Hunting	Game management
	Other products	Mushrooms, fruits, pharmaceutical molecules
Protection and restoration	Habitats	Naturalness as an ecological heritage (reserves) Protected habitats Microhabitats (ponds, peat bogs) Patches of senescent forests Deadwood material (large woody debris)
	Plant biodiversity	Endangered or rare species Ordinary biodiversity Genetic diversity
	Diversity of other taxa	Endangered or rare species Hunting and fishing Wildlife, birds, insects, etc. Microorganisms (e.g., soil microbes)
	Carbon storage	
	Water quality	Chemical (avoid nitrates, xenobiotics, raise up pH) Ecological (microbial and vertebrate diversity in streams)
	Soil protection	Chemical (maintenance of soil fertility) Textural (prevention of compaction) Integrity (prevention of erosion)
	Forest health	Limit sensitivity to diseases and disturbances
	Human protection	Use forest to mitigate landslides, avalanches, falling stones
Social function	Landscape quality	Meso-scale (forests in landscapes) Microscale (managing hedges for scenery) Landscape diversity (patchiness, mixtures, canopy texture)
	Naturalness as a cultural value	Forest reserves, botanical gardens, arboreta Undisturbed or low-impacted landscapes
	Tourism	Hiking, bicycle paths
	Other cultural values	Trees, flowers, fruits, animals of high cultural relevance Religious holy sites
	Educational value	Forests as support for education (ecology, environment)

Characterize and understand the different functions and the potential incompatibilities between them

Enhance forest productivity in a sustainable manner. Understand how the social and natural environments create, reinforce and localize the tradeoffs between the multiple functions of forestry. These understandings can constrain and guide developments in the following areas of AKST.

- Breeding trees for the future (for specialized plantations): study molecular, biochemical and physiological processes determining wood and fiber properties, water and nutrition biology and interactions with insects and microorganisms; this could include the identification and functional analysis of relevant tree genes as well as the elucidation of signal pathways and components required for the expression of genes important in tree improvement. More effective breeding strategies could be developed using molecular genetics including genetic engineering in order to use them to generate new tree varieties with characteristics that fit the local multifunctional needs of forestry; these may include wood fiber characteristics that provide enhanced economic as well as environmental value (higher cellulose, lower lignin, fewer chemicals during paper manufacturing, stronger rot resistant wood (for construction), xenobiotic degraders for phytoremediation, biotic and abiotic stress tolerances to allow expansion of forests to harsher climates or more marginal lands, hypoallergenic pollen producers to enhance urban landscapes without jeopardizing health) (FTP, 2007);
- Developing tools to anticipate new invasive species problems in forestry and improve tools to manage existing invasive plants, insect pests and pathogens, which are one of the major threats to forest quality in the NAE (Pimentel et al., 2000; Allen and Humble, 2002);
- Enhancing the availability and use of forest biomass for products and energy and finding a balance between the increasing demand for forest biomass for energy production and an increasing demand for forest-based products; and
- Accentuating the environmental assets of wood (compared to other materials) by developing innovative products for changing markets and customer needs: intelligent and efficient manufacturing processes that require little or no chemical products, reduced energy consumption, etc. (Forest-based Sector Technology Platform, Vision 2030).

Provide environmental and social services. Important areas for the development of knowledge, science and technology in forest management include:

- Analyzing the role of biological diversity (both functional and heritage value) and other factors (soil, water) in maintaining the stability and primary production of forest ecosystems (UNECE-FAO, 2005);
- Forecasting future dynamics of forest biodiversity and productivity, especially in relation to environmental change;
- Exploring further the positive effects of forests on water quality and accordingly exploring the potential benefits of urban and periurban forests;
- Evaluating the impacts of exurban sprawl on forest fragmentation and forest quality; and
- Continuing research activities that focus on determining the effects, at various scales, of optional forest management strategies on environmental services such as carbon sequestration and social services such as amenities and recreation (UNECE-FAO, 2005).

Once these different functions are characterized a clear definition can be developed with the help of indicators. These indicators could help to better assess and quantify these incompatibilities and also help in deciding where certain functions can be compromised compared to the others.

Define optimal management methods that guarantee the provision of these multiple functions

Defining optimal management methods can help better address this issue of multifunctionality at the appropriate geographical scales. Currently forest management can be broadly divided into three types. The first two types are based on complete geographic segregation of the different functions: intensive production forests that are dedicated solely to production (intensively managed conifer monocultures, e.g., Southern pines in the USA, Sitka spruce in the UK and maritime pine in France) and natural reserves that are left untouched with little or no human intervention. The third type is that of semi intensive forests ensuring production, environmental and social services, often using trees more adapted to local conditions.

There are many ways of guaranteeing the multifunctionality of forests based on the above mentioned three forest types. One way would be to have all the three forest types in the same zone and the other way would be to have only semi intensive forests wherein depending on the needs one function would dominate slightly over the other.

More research can shed light on how to optimize the overall distribution of intensive, natural reserves and semi intensive forests in NAE and its sub-regions, keeping in mind that total forest stocks have remained relatively constant in most of North America and are increasing throughout Europe (Karjalainen et al., 1999).

As at the local level semi intensive forests could be viewed as a complex multifunctional system. More research can contribute to developing models of this system as a whole (one that includes the production, environmental and social services) based on a meaningful knowledge representation elaborated with the help of the different stakeholders involved.

Provide methods and tools for monitoring and improving the environmental sustainability of forests:

- Extending existing and promoting new and integrated forest inventory services: develop tools for monitoring forest health, nutrition, greenhouse gas absorption, evolution of populations and communities, in addition to the traditional growth and yield studies (Birot et al., 2005);
- Adapting forestry to climate change (European Forest Inventory) (European Forest Institute, 2007), particularly in drought prone regions e.g., Southern Europe and Western United States; development of adaptive forest management methods comprised of heterogeneous

species populations for improved resilience, improved adaptive capacity of the forest reproductive material, deciphering the buffering capacities of tree species and genetic diversity to climate change;

- Developing better risk assessment, risk management methods and improved risk sharing instruments to integrate risk and other environmental and economic changes into forest management (climate change, invasive species, exurban sprawl, fires, gales, floods, pest and disease outbreaks, uncertainties regarding the economic value and abrupt changes in the market, etc.); for example, assessing the vulnerability of various management strategies in regard to the different risks (FTP, 2007);
- Monitoring genetic diversity of natural forest populations to elaborate methods to keep genetic integrity during reforestation and other forest management events; selecting areas for forest genetic reserves (*in situ* preservation); and
- Better and more exhaustive mapping of forest resources in terms of quantity and quality through a wider use of present technologies (GIS, remote sensing, ground laser technique, etc.) as well as the use of satellite imagery and modeling as a decision support tool in forest planning and management.

The future of successful forest management rests on a revision of forestry concepts in the light of climate, invasive species and other environmental changes, a recognition of new concepts of sustainability in which risk management and forest resilience are prioritized and the encouragement of an improved dialogue among scientists, managers and the public that transcends national boundaries.

6.2.9 Developing sustainable systems for fisheries and aquaculture

6.2.9.1 Coastal capture fisheries in the NAE region

With the collapse of many fisheries and the expansion of fishing efforts, improving the sustainability of coastal capture fisheries and increasing their productivity have become acute problems (Pauly et al., 2002, 2003, 2005; Garcia and Grainger, 2005). Sustainability can be achieved by: (1) broadening the focus to include the entire food web and habitats that support the target species; (2) efficient management systems that take into account the ecosystem and (FAO, 2003); and (3) better fishing technologies that help in preventing overexploitation of all target species (De Alessi, 2003; EC, 2003). Productivity on the other hand can be increased by adopting new processing methods that add value to the current system, such as creating markets for by-catch (Christensen et al., 2003).

In the following text, options for research and technological development in the area of coastal fisheries have been explored within the different compartments of the system namely: sustainable marine ecosystem management, fishing technologies and fish processing. These compartments have various interactions among them. Only an integrated vision of the entire system and its different compartments as a whole would allow a fuller understanding of the functioning of the system.

Develop an ecosystems approach for a sustainable coastal marine ecosystem management

Up to now, most research has concentrated on the consequences of fishing activities on the target fish stocks. Future research is increasingly taking into account the social, economic and ecological consequences of fisheries not just at the local but also at a global level (Pauly, 2005). More research on the following may be useful to achieve this:

- Further analyze the impacts of coastal fishing drag-net methods that disturb the entire benthic community and harvest entire food webs. These methods disrupt entire communities and irreversibly alter benthic habitats, changing the reproductive potential of the target species and associated by-catch (Francis et al., 2007);
- Further develop the construction of mathematical models for complex systems that help understand and predict ecosystems behavior, by multidisciplinary approaches and considering biological, ecological, economic and social driving forces (Dame & Christian, 2007). Such models and toolboxes would facilitate the study of the different effects of fisheries management, regulation and policy;
- Collect long term observation data: identify a representative sample of long term observatories to collect data that can be used for the constitution of reliable and continuous series data (biological, economic and social) for use in present and future research, namely for the validation and adjustments of the above mentioned models; this activity exists but is presently inadequate because marine resources are difficult to access and not well known;
- Develop tools and indicators: appropriate tools and indicators that take into account an ecosystem as a whole reflecting resource health including its economic and social components as well as integrate global phenomena such as climate change do not yet exist and will be useful now and in the future;
- Develop experimental research on consequences of human activity on wild fish: effects of fishing and pollution on growth and reproduction;
- Develop specific multidisciplinary research on marine ecosystems: including ecological engineering, "ecological therapy" (how to cure, restore an ecosystem), environmental economy and sociology; and
- Evaluate the effect of enforcement of territorial waters (or lack thereof) on the sustainability of fisheries. Combating illegal fishing has been identified as a crucial element (COFI, 2007).

All the above mentioned research activities will require an integrated global research effort, coupled with stronger enforcement measures throughout NAE. Focusing on selective fishing and sustainable harvest levels can prevent the overexploitation of all species. By concentrating research efforts on overcoming the barriers mentioned below it may be possible to adapt "selective" adaptation levels to the renewal capacity of fish stocks:

- Develop innovative methods for direct evaluation of fish stocks, e.g., acoustics, buoys, AUV (autonomous underwater vehicles);
- Promote selective fishing that takes into account the

present and potentially renewable fish stocks: in this context a better quantification of the long term biological and economic benefits of selective fishing could convince the actors of its importance and urgency;
- Devise new fishing techniques that are highly selective with a minimum impact on the ecosystem in coastal and high sea fisheries and/or that have a smaller ecological effect on the habitats that sustain the fisheries; and
- Improve existing fishing technologies: to obtain a higher quality of fished products while simultaneously minimizing by-catch.

Fish processing for food:
- Focus on processing and adding value to small pelagic fishes for human consumption, usually fished to make fish food and animal foods (yearly around 18 million tonnes in the world); and
- Improve processing methods and quality of the existing processing units or build new processing units that have the least environmental impact feasible.

6.2.9.2 Aquaculture

Developing countries such as China, others in South East Asia and some CEC countries will continue to expand production of low-valued fish, such as carp, and will greatly expand production of some high-valued fish, such as shrimp and salmonids (Delgado et al., 2003). The NAE region will continue relatively constant production of high-valued fish, such as salmonids. Trade in aquaculture is likely to increase or at least stay the same and perhaps South-South trade will increase. However, with increasing affluence in some developing countries, this trade dynamic is likely to change with a reduction in trade due to increased home consumption. Some of the main areas with identified research gaps are listed below.

Disease and water quality. An important aspect of aquaculture is combating viruses, bacterial and parasitic diseases. It is however difficult to guarantee a strict sanitary isolation of aquaculture sites. Therapeutic options are limited and often lead to environmental problems. Research on the fish immune system can provide more therapeutic options and allow an understanding of the environmental determinants of immunity. These are necessary prerequisites for the sustainable use of genetic resources for disease resistance.

The aquatic environment is subjected to pollution from both human activities (accumulation of pollutants: e.g., heavy metals, pesticides), nature (e.g., heavy metals and acids from volcanoes) as well as aquatic microorganisms (toxin production by microalgae). More research on ecotoxicology and ecopathology will provide a better understanding of these impacts on fish quality and production. It would be advantageous if such research were done in close collaboration with research in physical sciences (hydrodynamics, modeling of the pollutant flux, etc.). Also research to improve the quality of the aquatic environment could be done, through the optimization of physicochemical and microbiological quality of the aquatic environment.

Environmental impacts of aquaculture. Aquaculture can have negative impacts on the environment (SOFIA, 2006). Firstly, aquaculture activities in general can perturb and alter the surrounding environment. Secondly, wastes from intensive aquaculture often have adverse effects on the environment. Thirdly, escaped fish can destabilize nearby native fish communities. The impact of the wastes of intensive aquaculture is due, primarily, to the quality and the quantity of fish diet (concentrated fish feed). Some of the ways to combat this is are, for example, to (1) increase the efficiency of fish feed to reduce the quantity of the overall diet; (2) substitute fish meals that are classically made of fish oil and fish meal with plant products and (3) in closed systems, develop biofilters using biofilms that recycle wastes back to fish feed. Aquaculture can have positive effects on the environment by reducing the pressures on native fisheries, as well as increasing the fertility of the water though wastes.

Reduce dependence of high value fish farming on fish meal derived from coastal capture fisheries. To reduce impacts on fisheries, fish meal made from plant products might be substituted for fish meals that are classically made of fish oil derived from coastal capture fisheries. One of the areas of research that might reduce dependence on coastal fisheries is to produce feed crops with high levels of oils and proteins required in aquaculture.

Labeling or certification for responsible fish farming: Aquaculture's future is determined not only by the market price but also by consumers' acceptance of its products. Aquaculture may find it useful to expand linkages of its impact (or lack thereof) on the environment with the type of products produced and production practices with marketing. Initiatives that integrate these aspects have already been adopted in certain areas (the global aquaculture alliance—created by shrimp farm producers—which proposes a code of good practices that would help in reducing the environmental impacts of their activities; organic aquaculture; labels certifying the quality like red label, etc.) but more research is necessary to define appropriate management strategies and establish the relevant criteria that would help in the evaluation of the efficiency of these strategies and lead to an eventual labeling or certification of the product.

Moderate intensification of extensive systems: Intensification of aquaculture seems inevitable due to increasing reduction in the area available for these activities (SOFIA, 2006). The most desired solution may be to opt for moderate intensification of current extensive aquaculture systems (often polyculture systems comprising different species). This transition can only be successful following more research on:
- Identifying optimum conditions for the different types of polyculture systems as the different species involved in polyculture systems have different ecological roles;
- The criteria for the amelioration of environmental impact of a multi-specific population under trophic constraints (e.g., increase in the growth rate, nutritional behavior). Amelioration of the impact of one species in a polyculture could be done to the detriment of others and may not result in the overall amelioration of impacts; and
- Better understanding of the integration of these systems in rural areas: eventual constraints (e.g., water management), opportunities, complementarities (e.g., use of ag-

ricultural by-products in aquaculture and effluents from aquaculture in agriculture), etc.

Introduction and naturalization of species: Aquaculture is currently based on a limited number of species that have been disseminated all over the world. Aquaculture populations commonly escape establishing feral populations that can adversely effect the population density, health (e.g., when one population is a pathogen carrier), or genetic diversity of native species. More research evaluating and quantifying the impacts of introductions of aquaculture species on natural populations can ensure better integration of these species in the ecosystem while avoiding harmful effects on the surrounding environment. The culture of triploid fish is one way to ensure the non-reproduction of escaped fish.

6.2.10 Ensuring socioeconomic viability of the systems and improving rural livelihoods

Changing priorities and the reform of agricultural policies recently have reduced the financial rewards for farm production in NAE with economic and social consequences for those whose livelihoods depend on it. Simultaneously, concerns have grown about the high, yet hidden, social and environmental costs of intensive agricultural systems. It is critical that, drawing on lessons from the past, socio-economic mechanisms are harnessed to help achieve the new paradigm of multifunctional agriculture, securing the incentives and benefits to those engaged in its delivery and maximizing overall welfare. Doing this has major implications for the types of AKST required and how AKST can best be mobilized to meet new expectations.

6.2.10.1 Social issues

Development of AKST in agriculture strongly affects and is strongly affected by the multiple societal issues related to rural society. Ensuring social sustainability of locally dynamic economies will require AKST research on the necessary social relationships that could be reinforced or developed to meet goals for NAE (Narayan, 1999; Flora and Flora, 2004).

Social institution building

Because many of the institutions in rural NAE have been developed and maintained to support national agricultural commodities and commodity prices, it is likely that new institutions will be required to support rural economies that have a strong local component. Research can help determine how the present institutions can support and maintain a focus on a local economy and the institutional changes required. Several measures can be considered including:

- Providing appropriate training and new credit systems to enable rural workers to become farm owners and operators;
- Establishing locally-based market linkages between farm products and consumers;
- Improving rural quality of life, including better schools, health care, recreation and food quality and availability;
- Identifying and encouraging institutions to facilitate transitions to a multifunctional agriculture;
- Developing instruments for the provision of new income, in particular for goods and services that are not marketed today; and
- The new paradigm of multifunctional agriculture emphasizes environmental sustainability and the provision of public goods. There will be increasing demand for collective (community) rather than individual actions (Ostrom, 2003), encouraging a new "moral" economy in which people constrain their immediate individual freedoms in order to achieve improved common and subsequently individual, welfare (Trawick, 2004). There is a role here for AKST to devise new mechanisms for joint action especially concerning the management of scarce natural resources (Trawick et al., 2005).

Farmer organizations. Building producer capacities, an important objective of AKST, could be facilitated through professional and inter-professional organizations. To be effective, agricultural development requires the participation of the farmers and their organizations in domains such as elaboration of agricultural policies, extension and training systems, organization of the markets and the supply chains, rural credit, land policies. The roles that producer organizations play are diverse and can cover various topics such as:

- Policy representation and defense of the interests of the members ;
- Economic services through the supply chain organization and the collective setting in markets;
- Development of technical services such as economic and technical advice, training, the use of materials owned jointly; and
- Provision of public services, for instance the elimination of illiteracy, infrastructure maintenance, etc.

The current debate on the place and the role of the agricultural organizations in supporting family farms revolves around three themes (Mercoiret et al., 2001): producer support mechanisms; creating new forms of coordination between actors; building and strengthening the capacity to face global phenomena.

Supporting these professional and nonprofessional organizations could lead to the building of new relations between the different actors, based on the partnership, dialogue and negotiation.

On the economic side, strengthening the economic organization of agriculture is essential to ensure decent incomes through economic market management (MAP, 2006). It is important to accord a specific place to inter-professional organizations. They are private organizations bringing together the partners upstream and downstream of an agri-food network related to a product or a group of products. Their goal is to sign inter-professional agreements which define and promote contracting policies between members, contribute to market management (improved product adaptation and promotion) and reinforce food safety.

Organization of workforce. Demand for agricultural labor remains high in those regions in NAE that fill the increased consumer demand in domestic and export markets for vegetable, fruits, nuts, wines and juice products. Many tasks in this agricultural sector including planting, pruning, cultivating and harvest, remain labor intensive. The tendency towards more elaborate processing and packaging for all crops, including grain-based and meat products, creates continuing strong demand for workers in these agriculturally based industries. Organic and alternative agricultural practices also typically increase labor demand. The strength of the demand for agricultural hired labor in some regions and crops is often disguised by the longer term and more general decline in agriculture overall. Both changing markets and the prospect of climate change will create new needs for knowledge and skills in the agricultural labor force. Although the demand for agricultural labor remains strong in much of NAE, research shows that the inequalities created by low-paid farm labor constitute a significant share of overall income inequality in most of NAE (Alderson and Nielsen, 2002; Martin, 2003; ERS, 2007).

Development and sustainability goals have important and unresolved implications for required improvements in the welfare of agricultural workers and farm families in terms of wages, overall working conditions, health and safety problems, health insurance, job security and housing. Meeting development and sustainability goals also requires that a healthy and stable rural work force be available to agricultural employers.

Addressing the problems raised by farm labor will require broad public policy initiatives over the long term in farm subsidy programs, immigration law, labor law, health policy, regulatory law, housing policy, regional planning, governmental budgeting and other complex areas. However, existing research indicates that there are measures specifically within the agricultural sector and short of the broader and deeper reforms that are required, that can work to stabilize and improve the welfare of farm workers and families, yielding numerous advantages to society and the environment. The applicability of such measures will obviously vary by region. Among them are (Findeis, 2002; Martin, 2003; Strochlic and Hamerschlag, 2006):

- Value-added, on-farm activities and product diversification that allow for a more stable stream of farm and labor income while providing year-round employment, creating incentives for improvement in farm worker skills, in turn improving worker productivity and morale;
- Frequent consultation with farm workers, their families and rural residents to address both the nature of rural work and worker welfare; and
- Improved working conditions and higher wages, recognizing these as fundamental to maintaining a stable and skilled work force and often contributing to farm profitability.

Gender issues. The role of gender in North America and Europe is extremely varied from country to country and regionally within countries. NAE researchers and institutions have done a great deal of work on gender inequities in agriculture outside of NAE, but relatively little attention has been paid to gender issues within NAE agriculture. Discussion of gender inequities within families on what has been termed "the discourse of the family farm" initially focused on the masculine dominance of the farm family and inequities of power and welfare as a consequence. Later research has focused on the recognition and development of more complex familial relationships in terms of ownership, work roles, decision-making and welfare outcomes. In analyzing these more complex relationships researchers have more recently seen the way in which women play active roles within the family, more typically working with male family members to deal with difficulties imposed on the farm enterprise from outside the family structure (Brandth, 2002).

Gender inequities within the professions of agricultural research, education and extension are striking in much of the region and particularly in the higher reaches of academic research and teaching. Researchers have focused on the factors that lead women to choose other kinds of work and that determine the relative lack of women in agricultural research. Much of this analysis focused on gender inequities within the AKST profession a decade or more ago, setting an agenda for change. There seems to be an opportunity for reassessment of the prevailing situation and of future prospects (Van Crowder, 1997; Foster, 2001).

The last twenty years have seen a striking emphasis within rural and agricultural development work done outside NAE on gender analysis and appropriate policy responses. Much of this work is performed and/or directed by institutions based in NAE (including the World Bank and NAE national foreign assistance programs). This makes it urgent that gender imbalances among the professionals engaged in such work not undermine the quality and effectiveness of research and policy carried on abroad, as well as at home. Farm workers in NAE experience a variety of work situations involving gender that create hazards, inequities and significant stresses (Barndt, 2002; Nevins, 2002; Fox and Rivera-Salgado, 2004; VanWey et al., 2005). Among these are:

- Legal and illegal immigration across international borders often makes it difficult for families to remain together, posing high levels of insecurity and resulting in large economic costs. Most typically, men migrate internationally without their families when there are high risks and/or costs associated with border crossing and residence without legal documentation; this is particularly the case for some hundreds of thousands of migrants from the Caribbean, Mexico and Central America who work in agriculture in the United States and Canada;
- Gendered employment patterns, as for example women working in poultry processing plants while men work in slaughterhouses for pigs and beef cattle, often with significant gendered differences in pay and often resulting in family separation and inequities;
- sexual harassment and exploitation associated with women separated from families by gendered work situations; and
- Failure to exclude women, pregnant women and children from farm chemical exposure that have in some cases been shown to pose particular risks to women, fetuses and children. Serious toxicological issues remain in the analysis of this problem and while regulatory schemes in most of NAE have attempted to address the issue, problems of measurement, accountability and enforcement remain (Castorina, 2003; Bradman, 2005, 2007; Young et al., 2005; Eskanazi, 2006; Holland, 2006).

Equity in opportunity, participation and rewards for similar work ensure the sustainability of society. Rewards to farmers in NAE vary substantially as a function of land tenure. For farmers and employees in the rural economy, rewards also depend strongly on skill levels and on access to training and education. Security of employment varies widely and is critical for workers in an agricultural economy. In some parts of NAE, rewards and entitlements vary considerable according to social class, gender, ethnicity, age and formal education, often with adverse consequences for social and economic outcomes. More attention can be paid to the potential positive role of agriculture in promoting equal opportunities as a basis for social inclusion and sustainable development.

- Local, regional, national and international institutions to promote equity could be reinforced and improved;
- Local knowledge and knowledge of disadvantaged communities can be incorporated interactively into the AKST system; and
- Current trends are promoting further liberalization and reform of agricultural commodity markets, increased connectivity between people and food, as well as new "markets" for services previously considered un-traded and un-priced, such as water supply and access to the countryside. Market mechanisms require information and skills of negotiation and transaction to work properly. This has implications for the role of AKST in supporting economic efficiency and fairness.

6.2.10.2 Economic issues

In theory, economic sustainability requires that the most efficient means of production and consumption of goods and services be used and overall long term net welfare (benefits less costs) maximized (Begg, 2003). Sustainability also requires that agents engaging in economic activity, from farmers to plant breeders, who commit resources now with a view to enhanced benefit in the future, are justly rewarded in terms of incomes and returns on investment, creating the enabling conditions for risk-taking. This applies whether the processes are entirely driven by market forces or interventions by government. Underlying this, there must be clear signals indicating what society wants of its agricultural and rural sector, whether this is in the form of market prices for agricultural commodities, or payments for environmental services such as landscape management and flood storage.

In the past, economic growth (defined in terms of gross incomes or expenditure in the economy) was used as the dominant indicator of development. Sustainable development now requires that economic indicators be balanced with social factors associated with distributional, quality of life and ethical considerations, as well as environmental factors that reflect the state of natural systems and the beneficial services they provide (Pearce and Barbier, 2000; Hanley et al., 2001; Tietenberg, 2003). Many of these broader societal impacts are now included in so-called "extended" cost/benefit analysis and sustainability appraisal of development options, including AKST (DEFRA, 2005b).

In this context, an economic perspective has three particular contributions to make: (1) Assessment of the economic and institutional performance of food and fiber value chains.

A more complete understanding of the process by which value is added in food and fiber supply chains is necessary as this affects the efficiency of resource use, incentives, rewards, technology change, the sharing of risks among supply chain agents and end-user choice and welfare.

Further developments in AKST in the following areas, could be helpful to:

- Identify opportunities for adding value through market orientation, quality assurance, product differentiation, including the promotion of sustainable production and consumption;
- Evaluate the life cycle performance of alternative value chains, developing appropriate data bases and analytical methods (such as LCA) and decision support tools;
- Develop tools such as multi-agent modeling to help improve supply chain performance;
- Conduct value chain analysis that can help evaluate the total contribution of agriculture. This includes analysis of the competitiveness of the whole food and non-food chain and the economics of quality;
- Justify and guide investments in supply chain and logistics to improve economic efficiency; and
- Develop efficient supply chains for new products and markets, such as biofuels and medicinal crops.

Specifically, there may be significant opportunities to increase the competitiveness, economic viability and contribution to economic welfare of the forestry/wood chain:

- Optimization of the value chain from the forest to the end product, including recycling;
- A stronger coupling between wood producers and industrial consumers;
- Analysis of all sections of forestry-wood chain (silviculture operations, sales procedures and transaction costs, harvesting and logistics costs, etc.) in order to determine how to best improve competitiveness and economic viability;
- Diversification of wood and fiber-based products through technological innovations: this could apply to packaging with new functionalities (embedded information technologies); advanced hygiene and healthcare products; "green chemicals"; new generation of composites; and
- Development of logistic and decision support systems for optimized supply chain management.

(2) Identifying and valuing the costs and benefits of goods and services produced by agriculture. This requires valuation not only of marketed crop and livestock commodities but also of nonmarket outputs that have consequences for economic welfare. These include the "public good" or "external benefits" (e.g., food security, diets and nutrition, watershed protection, landscape management, access to the countryside, sustenance of vulnerable human communities) as well as the "public bad" or "external costs" (e.g., diffuse pollution, soil loss, habitat loss, displacement of people, health and safety risks) of agriculture and an understanding of how these are distributed spatially and over time (Costanza, 1997; Brouwer et al., 1999; Pretty et al., 2000; Hartridge and Pearce, 2001; Environment Agency, 2002; EFTEC, 2005). AKST could be developed to:

- Identify indicators that reflect or give an idea about the evolution of these external costs and benefits over time;
- Identify the scale at which these external costs and benefits can be studied: farm level (identification of the individual farmer's contribution to a specific externality), landscape level;
- Design policies that take into account the external costs and benefits associated with agriculture. Policies adopted to promote some public goods could worsen, or at least fail to alleviate some external costs (Sutherland et al., 2006). Specifically, the national (or larger scale) performance of the agricultural sector can be evaluated and the consequences analyzed at the local and farm level so that local policies do not contradict national ones; and
- Develop and promote innovative entrepreneurship initiatives, such as safe water production, eco- and nature tourism, recreation, hunting, including forest, upland and wetland systems.

Estimation of the contribution of agriculture and rural services to economic welfare is inefficient and can be improved. This requires redefinition of economic efficiency beyond conventional measures of tradeable inputs and outputs, which is "internalizing the externalities" of agriculture to obtain a comprehensive measure of the social and environmental "footprint" of the sector and its contribution to long term welfare (Barnes, 2002; UN SEEA, 2003; EFTEC, 2005).

(3) Design economic instruments to help achieve sustainability. Design economic instruments such as fiscal measures, compensatory and incentive regimes, market support and trading systems that can help achieve sustainable development, promoting the appropriate balance of private and public goods. Examples include capital and maintenance grants for organic farming, agroforestry projects, extensive livestock systems in less favored areas, farm diversification schemes, voluntary schemes to pay farmers for environmental services, grants and subsidies for cleaner, welfare oriented technologies and tradable permits for water licenses (OECD, 2000; DEFRA, 2002ab). For example, further research can examine how the supply of land to agriculture might respond to the fall in output prices as a result of the elimination of farm price and income support policies in many countries. Also, research could better examine how the supply of public goods associated with agricultural land responds to payments based on land area. Specific topics include:

- Cause and effects of price instabilities, including consequences for production and investments;
- Effects of the different public instruments in terms of market distortions, price stabilization (e.g., intervention prices, quotas, decoupled payments);
- Role of market mechanisms such as stock markets and commodity futures markets to face price risks; and
- Importance and role of contracts and conventions between the players of a sector (farmers, agribusiness, retailers).

6.2.10.3 Sustainable rural livelihoods

There is continuing concern about persistent poverty and the vulnerability of individuals and families in some rural populations in NAE, whether due to increased pressure on land and water resources or economic factors associated with structural change. The concept of sustainable livelihoods is used to analyze the social and economic viability of agricultural and rural systems (Chambers and Conway, 1992; Carswell, 1997; Hussein and Nelson, 1998; Scoones, 1998; Ellis 2000; Turner et al., 2001). Whereas the term "livelihood" focuses on productivity, income and poverty reduction, the term "sustainability" refers to the resilience of livelihoods and the maintenance of natural resources on which they depend. This analytical framework can help to understand how households and communities cope with shocks and stresses, such as those associated with policy or climate change.

The sustainable livelihood framework concept has considerable relevance for understanding the social and economic aspects of farming systems in the NAE region (Pretty, 1998) (Figure 6-1). It emphasizes the critical relationships between high level drivers and contextual factors, resources and assets, institutional processes, farmer motivation and coping strategies and resultant welfare (Scoones, 1998; DFID, 1999).

It is important to better understand the diversity of livelihoods within rural households and communities as a whole and the critical synergy between rural and urban dimensions of livelihoods, especially as these affect the transfer of assets, knowledge, goods and services between the rural and urban sectors, with consequences for welfare. The critical influence of local and distant institutions (e.g., local customs regarding access to common property resources, local and national land tenure rules), social relations (e.g., based on gender, kinships, tenure) and economic, value-adding opportunities are also recognized.

In the context of meeting development and sustainability goals in the NAE region, there is considerable merit in applying the livelihoods framework to guide future development of AKST, particularly to address the needs of the most vulnerable farming and rural communities. AKST clearly interacts with and is shaped by, the factors that describe the context for rural livelihoods, such as the policy and market drivers. As these change, so will the requirement for additional AKST as it is clearly embedded within the assets of households and communities. These include the products, tools, equipment and processes (physical assets), the knowledge and skills available (human capital) and the systems of governance (social capital) available to a farming community. Changing circumstances, whether induced by global or local factors, have implications for AKST in its widest sense.

AKST is closely linked with the availability, use and productivity of natural capital such as land and water resources and financial capital as this determines access to farming and other inputs. The livelihoods framework confirms the importance of governance systems as these influence patterns of resource use and rural development, in turn shaping the development and dissemination of AKST. Hence, AKST is central to the livelihood strategies evident in farming sys-

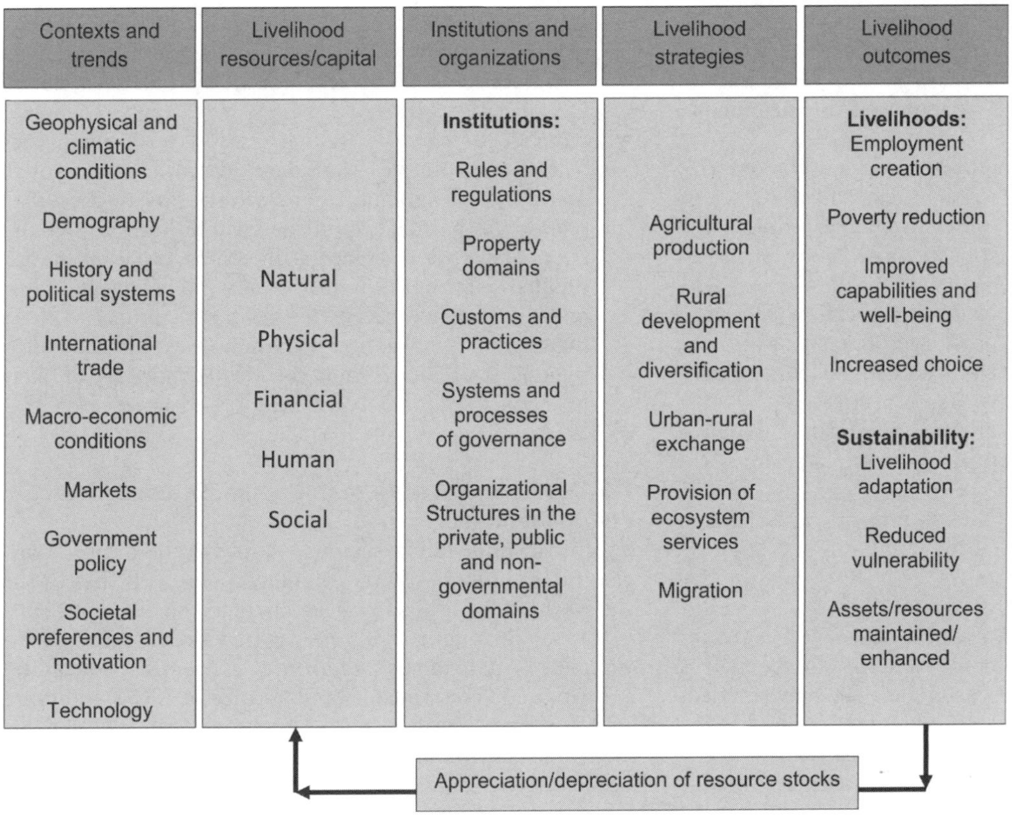

Contexts and trends	Livelihood resources/capital	Institutions and organizations	Livelihood strategies	Livelihood outcomes
Geophysical and climatic conditions Demography History and political systems International trade Macro-economic conditions Markets Government policy Societal preferences and motivation Technology	Natural Physical Financial Human Social	**Institutions:** Rules and regulations Property domains Customs and practices Systems and processes of governance Organizational Structures in the private, public and non-governmental domains	Agricultural production Rural development and diversification Urban-rural exchange Provision of ecosystem services Migration	**Livelihoods:** Employment creation Poverty reduction Improved capabilities and well-being Increased choice **Sustainability:** Livelihood adaptation Reduced vulnerability Assets/resources maintained/ enhanced

Appreciation/depreciation of resource stocks

Figure 6-1. *Framework for the analysis of sustainability.* Source: Adapted from Scoones, 1998, DFID, 1999.

tems and management practices, as well as the social and economic outcomes for farming families and communities.

There are critical synergies between livelihood outcomes and the stock of "capitals" on which livelihoods are based. Uncertain and declining livelihoods often result in depreciation of the capital stock, especially natural capital, further increasing vulnerability. By comparison, secure and improving livelihoods can support investment and enhancement of capital stocks, such as improved land management skills and practices.

AKST is a critical component of the stock of capital in the livelihoods framework. It is recommended that the sustainable livelihood framework, adapted to accommodate local conditions, is used to inform future development of AKST to meet the social and economic needs of farming households and communities, especially targeting the needs of the most vulnerable groups.

6.2.10.4 Understanding farmer attitudes and behavior

The development and successful application of AKST depends on the attitudes, motivation and behavior of the potential user community, especially land managers. An understanding of the processes by which land managers learn about, evaluate and adopt or reject new technologies is essential for the management of technology change and the design of appropriate AKST.

Innovation-decision models have long been used to explain technology adoption behavior among rural communities (Ryan and Goss, 1943; Rogers, 2003). Prior conditions, such as policy drivers or perceived needs, shape the disposition of potential adopters towards a new product or practice. This process is influenced by characteristics of decision makers (such as personal and contextual social, economic and cultural factors) and characteristics of innovations (such as relative advantage, compatibility with values and preferences, simplicity and ability to trial and observe benefits). These models also confirm the importance of communication channels, agents of change and contextual and cultural factors, including the relative balance of individual and collective decision making. These models have however been criticized as too rigid, seeing adoption as an externally driven, linear process. Alternative models emphasize different elements of the decision process, namely systems models, information models, models of reasoned action and learning and knowledge transfer models (Garforth and Usher, 1997; Beedell and Rehman, 1999; Morris et al., 2000; Phillipson and Liddon, 2007).

In this context, there is an urgent call for improvement in the understanding of technology change and adoption behavior, in particular to:

- Improve the understanding of variation in farmer motivation and behavior with respect to new technologies and how this is shaped by policy and market drivers, personal circumstances, common practices, local and distant institutions, issues of gender and ethnicity and perceptions of risk;

- Develop and appraise empirically based models of knowledge "exchange" suited to the new agricultural paradigm, combining indigenous and new knowledge sources and linked to concepts of sustainable livelihoods;
- Develop participatory methods for identifying criteria for AKST designs that meet the needs and resources of different target groups, especially as this informs the advantage, acceptability, robustness and convenience of AKST offerings to users;
- Develop and mobilize new communication channels, agents of change and "knowledge brokers" where appropriate, including web based sources, machinery contractors and specialist advisors, respectively;
- Develop a framework for the analysis and design of programs of collective action, for example in water management; and
- Integrate social science research into other sciences to ensure relevance of AKST products.

6.2.10.5 Rural development

Research and development can be undertaken with a greater concern for its role in sustainable rural development. It is important to factor in differences in social and environmental contexts as well as farmers' livelihood strategies and the diverse range of stakeholder interests. A key question concerns the roles that agriculture can assume in the sustainable development of rural areas. Agricultural research can and should play an important role in the collective efforts aiming at sustainable rural development:

- The contribution of agricultural research can address the challenges of a more complex countryside. Farmers follow many different and new livelihood strategies and an increasingly diverse range of stakeholders need to be taken into account; an improved understanding of the dynamics and multifaceted nature of rural development and of the roles that agriculture can assume in a more comprehensive process of sustainable development is necessary (FAO, 2003; Knickel, 2003);
- The more recent emphasis on countryside stewardship has at least three driving forces, all related to consumption: first, the rising environmental movement; second, increasing interest in recreation in the countryside; and third, a great residential shift out from the cities to small towns and villages. Use of labor in stewardship tasks consistent with the concept and financing structures of a policy of multifunctionality in agriculture can greatly increase the quality of community life in rural areas. A key question is how to balance the often-diverging interests or the occurrence of "clusters of compatible and mutually reinforcing activities" (Van der Ploeg and Renting, 2000). The active construction of synergies at farm household, farm and regional level could be better understood and promoted (Knickel and Renting, 2000);
- The multifunctionality concept effectively changed the understanding of the relationship between agriculture and society in more integrative ways; it recognizes that a strict segregation of different functions (living, producing, nature conservation, etc.) is less and less realistic; research approaches can be adapted accordingly (Marsden, 1995; Saccomandi and Van der Ploeg, 1995; Van Depoele, 2000; Knickel et al., 2001; Hervieu, 2003; Cairol et al., 2005);
- Sufficient research is lacking on how to optimally facilitate and ease the future development of less-favored areas and of agriculture and rural areas in the NAE region and particularly in the Eastern European countries. The latter are faced with a substantial fall in the number of farms due to historical trend of consolidation and a particularly severe decline in agricultural employment. A marginalization of farm households and entire regions is predicted, and the related impacts of such on rural livelihoods can be addressed in research and policy.

6.3 Development of Human Capital, Organizations and Institutions

Paradigm shifts and key issues relating to the future of agriculture within NAE and its interactions with the rest of the world, as explored in earlier chapters, have not just simply arisen overnight. Over the past few decades, increasing numbers of individuals and groups of scientists, educators, practitioners, policymakers and a range of AKST end-users in NAE have already been identifying, exploring and increasing their understanding of multifunctionality and its implications for design and delivery of AKST. In this regard, a number of individuals, groups and organizations in some of the countries of the NAE region have initiated changes that facilitate the development of human capital and associated institutional arrangements necessary for generating, providing access to and promoting the uptake of the newer and wider forms of AKST (OECD, 1995a; Lucey, 2000). A process of change has begun but it is still in the hands of the innovators and early adopters. A number of governments have encouraged the process. There have been some individual success stories but most of the newer approaches are hardly yet mainstream or sustainable; the rhetoric exists, but the reality lags well behind. It appears that there are many barriers, not only human, but also organizational, institutional or systemic (EURAGRI, 2005).

It is proposed that the process of reconfiguring AKST activities, both within NAE and in their partnerships with other regions, be dramatically accelerated so that they are jointly enabled to contribute most effectively to meeting sustainable development goals (Schneider, 2004).

The following sections explore some of the options, on a range of fronts, for this desired development, based in part on the experiences in NAE countries and analyses conducted to date by the OECD, governments, AKST agencies and individual scholars.

6.3.1 Towards interactive knowledge networks

Agricultural Knowledge Systems or AKS (long-standing OECD-adopted term) span the three main components of research, education and extension (OECD, 1995a). There are close links between these three elements of the "knowledge system," which now require more of a "network approach" and the development of substantially greater synergy. There is an increasing shift from a unidirectional paradigm of knowledge generation and transfer (knowledge production—enlightenment—adoption) towards a paradigm of interactive knowledge networks involving multiple stakeholders who contribute to problem definition, research conception, execution and provision of results to a range of end-users for whom the research is in some way deemed to be relevant. In this way AKS can contribute better to society's wider agenda (e.g., increasing concern with aspects of nutritional policy, food safety, animal welfare and other ethical aspects of food production and natural resource use). It is therefore essential that providers of advisory, higher education and research services become more engaged in building networks and coalitions to address newer objectives in such areas as global competitiveness, agricultural sustainability, rural development and multifunctional systems. Moreover, governments can help ensure that organizational and structural arrangements do not impede but rather encourage these cooperative efforts among components of the AKS (OECD, 1995b).

The AKS concept was further developed in collaborative work undertaken by the FAO and the World Bank which stressed the integrative nature of Agricultural Knowledge and Information Systems (AKIS), linking people and organizations to promote mutual learning and to generate, share and use agriculture-related technology, knowledge and information (FAO and World Bank, 2000). More recently, there have been noteworthy advances in applying an "Innovation Systems Concept" to agriculture, especially in approaching hunger and poverty issues in developing countries. Like AKS/AKIS, it stresses interactive knowledge networks, but recognizes an even broader range of actors/stakeholders and disciplines in a wider set of relationships that can potentially foster innovation. Innovation system analysis recognizes that creating an enabling environment to support the use of knowledge is as important as making that knowledge available through research and dissemination mechanisms (World Bank, 2007)

6.3.1.1 Promote stakeholder interaction

Stakeholder interaction in AKST is required to reinforce two recent trends: a shift from stakeholder management strategies to stakeholder involvement strategies; and a broadening of the types of stakeholders involved. Stakeholder management strategies are aimed at recognizing ways stakeholders can influence decisions and at limiting their ability to affect the process in ways contrary to the interests of the decision-makers (Eden and Ackermann, 1998). For the public sector, stakeholder management strategies have the long-term effect of alienating stakeholders as they come to recognize that their voice is not being heard and their input ignored, further isolating decision-makers. Even in the private sector, where stakeholder management is the norm, this can have similar adverse effects (Daft, 1998). When faced with a novel, complex problem, decision-makers are often unable to assess reliably the states of consensus in disciplines, incompetent in the face of burgeoning literature and prone to mistaken agreements (Fischer, 2005). Broader stakeholder involvement reflecting the multiple functions of agriculture can help improve the decision-making process.

NAE-AKST has been particularly successful at involving the dominant pre farm-gate and farm interests within the prioritization process and in recent decades the dominant post farm-gate food processing interests have also become effectively involved. Some, in fact, would argue that farmers and their organizations have possibly been heard too well. In the development of AKST, NAE has however been less successful in involving other interests. Traditionally, stakeholders are classified into eight kinds based on the legitimacy of their claims, their power and the urgency of their claims (Grimble and Wellard, 1997). Legitimacy refers to the perceived validity of the stakeholder's claim to a stake. Power refers to the ability or capacity of a stakeholder to produce an effect. Urgency refers to the degree to which the stakeholder's claim demands immediate attention. The stakeholders successfully involved in NAE-AKST are ones with legitimacy, power and urgency and these are sometimes referred to as definitive stakeholders. This kind of stakeholder is the easiest to involve and maintain. NAE-AKST has been less effective at involving stakeholders with little power to assert their interests when the definitive stakeholders and the AKST system do not recognize their legitimacy or urgency. For many years organic farmers felt they were in this category and many other stakeholder groups in society still feel as though they are. New stakeholder involvement methods could assist in developing methods to establish standards for legitimacy for inclusion in the development of NAE AKST, especially given the increasingly multifunctional importance of agriculture and the diversity of interests that must be serviced by rural areas (De Groot et al., 2002; Chiesura and de Groot, 2003).

Stakeholder involvement strategies aim to engage stakeholders in the decision-making process, either through representative or participatory processes (Grimble and Wellard, 1997). Stakeholder involvement processes can be costly and ineffective unless appropriately focused. The use of representative or participatory processes during stakeholder analysis depends on the cultural context and specific circumstances. A participatory process is one where the relevant stakeholders are involved directly, without the assumptions or structures to ensure that they are representing a broader group of like-minded stakeholders. While participatory processes are used when there are small numbers and types of stakeholders, a representative process is generally used when the number of stakeholders is large. Cost-effective participatory processes at larger scales as well as smaller scales of aggregation can be developed. The Danish Consensus Conferences and its variants (e.g., Joss, 1998; Einsiedel and Easlick, 2000), are examples of such cost-effective, large-scale participatory processes that have been successfully exported to other places. Much can be learned from these experiences.

6.3.1.2 Recognize the importance of indigenous and traditional knowledge

In recent decades, the importance of traditional and indigenous knowledge in agriculture has been newly recognized for its present and potential value. In a sense, all agriculture and AKST is built upon the traditional and indigenous systems that developed through the domestication and development of crop varieties and through the development of myriad cultivation techniques integrated within society and culture. In this sense, the "science of agriculture" of the last two centuries or so represents innovation based on a continuing and dynamic relationship with a foundation of older knowledge, even when the consequence of innovation is to replace older practices or knowledge. Researchers have pointed out that the categorical distinctions between "scientific" or "Western" knowledge and technologies on one hand and "traditional" or "indigenous" knowledge can thus be arbitrary and confusing and more recent research usually attempts to avoid overly dichotomous categorizations (Inglis, 1993; Agrawal, 1995; Tyler, 2006).

Important new developments in the study of traditional and indigenous knowledge have led to a heightened and more sophisticated recognition of:

- The fact that indigenous knowledge in agriculture is sometimes a vital element for the physical and cultural survival of indigenous groups, including some within North America and Europe (Berkes, 1999; Berkes et al., 2000; World Bank, 2004b);
- The role that indigenous and traditional knowledge plays in "adaptive management," that is, the way in which long evolved agricultural knowledge sometimes represent advantageous adaptation to specific local conditions and response to stresses such as lack of capital, lack of reliable access to water, flood, poor soil and pest invasion (Altieri, 1995; Berkes et al., 2000; World Bank, 2004b; Tyler, 2006);
- The potential for a deeper understanding of traditional and indigenous knowledge to contribute to innovation in AKST, in areas ranging from plant breeding to water and soil management (Inglis, 1993; Berkes, 2000; Tyler, 2006);
- The important role of traditional and indigenous knowledge in biodiversity conservation and the in situ conservation of genetic resources (Mauro and Hardison, 2000);
- An understanding that valuable indigenous and traditional knowledge cannot be maintained or developed without access to land and other agricultural resources by those who use it, whether they be indigenous people or commercial farmers; the situation is especially critical for indigenous people, including some portion of the several million indigenous people of Europe and North America (Tyler, 2006);
- The difficulties that sometimes exist for those trained in the AKST academic disciplines in recognizing or understanding the existence or the underlying rationales of traditional and indigenous knowledge; such difficulties can be exacerbated by cultural or socioeconomic distance between practitioners and researchers; interdisciplinarity and special training have proven important in overcoming these difficulties (Grenier, 1998; Stephen, 2006);
- The complexity and sensitivity of intellectual property rights with regard to the actual and potential products of traditional and indigenous knowledge and practitioners (Brush and Stabinsky, 1995; Mauro and Hardison, 2000); and
- A more highly developed framework for researching and evaluating the potential of traditional and indigenous knowledge, recognizing that traditional and indigenous knowledge may have either positive or negative social or environmental consequences (Stephen, 2006).

This knowledge, as other forms of AKST, is necessarily context dependent with regard to its consequences. The new framework does not make a priori assumptions about the positive or negative value of traditional or indigenous knowledge, except in recognizing the positive value of preserving all knowledge, whether or not it forms the basis for present or future practices. In rejecting such a priori assumptions, respect among AKST researchers and educators and practitioners of traditional and indigenous knowledge widens opportunities for mutual learning and improved practices (Agrawal, 1995; Berkes et al., 2000; Tyler, 2006).

6.3.2 Toward meaningful interdisciplinarity

6.3.2.1 Enlarge the scope of agricultural knowledge systems

Improvement of AKS has the capability to make powerful contributions to newer and wider issues and, in many cases, new partnerships would benefit the general scientific community. Interrelationships are required with the life sciences and in the economic and social sciences in terms of research, educational and extension/development work. The issue of developing successful linkages is important and can be addressed across the NAE region. Moving beyond "science versus humanities" dichotomies in many national education systems and developing skills in complex systems sciences is essential. Effective interdisciplinarity should not compromise disciplinary excellence, the base from which high quality interdisciplinary approaches to AKST issues can be developed. Meaningful interdisciplinary approaches are widely recognized as essential. Systemic barriers to their implementation can be addressed and overcome (Box 6-11).

If interdisciplinary approaches are to reach the required critical mass to become a centrally effective feature of AKST, it is clear that more is required than the development of individual talent or the mere allocation of extra funding. Governments and stakeholders at local, national and transnational levels could identify inhibitors and design corrective measures appropriate to their particular contexts. It would be wise for research funding bodies to further develop procedures to encourage rather than inhibit interdisciplinarity. Educational and research providers could bring their internal incentive, resource allocation and reward systems (including promotion procedures and criteria) as well as their program approval procedures to be more consistent and better reflect the broader AKST aims. Substantially enhanced funding is necessary to promote interdisciplinarity and interactive knowledge networking among AKST stakeholders. However, it is important that the systemic inhibitors to interdisciplinarity be simultaneously countered so that funding accelerates the "mainstreaming" and sustainability of the required new approach and drives it towards the "Tipping Point". In the short run it is recommended that NAE governments, AKST providers and funding agencies take steps to identify the variety of barriers to interdisciplinarity/networking at local, national and transnational levels. It is then vital to collate and analyze examples of "good

Box 6-11. Systemic barriers to interdisciplinarity

The rhetoric of interdisciplinarity has not yet been matched by the reality. In Europe, for example, the President of EURAGRI, at their 2002 Conference on "Placing Agricultural Research at the Heart of Society," identified some key systemic barriers to interdisciplinary work in research:

Interdisciplinary work and professional reality: Interdisciplinary agricultural research is essential, but there are major obstacles. First, the organization, funding and evaluation of research are biased towards work in specific disciplines. Second, co-operative research is time-consuming. In order to climb the career ladder and to receive peer recognition and funding for their research, scientists are often forced to "publish or perish" and to focus their activities on a relatively narrow field. To overcome these obstacles, it is important to address issues such as language, culture, values, and also the methods and traditions of scientific disciplines. It is also essential to remove legal and organizational constraints that hinder EU-scale co-operation.

Innovative research and research funding: Breakthroughs in science occur more often at the edge of disciplines than in the centre, and the scientists most willing to question traditional approaches and theories are often quite young. Unfortunately in some areas of NAE, their research proposals are rarely ranked high enough to receive funding, because the peers chosen to evaluate research proposals mainly represent the mainstream. This is an obstacle to innovative, more risky research and in the long term it may undermine economic competitiveness. We therefore need to examine how to correct these inbuilt shortcomings within the system.

Analogous difficulties exist in relation to interdisciplinary course design and course approval processes in educational institutions as well as subsequent course delivery mechanisms and learner assessment procedures. Promotion of many such initiatives is almost completely dependent on a "champion" who has the vision to catalyze a team to design the program proposal. The "champion" is usually sufficiently senior or influential to "guide" the proposal through the approval/funding processes and who is sufficiently well placed to "protect" the delivery team during the early cycles of the program until its (hoped-for) success. Earlier obstructionists who later acquiesce sometimes even claim that the success was due to the rigorous assessment procedures through which they had forced the original program proposal to pass! The sustainability of such initiatives (no matter how successful in the minds of the beneficiaries) after the well placed champion moves on or retires is often quite doubtful, in the absence of a pro-active institutional culture oriented to the fostering and "active mainstreaming" of such initiatives. Where multiple institutions are involved, the problems and difficulties are greater, often more than proportionately. For younger staff, the personal risks are often high relative to the potential for career advancement. This problem could be rectified as was demonstrated in cases of successful collaboration where the young researcher gets his/her name on far more papers than he/she would otherwise, and is typically lead author on the papers where he/she did the most work. Many leading journals now list the contribution of each author to a paper, which facilitates faculty advancement boards. This practice could be broadened to encourage more such collaborations.

Similar situations exist in the areas of extension/outreach/development activities, where the successful promotion of interdisciplinary teamwork, especially involving personnel from different agencies, is often due to the commitment and dedication of mid-level personnel at local level with the courage to act without formal approval from the top levels of their agencies.

It is clear, therefore, that a significantly greater level of level of institutional capacity development is necessary whereby AKST institutions acquire/develop an organizational ethos that facilitates/encourages/promotes various networking developments and encourages active participation of its personnel in such networks, as part of "mainstream" institutional activity attracting parity of esteem for professional recognition and career progression prospects. The "transactions costs" involved in establishing, operating and evaluating partnerships need to be kept reasonable, so that the barriers/obstacles to desirable co-operation can be reasonably surmounted. There is considerable evidence that crossing institutional boundaries can be quite difficult, especially if it also involves crossing ministerial boundaries.

practice" designed to overcome them with a view to promoting more rapid development and wider adoption of the desired AKST interdisciplinary and networking approaches. This work could be undertaken multilaterally or could build on the earlier OECD activities in this area.

6.3.2.2 New skills for AKST personnel

In order to enable these developments, newly arisen capacity building needs for existing and future AKST personnel should be addressed so that they can understand and function more comfortably in the context of the wider vision and provide AKST services to the wider range of practitioners who will engage themselves in the enlarged vision of agriculture in NAE. Major implications arise both for providing initial education and lifelong learning opportunities for AKST personnel and for their various clients, whether "traditional" or "potentially new" groups. In addition to the "content" knowledge demanded by the wider vision, the increasingly interactive networking activities will require enhanced "process" skills on the part of participants, as they adjust from the earlier unidirectional flow-of-knowledge paradigm and learn how to build new relationships and work smoothly with various new types of partners. In this regard, the European Parliament, in the Explanatory Statement accompanying a recent report on agriculture and agricultural research, highlights the need to safeguard inter- and trans-disciplinary research in the long term and to integrate in the teaching curriculum the ability to cooperate on an interdisciplinary basis. Additional teaching posts might be created by colleges and universities in order to promote the new approach to teaching and research which this would entail (European Parliament, 2004).

Traditionally, NAE agricultural higher education has been broadly based on the multidisciplinary study of a range of sciences/technologies focused on agriculture, often with a production orientation. Disciplinary specialization tended to occur at a subsequent stage via postgraduate studies. For the future, in order to enhance the pool of persons capable of making interdisciplinary contributions, it could be advisable to promote multiple entry into the agricultural education system, such that persons with initial specialized study in various other disciplines could undertake postgraduate studies (e.g., academic master's) providing understanding of the wider agricultural context in which they would hope to apply their particular disciplinary education/training. Such could be fulltime (oriented to younger graduates or those who can take time out for full time studies) or part-time (oriented to mid career personnel in a range of occupations as part of lifelong learning or continuing professional development). Some tertiary educational institutions have experienced high growth in demand for such programs, which are expected to become increasingly important if the wider contextual understanding of agriculture is to be realized.

6.3.2.3 Need for new learning opportunities

Promotion of a wider understanding of the multiple functions of agriculture has to extend far beyond the AKST personnel themselves and the universities and colleges that educate them. Learning opportunities for understanding, participation, contextualization and adaptation could be fostered for a range of stakeholders. Options could be developed in initial education/training and ongoing adult learning to promote better understanding of various levels of complexity in interpreting and responding in a sustainable way to the needs of the future. In particular, learning materials readily available via internet and new modes of interactive learning could be developed that could build on the experiential learning of various groups, enhance their mutual understanding and enhance their skills for developing sustainable provision of the multiple functions of agriculture in their particular contextual situations. Appropriate educational bodies could often accredit these learning opportunities, with credit accumulation and possible progression to suitable adult learning awards. Specific examples of target groups could include:

- All the players participating in the agriculture and food chain;
- Environmental interest groups;
- People engaged in a range of rurally located enterprises/occupations;
- Community development groups;
- Local public officials (both career and elected); and
- Interested local residents

6.3.2.4 Interactions with policy makers and political leaders

While agricultural, food and environmental issues have become wider and more complex throughout OECD countries, government has, in a sense, become but one of several clients for AKS services, albeit the client who has the important responsibility for the public good (OECD, 2000). Policy makers, meanwhile, are often torn between scientific evidence on the one hand and often emotionally charged consumer/interest group concerns on the other. The urgency of promoting more open and enhanced two-way communication among AKS, the public and policy makers was of major concern to the 2000 OECD AKS Conference, which recommended that effective steps be taken as a matter of urgency to develop an ongoing two-way dialogue among those three parties not only at national level but also under the auspices of OECD on an OECD-wide basis. Two-way learning opportunities for policy makers and AKST personnel are in urgent need of enhancement and a range of professional development policy-oriented learning could be developed which would enhance more productive interactions. These could involve policy makers from the Ministry of Agriculture but also other sectors like Industry, Environment, Health, Economy, etc., as well as personnel from various state agencies and AKST leaders. This would facilitate a more two-way communication between AKS and the policy makers. Also, as people are increasingly suspicious of scientists and science, it is important to consolidate an independent, trustworthy agricultural research community capable of guiding complex decision-making; this is particularly crucial when it comes to integrating the sustainability concept into policy (EURAGRI, 2000).

6.3.2.5 Public understanding of the multiple roles of agriculture

If citizens are to participate adequately in decisions about research, development and new technologies, they must have the capacity to understand the scientific issues. Conversely, scientists require communication skills and an awareness of society's needs and demands. They must take time to explain what they are doing, what they hope to achieve and how their work could benefit society. The development and delivery of messages and materials designed to enhance public understanding of the multiple functions of agriculture and to promote awareness of the related complexities and trade-offs that may be involved will become an increasing responsibility of educational research and outreach components of the AKST system. For the general public, this could lead to the promotion of a new concept of "agricultural literacy" that can be summed up as the goal of education about the new vision of agriculture. Achieving the goal of "agricultural literacy" would help to produce informed citizens able to participate in establishing the policies that will support a competitive and sustainable agricultural industry in the NAE region. Options to be considered include the development of adult learning materials and the development of material suitable for developing elements of the wider understanding during pre-kindergarten through 12th grade communities, thereby recognizing the importance of early-childhood development and creating organized ways to enhance child development.

6.3.2.6 Initial education/training for farmers

In many NAE countries, initial education/training of farmers has been conducted in specialized institutions under the aegis of their ministries of agriculture, as part of a general pattern in which sector training was the responsibility of the relevant sector ministry. In other countries, vocational agriculture courses were offered as part of general second level education. In both cases, these have been largely production oriented, for which demand has been declining in many cases in line with the decline in NAE farm employment. Many NAE countries are reviewing these arrangements. In France, for example, there have been proposals for radical reform aimed at developing wider suites of programs oriented to a broad range of rurally based occupations. In Ireland, steps have been taken to integrate the specialized agricultural colleges with the national system of higher education and training awards and an increasing provision of rural development or agribusiness programs leading to these qualifications, in addition to traditional programs which are now set in a wider environmental and livelihoods context.

6.3.2.7 *Stimulate links between higher education and research and facilitate the harmonization of the different education systems*

The links between higher education and research could be strengthened as a key component of human capital development for the agriculture, food and rural sectors. A crucial interface between the research and education areas lies in the development of significantly expanded doctoral level studies in NAE higher education institutions that would be essential for expanding the training of adequate numbers of future researchers and higher level educators who will educate the next waves of AKST personnel. NAE higher education could develop far-reaching programs at the doctoral level producing a cadre of scholars capable of seriously addressing the wider issues and new paradigms associated with the enlarged vision of agriculture in appropriate interactive knowledge networks. One example of strengthening the links between higher education and research in a European country is the promotion of special cooperative centers that must include a university (under aegis of Ministry of Education) and an agricultural research centre (under aegis of Ministry of Agriculture). This is a brave attempt to cross ministerial boundaries in an attempt to rectify the excessive compartmentalization of research and higher education when research becomes concentrated in National Agricultural Research Institutes (NARIs), to the detriment of developing a research base at the university/college level. Another such example in the US is that of the many researchers and extension personnel of the USDA who are based on university campuses, embedded within the appropriate academic departments, with adjunct university appointments and benefit from both worlds.

Another important issue is the development of greater harmonization among the various widely differing national education systems across NAE that will have enormous implications for curriculum design and delivery, articulation and transfer arrangements, institutional "niche marketing", international student and staff mobility arrangements and potential development of transnational program delivery not just for initial higher education but also for lifelong learning. Greater harmonization does not of course imply uniformity. The challenge is to encourage articulation and mobility, without compromising academic freedom and organizational diversity.

6.3.2.8 *Promote lifelong learning and create a learning society*

There is a need to ensure that the remarkable growth in demand for education throughout the lifetime of every citizen can be satisfied and to demonstrate that this demand can be filled at the highest level of quality imaginable, along with the greatest efficiency possible. More universities and colleges could consider making continuing learning a part of their core mission. This could lead to the creation of a learning society that values and fosters habits of lifelong learning, ensures that there are responsive and flexible learning programs and that learning networks are available to address all student needs. (Kellogg Commission, 2000). Such a development of a learning society could have enormous value in promoting more widespread understanding of the issues and opportunities associated with multifunctionality among a wide range of rural and urban residents. It could also enhance a wider set of skills necessary for functioning with various parts of a multifunctional agriculture. It could also stimulate the creation of new knowledge through research and other means of discovery and use that knowledge for the benefit of society and as a result could promote the wider recognition that investments in learning contribute to overall competitiveness and the economic and social well-being of nations. It is recommended that greater effort be expended on accreditation of lifelong learning courses within national or even wider mutual recognition systems so that proper credit accumulation procedures could more easily enable adult learners to progress to more advanced courses with organizations other than their original providers. Such credit accumulation and articulation arrangements could make it easier for rural residents to widen their knowledge/skills to work with the new paradigm and also to deepen their knowledge in specific areas, now set in the wider context. It would also make it easier for potential learning providers to identify opportunities for program design, learner recruitment and program provision.

6.3.3 Strengthening information and knowledge-based systems

Currently, we remain in the throes of an information technology (IT) boom that began over 30 years ago. The speed and quantity of information is still rapidly increasing and the modes of information acquisition are becoming increasingly more convenient and inexpensive. The conversion of this information into knowledge is a process that lags considerably behind (Hassell, 2007). It is expected that these trends will continue at least for the next two decades, ushering in unprecedented flows of information. The policy framework surrounding agriculture will also lead to the delivery of standardized information to various public authorities.

These changes, when allied to the paradigm shift developed earlier in this chapter, will create several significant challenges for the NAE AKST system that will also require adjustments in institutional arrangements. If the paradigm shift is to lead to really meaningful developments, the NAE AKST information and knowledge-based systems will need to be expanded and strengthened to enable rapid flow of information both to and from the various agricultural sectors and the AKST system, including those parts of the system involved in the policy framework. For example, information-based systems have enhanced the value of literacy in the agricultural sector (Warschauer, 1999) and this trend will probably continue in the future. Some of the options to strengthen these systems are described below.

6.3.3.1 Reducing the "Digital-divide"

Currently, the availability and use of IT in AKST in the NAE is uneven among countries and sectors. Some countries, such as those in Eastern Europe and to a lesser extent, Central Europe, have lower access to the technologies (Chinn and Fairlie, 2007). In comparison to Western Europe, availability in Eastern Europe is about 20-30%. The present uneven distribution of IT sets up some short term scenarios that might be useful to avoid, as they could create conditions that favor the persistence of long-term inequities. Some of the ways to counteract this digital divide are by:

- Using data and information sources that can improve production;
- Increasing access to software products that assist production (expert systems) in the production sectors in Central and Eastern Europe;
- Encouraging investments both by the private and state sectors in capitalization of the production sector, IT maintenance and repair infrastructure and software development to help meet production goals; and
- Providing education to be able to manage these IT systems in production.

6.3.3.2 Reconfiguration of information systems

If IT development progresses as expected, in the future vastly greater quantities of more detailed information will become available by faster and more convenient means for use by the AKST system and the wider range of stakeholders and clients with whom it will need to interact. If access to the hardware, software and information continues to increase, there will be too much information to be useful. Some specific challenges will be problems associated with temporal and spatial scale matching and extraction of useful knowledge from the dense and numerous sources of information. In the future, information systems will be necessary to identify and control emerging threats all at pertinent spatial and temporal scales. To avoid potential problems associated with information overload, several changes in the NAE AKST may be required, as mentioned below:

- Define collectively (active participation by farmers, extension services, etc.) what information is necessary and would be efficient for better farm and landscape management of resources (biophysical and economic) at the different pertinent scales;
- Promote, as far as possible, consistencies among data formats to be supplied for regulation purposes (control, follow-up...) and data used for farm, land and environmental management;
- Reconfigure information flow and information management practices to prioritize environmental land management goals in agricultural practice, in environmental practice and in government support policies, incorporating a cross-compliance approach to agricultural land management; and
- Develop specialized software and data management programs that can access and use the high volume of information.

6.3.3.3 From information systems to knowledge based systems

Information systems have been widely developed to the point that many people have access to so much information that they cannot use it effectively. In the NAE, the primary focus of knowledge generation (integration of information so that it is useful in making decisions and taking actions) in the AKST system has been educational and research institutions (Leeuwis, 2004). It is essential to promote the development of multiple loci of knowledge generation so that it will be possible to harness the vast flows of information to improve site-specific and temporally dynamic management (Hassell, 2007).

- Encourage land managers to become sources of knowledge production and facilitate multi-directional flows of knowledge by the education and lifelong learning systems;
- Expand the sources of knowledge-generation of AKST to go well beyond the institutional boundaries of educational institutions, especially with electronic and other distance learning systems in a lifelong learning context; and
- Develop several new and structurally innovative models for turning information into knowledge.

Similarly, many developing countries will probably experience a rising flood of information, although it is likely to be more uneven and lag behind the NAE (Chinn and Fairlie, 2007). It is also probable that the availability of IT and the AKST demands for its products will vary from region to region. It will be important to evaluate these regional needs and evaluate the relevance of the NAE experience so that IT is appropriately contextualized in the development strategy.

6.3.4 Promoting appropriate institutional and organizational arrangements

6.3.4.1 Towards new and "engaged" public institutes

A new kind of public institution is one that is as much a first-rate student university as it is a first-rate research university, one that provides access to success to a more diverse student population as easily as it reaches out to "engage" the larger community. Perhaps most significantly, this new type of university will be the engine of lifelong learning in the NAE region, because it will have reinvented its organizational structures and re-examined its cultural norms in pursuit of a learning society.

Engagement, on the other hand, goes well beyond extension, conventional outreach and even most conceptions of public service. Inherited concepts emphasize a one-way process in which the university transfers its expertise to key constituents. Embedded in the engagement ideal is a commitment to sharing and reciprocity. Engagement could give rise to partnerships, two-way streets defined by mutual respect among the partners for what each brings to the table. The engaged institution can:

- Be organized to respond to the needs of today's students and tomorrow's;
- Bring research and engagement into the curriculum and offer practical opportunities for students to prepare for the world; and
- Put its resources—knowledge and expertise—to work on problems that face the communities it serves.

Engagement, two-way outreach and civic service are all critical elements of public university missions, whether specifically included in the mission statement or not and are defining characteristics of the public university of today and tomorrow (Kellogg Commission, 1998)

6.3.4.2 Innovative education and research models
It was noted earlier that there are many obstacles, both personal and institutional, to the achievement of greater and more genuine interdisciplinary research and education in the AKST fields (Box 6-12). Similarly, there are major learning experiences, both personal and institutional, to be undertaken within AKST institutions that adopt or profess a commitment to become more "engaged," if we are to ensure that really interactive two-way knowledge exchange and development actually occurs. The potential partners in the "engagement" process also require support in learning to develop their skills to participate in, contribute to and benefit optimally from the new interactive knowledge networking with the engaged institutions.

There are already numerous examples of establishing such networks, some quite formal, some informal, which can have their origins either from AKST invitations to engage, from farmers who share a common problem or from a local NGO that identifies a local public good or environmental issue, for example. Indeed, networks can arise in the context of frustrations by farmers and by researchers/educators with the more traditional unidirectional delivery of research and extension services under existing institutional arrangements.

Many of these innovative education and research models show that successful development and application of innovative agricultural knowledge, science and technologies can be significantly improved by introducing more active collaboration between farmers, researchers, extension agents and other educators. Such collaboration, if it is to be most successful, begins by dispensing with the assumption that formal researchers and educators necessarily already hold the most useful and important knowledge. There is recognition that farmers and other practitioners not only have useful knowledge but that they can participate actively in formal, scientific research. The mutual learning that can occur among groups of farmers, researchers, extension agents and teachers can result in important innovations that are more readily accepted and applied by practitioners and that form a firm basis for further research (Box 6-12).Through participation by well-qualified researchers, farmers are able to sponsor and actively participate in producing rigorously scientific research results publishable in peer-reviewed journals.

It is essential that a greater level of support be provided for the more active and widespread promotion of a variety of innovative education and research models of this kind so that genuinely interactive knowledge networks can emerge. Such networks could be adapted to contextual issues and needs and to be effective they could receive the support from key people in relevant institutions required for them to become successful and sustainable relative to their purpose. It is essential that the networks always have the capacity to evolve as the needs and issues change. This could involve dissolution if their goals are reached or reconfiguring themselves into new or transformed networks as new needs and issues emerge in their spheres of influence.

Experiences of the variety of new and innovative education and research models which have been tried in NAE AKST could, in the short term, be collated and analyzed so as to identify success and failure elements, risk factors,

Box 6-12. An example of innovative education and research model: BIFS

Innovative models can range from informally organized "farmer circles", (which invite academic and/or extension personnel as resource persons), to a variety of more formally organized and funded programs such as the Biologically Integrated Farming Systems (BIFS) Program in California, whose projects involve farmers, University of California Cooperative Extension researchers, federally funded research staff, conservation organization staff, and private sector consultants. Originally begun to attempt to solve some of the seemingly intractable problems of heavy pesticide dependence in some orchard crops, the program has been extended to a wide variety of other crops, including row crops, ranging from cotton to melons. The program has developed innovative solutions that have reduced dependence on pesticides and synthetic fertilizers, reduced environmental impacts, and improved farm profitability. It has also revitalized the relationship among farmers and research and extension staff and has improved positive interactions among farmers themselves. Projects have been successful among both small and large-scale producers.

Key elements of the BIFS approach include, in the slightly abbreviated words of BIFS evaluators (Mitchell et al., 2001):

- Experienced farmers who voluntarily share information about their production systems with other farmer participants, consultants, and researchers;
- On-farm side-by-side demonstration evaluations of conventional and alternative management practices;
- A small management team that provides technical assistance and project leadership made up of farmers, consultants, and academic researchers;
- Customized information support to facilitate evaluation of alternative production practices; and
- An emphasis on providing opportunities for "co-learning" environments in which farmers, researchers, and consultants share insights.

sustainability factors, effects of differential support mechanisms and elements of "good practice" so as to inform and guide the introduction of and to promote more widespread adoption of practically oriented interactive networking with a range of end users of AKST services.

6.3.4.3 Setting up institutional and organizational arrangements for knowledge based systems

A continual accumulation and application of agricultural knowledge, science and technology, broadly defined as AKST, has been the necessary factor making possible the development of a global food and agriculture system. Several major changes are affecting the way this AKST is and will be made available in the future. Firstly, the political base for public food and agriculture support systems is eroding as rural populations change. Institutions once uninterested in food and agriculture are now devoting resources to food and agriculture. Secondly, there is a major shift in the generation of AKST toward private rather than public funding. Further complicating matters, the above information on any subject is now easily available on the web and elsewhere, unrestrained by quality standards.

Considering these elements, some of the options for action that would ensure the right dissemination and adoption of AKST would be to

- Set up new forms of local innovation networks and efficient "value-chains" associating all concerned actors to turn science into practice. For example, review the current link between science/extension/farmers to make it more efficient and, widen the more effective involvement of end-users (e.g., private sector, suppliers of goods and services, consumers, processing) and their potential benefits; and
- Set up information systems that would aid AKST users in accessing information that is clear, transparent and reliable even if this means that some categories of users will have to pay a fee for it. For certain areas of a public good where public intervention is legitimate/desirable such as food security, impacts of climate change, the long term sustainability of agricultural systems, the protection of natural resources and the environment and the livelihoods of vulnerable rural communities, large diffusion systems can be strengthened. In these areas public funding could support open and user-friendly information systems.

6.3.4.4 AKST interactions between NAE and other regions

The development of AKST in North America and Europe has had both positive and negative effects on human welfare, independence, security and environmental quality in other regions of the world. It is important that the further development of AKST in NAE serve development and sustainability goals to reduce hunger and poverty, improve rural livelihoods and health, increase incomes and facilitate equitable environmentally, socially and economically sustainable development in all world regions. The first and essential element in serving the purpose of empowering people and nations outside NAE in gaining new power to improve their own situation is the recognition that it is possible to improve the nature of the interactions of NAE with other regions. It is therefore strongly recommended that the guiding principles for people and institutions in NAE be reexamined.

The next fifty years of NAE AKST interactions with other regions could be approached from a different point of view; that of two-way sharing rather than the predominant unidirectional view in which one part of the world helps another, less fortunate part of the world.

The contributions of AKST to NAE have been partly documented in earlier chapters. New developments in AKST have the potential to play a key role in assisting other world regions to achieve higher levels of self-sufficiency and meet the challenges that will develop in the whole world over the next fifty years as we address the IAASTD question. Sustainability issues in particular will require an increase in international cooperation and coordination.

The agriculture and food sector is the basis of economic livelihood for most developing countries and its health lies at the heart of the development process. Food security is more than food production. It is the efficient, reliable combination of access to needed food supplies (directly or through markets) and the ability to pay for them. Consequently, while agricultural development is a critical starting block for the economic development process, more is needed. No country has successfully ended rural poverty on the back of agriculture alone. However, the converse also applies: for the poorest countries, economic growth and sustained poverty reduction are unlikely to be achieved without initially stimulating sustained agricultural production growth. As agricultural development takes hold, its growth in productivity releases labor that needs to find alternative productive uses. This is both an opportunity and a challenge for development because uncontrolled migration to already overcrowded urban centers in many developing countries is equally problematic.

More effort is called for in planning and funding effective rural development strategies, including the investments in physical infrastructure and human capital that will connect a more diversified rural economy efficiently, through local and national markets, to the emerging global economy.

AKST institutions in NAE need to be ready to participate actively with AKST institutions in other regions to address the IAASTD question. It is suggested that the issues associated with interdisciplinarity and interactive knowledge networks developed in this section may also be of fundamental importance in facilitating the development of the most appropriate working relationships between AKST in NAE and other regions. Previously articulated principles and issues could be used for developing different types of interactions between NAE and partners in other regions. Three examples of interactions are discussed more specifically below, one of them in SSA where the hunger and poverty issues are most stark and the two others through international agricultural research organizations and forums.

The Framework for African Agricultural Productivity (FAAP)

Africa is a region in critical need of new directions in agricultural research and development. Africa's leaders see agriculture as an engine for overall economic development. Sustained agricultural growth at a higher rate than in the past is crucial for reducing hunger and poverty across the continent, in line with Millennium Development Goals. The African Union's (AU) New Partnerships for African Development (NEPAD) has issued a Comprehensive African Agriculture Development Programme (CAADP) that describes African leaders' collective vision for how this can be achieved. It sets a goal of 6% per annum growth for the sector.

A key component of the vision calls for improving agricultural productivity through enabling and accelerating innovation. CAADP Pillar IV constitutes NEPAD's strategy for revitalizing, expanding and reforming Africa's agricultural research, technology dissemination and adoption efforts. Currently, chronic shortcomings afflict many of the continent's agricultural productivity programs. This explains the historical underperformance of the sector and the current plight of African farmers. Consultations with agricultural leaders, agricultural professionals, agribusiness and farmers shows substantial agreement that institutional issues such as, capacity weaknesses, insufficient end user and private sector involvement and ineffective farmer support systems persist in most of Africa's agricultural productivity programs and organizations, hampering progress in the sector. These problems are compounded by the fragmented nature of support and by inadequate total investment in agricultural research and technology dissemination and adoption. So, restoring and expanding Africa's agricultural innovation capacities requires radical modifications and changes in human and institutional capacity building (Youdeowei, 2007).

Despite the enormous challenges facing African agriculture, there are reasons for optimism. The African Union (AU), in establishing NEPAD and formulating CAADP, has given its unequivocal political backing for this effort. In setting up the Forum for Agricultural Research in Africa (FARA/AU/NEPAD, 2006), Africa has created a way of bringing technical leadership into play.

The Framework for African Agricultural Productivity (FAAP) brings together the essential ingredients suggested for the evolution of African national agricultural productivity programs. A number of guiding principles have been derived from consultation with Africa's agricultural people and with their development partners. The FAAP indicates how such best practice can be employed to improve the performance of agricultural productivity in Africa. Beyond improving the performance of individual initiatives, the FAAP also highlights the importance of replicating and expanding such programs through increased levels of investment. It also stresses how increased funding must be made available through much less fragmented mechanisms than has been the case in the past. If these efforts are to have their desired effect, the harmonization of Africa's own resources with those of development partners will therefore need to be placed high on the agenda.

The FAAP has been developed as a tool to help stakeholders come together to bring these political, financial and technical resources to bear in addressing problems and strengthening Africa's capacity for agricultural innovation. The Heads of State and Governments of the African Union (AU) endorsed the FAAP at its Heads of State Summit in Banjul in June/July 2006. Specifically, the AU urges regional economic communities and member states to realign their regional and national research priorities to the FAAP with the support of the FARA.

The FAAP, in its detailed discussion of the evolution and reform of agricultural institutions and services, has several proposals regarding the future strengthening of extension, research training and education, several of which resonate loudly with the proposals of this section:

- End-users should be actively engaged in the processes of agricultural research priority setting, planning and work program management;
- The quality of tertiary agricultural education is critical because it determines the expertise and competencies of scientists, professionals, technicians, teachers and civil service and business leaders in all aspects of agriculture and related industries. It raises their capacities to access knowledge and adapt it to the prevailing circumstance and to generate new knowledge and impart it to others; there is a consensus among recent studies, such as those by the Inter-Academy Council and the Commission for Africa, that urgent action must be taken to restore the quality of graduate and postgraduate education in Africa;
- Establishment of national agricultural research strategies through participatory and multidisciplinary processes and the endorsement of these at national level through inclusion in the poverty reduction strategies;
- Breakdown of the institutional and programmatic separation between universities and NARIs which results in inefficient use of capacity and unproductive competition; and
- Create synergies among institutions and curricula in education, research and extension.

The FAAP document suggests that international contributions could be in the following principal areas, among others:

- Bringing best practices, data, knowledge and expertise from other regions of the world to bear on African issues;
- Providing research-based, relevant information and data for training and curricula and course development;
- Providing specialized expertise in cutting-edge sciences including biosciences, social sciences and policy analysis;
- Creating critical mass and building capacity through collaborative research; and
- Enabling cross-country and cross-continent replications and comparisons to inform African research and development.

Already then, at this stage, there is an articulated set of measures to which NAE AKST institutions can be enabled to respond, not solely through the International Organizations/Institutes, but also through national and international consortia or networks of NAE AKST institutions that could link with similar networks of AKST institutions in other regions or sub regions. One such European Network is NATURA, a network of about 30 European universities and research complexes which have agricultural partnership links with developing countries. In the US, the National Association of State Universities and Land-Grant Colleges (NASULGC), is similarly placed for appropriate networking.

It is recommended that initially, development funding could be made available for a number of pilot partnerships involving networks of NAE institutions and AKST institutions in developing countries in order to address the issues of generating, providing access to and promoting the uptake of AKST to address the IAASTD question. In the medium term the results from such pilots could be scaled up and outwards to regional level such as those visualized in the FAAP and the BASIC program aimed at Building African Scientific and Institutional Capacity (FARA/ANAFE, 2005).

The contribution of NAE to the CGIAR

Guided by NAE countries and the Green Revolution concept as the general horizon for research in the 1960s, the CGIAR agenda initially focused on food supply, mostly through the breeding of high yielding cultivars that were highly responsive to agrochemical inputs and could express their full potential only when provided with sufficient fertilizer and water. Because 70% of the poor are living in rural areas, reducing poverty in developing countries will require that more food be produced by the poor and thus should consider the present context of their socioeconomic and ecological environments (It will also be necessary to reduce poverty among the growing numbers of urban poor). As poor farmers have limited access to inputs, sustainable improvements of their farming systems and family incomes will be achieved (a "doubly Green Revolution") through (1) better use of locally available resources like biological diversity, ecosystem services and diversification of income-generating products, (2) increased access to credit, agricultural inputs as well as empowerment through training and capacity building in ways that do not jeopardize the livelihood of the poor, (3) decreased food costs, especially of staples, (4) overall economic development in nonagricultural sectors that stimulates the agricultural sector, or (5) some combination of these. Without question, there have been important contributions such as new maize, cassava and rice varieties. However, the CGIAR/NARS relationships have, for some time, been "festering" (Eicher, 2001) and there exists a great challenge for the CGIAR to build genuine partnerships with developing-country NARS (World Bank, 2004).

Despite the fact that NAE countries have the major part of the AKST resources of the world, their research and educational agendas scarcely consider major technological spillovers and the ecological, social and economic footprints produced by agrifood systems with regard to development and sustainability goals. This suggests a strong awareness effort is required to encourage politicians to accept that poverty will not disappear without a strong financial commitment of NAE AKST to agricultural development. This must be based on a wide societal and global view about the role of agrifood systems and the scope of AKST and on a strong, concerted research and educational effort to find and implement solutions that are well adapted to the conditions of the poor and take into account the many impacts of technological change.

The CGIAR centers have a unique position and enormous challenges. Taking the relatively low proportion of world R&D resources CGIAR centers directly use, even if it were substantially increased, the most effective option to use these resources is as a mediator affecting and utilizing NAE AKST, thereby simultaneously supporting the human and organizational capacity building in developing country NARS, including universities. CGIAR centers, which are research organizations, could evolve to assume an additional role as facilitators or honest brokers to support development networks that will bring together the key decision makers at different levels of public and private AKST organizations. The different stakeholders from national and regional systems include research, education, development, socioeconomic actors, including farmers' organizations, local and national authorities, NGOs and civil society as well as the best and most useful parts of the upstream science conducted in and outside the NAE countries. Summing up, partners from the NAE countries can help the CGIAR better contribute to the IAASTD agenda by:

- Raising public awareness (particularly among youth, politicians, donors) and strong financial support of both development and sustainability goals and the role of research, education and innovation to address the issues (such as the Davos Economic Forum that is organized every year. "Research for Development [RforD]", under CGIAR coordination, can have an annual forum putting RforD high in the news on a regular basis);
- Including a global perspective on agriculture and food systems as part of common basic education of all agricultural, food and environmental university programs utilizing expertise from developed and developing countries mediated by the CGIAR system: (1) encouraging youth in industrialized countries to work in agricultural research and for developing countries; from regular lecture programs in high schools and universities to increasing attractive scholarship and fellowship programs to encourage young scientists to do their thesis or postdoctorate work in developing countries; (2) encouraging CG scientists to co-advise more students in NAE institutions and even participate in their teaching programs, while encouraging university personnel to participate more fully in the design and implementation of CGIAR activities;
- Allocating special financial resources to the intensification of agricultural education and knowledge systems in developing countries;
- Working together to build a concerted, global effort for training and capacity building in poor countries; these programs can aim to strengthen the capacity of NARS (including universities) to undertake collaborative scientific research and educational activities to realize development and sustainability goals; this could include more targeted training with policy makers, intensified training partnerships of CGIAR centers with local universities and recognizing the importance of informal learning which takes place in the course of joint activities, seminars and other events (Stern et al., 2006);
- Developing more efficient ways to group experts, intermediaries and end-users in different regions, so more aid money goes directly to improvement rather than administration; and
- Continuing work on targeted research programs that have a strong impact on development and sustainability goals (for e.g., challenge programs (Box 6-13) and which call for new patterns of interaction; this leads to the development of wider networks and consortia with

Box 6-13. The Challenge Programs in the CGIAR

Recently the CGIAR (Consultative Group in International Agricultural Research) system launched challenge programs (CPs), with a double objective of encouraging the centers to work better together and mobilizing other research institutions around common development objectives. Four pilot CPs have been started. Although the networking role of this approach has already proved extremely successful, these programs are still too young to show any real impact on resource-poor farmers in developing countries. CPs have significantly increased the overall budget of the CGIAR and mobilized scientists and institutions that were not previously working on development issues. The CPs were criticized for not being sufficiently inclusive of national programs and development stakeholders. Additional CPs, or similar types of collective actions, could be launched, involving partners from NAE and developing countries together. Oriented towards farmers and building practical solutions, these new collective actions may address:

- The forecasted impact of climate change on crop and animal productions in poor countries;
- The forecasted reduction of renewable and nonrenewable resources, mostly water and fossil energy, and the potential of diversity and diversification;
- The relation between new, emerging illness in poor countries and agricultural development;
- The growing urbanization and the role for agricultural intensification in favorable and non favorable environments;
- The potential conflicts in land use arising, for example, between biofuels and food, between exports and domestic consumption; the development of stronger food supply chains and more efficiently functioning marketing arrangements; and
- The development of rural innovation and raising rural incomes.

members from the other CG centers or NARS—including universities—private sector and the NGOs.

- Working on common research issues, among them food diversification and its role in reducing malnutrition, plant adaptation to climate change, or more specifically plant tolerance to drought and other biotic and abiotic stresses, sustainable farming systems and practices to provide niche products for solvent markets and staples for local markets, relying on local resources and ecosystem services, developing new environmentally friendly agricultural technologies, developing more effective post harvesting market arrangements; and
- Ensuring that international funding for AKST does not perpetuate donor dependence and undermine efforts to develop domestic political support for sustainable funding, especially for the smallholder sector.

Making international agricultural research work better for the poor implies developing well targeted research activities, but this research must, more so than in the past, be able to promote appropriate research carried out in NAE countries. Hence a major question for the CGIAR is how to optimize this evolution, or how to initiate, sustain and mobilize appropriate research in NAE that contributes to the international efforts of the CGIAR centers, which are now trying to orchestrate and strengthen the sustainable cooperative capacity of NARS—including universities—in developing countries.

This new way of working can mark a shift in how research for development activities is designed, monitored and evaluated in CGIAR centers and NAE country institutions altogether. All contributors, from upstream science to delivery systems and impact assessment must work effectively together from day one to ensure that the expected outcomes and impacts on food security and poverty alleviation are oriented to poor communities, farmers and other relevant food system actors with less voice and that practical solutions are developed that can be realized and sustained for generations to come.

The Global Forum on Agricultural Research (GFAR). GFAR is a joint undertaking of all agricultural research stakeholders at the global level built through a bottom-up process from the National Agricultural Research Systems (NARS) through sub-Regional and Regional Fora (SRF/RF) in the different geographical regions of the world. The GFAR goals are to:

- Facilitate the exchange of information and knowledge in all agricultural research sectors: crop and animal production, fisheries, forestry and natural resources management;
- Promote the integration of NARS from the south and enhance their capacity to produce and transfer technology that responds to users' needs;
- Foster cost-effective, collaborative partnerships among the stakeholders in agricultural research and sustainable development;
- Facilitate the participation of all stakeholders in the formulation of a truly global framework for development-oriented agricultural research; and
- Increase awareness among policymakers and donors of the need for long-term commitment to and investment in, agricultural research.

In the NAE region, the stakeholders involved in Agriculture Research for Development (ARD) have organized themselves in different ways:

- In Europe, EFARD[18] provides a platform for strategic dialogue among European stakeholder groups in order to promote research partnerships between European and

[18] EFARD, the European Forum on Agricultural Research for Development, represents the various stakeholders through National Fora on ARD in European Union (EU) Member States and applicant countries, as well as Norway and Switzerland. EFARD's mission is to strengthen the contribution of European ARD to three major worldwide challenges: (1) alleviating poverty and hunger, (2) achieving food security, and (3) assuring sustainable development.

Southern research communities; up to now, EFARD has developed a strategic research agenda, set up an ERA-ARD and established a strategic alliance with the Forum for Agricultural Research in Africa (FARA); and
- In North America, progress is still underway to link NA Agricultural Research Institutions (ARIs) with a vested interest in Agricultural Research for Development to GFAR; also, the PROCINORTE[19] cooperative program could join the other "PROCIs" (PROCIANDINO, PROCISUR, PROCITROPICOS and PROCICARIBE) under the umbrella of the Latin American and Caribbean Forum: FORAGRO (http://www.iica.int/foragro/).

Therefore, GFAR provides an ideal platform for addressing issues of global concern, where the participation of a broad and diverse set of actors is required. One of its obvious added values is the increased exchange of information, experience and best practices between regions.

This relatively recent initiative can rightly claim significant results in AKST (identification of knowledge needs; knowledge generation, dissemination, access, adoption and use) within and between the less developed regions in the world. The best evidence of GFAR success was the official support it received at the G-8 Summit of Evian in 2003, seven years after its official launching.

However, despite previous efforts in NAE, it seems more difficult and challenging to mobilize the different categories of stakeholders in the NAE region:
- In Europe, EFARD has succeeded in mobilizing the different stakeholders for some specific tasks but its legitimacy is based on the existence of an active and truly representative national forum; however the situation varies greatly from one European Member State to another: for example, Denmark and Switzerland have established active and successful national fora (http://www.sfiar.ch/ and http://www.netard.dk/) whereas Germany, after a strong launching phase, could not maintain its national ARD forum; France, in spite of being the first ARD contributor in Europe, has yet to establish its national ARD forum; and
- In North America, the different categories of ARD stakeholders seem to be working even more in isolation than in other regions, particularly universities. So far, NAFAR has not succeeded in convincing stakeholders of its added value and PROCINORTE is currently more a research program than a multistakeholder forum.

Lessons have to be drawn from this innovative, bottom-up, highly participative multistakeholder mechanism and its impact on AKST after 10 years of existence. The second external evaluation was completed in February 2007. Options for action can be discussed in the light of this last evaluation and focused on two major issues:
- The building up of two or three strong and active ARD forums in the NAE region (North America, Western Europe, Eastern Europe including Russia) to significantly help the work at the global level in collaboration with other regions; and
- The analysis of strength and weaknesses of the major past projects to identify the conditions for success of the future projects supported by GFAR, both at the regional and global level and taking into account regional specificities and diversities in the analysis.

It would be worthwhile if cooperation at the academic level be made between NAE and south AKST with strong political support but without political interference, in an effort to gain mutually useful knowledge firmly oriented towards development and sustainability goals.

6.4 Reshaping Policy Environment and Governance Systems

The agenda for agricultural and rural development policies nowadays is broader than in previous decades. The agricultural sector is being exposed to a more diversified set of demands, not only from consumers, who are increasingly concerned over issues such as food quality and safety, but from wider society, whose expectations increasingly involve territorial, social, environmental and cultural matters. This may require a wider and more coherent policy framework, the establishment of new proprietary regimes as well as the reshaping of IPR. In addition, governance options particularly at the local level can also be reconsidered.

6.4.1 Developing a coherent policy framework

The intricate complexity of the development of agriculture and rural areas, the multifaceted linkages with policy, the diversity in agricultural and rural systems and the important dynamics of changes in the overall system mean that policies are typically formulated on the basis of a partial knowledge of the overall situation. The guiding principles in any intervention and in the supporting research simultaneously consider the economic, social and environment dimensions of sustainability (FAO, 2005b; Martin, 2005):
- Economic: implies that production is profitable and demand-driven and contributes to the livelihoods of the citizens;
- Social: implies that production concentrates on product safety and quality, contributes to better health of all the citizens and is transparent and responsible, etc.; and
- Environmental: implies that production processes should respect environmental carrying capacity, respond to climate change, participate in improving the energy policies, etc.

The adjustments in policy issues and regulatory frameworks have implications for research to tackle some of the main challenges, such as:
- To provide a trans-ministerial/interagency approach for better coherence of the complex overall framework (e.g., between agricultural, economic and health ministries that would result in the production of diversi-

[19] PROCINORTE is a cooperative programme in research and technology for North American countries (Canada, United States and Mexico) that aims to strengthen the capacity of the three countries to carry out agricultural research and technology transfer through exchanges and partnership in a cost effective way. This program is under the leadership of the Inter-American Institute for Cooperation on Agriculture (IICA).

fied and healthy foods, at the local, national and global levels) as well as collaborations between governmental departments and private sectors and NGO actors;

- Define the criteria for balancing between these different policies; and
- Identify ways that ensure the articulation of these policies at the local, regional and global levels.

For example formulating new policies to improve current policies in order to integrate the sustainability and multifunctional aspect of agriculture and facilitate the development of sustainable food and farming systems. Such possibilities include:

- Adoption of policies that facilitate rapid uptake of technologies that maintain or increase productivity and have a smaller environmental footprint than current technologies;
- Elaboration of policies that consider the holistic approach to agriculture (Bryden, 2001). This would lead to more encompassing policy instruments for the achievement of multiple objectives that are more efficient than separate policies for each of the multifunctional attributes of agriculture;
- Development of new policies while keeping in mind the transaction costs (other than administrative costs) involved and determining if these costs could be reduced through policies aimed at selective targeting of farms subject to the programs, by using agricultural price and income support programs, etc. (Abler, 2004); and
- Elaboration of policies to reduce the negative and promote the positive externalities or public goods at the farm level (identification of the individual farmer's contribution to a specific externality) and the landscape level (INRA et al., 2004).

6.4.2 Developing AKST in response to international agreements

International cooperation in making development and sustainability goals is critical in order to facilitate the implementation and development of international treaties and conventions. In many cases, successful implementation of international agreements will require changes in the use of agricultural technologies. Many agreements offer opportunities for scientific, technological and policy innovation. Policy responses to international agreements can often be made practicable or facilitated by accompanying scientific and technological change (Kiss, 2003; Mitchell, 2003; Porter et al., 2006; Anton et al., 2007).

The Kyoto Protocol signed in 1997 and the subject of continuing development since that time requires the reduction of Greenhouse Gas Emissions (GHE) in NAE and presents challenges in all sectors, including agriculture. As discussed elsewhere in this report, methane production deriving from agriculture, the use of agricultural chemicals that contribute to the greenhouse effect, energy use in agriculture and deforestation for agricultural production are among the most important areas where innovations may help in meeting the goals of the agreement. Agricultural techniques that improve rates of carbon sequestration may prove important and the development of biofuels that meet other environmental and social criteria and result in net reduction of GHE may also make an important contribution.

The fact that the United States is not signer of the Kyoto Protocol presents its own set of policy challenges and by implication, challenges for both policy and science among all nations. Adherence of the United States to the goals of Kyoto and/or the creation of new international agreements that adequately address global climate change and that enlist the commitment of all nations in NAE is of great importance. The development of existing and new agricultural technologies and policies or their novel application should be considered important tools not only in achieving compliance with existing agreements but also in removing obstacles to the design of effective new agreements (Tamara, 2006; Eyckmans and Finus, 2007).

Similar issues arise with respect to the non-ratification by the United States of the Convention on Biological Diversity and the Law of the Sea Treaty. Both of these agreements have important implications for agricultural production and for the development of AKST. The fact that the United States remains one of the few nations in the world that have not ratified them presents challenges to further developments in AKST and related policies, with particular significance within NAE. Resolution of the problems standing in the way of ratification by all NAE countries could clear the way for more straightforward and coordinated responses to the efforts needed to meet the goals and provisions of the agreements at the global scale.

The Montreal Protocol, first adopted in 1987, continues to challenge agriculture and other sectors to achieve full compliance. For example, the full phase-out of methyl bromide as an agricultural fumigant has yet to be achieved and has led to the search for and/or development of substitute technologies and practices (for current regulatory status and actions, see http:www.epa.gov/ozone/mpr).

Similarly, the 1998 Rotterdam Convention for the Application of Prior Informed Consent for Trade in Hazardous Chemicals and Pesticides and the 2001 Stockholm Convention on Persistent Organic Pollutants create the demand for technologies to substitute for those that may be used less widely or eliminated by implementation of these treaties. These treaties in some of their implications may create a potential imbalance between wealthier and poorer nations, as poorer nations may have a more difficult time finding substitutes for older and cheaper technologies disfavored by the treaties, which in many cases include tools or techniques with expired patents. The higher cost of newer substitutes with patent protection may make it difficult for many countries to comply with the provision of the treaties. Research and innovation in NAE, possible compensatory schemes and the application of features of intellectual property rights may be critical in identifying viable alternatives that can be made practical and affordable to all (Nakada, 2006).

International agreements are not always entirely consistent with one another, as appears to be the case with provisions of the Convention on Biological Diversity and the World Trade Organization's TRIPs (trade-related intellectual property rights) agreement. It will be important to forge agreements that create clear and consistent property rules, or that create ways in which inconsistencies among international agreements and among international agreements and national legal regimes, may be mediated (Chiarolla, 2006; Rosendal, 2006).

Many other treaties have implications for agriculture. In general, the credibility and effectiveness of international efforts to improve agricultural knowledge, science and technologies for development and meet development and sustainability goals will partially depend on the consistency and effectiveness of international conventions and agreements. Increasingly, agricultural scientists will find their own work shaped by such agreements and will find opportunities in the ability to provide innovations that facilitate their implementation.

6.4.3 Enlarging the range of proprietary regimes

6.4.3.1 General issues concerning proprietary regimes and IPR

A continuing reconsideration of the legal and cultural definitions of property is necessary as agriculture faces the challenges of a changing world. In the late 20th century, international institutions and most national governments promoted relatively simple property rules based on either the private ownership of goods or public ownership (goods that were considered as a public utility and were either publicly owned or heavily regulated by the government). There have been counter-trends in the definition of property that have been more compelling and many of them are likely to become critical pieces of the response agriculture will have to make to global economic, social and environmental challenges over the next half-century. At a minimum, a critical re-assessment is advised while allowing for more research and experimentation in the area of property regimes. In order to better understand the different property regimes a quick review of the classification of the different goods that determine their property regime, based on their consumption and access, is essential (Table 6-3).

As mentioned above, there has been a tendency so far to simplify the concept and attribute only two kinds of regimes: public or private. In reality of course not all goods can be classified under these two categories as there are few goods that are purely public or purely private. For example, air used to be thought of as a public good, but as a result of pollution, this has come to be considered as somewhat of a hybrid public good, because its erstwhile non-rival nature has been eroded due to technology and policy (Box 6-14).

Table 6-3. **Property regimes by levels of consumption and access.**

Consumption	Access	
	Exclusive	**Non-exclusive**[1]
Rival	*Private* (e.g., food, clothing, cars)	*Common pool* (e.g., air, water, soil and ocean fisheries,[2] landscapes)
Non-rival[3]	*Club/Toll* (e.g., toll-roads INTELSAT, Suez Canal, Panama Canal, private schools, theatres, professional associations)	*Public* (e.g., public roads, sunshine, national defense)

[1]Non-exclusive: once available, it is not possible to prevent free access to it by all.

[2]In some cases, soil and ocean fisheries access may also be viewed as exclusive.

[3]Non-rival: one person's consumption does not diminish its availability to others.

This has lead to the emergence of a new category of goods called "impure public/private goods," which can be further divided into club/toll or common pool goods. These goods may call for a double approach: partly legitimizing privatization of these goods and partly seen as a global common good by the society. A new proprietary regime can be established for these "hybrid" goods that would do more justice than either purely public or private ownership. This type of regime could allow a sustainable management of the commons and avoid over-exploitation or loss of associated resources as is expected in the "tragedy of the commons" (Hardin, 1968).

Such a vision of "hybrid" goods has been established with the concept of "common property regimes," developed for natural resource management projects. Common property regimes can be defined as those resource management systems in which resources or facilities are subject to individual use but not to individual possession or disposal, where access is controlled and the total rate of consumption varies according to the number of users and the type of use (Forni, 2000).

Thus, proprietary questions undoubtedly raise many complex issues of which more research would allow a better understanding, so that they could be used to maximize ben-

Box 6-14. The complexity of property questions illustrated with water law reform or species and genetic resource protection

For various reasons, throughout Europe and North America, and much of the rest of the world, water has historically been to a large degree considered a public good to be owned and traded outside the market, and/or with strong restrictions on market transactions. There are arguments that promote the creation of water markets. It has been shown that in many circumstances water markets can be created that provide efficiencies so convincing that difficulties can be overcome while meeting reasonable concerns for quality, access, and equity. But the creation of water markets raises other important questions such as the ownership claims (is a water right held by a landowner or by the legally constituted water district of which the landowner is a member?), varied and complicated market rules (different legal and geographic conditions prevailing in the different regions), etc. (Roth et al., 2005).

Property rules and policy with regard to such fundamental resources as water can have critical impacts on such clearly nonmarket issues as the survival of endangered species. The effort to protect species has already created highly charged conflicts regarding private and public claims on land and resources. These conflicts involve matters that clearly cannot be addressed simply through market mechanisms; they are in fact claims that are based on a universal human interest in the protection of species in conflict with private property interests (Fairfax and Guenzler, 2001).

efits. Such research might concentrate on the identification and analysis of the factors conducive to the organization of common-property regimes as opposed to a private-property regime (Orstrom, 2003). The greatest value might be gained from such research if it associates agricultural scientists with social scientists, philosophers, ethicists, public policy practitioners and lawyers, with the participation of the public at large and of public institutions in the evaluation and implementation of research results. Significant and sustained promotion of interdisciplinarity, public forums and public policy discussions of the nature and implications of property and property law for the future could be useful.

6.4.3.2 *Intellectual property rights*

Intellectual property rights (IPR) have clearly benefited agriculture and the environment. Much of the harmful and contaminating pesticides, insecticides or herbicides have been replaced by generation after generation of proprietary, IP-protected products. Each generation was safer than the previous, both to humans and to the environment and all generations safer than the materials initially used. This enhanced safety was due to stiffer regulation coupled with the knowledge that there would be IP (Intellectual Property) protection that would cover investments in producing subsequent generations. But the downside to this IP protection, due to the pioneering nature of the applications, is the broad coverage that the patent offices granted that often extended beyond the enabling information in the applications. This has led to a few companies obtaining broad coverage, to the point of cornering areas and making it exceedingly hard for others to have freedom to innovate. While patents are most important to reward the discoveries of astute inventors, there can indeed be problems in getting wanted and novel products to market, especially from the public sector. For instance, the inability to obtain the license on any one element in developing a transgenic crop can prevent a crop from getting to market, which can be to the detriment of agriculture. Also, patents controlled by large agricultural companies have protected certain enabling technologies essential to agricultural sciences, such as transformation methods, constitutive promoters and selectable markers.

Reshaping IPR and its associated regulatory environment can facilitate the generation, dissemination, access and use of AKST. Some of the options are listed below:

- The patent offices can continue the trend to issue narrower patents even on pioneering technologies;
- University groups and the private sector could be encouraged to pool patents through cross licensing (and free licensing to the developing world). Several interesting public initiatives are now coordinating collective networks for the management of patents and other exploitable assets (know-how, software, etc.) held by public research organizations in the field of agricultural biotechnologies (e.g., CAMBIA[20] in Australia, PIPRA[21] in USA, EPIPAGRI[22] in Europe). They may also ensure common development and patenting of novel biotechnological techniques, vectors, genes, etc. They make them available by royalty-free license on the proviso that improvements be immediately made available to all other licensees. Having the technologies as "open source" leads to what they call "collaborative invention", as all the different players working with the open source material further develop it for all and innovations are quickly disseminated, instead of remaining proprietary knowledge within a company. This is an excellent rationalization of the system, for the common good, with adequate economic incentives for the developer.
- Consider legislation to allow compulsory licensing, if and when necessary for agriculture and food security (So far this measure has been used or threatened only for pharmaceuticals in relation to critical health issues); and
- Address the trend wherein patent offices are limiting what had been known as the "American unwritten exemption for not for profit research" on using patented intellectual property, which is being eroded by the courts. As patent law was written to optimize the acquisition of new knowledge and its being put to use while rewarding inventors, it is time for the legislators to codify research exemptions so as not to stifle research.

In conclusion, it would be advisable to have more uniformly accepted and coherent IPR regimes in order to encourage research and other endeavors that would facilitate the achievement of the MDGs.

6.4.3.3 *Access to genetic resources for food and agriculture*

The current international basis for the exchange of genetic resources was established in 1992-1994, with the quasi-universal adoption of two international agreements: the Convention on Biological Diversity (CBD) and the Trade Related Intellectual Property Rights agreements (TRIPs) of the World Trade Organization. The former qualification of genetic resources as human heritage was then replaced by the principle of national sovereignty over natural resources and patentability extended to any domain of invention applied to living organisms. This framework has been further developed with the adoption in 1994 of the International Treaty on Plant Genetic Resources for Food and Agriculture (PGRFA), which adapts the CBD principles to the specific field of agricultural plant genetic resources and establishes a multilateral system for facilitated access and benefit-sharing arising from the use of resources: it allows the use of PGRFA

[20] CAMBIA is an independent, international nonprofit institute that has been creating new technologies, tools and paradigms to foster collaboration and life-sciences enabled innovation: www.cambia.org.

[21] PIPRA (Public Intellectual Property Resource for Agriculture) is a nonprofit initiative that brings together intellectual property from over 40 universities, public agencies, and nonprofit institutes to help make their technologies available to innovators around the world: www.pipra.org.

[22] EPIPAGRI is a European project (specific support action) that aims to set up a collective network for the management of patents and other exploitable assets (know-how, software, etc.) held by European public research organizations in the field of agricultural biotechnologies.

in a multilateral way through the conclusion of a "standard" material transfer agreement, but there are shortfalls in funding. The PGRFA treaty also states that the responsibility for realizing farmers' rights (including, when appropriate, the protection of traditional knowledge, the sharing of benefits arising from the use of PGRFA, the right to participate in making decisions at national level on matters related to PGRFA) rests with national governments. Thus, a reflection is also going on since 2001 at the World Intellectual Property Organization on the relationship between intellectual property, genetic resources and traditional knowledge.

Stakeholders are being increasingly challenged by this continuously evolving and complex legal framework (Visser et al., 2000). There are several options to consider for AKST to contribute to the clarification of the regulatory framework in the context of more systemic governance approaches linking global, national and local levels on the one hand and conservation, knowledge, utilization on the other. Among others:

- Ensure a better coherence of national, community and private rights systems over genetic resources and traditional knowledge relevant to agriculture, while encouraging the implementation of effective dialogues between agricultural communities and governments; and
- Encourage intellectual or other property rights that increase easy access to genetic variability and associated knowledge, while ensuring that the royalties associated to such rights will effectively permit the maintenance and regeneration of the genetic resources and their trustees.

6.4.4 Setting up new modes of governance

6.4.4.1 General governance issues in food and farming systems

New modes of governance can contribute to the sustainable development of food and farming systems. This calls for the development of innovative networks at the local level (both terrestrial and marine). It is advised that some of the research concentrate on:

- Area required to ensure a good balance between diagnostic and action as well as between action and needed resource mobilization: in most cases, the size of an environmental space (e.g., a watershed) will not fit either the economic or policy space of action, suggesting compromise as a tool to define the optimal boundaries;
- Development of methods and processes to create innovative networks at local levels to solve problems: mobilization of stakeholders to be part of the network, collective identification of potential conflicts among stakeholders to face and solve the problems, relevant collective organization and resource mobilization for action and follow-up; and
- Development of common tools to facilitate local governance: local databases, easy to use integrated software packages to model complex systems and build up indicators to compare response strategies.

A systematic exploration and scientifically sound examination of practical experience could facilitate in fulfilling these needs. Such research would be effective if it is transdisciplinary, i.e., also involving stakeholders using suitable participatory approaches (focus groups, expert panels, etc.). Stakeholders are the farming sector, consumers, taxpayers, citizens with food safety, environment and animal welfare interests, the food industry as well as regional level decision-makers and administrators (World Bank, 2008). A challenging question is how to combine qualitative and quantitative research to effectively support the related decision processes. The aim must be to effectively bridge different research paradigms and to embed the analyses within a process of stakeholder interactions.

Examples of such new modes of governance include:

(1) Food policy councils. A food policy council is a coalition of food system stakeholders who advise a city, county, or state government on policies related to agriculture and food. These councils focus on areas such as using agriculture and food systems as an economic development tool, protection for farmland and farming, prevention of hunger, fostering the processing and local marketing of food and agricultural products, reducing producer risk, enhancing food safety and promoting nutrition education. They develop legislation, recommendations to departments of agriculture and other policymakers, support and promote state and regional food marketing programs and promote education about local food issues. One of the key functions and benefits of these councils is the increased coordination between state agencies. They also serve as a venue for communication between food and agricultural businesses, consumers and policymakers. The work of Food Policy Councils across the United States of America has so far engaged a large number of stakeholders from food businesses, agriculture, government, consumer groups, non-governmental advocates, nutritionists and institutions in a dialogue about how to promote food and farm businesses for the well-being of the current and future residents of their respective states (Lipstreu, 2007).

Food policy councils could provide a crucial forum to encourage more creative and lasting solutions to food system issues. Based on their ability to bring together diverse organizations and interests to develop win-win solutions, food policy councils can have a significant influence, even with modest resources. They have proven to be a voice for the critical role of food issues in public policy, both at the municipal and state level. Food policy councils can help put healthy food on the radar screens of local and state governments (Food Security Learning Center, 2007).

(2) European Water Framework Directive: Integrated river basin management for Europe. This is the most substantial piece of water legislation ever produced by the European Commission and will provide the major driver for achieving sustainable management of water in the Member States for many years to come. It requires that all inland and coastal waters within defined river basin districts must reach at least good status by 2015 and defines how this should be achieved through the establishment of environmental objectives and ecological targets for surface waters.

Success will depend on close cooperation and coherent action at community, member state and local level as well as on information, consultation and involvement of the public.

Public participation in "River Basin Management" projects is deemed to be crucial and the framework states clearly that caring for Europe's waters will require more involvement of citizens, interested parties and non-governmental organizations. To that end the Water Framework Directive will require information and consultation when river basin management plans are established: the river basin management plan must be issued in draft and the background documentation on which the decisions are based must be made accessible. Furthermore, a biannual conference is said to be important to provide for a regular exchange of views and experiences in implementation. The Framework Directive underlines the need for establishing very early on a network for the exchange of information and experience between water professionals throughout the community (Directive 2000/60/EC of the European Parliament and of the Council, 2007).

To facilitate the implementation of the EU Water Framework Directive, the Department of the Environment, Heritage and Local Government in Ireland as well as the Environment Agency in England and the Scottish Environment Protection Agency (SEPA) are promoting the establishment of river basin management projects by local authorities for River Basin Districts in relation to all inland and coastal waters that will facilitate participation by all stakeholders and lead to the identification and implementation of effective measures for improved water management. The overall objective of these projects is to develop a River Basin Management System, including a program of measures designed to maintain and/or achieve at least good water status for all waters and to facilitate the preparations of River Basin Management Plans. In order to implement this Directive the government of France has for instance established the Rhine Network for a better participatory management of the Rhine River. This network's primary role is to identify and encourage water management based on local participatory practices as well as reinforce European cooperation at the watershed scale.

6.4.4.2 Fisheries issues

One of the main areas where research can be developed in this domain concerns the regulation of the access to marine resources and their exploitation, for instance:

- Better define the rights of use and the rights of property of marine ecosystems: the regulation of the access to marine resources and their exploitation leads to the separation of the rights of property and the rights of use. The inadequacies of many present regimes and particularly where the property of the resources is declared common, results mostly from the absence of a clear regime of access and rights of usage. The evolution of access and property regimes is an essential condition for the establishment of a sustainable exploitation of fisheries resources. Alternatives for resource management could be built on scenarios allowing for the testing of various property regimes (e.g., private/collective), various systems of rights exchange, at various resources levels (stock/ecosystems).

In addition, in this area the importance of local governance and the integration of stakeholders' advice in governance could be underlined. Models where stakeholders' advice is taken into account could be developed to help build scenarios of sustainable fisheries management. This could be done by either strengthening or improving existing institutions (e.g., Regional Advisory Councils in the EU). In this context, the creation of localized Territorial Use Rights in Fisheries (TURF) and the granting of the TURFs to fishing communities offer new opportunities to provide local control over the resources within a territory with local determination of the objectives to be derived (Christy, 1982). The community would be in a position to choose whether it wishes to extract resource rents, to increase the income levels of its fishermen, to increase employment opportunities, or to achieve some combination of these goals. It could also determine the kind of gear to be used, the technological innovations to adopt, the time and seasons of fishing and other management measures. Exclusive territorial rights could be a strong incentive for ensuring that the management measures are respected. Further studies are necessary to develop this TURF concept such as (1) detailed examinations of the conditions permitting the creation of localized TURFs or the maintenance and enhancement of traditional territorial rights; (2) defining the ways in which the benefits of traditional systems are shared or distributed and identify the kinds of controls over newly created TURFs that would ensure equitable distribution of benefits both within communities acquiring the rights and among neighboring communities of fishermen.

6.4.4.3 Forestry issues

The forest sector has been affected by important changes in terms of modes of governance and management since the beginning of the 1990s (Tikkanen et al., 1997). The conventional mode of decision-making in the forestry sector is basically a top-down, command-and-control, centralized system, where the technical expertise of the state forest administration staff is exclusive. With time, this framework has slightly moved towards new modes of governance and management, where participation (in fact consultation in most of the cases) and deliberation among stakeholders (mostly production-based ones) are becoming more prominent (FAO/ECE/ILO, 1997; GoFOR, 2007).

The main changes identified in the forest sector are highlighted below:

Schemes of certification. Under strong pressure from some major environmental NGOs, the idea has been introduced that the evaluation of the sustainability of forest management could work completely differently from what has been the case, where forest managers were more or less their own evaluators. It is admitted today that a certification procedure carried out by neutral actors is the only way to ensure a label of sustainable forest management (Viana et al., 1997).

Three main certification schemes are coexisting today. The FSC (Forest Stewardship Council): promoted by environmentalists (mainly WWF) and based mainly on performance indicators; the PEFC (Program of Endorsement of Forest Certification schemes), promoted by producers, including private forest owners in Europe and based on system indicators; and Smartwood, basically a North American joint initiative of an environmental NGO (the Rainforest Alliance) and industry. The effectiveness of those certification schemes is still questioned: multiplicity of labels creates confusion (still a battle among certifiers); chain of custody controls are unclear in some cases (example of PEFC); labels are used as market instruments more than tools promoting sustainability; and labels are unknown to customers (Burger et al., 2005). But the changes in the forest sector are promising, because the introduction of certification schemes has boosted both the participation among actors and the development of a more accountable expertise. An increasing share of the wood products on the market is certified through these various schemes (Rametsteiner, 2000).

National Forest Programs (NFPs): An NFP is a strategic document, established within a timeframe of 10-15 years and provides the rationale and directive for public action in the forest sector. It is established through a formal participatory process associating the stakeholders and the public and gives guidance on the establishment of partnerships and share of responsibilities in carrying out the activities. This new way of formulating forest strategies has replaced the conventional, technical top-down mode of planning of the forest administration (Glück et al., 1999; COST E19, 2000). The NFPs are based on several elements: participation—all stakeholders and sometimes the public are strongly invited to be involved in the designing and implementation of forest activities (FAO/ECE/ILO, 2000, 2003; IUCN, 2001); links across sectors—programming of forest activities is elaborated in connection with other sectors, especially environment and land use (Tikkanen et al., 2002); coordination between various levels of governance to ensure comprehension between international, national and local actions (Niskanen and Väyrynen, 1999; Slee and Wiersum, 2001); accountable expertise from various sources and subject to public debate; and iterative processes to promote adaptive management based on collaborative learning.

During the last 10 years, there has been a significant increase of NFPs, especially in Central and Eastern Europe, as this was an informal requirement before becoming integrated within the European Union (Glück and Humphreys, 2002; Glück et al., 2003). As for Western European countries, the NFPs elaborated are more formal documents, except in Finland and Scotland (Gislerud and Neven, 2002; Humphreys, 2004). In the US, where decision makers are more results, than process-oriented, NFPs are not yet part of the culture.

Although those changes are significant in a very traditional sphere such as the forestry sector, all the characteristics of these new modes of governance and management are far from being present in all national frameworks: participation is used as an alibi, cross-sector approaches are advocated only when they reinforce the sector and the accountability of expertise is diverted by the conventional experts (Buttoud et al., 2004).

6.5 Funding Investments in Research and Development (R&D) for Agriculture

Research and Development are key elements of technology change. R&D needs and priorities will vary among plausible futures. From an investment viewpoint, decisions are required on how much to spend on R&D, who should pay for it, who should do it and on issues of governance.

6.5.1 Spending on R&D

Spending on R&D is worthwhile if it gives a satisfactory return in absolute terms (extra benefits are greater than or equal to extra costs) and relative to other investment opportunities. The benefits of R&D are multiple and diverse, some being immediate and others long term. Some benefits of agricultural R&D accrue directly and exclusively to the users of research products (private benefits/goods) while others generate indirect benefits for society at large (public benefits/goods). Investment in crop genetics, for example, can deliver private benefits to farmers who use the research products, as well as public benefits associated with increased food security. Extra costs include the costs of resources committed to R&D activities. They may also include public costs associated, for example, with unwanted social or environmental side effects.

Methodologies to evaluate investments in R&D are available but there are theoretical and practical challenges (Alston et al., 1995). Previous investments in agricultural R&D in NAE have shown relatively high rates of return (Alston et al., 1999, 2000, 2001; Thirtle, 1999, 2003; ADAS, 2002; Marra et al., 2002; Sylvester-Bradley and Wiseman, 2005), perhaps suggesting a degree of under-investment. However, some estimates are liable to errors associated with overestimation, double counting, over-attribution of benefits to individual R&D programs (Alston et al., 2001) and possible omission of some of the negative social and environmental effects of improvements in productivity due to use of R&D products (Julian and Pardey, 2001; Barnes, 2002; Koeijer et al., 2002). In some cases, R&D may not have been the best way of achieving the desired outcome.

6.5.2 Paying for agricultural R&D

Generally, the criteria for payment is that the beneficiary pays, moderated by ability to pay. Broadly, providers of R&D products should be remunerated by those who derive private benefits from their use. In this way, the researchers recover their costs and are given incentives to invest in R&D. Where potential users of R&D products cannot afford to pay and yet it is considered that overall social welfare is enhanced if they use R&D products, there may be a case for public funding. This could be to help finance the R&D process or the acquisition of research products by users. Here, government intervention is addressing the failure of R&D markets to deliver a socially optimum R&D spend. Some R&D may also provide public goods by reducing the negative externalities of agriculture such as diffuse pollution; in this case, it is sometimes questionable whether remedial R&D of this kind constitutes the most economically efficient approach: other policy interventions might be better, including removing incentives which cause the externalities in the first place, such as production subsidies.

Definitions of public goods associated with agricultural R&D vary over time and space, as does the justification for government funding. In postwar Europe, food production to feed nations was regarded as a public good, justifying major commitments of public funds for crop, livestock and agricultural engineering research. Much of the fundamental R&D that underpinned the gains in agricultural productivity in NAE was publicly funded between the 1950s and early 1980s. Much of this stock of R&D knowledge and the research capability that provided it has now either been used up or depleted. Less is available for future needs.

Government funded research is now largely confined to addressing non-market social, environmental and strategic issues as well as supporting fundamental research that would not attract private funding. Examples include research into livestock systems to reduce environmental burdens and research programs to assist farmers in disadvantaged, upland areas. Under the new paradigm of multifunctional, sustainable agriculture, it is clear that there will be a continued need for government funding of R&D in the public interest.

Spending on private sector agricultural research has grown relatively rapidly over the last 25 years and now exceeds public spending in many developed countries (Alston, 2000). An additional and growing source of funding is that of nongovernmental, not-for-profit organizations that sponsor research in pursuit of an organizational agenda, mostly associated with social, environmental, ethical or political objectives. They also provide a conduit for private or government funds. Private benefactors also channel funds into trusts that pursue selected themes. In the future, increased emphasis on the multiple functions of agriculture is likely to call on a greater range of funding sources.

The paradigm shift in agriculture towards multifunctionality and the concomitant shift in AKST have major implications for the provisioning of R&D in terms of priority setting, funding and delivery mechanisms. Continuing reform of agricultural policy throughout the NAE region is likely to promote greater market orientation for agriculture, implying that governments will further retract from R&D that is "near market", leaving this largely to the private sector. Government funding of R&D is likely to focus on aspects of public good, addressing strategic issues such as food security, impacts of climate change, the long-term sustainability of agricultural systems and the protection of natural resources, the environment and the livelihoods of vulnerable rural communities.

Where private R&D initiatives fail to respond to market potential because of high costs, high risks or long investment periods, governments can collaborate with private partners to underwrite commercial risks if they perceive a potential net gain in social and economic welfare. Collaborative funding of R&D to support bioenergy cropping and processing systems is a case in point. Where technology change is policy-induced, there is a strong case for collaborative public–private funding mechanisms for R&D. The EU Integrated Pollution Control Regulations and the Water Framework Directive are cases in point, justifying collaborative R&D ventures that share the burden of costs associated with new regulations.

6.5.3 Undertaking agricultural R&D

There has been a recent tendency to separate R&D funding and delivery mechanisms, with increased "contracting out" of government-funded research, diversifying the range of organizations engaged in research. As agricultural enhancement has become less important as a policy goal, direct government involvement in R&D to improve productivity has declined: it is now regarded as too "near market." As a result, the number and size of government research institutes in some parts of NAE have decreased. In other cases, specialist government research institutes and universities have increased their share of private and NGO funded research, utilizing specialist skills and facilities. Funding and delivery regimes come together to provide a range of options for R&D management, considering the three main players: government, NGOs and private organizations (Table 6-4). Universities are a key delivery agent.

There is now much greater diversity in the provisioning of agricultural research, especially regarding biosciences, with a growth in public-private funding partnerships among industry, NGOs and government, often involving universities as research contractors. Potential complementarities include joint funding, pooling of facilities and expertise (including research management), economies of scale and learning, risk sharing and dissemination and commercialization of research products into research outcomes.

6.5.4 Deciding on R&D

Regarding public goods, this is clearly a role for government and intergovernmental development agencies, engaging the key stakeholders in the process. Most NAE governments have R&D priority setting regimes in place and these will be increasingly linked to strategies to promote sustainable agriculture (e.g., DEFRA, 2002b, 2005b).

Regarding private good aspects, decisions rest on the commercial considerations of business enterprises. Public–private partnerships can help to underwrite private R&D investment costs where risks are high and the development of successful research capabilities and products can significantly enhance the public good. At a national level, this may include improving international competitive advantage (Gopinath et al., 1997; Ball et al., 2001).

6.5.5 Institutional arrangements and collaboration

The organizational arrangements for identifying, prioritizing, funding and carrying out R&D and, not least, the transposition of research outputs into knowledge, products and processes for adoption by end users, are critical to the overall successful outcomes of R&D. Again, this will reflect the dominant purposes to be achieved and, as far as serving the public good is concerned, it is a responsibility of Government to provide an institutional framework within which various stakeholders can interact.

R&D management includes arrangements for identification of needs, priority setting, pre-investment appraisal, research procurement, dissemination and follow-up. As mentioned in earlier sections, it is imperative that key stakeholders are involved throughout this process to ensure R&D is relevant to end-user constituencies. The latter also implies full integration with the processes of "knowledge exchange", including those of advisory and extension services. There are also important links with other services that affect technology adoption, notably credit and marketing.

The further development of funding arrangements could be designed to promote enhanced cooperation not only

Table 6-4. **Agricultural research: Who pays and who delivers?**

		Who Pays?		
		Government including parastatal organizations	**Nongovernmental, organizations**	**Private, commercial**
Research Objectives		Public good, e.g., food security, environmental protection	Organizational Agenda, e.g., poverty alleviation, animal welfare	Private good, e.g., profits, increased utility of consumers
Who delivers?	**Government, including parastatal organizations**	Government funded research institutes	Government research institutes conducting external research contract	Government research institutes conducting external research contract
	Universities*	Govt funded research programs in Universities	University research under contract to NGOs	University research under contracts to commercial companies
	Nongovernmental, organizations	NGOs undertaking research on contract to Government	NGOs funding and operating own research programs	NGOs conducting research on contract to private companies
	Private, commercial	Commercial research organizations on contract to Government	Commercial research organizations on contract to NGOs	Market driven, research for competitive advantage, conducted "in house" or contracted out

*Universities also fund their own research programs, but this usually draws indirectly on external funding sources, such as trust funds.

among AKS components but also between AKS components and the more general scientific and higher education communities (life sciences and economic/social disciplines) as well as policy makers, stakeholders and the general public.

Funding for cooperation across AKS institutions has led to the emergence of new partnerships and networks providing cross institutional and cross-disciplinary synergy in some NAE countries. Open dialogue, joint planning and fair sharing of credit are key success features in the promotion of these partnerships. It is now vital to design mechanisms for scaling these partnerships up and outwards, not only nationally, but also across the NAE region, both in research and in human capital development.

It is noteworthy that a recently published OECD study on Human Capital Investment concluded that "human capital seems to offer rates of return comparable to those available for business capital" (OECD/CERI, 1998). Allied to this conclusion is the increasing acceptance by many OECD governments that investment in the development of a knowledge based society can be a powerful stimulant to promoting innovation and competitiveness. Many governments have increasingly been prepared to give extra public funding for innovative research (especially interdisciplinary ones) to promote competitiveness and for human capital development through higher education designed to achieve competitiveness and life-long learning/re-education to maintain competitiveness. The experiences over several decades of AKST institutions linking research, higher education and extension/development in an integrated manner offers a valuable model for AKS to play a central role in addressing the wider societal issues including food safety, the food chain, sustainability of natural resource use and rural development. New partnerships, networks and relationships will be required for this potential to be realized and AKS institutions could be encouraged to take action accordingly.

6.5.6 Funding mechanisms and enablers of AKST

The new paradigm of multifunctional sustainable agriculture calls for new, increased and more diverse funding and delivery mechanisms for agricultural R&D and human capital development. There is continuing need for public funding to serve the public interest, as well as new investments by private organizations responding to market needs and opportunities. Funding arrangements can promote cooperation among all stakeholders. Open dialogue, joint planning and fair sharing of rewards are key success features in the promotion of these partnerships. Depending on circumstances, the following will be required:

- Public investment in R&D to serve the public good, addressing strategic issues such as food security and safety, impacts of climate change, long-term environmental sustainability of the system, social viability, protection of biodiversity, achieving strategic balance between land use for food and bioenergy, as well as other non-market issues that do not attract private funding;
- Public investment in human capital development to achieve widespread understanding of the complexities of multifunctionality and to develop the knowledge and skill sets necessary for effective decision making by all stakeholders. These developments will encompass initial education, professional formation and lifelong learning for AKST personnel as well as for a much wider range of clients, including civil society and public policy makers as well as farmers and others (especially women) involved in rural livelihoods;
- Public investment to support the development of multi-

Box 6-15. Enablers of AKST

Policy drivers providing high level commitment to multi-functional agriculture within the broader context of sustainable development:

- A knowledge and science culture at all levels of governance in society, supported by an informed science and society discourse, including aspects of welfare and ethics.
- Incentives, rewards and risk sharing using an appropriate balance of public and private involvement.
- Institutional frameworks governing the rules, regulations and ways of doings things, including regulation of intellectual property rights, patents, and fair trading.
- Stakeholder engagement and exchange amongst providers, brokers and users of AKST, including joint and collaborative working.
- Experimentation, testing and demonstration of new forms of AKST in real world conditions as a precursor to adoption and diffusion.
- Funding and delivery mechanisms suited to the wide range AKST products and services, including public, private and joint public-private partnerships.

disciplinary research and education programs that promote an articulation between research and educational goals consistent with development and sustainability goals and judge the research and educational outcomes against attainment of these goals;

- Private investments in R&D, made in response to market opportunities and potential private gain by those supplying and using new technologies, as an important and growing source of new AKST. It is critical that private suppliers of ASKT are given the necessary incentives and rewards to make new investments in R&D and have access to essential commercial services such as market information and credit;
- Public–private partnerships to provide technical assistance and joint funding of R&D investments, especially where risks are high and where development of successful research capabilities and products/services in the private sector can significantly enhance the public good. Various forms of public-private partnerships will be relevant for advisory/information services of a near market nature. There are significant public good aspects to the development of human capital and skills relating to many pre-market, quasi-market or non-market multifunctional services, some of which may transfer to private funding at a later date; and
- Nongovernmental organizations to act as channels for public and private funding of technical assistance, knowledge transfer and applied research, especially at the local scale. Further support will be required to facilitate this.

With respect to the above, there is need for a framework that can support the cost-benefit analysis of future AKST investments under the new paradigm for agriculture.

It is clear that a range of enabling conditions are required to support new forms of AKST investment in support of development and sustainability goals (Box 6-15).

References

Abler, D. 2004. Multifunctionality, agricultural policy and environmental policy. Agric. Res. Econ. Rev. 33:8-17.

ACRE. 2006. Managing the footprint of agriculture: Towards a comparative assessment of risks and benefits for novel agricultural systems (Consultation draft). Available at http://www.defra.gov.uk/environment/acre/fsewiderissues/pdf/acre-wi-final.pdf. ACRE Secretariat, DEFRA, London.

Action Aid International. 2006. Power hungry: Six reasons to regulate global food corporations. Johannesburg, South Africa.

ADAS. 2002. The role of future public research investment in the genetic improvement of UK grown crops. DEFRA, London.

Aebersold, R. 2005. Molecular systems biology: A new journal for a new biology? Mol. Systems Biol. 1:1.

AFES (French Assoc. Soil Sci.). 1998. The aims of soil science, challenges to be met by soil science, the services soil science can render. 16th World Cong. Soil Science: Introductory conferences and debate. AFES, Montpellier.

Afman, M., and M. Muller. 2006. Nutrigenomics: From molecular nutrition to prevention of disease. J. Am. Diet Assoc. 106:569-576.

Agrawal, A. 1995. Indigenous and scientific knowledge: Some critical comments. Article and exchange. Indigen. Knowl. Dev. Mon. 3:3-4, 4:1-2.

Aigner, S.M., C.B. Flora, and J.M. Hernandez. 2001. The premise and promise of citizenship and civil society for renewing democracies and empowering sustainable communities. Soc. Inquiry 71:493-507.

Aitken, R.J., K.S. Creely, and C.L. Tran. 2004. Nanoparticles: An occupational hygiene review. Res. Rep. 274. Health Safety Exec. London.

Alderson, A.S., and F. Nielsen. 2002. Globalization and the great u-turn: Income inequality trends in 16 OECD countries. Am. J. Soc. 107:1244.

Allaire, G. 2002. Economy of quality, its chains, territories and myths. Géographie, Econ. Société 4:155-180.

Allen, E.A., and L.M. Humble. 2002. Nonindigenous species introductions: A threat to Canada's forests and forest economy. Can. J. Plant Path. 24:103-110.

Alston, J.M. 2000. Agricultural R&D, technological change, and food security. Dep. Agric. Res. Econ., UC Davis CA.

Alston, J., M.C. Marra, P.G. Pardey, and T.J. Wyatt. 2000. Research returns redux: A meta-analysis of agricultural R&D evaluations. Aust. J. Agric. Res. Econ. 44:185-216.

Alston, J., P. Pardey, and V Smith. 1999. Paying for agricultural productivity. Johns Hopkins Univ. Press, Baltimore.

Alston, J.M., and P.G. Pardey. 2001. Attribution and other problems in assessing the returns to agricultural R&D. Agric. Econ. 25:141-152.

Alston, J.M., G.W. Norton, and P.G. Pardey.1995. Science under scarcity: Principles and practice for agricultural research evaluation and priority setting. Cornell Univ. Press, Ithaca, NY.

Altieri, M.A., and C.I. Nicholls. 1999. Biodiversity, ecosystem function and insect pest management in agricultural systems. p. 69-84 *In* W.W. Collins and C.O. Qualset

(ed), Biodiversity in agroecosystems. CRC Press, Boca Raton.
Altieri, M.A. 1995. Agroecology: The science of sustainable agriculture. Westview Press, Boulder CO.
Altieri, M.A. 2005. Agroecology: Principles and strategies for designing sustainable farming systems. p. 291. *In* UNEP et al. (ed), Agroecology and the search for a truly sustainable agriculture. Univ. California, Berkeley and UNEP, Mexico.
ANAFE, 2007. African network for agriculture, agroforestry and natural resources education [Online]. Available at http://www.anafeafrica.org/. ANAFE, Nairobi.
Anderson, K. 2004. Subsidies and trade barriers. *In* B. Lomborg (ed), Global crises, global solutions. Cambridge Univ. Press, UK.
Anderson, P.K., A.A.. Cunningham, N.G. Patel, F.J. Morales, P.R. Epstein, and P. Daszak. 2004. Emerging infectious diseases of plants: Pathogen pollution, climate change and agrotechnology drivers. Trends Ecol. Evol. 19:535-44.
Andow, D.A., and C. Zwahlen. 2006. Assessing environmental risks of transgenic plants. Ecol. Letters 9:196-214.
Anton, D., P. Mathew, W. Morgan. 2007. International law. Cases and materials. Oxford, New York.
Attieh, J.M., R. Djiana, P. Koonjul, C. Ettiene, S. Sparace, H.S. Saini. 2002. Cloning and functional expression of two plant thiolmethyltransferases, a new class of enzymes involved in the biosynthesis of sulfur volatiles. Plant Mol. Biol. 50:511-521.
Azais-Braesco, V., C. Goffi, and E. Labouze. 2006. Nutrient profiling: Comparison and critical analysis of existing systems. Public Health Nutr. 9:613-622.
Ball, V.E., J.C. Bureau, J.P. Butault, and R. Nehring. 2001. Levels of farm sector productivity: An international comparison. J. Productivity Anal. 15:5-29.
Barndt, D. 2002. Tangled routes: Women, work, and globalization on the tomato trail. Rowman and Littlefield, Lanham, MD.
Barnes, A.P. 2002. Publicly-funded UK agricultural R&D and "social" total factor productivity. Agric. Econ. 27:65-74.
Beedell, J.D.C., and T. Rehman. 1999. Explaining farmers' conservation behaviour: Why do farmers behave the way they do? J. Environ. Manage. 57.
Begg, D., Fisher, S. and Dornbusch, R. 2003. Economics: 7th ed. McGraw-Hill, London.
Bellisle, F. 2003. Why should we study human food intake behaviour? Nutr. Metab. Cardiovasc. Dis. 13:189-193.
Bérard, L., and P. Marchenay. 2004. Products of "terroir" between cultures and regulations. (In French). CNRS Ed., Paris.
Bérard, L., and P. Marchenay. 2006. Local products and geographical indications: Taking account of local knowledge and biodiversity. Int. Soc. Sci. J. 58:109-116.
Bérard, L. and P. Marchenay. 2007. Localized products in France: Definition, protection and value-adding. *In* V. Amilien, and G. Holt (ed), From local food to localized food. Anthropol. Food, S2, Mars, special issue.
Berbel, J. and C.G. Martin. 2004. The sustainability of European irrigation under water framework directive and agenda 2000: EC DG Res., Luxembourg.
Berkes, F. 1999 Sacred ecology: Traditional ecological knowledge and resource management. Taylor and Francis, Philadelphia, PA.
Berkes, F., C. Folke, J. Colding. 2000. Linking social and ecological systems: Management practices and social mechanisms for building resilience. Cambridge Univ. Press, Cambridge.
Betts, R.A., P.D. Falloon, K.K. Goldewijk and N. Ramankutty, 2007. Biogeophysical effects of land use on climate: Model simulations of radiative forcings and large-scale temperature change. Agric. For. Meteorol. 142:216-233
Bicknell, K.B, R.J. Ball, R. Cullen and H.R. Bigsby. 1998. New methodology for the ecological footprint with an application to the New Zealand economy. Ecol. Econ. 27:149-160.
Birot, Y., and R. Päivinen. 2005. Report from the expert group on the vision and strategic objectives for the EU forest action plan. Eur. Community, Luxembourg.
Bonnal, P. (ed). 2004. European series on multifunctionality 4: Public policies and international comparisons. (In French). Available at http://www.inra.fr/sed/multifonction/textes/CAHIERMF4.pdf. INRA, CIRAD, CEMAGREF, France.
Bonny, S. 2005. Agricultural systems and agribusiness. (In French). Economie Rurale 288.
Bourgine, P., and J. Johnson. 2005. Living roadmap for complex systems science. IST-FET, Coordination action. Open network of centres of excellence in complex systems. Project FP6-IST 29814.ONCE-CS. Paris.
Boussard, J.M., F. Gérard, and M.G. Piketty. 2005. Evaluating the benefits from liberalization in agriculture: Are standard Walrasian models relevant? Available at: http://www.economicswebinstitute.org/essays/liberalization.htm. Economics Web Institute.
Bowes, M.J., et al. 2005. The relative contribution of sewage and diffuse phosphorus sources in the river Avon catchment, southern England: Implications for nutrient management. Sci. Total Environ. 344:67-81.
Bradman, A., B. Eskenazi, D.B. Barr, R. Bravo, R. Castorina, J. Chevrier et al. 2005. Organophosphate urinary metabolite levels during pregnancy and after delivery in women living in an agricultural community. Environ. Health Perspect. 113:1802-1807.
Bradman, A., L. Fenster, A. Sjodin, R.S. Jones, D.G. Patterson, and B. Eskenazi. 2007. Polybrominated diphenyl ether levels in the blood of pregnant women living in an agriculture community in California. Environ. Health Perspect. 115:71-74.
Brandth, B. 2002. Gender identity in European family farming: A literature review. Soc. Ruralis. 42:181-200.
Brouwer, R., I.H. Langford, I.J. Bateman, and R.K. Turner. 1999. A meta-analysis of contingent valuation studies. Reg. Environ. Change 1:47-57.
Brovkin, V., S. Sitch, W. Von Bloh, M. Claussen, E. Bauer, W. Cramer. 2004. Role of land cover changes for atmospheric CO2 increase and climate change during the last 150 years. Glob. Change Biol. 10:1-14.
Brown, L.K., and K.G. Zeiler. 1993. Aquatic biomass and carbon dioxide trapping. Energy Conserv. Manage. 34:1005-1013.
Brown, R.J. 2002. Thinking globally, working locally. Pres. Nat. Conf. Food Safety Educators, Orlando, Florida. 18 Sept. 2002.
Bruin, S., and Th.R.G. Jongen. 2003. Food process engineering: The last 25 years and challenges ahead. Comp. Rev. Food Sci. Tech. 2:42.
Brush, S.B., and D. Stabinsky. 1995. Valuing local knowledge: Indigenous people and intellectual property rights. Island Press, Washington, DC.
Bryden, J. 2001. Section 3: Rural development. *In* Landsis g.e.i.e. proposal on agri-environmental indicators PAIS. Luxembourg.
BSI (British Standards Institution). 2006. NTI/1 Nanotechnologies: Committee activities. Available at http://www.bsi-global.com/en/Standards-and-Publications/Industry-Sectors/Nanotechnologies/Committee-Activities. BSI, London.
Bureau, J.C., E. Gozlan, S. Jean. 2005. Globalization of agricultural markets, an opportunity for developing countries? (In French). Rev. Française d'Econ. 20:109-145.
Burger, D., J. Hess, and B. Lang (ed). 2005. Forest certification: An innovative instrument in the service of sustainable development? Available at http://www.gtz.de/de/dokumente/en-forest-certification-introduction-and-summary.pdf. GTZ, Eschborn.
Buttoud, G., B. Solberg, I. Tikkanen, and B. Pajari (ed). 2004. The evaluation of forest policies and programmes: EFI Proc. 52. Available at http://www.efi.int/portal/virtual_library/publications/proceedings/52/. EFI, Finland.
Cairol, D., E. Coudel, D. Barthélémy, P. Caron, E. Cudlinova, P. Zander, et al. 2005. Multifunctionality of agriculture and rural areas: From trade negotiations to contributing to sustainable development: New challenges for research. EU Project MULTAGRI.
Caponera, D.A., and M. Nanni (ed). 2007. Principles of water law and administration: 2nd ed. Taylor and Francis, The Netherlands.
Carswell, G. 1997. Agricultural intensification and rural sustainable livelihoods: A think piece. IDS Working Pap. 64. Inst. Dev. Studies, Univ. Sussex, Brighton.
Castorina, R., A. Bradman, T.E. McKone, D.B. Barr, M.E. Harnly, and B. Eskanazi. 2003. Organophosphate pesticide exposure and risk assessment among pregnant women

living in an agricultural community: A case study from the CHAMACOS Cohort. Environ. Health Perspect. 111:1640-1648.

Caughley, G., and A. Gunn. 1996. Conservation biology in theory and practice. Blackwell Science, Cambridge.

CBD (Convention on Biological Diversity). 2005. Agricultural biodiversity: Further development of the International initiative for the conservation and sustainable use of soil biodiversity. SBSTA Tenth meeting, Bangkok, 7-11 Feb 2005.

Chabe-Ferret, S., J. Gourdon, M.A. Marouani, and T. Voituriez. 2005. Trade-induced changes in Economic inequalities: Assessment issues and policy implications for developing countries. Available at http://www.iddri.org/Activites/Ateliers/abcde_tokyo_06-05_iddri_paper.pdf. World Bank, Washington, DC.

Chiao, M., K.B. Lam, and L.W. Lin. 2006. Micromachined microbial and photosynthetic fuel cells. J. Micromech. Microeng. 16: 2547-2553.

Chiarolla, C. 2006. Commodifying agricultural biodiversity and development-related issues. J. World Intellectual Prop. 9:1.

Chiesura, A. and R. de Groot. 2003. Critical natural capital: A sociocultural perspective. Ecol. Econ. 44:219-231.

Chinn, M.D. and R.W. Fairlie. 2007. The determinants of the global digital divide: A cross-country analysis of computer and internet penetration. Oxford Econ. Papers 59:16-44.

Chisti, Y. 2007. Biodiesel from microalgae. Biotech. Adv. 25:294-306.

Christensen, V.S., J.J. Guénette, C. Heymans, R. Walters, R. Watson, D. Zeller et al. 2003. Hundred year decline of north atlantic predatory fishes. Fish Fisheries 4:1-24.

Christy, F.T.J. 1982. Territorial use rights in marine fisheries: definitions and conditions, FAO Fish. Tech. Pap. 227:1-10.

COFI (Committee on Fisheries). 2007. 27th meeting, Rome. 5-9 Mar., 2007. UN-FAO, Rome.

Colvin, V.L. 2003. The potential environmental impact of engineered nanoparticles. Nature Biotechnol. 21:1166-1170.

COST E19. 2000. National forest programmes: Social and political context. Proc. Cost Action E19 Sem., Madrid. 18-21, Oct 2000. Min. Environ., Spain.

Costanza, R., S.C. Farber, and J. Maxwell. 1989. Valuation and management of wetland ecosystems, Ecol. Econ. 1:335-361.

Costanza, R., R. d'Arge, R. de Groot, S. Farber, M. Grasso et al. 1997. The value of the world's ecosystem services and natural capital. Nature 387:253-260.

Daft, R.L. 1998. Organization theory and design. 6th ed. South-Western College Publ., Cincinnati, OH.

Dame, J.K., and R.R. Christian. 2007. Uncertainty and the use of network analysis for ecosystem-based fishery management. Fisheries 31:331-341.

Danbom, D.B. 2006. Born in the country: A history of rural America, 2nd ed. Johns Hopkins Univ. Press, Baltimore.

Daszak, P., G.M. Tabor et al. 2004. Conservation medicine and a new agenda for emerging diseases. Annals NY Acad. Sci. 1026:1-11.

De Alessi, M. 2003. Fishing for solutions. Inst. Econ. Affairs, London.

De Groot, R., W. Wilson, and R. Boumans. 2002. A typology for the classification, description and valuation of ecosystem functions, goods and services. Ecol. Econ. 41:393-408.

DEFRA. 2002a. Farming and foods' contribution to sustainable development: economic and statistical analysis.DEFRA, London.

DEFRA. 2002b. The strategy for sustainable farming and food: Facing the future. DEFRA, London.

DEFRA. 2005a. Characterizing the potential risks posed by engineered nanoparticles: A first UK government research report. DEFRA, London.

DEFRA. 2005b. The first report of the sustainable farming and food research priorities group. DEFRA, London.

Dekker , M., E. Turnhout, B. M. S. D. L. Bauwens and G.M.J. Mohren. 2007. Interpretation and implementation of ecosystem management in international and national forest policy. Forest Policy Econ. 9:546-557.

Delgado, C. L., N. Wada, M.W. Rosegrant, S. Meijer, and M. Ahmed. 2003. Outlook for fish to 2020. World Fish Center and IFPRI, Washington, DC.

Desenclos, J.C., and H. de Valk. 2005. Emerging infectious diseases: Importance in public health, epidemiological aspcets, determinants and prevention (In French). Médecine et maladies infectieuses 35:49-61.

DFID (Department for International Development). 1999. Sustainable Livelihood Guidance Sheets [Online]. Available at http://www.livelihoods.org/info/guidance_sheets_pdfs/section1.pdf. DFID, London.

Dobrowolski, J.P., and M.P. O'Neill (ed). 2005. Agricultural water security listening session final report. USDA REE, Washington, DC.

Dochain D., W. Marquardt, S. C. Won, O. Malik and M. Kinnaert. 2005. Monitoring and control of process and power systems: Towards new paradigms. Ann. Rev. Control 30:1.

Dupouey, J.L. et al. 2006. The role of agriculture and forests in greenhouse effect (In French.) *In* Paul Colonna (ed), La chimie verte. Éditions Tech. Doc. Lavoisier, Cachan, France

Easterling, W.E. 1996. Adapting North American agriculture to climate change. Agric. Forest Meteorol. 80:1-53.

EC (European Commission). 2003. European commission fisheries yearbook 1993-2002. Off. Official Publ. European communities, Luxembourg.

EC. 2005. Promoting healthy diets and physical activity: A European dimension for the prevention of overweight, obesity and chronic diseases. Green Pap. Available at http://ec.europa.eu/health/ph_determinants/life_style/nutrition/documents/nutrition_gp_en.pdf. European Commission, Luxembourg.

EC. 2006. Framework program VI Information society technology 2005-2006 (Good food project): Strategies for leadership. Available at http://www.goodfood-project.org/. Eur. Commission, Luxembourg.

Economic Research Service. 2007. Briefing Room: Farm Labor. Available at www.ers.usda.gov/Briefing/FarmLabor. USDA, Washington, DC.

Eden, C. and F. Ackermann (ed). 1998. Making strategy: The journey of strategic management. Sage Publ., London.

EEA. 1999. Sustainable water use in Europe Part 1: Sectoral use. European Environ. Agency, Copenhagen.

EEA. 2001a. Sustainable water use in Europe Part 2: Demand management. European Environ. Agency, Copenhagen.

EEA. 2001b. Sustainable water use in Europe Part 3: Hydrological events: Floods and droughts. European Environ. Agency, Copenhagen.

EFTEC. 2005. The economic, social and ecological value of ecosystem services: A literature review. DEFRA, London.

EFTEC and IEEP. 2004. Framework for environmental accounts for agriculture. Economics for Environment Consultancy (EFTEC) and Institute for European Environmental Policy IEEP). DEFRA, London

Ehler, L.E. and E. Lester. 2005. Integrated pest management: A national goal? The history of federal initiatives in IPM has been one of redefining the mission rather than accomplishing it. Issues Sci. Tech. 22:25-26.

Eicher, C.K. 2001. Africa's unfinished business: Building sustainable agricultural research systems. Available at http://agecon.lib.umn.edu/cgi-bin/pdf_view.pl?paperid=2820&ftype=.pdf. Michigan State Univ., East Lansing.

Einsiedel, E.F., and D.L. Easlick. 2000. Consensus conferences as deliberative democracy. Sci. Commun. 21:323-343.

Elliot, N.C., and D. Dent. 1995. Integrated pest management. Springer, New York.

Ellis, F. 2000. Rural livelihoods and diversity in developing countries. Oxford Univ. Press, Oxford.

Engberg, R.A. (ed). 2005. Proceedings of the second water resources policy dialogue, Tucson. 14-15 Feb. 2005. Amer. Water Res. Assoc., Tucson, AZ.

English Nature. 2002. Policy mechanisms for the control of diffuse agricultural pollution. English Nature Res. 455. English Nature, Peterborough.

EA (Environmental Agency). 2002. Agriculture and natural resources: Benefits, costs and potential solutions. Environ. Agency for England and Wales, Bristol.

Eskanazi, B., A.R. Marks, A. Bradman, L. Fenster, C. Johnson, D.B. Barr et al. 2006. Organophosphate pesticide exposure and neurodevelopment in young Mexican-American children. Pediatrics 118:233-241.

EU. 2007. Organic farming: Agriculture and rural development [Online]. Available at http://ec.europa.eu/agriculture/qual/organic/index_en.htm. European Commission, Brussels.

EURAGRI. 2005. Science for society, science with society. Conf. Proc. Brussels. 3-4 Feb 2005. Available at http://www.ec.europa.eu/research/agriculture/pdf/summary_euragri_con_brussels2005_en.pdf. European Commission, Brussels.

European Forest Institute, 2007. News 1:15. European Forest Inst., Brussels.

European Parliament. 2004. Report on agriculture and agricultural research in the framework of CAP Reform: Committee on agriculture and rural development: A5-0018/2004. Eur. Parliament, Brussels.

European Parliament 2007. Directive 2000/60/EC of the European Parliament and of the Council establishing a framework for community action in the field of water policy. Eur. Parliament, Brussels.

Evans, R. 1996. Soil erosion and its impacts in England and Wales. Friends of the Earth Trust. London.

Eyckmans, J., and M. Finus. 2007. Measures to enhance the success of global climate treaties. Available at http://www.springerlink.com/content/m22v33551h1j5123/. Int. Environ. Agree. Polit. Law Econ. 7:73-97.

FABRE Technology Platform. 2006. Sustainable farm animal breeding and reproduction: A vision for 2025 [Online]. Available at http://www.fabretp.org/. Eur. Commission, Brussels.

Fairfax, S.K., and D. Guenzler. 2001. Conservation trusts. Univ. Kansas Press, Lawrence.

FAO. 2002. World agriculture: Towards 2015-2030: An FA0 perspective. FAO, Rome.

FAO. 2003. The ecosystem approach to fisheries: Fisheries Tech. Pap. 443. FAO, Rome.

FAO. 2004a. Economic valuation of water resources in agriculture: Water Res. Rep. 27. FAO, Rome.

FAO. 2004b. The state of food and agriculture 2003-04: Agricultural biotechnology: Meeting the needs of the poor? FAO, Rome.

FAO. 2005a. Urban and peri-urban agriculture (UPA). Available at http://www.rlc.fao.org/en/prioridades/aup/. FAO, Rome.

FAO. 2005b. An approach to rural development: Participatory and negotiated territorial development (PNTD). Available at http://www.fao.org/sd/dim_pe2/docs/pe2_050402d1_en.pdf. FAO Rural Dev. Div., Rome.

FAO/ECE/ILO. 1997. People, forests and sustainability. ILO, Geneva.

FAO/ECE/ILO. 2000. Public participation in forestry. ILO, Geneva.

FAO/ECE/ILO. 2003. Raising awareness of forests and forestry. ILO, Geneva.

FAO and World Bank. 2000. Agricultural knowledge and information systems for rural development: Strategic vision and guiding principles. FAO, Rome and World Bank. Washington DC.

FAO/WHO. 2006. Food standards. Available at http://www.codexalimentarius.net/web/index_en.jsp. Codex Alimentarius Commission, FAO, Rome.

FARA/ANAFE. 2005. Building Africa's scientific and institutional capacity for agriculture and natural resources (BASIC). FARA, Accra and ANAFE, Nairobi.

FARA/AU/NEPAD. 2006. Framework for african agricultural productivity (FAAP). FARA, Accra.

Farrell, A.E., R.J. Plevin, B.T. Turner, A.D. Jones, M. O'Hare, and D.M. Kammen. 2006. Ethanol can contribute to energy and environment goals. Science 311:506-508.

Fearne, A. 1994. Strategic alliances in the European food industry. Eur. Bus. Rev. 94:30-36.

Findeis, J.L. (ed). 2002. The Dynamics of hired farm labour: Constraints and community responses. CABI, UK.

Fischer. G. 2005. Socioeconomic and climate impacts on agriculture: An integrated assessment. Philos. Trans. R. Soc. Lond. Ser. B: Biol. Sci. 360:2067-2083.

Flora, C.B., and J. Flora (ed). 2004. Rural communities: Legacy and challenges. 2nd ed. Westview, Boulder, CO.

Fluck, R.C. 1992. Energy of agricultural production. p. 13-28. *In* R.C. Fluck (ed), Energy analysis in agricultural systems. Vol. 6 of Energy in world agriculture. Elsevier, Amsterdam.

Food Security Learning Center. 2007. Food policy councils. Available at http://www.worldhungeryear.org/fslc. World Hunger Year, New York.

FTP. 2007. Forest-based sector technology platform vision 2030: A European technology platform initiative by the European forest based sector. Available at: www.forestplatform.org. FTP, Luxembourg.

Forni, N. (ed). 2000. Common property regimes: Origins and implications of the theoretical debate: Land reforms: Land settlements and cooperatives. FAO, Rome.

Foster, B. 2001. Women in agricultural education. Proc. Ann. Nat. Agric. Educ. Res. Conf. 28th, New Orleans. 12 Dec. 2001. Available at http://aaae.okstate.edu/proceedings/2001/program.HTM. Am. Assoc. Agric. Ed., Oklahoma State Univ.

Fox, J., and G. Rivera-Salgado. 2004. Indigenous Mexican migrants in the United States. Center for U.S.-Mexican Studies, UC San Diego, CA.

Francis, R. C., M. A. Hixon, M. E. Clarke, S. A. Murawski, and S. Ralston. 2007. Ten commandments for ecosystem-based fisheries scientists. Fisheries 32:217-233.

Friedman, T. 1999. The lexus and the olive tree. Harper Collins, London.

Fry, W.E., and S.B. Goodwin. 1997. Resurgence of the Irish potato famine fungus. Bioscience 47:363-371.

Gan, J., S.R. Yates, H.D. Ohr, and J.J. Sims. 1998. Production of methyl bromide by terrestrial higher plants. Geophys. Res. Letters 25:3595-3598.

Garbellotto, M., and D.M. Rizzo. 2005. A California-based chronological review (1995-2004) of research on Phytophthora ramorum: The causal agent of sudden oak death. Phytopathol. Mediterr. 44:127-143.

Garcia, S. M., and R.J.R. Grainger. 2005. Gloom and doom? The future of marine capture fisheries, Phil. Trans. R. Soc. B. 360:21-46.

Garforth, C., and R. Usher. 1997. Promotion and uptake pathways for research output: A review of analytical frameworks and communication channels. Agric. Syst. 55:301-322.

Giampietro, M., G. Cerretelli, and D. Pimentel. 1992. Energy analysis of agricultural ecosystem management: Human return and sustainability. Agric. Ecosyst. Environ. 38:219-244.

Gislerud, O., and I. Neven (ed). 2002. National forest programmes in a European context: EFI Proc. 44. EFI, Finland.

Gliessman, S.R. 1997. Agroecology: Ecological processes in sustainable agriculture. Ann Arbor Press MI.

Glück, P., and D. Humphreys (ed). 2002. National forest programmes in a European context. Forest Policy Econ. 4:253-358.

Glück, P., A.C. Mendes, and I. Neven. 2003. Making NFPs Work: Publication series 48. Inst. For. Sect. Policy and Econ. BOKU, COST, Vienna.

Glück, P., G. Oesten, H. Schanz, and K.R Volz (ed). 1999. Formulation and implementation of national forest programmes: EFI Proc. 30:3. EFI, Finland.

GoFOR. 2007. New modes of governance for sustainable forestry in Europe [Online]. Available at www.boku.ac.at/GoFOR/. GoFOR, Vienna.

Gollehon, N., L. Hansen, R. Johansson, W. Quinby, and M. Ribaudo. 2006. Water and wetland resources: AREI. ERS, USDA, Washington, DC.

Gopinath, M., C. Arnade, M. Shane, and T.Roe. 1997. Agricultural competitiveness: The case of the United States and major EU countries. Agric. Econ. 16:99-109.

Graf, W.L. (ed). 2003. Dam removal research: Status and prospects. H. John Heinz III Center, Washington DC.

Grandt-Dowton, R.T., and H.G. Dickinson. 2005. Epigenetics and its implications for plant biology: The Epigenetic network in plants. Ann. Bot. 96:1143-1164.

Grandt-Dowton, R.T., and H.G. Dickinson.

2006. Epigenetics and its implications for plant biology: The epigenetic epiphany: Epigenetics, evolution and beyond. Ann. Bot. 97:11-27.
Grenier, L. 1998. Working with indigenous knowledge: A guide for researchers. IDRC, Ottawa.
Gressel, J., and A. Zilberstein. 2003. Let them eat (GM) straw. Trends Biotech. 21:525-530.
Griffon M., and J. Weber. 1996. Doubly green revolution, economy and institutions (In French). Agric. 5:239-242. CIRAD, Paris.
Grimble, R., and K. Wellard. 1997. Stakeholder methodologies in natural resources management: A review of principles, contexts, experiences and opportunities. Agric. Syst. 55:173-193.
Guzman, K.A., M.R.Taylor and J.F. Banfield. 2006. Environmental risks of nanotechnology: National nanotechnology initiative funding 2000-2004. Environ. Sci. Tech. 40:1401-1407.
Hallman, W K., N.D. Weinstein, S.S. Kada'Kia, and C. Chess. 1995. Precautions taken against Lyme disease at three recreational parks in endemic areas of New Jersey. Environ. Behavior 27:437-453.
Hamdy, A., R. Ragab, and M.E. Scarascia. 2003. Coping with water scarcity: Water saving and increasing water productivity. Irrig. Drain. 52:3-20.
Hanley, N., J. Shogren, and B. White. 2001. Introduction to environmental economics. Oxford Univ. Press, Oxford.
Hardin, G. 1968. The tragedy of the commons. Science 162:1243-1248.
Harrus S., and G. Baneth. 2005. Drivers for the emergence and re-emergence of vector-borne 16 protozoal and bacterial diseases. Int. J. Parasitol. 35:1309-1318.
Hartridge.O., and D.W. Pearce.2001. Is UK agriculture sustainable? Environmentally adjusted economic accounts for UK agriculture. CSERGE-Economics, Univ. College, London.
Hassell, L. 2007. A continental philosophy perspective on knowledge management. Infor. Syst. J. 17:185-195.
Hatch, D.L., and Z.I. Imam. 1996. Collaboration: The key to investigations of emerging and re-emerging diseases. E. Mediter. Health J. 2:1:30-36.
Health and Safety Executive (HSE). 2004. Nanoparticles: An occupational hygiene review. Inst. Occup. Med. Health Safety Exec., London.
Heaton, E., T. Voigt, and S.P. Long. 2004. A quantitative review comparing the yields of two candidate C-4 perennial biomass crops in relation to nitrogen, temperature and water. Biomass Bioenergy 27:21-30.
Hector, A., B. Schmid, C. Beierkuhnlein, M.C. Caldeira, M. Diemer, P.G. Dimitrakopoulos et al. 1999. Plant diversity and productivity experiments in European Grasslands. Science 286:1123-1127.
Heller, M.C., and G.A. Keoleian. 2000. Life cycle-based sustainability indicators for assessment of the U.S. food system. School Nat. Res. Environ., Univ. Michigan.
Heller, M.C., and G.A. Keoleian. 2003. Assessing the sustainability of the US food system: a life cycle perspective. Agric. Syst. 76:1007-1041.
Herman, A. 2001. How the Scots invented the modern world. Three Rivers Press, New York.
Hetherington, M.M. 2002. The physiological-psychological dichotomy in the study of food intake. Proc. Nutr. Soc. 61:497-507.
Holland, N., C. Furlong, M. Bastaki, R. Richter, A. Bradman, K. Huen et al. 2006. Paraoxonase polymorphisms, haplotypes, and enzyme activity in Latino mothers and newborns. Environ. Health Perspect. 114:985-91.
Horwitz, P., and B.A. Wilcox. 2005. Parasites, ecosystems and sustainability: Ecological and complex systems perspectives. Int. J. Parasitol. 35:725-732.
Houllier, F., J. Novotny, R. Päivinen, K. Rosén, G. Scarascia-Mugnozza, and K. Von Teuffel. 2005. Future forest research strategy for a knowledge based forest cluster: An asset for sustainable Europe. EFI Disc. Pap. 11. EFI, Finland.
Hubacek, K., and S. Giljum. 2003. Applying physical input-output analysis to estimate land appropriation (ecological footprints) of international trade activities. Ecol. Econ. 44:137-151.
Hubert, B., and D. Despréaux. 2007. Added value and territories (In French.) Valeur ajoutée et territoires. *In* Loyat J. (ed), Ecosystèmes et sociétés: Concevoir une recherche pour un développement durable. Cemagref, Cirad, Ifremer, INRA, IRD, MNHN, Paris.
Hughes, J.M. 2001. Emerging infectious diseases: A CDC perspective. Emerg. Infect. Dis. 7(3Supp):494-496.
Humphreys, D. (ed). 2004. Forests for the Future: National forest programmes in Europe: Country reports from COST Action E19. Off. Official Publ. Eur. Communities, Luxembourg.
Hussein, K., and J. Nelson.1998. Sustainable livelihoods and livelihoods diversification: IDS working paper 69. IDS, Univ. Sussex, Brighton.
Hutson, S.S., N.L. Barber, J.F. Kenny, K.S. Linsey, D.S. Lumia., and M.A. Maupin. 2005. Estimated use of water in the United States in 2000. Circular 1268. US Geological Survey, Washington, DC.
IEA. 2004. Analysis of the impact of high oil prices on the global economy. IEA, Paris.
IFAD. 2001. Rural poverty report 2001: The challenge of ending rural poverty. Oxford Univ. Press, Oxford.
Inglis, J. 1993.Traditional ecological knowledge: Concepts and cases. IDRC, Ottawa.
IPCC. 2001. Third Assessment Report of the Intergovernmental Panel on Climate Change. IPCC/WMO, Geneva.
IUCN, 2001. Communities and forest management in Western Europe. IUCN, Switzerland.
Jackson, W. 1980. New roots for agriculture. Univ. Nebraska Press, Lincoln.
John, W.B. Jr. 2001. Energy conservation for commercial greenhouses: NRAES [Online]. Available at http://www.nraes.org/publications/nraes3.html. Nat. Resource, Agric., and Engineering Service, Ithaca NY.
Johnson, K.H., K.A. Vogt, H.J. Clark, O.J. Schmitz and D.J. Vogt. 1996. Biodiversity and the productivity and stability of ecosystems. Trends Ecol. Evol. 11:9:372-377.
Johnson, C., G. Bentrop, and D. Rol. 1999. Conservation corridor planning at the landscape level: Management for wildlife. USDA and NRCS, Washington, DC.
Joseph, T., and M. Morrison. 2006. Nanotechnology in agriculture and food. Available at http://www.nanoforum.org/. Eur. Nanotechnology Gateway, Germany.
Joss, S. 1998. Danish consensus conferences as a model of participatory technology assessment: An impact study of consensus conferences on Danish parliament and Danish public debate. Sci. Public Policy 25:2-22.
Julian, M.A., and P.G. Pardey. 2001. Attribution and other problems in assessing the returns to agricultural R&D. Agric. Econ. 25:141-152.
Karjalainen, T., H. Spiecker, and O. Laroussinie (ed). 1999. Causes and consequences of accelerating tree growth in Europe. Eur. Forest Inst. Proc. 27. EFI, Finland.
Kasnakoglu, H. (ed) 2006. FAO statistical yearbook 2005-2006. FAO, Rome.
Kellogg Commission on the future of state and land-grant universities. 2000. Returning to our roots. NASULGC. Kellogg Commission, Washington DC.
Kerr, J.T., and D.J. Currie. 1995. Effects of human activity on global extinction risk. Conserv. Biol. 9:1528-1538.
Kiss, A.C. (ed). 2003. Economic globalization and compliance with environmental agreements. Int. Environ. Law Policy Ser. 63. Kluwer, The Hague.
Knickel, K., and H. Renting. 2000. Methodological and conceptual issues in the study of multifunctionality and rural development. Sociol. Ruralis 40:4:512-528.
Koeijer, T.J., G.A.A. de Wossink, J.A. Renkema, and P.C. Struik. 2002. Measuring agricultural sustainability in terms of efficiency: The case of sugar beet growers. J. Environ. Manage. 66:9-17.
Kuzma, J., and P. VerHage. 2006. Nanotechnology in agriculture and food production: Anticipated applications. Woodrow Wilson Center for Int. Scholars, Univ. Minnesota, Minneapolis.
Ladygina, N., E.g., Dedyukhina, and M.B. Vainshtein. 2006. A review on microbial synthesis of hydrocarbons. Process Biochem. 41:1001-1014.
Lahmar, R., M. Dosso, A. Ruellan, and

L. Montanarella (ed). 2000. Soils in central and eastern European countries, in the new independent states in central Asian countries and in Mongolia: Present situation and future perspectives. Eur. Soil Bureau, European Commission, Luxembourg.

Lal, R. (ed). 2002. Encyclopedia of soil science. Marcel Dekker, New York.

Lamont, J.T.J. 1992. Agricultural marketing systems: Horizontal and vertical integration in the seed potato industry. Brit. Food J. 94(8).

Lane, S.N., J. Morris, P.E. O'Connell, and P.E. Quinn. 2006. Managing the rural landscape. *In* C.R. Thorne et al. (ed) Future flood and coastal erosion risk. Thomas Telford, London.

Lassey, K. 2005. Livestock methane emissions: Measurements, methods, inventory estimation, and the global methane cycle. Agric. For. Meterol. 142:120-132.

Lazarova, V., and A. Bahri (ed). 2004. Water reuse for irrigation: Agriculture, landscapes, and turf grass. CRC Press, Boca Raton FL.

Leeuwis, C. 2004. Communication for rural innovation: Rethinking agricultural extension. 3rd ed. Blackwell Sci., Oxford.

Lenzen, M., and S.A. Murray. 2001. A modified ecological footprint method and its application to Australia. Ecol. Econ. 37:229-255.

Lenzen, M., and S.A. Murray. 2003. The ecological footprint: Issues and trends. Integrated Sustainability Analysis Res. Pap. 01-03. Univ. Sydney, Australia.

Lipstreu, A. 2007. A review of state food policy councils in the United States and opportunities for the state of Ohio [Online]. Available at http://www.thefarmlandcenter.org/documents/FoodPolicyBrief07.pdf. The Farmland Center, OH.

Littlefield, J. 1998. Greenhouses feature high-tech hydroponics, controlled environment agriculture. Available at http://cals.arizona.edu/pubs/general/resrpt2000/controlledenvironment.pdf. Univ. Arizona College Agri. Life Sci., AR.

Logan, B.E., and J.M. Regan. 2006. Electricity-producing bacterial communities in microbial fuel cells. Trends Microbiol. 14:512-518.

Loreau, M., N. Mouquet, and A. Gonzalez. 2003. Biodiversity as spatial insurance in heterogeneous landscapes. PNAS 100:12765-12770.

Loreau, M., S. Naeem, P. Inchausti, J. Bengtsson, J.P. Grime, A. Hector et al. 2001. Biodiversity and ecosystem functioning: Current knowledge and future challenges. Science 294:804-808.

Loyat, J. 2004. Agriculture and development, paradox and wrong solutions (In French). La Documentation Française, Questions Internationales 6.

Loyat, J. 2006. Striking a balance between economic efficiency and solidarity: Using agriculture to promote sustainable development [Online]. Available at http://www.cgiar.org/pdf/france_cgiar_ENG_chapter3.pdf. Washington, DC.

Lucey, D. 2000. Summary and evaluation of main developments and changes in organizational forms of and approaches by the AKS in OECD member countries. OECD, Paris.

Lynch, J.M., and A.J. Moffat. 2005. Bioremediation: Prospects for the future application of innovative applied biological research. Annals Appl. Biol. 146:217-221.

Lyson, T.A. and A.L.Raymer. 2000. Stalking the wily multinational: Power and control in the U.S. food system. Agric. Human Values 17:199-208.

Ma, W., D. Shi, Q. Wang, L. Wei, and H. Chen. 2005. Exogenous expression of the wheat chloroplastic fructose-1-6-bisphosphate gene enhances photosynthesis in the transgenic cyanobacterium Anabaena PCC7120. J. Appl. Phycology 17:273-280.

Mack, R.N., and M. Erneberg. 2002. The United States naturalized flora: Largely the product of deliberate introductions. Annals Missouri Bot. Garden 89:176-189.

Mackova, M., D.N. Dowling, and T. Macek. 2006. Phytoremediation and rhizoremediation: Focus on biotechnology 9A. Springer, Dordrecht.

MAP (French Ministry of Agriculture). 2006. Reinforce economic organization. (In French). Available at www.agriculture.gouv.fr/spip/actualites.loa_a6276.html. MAP, Paris.

Marra, M., P. Pardey, and J. Alston. 2002. The payoffs of agricultural biotechnology: An assessment of the evidence. IFPRI, Washington DC.

Martin, P. 2003. Managing labor migration: Temporary worker programs for the 21st century. ILO, Geneva.

Mattson, J.W., and W.W. Koo. 2006. Forces reshaping world agriculture: Agribusiness and applied economics report No. 582. Center for Agric. Policy and Trade Studies, North Dakota State Univ., ND.

Mauro, F., and P. Hardison. 2000. Traditional knowledge of indigenous and local communities: International debate and policy initiatives. Ecol. Appl. 10:1263-1269.

Mazoyer, M., and L. Roudart. 2006. A History of world agriculture. From the neolithic age to the current crisis. Earthscan, London.

McNeely, J. and S.J. Scherr. 2002. Ecoagriculture: Strategies to feed the world and save biodiversity. Island Press, Washington, DC.

MEA (Millennium Ecosystem Assessment). 2005. Ecosystems and human well-being: Current states and trends. vol. 1. Island Press, Washington, DC.

Mercoiret, M.R., B. Goudiaby, S. Marzaroli, D. Fall, S. Gueye, and J. Coulibaly. 2001. Empowering producer organizations: Issues, goals and ambiguities. p. 20-28. *In* P. Rondot, M-H. Collion (ed), Agricultural producer organizations: Their contribution to rural capacity building and poverty reduction. Workshop Rep., Washington, 28-30 June 1999. World Bank, Washington DC.

Miles, H.L., P.L. Hofman, and W.S. Cutfield. 2005. Fetal origins of adult disease: A paediatric perspective. Rev. Endocr. Metab. Disord. 6:261-268.

Minorsky, P.V. 2003. Achieving the in silico plant: System biology and the future of biological research. Plant Physiol. 132: 404-409.

Mitchell, J.P., P.B. Goodell, R. Krebill-Prather, T.S. Prather, K.J. Hembree, D.S. Munk et al. 2001. Innovative agricultural extension partnerships in California's central San Joaquin Valley. Available at http://www.joe.org/joe/2001december/rb7.html. J. Extension 39:6.

Mitchell, R. 2003. International environmental agreements: A survey of their features, formation, and effects. Ann. Rev. Environ. Resour. 28:429-461.

Morisset, J. 1997. Unfair trade? Empirical evidence in world commodity markets over the past 25 years: Poverty research wor. Pap. Ser. 1815. World Bank, Washington DC.

Morris, J. 2007. Water policy, economics and the EU framework Directive. *In* J. Pretty (ed) Handbook for environmental management. Sage Publ., London.

Morris, J., J. Mills, and I.M. Crawford. 2000. Promoting farmer uptake of agrienvironment schemes: The countryside stewardship arable options scheme. Land Use Policy 17:241-254.

Morris, J. and H. Wheater. 2006. Catchment land use. *In* C.R. Thorne et al. (ed), Future flood and coastal erosion risk. Thomas Telford, London.

Murray, D. 2005. Oil and food: A rising security challenge. Earth Policy Inst., Washington, DC.

NAFTA-CEC (Commission for Environmental Cooperation). 2004. Maize and biodiversity: The effects of transgenic maize in Mexico: Key findings and recommendations. Commission for Environ. Coop., Montreal.

Nakada, M. 2006. Distributional conflicts and the timing of environmental policy. Int. Environ. Agreements 6:1.

Napper, D. 2006. Hygiene in food factories of the future. Proc. (CD-ROM). Food Factory for the Future, Goteborg, Sweden.

Narayan, D. 1999. Bonds and bridges: Social capital and poverty. Rep. No. 2167. World Bank, Washington, DC.

National Academy of Sciences. 2000. The future role of pesticides in US agriculture: Report of the board on agriculture committee. Nat. Acad. Press, Washington DC.

Neal, C., W.A. House, H.P. Jarvie, M. Neal, L. Hill, and H. Wickham. 2005. Phosphorus concentrations in the river Dun, the Kennet and Avon Canal and the river Kennet, southern England. Sci. Total Environ. 344:107-128.

Nevins, J. 2002. Operation gatekeeper: The rise of the illegal alien and the remaking of the U.S. Mexico boundary. Routledge, London.

Niskanen, A., and J. Väyrynen (ed). 1999. Regional forest programmes. Proc. 32.240. EFI, Finland.

NRC (National Research Council). 2001. Envisioning the agenda for water resources research in the twenty-first century. Nat. Acad. Press, Washington, DC.

NRC. 2004. Confronting the nation's water problems: The role of research. Nat. Acad. Press, Washington DC.

NRC. 2007. The new science of metagenomics: Revealing the secrets of our microbial planet. Committee on Metagenomics. Nat. Acad. Sciences, Washington, DC.

O'Connell, P.E., K.J. Beven, J.N. Carney, R.O. Clements, J. Ewen, H. Fowler et al. 2004. Review of impacts of rural land use and management on flood generation, part A: Impact study report. DEFRA, London.

OACC (Organic Agriculture Centre of Canada). 2007. Available at http://www .organicagcentre.ca/index_e.asp. OACC, Truro, Nova Scotia.

OECD. 1995a. The state of agricultural knowledge systems in OECD member countries: A synthesis report. AGR/ REE(95)3. OECD, Paris.

OECD. 1995b. First Joint Conf. of Dir. and Reps. of Agric. Research, Agric. Advisory Services and Higher Ed. in Agriculture. Conf. Summary. Paris, 4-8 Sep 1995. AGR/ CA 95:22. OECD, Paris.

OECD. 2000. Second Conf. of Dir. and Reps. of Agric. Knowledge Systems (AKS): Summary Report, Paris. 10-13 Jan. 2000. AGR/CA 2000:1. OECD, Paris.

OECD. 2006. Environment, water resources and agricultural policies: Lessons from China and OECD countries. OECD, Paris.

OECD/CERI. 1998. Human capital investment. OECD, Paris.

OECD/FAO. 2007. Agricultural outlook 2007-2016. OECD, Paris.

OFT. 2006. The grocery market: The OFT's reasons for making a reference to the competition commission. Available at http://www.oft.gov.uk/shared_oft/reports/ comp_policy/oft845.pdf. Off. Fair Trading, London.

Olesen, J.E., and M. Bindi. 2002. Cosequences of climate change for European agricultural productivity, land use and policy. Eur. J. Agron. 16:239-262.

Opara, L.U. 2003. Traceability in agriculture and food supply chain: A review of basic concepts, technological implications, and future prospects. J. Food Agric. Environ. 1:101-106.

Ostrom, E. 2003. How types of goods and property rights jointly affect collective action. J. Theor. Politics 15:239-270.

Padel, S., and P. Midmore. 2005. The development of European market for organic products: Insights from a Delphi study. Brit. Food J. 107:626-647.

Parrott, L., and R. Kok. 2000. Use of an object-based model to represent complex features of ecosystems. Proc. Int. Conf. Complex Syst. 3rd. Nashua, NH. 21-26 May 2000. New England Complex Syst. Inst., NH.

Pauly, D., J. Alder, E. Bennett, V. Christensen, P. Tyedmers, and R. Watson. 2003. Viewpoint: The future for fisheries. Science 302:1359-1361.

Pauly, D., R. Watson, and J. Alder. 2005. Global trends in world fisheries: Impacts on marine ecosystems and food security. R. Soc. London Phil. Trans. Biol. Sci. 360:5-12.

Pauly, D., V. Christensen, S. Guénette, T.J. Pitcher, U.R. Sumaila, C.J. Walters et al. 2002. Towards sustainability in world fisheries. Nature 418:689-695.

PCFFF. 2002. Farming and food: A sustainable future. Rep. Policy Commission on the Future of Farming and Food (The Curry Report). [Online]. Available at http:// archive.cabinetoffice.gov.uk/farming/pdf/ PC%20Report2.pdf. Cabinet office, UK Gov., London.

Pearce, D., and E.D. Barbier. 2000. Blueprint for a sustainable economy. Earthscan, London.

Phillipson J., and A. Liddon. 2007. Common knowledge: An exploration of knowledge transfer. RELU briefing paper 6 [Online]. Available at http:www.relu.ac.uk/news/ briefings.htm. UK Res. Councils, London.

Pimentel, D., and M. Giampietro. 1994. Food, land and population and the US economy. Carrying Capacity Network, Washington, DC.

Pimentel, D., L. Lach, R. Zuniga, and D. Morrison. 2000. Environmental and economic costs of nonindigenous species in the United States. BioScience 50:53-65.

Plants for the Future (The European Technology Platform). 2005. 2025: A European vision for plants genomics and biotechnology. Available at http:// www.epsoweb.org/catalog/tp. European commission, Luxembourg.

Polaski, S. 2006. Winners and losers: Impact of the Doha round on developing countries. Carnegie Endowment Int. Peace, Washington, DC.

Porter, G., J.W. Brown, and P. Chasek. 2006. Global environmental politics. 4th ed. Westview, Boulder, CO.

Pretty, J.N. 1998. The living land: Agriculture, food and community regeneration in rural Europe. Earthscan, New York.

Pretty, J.N., A.S. Ball, T. Lang, and J.I.L. Morison. 2005. Food costs and food miles: An assessment of the full cost of the UK weekly food basket. Food Policy 30:1-19.

Pretty, J.N., D. Brett, D. Gee, R.E. Hine, C.F. Mason, J.I.L. Morison et al. 2000. An assessment of the total external costs of UK agriculture. Agric. Syst. 65:113-136.

Raison, R.J., A.G. Brown, and D.W. Flinn (ed). 2001. Criteria and indicators for sustainable forest management. CABI Publ., Wallingford.

Rametsteiner, E. 2000. Sustainable forest management certification. MCPFE, Vienna.

Rauscher, H.M. 1999. Ecosystem management decision support for federal forests in the United States: A review. Forest Ecol. Manage. 114:173-197.

Rees, W.E. 1992. Ecological footprints and appropriated carrying capacity: What urban economics leaves out. Environ. Urban 4:121-130.

Reich P.B., D. Tilman, S. Naeem, D.S. Ellsworth, J. Knops, J. Craine et al. 2004. Species and functional group diversity independently influence biomass accumulation and its response to CO_2 and N. PNAS 101: 10101-10106.

Reijntjes, C.B., B. Haverkort, and A. Waters-Bayer. 1992. Farming for the future. MacMillan Press, London.

Richardson, D.P., T. Affertsholt, N.-G. Asp, S. Bruce, R. Grossklaus, J. Howlett, et al. 2003. PASSCLAIM—Synthesis and review of existing processes. Eur. J. Nutr. 42(Sup1):96-111.

Rizzo, D.M., M. Garbelotto, and E.M. Hansen. 2005. Phytophthora ramorum: Integrative research and management of an emerging pathogen in California and Oregon forests. Ann. Rev. Phytopathol. 43:309-335.

Roberfroid, M.B. 2002. Global view on functional foods: European perspectives. Brit. J. Nutr. 88:S133-S138.

Rogers, E.M. 2003. Diffusion of innovations. 5th ed. Free Press, New York.

Rollins, B.E. 1995. The Frankenstein syndrome: Ethical and social issues in the genetic engineering of animals. Cambridge studies in philosophy and public policy. Cambridge Univ. Press, Cambridge.

Rosendal, G.K. 2006. The convention on biological diversity: Tensions with the WTO TRIPS agreement over access to genetic resources and the sharing of benefits. *In* S. Oberthur and T. Gehring (ed), Institutional interaction in global governance. MIT Press, Cambridge.

Rosegrant, M.W., X. Cai, and S.A. Cline. 2002. World water and food to 2025: Dealing with scarcity. IFPRI, Washington, DC.

Roth, D., R. Boelens, and M. Zwarteveen (ed) 2005. Liquid relations: Contested water rights and legal complexity. Rutgers Univ. Press, Piscataway, NJ.

RS and RAE (Royal Soc. Royal Acad. Engineering). 2004. Nanoscience and nanotechnologies: Opportunities and uncertainties. The Royal Society, London.

Ruellan, A. 2000. Morphology of the soil cover. p. 49-52. *In* I. Kheoruenromne and S. Theerawong (ed) Proc. Int. Symp. Soil Sci.: Accomplishments and changing paradigms toward the 21st century. Bangkok.

Ruellan, A. 2005. Classification of pedological systems: A challenge for the future of soil science. Ann. Agrarian Sci. 3:24-28.

Ryan, B., and N.C. Goss. 1943. The diffusion of hybrid seed corn in two Iowa communities. Rural Soc. 8:15-24.

Sachs, I. 2002. Development and ethics: Whither Latin America? National development strategies in the globalization age. Proc. IDB meeting on ethics and development. Buenos Aires. 5-6 Sep. 2002 [Online]. Available at http://www.iadb.org/Etica/Documentos/ar2_sac_desar-i.pdf. Inter-American Dev. Bank, Washington DC.

Salamanca-Buentello, F., D.L. Persad, E.B. Court, D.K. Martin, A.S. Daar, and P.A. Singer. 2005. Nanotechnology and the developing world [Online]. Available at http://www.issues.org/21.4/singer.html. PLoS Med 2:4:e97.

Salman, M.D. 2004. Controlling emerging diseases in the 21st century. Prev. Vet. Med. 62:177-184.

Samson, R., M. Sudhagar, R. Boddey, S. Shahab, D. Quesada, S. Urquiaga et al. 2005. The potential of C-4 perennial grasses for developing global BIOHEAT industry. Critical Rev. Plant Sci. 24:461-495.

Sankoff, D., and J.H. Nadeau (ed) 2000. Comparative genomics: Empirical and analytical approaches to gene order dynamics, map alignment and the evolution of gene families. Kluwer Acad. Press, Dordrecht.

Saunders, C., A. Barber, and G. Taylor. 2006. Food miles: Comparative energy/emissions: Performance of New Zealand's agriculture industry. Res. Rep. 258. Lincoln Univ. Agribusiness Econ. Res. Unit, New Zealand.

Savvas, D., and H. Passam. 2002. Hydroponics production of vegetables and ornamentals. Embryo Publ., Athens.

SCENIHR (Sci. Committee on Emerging and Newly Identified Health Risks). 2005. Opinion on the appropriateness of existing methodologies to assess the potential risk associated with engineered and adventitious products of nano-technologies [Online]. Available at http://ec.europa.eu/health/ph_risk/documents/synth_report.pdf. European Commission, Luxembourg.

Schmer, M.R., K.P. Vogel, R.B. Mitchell, L.E. Moser, K.M. Eskridge, and R.K. Perrind. 2006. Establishment stand thresholds for switchgrass grown as a bioenergy crop. Crop Sci. 46:157-161.

Schneider, J. 2004. The role of social capital in building healthy communities. Casey Foundation, Baltimore MD.

Scoones, I. 1998. Sustainable rural livelihoods: A framework for analysis. IDS Working Pap. 72. Univ. Sussex, Inst. Dev. Studies, Brighton, UK.

SEENET (South and East European Ecological Network). 2007. [Online]. Available at http://www.seenet.info/. SEENET, the Netherlands.

Seguin B., N. Brisson, D. Loustau, and J.L. Dupouey. 2005. Impact du changement climatique sur l'agriculture et la forêt. (In French). *In* L'homme face au climat? Actes du symposium du Collège de France, Paris. 12-13 Oct. 2004. Odile Jacob, Paris.

Seguin, <B>[[INITIAL?]]<B> et al. 2005. Moderating the impact of agriculture on climate. Agric. For. Meteorol. 142:2-4.

Sen, A. 1981. Poverty and famines, an essay on entitlement and deprivation. Clarendon Press, Oxford.

Shani Z., M. Dekel, G. Tzbary, C.S. Jensen, T. Tzfira, R. Goren et al. 1999. Expression of Arabidopsis thaliana endo-1,4-beta-glucanase (cel1) in transgenic plants. *In* A. Altman et al (ed), Plant biotechnology and in vitro biology in the 21st century. Kluwer Acad. Publ., Dordrecht.

Shlyk-Kerner, O., I. Samish, D. Kaftan, N. Holland, P.S. Sai, H. Kless et al. 2006. Protein flexibility acclimatizes photosynthetic energy conversion to the ambient temperature. Nature 442:827-830.

Slee, B., and F. Wiersum (ed) 2001. New opportunities for forest-related rural development. Forest Policy Econ. 3:1-4

Smith, A., P. Watkiss, G. Tweddle, A. McKinnon, M. Browne, A. Hunt et al. 2005. The validity of food miles as an indicator of sustainable development. Report to UK Gov. DEFRA. AEA Tech. Environ. Didcot, Oxford.

Snow, A.A., D.A. Andow, P. Gepts, E.M. Hallerman, A. Power, J.M. Tiedje et al. 2005. Genetically engineered organisms and the environment: Current status and recommendations. Ecol. Appl. 15:377-404.

Sokratous, G. 2003. Integrated water resources planning Cyprus: Conf. Int. Water Manage. Policy. Nicosia-Cyprus. 19-21 June 2003. Agric. Res. Inst., Cyprus.

SOFIA. 2006. State of fisheries and aquaculture. FAO, Rome.

SOMO (Centre for Res. Multinational Corporations). 2006. Who reaps the fruit? Critical issues in the fresh fruit and vegetable chain. SOMO, Amsterdam.

Soule, J., and J.K. Piper. 1992. Farming in nature's image: An ecological approach to agriculture. Island Press, Washington, DC.

Sourie, J.C., D. Tréguer, S. Rozakis. 2005. L'ambivalence des filières biocarburants (In French). *In* INRA Sciences Sociales 2. Ivry-sur-Seine. INRA, Paris.

Soussana, J.F., S. Saletes, P. Smith, R. Schils, S. Ogle, M.C. Amezquita et al. 2004. Greenhouse gas emissions from European grasslands. CarboEurope-Greenhouse Gases. Specific study 3. INRA, Clermont-Ferrand, France.

Srivastava, N., J.H. Smith, and D.A. Forno (ed) 1996. Biodiversity and agricultural intensification: Partners for development and conservation. World Bank, Washington, DC.

Stephen, R.T. 2006. Comanagement of natural resources: Local learning for poverty reduction. IDRC, Ottawa.

Stern, N. 2006. The Stern review: The economics of climate change. HM Treasury, UK Gov., London.

Stiglitz, J. 2006. Political organization of the world, will it contribute to the general interest of the planet (In French.) *In* M. Albin (ed) L'avancée des biens publics: Politique de l'intérêt général et mondialisation. France.

Stolze, M., A. Piorr, and S. Dabbert. 2000. The environmental impacts of organic farming in Europe: Economics and Policy, 4. Univ. Hoheneim, Germany.

Stout, B.A. 1991. Handbook of energy for world agriculture. Elsevier, New York.

Straughan, R. 1999. Ethics, morality and animal biotechnology. Biotech. Biol. Sci. Res. Council, Swindon.

Stringham, G.E., and W.R. Walker. 1987. Surge flow: Automation of surface irrigation, at last. *In* L.G. James and M.J. English (ed) Irrigation systems for the 21st Century. Am. Soc. Civil Engineers Press, New York.

Strochlic, R. and K. Hamerschlag. 2006. Best labor practices on twelve California farms: Toward a more sustainable food system. California Inst. Rural Studies, Davis, CA.

Sutherland, W.J., S. Armstrong-Brown, P.R. Armsworth, T.Brereton, J. Brickland, C.D. Campbell et al. 2006. The identification of 100 ecological questions of high policy relevance in the UK. J. Applied Ecol. 43(4):617-627.

Sylvester-Bradley, R., and J. Wiseman (ed) 2005. Yields of farmed species, constraints and opportunities in the 21st century. Nottingham Univ. Press, UK.

Tabary, J.C. 1993. Théorie de la connaissance et autonomie biologique [Online] (in French). Available at http://cerveau.pensee.free.fr/these/Tabary-these.pdf.

Tamara, K. 2006. Climate change and the credibility of international commitments: What is necessary for the U.S. to deliver on such commitments? Int. Environ. Agreements 6:3.

Tetali, S.D., M. Mitra, and A. Melis. 2006. Development of the light-harvesting chlorophyll antenna in the green alga *chlamydomonas reinhartii* is regulated by the novel Ta1 gene. Planta 225:813-829.

Thies, C., and T. Tscharntke. 1999. Landscape structure, and biological control in agroecosystems. Science 285:893-895.

Thirtle, C., and J. Holding. 2003. Productivity in UK agriculture: Causes and constraints. DEFRA, and Univ. London, Wye, Kent.

Thirtle, C., J. Piesse, and V. Smith (ed). 1999. Agricultural R&D policy in the United Kingdom. *In* J. Alston et al. (ed) Paying for productivity: Financing R&D in the rich countries. Johns Hopkins Univ. Press, Baltimore MD.

Thompson, R.C.A. 2000. Molecular

epidemiolmogy: Applications to problems of infectious disease. p. 1-4. *In* R.C.A. Thompson (ed) Molecular epidemiology of infectious diseases. Arnold, London.

Thompson, R.C.A. 2001.The future impact of societal and cultural factors on parasitic disease: Some emerging issues. Int. J. Parasitol. 31:949-959.

Thorne, C.R., E.P. Evans, and E. Penning-Rowsell. 2006. Future flooding and coastal erosion risks. Thomas Telford, London.

Tietenberg, T. 2003. Environmental and natural resource economics. 6th ed. Addison Wesley, NY.

Tikkanen, I., P. Glück and B. Solberg (ed). 1997. Review of forest policy issues and policy processes. EFI Proc.12. EFI, Finland.

Tikkanen, I., P. Glück and H. Pajuoja (ed) 2002. Cross-sectoral policy impacts on forests: EFI Proc. 46. EFI, Finland.

Tracy, B.F., and D.B. Faulkner. 2006. Pasture and cattle responses in rotationally stocked grazing systems sown with differing levels of species richness. Crop Sci. 46: 2062-2068.

Trawick, P. 2006. Sustainable development as a "collective choice" problem: Theoretical and practical implications. Rep. SD14003. Defra, London, UK.

Trujillo, E., C. Davis and J. Milner. 2006. Nutrigenomics, proteomics, metabolomics, and the practice of dietetics. J. Am. Diet Assoc. 106:403-413.

Tsujimura, S., A. Wadano, K. Kano, and T. Ikeda. 2001. Photosynthetic bioelectrochemical cell utilizing cyanobacteria and water-generating oxidase. Enzyme and microbial technology 29:225-231.

Turner, R.K., R. Brouwer, and S. Georgiou. 2001. Ecosystems functions and the implications for economic valuation: Res. Rep. 441. English Nature, Peterborough, UK.

Tyler, S.R. 2006. Comanagement of natural resources: Local learning for poverty reduction. IDRC, Ottawa.

UK Food Group. 2005. EU competition rules and future developments from the perspective of farmers and small suppliers. UK Food Group, London.

UN SEEA (UN System Econ. Environ. Accounting). 2003. Integrated environmental and economic accounting. Available at http://unstats.un.org/unsd/envaccounting/seea.asp. UN Stat. Division, New York.

UNCTAD. 2006. Tracking the trend towards market concentration: The case of the agricultural input industry. Proc. UN Conf. Trade and Dev., Geneva. 20 Apr 2006. Available at http://www.unctad.org/en/docs/ditccom200516_en.pdf. UN, NY.

UNECA (United Nations Economic Commission for Africa). 2002. Harnessing technology for sustainable development. Available at http://www.uneca.org/harnessing/. UNECA, Addis Ababa.

UNECE-FAO. 2005. European forest sector outlook study 1960-2000-2020. UNECE-FAO, Geneva.

USHHS and USDA. 2005. Dietary guidelines for Americans. 6th ed. Available at http://www.healthierus.gov/dietaryguidelines/. USHHS and USDA, Washington DC.

USDA. 2007. National organic program in United States of America. Available at http://www.ams.usda.gov/nop/indexIE.htm. USDA, Washington, DC.

Van Crowder, L. 1997. Women in agricultural extension, education and communication service (SDRE). FAO, Geneva.

Van der Ploeg, J.A., and H. Renting. 2000. Impact and potential: A comparative review of European rural development practices. Sociol. Ruralis 40:4:529-543.

Van Ruijven, J., and F. Berendse. 2005. Diversity-productivity relationships: Initial effects, long-term patterns, and underlying mechanisms. PNAS 102:695-700.

Van Camp, L., B. Bujarrabal, A.R. Gentile, R.J.A. Jones, L. Montanarella, C. Olazabal et al. 2004. Reports of the technical working groups established under the thematic strategy for soil protection. Off. Official Publ. European Communities, Luxembourg.

Vandermeer, J. 1995. The ecological basis of alternative agriculture. Ann. Rev. Ecol. Syst. 26:201-224.

VanWey, L.K., C. Tucker, and E.D. McConnell. 2005. Community organization, migration, and remittances in Oaxaca. Latin Am. Res. Rev. 40(1):83-106.

Viana, V., H. Gholz, and R. Donovan (ed) 1996. Certification of forest products: Issues and perspectives. Island Press, Washington, DC.

Visser, B., D. Eaton, N. Louwaars and J. Engels. 2000. Transaction costs of germplasm exchange under bilateral agreements: Background Study Pap. 14. Available at ftp://ftp.fao.org/ag/cgrfa/BSP/bsp14e.pdf. FAO, Geneva.

Vorley, B., 2003. Food, Inc.: Corporate concentration from farm to consumer. UK Food Group, London.

Wackernagel, M., and W.E. Rees. 1995. Our ecological footprint: Reducing human impact on the Earth. New Society Publ., Philadelphia, PA.

Warheit, D.B. 2004. Nanoparticles: Health impacts? Materials Today 7:32-35.

Warschauer, M. 1999. Electronic literacies: Language, culture, and power in online education. Lawrence Erlbaum Assoc., Mahwah, NJ.

Waterland, R.A., and C. Garza. 1999. Potential mechanisms of metabolic imprinting that lead to chronic disease. Am. J. Clin. Nutr. 69:179-197.

Watson, R.T., M.C. Zinyowera, and R.H. Moss (ed) 1997. The regional impacts of climate change: An assessment of vulnerability. A special report of IPCC working group II. Cambridge Univ. Press, Cambridge.

WHO. 1998. The world health report 1998: Health futures: Life in the 21st century. WHO, Geneva.

WHO. 2003. Diet, nutrition and the prevention of chronic diseases. Tech. Rep. Ser. 916. WHO, Geneva.

WHO. 2005. Modern food biotechnology, human health and development: An evidence based study. WHO, Geneva.

Wiley, H.S. 2006. Systems biology: Beyond the buzz. The Scientist 2006:53-57.

Williams, A.G., E. Audsley and D.L. Sandars. 2006. Determining the environmental burdens and resource use in the production of agricultural and horticultural commodities: Main Rep. Defra Res. Project IS0205. Bedford, UK.

Windhab, E. 2006. Engineering concept for TMF. Lufost, Nantes, France.

Wolfensohn, J.D. and F. Bourguignon. 2004. Development and poverty reduction: Looking back, looking ahead. Prep. 2004 World Bank, IMF Meeting. Available at http://www.worldbank.org/ambc/lookingbacklookingahead.pdf. World Bank, Washington, DC.

World Bank. 2004a. The CGIAR at 31. World Bank, Washington DC.

World Bank. 2004b. IK (Indigenous Knowledge) Notes, 2004: Marking five years of the World Bank indigenous knowledge for development program. Washington, DC.

World Bank. 2007. Enhancing agricultural innovation: How to go beyond the strengthening of research systems. World Bank, Washington, DC.

World Bank. 2008. World development report: Agriculture for development. World Bank, Washington, DC.

Youdeowei, A. 2007. Building Africa's human and institutional capacity for the agricultural industry to meet its potential to contribute to the achievement of the MDGs. *In* Proc. FARA 4th Gen. Assembly. Johannesburg. 10-16 Jun 2007. FARA, Accra.

Young, J.G., B. Eskenazi, E.A. Gladstone, A. Bradman, L. Pedersen, C. Johnson et al. 2005. Association between in utero organophosphate pesticide exposure and abnormal reflexes in neonates. Neurotoxicology 26(2):199-209.

Young, V.R. 2002. Human nutrient requirements: The challenge of the post-genome era. J. Nutr. 132:621-629.

Zang, W.X. 2003. Nanoscale iron particles for environmental remediation: An overview. J. Nanoparticle Res. 5:323-332.

Annex A
NAE Authors and Review Editors

Canada
Guy Debailleul • Laval University
John M.R. Stone • Carleton University

Cyprus
Georges Eliades • Agricultural Research Institute (ARI)
Costas Gregoriou • Agricultural Research Institute (ARI)
Christoph Metochis • Agricultural Research Institute (ARI)

Czech Republic
Miloslava Navrátilová • State Phytosaniţary Administration

Finland
Riina Antikainen • Finnish Environment Institute
Henrik Bruun • Helsinki University of Technology
Helena Kahiluoto • MTT Agrifood Research
Jyrki Niemi • MTT Agrifood Research
Reimund Roetter • MTT Agrifood Research
Timo Sipiläinen • MTT Agrifood Research
Markku Yli-Halla • University of Helsinki

France
Loïc Antoine • IFREMER
Gilles Aumont • Institut National de la Recherche Agronomique (INRA)
Yves Birot • Institut National de la Recherche Agronomique (INRA)
Gérard Buttoud • Institut National de la Recherche Agronomique (INRA)
Bernard Chevassus • French Ministry of Agriculture and Fisheries
Béatrice Darcy-Vrillon • Institut National de la Recherche Agronomique (INRA)
Jean-François Dhôte • Institut National de la Recherche Agronomique (INRA)
Tilly Gaillard • Independent
Ghislain Gosse • Institut National de la Recherche Agronomique (INRA)
Jean-Marc Guehl • Institut National de la Recherche Agronomique (INRA)
Hugues de Jouvenel • Futuribles
Trish Kammili • Institut National de la Recherche Agronomique and FI4IAR
Véronique Lamblin • Futuribles
Marie de Lattre-Gasquet • CIRAD
Marianne Lefort • Institut National de la Recherche Agronomique and AgroParisTech
Jacques Loyat • French Ministry of Agriculture and Fisheries
Jean-Luc Peyron • GIP ECOFOR
Pierre Ricci • Institut National de la Recherche Agronomique (INRA)
Alain Ruellan • Agrocampus Rennes
Yves Savidan • AGROPOLIS
Bernard Seguin • Institut National de la Recherche Agronomique (INRA)
Andrée Sontot • Bureau de Ressources Genetiques

Germany
Tanja H. Schuler • Independent

Ireland
Denis Lucey • University College Cork – National University of Ireland

Italy
Maria Fonte • University of Naples
Francesco Vanni • Pisa University

Latvia
Rashal Isaak • University of Latvia

Netherlands
Willem A. Rienks • Wageningen University and Research Centre

Poland
Dariusz Jacek Szwed • Independent
Dorota Metera • IUCN – Poland

Russia
Sergey Alexanian • N.I. Vavilov Research Institute of Plant Industry

Slovakia
Pavol Bielek • Soil Science and Conservation Research Institute

Spain
Maria del Mar Delgado • University of Córdoba
Luciano Mateos • Instituto de Agricultura Sostenible, CSIC

Sweden
Susanne Johansson • Swedish University of Agricultural Sciences
Richard Langlais • Nordregio, Nordic Center for Spatial Devleopment
Veli-Matti Loiske • Södertörns University College
Fred Saunders • Södertörns University College

Ukraine

Yuriy Nesterov • Heifer International

United Kingdom

Michael Appleby • World Society for the Protection of Animals, London
Joanna Chataway • Open University
Janet Cotter • Greenpeace International, University of Exeter
Barbara Dinham • Pesticide Action Network
Les Firbank • North Wyke Research
Anil Graves • Cranfield University
Andrea Grundy • National Farmers' Union
Brian Johnson • Independent
Peter Lutman • Rothamsted Research
John Marsh • Independent
Mara Miele • Cardiff University
Selyf Morgan • Cardiff University
Joe Morris • Cranfield University
Gerard Porter • University of Edinburgh
Paresh Shah • London Higher
Joyce Tait • University of Edinburgh
K.J. Thomson • University of Aberdeen
Bill Vorley • International Institute for Environment and Development

United States

Molly D. Anderson • Food Systems Integrity
David Andow • University of Minnesota
Dave Bjorneberg • U.S. Department of Agriculture
Rodney Brown • Brigham Young University
Rebecca Burt • U.S. Department of Agriculture
Randy L. Davis • U.S. Department of Agriculture
Denis Ebodaghe • U.S. Department of Agriculture
Paul Guillebeau • University of Georgia
Mary Hendrickson • University of Missouri
William Heffernan • University of Missouri
Kenneth Hinga • U.S. Department of Agriculture
Uford Madden • Florida A&M University
Elizabeth Ransom • University of Richmond
Peter Reich • University of Minnesota
Michael Schechtman • U.S. Department of Agriculture
Leonid Sharashkin • Independent
Pai-Yei Whung • U.S. Department of Agriculture
Angus Wright • California State University, Sacramento

Annex B

Peer Reviewers

Benin

Peter Neuenschwander • IITA

Canada

Brad Fraleigh • Agriculture and Agri-Food Canada
Edward Gregorich • Agriculture and Agri-Food Canada
H. Henry Janzen • Agriculture and Agri-Food Canada
Robert MacGregor • Agriculture and Agri-Food Canada
Priyadarshini Mir • Agriculture and Agri-Food Canada
Radhey Pandeya • Agriculture and Agri-Food Canada

Denmark

Mette Stjernholm Meldgaard • IFOAM and The Danish Association for Organic Farming

Egypt

Malcolm Beveridge • WorldFish Center

Finland

Jukka Peltola • MTT
Riikka Rajalahti • Ministry of Foreign Affairs

Germany

Jan van Aken • Greenpeace International

Italy

Piero Morandini • University of Milan

Mexico

Armando Paredes • Consejo Coordinador Empresarial

Netherlands

Johan C. van Lenteren • IOBC Global

Russia

Eugenia Serova • IET, Center AFE

Sweden

Gunnela Gustafson • Swedish University of Agricultural Sciences

Switzerland

Alain Gaume • AGROSCOPE

United Kingdom

Michael Abberton • Institute for Grasslands and Environmental Research
UK Department of Environment, Food and Rural Affairs
UK Department for International Development
Emma Hennesey • DEFRA
Mervyn Humphreys • Institute for Grasslands and Environmental Research
Patrick Mulvaney • Practical Action National Family Farm Coalition
Patience Purdy • National Council of Women of Great Britain [NCW]
John Reynolds • Northeast BioFuels, Ltd.
Jerry Rider • TFK
Pete Riley • GM Freeze
Reyes Tirado • Greenpeace International
Stephanie Williamson • Pesticide Action Network, UK

United States

Patrick Avato • The World Bank
Philip L. Bereano • University of Washington
Charles Bertsch • US Department of Agriculture
Lynn Brown • The World Bank
Marilyn Buford • U.S. Forest Service
Tom Buis • National Farmers Union
Cheryl Christensen • US Department of Agriculture
Douglas Constance • Sam Houston State University
Tom Crow • US Forest Service
Kenneth A. Dahlberg • Western Michigan University
David Darr • US Forest Service
Mary Dix • US Department of Agriculture
Norman Ellstrand • University of California, Riverside
Karin Ferriter • US Patent and Trademark Office
Steven Finch • U.S. Department of Agriculture
Jaeda Harmon • Oxfam America
Gregory Jaffe • Center for Science in the Public Interest
Willem Janssen • World Bank
John Jefferies • US Department of Agriculture
Randy Johnson • US Forest Service
Susan Koehler • US Department of Agriculture
World Nieh • US Forest Service
Susan J. Owens • US Department of Agriculture
Margaret Reeves • Pesticide Action Network North America
Al Riebau • US Forest Service
Jill Roland • US Department of Agriculture
Marc Safley • US Department of Agriculture
Sara Scherr • Ecoagriculture Partners
Seth Shames • Ecoagriculture Partners
Jimmy Smith • World Bank
Doreen Stabinsky • College of the Atlantic
Kitisri Sukhapinda • US Patent and Trademark Office
Bea VanHorne • US Forest Service
Pai-Yei Whung • US Department of Agriculture

Annex C

Glossary

Agriculture A linked, dynamic social-ecological system based on the extraction of biological products and services from an ecosystem, innovated and managed by people. It thus includes cropping, animal husbandry, fishing, forestry, biofuel and bioproducts industries, and the production of pharmaceuticals or tissue for transplant in crops and livestock through genetic engineering. It encompasses all stages of production, processing, distribution, marketing, retail, consumption and waste disposal.

Agricultural biodiversity Encompasses the variety and variability of animals, plants and microorganisms necessary to sustain key functions of the agroecosystem, its structure and processes for, and in support of, food production and food security.

Agricultural extension Agricultural extension deals with the creation, transmission and application of knowledge and skills designed to bring desirable behavioral changes among people so that they improve their agricultural vocations and enterprises and, therefore, realize higher incomes and better standards of living.

Agricultural innovation Agricultural innovation is a socially constructed process. Innovation is the result of the interaction of a multitude of actors, agents and stakeholders within particular institutional contexts. If agricultural research and extension are important to agricultural innovation, so are markets, systems of government, relations along entire value chains, social norms, and, in general, a host of factors that create the incentives for a farmer to decide to change the way in which he or she works, and that reward or frustrate his or her decision.

Agricultural population The agricultural population is defined as all persons depending for their livelihood on agriculture, hunting, fishing or forestry. This estimate comprises all persons actively engaged in agriculture and their non-working dependants.

Agricultural subsidies Agricultural subsidies can take many forms, but a common feature is an economic transfer, often in direct cash form, from government to farmers. These transfers may aim to reduce the costs of production in the form of an input subsidy, e.g., for inorganic fertilizers or pesticides, or to make up the difference between the actual market price for farm output and a higher guaranteed price. Subsidies shield sectors or products from international competition.

Agricultural waste Farming wastes, including runoff and leaching of pesticides and fertilizers, erosion and dust from plowing, improper disposal of animal manure and carcasses, crop residues and debris.

Agroecological Zone A geographically delimited area with similar climatic and ecological characteristics suitable for specific agricultural uses.

Agroecology The science of applying ecological concepts and principles to the design and management of sustainable agroecosystems. It includes the study of the ecological processes in farming systems and processes such as nutrient cycling, carbon cycling/sequestration, water cycling, food chains within and between trophic groups (microbes to top predators), lifecycles, herbivore/predator/prey/host interactions, pollination, etc. Agroecological functions are generally maximized when there is high species diversity/perennial forest-like habitats.

Agroecosystem A biological and biophysical natural resource system managed by humans for the primary purpose of producing food as well as other socially valuable nonfood goods and environmental services. Agroecosystem function can be enhanced by increasing the planned biodiversity (mixed species and mosaics), which creates niches for unplanned biodiversity.

Agroforestry A dynamic, ecologically based, natural resources management system that through the integration of trees in farms and in the landscape diversifies and sustains production for increased social, economic and environmental benefits for land users at all levels. Agroforestry focuses on the wide range of work with trees grown on farms and in rural landscapes. Among these are fertilizer trees for land regeneration, soil health and food security; fruit trees for nutrition; fodder trees that improve smallholder livestock production; timber and fuelwood trees for shelter and energy; medicinal trees to combat disease; and trees that produce gums, resins or latex products. Many of these trees are multipurpose, providing a range of social, economic and environmental benefits.

AKST Agricultural Knowledge, Science and Technology (AKST) is a term encompassing the ways and means used to practice the different types of agricultural activities, and including both formal and informal knowledge and technology.

Alien Species A species occurring in an area outside of its historically known natural range as a result of intentional or accidental dispersal by human activities. Also referred to as introduced species or exotic species.

Aquaculture The farming of aquatic organisms in inland and coastal areas, involving intervention in the rearing process to enhance production and the individual or corporate ownership of the stock being cultivated. Aquaculture practiced in a marine environment is called mariculture.

Average Rate of Return Average rate of return takes the whole expenditure as given and calculates the rate of return to the global set of expenditures. It indicates whether or not the entire investment package was successful, but it does not indicate whether the allocation of resources between investment components was optimal.

Biodiversity The variability among living organisms from all sources including, inter alia, terrestrial, marine and other aquatic ecosystems and the ecological complexes of which they are part; including diversity within species and gene diversity among species, between species and of ecosystems.

Bioelectricity Electricity derived from the combustion of biomass, either directly or co-fired with fossil fuels such as coal and natural gas. Higher levels of conversion efficiency can be attained when biomass is gasified before combustion.

Bioenergy (biomass energy) Bioenergy is comprised of bioelectricity, bioheat and biofuels. Such energy carriers can be produced from energy crops (e.g., sugar cane, maize, oil palm), natural vegetation (e.g., woods, grasses) and organic wastes and residues (e.g., from forestry and agriculture). Bioenergy refers also to the direct combustion of biomass, mostly for heating and cooking purposes.

Biofuel Liquid fuels derived from biomass and predominantly used in transportation. The dominant biofuels are ethanol and biodiesel. Ethanol is produced by fermenting starch contained in plants such as sugar cane, sugar beet, maize, cassava, sweet sorghum or beetroot. Biodiesel is typically produced through a chemical process called trans-esterification, whereby oily biomass such as rapeseed, soybeans, palm oil, jatropha seeds, waste cooking oils or vegetable oils is combined with methanol to form methyl esters (sometimes called "fatty acid methyl ester" or FAME).

Bioheat Heat produced from the combustion of biomass, mostly as industrial process heat and heating for buildings.

Biological Control The use of living organisms as control agents for pests, (arthropods, nematodes mammals, weeds and pathogens) in agriculture. There are three types of biological control:

Conservation biocontrol: The protection and encouragement of local natural enemy populations by crop and habitat management measures that enhance their survival, efficiency and growth.

Augmentative biocontrol: The release of natural enemies into crops to suppress specific populations of pests over one or a few generations, often involving the mass production and regular release of natural enemies.

Classical biocontrol: The local introduction of new species of natural enemies with the intention that they establish and build populations that suppress particular pests, often introduced alien pests to which they are specific.

Biological Resources Include genetic resources, organisms or parts thereof, populations, or any other biotic component of ecosystems with actual or potential use or value for humanity.

Biotechnology The IAASTD definition of biotechnology is based on that in the Convention on Biological Diversity and the Cartagena Protocol on Biosafety. It is a broad term embracing the manipulation of living organisms and spans the large range of activities from conventional techniques for fermentation and plant and animal breeding to recent innovations in tissue culture, irradiation, genomics and marker-assisted breeding (MAB) or marker assisted selection (MAS) to augment natural breeding. Some of the latest biotechnologies, called "modern biotechnology", include the use of *in vitro* modified DNA or RNA and the fusion of cells from different taxonomic families, techniques that overcome natural physiological reproductive or recombination barriers.

Biosafety Referring to the avoidance of risk to human health and safety, and to the conservation of the environment, as a result of the use for research and commerce of infectious or genetically modified organisms.

Blue Water The water in rivers, lakes, reservoirs, ponds and aquifers. Dryland production only uses green water, while irrigated production uses blue water in addition to green water.

BLCAs Brokered Long-term Contractual Arrangements (BLCAs) are institutional arrangements often involving a farmer cooperative, or a private commercial, parastatal or a state trading enterprise and a package (inputs, services, credit, knowledge) that allows small-scale farmers to engage in the production of a marketable commodity, such as cocoa or other product that farmers cannot easily sell elsewhere.

Catchment An area that collects and drains rainwater.

Capacity Development Any action or process which assists individuals, groups, organizations and communities in strengthening or developing their resources.

Capture Fisheries The sum (or range) of all activities to harvest a given fish resource from the "wild". It may refer to the location (e.g., Morocco, Gearges Bank), the target resource (e.g., hake), the technology used (e.g., trawl or beach seine), the social characteristics (e.g., artisanal, industrial), the purpose (e.g., commercial, subsistence, or recreational) as well as the season (e.g., winter).

Carbon Sequestration The process that removes carbon dioxide from the atmosphere.

Cellulosic Ethanol Next generation biofuel that allows converting not only glucose but also cellulose and hemicellulose—the main building blocks of most biomass—into ethanol, usually using acid-based catalysis or enzyme-based reactions to break down plant fibers into sugar, which is then fermented into ethanol.

Climate Change Refers to a statistically significant variation in either the mean state of the climate or in its variability, persisting for an extended period (typically decades or longer). Climate change may be due to natural internal processes or external forcing, or to persistent anthropogenic changes in the composition of the atmosphere or in land use.

Clone A group of genetically identical cells or individuals that are all derived from one selected individual by vegetative propagation or by asexual reproduction, breeding of completely inbred organisms, or forming genetically identical organisms by nuclear transplantation.

Commercialization The process of increasing the share of income that is earned in cash (e.g., wage income, surplus production for marketing) and reducing the share that is earned in kind (e.g., growing food for consumption by the same household).

Cultivar A cultivated variety, a population of plants within a species of plant. Each cultivar or variety is genetically different.

Deforestation The action or process of changing forest land to non-forested land uses.

Degradation The result of processes that alter the ecological characteristics of terrestrial or aquatic (agro)ecosystems so that the net services that they provide are reduced. Continued degradation leads to zero or negative economic agricultural productivity.

For loss of *land* in quantitative or qualitative ways, the term *degradation* is used. For water resources rendered unavailable for agricultural and nonagricultural uses, we employ the terms *depletion* and *pollution*. *Soil* degradation refers to the processes that reduce the capacity of the soil to support agriculture.

Desertification Land degradation in drylands resulting from various factors, including climatic variations and human activities.

Domesticated or Cultivated Species Species in which the evolutionary process has been influenced by humans to meet their needs.

Domestication The process to accustom animals to live with people as well as to selectively cultivate plants or raise animals in order to increase their suitability and compatibility to human requirements.

Driver Any natural or human-induced factor that directly or indirectly causes a change in a system.

Driver, direct A driver that unequivocally influences ecosystem processes and can therefore be identified and measured to different degrees of accuracy.

Driver, endogenous A driver whose magnitude can be influenced by the decision-maker. The endogenous or exogenous characteristic of a driver depends on the organizational scale. Some drivers (e.g., prices) are exogenous to a decision-maker at one level (a farmer) but endogenous at other levels (the nation-state).

Driver, exogenous A driver that cannot be altered by the decision-maker.

Driver, indirect A driver that operates by altering the level or rate of change of one or more direct drivers.

Ecoagriculture A management approach that provides fair balance between production of food, feed, fuel, fiber, and biodiversity conservation or protection of the ecosystem.

Ecological Pest Management (EPM) A strategy to manage pests that focuses on strengthening the health and resilience of the entire agro-ecosystem. EPM relies on scientific advances in the ecological and entomological fields of population dynamics, community and landscape ecology, multi-trophic interactions, and plant and habitat diversity.

Economic Rate of Return The net benefits to all members of society as a percentage of cost, taking into account externalities and other market imperfections.

Ecosystem A dynamic complex of plant, animal, and microorganism communities and their nonliving environment interacting as a functional unit.

Ecosystem Approach A strategy for the integrated management of land, water, and living resources that promotes conservation and sustainable use in an equitable way.

An ecosystem approach is based on the application of appropriate scientific methodologies focused on levels of biological organization, which encompass the essential structure, processes, functions, and interactions among organisms and their environment. It recognizes that humans, with their cultural diversity, are an integral component and managers of many ecosystems.

Ecosystem Function An intrinsic ecosystem characteristic related to the set of conditions and processes whereby an ecosystem maintains its integrity (such as primary productivity, food chain biogeochemical cycles). Ecosystem functions include such processes as decomposition, production, pollination, predation, parasitism, nutrient cycling, and fluxes of nutrients and energy.

Ecosystem Management An approach to maintaining or restoring the composition, structure, function, and delivery of services of natural and modified ecosystems for the goal of achieving sustainability. It is based on an adaptive, collaboratively developed vision of desired future conditions that integrates ecological, socioeconomic, and institutional perspectives, applied within a geographic framework, and defined primarily by natural ecological boundaries.

Ecosystem Properties The size, biodiversity, stability, degree of organization, internal exchanges of material and energy among different pools, and other properties that characterize an ecosystem.

Ecosystem Services The benefits people obtain from ecosystems. These include provisioning services such as food and water; regulating services such as flood and disease control; cultural services such as spiritual, recreational, and cultural benefits; and supporting services such as nutrient cycling that maintain the conditions for life on Earth. The concept "ecosystem goods and services" is synonymous with ecosystem services.

Ecosystem Stability A description of the dynamic properties of an ecosystem. An ecosystem is considered stable if it returns to its original state shortly after a perturbation (resilience), exhibits low temporal variability (constancy), or does not change dramatically in the face of a perturbation (resistance).

Eutrophication Excessive enrichment of waters with nutrients, and the associated adverse biological effects.

Ex-ante The analysis of the effects of a policy or a project based only on information available before the policy or project is undertaken.

Ex-post The analysis of the effects of a policy or project based on information available after the policy or project has been implemented and its performance is observed.

Ex-situ Conservation The conservation of components of biological diversity outside their natural habitats.

Externalities Effects of a person's or firm's activities on others which are not compensated. Externalities can either hurt or benefit others—they can be negative or positive. One negative externality arises when a company pollutes the local environment to produce its goods and does not compensate the negatively affected local residents. Positive externalities can be produced through primary education—which benefits not only primary school students but also society at large. Governments can reduce negative externalities by regulating and taxing goods with negative externalities. Governments can increase positive externalities by subsidizing goods with positive externalities or by directly providing those goods.

Fallow Cropland left idle from harvest to planting or during the growing season.

Farmer-led Participatory Plant Breeding Researchers and/or development workers interact with farmer-controlled, managed and executed PPB activities, and build on farmers' own varietal development and seed systems.

Feminization The increase in the share of women in an activity, sector or process.

Fishery Generally, a fishery is an activity leading to harvesting of fish. It may involve capture of wild fish or the raising of fish through aquaculture.

Food Security Food security exists when all people of a given spatial unit, at all times, have physical and economic access to sufficient, safe and nutritious food to meet their dietary needs and food preferences for an active and healthy life, and that is obtained in a socially acceptable and ecologically sustainable manner.

Food Sovereignty The right of peoples and sovereign states to democratically determine their own agricultural and food policies.

Food System A food system encompasses the whole range of food production and consumption activities. The food system includes farm input supply, farm production, food processing, wholesale and retail distribution, marketing, and consumption.

Forestry The human utilization of a piece of forest for a certain purpose, such as timber or recreation.

Forest Systems Forest systems are lands dominated by trees; they are often used for timber, fuelwood, and non-wood forest products.

Gender Refers to the socially constructed roles and behaviors of, and relations between, men and women, as opposed to sex, which refers to biological differences. Societies assign specific entitlements, responsibilities and values to men and women of different social strata and subgroups.

Worldwide, systems of relation between men and women tend to disadvantage women, within the family as well as in public life. Like the hierarchical framework of a society, gender roles and relations vary according to context and are constantly subject to changes.

Genetic Engineering Modifying genotype, and hence phenotype, by transgenesis.

Genetic Material Any material of plant, animal, microbial or other origin containing functional units of heredity.

Genomics The research strategy that uses molecular characterization and cloning of whole genomes to understand the structure, function and evolution of genes and to answer fundamental biological questions.

Globalization Increasing interlinking of political, economic, institutional, social, cultural, technical, and ecological issues at the global level.

GMO (Genetically Modified Organism) An organism in which the genetic material has been altered anthropogenically by means of gene or cell technologies.

Governance The framework of social and economic systems and legal and political structures through which humanity manages itself. In general, governance comprises the traditions, institutions and processes that determine how power is exercised, how citizens are given a voice, and how decisions are made on issues of public concern.

Global Environmental Governance The global biosphere behaves as a single system, where the environmental impacts of each nation ultimately affect the whole. That makes a coordinated response from the community of nations a necessity for reversing today's environmental decline.

Global Warming Refers to an increase in the globally-averaged surface temperature in response to the increase of well-mixed greenhouse gases, particularly CO_2.

Global Warming Potential An index, describing the radiative characteristics of well-mixed greenhouse gases, that represents the combined effect of the differing times these gases remain in the atmosphere and their relative effectiveness in absorbing outgoing infrared radiation. This index approximates the time-integrated warming effect of a unit mass of a given greenhouse gas in today's atmosphere, relative to that of carbon dioxide.

Green Revolution An aggressive effort since 1950 in which agricultural researchers applied scientific principles of genetics and breeding to improve crops grown primarily in less-developed countries. The effort typically was accompanied by collateral investments to develop or strengthen the delivery of extension services, production inputs and markets and develop physical infrastructures such as roads and irrigation.

Green Water Green water refers to the water that comes from precipitation and is stored in unsaturated soil. Green water is typically taken up by plants as evapotranspiration.

Ground Water Water stored underground in rock crevices and in the pores of geologic materials that make up the Earth's crust. The upper surface of the saturate zone is called the water table.

Growth Rate The change (increase, decrease, or no change) in an indicator over a period of time, expressed as a percentage of the indicator at the start of the period. Growth rates contain several sets of information. The first is whether there is any change at all; the second is what direction the change is going in (increasing or decreasing); and the third is how rapidly that change is occurring.

Habitat Area occupied by and supporting living organisms. It is also used to mean the environmental attributes required by a particular species or its ecological niche.

Hazard A potentially damaging physical event, phenomenon and/or human activity, which my cause injury, property damage, social and economic disruption or environmental degradation.

Hazards can include latent conditions that may represent future threats and can have different origins.

Household All the persons, kin and non-kin, who live in the same or in a series of related dwellings and who share income, expenses and daily subsistence tasks. A basic unit for socio-cultural and economic analysis, a household may consist of persons (sometimes one but generally two or more) living together and jointly making provision for food or other essential elements of the livelihood.

Industrial Agriculture Form of agriculture that is capital-intensive, substituting machinery and purchased inputs for human and animal labor.

Infrastructure The facilities, structures, and associated equipment and services that facilitate the flows of goods and services between individuals, firms, and governments. It includes public utilities (electric power, telecommunications, water supply, sanitation and sewerage, and waste disposal); public works (irrigation systems, schools, housing, and hospitals); transport services (roads, railways, ports, waterways, and airports); and R&D facilities.

Innovation The use of a new idea, social process or institutional arrangement, material, or technology to change an activity, development, good, or service or the way goods and services are produced, distributed, or disposed of.

Innovation System Institutions, enterprises, and individuals that together demand and supply information and technology, and the rules and mechanisms by which these different agents interact.

In recent development discourse agricultural innovation is conceptualized as part and parcel of social and ecological organization, drawing on disciplinary evidence and understanding of how knowledge is generated and innovations occur.

In-situ Conservation The conservation of ecosystems and natural habitats and the maintenance and recovery of viable populations of species in their natural habitats and surroundings and, in the case of domesticated or cultivated species, in the surroundings where they have developed their distinctive properties and were managed by local groups of farmers, fishers or foresters.

Institutions The rules, norms and procedures that guide how people within societies live, work, and interact with each other. Formal institutions are written or codified rules, norms and procedures. Examples of formal institutions are the Constitution, the judiciary laws, the organized market, and property rights. Informal institutions are rules governed by social and behavioral norms of the society, family, or community. Cf. Organization.

Integrated Approaches Approaches that search for the best use of the functional relations among living organisms in relation to the environment without excluding the use of external inputs. Integrated approaches aim at the achievement of multiple goals (productivity increase, environmental sustainability and social welfare) using a variety of methods.

Integrated Assessment A method of analysis that combines results and models from the physical, biological, economic, and social sciences, and the interactions between these components in a consistent framework to evaluate the status and the consequences of environmental change and the policy responses to it.

Integrated Natural Resources Management (INRM) An approach that integrates research of different types of natural resources into stakeholder-driven processes of adaptive management and innovation to improve livelihoods, agroecosystem resilience, agricultural productivity and environmental services at community, eco-regional and global scales of intervention and impact. INRM thus aims to help to solve complex real-world problems affecting natural resources in agroecosystems.

Integrated Pest Management The procedure of integrating and applying practical management methods to manage insect populations so as to keep pest species from reaching damaging levels while avoiding or minimizing the potentially harmful effects of pest management measures on humans, non-target species, and the environment. IPM tends to incorporate assessment methods to guide management decisions.

Intellectual Property Rights (IPRs) Legal rights granted by governmental authorities to control and reward certain products of human intellectual effort and ingenuity.

Internal Rate of Return The discount rate that sets the net present value of the stream of the net benefits equal to zero. The internal rate of return may have multiple values when the stream of net benefits alternates from negative to positive more than once.

International Dollars Agricultural R&D investments in local currency units have been converted into international dollars by deflating the local currency amounts with each country's inflation ration (GDP deflator) of base year 2000. Next, they were converted to US dollars with a 2000 purchasing power parity (PPP) index. PPPs are synthetic exchange rates used to reflect the purchasing power of currencies.

Knowledge The way people understand the world, the way in which they interpret and apply meaning to their experiences. Knowledge is not about the discovery of some finale objective "truth" but about the grasping of subjective culturally-conditioned products emerging from complex and ongoing processes involving selection, rejection, creation, development and transformation of information. These processes, and hence knowledge, are inextricably linked to the social, environmental and institutional context within which they are found.

Scientific knowledge: Knowledge that has been legitimized and validated by a formalized process of data gathering, analysis and documentation.

Explicit knowledge: Information about knowledge that has been or can be articulated, codified, and stored and exchanged. The most common forms of explicit knowledge are manuals, documents, procedures, cultural artifacts and stories. The information about explicit knowledge also can be audiovisual. Works of art and product design can be seen as other forms of explicit knowledge where human skills, motives and knowledge are externalized.

Empirical knowledge: Knowledge derived from and constituted in interaction with a person's environment. Modern communication and information technologies, and scientific instrumentation, can extend the "empirical environment" in which empirical knowledge is generated.

Local knowledge: The knowledge that is constituted in a given culture or society.

Traditional (ecological) knowledge: The cumulative body of knowledge, practices, and beliefs evolved by adaptive processes and handed down through generations. It may not be indigenous or local, but it is distinguished by the way in which it is acquired and used, through the social process of learning and sharing knowledge.

Knowledge Management A systematic discipline of policies, processes, and activities for the management of all processes of knowledge generation, codification, application and sharing of information about knowledge.

Knowledge Society A society in which the production and dissemination of scientific information and knowledge function well, and in which the transmission and use of valuable experiential knowledge is optimized; a society in which the information of those with experiential knowledge is used together with that of scientific and technical experts to inform decision-making.

Land Cover The physical coverage of land, usually expressed in terms of vegetation cover or lack of it. Influenced by but non synonymous with land use.

Land Degradation The reduction in the capability of the land to produce benefits from a particular land use under a specific form of land management.

Landscape An area of land that contains a mosaic of ecosystems, including human-dominated ecosystems. The term cultural landscape is often used when referring to landscapes with characteristic form and uses, often traditional.

Land Tenure The relationship, whether legally or customarily defined, among people, as individuals or groups, with respect to land and associated natural resources (water, trees, minerals, wildlife, and so on).

Rules of tenure define how property rights in land are to be allocated within societies. Land tenure systems determine who can use what resources for how long, and under what conditions.

Land Use The human utilization of a piece of land for a certain purpose (such as irrigated agriculture or recreation). Land use is influenced by, but not synonymous with, land cover.

Leguminous Cultivated or spontaneous plants which fix atmospheric nitrogen.

Malnutrition Failure to achieve nutrient requirements, which can impair physical and/or mental health. It may result from consuming too little food or a shortage or imbalance of key nutrients (e.g., micronutrient deficiencies or excess consumption of refined sugar and fat).

Marginal Rates of Return Calculates the returns to the last dollar invested on a certain activity. It is usually estimated through econometric estimation.

Marker Assisted Selection (MAS) The use of DNA markers to improve response to selection in a population. The markers will be closely linked to one or more target loci, which may often be quantitative trait loci.

Minimum Tillage The least amount possible of cultivation or soil disturbance done to prepare a suitable seedbed. The main purposes of minimum tillage are to reduce tillage energy consumption, to conserve moisture, and to retain plant cover to minimize erosion.

Model A simplified representation of reality used to simulate a process, understand a situation, predict an outcome or analyze a problem. A model can be viewed as a selective approximation, which by elimination of incidental detail, allows hypothesized or quantified aspects of the real world to appear manipulated or tested.

Multifunctionality In IAASTD, multifunctionality is used solely to express the inescapable interconnectedness of agriculture's different roles and functions. The concept of multifunctionality recognizes agriculture as a multi-output activity producing not only commodities (food, feed, fibers, agrofuels, medicinal products and ornamentals), but also non-commodity outputs such as environmental services, landscape amenities and cultural heritages (See Global SDM Text Box)

Natural Resources Management Includes all functions and services of nature that are directly or indirectly significant to humankind, i.e., economic functions, as well as other cultural and ecological functions or social services that are not taken into account in economic models or not entirely known.

Nanotechnology The engineering of functional systems at the atomic or molecular scale.

Net Present Value (NPV) Net present value is used to analyze the profitability of an investment or project, representing the difference between the discounted present value of benefits and the discounted present value of costs. If NPV of a prospective project is positive, then the project should be accepted. The analysis of NPV is sensitive to the reliability of future cash inflows that an investment or project will yield.

No-Till Planting without tillage. In most systems, planter-mounted coulters till a narrow seedbed assisting in the placement of fertilizer and seed. The tillage effect on weed control is replaced by herbicide use.

Obesity A chronic physical condition characterized by too much body fat, which results in higher risk for health problems such as high blood pressure, high blood cholesterol, diabetes, heart disease and stroke. Commonly it is defined as a Body Mass Index (BMI) equal to or more than 30, while overweight is equal to or more than 25. The BMI is an index of weight-for-height and is defined as the weight in kilograms divided by the square of the height in meters (kg/m^2).

Organic Agriculture An ecological production management system that promotes and enhances biological cycles and soil biological activity. It is based on minimal use of off-farm inputs and on management practices that restore, maintain and enhance ecological harmony.

Organization Organizations can be formal or informal. Examples of organizations are government agencies (e.g., police force, ministries, etc.), administrative bodies (e.g., local government), non governmental organizations, associations (e.g., farmers' associations) and private companies (firms). Cf. with Institutions.

Orphan Crops Crops such as teff, finger millet, yam, roots and tubers that tend to be regionally or locally important for income and nutrition, but which are not traded globally and receive minimal attention by research networks.

Participatory Development A process that involves people (population groups, organizations, associations, political parties) actively and significantly in all decisions affecting their lives.

Participatory Domestication The process of domestication that involves agriculturalists and other community members actively and significantly in making decisions, taking action and sharing benefits.

Participatory Plant Breeding (PPB) Involvement of a range of actors, including scientists, farmers, consumers, extension agents, vendors, processors and other industry stakeholders—as well as farmer and community-based organizations and non-government organization (NGOs) in plant breeding research and development.

Participatory Varietal Selection (PVS) A process by which farmers and other stakeholders along the food chain are involved with researchers in the selection of varieties from formal and farmer-based collections and trials, to determine which are best suited to their own agroecosystems' needs, uses and preferences, and which should go ahead for finishing, wider release and dissemination. The information gathered may in turn be fed back into formal-led breeding programs.

Pesticide A toxic chemical or biological product that kills organisms (e.g., insecticides, fungicides, weedicides, rodenticides).

Poverty There are many definitions of poverty.

Absolute Poverty: According to a UN declaration that resulted from the World Summit on Social Development in 1995, absolute poverty is a condition characterized by severe deprivation of basic human needs, including food, safe drinking water, sanitation facilities, health, shelter, education and information. It depends not only on income but also on access to services.

Dimensions of Poverty: The individual and social characteristics of poverty such as lack of access to health and education, powerlessness or lack of dignity. Such aspects of deprivation experienced by the individual or group are not captured by measures of income or expenditure.

Extreme Poverty: Persons who fall below the defined poverty line of US$1 income per day. The measure is converted into local currencies using purchasing power parity (PPP) exchange rates. Other definitions of this concept have identified minimum subsistence requirements, the denial of basic human rights or the experience of exclusion.

Poverty Line: A minimum requirement of welfare, usually defined in relation to income or expenditure, used to identify the poor. Individuals or households with incomes or expenditure below the poverty line are poor. Those with incomes or expenditure equal to or above the line are not poor. It is common practice to draw more than one poverty line to distinguish different categories of poor, for example, the extreme poor.

Private Rate of Return The gain in net revenue to the private firm/business divided by the cost of an investment expressed in percentage.

Processes A series of actions, motions, occurrences, a method, mode, or operation, whereby a result or effect is produced.

Production Technology All methods that farmers, market agents and consumers use to cultivate, harvest, store, process, handle, transport and prepare food crops, cash crops, livestock, etc. for consumption.

Protected Area A geographically defined area which is designated or regulated and managed to achieve specific conservation objectives as defined by society.

Public Goods A good or service in which the benefit received by any one party does not diminish the availability of the benefits to others, and/or where access to the good cannot be restricted. Public goods have the properties of non-rivalry in consumption and non-excludability.

Public R&D Investment Includes R&D investments done by government agencies, nonprofit institutions, and higher-education agencies. It excludes the private for-profit enterprises.

Research and Development (R&D) Organizational strategies and methods used by research and extension program to conduct their work including scientific procedures, organizational modes, institutional strategies, interdisciplinary team research, etc.

Scenario A plausible and often simplified description of how the future may develop based on explicit and coherent and internally consistent set of assumptions about key driving forces (e.g., rate of technology change, prices) and relationships. Scenarios are neither predictions nor projections and sometimes may be based on a "narrative storyline". Scenarios may be derived from projections but are often based on additional information from other sources.

Science, Technology and Innovation Includes all forms of useful knowledge (codified and tacit) derived from diverse branches of learning and practice, ranging from basic scientific research to engineering to local knowledge. It also includes the policies used to promote scientific advance, technology development, and the commercialization of products, as well as the associated institutional innovations. *Science* refers to both basic and applied sciences. *Technology* refers to the application of science, engineering, and other fields, such as medicine. *Innovation* includes all of the processes, including business activities that bring a technology to market.

Shifting Cultivation Found mainly in the tropics, especially in humid and subhumid regions. There are different kinds; for example, in some cases a settlement is permanent, but certain fields are fallowed and cropped alternately ("rotational agriculture"). In other cases, new land is cleared when the old is no longer productive.

Slash and Burn Agriculture A pattern of agriculture in which existing vegetation is cleared and burned to provide space and nutrients for cropping.

Social Rate of Return The gain to society of a project or investment in net revenue divided by cost of the investment, expressed by percentage.

Soil and Water Conservation (SWC) A combination of appropriate technology and successful approach. Technologies promote the sustainable use of agricultural soils by minimizing soil erosion, maintaining and/or enhancing soil properties, managing water, and controlling temperature. Approaches explain the ways and means which are used to realize SWC in a given ecological and socio-economic environment.

Soil Erosion The detachment and movement of soil from the land surface by wind and water in conditions influenced by human activities.

Soil Function Any service, role, or task that a soil performs, especially: (a) sustaining biological activity, diversity, and productivity; (b) regulating and partitioning water and solute flow; (c) filtering, buffering, degrading, and detoxifying potential pollutants; (d) storing and cycling nutrients; (e) providing support for buildings and other structures and to protect archaeological treasures.

Staple Food (Crops) Food that is eaten as daily diet.

Soil Quality The capacity of a specific kind of soil to function, within natural or managed ecosystem boundaries, to sustain plant and animal productivity, maintain or enhance water and air quality, and support human health and habitation. In short, the capacity of the soil to function.

Subsidy Transfer of resources to an entity, which either reduces the operating costs or increases the revenues of such entity for the purpose of achieving some objective.

Subsistence Agriculture Agriculture carried out for the use of the individual person or their family with few or no outputs available for sale.

Sustainable Development Development that meets the needs of the present without compromising the ability of future generations to meet their own needs.

Sustainable Land Management (SLM) A system of technologies and/or planning that aims to integrate ecological with socio-economic and political principles in the management of land for agricultural and other purposes to achieve intra- and intergenerational equity.

Sustainable Use of Natural Resources Natural resource use is sustainable if specific types of use in a particular ecosystem are considered reasonable in the light of both the internal and the external perspective on natural resources. "Reasonable" in this context means that all actors agree that resource use fulfils productive, physical, and cultural functions in ways that will meet the long-term needs of the affected population.

Technology Transfer The broad set of deliberate and spontaneous processes that give rise to the exchange and dissemination of information and technologies among different stakeholders. As a generic concept, the term is used to encompass both diffusion of technologies and technological cooperation across and within countries.

Terms of Trade The *international terms* of trade measures a relationship between the prices of exports and the prices of imports, this being known strictly as the barter terms of trade. In this sense, deterioration in the terms of trade could have resulted if unit prices of exports had risen less than unit prices for imports. The *inter-sectoral terms of trade* refers to the terms of trade between sectors of the economy, e.g., rural & urban, agriculture and industry.

Total Factor Productivity A measure of the increase in total output which is not accounted for by increases in total inputs. The total factor productivity index is computed as the ratio of an index of aggregate output to an index of aggregate inputs.

Tradeoff Management choices that intentionally or otherwise change the type, magnitude, and relative mix of services provided by ecosystems.

Transgene An isolated gene sequence used to transform an organism. Often, but not always, the transgene has been derived from a different species than that of the recipient.

Transgenic An organism that has incorporated a functional foreign gene through recombinant DNA technology. The novel gene exists in all of its cells and is passed through to progeny.

Undernourishment Food intake that is continuously inadequate to meet dietary energy requirement.

Undernutrition The result of food intake that is insufficient to meet dietary energy requirements continuously, poor absorption, and/or poor biological use of nutrients consumed.

Urban and Peri-Urban Agriculture Agriculture occurring within and surrounding the boundaries of cities throughout the world and includes crop and livestock production, fisheries and forestry, as well as the ecological services they provide. Often multiple farming and gardening systems exist in and near a single city.

Value Chain A set of value-adding activities through which a product passes from the initial production or design stage to final delivery to the consumer.

Virtual Water The volume of water used to produce a commodity. The adjective "virtual" refers to the fact that most of the water used to produce a product is not contained in the product. In accounting virtual water flows we keep track of which parts of these flows refer to green, blue and grey water, respectively.

The real-water content of products is generally negligible if compared to the virtual-water content.

Waste Water "Grey" water that has been used in homes, agriculture, industries and businesses that is not for reuse unless it is treated.

Watershed The area which supplies water by surface and subsurface flow from precipitation to a given point in the drainage system.

Watershed Management Use, regulation and treatment of water and land resources of a watershed to accomplish stated objectives.

Water Productivity An efficiency term quantified as a ration of product output (goods and services) over water input.

Expressions of water productivity. Three major expressions of water productivity can be identified: 1) the amount of carbon gain per unit of water transpired by the leaf or by the canopy (photosynthetic water productivity); 2) the amount of water transpired by the crop (biomass water productivity); or 3) the yield obtained per unit amount of water transpired by the crop (yield water productivity).

Agricultural water productivity relates net benefits gained through the use of water in crop, forestry, fishery, livestock and mixed agricultural systems. In its broadest sense, it reflects the objectives of producing more food, income, livelihood and ecological benefits at less social and environmental cost per unit of water in agriculture.

Physical water productivity relates agricultural production to water use—more crop per drop. Water use is expressed either in terms of delivery to a use, or depletion by a use through evapotranspiration, pollution, or directing water to a sink where it cannot be reused. Improving physical water productivity is important to reduce future water needs in agriculture.

Economic water productivity relates the value of agricultural production to agricultural water use. A holistic assessment should account for the benefits and costs of water, including less tangible livelihood benefits, but this is rarely done. Improving economic water productivity is important for economic growth and poverty reduction.

Annex D
Acronyms, Abbreviations and Units

ACP African, Caribbean and Pacific
AIDS Acquired immune deficiency syndrome
AKST Agricultural knowledge, science and technology
ARI agricultural research institute
AST Agricultural science and technology
ATEAM Advanced Terrestrial Ecosystem Analysis and Modeling
AUV autonomous underwater vehicles
billion one thousand million
BSE Bovine spongiform encephalopathy
Bt soil bacterium *Bacillus thuringiensis* (usually refers to plants made insecticidal using a variant of various *cry* toxin genes sourced from plasmids of these bacteria)
C carbon
CAFTA Central America Free Trade Agreement
CAP Common Agricultural Policy
CBD Convention on Biological Diversity
CDM Clean Development Mechanism
CEC Commission of the European Community
CEE Central and Eastern Europe
CERN European Organization for Nuclear Research
CGE computable general equilibrium
CGIAR Consultative Group on International Agricultural Research
CH_4 methane
CIA US Central Intelligence Agency
CIAT International Center for Tropical Agriculture
CIFOR Center for International Forestry Research
CIMMYT International Maize and Wheat Improvement Center
CIP International Potato Center
CIS Commonwealth of Independent States
CO_2 carbon dioxide
COA certified organic agriculture
Codex Codex Alimentarius
CSFP Commonwealth Scholarship and Fellowship Plan
CSO civil society organization
CWANA Central and West Asia and North Africa
D&S development and sustainability
Defra UK Department of Environment, Food and Rural Affairs
DFID UK Department of International Development
DNA deoxyribonucleic acid
EBPM ecologically-based pest management
EC European Commission
EFMN European Foresight Monitoring Network
EFSA European Food Safety Authority
EJ Exajoules
EMBL European Molecular Biology Laboratory
EPA US Environmental Protection Agency
EPTA European Parliamentary Technology Assessment
ERS Economic Research Service of USDA
ESF European Science Foundation
ESAP East and South Asia and the Pacific
ETP European Technology Pleatform
EU European Union
EUFO European Futures Observatory
FAO Food and Agriculture Organization of the United Nations
FAPRI Food and Agricultural Policy Research Institute
FDA US Food and Drug Administration
FFS farmer field school
FLO Fair Trade Labeling Organization
FMD foot and mouth disease
FSR Farming systems research
FSRE Farming systems research and extension
FQPA US Food Quality Protection Act
FSC Forest Stewardship Council
g gram (10^{-3} kg)
GBA Global Biodiversity Assessment
GCM general circulation model
GDP Gross domestic product
GE genetic engineering/genetically engineered
GEF Global Environment Facility
GEO Global Environment Outlook
GFAR Global Forum for Agricultural Research
GFSI Global Food Safety Initiative

Gg gigagram (10^6 kg)
GHG greenhouse gas
GIS geographic information system
GLASOD Global assessment of human-induced soil degradation
GM genetically modified/genetic modification
GMO genetically modified organism
GNP Gross National Product
GSG Global Scenarios Group
Gt gigaton/gigatonne; 10^{19} tonnes
GTAP Global Trade Analysis Project
GURT Genetic Use Restriction Technologies
ha hectare (10^4 m^2)
HACCP Hazard Analysis Critical Control Point
HIV Human immunodeficiency virus
HR herbicide resistant
HT herbicide tolerant
HYV High yielding variety
IAASTD International Assessment of Agricultural Knowledge, Science and Technology for Development
IARC International Agricultural Research Center
ICARDA International Center for Agricultural Research in the Dry Areas
ICRAF World Agroforestry Center
ICRISAT International Crops Research Institute for Semi-arid Tropics
ICT information and communication technologies
IEA International Energy Agency
IFAD International Fund for Agricultural Development
IFC International Finance Corporation
IFI international financial institution
IFOAM International Federation of Organic Agriculture Movements
IFPRI International Food Policy Research Institute
IFS integrated farming systems
IIASA International Institute for Applied System Analysis
IITA International Institute for Tropical Agriculture
IK Indigenous knowledge
ILO International Labour Organisation
ILRI International Livestock Research Institute
IMF International Monetary Fund
INM Integrated Nutrient Management
INRA Institut National de la Recherche Agronomique (France)
INRM Integrated Natural Resources Management
IP intellectual property
IPCC Intergovernmental Panel on Climate Change
IPGRI Bioversity International
IPM Integrated pest management
IPNS Integrated plant nutrient systems
IPPC International Plant Protection Convention
IPR intellectual property rights
IPTS Institute for Prospective Technological Studies
IR insect resistant
IRR internal rate of return
IRRI International Rice Research Institute
IS innovation systems
ISNM Integrated soil and nutrient management
ISO International Organization for Standardization
ISPM International sanitary and phytosanitary measure
ITU International Telecommunications Union
IWM Integrated Weed Management
IWMI International Water Management Institute
IWRM Integrated water resources management
K potassium
kcal kilocalorie
kg kilogram, 10^3 grams
km kilometer
kWh kilowatt hour
LAC Latin America and the Caribbean
LDC least developed countries
LEISA Low-External Input Sustainable Agriculture
LIC low income country
LTSP Long-Term Soil Productivity
LUC land use change
m 10^2 cm
MA Millennium Ecosystem Assessment
MAB/S marker assisted breeding/selection
MDG Millennium Development Goals
Mg magnesium
mg milligram (10^{-3} grams)
MIGA Multilateral Investment Agency
MJ megajoule
MNC multinational corporation
MNP Netherlands Environmental Assessment Agency
MRL maximum residue level
MSA mean species abundance
MTA material transfer agreement
MV Modern variety
N nitrogen
NA North America
NAE North America and Europe
NAFTA North American Free Trade Agreement
NARS national agricultural research systems
NBF National Biosafety Frameworks
NFMA US National Forest Management Act
NFP national forest program
ng nanogram (10^{-9} grams)

NGO nongovernmental organization
NIAS US National Institute for Agricultural Security
N_2O nitrous oxide
NPK nitrogen, phosphorus, potassium
NRM Natural resource management
NTFP non-timber forest product
OA organic agriculture
OECD Organization of Economic Cooperation and Development
OF organic food/farming
OIE World Organization for Animal Health
P phosphorus
PBR plant breeders rights
PCA Partnership and Cooperation Agreement
PE partial equilibrium
PEFC Program of Endorsement of Forest Certification
PES Payments for environmental services
PGRFA Plant Genetic Resources for Food and Agriculture
PIPRA Public-Sector Intellectual Property Resource for Agriculture
ppm parts per million
ppmv parts per million by volume
PPA participatory poverty assessment
PPP Purchasing Power Parity
PRA participatory rural appraisal
R&D research and development
rBST recombinant bovine somatotropin
RELU Rural Economy and Land Use Programme
RFID radio frequency identification
RNA ribonucleic acid
RNAi RNA interference
ROR rates of return
RRA rapid rural appraisal
RTO Research and technology organization
S&E science and engineering
S&T science and technology
SEPA Scottish Environment Protection Agency
SFP Single farm payment
SME small or medium size enterprise
SPIA Standing Panel on Impact Assessment
SPLT Substantive Patent Law Treaty
SPS Sanitary and Phytosanitary
SRES Special Report on Emission Scenarios
SRL Sustainable Rural Livelihoods
SSA Sub-Saharan Africa
TAC Technical Advisory Committee of the CGIAR
TFP Total Factor Productivity
Tg teragram, unit of mass equal to one megatonne
TNC transnational corporation
tonne 10^3 kg (metric ton)
TRIPS Trade-Related Aspects of Intellectual Property Rights
TURF Territorial Use Rights in Fisheries
TWAS Third World Academy of Scientists
UN United Nations
UNCBD UN Convention on Biodiversity
UNCCD United Nations Convention to Combat Desertification
UNCED UN Conference on Environment and Development
UNCTAD UN Conference on Trade and Development
UNDP United Nations Development Program
UNEP United Nations Environment Programme
UNESCO United Nations Educational, Scientific and Cultural Organization
UNFCCC United Nations Framework Convention on Climate Change
UNICEF United Nations Children's Fund
UPOV International Union for the Protection of New Varieties of Plants
US United States
USDA US Department of Agriculture
vCJD variant Creutzfeldt-Jakob Disease
WARDA Africa Rice Center
WHO World Health Organization
WIPO World Intellectual Property Organization
WMO World Meteorological Organization
WRI World Resources Institute
WSSD World Summit on Sustainable Development
WTO World Trade Organization
WWF World Wildlife Fund
yr year
μg microgram

Annex E
Secretariat and Cosponsor Focal Points

Secretariat

World Bank
Marianne Cabraal, Leonila Castillo, Jodi Horton, Betsi Isay, Pekka Jamsen, Pedro Marques, Beverly McIntyre, Wubi Mekonnen, June Remy

UNEP
Marcus Lee, Nalini Sharma, Anna Stabrawa

UNESCO
Guillen Calvo

With special thanks to the Publications team: Audrey Ringler (logo design), Pedro Marques (proofing and graphics), Ketill Berger and Eric Fuller (graphic design)

Regional Institutes

Sub-Saharan Africa—African Centre for Technology Studies (ACTS)
Ronald Ajengo, Elvin Nyukuri, Judi Wakhungu

Central and West Asia and North Africa – International Center for Agricultural Research in the Dry Areas (ICARDA)
Mustapha Guellouz, Lamis Makhoul, Caroline Msrieh-Seropian, Ahmed Sidahmed, Cathy Farnworth

Latin America and the Caribbean – Inter-American Institute for Cooperation on Agriculture (IICA)
Enrique Alarcon, Jorge Ardila Vásquez, Viviana Chacon, Johana Rodríguez, Gustavo Sain

East and South Asia and the Pacific – WorldFish Center
Karen Khoo, Siew Hua Koh, Li Ping Ng, Jamie Oliver, Prem Chandran Venugopalan

Cosponsor Focal Points

GEF	Mark Zimsky
UNDP	Philip Dobie
UNEP	Ivar Baste
UNESCO	Salvatore Arico, Walter Erdelen
WHO	Jorgen Schlundt
World Bank	Mark Cackler, Kevin Cleaver, Eija Pehu, Juergen Voegele

Annex F
Steering Committee for Consultative Process and Advisory Bureau for Assessment

Steering Committee

The Steering Committee was established to oversee the consultative process and recommend whether an international assessment was needed, and if so, what were the goals, the scope, the expected outputs and outcomes, governance and management structure, location of the secretariat and funding strategy.

Co-chairs

Louise Fresco, Assistant Director General for Agriculture, FAO
Seyfu Ketema, Executive Secretary, Association for Strengthening Agricultural Research in East and Central Africa (ASARECA)
Claudia Martinez Zuleta, Former Deputy Minister of the Environment, Colombia
Rita Sharma, Principal Secretary and Rural Infrastructure Commissioner, Government of Uttar Pradesh, India
Robert T. Watson, Chief Scientist, The World Bank

Nongovernmental Organizations

Benny Haerlin, Advisor, Greenpeace International
Marcia Ishii-Eiteman, Senior Scientist, Pesticide Action Network North America Regional Center (PANNA)
Monica Kapiriri, Regional Program Officer for NGO Enhancement and Rural Development, Aga Khan
Raymond C. Offenheiser, President, Oxfam America
Daniel Rodriguez, International Technology Development Group (ITDG), Latin America Regional Office, Peru

UN Bodies

Ivar Baste, Chief, Environment Assessment Branch, UN Environment Programme
Wim van Eck, Senior Advisor, Sustainable Development and Healthy Environments, World Health Organization
Joke Waller-Hunter, Executive Secretary, UN Framework Convention on Climate Change
Hamdallah Zedan, Executive Secretary, UN Convention on Biological Diversity

At-large Scientists

Adrienne Clarke, Laureate Professor, School of Botany, University of Melbourne, Australia
Denis Lucey, Professor of Food Economics, Dept. of Food Business & Development, University College Cork, Ireland, and Vice-President NATURA
Vo-tong Xuan, Rector, Angiang University, Vietnam

Private Sector

Momtaz Faruki Chowdhury, Director, Agribusiness Center for Competitiveness and Enterprise Development, Bangladesh
Sam Dryden, Managing Director, Emergent Genetics
David Evans, Former Head of Research and Technology, Syngenta International
Steve Parry, Sustainable Agriculture Research and Development Program Leader, Unilever
Mumeka M. Wright, Director, Bimzi Ltd., Zambia

Consumer Groups

Michael Hansen, Consumers International
Greg Jaffe, Director, Biotechnology Project, Center for Science in the Public Interest
Samuel Ochieng, Chief Executive, Consumer Information Network

Producer Groups

Mercy Karanja, Chief Executive Officer, Kenya National Farmers' Union
Prabha Mahale, World Board, International Federation Organic Agriculture Movements (IFOAM)
Tsakani Ngomane, Director Agricultural Extension Services, Department of Agriculture, Limpopo Province, Republic of South Africa
Armando Paredes, Presidente, Consejo Nacional Agropecuario (CNA)

Scientific Organizations

Jorge Ardila Vásquez, Director Area of Technology and Innovation, Inter-American Institute for Cooperation on Agriculture (IICA)
Samuel Bruce-Oliver, NARS Senior Fellow, Global Forum for Agricultural Research Secretariat
Adel El-Beltagy, Chair, Center Directors Committee, Consultative Group on International Agricultural Research (CGIAR)
Carl Greenidge, Director, Center for Rural and Technical Cooperation, Netherlands
Mohamed Hassan, Executive Director, Third World Academy of Sciences (TWAS)
Mark Holderness, Head Crop and Pest Management, CAB International
Charlotte Johnson-Welch, Public Health and Gender Specialist and Nata Duvvury, Director Social Conflict and Transformation Team, International Center for Research on Women (ICRW)
Thomas Rosswall, Executive Director, International Council for Science (ICSU)
Judi Wakhungu, Executive Director, African Center for Technology Studies

Governments

Australia: Peter Core, Director, Australian Centre for International Agricultural Research

China: Keming Qian, Director General Inst. Agricultural Economics, Dept. of International Cooperation, Chinese Academy of Agricultural Science

Finland: Tiina Huvio, Senior Advisor, Agriculture and Rural Development, Ministry of Foreign Affairs

France: Alain Derevier, Senior Advisor, Research for Sustainable Development, Ministry of Foreign Affairs

Germany: Hans-Jochen de Haas, Head, Agricultural and Rural Development, Federal Ministry of Economic Cooperation and Development (BMZ)

Hungary: Zoltan Bedo, Director, Agricultural Research Institute, Hungarian Academy of Sciences

Ireland: Aidan O'Driscoll, Assistant Secretary General, Department of Agriculture and Food

Morocco: Hamid Narjisse, Director General, INRA

Russia: Eugenia Serova, Head, Agrarian Policy Division, Institute for Economy in Transition

Uganda: Grace Akello, Minister of State for Northern Uganda Rehabilitation

United Kingdom Paul Spray, Head of Research, DFID

United States: Rodney Brown, Deputy Under Secretary of Agriculture and Hans Klemm, Director of the Office of Agriculture, Biotechnology and Textile Trade Affairs, Department of State

Foundations and Unions

Susan Sechler, Senior Advisor on Biotechnology Policy, Rockefeller Foundation

Achim Steiner, Director General, The World Conservation Union (IUCN)

Eugene Terry, Director, African Agricultural Technology Foundation

Advisory Bureau

Non-government Representatives

Consumer Groups

Jaime Delgado • Asociación Peruana de Consumidores y Usuarios
Greg Jaffe • Center for Science in the Public Interest
Catherine Rutivi • Consumers International
Indrani Thuraisingham • Southeast Asia Council for Food Security and Trade
Jose Vargas Niello • Consumers International Chile

International organizations

Nata Duvvury • International Center for Research on Women
Emile Frison • CGIAR
Mohamed Hassan • Third World Academy of Sciences
Jeffrey McNeely • World Conservation Union (IUCN)
Dennis Rangi • CAB International
John Stewart • International Council of Science (ICSU)
Mark Holderness • Global Forum on Agricultural Research

NGOs

Kevin Akoyi • Vredeseilanden
Hedia Baccar • Association pour la Protection de l'Environment de Kairouan
Benedikt Haerlin • Greenpeace International
Juan Lopez • Friends of the Earth International
Khadouja Mellouli • Women for Sustainable Development
Patrick Mulvaney • Practical Action
Romeo Quihano • Pesticide Action Network
Maryam Rahmaniam • CENESTA
Daniel Rodriguez • International Technology Development Group

Private Sector

Momtaz Chowdhury • Agrobased Technology and Industry Development
Giselle L. D'Almeida • Interface
Eva Maria Erisgen • BASF
Armando Paredes • Consejo Nacional Agropecuario
Steve Parry • Unilever
Harry Swaine • Syngenta (resigned)

Producer Groups

Shoaib Aziz • Sustainable Agriculture Action Group of Pakistan
Philip Kiriro • East African Farmers Federation
Kristie Knoll • Knoll Farms
Prabha Mahale • International Federation of Organic Agriculture Movements
Anita Morales • Apit Tako
Nizam Selim • Pioneer Hatchery

Government Representatives

Central and West Asia and North Africa

Egypt • Ahlam Al Naggar
Iran • Hossein Askari
Kyrgyz Republic • Djamin Akimaliev
Saudi Arabia • Abdu Al Assiri, Taqi Elldeen Adar, Khalid Al Ghamedi
Turkey • Yalcin Kaya, Mesut Keser

East and South Asia and the Pacific

Australia • Simon Hearn
China • Puyun Yang
India • P.K. Joshi
Japan • Ryuko Inoue
Philippines • William Medrano

Latin America and Caribbean

Brazil • Sebastiao Barbosa, Alexandre Cardoso, Paulo Roberto Galerani, Rubens Nodari
Dominican Republic • Rafael Perez Duvergé
Honduras • Arturo Galo, Roberto Villeda Toledo
Uruguay • Mario Allegri

North America and Europe

Austria • Hedwig Woegerbauer
Canada • Iain MacGillivray
Finland • Marja-Liisa Tapio-Bistrom
France • Michel Dodet
Ireland • Aidan O'Driscoll, Tony Smith
Russia • Eugenia Serova, Sergey Alexanian
United Kingdom • Jim Harvey, David Howlett, John Barret
United States • Christian Foster

Sub-Saharan Africa

Benin • Jean Claude Codjia
Gambia • Sulayman Trawally
Kenya • Evans Mwangi
Mozambique • Alsácia Atanásio, Júlio Mchola
Namibia • Gillian Maggs-Kölling
Senegal • Ibrahim Diouck

Annex G

Reservations on NAE Report

Canada: In recognizing the important and significant work undertaken by IAASTD authors, Secretariat and stakeholders on the background Reports, the Canadian Government notes these documents as a valuable and important contribution to policy debate which needs to continue in national and international processes. While acknowledging the valuable contribution these Reports provide to our understanding on agricultural knowledge, science and technology for development, there remain numerous areas of concern in terms of balanced presentation, policy suggestions and other assertions and ambiguities. Nonetheless, the Canadian Government advocates these reports be drawn to the attention of governments for consideration in addressing the importance of AKST and its large potential to contribute to economic growth and the reduction of hunger and poverty.

United States of America: The United States joins consensus with other governments in the critical importance of AKST to meet the goals of the IAASTD. We commend the tireless efforts of the authors, editors, Co-Chairs and the Secretariat. We welcome the IAASTD for bringing together the widest array of stakeholders for the first time in an initiative of this magnitude. We respect the wide diversity of views and healthy debate that took place.

As we have specific and substantive concerns in each of the reports, the United States is unable to provide unqualified endorsement of the reports, and we have noted them.

The United States believes the Assessment has potential for stimulating further deliberation and research. Further, we acknowledge the reports are a useful contribution for consideration by governments of the role of AKST in raising sustainable economic growth and alleviating hunger and poverty.

Annex H
Additional NAE Figures and Tables

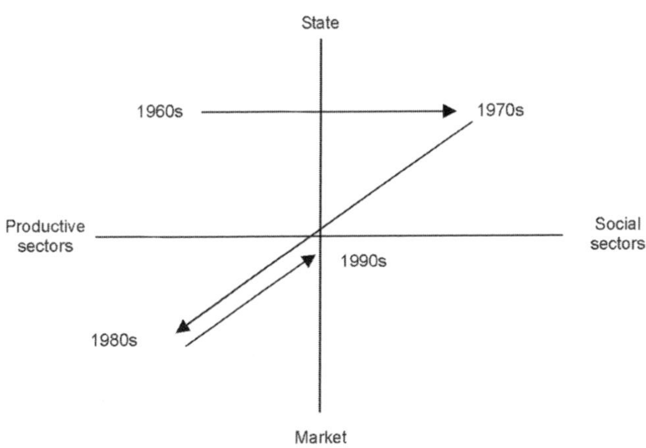

Figure 4-2.

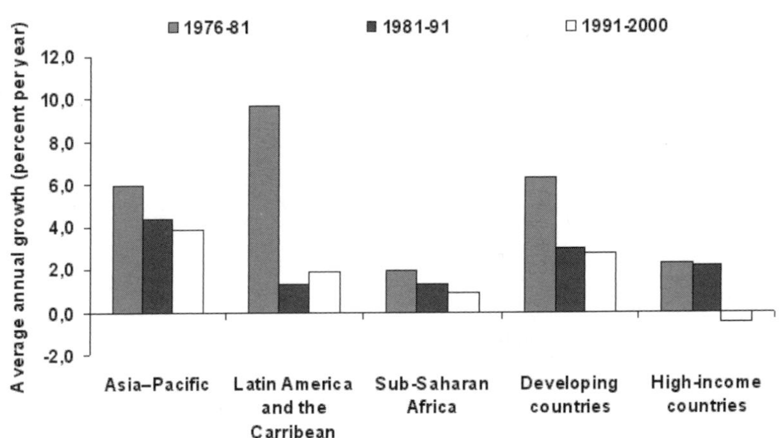

Figure 4-3.

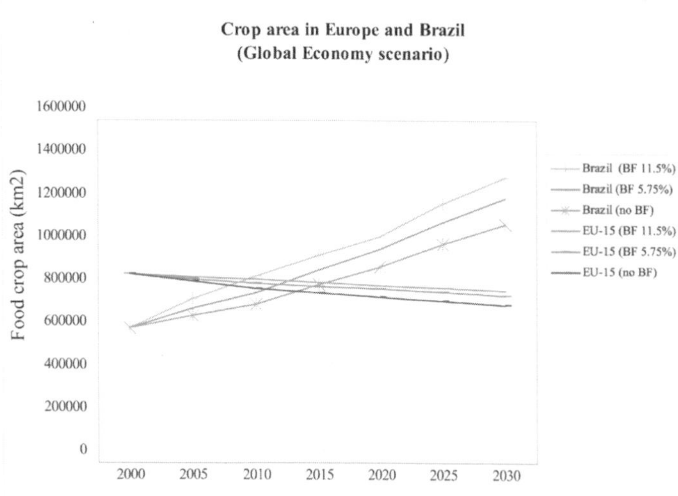

Graphic for Box 5.1

Table 4-4. **Total public and private agricultural R&D expenditures in selected OECD countries 1981 to 2000. Source: Pardey et al., 2006**

	Private share of total (percent)			(percent per year)	
Country	1981	1991	2000	1981-91	1991-2000
Australia	5.9	22.0	24.8	15.3	4.0
Canada	17.3	21.5	34.0	2.5	5.5
France	44.1	52.0	74.7	8.2	2.7
Germany	56.2	43.6	53.6	2.4	0.7
Japan	36.6	48.4	58.6	7.5	1.8
The Netherlands	44.8	56.1	57.7	9.3	1.1
United Kingdom	55.9	66.8	71.5	6.0	1.7
United States	49.3	51.0	51.5	3.6	2.4
OECD total (22)	43.6	48.5	54.3	5.2	2.1

Note: 2000 international dollars.
Average annual growth rates calculated using the least-squares regression method, as described by the World Bank, 2006). In 1981, private sector agricultural R&D spending was estimated to be $6,422 million, $9,930 million in 1991, and $12, 086 million in 2000.

Table 4-6. **Aid to agriculture 1970 to 2004. Source: Pardey et al., 2006, adapted from Alston, Dehmer and Pardey, 2006.**

		Bilateral aid	
Year	**Total official development assistance (ODA) (million 2000 U.S. dollars)**	**Amount (million 2000 U.S. dollars)**	**Share to agriculture (percent)**
1970	24, 719	20, 886	4.91
1975	35, 448	26, 233	11.13
1980	49, 166	31, 875	16.63
1985	41, 773	30, 782	15.93
1990	67, 071	47, 540	11.39
1995	64, 077	44, 129	9.82
2000	53, 749	36, 064	6.36
2003	65, 502	47, 222	4.22
2004	74, 483 [a]	50, 700[a]	n/a

Note: n/a indicates not available
[a] preliminary estimate

Index

Page references followed by *f*, *t*, and *b* indicate figures, tables, and boxes, respectively.

A

ABARE (Australian Bureau of Agricultural and Resource Economics), 155
Access, control and distribution of AKST, 196–98
Advanced Terrestrial Ecosystem Analysis and Modeling (ATEAM), 190
AEA (agroecosystem analysis), 124
African Union (AU), 255
Agreements. *See* International treaties and agreements
Agricultural industry
 bipolarisation of demand, *224b*
 and climate change, 186–88
 direct drivers for, 4, 172–92, *177t, 183f, 184ff, 187f*
 environmental consequences of changes in, 90–92, *91f*
 food consumption and distribution trends and consequences, 172–75
 and natural resources, 180–86, *183f, 184f*
 vulnerability to climate change, 215–16
 See also Food and farming; Seed industry; Specialization of agriculture
Agricultural knowledge, science and technology. *See* AKST
Agricultural knowledge systems, 248–51, *252b*
Agricultural policies
 post-World War II, 15, 23, 30–34
 responding to climate change, 214–16, *215b*
 trends and uncertainties, 175–77, *177t*
Agricultural research. *See* Research and development
Agricultural Research for Development (ARD), 257–58
Agricultural Research Institutions (ARIs), 134, 258
Agrifood systems
 actors in, 2
 AKST options for improving, 210–11
 alternative systems, 104
 and application of AKST, 9
 development of, 14–16
 overview, 14–16, 22–30, *23f, 23t, 24f, 25f, 26t*
 post-production sectors, 15
 public control of, 141–44
 and public policy, 16
 quality turn, 2–3, 8
 reducing detrimental effects of, 209
 sustainability as goal, 8
 wageworkers vs. management incomes, 7
 See also Food and farming
Agroecosystem analysis (AEA), 124
Agroecosystems, 16, 127, 180, 210, 226–27, 229, 235
AKST (agricultural knowledge, science and technology)
 agenda and funding, 2, 133–38, *133f, 134f, 135f, 136t*
 agriculture R&D funding, 264–67, *266t, 267b*
 and aquaculture, 61
 challenges for, 16–17
 current application of, 2–3, 8–10
 direct drivers for agriculture, 4, 172–92, *177t, 183f, 184ff, 187f*
 direct drivers for KST and, 165–72, *171t*
 economic impacts of agriculture and, 92–99, *93f, 94ff, 94t, 95f, 96f, 97t, 98ff, 99f, 100t*
 enablers of, 266–67, *267b*
 environmental impacts of agriculture and, 81–92, *83f, 83t, 84t, 89f, 89t, 91f*
 in forestry, 55
 forms and permutations, 4–5
 future organization and funding, 192–95, 211
 future systems and sustainable development goals, 198–201
 IAASTD reports' focus on, 3–5, *3f, 4f, 5f*
 indirect drivers, 161–65, *163t, 164tt*
 integration of KST and, 117, 126, 138, 153, 192–96, 197, 200
 integration of perspectives, 117, 126–30, *127b, 128t*, 197, 211
 interactions between NAE and other regions, 254
 interdisciplinarity, 248–51, *249b*
 key drivers for AKST and research/innovation systems, 192–98
 lessons learned, 118–20
 and livestock productivity, 50–52
 options for funding, 212
 and pesticides, 45–46
 social impacts of agriculture and, 99–104, *100t, 101f, 101t, 102t, 105tt, 106f*
 structures, funding and agenda development, 119, 130–42, *131t, 133f, 134f, 135f, 136t*
 See also Future for KST, agriculture, and AKST; Innovative models and systems
Algal/cyanobacterial biofuels, 218–20, *220t*
Alternative resource management strategies, 236
Animal production, environmental consequences of changes in, 87–88
Animal welfare issues
 desirability of high standards, 15–16
 EU standards and compliance requirements, 32, 52
 intensive livestock production, 48, 52–53, 69, 88, 100–101
 vegetarian movement in response to, 15
Antibiotic usage
 in aquaculture, 89, *89f*
 in livestock production, 53, 80, 88, 101, 172
Aquaculture systems
 antibiotic usage, 89, *89f*
 development and sustainability, 186, 210, 240–41
 environmental impacts of, 80, 88–90, *89f, 89t*
 fish farming techniques, 57–61, *57f, 58ff, 59f*
 fish meal from coastal capture fisheries, 240–41
 overview, 57–61, *57f, 58ff, 59f*
 sustainability improvements, 210
 uncertainties and consequences for AKST, 185–86
 See also Fisheries
ARD (Agricultural Research for Development), 257–58
ARIs (Agricultural Research Institutions), 134, 258
ATEAM (Advanced Terrestrial Ecosystem Analysis and Modeling), 190
AU (African Union), 255
Australian Bureau of Agricultural and Resource Economics (ABARE), 155

B

Beneficiaries, influence of, 138–139, 197
BIFS (Biologically Integrated Farming Systems), *253b*
Bilateral agreements, 144
Bimodal distribution of farms in US, 23–24
Bioassays, 170
Biodiesel, 218*n*
Biodiversity
 access to seed varieties, 213–14, *233b*
 and AKST, 228–30
 of crops in NAE countries, 2
 development of new crops, 216
 and ecosystem approach, 124, *125b*, 226
 foresight exercises on, 155–56
 in forests, 54–55, 90, 235, *235t*, 238, 239
 future of, 185, 238
 loss of, 8, 15, 52, 57, 81–82, 91–92
 mechanization vs., 80
 and plant variety protection, 42, 144, 195–96

Biodiversity (*continued*)
as public good not recognized by markets, 222
research on, 56, 155–56, 158, 210, 265, 266, *266t*
strengthening breeding activities, 210
and sustainability, 124
and transhumance, 13
and water law, *260t*
of weeds, 217, 218
See also Environmental quality
Bioenergy trends and issues, 191–92
Bioethanol, 218*n*
Biofuels
biomass and policy changes, 22
biomass feedstock used for producing, 191, 192
cellulosic, 191, 192, 218–20, 219–20
diversion of food production, 10, 13, 219
environmental consequences of growing crops for, 80, 87
and Eururalis foresight exercise, 157, 158, *158b, 159t*
and global carbon balance, 220–21
and innovations in AKST, 209
overview, 47, 218–20
See also Biomass/bioenergy crops
Biohydrogen, 220
Biologically Integrated Farming Systems (BIFS), *253b*
Biology of animals and plants
knowledge as intellectual property of TNCs, 10, 21, 194
plant breeding, seeds and genetics, 40–43, *42f*
productivity increases from, 21
See also Genetic engineering
Biomass/bioenergy crops
from agriculture and forestry, 13, 183, 238
as animal fodder, 88
biomass production for biofuels, 47, *158b*
carbon in, 87
as fuel source, 22, 47, 80
and GHG emissions, 188
liquid fuels from, 153, 191
and natural resource availability, 181
recycling of, 226
and soil-related technology, 227
See also Biofuels
Biotechnology, 210, 217, 231–34, *232b, 234b*
Bipolarisation of agricultural demand, *224b*
Blue Baby syndrome, 6
Border protection
EU's Global Economy scenario vs., 158, 165, 175
and immigration, 27, 242
multinational food companies vs., 180, 226
pests and disease-causing pathogen issues, 218
as supply management technique, 31, 32
Bovine tuberculosis management costs, 99, *100t*
Brundtland Commission, 189–90
Business interests. *See* Transnational and multinational corporations

C

CAADP (Comprehensive African Agriculture Development Programme), 255
CAMBIA, 261*n*
Canada
aquaculture in, 57, *58f*, 59, *59f*, 61
bipolar farm policy, 31
economic development in, 11
Food Systems 2002 (1987), 45
genetic engineering regulations, 143–44
Market Net Income of farmers, 6
number of farms and farm size, *23f*
protection of aboriginal rights, 12
rural and farm populations, *27f*
voluntary organic standard, 48
See also entries under "NAE" *and* "North America"
Canadian Community Healthy Survey, 6
CAP. *See* Common Agricultural Policy
Carbon
global strategy for a low carbon economy, 218–21, *220t*
sequestration methods, 54, 82, 90, 161, 214–15, 259
Carbon dioxide (CO_2), 43, 53, 87, 187–88, *187f*, 218, 220
CBD (Convention on Biological Diversity), *125b*, 175, 195, 261–62
CEC (Commission of the European Community), 137, 182. *See also* European Commission
Cellulosic material biofuels, 192, 218–20, 219–20, *220t*
Central and Eastern European Countries (CEECs), 33–34, 34*n*
Central and Eastern Europe (CEE)
agricultural policies, 33–34
agriculture and food production changes, 29, 38, 67, 68
AKST organization, 131, 138, 196
and European Union, 28–29, 38
farm structure, 24–26, 40
food retail sector, 61–62
forestry and forest land, 54–55, 56
livestock in, 52, 53
NFPs in, 264
trade with other countries, 106, *106t*
transition to market economy, 49
See also Europe; European Union; NAE region
Centralization and consolidation, 195
CGE (computable general equilibrium) models, 155, 156, *223b*
CGIAR. *See* Consultative Group on International Agricultural Research
Challenge programs (CPs), *257b*
CIS. *See* Commonwealth of Independent States
Climate change
adapting to, 189–91
as AKST driver, 4–5
carbon sequestration methods, 54, 82, 90, 161, 214–15, 259
effect in NAE vs. developing regions, 9
methane, environmental effects of, 88
variability, and technologies, 186–91, *187f, 188f*
See also Greenhouse gas emissions
CLUE-s (Conversion of Land Use and its Effects) model, 157–58
Coastal capture fisheries, 185, 210, 239–40
Codex Alimentarius Commission, 66, *236b*
Collaboration. *See* Integration
Collective intelligence knowledge production, 165, 166
Commission of the European Community (CEC), 137, 182. *See also* European Commission
Common Agricultural Policy (CAP) in Europe
domestic price supports, 32, 94
and European Union, 32–33
and farm specialization, 48
founding and implementation of, 11, 31–32, 123
future of, 157, 175
organic cropping systems, 47–48
output and productivity trends, 49, 53, 108, *108f*
rural development support, 16, 28
single farm payments, 32
tariffs and import quotas, 32
Common products, *224b*
Commonwealth of Independent States (CIS)
agricultural policy, 44, 54, *54f*
economics and international trade, 164, 176, *177t*
fertility rate projection, 163, *163t*
food insecurity in, 6, *7f*, 104
future of, 165
Communication technologies, 152, 157, *159t*, 169, 170, 200, 246–48
Communities
Canadian health survey of, 6
CAP's support of, 16, 28
effect of migration from rural areas, 102–3
farming and rural population changes, 26–29, *27f, 28f*, 97–98, 100–102, *101t, 102t*
in forest areas of Europe, 56
and market structure changes, 34
rural, pre-WWII, 14
separation from food production, 92, 104
See also Human health
Competition Commission (UK), 64, 96
Complex systems, 214*n*, *215b*
Comprehensive African Agriculture Development Programme (CAADP), 255
Computable general equilibrium (CGE) models, 155, 156, *223b*
Concept of Economic Rent on land, *223b*
Consequences for AKST
in access, control and distribution of AKST, 197–98
agricultural labor and organizations, 180
in agricultural policies, trade and markets, 177
climate change and variability, 190–91
energy and bioenergy, 192
farming systems and farm structures, 178
in food consumption and distribution, 174–75
natural resource availability and management, 185–86
organizations, 194–95
in science and technology funding, 195
Consultative Group on International Agricultural Research (CGIAR)
challenge programs of, *257b*
collaboration with others, 127, 128

founding of, 5
funding and budget, 137, 142
NAE contributions to, 256–58
overview, 132, 193
and strategy to ensure food sufficiency, 123–24
Consumers, 2, *98f,* 104, 160, 226–27, 243
Consumption footprint, 10
Consumption systems, changes in, 70–71
Contextualization of AKST organizations, 195
Control and distribution of AKST, 196–98
Control and influence, equity in, 81, 103
Convention on Biological Diversity (CBD), *125b,* 175, 195, 261–62
Conversion of Land Use and its Effects (CLUE-s) model, 157–58
Cooperative Agricultural Extension Service, 131
Cooperatives
in CEE, 26
centers with a research center and an educational institution, 251
dairy cooperatives in US, 35
extensive service of land-grant universities, 40
PROCINORTE cooperative program, 258
as response to market power, 96–97, *97t, 98f*
Cooperative Wheat Production Program, 123–24
CPs (challenge programs), *257b*
Cropping systems
and biofuel/biomass production, 47, *158b*
environmental consequences of, 80, 82–87, *83f, 83t, 84t*
increasing productivity, 40–46, *42f, 44f, 45f, 46f*
and marketing contracts, 37
organic, 47–48
overview, 38–40, *39ff,* 48
and soil AKST, 37–38
See also Plants
Cross-sectoral public-private governance platforms, 154, 221
Cultural developments and AKST, 2
Cyanobacterial biofuels, 218–20, *220t*

D

De/centralization and consolidation, 195
Demographic drivers of AKST, 161–63, *163t, 164tt*
Demography and AKST, 2
Developing countries or regions
agricultural component of economics of, 125–26
and CGIAR, 256–57
and climate change, 9
and digital divide, 130, 197
effect of trade agreements, 9, 53, 152
impact of NAE paradigms, 123–24
labor from, 10
NAE influence on, 2, 8–10, 81, 117–18, 119
undernourishment in, 6–7, *7f*
See also Consultative Group on International Agricultural Research
Development and sustainability
and agroecosystems, 16, 127, 180, 210, 226–27, 229, 235
of aquaculture systems, 240–41
for forestry, 236–39, *237t*
future AKST systems and, 198–201
as goals, 5–8, *5f, 7f,* 16, 118, 119–20, 198–201
lessons learned, 118–20
meeting goals, 212–14
NAE control of resources, 9–10
overview, 117–18
paradigm shift in process, 209
TNC efforts, 16
treaties and conventions that promote, 212
Diet/consumption changes, 69–71, *69t, 70tt*
Diet-related health problems, 2, 6, 8, 9. *See also* Human health
Digital divide, 130, 197
Direct payments to farmers, 31–33, 47, 93, 175
Diseases, new and emerging, 209, 216–18. *See also* Human health
Dis/integration of organizations, 194–95
DNA-based techniques, 41, 217, 225, 232–34. *See also* Genetic engineering
Doubly Green Revolution, 124, 256

E

Eastern Europe
and access to capital, 2
effect of communism, 11, 12
fertility rate projection, 163, *163t*
food insecurity and subsistence farming in, 5–6, *7f*
foresight exercises for, 158
See also Central and Eastern Europe; Europe; European Union; NAE region; *entries under* "NAE" *and* "Europe"
Ecologically-based pest management (EBPM), 235
Ecological management
agroecosystem analysis, 124
agroecosystems, 16, 127, 180, 210, 226–27, 229, 235
ecosystem approach to natural resources management, 124–25, *125b,* 239
ecosystem-oriented AKST innovation system, 154, 198–99, 210
environmental sustainability via, 226–46, *230b, 232b, 233b, 234b, 236b, 237t, 245f*
and evolutionary sciences, 210
Economic development
and agrifood systems, 14–16
as AKST driver, 164
environmental costs of, 8
and farm structure, 22
NAE concepts and ideas about, 10
and natural resources, 12–14
overview, 8–9, 10–12
social issues, 241–46
Economic migration, 12
Economic modeling, *223b*
Economic policies
agrifood systems and economic signals, 14–15
and AKST, 2
global strategy for a low carbon economy, 218–21, *220t*
impact in NAE, 80–81, 92–99, *93f, 94ff, 94t, 95f, 96f, 97t, 98ff, 99f, 100t*
price supports, 30, 31–33, 93
supply control, 30, 31, 32, 81, 145
tariffs, 5, 16, 30, 32, 175, *222b*
See also Subsidies
Economic Research Service (ERS) of the USDA, 160
Ecosystem approach to natural resources management, 124–25, *125b,* 239. *See also* Ecological management
Ecosystem-oriented AKST innovation system, 154, 198–99, 210
Educational institutions
for farmers, 250–51
links between research and, 251
research approaches for, 213
student interests, 137–138
trends and uncertainties, 172
university trends in research, 153, 167–68, 179–80
See also Research and development
EEC (European Economic Community), 11
EFMN (European Foresight Monitoring Network), 157
Energy
as AKST driver, 4–5
and bioenergy trends and issues, 191–92
resources in NAE region, 13
wind energy, 47
See also Biofuels
Energy efficiency of supply chains, 214, 220–21, 231
Energy efficient food and farming systems, 230–31, *230b*
Environmental impacts
agricultural industry changes, 90–92, *91f*
agriculture and AKST, 80, 81–92, *83f, 83t, 84t, 89f, 89t, 91f,* 99, *100t,* 210
animal production changes, 87–88
aquaculture systems, 80, 88–90, *89f, 89t*
cropping systems, 80, 82–87, *83f, 83t, 84t*
economic development, 8
farm size and structure changes, 87
fertilizers, 84
field drainage, 84–85
food and farming, 81–92, *83f, 83t, 84t, 89f, 89t, 91f*
forest management changes, 90
genetic engineering, 80, 85–86
grassland loss, 82, 84–85, 91, 188, *188f*
growing biofuel crops, 80, 87
irrigation, 85
mechanization, 80, 86–87
methane, 88
pesticide usage, 84
soil management, 82–84, *83f, 83t, 84t*
veterinary medicine use, 88
Environmental quality
and aquaculture, 60–61
awareness of, 22
certification of forest management, 263–64
degradation of, 2, 8, 9
from ecological management, 226–46, *230b, 232b, 233b, 234b, 236b, 237t, 245f*
enhancement goals, 8
environmental sustainability via ecological management, 226–46, *230b, 232b, 233b, 234b, 236b, 237t, 245f*

Environmental quality (*continued*)
 environment and governance system policies, 258–64, *260b, 260t*
 and EU agricultural policy, 32
 global assessments, 155–56
 NAE's footprint, 10
 in pre-1945 agrifood systems, 14
 water pollution vs., 46
 See also Biodiversity; Development and sustainability; Natural resources; Sustainability
EPIPAGRI, 261*n*
Equity of food systems, 102–4, 212. *See also* Wealth and asset inequity
ERS (Economic Research Service) of the USDA, 160
ESF (European Science Foundation), 157
ESIM (European Simulation Model), 157–58
ETPs (European Technology Platforms), 156–57
EU27 countries, 3*n*
EUFO (European Futures Observatory), 157
Europe
 aquaculture, 59–60
 concentration of retail, 61
 diet-related health problems, 6
 economics and international trade, 164, 176, *177t*
 genetic engineering regulations, 143–44
 grassland-based cattle systems, 52
 livestock systems, 48–53, *49f, 50f, 51f*
 organic farming and market, 47–48, 67, *68t*
 pesticide use in, 44–45, *45f*
 population projections, 163, *164t*
 retail sector, 62–63, *63t*
 rural and farm populations, 27–29, *28f*
 Western Europe, 2, 11, 12–14, 163, *163t*
 See also Central and Eastern Europe; Common Agricultural Policy; Eastern Europe; European Union; NAE region
European Commission
 and CAP reform, 175
 foresight exercises, 156–58, *158b, 159t*, 165
 on genetic engineering, *233b*
 and international AKST, 136–137
 on soil degradation, 182
 Standing Committee on Agriculture Research, 158
 water legislation, *262–63*
European Economic Community (EEC), 11
European Foresight Monitoring Network (EFMN), 157
European Forum on Agricultural Research for Development, 257–58
European Futures Observatory (EUFO), 157
European Science Foundation (ESF), 157
European Simulation Model (ESIM), 157–58
European Technology Platforms (ETPs), 156–57
European Union (EU)
 agricultural imports and exports, 106, *107f*
 alternative energy sources, 47
 animal husbandry restrictions, 88
 biofuels directive, 47
 forests and livelihoods, 56
 founding of, 11
 frontier research concept of knowledge production, 165, 166
 gender pay gap, 12
 livestock production and trade, 108–9, *109t*
 River Basin Management programs, 263
 See also Common Agricultural Policy; Europe
European Water Framework Directive, 262–63
Eururalis foresight exercise, 157, 158, *158b, 159t*
Evolutionary and ecological sciences, 210
Evolution of AKST, 126–28, *127b, 128t*
Export subsidies, 93, 106

F

FAAP (Framework for African Agricultural Productivity), 254–56
Fair trade policies, 68–69, *68t, 69t*, 210, 221–22
FAO. *See* Food and Agriculture Organization
Farmer associations, 179
Farmer attitudes and behavior, 245–46
Farming systems and farm structures, 177–78. *See also* Agricultural industry; Food and farming
Farming systems research (FSR), 124, *127b*
Farmworkers, 7, 10, 27, 152–53, 178–79, 244–45, *245f*
Fertilizers
 corporate control of production, 15, 34
 and crop production, 21
 environmental consequences of chemicals, 84
 organic, 38
 and precision agriculture, 38
 run-off issues, 6, 8, 38
 and soil testing, 37–38
FFRAF (foresight food, rural and agrifutures), 157, 158, *159t*
Field drainage, environmental consequences of, 84–85
Fisheries
 availability and management, 183, 185, 186
 and climate change, 186–88
 coastal capture, 185, 210, 239–40
 governance issues, 263
 marine fishery stocks, 8
 in NAE region, 13
 overview, 13
 sustainability, 210, 239–40
 See also Aquaculture systems
Fisheries and Oceans Canada, 13
Fish farming techniques, 57–61, *57f, 58ff, 59f*
Fish meal from coastal capture fisheries, 240–41
Food and Agriculture Organization (FAO) of the United Nations
 FAO/IAEA mutation varieties database, 41
 FAO/OECD global foresight model, 155–56, *156t*
 FAO/WHO Codex Alimentarius, 66, *236b*
 on food insecurity, 5–6, 15, 38
 International Treaty on Plant Genetic Resources for Food and Agriculture, 5, 175, 195–96, 261–62
 on livestock development practices, 88
 World Bank and AKS concept development, 247
 World Food Model, 156
Food and farming
 cooperative responses to market power, 96–97, *97t, 98f*
 effect of standards on small-scale producers, 152
 energy efficient systems, 230–31, *230b*
 environmental impact, 81–92, *83f, 83t, 84t, 89f, 89t, 91f*
 in Europe, 5–6, *7f*
 farm size and wealth inequity, 9
 farmworkers incomes, 7, 10
 food safety issues, 223–26, *224b, 225b*
 governance issues, 262–63
 mixed systems, 14, 48, 80, 87–88, 92
 precision agriculture, 40, 169, 178, 210, 236
 rural and farm populations, 26–29, *27f, 28f*, 97–98, 100–102, *101t, 102t*, 152–53
 rural development, 246
 socioeconomic viability of systems, 241–44
 sustainability of rural livelihoods, 244–45, *245f*
 systems and structures, 177–78
 wireless communications for, 170
 by women, 7–8, 29, 180, 211, 242
 women as farm operators, 7–8
 See also Cropping systems; Food safety issues; Labor
Food consumption and distribution trends, uncertainties and consequences for AKST, 172–75
Food demand and consumption as AKST driver, 4–5
Food insecurity, 5–6, *7f*, 81, 104
Food manufacturing and processing, 65, *65t*
Food miles, 92, 195, 220
Food policy councils, 262
Food quality improvements, 224–25, *225b*
Food retail sector, 21–22, 61–65, *62f, 62t, 63tt, 64f, 64t*, 80, 103
Food safety issues
 AKST options for improving, 210
 in diverse food and farming systems, 223–26, *224b, 225b*
 future of, 174, 177, 185
 global demand for, 99
 in meat industry, 52–53, 81, 173
 and modes of governance, 81, 175–76, 262
 post-World War II solutions, 117
 research resources for, 135, 139, 142
 retail sector provisions, 173
 SFP tied to, in EU policy, 32
 standards for, 66–67
 systems for controlling, 210
 See also Food security
Food Safety Management Systems Standard (ISO), 67, *67t*
Food security
 and development and sustainability goals, 5–6, 22, 137
 and EFARD, 257*n*
 ethical considerations, 219
 inequities in, 81, 104, 139
 and local production systems, 221
 options for action, 210
 and organic agriculture, *236b*
 post-World War II achievements, 21, 117, 139
 as public good, 194, *222b*, 243, 254, 264, 266, *266t*

in Soviet Union, 21, 24–25
and sustained poverty reduction, 254
USDA food security assessment, *162t*
and water demands and land scarcity, 156
See also Food safety issues
Food sovereignty, 9
Food systems approaches, 127–28, *127b*
Food system specialization, 22–26, *23f, 23t, 24f, 25f, 26t*. *See also* Supply chains
Foresight food, rural and agrifutures (FFRAF), 157, 158, *159t*
Forest institutions, 55–56
Forest management
and biodiversity, 54–55, 90, 235, *235t*, 238, 239
certification of, 263–64
development and sustainability, 210, 236–39, *237t*
environmental consequences of changes in, 90
governance issues, 263–64
mechanization of, 101, *101f*
overview, 53–57, *54f*, 210, 236–39, *237t*
Forest productivity, 238
Forest resources
and biomass crops, 47
and climate change, 186–88
mapping of, 239
in NAE region, 13–14
overview, 183
quality of stands, 54, 90
Forest Stewardship Council (FSC), 263–64
ForSociety, 157
Framework for African Agricultural Productivity (FAAP), 254–56
Free-trade agreements, 31, 144
Freshwater resources in NAE region, 13. *See also entries beginning with* "Water"
Frontier research concept of knowledge production, 165, 166
FSC (Forest Stewardship Council), 263–64
FSR (farming systems research), 124, *127b*
Funding. *See* Privatization; Public funding
Fungicides. *See* Pesticide usage
Future for KST, agriculture, and AKST
access, control and distribution of AKST, 196–98
agricultural labor and organizations, 178–80
climate change and variability, 186–91, *187f, 188f*
context, 154–61, *156t, 158b, 159t*
direct drivers for agriculture, 4, 172–92, *177t, 183f, 184ff, 187f*
direct drivers for KST, 165–72, *171t*
energy and bioenergy trends and issues, 191–92
farm systems and structures, 177–78
and GHG emissions, 188–91
indirect drivers for AKST, 161–65, *163t, 164tt*
international trade and, 176–77, *177t*
key drivers for AKST and research/innovation systems, 192–98
natural resources availability and management, 180–86, *183f, 184ff*
organization and funding of AKST, 192–95
overview, 152–54
proprietary regimes, 195–96
and sustainable development goals, 198–201
trade, markets and agricultural policies, 221–23, *222b, 223b*
Future needs and priorities for AKST. *See* Options for action

G

GE. *See* Genetic engineering
Gender gaps, 7–8, 12, 242–43. *See also* Women
General Agreement on Tariffs and Trade (GATT), 5
Genetic engineering (GE)
in animal breeding, 51–52
consumer concerns, 100–101
and development and sustainability goals, *233b, 234b*
DNA-based techniques, 41, 217, 225, 232–34
environmental consequences of, 80, 85–86
ethical issues, 100–101
of livestock, 50–51
and nutritional value, 170
overview, 41–43, *42f*, 80, 85–86
potential of, 231–34, *232b*
and productivity increases in crops, 40–43, *42f*
for reducing drought stress, 228
regulations in Europe and Canada, 143–44
resource accessibility, 261–62
traditional practices vs., 4
Genetic resources collections, 2, 229–30
GEO (Global Environment Outlook), 155–56, *156t*
Geographic scope of NAE region, 3, 133*n*, 175
GFAR (Global Forum on Agricultural Research), 257–58
GHG. *See* Greenhouse gas emissions
Global assessment, aspects of, 3–4, 161, 181
Global Environment Outlook (GEO), 155–56, *156t*
Global foresight exercises, 155–56
Global Forum on Agricultural Research (GFAR), 257–58
Global issues, AKST options for addressing, 209–10
Global population growth, 12
Global Scenario Group (GSG), 155, 156
Global strategy for a low carbon economy, 218–21, *220t*
Global Technology Revolution 2020 (Rand Corp.), 170
Global Trade Analysis Project (GTAP), 157, 160
Goals
environmental enhancement, 8
for food security, 5–6, 22, 137
for genetic engineering, *233b, 234b*
meeting sustainability goals, 212–14
paradigm shift to meet development and sustainability goals, 212–14
poverty reduction, 6–7, *7f*, 211
for public policies, 144–45
and scope of AKST, 118
sustainability, 5–8, *5f, 7f*, 8, 16, 118, 119–20, 198–201
Governance systems and policy environment redesign, 258–64, *260b, 260t*
Grameen Bank, 195
Grasslands
and biodiversity, 229
crop and livestock takeovers, 52, 81–82
ecosystem-based management, 57
environmental effect of losing, 82, 84–85, 91, 188, *188f*
grazing systems, 87
recovering to counter climate change, 215
Great Depression, 30–31
Greenhouse gases, 187*n*
Greenhouse gas (GHG) emissions
from agriculture, 80
carbon dioxide (CO_2), 43, 53, 87, 187–88, *187f*, 218, 220
and climate change, 186–91, *187f, 188f*
effect on consumer choices, 8
future for KST, agriculture, and AKST, 188–91
methane, 13, 80, 88, *89f, 187f*, 187*n*, *230b*
nitrous oxide (N_2O), 80, 83, 187, *187f*, 230
ozone, 90, 187
reducing emissions, 209
See also Climate change
Green manufacturing, 170
Green Revolution, 123–24, 127, 256
Gross domestic product (GDP)
contribution of agricultural sector, 92
education funding as percent of, 12
influence on agriculture and AKST, 164
and penetration of "modern" retail, 61, *62f*
and percentage of income spent on food, 69
per person engaged in agriculture, 93, *94f*
research and development percentage, 122, 136, 170–71, *171t*, 193
GSG (Global Scenario Group), 155, 156
GTAP (Global Trade Analysis Project), 157, 160

H

Hazard Analysis at Critical Control Point (HACCP), 53, 66–67
Health. *See* Human health
Herbicides. *See* Pesticide usage
Herbicide tolerant (HT) crops, 41, 85–86
Housing for animals, 51, 52, 231, 235
Housing for humans, 69, *69t, 70tt*, 170, 242
HT (herbicide tolerant), 41, 85–86
Human capital. *See* Labor
Human health
and AKST development choices, 2
and AKST policies and investment, 6, 9
diet/consumption changes, 69–71, *69t, 70tt*
diet-related health problems, 2, 6, 8, 9
increasing awareness of food and, 22, 152
new and emerging diseases, 216–17
and nutrients in food, 21, 69–70, *70t*, 104, *105tt, 106f*, 170, 210
obesity and overweight, 6, 81, 104, *105tt, 106f*, 173
pesticide poisonings, 6
and targeted drug deliveries, 170
veterinary and public health infrastructure integration, 217–18
Hungarian exports, 61–62

Hunger and food insecurity eradication, 5–6
Hybrid public goods, 141, 260
Hybrid seeds and plants, 34, 40–41, 42, 123, 144, 195
Hybrid vehicles, 170

I

IAASTD conceptual diagram, *4f*
ICT (information and communication technologies), 152, 157, *159t*, 169, 170, 200
Identified products, *224b*
IFPRI (International Food Policy Research Institute), 155, *156t*
Illegal immigrants, 7, 152–53, 180, 242
Immigrant agricultural populations, 22, 27
Immigration, 27, 163, 242
IMP (Integrated Pest Management), 124, 235
Indigenous knowledge, 4, 9, 14, 247–48
Indigenous peoples, 12, 13, 14
Influence
- of beneficiaries, 138–139, 197
- equity in control and, 81, 103
- of media on consumer preferences, 179
- of NAE on developing countries, 2, 8–10, 81, 117–18, 119

Information and communication technologies (ICT), 152, 157, *159t*, 169, 170, 200
Information systems, reconfiguration of, 252
Information technology (IT)
- and communication, 170
- and digital divide, 130, 197
- and food supply chain, 95
- gleaning knowledge from, 251–52
- and precision farming, 199
- trends and uncertainties, 169
- and value chains, 126

Innovative models and systems
- for AKST dissemination, 254
- Biologically Integrated Farming Systems program, 253
- for crops and livestock food and farming, 231–36, *232b, 233b, 234b, 236b*
- education and research model, 253–54, *253b*
- future of, 154–55
- normative agriculture innovation systems, 198–201
- overview, 153, 155–61, 192–98
- transformation in, 166–69
- trends and uncertainties, 165–69

Inputs enterprises, 179
INRA (Institut National de la Recherche Agronomique), 157
Insecticides. *See* Pesticide usage
Insect resistant (IR) crops, 41, 85–86
Institute for Prospective Technological Studies (IPTS), 157, *159*
Institutions
- effect of organizations and, 10
- forest institutions, 55–56
- institutional arrangements and collaboration, 265–66
- lessons learned, 118–20
- organizational and institutional arrangements, 252–58, *253b, 257b*
- public institutions, 42, 43, 53, 160, 198, 252–53, 261
- *See also* Educational institutions

Institut National de la Recherche Agronomique (INRA), 157
Integrated Pest Management (IMP), 124, 235
Integration
- of AKST and KST, 117, 126, 138, 153, 192–96, 197, 200
- of education and research programs, 137–39, 211
- interdisciplinarity overview, 248–51, *252b*
- paradigms as barriers to, 129, *249b*
- of perspectives within AKST, 117, 126–30, *127b, 128t*, 197, 211
- of research on coastal fisheries, 239–40
- for studying ecology of emerging diseases, 216–17

Intellectual Property Rights (IPR), 10, 21, 42, 143, *144b*, 194, 195–96, 212
Intensive agriculture, 80, 90–92, *91f*
Intensive livestock production vs., 48, 52–53, 69, 88, 100–101
Interactive knowledge networks, 246–48
Intergovernmental Panel on Climate Change (IPCC), 155, *156t*, 188, 216
International agreements. *See* International treaties and agreements
International AKST, 136–38
International Food Policy Research Institute (IFPRI), 155, *156t*
International Obesity Task Force (IOTF), 104, *105t*
International Organization for Standardization (ISO), 66–67, *67t*
International purchasing and marketing organizations, 62–63, 63t
International quality standards, 5
International R&D, 117–18, 119, 137, 168. *See also* Consultative Group on International Agricultural Research; Research and development
International trade
- agricultural policies, markets and, 175–77, *177t*
- AKST and production changes, 98–99, *98f, 99f*
- CEE trade with other countries, 106, *106t*
- CIS economics and international trade, 164, 176, *177t*
- European economics and, 108–9, *109t*, 164, 176, *177t*
- export subsidies, 93, 106
- Global Trade Analysis Project, 157, 160
- intellectual property rights, 144, *144b*
- of meat and dairy products, 53, 108–9, *109t*
- NA economics and, 164, 176, *177t*
- overview, 2, 104, 106–9, *106t, 107ff, 108ff, 109t*, 164–65
- promoting fair trade and market reform, 210, 211, 221–23, *222b, 223b*
- and tariffs, 5, 16, 30, 32, 175, *222b*
- trade agreements, 9, 16, 53, 144, *144b*, 152
- trends, uncertainties and consequences, 176–77, *177t*

International treaties and agreements
- Convention on Biological Diversity, *125b*, 175, 195, 261–62
- developing AKST in response to, 259–60
- and development goals of other countries, 152
- on intellectual property rights, 10, 21, 42, 144, *144b*, 194, 195–96, 212
- North American Free Trade Agreement (NAFTA), 16, 27, 53, 106
- Partnership and Cooperation Agreement, 165
- plant breeders rights, 42, 144, 195
- Plant Genetic Resources for Food and Agriculture, 5, 175, 195–96, 261–62
- Trade-Related Aspects of Intellectual Property Rights, 42, 144, *144b*, 195, 261
- *See also* European Union

International Union for the Protection of New Varieties of Plants (UPOV), 42, 195
International Water Management Institute (IWMI), 155
IOTF (International Obesity Task Force), 104, *105t*
IPCC (Intergovernmental Panel on Climate Change), 155, *156t*, 188, 216
IPR (Intellectual Property Rights), 10, 21, 42, 143, *144b*, 194, 195–96, 212
IPTS (Institute for Prospective Technological Studies), 157, *159*
IR (insect resistant) crops, 41, 85–86
Irrigation
- environmental consequences of, 85
- overview, 13, 46
- traditional knowledge on, 9

ISO (International Organization for Standardization), 66–67, *67t*
Israel, 2, 12
IT. *See* Information technology
IWMI (International Water Management Institute), 155

K

Knowledge. *See* AKST
Knowledge, science and technology. *See* Future for KST, agriculture, and AKST; KST
Knowledge-based systems, 241–42, 251–58, *253b, 257b*
Knowledge networks, 246–48
Knowledge production models and trends, 165–66
KST (knowledge, science, and technology)
- advancement in, 118
- as AKST driver, 119–20
- direct drivers, 165–72, *171t*
- export from NAE, 9–10
- integration of AKST and, 117, 126, 138, 153, 192–96, 197, 200
- uncertainties and consequences for AKST, 166, 168–69, 171, 172
- veterinary and public health infrastructure integration, 217–18
- *See also* Future for KST, agriculture, and AKST

L

Labor
- as AKST driver, 4–5
- AKST options for strengthening, 211
- CAP focus on equitable standards, 31
- decrease in need for, 22–23, *23t*, 40, 48
- from developing countries, 10
- development of organizations, institutions and, 246–58

farming and rural population changes, 26–29, *27f, 28f,* 97–98, 100–103, *101t, 102t*
farmworkers, 7, 10, 27, 152–53, 178–79, 244–45, *245f*
future of farmworkers, 178–80
GDP per person engaged in agriculture, 93, *94f*
and gender dynamics, 178–79
improving welfare of, 241–42
livestock and, 52–53
on Soviet collectivized farms, 24
Land use and management
as AKST driver, 4–5
biodiversity enhancement, 229
for biofuels, 10
and climate change, 186–87, 190, 214–15
competition for land use, 181, 237, *257b*
effect intensive agriculture, 90–92, *91f*
and integrated farming systems, 125
and mechanization, 40
models of, 157–58, *158b*
multiple development objectives, 186, 190
and soil classification, 38
soil degradation from, 80, 82
uncertainties about, 185
See also Forest management; Grasslands
LEITAP model, 157–58
Lignocellulosics, 220
Literacy rates, 12
Livestock processing, 34–35, *35t,* 89
Livestock production
antibiotic use in, 53, 80, 88, 101, 172
environmental consequences of veterinary medicines, 88
environmental impacts of differing systems, 87–88
intensive, 48, 52–53, 69, 80, 88, 100–101
and methane gas, 80, 88–89, 215, 232
new and emerging diseases, 216–17
and trade, European, 108–9, *109t*
Livestock production contracts, 36–37
Livestock systems, 48–53, *49f, 50f, 51f*
Local food-supply led AKST innovation system, 154, 199, 221
Local-learning AKST innovation system, 200–201
Low-income countries. *See* Developing countries or regions

M

Maize development, 42, 44
Manures, environmental effects of, 88
Marginal lands in NAE region, 13
Marker assisted selection, 41, 232–34
Marketing contracts, 37
Marketing/processing enterprises, 179
Market-led AKST innovation system, 153–54, 198
Market-led subsidies, 53, 109, 145
Market power, AKST impacts on, 95–97, *97t, 98f*
Market structure
and forestry production, 56
and information dissemination, 197
inputs and outputs, 34–37, *35f, 35t, 36f, 36t*
market driven policies, 144
niche markets, 67
organic and locally-produced goods, 15–16, 67, *68t*
post-harvest and consumption systems, 61–70, *62f, 62t, 63tt, 64f, 64t, 65t, 67t, 68tt, 69tt, 70tt*
segmentation in food markets, 65–66
trade, agricultural policies, and, 221–23, *222b, 223b*
See also Agrifood systems; Supply and demand; Supply chains
Mechanization, 40, 80, 86–87, 101, *101f*
Media influence on consumer preferences, 179
MEGAAF model, 157
Methane, 13, 80, 88, *89f, 187f,* 187*n, 230b*
Migrant labor, 2, 152–53, 178–79, 211, 242
Millennium Ecosystem Assessment (2003), 14
MNCs. *See* Transnational and multinational corporations
MNP (Netherlands Environmental Assessment Agency), 155, 157
Mode 1 and Mode 2 knowledge production, 165, 166
Montreal Protocol, 259
Multifunctional agriculture systems
and AKST, 194–95, 246, 265
drivers for, 8, 160–61, 197
and ecosystem-oriented AKST, 124–26, *125b,* 177, 198
farmers' challenges, 217
funding for R&D, 265, 266–67
future of, 209, 210, 211, 212, 246–47, 259
as new paradigm, 209
in rural areas, 190–91, 241, 246
training programs on, 251
Multifunctional economic signals, 15–16
Multifunctional forestry system, 238–39
Multifunctional sustainability of forests, 55, 57, 236–39, *237t*
Multinational companies. *See* Transnational and multinational corporations
Mutagenesis, 41, 233

N

NAE assessment roadmap, *3f*
NAE region
agrifood systems, 14–16
cropping system changes, 37–48
economic policy impacts, 92–99, *93f, 94ff, 94t, 95f, 96f, 97t, 98ff, 99f, 100t*
fair trade sales, 68–69, *68t, 69t*
farm policies, 33
forestry systems, 53–57, *54f*
geographic scope, 3, 133*n,* 175
influence on other countries, 2, 8–10, 81, 117–18, 119
Israel, 2, 12
natural resources and their exploitation, 12–14
overview, 10–12
post-harvest and consumption systems, 21–22, 61–70, *62f, 62t, 63tt, 64f, 64t, 65t, 67t, 68tt, 69tt, 70tt*
transferring wealth from urban to rural areas, 16, 28
See also Europe; North America
Nanotechnology
for livestock vaccinations, 217
potential of, *163t,* 234–35
and precision agriculture, 160
public response to, 170
sensors for monitoring soil health, 227
for traceability from raw materials to marketed products, 225
National Academy of Science (NAS), 66
National Agricultural Research Institutes (NARIs), 251, 255
National Agricultural Research Systems (NARS), 132, 256–57
National Forest Programs (NFPs), 264
National Intelligence Council (NIC), 160
Natural resources
alternative resource management strategies, 236, 260–61
availability and management, 4–5, 180–86, *183f, 184ff*
exploitation by external peoples, 13
in NAE region, 12–14
reducing pressure on, 231
See also Ecological management; Environmental quality
Netherlands Environmental Assessment Agency (MNP), 155, 157
New Partnerships for African Development (NEPAD), 255
NFPs (National Forest Programs), 264
Niche markets, 15–16, 67, *68t*
NIC (National Intelligence Council), 160
Nitrogen (N) fertilizers, 8, 41, 43–44, *45f,* 88
Nitrogen (N) in soil, 43
Nitrous oxide (N_2O), 80, 83, 187, *187f,* 230
Nonagricultural domains and AKST, 2
North America
and access to capital, 2
aquaculture, 57–59, *57f, 58ff, 59f*
economics and international trade, 164, 176, *177t*
farm size and number of farms, 22–26, *23f, 23t, 24f, 25f*
fertility rate projection, 163, *163t*
foresight exercises, 160–61, *162–63t*
livestock systems, 48–53, *49f, 50f, 51f*
natural resources and their exploitation, 12–14
organic farming in, 48
organic market in, 67, *68t*
pesticide usage in, *44f,* 45–46
population projections, 163, *164t*
range management, 52
See also Canada; United States
North American Free Trade Agreement (NAFTA), 16, 27, 53, 106
Nutrients
in cropping systems, 43–44, *44f, 45f*
in food, 21, 69–70, *70t,* 104, *105tt, 106f,* 170, 210

O

Obesity and overweight, 6, 81, 104, *105tt, 106f,* 173
OCED/FAO global foresight model, 155–56, *156t*
OECD (Organization of Economic Cooperation and Development), 104, *105t,* 132–33, 133*n,* 138

Options for action
and climate change, 214–18, *215b*
environmental sustainability via ecological management, 226–46, *230b, 232b, 233b, 234b, 236b, 237t, 245f*
environment and governance system policies, 258–64, *260b, 260t*
food quality and safety, 223–26, *224b, 225b*
global strategy for a low carbon economy, 218–21, *220t*
human capital, organization and institution development, 246–58, *249b, 253b, 257b*
investments in R&D for agriculture, 264–67, *266t, 267b*
new and emerging diseases, 209, 216–18
overview, 209–12, 214
paradigm shift to meet development and sustainability goals, 212–14
Organic agriculture, 47–48, 236, *236b*
Organic and locally-produced goods, 8, 15–16, 67, *68t*
Organizations
dis/integration of, 194–95
institutional and organizational arrangements, 252–58, *253b, 257b*
lessons learned, 118–20
overview, 178, 179–80
private AKST organizations, 194
See also Institutions
Other wooded land (OWL), 54
Output/outputs
advances in AKST and, 94–95, *96f*
market structure, inputs and, 34–37, *35f, 35t, 36f, 36t*
and productivity trends of CAP, 49, 53, 108, *108f*
in Russia, 25, *26t*
OWL (other wooded land), 54
Ozone-sensitive trees, 90

P

Paradigms in NAE AKST
as barriers to integration, 129
and CAMBIA, 261
choices based on, 152, 201
and development and sustainability goals, 119–20, 212–14
impact of, in developing countries, 123–24
policies, demand and, 195
for research and action, 212–13
in science, 141, 200, 262
and shift into privatization, 196–97
in societal context, 118, 122–26, *125b*
Pareto optimality, *223b*
Partial equilibrium (PE) models, 155
Partnership and Cooperation Agreement (PCA), 165
PBRs (plant breeders rights), 42, 144, 195
PEFC (Program of Endorsement of Forest Certification), 263–64
PE (partial equilibrium) models, 155
Pervasive sensors for real-time surveillance, 170
Pesticide poisonings, 6
Pesticide usage
consumers' aversion to, 15
corporate control of production, 15
in cropping systems, 44–46, *46f*
in Eastern Europe, 152
environmental consequences of, 84
plant breeding vs., 41–42
regulation in Europe and US, 142–43
as substitute for knowledge, 40
and US farm policy, 31
Pest management, 218, 235–36
Pests, new and emerging, 217–18
PGRFA (International Treaty on Plant Genetic Resources for Food and Agriculture), 5, 175, 195–96, 261–62
Phosphorus (P), 43–44, 88
PIPRA (Public Intellectual Property Resource for Agriculture), 261*n*
Plant breeders rights (PBRs), 42, 144, 195
Plants
hybrid seeds and plants, 34, 40–41, 42, 123, 144, 195
International Treaty on Plant Genetic Resources for Food and Agriculture (PGRFA), 5, 175, 195–96, 261–62
new and emerging diseases, 217–18
seeds and genetics, 40–43, *42f*
See also Seed industry
Platform model of knowledge production, 165, 166
Policies
developing a framework, 258–59
of environment and governance system, 258–64, *260b, 260t*
fair trade, 68–69, *68t, 69t,* 210, 221–22
interdisciplinary, 250
market driven, 144
options for improving, 211–12
trade, markets, and agricultural, 221–23, *222b, 223b*
See also Agricultural policies; Economic policies; Public policies
Policy environment and governance systems redesign, 258–64, *260b, 260t*
Political development, 10–12, 164–65
Post-harvest and consumption systems, 61–70, *62f, 62t, 63tt, 64f, 64t, 65t, 67t, 68tt, 69tt, 70tt*
Post-World War II/1945
agricultural policies, 15, 23, 30–34
agricultural workers, 26–27, *27f*
agrifood systems, 14–16
conditions in Soviet Union, 24–25, 38
food security achievements, 117
increase in productivity, 21
social, political and economic development, 11–12
Potassium (K), 43–44
Poverty
in European Union, 29
and food insecurity, 5–6, *7f,* 81, 104
goals and methods for reducing, 6–7, *7f,* 211
Grameen Bank vs., 195
hunger and food insecurity eradication goal, 5–6
in post-WWII United States, 11–12
See also Wealth and asset inequity
Precision agriculture, 40, 160, 169, 178, 186, 210, 236
Precision application of fertilizers and pesticides, 230, 235
Price supports, 30, 31–33, 93. *See also* Subsidies
Principal-agent model for agricultural research incentives, 140–41
Private labels, 62, 63, *63t,* 153–54
Private sector role in AKST development, 9, 153–54, 194, 213–14. *See also* Transnational and multinational corporations
Privatization
of agricultural land, 25, 34
future of, 197, 213
and influence of beneficiaries, 138–139, 197
of research and development, 119, 120–21, 134–35, 140–41, 194, 196–97, 198
Processing/marketing enterprises, 179
PROCINORTE cooperative program, 258
Production
and AKST advances, 92–93, *93f, 94f,* 98–99, *98f, 99f,* 117
AKST and production changes, 98–99, *98f, 99f*
biofuels as diversion from food production, 10, 13, 219
biomass, for biofuels, 47, *158b*
CEE agriculture and food production changes, 29, 38, 67, 68
Cooperative Wheat Production Program, 123–24
ethical dimensions of food production, 22
forestry production and market structure, 56
knowledge production models and trends, 165–66
See also Livestock production
Production and productivity paradigm, 2
Production contracts, 36–37
Productivity analysis, 123*n*
Program of Endorsement of Forest Certification (PEFC), 263–64
Property questions, complexity of, *260b*
Proprietary regimes, 195–96, 212, 260–61, *260b, 260t*
Public control of agrifood systems, 141–44
Public funding
of AKST organizations, 194
competitive grants and short-term contracts, 117, 119, 139–41
of R&D, 117, 123, 132–34, *133f, 136t,* 137–38, 265
of science and technology, 153
Public goods
and agricultural R&D, 264–65
food security, 194, *222b,* 243, 254, 266, *266t*
hybrid, 141, 260
Public institutions, 42, 43, 53, 160, 198, 252–53, 261. *See also entries beginning with* "United Nations"
Public Intellectual Property Resource for Agriculture (PIPRA), 261*n*
Public policies
and agrifood systems, 16
biofuels, biomass and policy changes, 22
bipolar farm policy in Canada, 31
changes in goals, 144–45
and Great Depression in US, 30–31

options for improving, 211–12
See also Common Agricultural Policy; Economic policies

Q

Quality standards for food, 65–66
Quality turn, 2–3, 8
Quantitative models, 155–56, *156t*, 223, 262

R

Radio frequency identification (RFID), 152, 169, 170
Rainforest Alliance, 69, 264
Reforestation, 53–54
Regional treaties and trade agreements, 144, *144b*
Remedial R&D, 264
Renewable energy, 13
Research and development
Asia's results, 153
on biodiversity, 56, 155–56, 158, 210, 265, 266, *266t*
on biofuels, 219–20, *220t*
on coastal fisheries, 239–40
in ecological and evolutionary sciences, 210
and education, 251
funding for agriculture, 264–67, *266t*, *267b*
international R&D, 117–18, 119, 137, 168
key drivers for AKST and, 192–98
on market mechanisms, 222–23
of multifunctional agriculture systems, 265, 266–67
overview, 117, 119
as percentage of GDP, 122, 136, 170–71, *171t*, 193
principal-agent model for agricultural research incentives, 140–41
privatization of, 119, 120–21, 134–35, *136t*, 140–41, 194, 196–97, 198
public funding of, 117, 123, 132–34, *133f*, *136t*, 137–38
public to private investment ratio, 10
review of foresight exercises, 155–61, *158b*, *159t*, *162–63t*
structure and policy changes, 117, 132–33
TNC/MNC competition, 168
trends, 166–69
See also AKST; Innovative models and systems
Research and technology organizations (RTOs), 167
Research definition and execution, 153, 165
Researchers, 166–67, *171t*
Research resources, reallocation of, 141
Research structures and management, 137–38
Resources, equity in access to, 9–10, 103
Response network for infectious diseases, 217
RFID (radio frequency identification), 152, 169, 170
Ricardo, David, *223b*
Risk regulation in agrifood systems, 141–43
River Basin Management programs, 263
RTOs (research and technology organizations), 167
Rules and norms, 5. *See also* International treaties and agreements; Standards
Rural development, 246
Rural development support, 16, 28, 33
Rural Economy and Land Use Programme (RELU), 135
Rural livelihoods. *See* Food and farming
Rural populations, 26–29, *27f*, *28f*, 97–98, 100–103, *101t*, *102t*
Russia, 21, 25, *26t*, 33–34. *See also* Soviet Union

S

SAPARD (Special Accession Programme for Agriculture and Rural Development), 33
Scenar 2020 foresight exercise, 157–58, *159t*
Science and technology
in aquaculture, 60–61
attitudes toward, 171–72
financial resources devoted to, 170–71, *171t*
funding for, 153, 170–71, *171t*, 192–95
integration of perspectives, 117, 126–30, *127b*, *128t*, 197, 211
student interest, 153
See also AKST
Science and Technology Foresight Unit, 157
Science of complex systems, *214n*, *215b*
Scientific knowledge
global spending, 136–38
organization of, 120–22, *121b*, *122f*, *122t*
Seedbanks, 230
Seed industry
effect of plant breeding technologies, 21
hybrid seeds and plants, 34, 40–41, 42, 123, 144, 195
market structure, 34, *36t*
oilseed crops, 22, 31–32, 38, 47, 218–19, 220, *220t*
organic seeds, 67
plant breeding, seeds and genetics, 40–43, *42f*
TNC domination of, 10, 16, 21, 34, 179
in United States, 34, *36t*
Seed varieties, access to, 213–14
Short-term grant contracts, 117, 139–140
Single farm payments (SFPs), 32–33
Smartwood certification scheme, 263–64
Social development, 7–8, 11–12, 22, 216, 241–43
Social gaps, 7–8. *See also* Gender gaps; Women
Societal context
cultural aspects of AKST, 5, 9
impacts of agriculture and AKST, 81, 99–104, *100t*, *101f*, *101t*, *102t*, *105tt*, *106f*
impacts of increased mechanization, 101, *101f*
trends of paradigms, 122–26, *125b*
See also Communities; Women
Socioeconomic viability of systems, 241–44
Sociopolitical drivers of AKST, 164–65
Soil management
AKST analytical techniques, 37–38
environmental effects of, 82–84, *83f*, *83t*, *84t*
European Commission on soil degradation, 182
improving standards for, 210, 226–27
Solar energy, 170, 180, 192, 220, 230, *230b*
Soviet Union
agrifood systems, 14, 15, 21, 24, 40
collapse and breakup of, 11, 98, 138
economic development of, 11, 12
post-WWII agricultural conditions, 38
See also Russia
Special Accession Programme for Agriculture and Rural Development (SAPARD), 33
Specialization of agriculture
bipolarisation of demand, *224b*
crop or livestock production vs. mixed systems, 92
as effect of AKST, 97–98, 126–27
and equity of food systems, 102–4
and food miles, 92, 195, 220
food system specialization, 22–26, *23f*, *23t*, *24f*, *25f*, *26t*
industrialization and, 123
overview, 22–26, *23f*, *23t*, *24f*, *25f*, *26t*, 100
potential problems, 152
precision agriculture, 40, 160, 186
quality products distinguished by origin, 225
social impacts, 100
See also Supply chains
Specialization of livestock agriculture, 48–49, *49t*, 52–53, 87–88, 123, 194
Standards
for animal welfare, 15–16
effect of, on small-scale producers, 152
ISO and HACCP, 66–67, *67t*
overview, 65–66, *65n*
Structures
AKST funding, agenda development and, 119, 130–42, *131t*, *133f*, *134f*, *135f*, *136t*
CEE farm structure, 24–26, 40
changes in, as effect of AKST, 97–98
changes in funding and, 140–42
development/establishment of, 130–32, *131t*
food and farming systems and, 177–78
rationalization of, 141
research and development structure and policy changes, 117, 132–33
research structures and management, 138-39
in US agriculture, 23–24, *23t*
See also Market structure
Subsidies
of biofuels, 22
for cotton producers in the US, 176
direct payments in EU, 31–33, 47, 93, 175
discrimination in distribution of, 7
elimination of, 198, *222b*
in Europe, 11, 16, 32, 48, 60, 175, 196
and excess productivity, 98
for exports, 93, 106
for forestry science development, 56
market-led policies, 53, 109, 145
NAE undermining developing countries with, 152, 175
in North America, 6–7, 30–31
for positive ecological, sociocultural and economic externalities, 200, 244
and private sector, 197
for productivity increases, 50, 53, 145, 176, 177
Subsistence or semi-subsistence growers, 2
Supermarkets, 15. *See also* Food retail sector
Supermarkets Code of Practice, 64
Supply and demand
in foresight exercises, 155, 223
impact of AKST on, 93, *94f*, *94t*, *95f*
of labor, 26–27, *27f*
supply driven policies, 144

Supply chains
agrifood chains, 16
and AKST, 224, *224b*
anti-competitive practices, 65, 81
economic issues, 241, 243
energy efficiency of, 214, 220–21, 231
in Europe, *64t*
in food system, 36–37, 66, 95, 96
of multinational food companies, 180
mutually-beneficial aspects of, 211, 225–26
and nanotechnology, 234–35
of supermarkets, 80, 92
See also Value chains
Supply control, 30, 31, 32, 81, 145
Surveillance networks, 217, 218
Sustainability
and adaptive capacity for climate change, 189–90
from ecological management, 226–46, *230b, 232b, 233b, 234b, 236b, 237t, 245f*
as goal, 8, 16
and innovation systems, 153–54
and R&D decisions, 265
of rural livelihoods, 244–45, *245f*
See also Development and sustainability
Synthetic pesticides, 44
Systems approaches, 124, 127–28, *127b*

T

Tariffs, 5, 16, 30, 32, 175, *222b*
Technology
assessing, 210
biotechnology, 210, 217, 231–34, *232b, 234b*
for communication, 152, 157, *159t*, 169, 170, 200, 246–48
and evolution of KST, 169–70
and role of private AKST organizations, 194
soil-related, 227
for water policies and water reuse, 228
See also AKST; Information technology; KST; Nanotechnology; Science and technology
Territorial Use Rights in Fisheries (TURF), 263
Tissue engineering, 170
TNCs. *See* Transnational and multinational corporations
Tools and approaches of AKST, 118–19
Trade agreements. *See* International trade; International treaties and agreements
Trade-Related Aspects of Intellectual Property Rights (TRIPs), 42, 144, 195, 261
Trade spending, 62
Traditional practices, 4, 9, 14, 247–48
Transgenesis, 233
Transhumance, 13, 48
Transnational and multinational corporations (TNCs and MNCs)
agrochemical companies, 43, 45
border protection vs., 180, 226
cooperatives or consolidation vs., 35, 96–97, *97t, 98f*, 168
development of, 15
dominance of, 10, 21, 34–37, *35t, 37t*, 153–54
in European aquaculture, 59
in food retail sector, 21–22, 61–65, *62f, 62t, 63tt, 64f, 64t, 65f*, 103
growth related to AKST application, 9
and organic foods, 67
purchasing and marketing organizations, 62–63, *63t*
supply chains of, 180
sustainability goals, 16
and wealth inequity, 9
Trends
in access, control and distribution of AKST, 196–97
in agricultural policies, trade and markets, 175–76, *177t*
in attitudes towards science and technology, 171
climate change and variability, 186–88, *187f, 188f*
consolidation in US food industry, 34, *35f*
emerging trends and AKST response, 152–53
emerging trends and uncertainties in KST, 153
energy and bioenergy, 191
farming systems and farm structures, 177–78
in food consumption and distribution, 172–74
in forests and forestry production, 53–54, *54f*, 56–57
historical, of scientific knowledge generation, 120–22, *121b, 122f, 122t*
identifying with foresight activities, 157–58, *158b*
information technology, 169
labor and gender dynamics, 178–79
natural resource availability and management, 180–85, *183f, 184ff*
niche markets, 15–16, 67, *68t*
obesity and associated diseases, 81
and options for action, 201
organizations, 179–80, 192–95
output and productivity, 49–53, *51f*
output and productivity in Europe, 53, 108, *108f*
of paradigms in societal context, 122–26, *125b*
pesticide usage, 44, *46f*
in post-harvest and consumption systems, 61, 63–65, *64t*
proprietary regimes, 195–96
in R&D spending, 133–34, *133f, 134t*
in research and development, 166–69
in science and technology funding, 170–71, *171t*, 192–94
in science education, 172
and small farms, 99, 101–2, *101t, 102t*
in stakeholder interaction, 247
transformation in KST models, 165–68
Triple helix model of knowledge production, 165, 166
TRIPs (Trade-Related Aspects of Intellectual Property Rights), 42, 144, 195, 261
TURF (Territorial Use Rights in Fisheries), 263

U

Uncertainties
in access, control and distribution of AKST, 197
agricultural labor and organizations, 180
in agricultural policies, trade and markets, 176, *177t*
in attitudes towards science and technology, 171–72
climate change and variability, 188–90
energy and bioenergy, 191–92
farming systems and farm structures, 178
in food consumption and distribution, 174
information technology, 169
natural resource availability and management, 185
organizations, 194–95
proprietary regimes, 195–96
in science and technology funding, 171, 194–95
in science education, 172
transformation in KST models, 166, 168–69
Undernourishment in countries in transition, 6–7, *7f*
Union for the Protection of New Varieties of Plants (UPOV), 42, 195
United Nations, founding of, 5
United Nations Children's Fund (UNICEF), 5
United Nations Convention on Biological Diversity, *125b*, 175, 195, 261–62
United Nations Environment Programme (UNEP), 155–56, *156t*
United States
agricultural imports and exports, 106, *108ff*
agriculture subsidies in, 6–7
American Indian to white economic disparity, 12
changes in livestock agriculture, *50f, 51f*
diet-related health problems, 6
fertility rate projection, 163, *163t*
food insecurity in, 5
funding for agricultural research, 134–35, *135f*
and Great Depression, 30–31
post-WWII economic recovery, 11–12
seed industry, 34, *36t*
See also NAE region; North America
University trends in research, 153, 167–68, 179–80. *See also* Educational institutions; Research and development
UPOV (International Union for the Protection of New Varieties of Plants), 42, 195
Urbanization, 12, 177
US Department of Agriculture (USDA), 42, 53, 160
US Energy Policy Act (2005), 47

V

Value chains
agrifood systems and, 14, 15
consolidation of, 8–9, 16, 195
development of, 4, 15–16, 52, 166, 243, 254
and information technology, 126
See also Supply chains
Vegetarian movement, 15
Veterinary and public health infrastructure integration, 217–18
Veterinary medicines, environmental consequences of, 88

W

Water control in cropping systems, 46
Water footprints, 10, 210
Water law reform, *260t*
Water management with AKST, 227–28
Water quality, 228
Water resources, 182–83, *183f, 184f*
Wealth, consumer awareness from, 153–54, 160
Wealth and asset inequity
  and AKST, 8–9, 118, 259
  birth/death rate and, 12
  and Caribbean banana producers, 64
  counteractive measures, 16, 28
  and demand for biofuel, 52
  farm size and, 9
  NAE vs. developing countries, 108, 120
  overview, 2, 81
  political power shifts related to AKST, 10
  in research and development, 124
  resources and, 9–10, 103
  TNCs and, 7, 10, 15
Western Europe, 2, 11, 12–14, 163, *163t. See also* Europe; European Union; NAE region
WHO (World Health Organization), 5, 66, *236b*
Wind energy, 47
WIPO (World Intellectual Property Organization), *144b,* 196
Women
  in agriculture workforce, 29
  in European Union, 29
  as farm operators, 7–8, 180
  fertility rates, 163, *163t,* 242
  future of, 180, 211, 242
  gender gaps, 7–8, 12, 242–43
  as higher education students, 12
  influence via civil society groups, 139
  rural livelihood issues, 242–43, 266
  in science, 121, *121b, 122f, 122t*
World Bank, 5
World Health Organization (WHO), 5, 66, *236b*
World Intellectual Property Organization (WIPO), *144b,* 196
World Technology Evaluation Center (WTEC), 160
World Trade Organization (WTO), 5
World War II, 24. *See also* Post-World War II

About Island Press

Since 1984, the nonprofit Island Press has been stimulating, shaping, and communicating the ideas that are essential for solving environmental problems worldwide. With more than 800 titles in print and some 40 new releases each year, we are the nation's leading publisher on environmental issues. We identify innovative thinkers and emerging trends in the environmental field. We work with world-renowned experts and authors to develop cross-disciplinary solutions to environmental challenges.

Island Press designs and implements coordinated book publication campaigns in order to communicate our critical messages in print, in person, and online using the latest technologies, programs, and the media. Our goal: to reach targeted audiences—scientists, policymakers, environmental advocates, the media, and concerned citizens—who can and will take action to protect the plants and animals that enrich our world, the ecosystems we need to survive, the water we drink, and the air we breathe.

Island Press gratefully acknowledges the support of its work by the Agua Fund, Inc., Annenberg Foundation, The Christensen Fund, The Nathan Cummings Foundation, The Geraldine R. Dodge Foundation, Doris Duke Charitable Foundation, The Educational Foundation of America, Betsy and Jesse Fink Foundation, The William and Flora Hewlett Foundation, The Kendeda Fund, The Andrew W. Mellon Foundation, The Curtis and Edith Munson Foundation, Oak Foundation, The Overbrook Foundation, the David and Lucile Packard Foundation, The Summit Fund of Washington, Trust for Architectural Easements, Wallace Global Fund, The Winslow Foundation, and other generous donors.